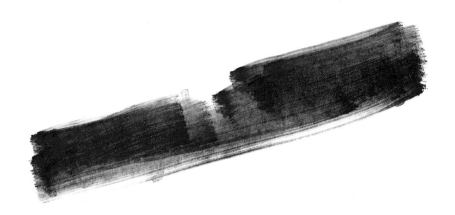

BACTERIAL CELL WALL

New Comprehensive Biochemistry

Volume 27

General Editors

A. NEUBERGER
London

L.L.M. van DEENEN
Utrecht

ELSEVIER
Amsterdam – London – New York – Tokyo

Bacterial Cell Wall

Editors

J.-M. GHUYSEN
*Centre d'Ingénierie des Protéines, Université de Liège, Institut de Chimie, B6,
B-4000 Sart Tilman (Liège 1), Belgium*

R. HAKENBECK
*Max-Planck Institut für Molekulare Genetik, Ihnestrasse 73,
D-14195 Berlin 33, Germany*

1994
ELSEVIER
Amsterdam – London – New York – Tokyo

Elsevier Science B.V.
P.O. Box 211
1000 AE Amsterdam
The Netherlands

ISBN 0 444 88094-1 (volume)
ISSN 0 444 80303-3 (series)

Library of Congress Cataloging-in-Publication Data

Bacterial cell wall / editors, J.-M. Ghuysen, R. Hakenbeck.
    p.   cm. -- (New comprehensive biochemistry ; v. 27)
Includes bibliographical references and index.
ISBN 0-444-88094-1 (acid-free paper)
1. Bacterial cell walls. 2. Peptidoglycans. 3. Bacterial proteins. I. Ghuysen, J.M. II. Hakenbeck, R. (Regine), 1948-. III. Series.
QD415.N48 vol. 27
[QR77.3]
574.19'2 s--dc20
[589.9'0875]
                                                          93-41268
                                                             CIP

Preface

Studies of the bacterial cell wall emerged as a new field of research in the early 1950s and, literally, took off primarily thanks to the pioneering work of Milton R.J. Salton. Since then, it has flourished in a multitude of directions and has reached the stage where the preparation of a monograph as a 'one-man show' is no longer possible. This volume is the result of a joint venture involving numerous contributors. Following a historic perspective of the field (Chapter 1), it has been conceived as an attempt to single out topics of great conceptual importance related to those groups of macromolecules which are localized outside the permeability barrier, i.e. the cytoplasmic membrane, of the Eubacteria, with special emphasis on the genetic, molecular and, when possible, atomic levels.

Peptidoglycan, the basic matrix of the bacterial wall, is present in all vegetative cells, except the L-forms and Mycoplasma, and is an essential constituent of bacterial endospores (Chapters 2–8). Teichoic and teichuronic acids, lipoteichoic acids, lipoglycans, neutral complex polysaccharides and several specialized proteins are frequently unique wall-associated components of Gram-positive bacteria (Chapters 9–11). Lipopolysaccharides, lipoproteins and a multitude of proteins including porins, form the outer membrane of the Gram-negative bacteria (Chapters 12–20). The periplasm is a transshipment region localized between the plasma membrane and the outer membrane (Chapter 21).

Bacterial cells know how to adapt to changing environmental conditions and, for that purpose, they have evolved signal transduction pathways. These pathways generally are initiated by 'receptors' consisting of a sensory domain which is responsible for signal reception on the outer face of the plasma membrane and a cytosolic domain which is responsible for the generation of an intracellular signal. Examples of adaptative responses of this kind are discussed in Chapters 22–24.

Acquired resistance to known antibiotics by an increasing number of bacterial species has become such an utterly serious concern that, in terms of effective antibacterial chemotherapy, the 1990s are beginning to look like the 'pre-antibiotic era'. This situation is the grim harvest of ignorance and complacency. It demands urgent attention. Chapters 25–27 deal with the underlying mechanisms of bacterial resistance to the main chemotherapeutically useful 'cell wall' antibiotics, the β-lactams and the glycopeptides.

With the rapid growth of scientific literature in the field, it was impossible to condense all the accumulated knowledge in a book of the size of the present volume. There are gaps. In particular, no chapter is devoted to the walls of *Mycobacterium tuberculosis* and *M. leprae* which are plagues that massively affect much of the Third World. Inevitably also, the book will contain 'out of date' sections by the time it is published. In spite of these imperfections, we hope that it provides an integrated collection of contributions forming a fundamental reference for researchers and of general use to teachers and advanced students in the life sciences. We also hope that it can provide useful guidelines for those wishing to make an informed decision about whether research on the bacterial cell wall is worth funding in these times .

Research often begins with a question. This book witnesses the multifaceted pursuits of scientists engaged in bacterial cell wall research with many of the why's, how's and wherefore's that they strive, with increasing success, to answer. We thank them for their valuable contributions. We also express our gratitude to Professor Laurens L.M. van Deenen, general co-editor of the *New Comprehensive Biochemistry* series who initiated this venture, to Mrs Amanda Shipperbottom and Mrs Annette Leeuwendal, publishing editors, and Mr. Dirk de Heer, desk editor, of Elsevier Science B.V., who produced the book.

Regine Hakenbeck   *Jean-Marie Ghuysen*
Berlin   Liège

October 1993.

List of contributors

J.A. Ayala, 73
Centro de Biología Molecular, CSIC, Universidad Autónoma de Madrid, Facultad de Ciencias, Cantoblanco, 28049 Madrid, Spain

E. Bartowsky, 485
Department of Molecular Microbiology, Washington University, Medical School, Box 8230, 660 S. Euclid Ave., St. Louis, MO 63110, USA

M.E. Bayer, 447
Fox Chase Cancer Center, Institute for Cancer Research, 7701 Burholme Avenue, Philadelphia, PA 19111, USA

R. Benz, 397
Lehrstuhl für Biotechnologie, Biozentrum der Universität Würzburg, Am Hubland, D-97074 Würzburg, Germany

C. Bernegger-Egli, 263
Department of Microbiology, University of British Columbia, 300 – 6174 University Blvd., Vancouver, B.C., V6T 1Z3, Canada

V. Braun, 319
Mikrobiologie II, Universität Tübingen, Auf der Morgenstelle 28, D-72076 Tübingen, Germany

C.E. Buchanan, 167
Department of Biological Sciences, Southern Methodist University, Dallas, TX 75275 0376, USA

S.W. Cowan, 353
Department of Structural Biology, Biocentre, University of Basel, Klingelbergstrasse 70, CH-4056 Basel, Switzerland

G. Dive, 103
Centre d'Ingénierie des Protéines, Université de Liège, Institut de Chimie, B6, B-4000 Sart Tilman (Liège 1), Belgium

J. Erickson, 485
Department of Molecular Microbiology, Washington University, Medical School, Box 8230, 660 S. Euclid Ave., St. Louis, MO 63110, USA

W. Fischer, 199
Institut für Biochemie der Medizinischen Fakultät, Universität Erlangen-Nürnberg, Fahrstrasse 17, D-91054 Erlangen, Germany

T. Garrido, 73
Centro de Investigaciones Biológicas, CSIC, C/ Velázquez, 144, 28006 Madrid, Spain

J.-M. Ghuysen, 103, 505
Centre d'Ingénierie des Protéines, Université de Liège, Institut de Chimie, B6, B-4000 Sart Tilman (Liège 1), Belgium

R. Hakenbeck, 535
Max-Planck Institut für Molekulare Genetik, Ihnestrasse 73, D-14195 Berlin 33, Germany
R.E.W. Hancock, 263
Department of Microbiology, University of British Columbia, 300 – 6174 University Blvd., Vancouver, B.C., V6T 1Z3, Canada
K. Hardt, 505
Centre d'Ingénierie des Protéines, Université de Liège, Institut de Chimie, B6, B-4000 Sart Tilman (Liège 1), Belgium
J. van Heijenoort, 39
Unité de Recherche Associée 1131 du CNRS, Biochimie Moléculaire et Cellulaire, Université Paris-Sud, Bât. 432, 91405 Orsay, France
U. Henning, 381
Max-Planck-Institut für Biologie, Corrensstrasse 38, D-72076 Tübingen, Germany
A.O. Henriques, 167
Centro de Tecnologia Química e Biológica, Apartado 127, 2780 Oeiras, Portugal
J.-V. Höltje, 131
Max-Planck Institut für Entwicklungsbiologie, Spemannstrasse 35, D-72076 Tübingen, Germany
C. Hughes, 425
Cambridge University Department of Pathology, Tennis Court Road, Cambridge CB2 1QP, UK
C. Jacobs, 485
Centre d'Ingénierie des Protéines, Université de Liège, Institut de Chimie, B6, B-4000 Sart-Tilman (Liège 1), Belgium
D. Jeanteur, 363
European Molecular Biology Laboratory, Postfach 10.2209, Meyerhofstrasse 1, D-69012 Heidelberg, Germany
B. Joris, 505
Centre d'Ingénierie des Protéines, Université de Liège, Institut de Chimie, B6, B-4000 Sart Tilman (Liège 1), Belgium
D. Karamata, 187
Institut de génétique et de biologie microbiennes, Rue César-Roux 19, CH-1005 Lausanne, Switzerland
D.N. Karunaratne, 263
Department of Microbiology, University of British Columbia, 300 – 6174 University Blvd., Vancouver, B.C., V6T 1Z3, Canada
M.A. Kehoe, 217
Department of Microbiology, University of Newcastle upon Tyne, Medical School, Newcastle upon Tyne, NE2 4HH, UK
R. Koebnik, 381
Max-Planck-Institut für Biologie, Corrensstrasse 38, D-72076 Tübingen, Germany
V. Koronakis, 425
Cambridge University Department of Pathology, Tennis Court Road, Cambridge CB2 1QP, UK

H. Labischinski, 23
Bayer AG, Institut für Chemotherapie, D-42096 Wuppertal, Germany
J.H. Lakey, 363
European Molecular Biology Laboratory, Postfach 10.2209, Meyerhofstrasse 1, D-69012 Heidelberg, Germany
F. Lindberg, 485
Department of Medicine, Washington University, Medical School, Box 8230, 660 S. Euclid Ave., St. Louis, MO 63110, USA
S. Lindquist, 485
Department of Microbiology, Umeå University, Umeå, Sweden
H. Maidhof, 23
Robert Koch-Institute of the Federal Health Office, Nordufer 20, D-13353 Berlin 65, Germany
M. Matsuhashi, 55
Department of Biological Science and Technology, Tokai University, School of High Technology for Human Welfare, Nishino 317, Numazu-shi, Shizuoka-ken, 410-03 Japan
H. Nikaido, 547
Department of Molecular and Cell Biology, University of California, 229 Stanley Hall, Berkeley, CA 94720, USA
S. Normark, 485
Department of Molecular Microbiology, Washington University, Medical School, Box 8230, 660 S. Euclid Ave., St. Louis, MO 63110, USA
F. Pattus, 363
European Molecular Biology Laboratory, Postfach 10.2209, Meyerhofstrasse 1, D-69012 Heidelberg, Germany
M.A. de Pedro, 73
Centro de Biología Molecular, CSIC, Universidad Autónoma de Madrid, Facultad de Ciencias, Cantoblanco, 28049 Madrid, Spain
P.J. Piggot, 167
Department of Microbiology and Immunology, Temple University School of Medicine, Philadelphia, PA 19140, USA
H.M. Pooley, 187
Institut de génétique et de biologie microbiennes, Rue César-Roux 19, CH-1005 Lausanne, Switzerland
P. Reeves, 281
Microbiology Department, Bldg G08, University of Sydney, Australia NSW 2006
M.R.J. Salton, 1
Department of Microbiology, NYU School of Medicine, New York, NY 10016, USA
T. Schirmer, 353
Department of Structural Biology, Biocentre, University of Basel, Klingelbergstrasse 70, CH-4056 Basel, Switzerland
G.E. Schulz, 343
Institüt für Organische Chemie und Biochemie der Albert-Ludwigs-Universität, Albertstrasse 21, D-79104 Freiburg im Breisgau, Germany

G.D. Shockman, 131
Department of Microbiology and Immunology, Temple University, School of Medicine, Philadelphia, PA 19140, USA

B.G. Spratt, 517
Microbial Genetics Group, School of Biological Sciences, University of Sussex, Falmer, Brighton BN1 9QG, UK

J.B. Stock, 465
Department of Molecular Biology, Princeton University, Princeton, NJ 08544, USA

M.G. Surette, 465
Department of Molecular Biology, Princeton University, Princeton, NJ 08544, USA

M. Vicente, 73
Centro de Investigaciones Biológicas, CSIC, C/ Velázquez, 144, 28006 Madrid, Spain

K. Weston-Hafer, 485
Department of Molecular Microbiology, Washington University, Medical School, Box 8230, 660 S. Euclid Ave., St. Louis, MO 63110, USA

M. Wikström, 485
Department of Microbiology, Umeå University, Umeå, Sweden

H.C. Wu, 319
Department of Microbiology, Uniformed Services University of the Health Sciences, 4301 Jones Bridge Road, Bethesda, MD 20814-4799, USA

Contents

Chapter 12. Molecular organization and structural role of outer membrane macromolecules
R.E.W. Hancock, D.N. Karunaratne and C. Bernegger-Egli ....

Chapter 13. Biosynthesis and assembly of lipopolysaccharide
P. Reeves ....

Chapter 15. Structure-function relationships in porins as derived from a 1.8 Å resolution crystal structure

Chapter 16. Structures of non-specific diffusion pores from Escherichia coli

Chapter 17. The porin superfamily: diversity and common features

J.-M Ghuysen and R. Hakenbeck (Eds.), *Bacterial Cell Wall*

CHAPTER 1

The bacterial cell envelope – a historical perspective

MILTON R.J. SALTON

Department of Microbiology, Milton R.J. Salton, NYU School of Medicine, New York, NY 10016, USA

1. Historical introduction

The evaluation of ideas and knowledge of the nature of our universe has never ceased to fascinate and challenge physicists, astronomers and cosmologists for millenia and will without doubt continue into the forseeable future. Unravelling the complexities of biological systems has, by comparison, been a relatively recent event in human history but the 'microcosmic' has proved to be equally challenging and complex but amenable to a level of direct manipulation and observation within a fairly reasonable span of time. This is particularly true for the development of our knowledge of the nature of the bacterial surface, which emerged with the discovery of minute microbes by Leeuwenhoek in 1675 as a complete mystery and advanced to the sophisticated molecular level of our current detailed understanding of the bacterial cell envelope structure as exemplified in the chapters of this book. The challenge and how this all came about is nonetheless fascinating and represents an exciting event in understanding the significant differences between prokaryotic bacteria and eukaryotic cells of 'higher organisms' and ourselves. These important differences between prokaryotic and eukaryotic cells turned out to be not merely a question of size, shape or surface areas but rather more fundamental anatomical and biochemical distinctions. Electron microscopy of thin sections of cells clearly revealed some of the basic differences between eukaryotic cells with a membrane-enveloped nucleus and discrete golgi, mitochondrial, lysosomal and other intracellular organelles, and the prokaryotic cells with their chromatin structure lacking an enveloping membrane and the absence of separate mitochondrial and other organelles commonly found in 'higher' cells. Equally dramatic was the discovery that the surface structures of bacteria contained unique compounds (e.g. muramic acid, α,ε-diaminopimelic acid, D-amino acids, teichoic acids, etc.) that were absent in eukaryotic cells. It can be deduced that each of these specific wall components accounted for about 1–3% of the dry weight of the bacterial cell and thus occurred in significant amounts rather than mere traces. The emergence of this knowledge thus set the stage for the fundamental biochemical and structural differences between the prokaryotic and eukaryotic worlds. Subsequent investigations have defined and refined these differences and added new molecular and structural parameters giving us an insight into differences and similarities of the two basic cellular types and exciting details of the Archaebacteria.

The extent of these advances made over the past 40 years or so is abundantly illustrated by browsing through one of the first major Symposia on *The Nature of the Bacterial Surface* (A Symposium of the Society for General Microbiology, April, 1949, Edited by A.A. Miles and N.W. Pirie). In his introductory Preface, Alexander Fleming put it succinctly, 'Something of the nature of the bacterial surface has been known for as long as bacteria themselves have been studied. Gross characters like spores, capsules, granules and flagella were recognized, but it is only in recent years that the finer structure and still more the extraordinary enzymic activity of these minute bodies has attracted the general attention of microbiologists.' Pirie further added that 'It is no longer necessary to apologize for the intrusion of biochemists into what used to be a purely bacteriological field. There is now general recognition that staining reactions, the antigen-antibody reaction, phagocytosis and the phenomenon of lysis are all biochemical processes and can only be described fully in biochemical terms (Introduction by N.W. Pirie, 1949).' The 'intrusion' of the biochemist over the past four decades has, indeed, paid off handsomely as abundantly illustrated by the state of the art and exciting details of the contributions presented in this monograph.

In the light of our current knowledge, it is hard to believe that four decades ago, very little specific could be said about the bacterial cell wall at this 1949 Symposium. Perhaps the most specific biochemical comments were those of Pirie [1] when he pointed out that 'lysis is the most obvious index that a change has been brought about in the cell wall' and that 'Lysozyme is probably the best understood of the lytic agents, it digests an insoluble amino-polysaccharide, which appears to be present in all organisms susceptible to the lytic action of the enzyme (Epstein and Chain, 1940) and in organisms killed without lysis.' The 'insoluble amino-polysaccharide' [2] of course turned out to be the cell wall structure of a sensitive organism such as *Micrococcus lysodeikticus* as shown by Salton [3], and chemically characterized and classified as a wall peptidoglycan by Ghuysen [4]. Thus, despite the fact that there was little information about the precise chemical nature of bacterial wall or envelope structures in 1949, much was known about surface lipopolysaccharides of Gram-negative bacteria, bacteriophage receptors and indeed the physical appearance of rigid cell wall structures of disrupted *Staphylococcus aureus* as elegantly demonstrated in the electron micrograph of Dawson [5]. This latter observation was the forerunner to the isolation and purification of cell walls for chemical analysis and characterization by Salton and Horne [6] and Salton [7].

Thus, the late 1940s to mid-1950s ushered in an era of cell fractionation and chemical and biochemical characterization of some of the major morphological entities of bacteria and of course many other eukaryotic cells of diverse origins. Although bacterial capsules, especially those of the pneumococcus, had been well studied and characterized by Heidelberger et al. [8], and the O-somatic antigenic lipopolysaccharides of Gram-negative bacteria were defined serologically and biochemically by Morgan [9], the precise nature of many of the other bacterial structures (e.g. flagella, cell walls of Gram-positive bacteria, envelopes of Gram-negative organisms, cytoplasmic membranes, endospores) remained a matter of conjecture or total mystery as to chemical composition. Weibull's [10] classical investigation of the flagella isolated from *Proteus vulgaris* established the protein nature of these structures. This was soon followed by the development of suitable methods for the isolation of cell walls and envelopes [6,11] and cytoplasmic membranes by Weibull [12], Salton and Freer [13], and Osborne and Munson [14], and ultimately their chemical and biochemical characterization [15]. This was a highly productive era for cell fractionation, leading not only to new insights into walls and membranes but other key bacterial structures including ribosomes, chromatin-DNA, spore coats and contents and diverse granules, to name but a few. Although this era was a landmark achievement for microbiology, it is well to remember the earlier valiant efforts made under less optimal conditions, to probe and define the bacterial surface and some of these seminal 'seeds' are brought to the readers attention, below.

2. Early observations

The remarkable discovery and description of the microscopic morphology of bacteria and other microorganisms by Antony van Leeuwenhoek in 1675 appears to have raised the key question in his mind as to what 'held them together, or contained them'. Leeuwenhoek clearly recorded the characteristic coccal, rod and spiral shapes of bacteria and anticipated that surface structures were responsible for their shape, one of the principal features and functions of the cell wall still recognized today. Not surprisingly, Leeuwenhoek observed these 'clear globules, without being able to discern any film that hold them together, or contained them' (see [16]) and it took almost another 300 years before the 'rigid' cell surface structures of bacteria became clearly visible in the electron microscope on examination by Mudd et al. [17]. Indeed, the many biologists who studied bacteria after Leeuwenhoek considered that their minuteness of size was incompatible with any further structural or morphological differentiation. This concept was finally put to rest by the advances made in electron microscopy and its associated techniques. We now know that the majority of bacteria are complex, differentiated cells, many with complicated, layered surface structures, some with a discrete periplasmic compartment and most with organelle functions, such as those of mitochondria localized or 'condensed' into a cytoplasmic membrane system.

Once bacteria became amenable to manipulation as individual species, through the pioneering investigations of Pasteur, Koch and many others, some early attempts were made prior to the turn of the century to establish a cell wall in bacteria and to probe its chemical nature. Two types of experimental approaches were made, one being chemical analysis of resistant residues of bacterial cells, the other more 'physiological' relating to the changes observed on plasmolysis of the cells. Vincenzi [18] made attempts to chemically characterize the alkali-resistant rod-shaped structures of *Bacillus subtilis*. It was generally believed that the residues remaining after various extraction procedures represented the resistant wall structures. There was of course no refined technique such as electron microscopy available for monitoring the morphological homogeneity of such preparations, so an element of ambiguity inevitably plagued the results and their interpretation. Vincenzi's preparation which had been alkali extracted, treated with gastric juice and alcohol and ether extracted retained the form of the original *B. subtilis* cell and was referred to as the 'Hülle der Zellen'. Under no conditions could he demonstrate the presence of cellulose and he could draw no further conclusion other than the presence of

Nitrogen (%N varied from 6.24% to 11.15%) in the outer envelope. At least one thing seemed certain that cellulose could be ruled out as a constituent of the *B. subtilis* envelope.

Fischer's [19] experiments on plasmolysis of bacteria were less direct than those of Vincenzi and little could be inferred as to the chemical nature of the outer structure differentiated by the collapse and contraction of the protoplasmic matter and its shrinkage away from the surface. The outer envelope clearly differed in its physio-chemical properties from those of the bacterial protoplasm.

The true nature of the bacterial cell wall remained a mystery until a direct approach was made in the 1950s. Cytologists and bacteriologists subjected bacteria to the whole gamut of tests for various constituents such as cellulose, chitin and many types of staining reactions [20], all to no avail. Ambiguity, uncertainty and disagreement reigned supreme. Dubos [21] aptly summed up the situation as follows: 'Unfortunately, very little information is available concerning the chemical composition of the wall, and many claims that it consists of cellulose, or of chitin, have not been substantiated.'

2.1. Birth of cell surface studies

With the superior resolution of the then newly introduced electron microscope, the cell wall became clearly visualized upon examination of disrupted bacteria by the techniques of electron microscopy of biological materials. Thus, Mudd et al. [17] established the appearance of the 'rigid' cell wall structures as empty sacs retaining the rod-shape of the original bacterium in the disintegrated cell fractions they examined. Such observations were further extended by Dawson [5] when he applied the biochemists' cell disruption technique with minute glass beads in the Mickle tissue disintegrator and revealed the flat, empty coccal-shaped walls of *Staphylococcus aureus*. Salton and Horne [6] further extended this approach and other disruption procedures to prepare 'pure' cell wall fractions from Gram-positive and Gram-negative bacteria conforming to the stringent requirements needed for precise analytical characterization. This could only be achieved by carefully monitoring the procedure and the fractions by electron microscopy to establish the morphological homogeneity of the preparations and their freedom from cytoplasmic elements [6], rather analogous to what Weibull [10] had achieved with bacterial flagella. Thus, isolated bacterial cell wall preparations became available for chemical analysis by Salton [7]. Weidel [11] had approached the same general problem of defining the bacterial surface in a different manner by characterizing the bacterio-phage receptors of *E. coli* in preparations which had been digested with enzymes and examined by electron microscopy. These preparations were different in that they contained electron-dense residues (presumably undigested cytoplasmic materials) within the otherwise 'empty' sac-like structures. It was deemed that such preparations would not be suitable for precise chemical characterization, but they were, however, invaluable in defining the surface nature of bacteriophage receptors. In all of these early investigations, electron microscopy played a pivotal role in establishing the nature and homogeneity of bacterial cell wall structures. The appearance of the rod-shaped structure of *Bacillus megaterium* cell wall is illustrated in the electron micrograph in Fig. 1 [22].

Fig. 1. Cell wall of *Bacillus megaterium* (air-dried preparation). The electron-dense spheres are polystyrene latex indicator particles (0.25 mm diameter). From Salton and Williams [22]. Reproduced with permission from Elsevier, Amsterdam.

The exact nature of comparable 'wall' fractions obtained from Gram-negative bacteria remained a matter of uncertainty until the advent of electron microscopy of ultra-thin sections of bacterial cells as demonstrated by Ryter and Kellenberger [23]. The response of cells to the Gram stain procedure and the early chemical analysis of 'cell wall' preparations clearly pointed to some basic differences in the properties of the envelopes of Gram-positive and Gram-negative bacteria and the greater chemical complexity of the isolated fractions from the latter organisms [24]. The structural complexity of the Gram-negative cell envelope was clearly resolved in the thin sections of Ryter and Kellenberger [23] and established the presence of an outer membrane as a characteristic feature of the cell envelope. The mysteries of the structure of the Gram-negative cell envelope were finally solved with the demonstration of the thin layer (peptidoglycan) between the outer and inner (cytoplasmic) membranes by use of lanthanum-stained thin sections by Murray et al. [25], and the isolation and demonstration of the thin murein sacculus of *E. coli* by Weidel et al. [26]. Thus, the complex cell envelopes of Gram-negative bacteria were resolved as consisting of an outer membrane anchored to the underlying rigid, thin, peptidoglycan layer or sacculus. As we now know, the outer membrane with its integral lipopolysaccharide structures, Braun [27] lipoproteins and specialized porin proteins [28] are unique to Gram-negative organisms and do not have their exact counterparts in the walls of Gram-positive bacteria.

Electron microscopy of thin sections of a variety of Gram-positive and Gram-negative bacteria provided the early key observations pointing to a fundamental difference in the organization of the surface structures of the two groups leading to the typical membrane profiles of the Gram-negative organisms compared to the thick, amorphous walls of Gram-positive bacteria with their underlying cytoplasmic membranes. Early investigations of the differences in chemical composition correlated well with the structural studies [29] and led Salton [30] to propose that the basic difference between wall and envelope structures of the two groups (G+ and G–) could account for the differential response to the solvent extraction of the crystal violet-iodine complex in the Gram stain procedure.

In addition to its important role in monitoring cell fractionation and envelope structure, electron microscopy has added much more to our understanding of the diversity and complexity of bacterial cell surface structures. It has led to the recognition of additional external structured layers in both Gram-positive and Gram-negative bacteria by Sletyr and Messner [31] and in some instances their structural and chemical characterization. Moreover, it has been instrumental in resolving the highly complex surface layers of bacteria such as Lampropedia, Deinococcus and Acinetobacteria studied by Lancy and Murray [32] and Glauert and Thornley [33], and more recently, those of the Archaebacteria.

During this gestation period and birth events of the bacterial cell wall era, a number of key biochemical discoveries were made, establishing some fundamental differences between the prokaryotic and eukaryotic worlds. The discovery of the unique bacterial compounds, the Park nucleotides, diaminopimelic acid, muramic acid, D-amino acids and teichoic acids provided the biochemical background and basis for the modern era explosion of our knowledge of the biochemistry of the bacterial surface, antibiotic action and cell function, to name but a few areas covered more extensively in this monograph.

3. The modern era

3.1. Uridine nucleotides

The observations by Park and Johnson [35] that uridine nucleotides accumulated in *Staphylococcus aureus* inhibited by penicillin provided an early biochemical clue of components and a process unique to the bacterial cell. This was further reinforced by the isolation and chemical characterization of the nucleotides by Park [36], although the unidentified amino sugar remained unknown at that time but was subsequently identified as muramic acid by Park and Strominger [37]. The important discovery of these nucleotides under the action of penicillin by Park [36] led the way to the detection of *N*-acetylmuramic acid, D-alanine and D-glutamic acid in these uridine nucleotides and established a close biochemical relationship to the bacterial cell wall as emphasized by Park and Strominger [37]. The detection of such nucleotides was a landmark event providing strong evidence that they were biosynthetic intermediates involved in the assembly of a unique cell wall polymer. Thus, the background was set for the unravelling of the series of stages in the biosynthesis of bacterial wall peptidoglycan and the action of various antibiotics inhibiting the enzymatic steps [38] and ultimately the sophisticated knowledge of the molecular interactions involved as exemplified in the present contributions to this book.

3.2. Diaminopimelic acid

Another landmark event was the discovery of the diamino acid, diaminopimelic acid by Work [39,40] and its demonstration as a cell wall component by Holdsworth [41], Salton [24] and Cummins and Harris [42]. Detected initially in hydrolysates of whole cells of bacteria, this amino acid had not previously been found in proteins [39,40] and as subsequent investigations showed, it really was not a protein amino acid but an important component of the peptide structure of certain bacterial peptidoglycans. The discovery of this amino acid in many bacteria, especially Gram-negative organisms and its presence in cell wall preparations of certain species added further evidence for the existence of some unique or unusual components in the bacterial cell. Diaminopimelic acid turned out to be one of the several diamino acids, alternatively with lysine in some species, but each exclusively performing cross-linking functions between adjacent peptide chains of the cell wall peptidoglycans defined by Ghuysen [4]. In retrospect, it is interesting to note that diaminopimelic acid was the first cell wall component not discovered initially in nucleotide form. The biochemical ramifications of the discovery of diaminopimelic acid extended beyond its role as an important building block in the cell wall peptidoglycan and spore peptides, with the recognition by Meadow and Work [43] of its significance in lysine biosynthesis in prokaryotes. The elucidation of the enzymatic steps in the biosynthesis of diaminopimelic acid by Gilvarg [44] was thus not only a key step in the overall understanding of wall peptidoglycan synthesis but also for the formation of spore peptides during the developmental stages of bacterial endospore differentiation. The only diamino acid found so far in peptidoglycans of Gram-negative species is *meso*-diaminopimelic acid, whereas lysine or diaminopimelic acid are present in walls of Gram-positive bacteria [4,45].

3.3. Muramic acid

Of all the prokaryotic molecular markers, the impact of the discovery of muramic acid has probably been the most dramatic with its almost universal presence in the majority of both Gram-positive and Gram-negative bacteria and its absence only in Mycoplasma groups lacking a cell wall and certain Archaebacteria. Attention to the significance of this 'uniquely' bacterial amino sugar was drawn by Strange and Powell [46] when they detected an unknown amino sugar in hydrolysates of the peptides secreted by germinating spores. An unknown amino sugar had thus been detected in uridine nucleotides by Park [36], in the spore peptides and in cell wall hydrolysates [42]. Following the isolation of the 'unknown' amino sugar from spore peptides by Strange and Dark [47], its structure was established as 3-*O*-carboxyethyl glucosamine and named 'muramic acid' by Strange [48]. The structure of muramic acid was subsequently confirmed by chemical synthesis by Strange and Kent [49], Zilliken [50], Lambert and Zilliken [51], and its significance in cell wall structure and biosynthesis realized by Work [45] and Park and Strominger [37]. Moreover, the 'unknown' amino sugar was recognized as one of the two amino sugars of the disaccharide product released from the walls of *M. lysodeikticus* digested with ly-sozyme by Salton [52]. The discovery and chemical characterization of muramic acid thus provided the key to unlocking the puzzle of the cell wall structure, its biosynthesis, the mode of action of penicillin and the specificity of the enzyme lysozyme. Muramic acid, or its N-acetylated form, therefore provided the central link between the nucleotide inter-mediates and what subsequently became known as the cell wall 'peptidoglycan' structures [4]. Elucidation of the nature of this key, major wall component triggered a decade or more of intensive work providing the basis for our present understanding of the primary structure of peptidoglycans, their biosynthesis and assembly and ultimately the enzymatic stages and reactions specifically inhibited by various antibiotics including β-lactams, gly-copeptides such as vancomycin and teichoplanin, bacitracin and fosfomycin.

3.4. Isomers of amino acids in walls

It was generally recognized by protein chemists that all of the amino acid building blocks of proteins were in the form of L-isomers. The occurrence of substantial amounts of cer-tain D-amino acids in hydrolysates of bacterial cells reported by Stevens et al. [53] prob-ably came as something of a surprise, since the significance of their possible origins from the cell walls was not apparent at that time. The source of such D-isomers was not clear until Snell et al. [54] and Ikawa and Snell [55] discovered a large quantity of D-alanine in a TCA-insoluble fraction of *Streptococcus faecalis* and traced its origin to the bacterial cell wall. The existence of a high proportion of cell wall alanine and glutamic acid as the D-isomers was thus demonstrated by Ikawa and Snell [55] and became another of the dis-tinctive features of bacterial wall chemistry. This filled in yet another piece of the puzzle, providing a link between the uridine nucleotides characterized by Park [36], the action of penicillin [37] and cell wall structure [45].

The variety of D-amino acids in cell walls was further extended to aspartic acid by Toennies et al. [56] when they found that this amino acid occurred partly as the D-isomer in the cell wall of *Streptococcus faecalis*. Ikawa and Snell [57] subsequently found that D-aspartic acid accounted for about 75% of the total aspartic acid content of lactic acid bac-teria. Thus bacterial cell walls became recognized as a prime source of D-amino acids and Armstrong et al. [58] found that their then recently discovered teichoic acids were also a source of ester-linked D-alanine.

In addition to the elucidation of the isomeric forms of alanine, glutamic and aspartic acids in walls, Hoare and Work [59] established that α,ε-diaminopimelic acid can occur in bacteria as the LL-, meso(DL)-, or DD- isomers and occasionally as the LL- and meso isomers together. Indeed, the isomeric forms of diaminopimelic acid have been found to be of taxonomic significance with certain isomers characteristic of particular groups or sub-groups of bacteria (e.g. cell walls of actinomycetes, studied by Lechevalier and Lechevalier [60], and other species by Schleifer and Kandler [61]. Thus, diaminopimelic acid and its isomeric forms have proved to be valuable cell wall markers in bacteria, es-pecially as the detection of this amino acid in whole-cell hydrolysates is in close agree-ment with its occurrence in isolated walls as shown by Dewey and Work [62].

Although it is recognized that D-isomers of amino acids have been found in a variety of bacterial products (e.g. *B. anthracis* capsular substance, antibiotics, peptidoglycans, peptidoglycolipids), the emphasis here has been on the discoveries relevant to some of the major amino acid constituents of peptidoglycans. In this regard, it is of interest to note that apart from diaminopimelic acid, the other major peptidoglycan diamino acid, lysine, has only been found as the L-isomer.

3.5. Teichoic and teichuronic acids

As pointed out by Salton [29], it is a curious fact that many of the 'unusual' compounds now known to be integral parts of the cell wall structures were first found in cellular fractions from bacteria rather than by direct examination of isolated walls. The discovery of the teichoic acids resulted from the identification of ribitol in a cytidine nucleotide isolated and characterized by Baddiley and Mathias [63] and Baddiley et al. [64]. The isolation of CDP-ribitol suggested that it played a role in the biosynthesis of a bacterial polymer and led Armstrong et al. [65] to the isolation and characterization of ribitol teichoic acid, a new, unusual ribitol phosphate polymer of the bacterial cell wall of *Lactobacillus arabinosus* and other bacteria. Further studies of the presence of polyolphosphate polymers in bacteria established that teichoic acids of the glycerol type may also occur in the walls of some bacteria with the glycerol phosphate polymer possessing ester-linked alanine and usually a sugar moiety as reported by Baddiley [66], and Baddiley and Davison [67]. These ribitol and glycerol teichoic acids were distinct from the subsequently discovered lipoteichoic acids containing glycerol phosphate, localized in the cytoplasmic membranes of Gram-positive bacteria. The teichoic acids and the lipoteichoic acids appear to be more characteristic of the walls and membranes respectively of Gram-positive bacteria than of Gram-negative organisms. Moreover, their covalent attachment to the cell wall peptidoglycan does not appear to play a role in the maintenance of the shape or rigidity of the cell wall since their extraction with TCA leaves an essentially intact residual wall structure [68]. Being highly negatively charged polymers, it has been proposed by Heptinstall et al. [69] that teichoic acids play an important role in divalent cation sequestration in the bacterial surface, as well as acting as an inhibitor of the cell's autolytic enzymes [70], perhaps thereby performing a control function of the process of autolysis. In addition to these cell surface functions mediating cellular relationships between external and internal environments and ensuring the cell's integrity from autolysis, the teichoic acids were found to be important antigens and bacteriophage receptors by Knox and Wicken [71] and Archibald and Coapes [72], respectively.

Teichuronic acid, a new class of wall polymer, was first isolated from *Bacillus licheniformis* by Janczura et al. [73] and shown by these investigators to consist of equimolar proportions of *N*-acetylgalactosamine and D-glucuronic acid. Following the discovery of teichuronic acid in *B. lichenformis*, polysaccharides with repeating acidic groups have been found in the walls of other Gram-positive bacteria (see [74]). Some organisms such as *M. lysodeikticus* (*luteus*) are completely devoid of wall teichoic acid but have instead a teichuronic acid composed of D-glucose and *N*-acetyl-D-mannosaminuronic acid as discovered by Perkins [75], Hase and Matsushima [76] and Rohr et al. [77]. Although teichoic acid appears to be absent altogether in some Gram-positive cocci, particularly *Micrococcus* species [74], Ellwood and Tempest [78] made the interesting discovery that growth of *Bacillus subtilis* under conditions of phosphate limitation resulted in replacement of wall teichoic acid by a teichuronic acid polymer (i.e. the same polymer as found earlier in *B. licheniformis*). Thus certain bacteria possessed the ability to respond to environmental (media) conditions by altering the nature of their wall polymers. In addition to involvement in regulation of the divalent cation economy of the cell, the teichuronic acid also appeared to have a function in cell separation and modulation of autolysin activity [74].

Although much attention has been focussed on the extensive studies, discovery and characterization of structure, function, antigenic and phage receptor properties of the teichoic and teichuronic acids of bacterial walls, it is well to remember that other wall associated polymers including neutral polysaccharides and proteins have been investigated in Gram-positive bacteria, and for more extensive discussions the reader is referred to various chapters in the monograph by Rogers et al. [74]. All of these early studies have attested to the chemical complexity of the bacterial cell wall with all its associated components, but with a robust, generally thick basal structure of the peptidoglycan polymer in the majority of Gram-positive organisms. The added complexity of the Gram-negative cell envelope with its lipopolysaccharides, porin/matrix proteins, phospholipids, lipoproteins and peptidoglycan sacculus is discussed briefly below.

4. The Gram-negative cell envelope

Ever since the introduction of Christian Gram's staining procedure in 1884 [79], the bacterial world has been divided into the two broad groups of Gram-positive and Gram-negative organisms. Correlations with this division of Eubacteria have been numerous and have reinforced the fundamental differences between the two groups separated by their response to the Gram stain. Bartholomew and Mittwer [80] have documented many of these key differences in responses to dyes such as crystal violet, growth in the presence of tellurite and azide, the presence of O-somatic antigens in Gram-negative bacteria, to name just a few characteristics. As indicated earlier in this historical survey of the field, the development of suitable thin-sectioning techniques for electron microscopy was a pivotal event in beginning to resolve and define the differences in the surface structures of the organisms from the two groups. In short, the majority of Gram-positive organisms were seen to be bounded by a relatively thick, amorphous cell wall, in contrast to the Gram-negative bacteria with their outer membrane external to an electron transparent zone (periplasm?) between the outer and inner cytoplasmic membranes as revealed by Ryter and Kellenberger [23]. Ultimately, the thin peptidoglycan layer of the Gram-negative envelope was revealed by Murray et al. [25] as a continuous, intervening layer by suitable staining procedures. Thus the complexity of the Gram-negative cell envelope was resolved into its outer membrane anchored to a peptidoglycan layer and underlying cytoplasmic membrane, thereby leading to the development of suitable methods for separating and isolating the outer and inner membranes of Gram-negative bacteria first initiated by Birdsell and Cota-Robles [81] and further developed by Osborne and Munson [14] and Schnaitman [82]. Electron microscopy played a further role in visualizing the thin peptidoglycan sacculus [26] and the difference in appearance of inner and outer membranes observed by Pollock et al. [83]. The physical separation of the constituent layers of the

Gram-negative envelope then permitted a meaningful chemical and biochemical characterization of the individual components [14].

4.1. Lipopolysaccharides

Immunologists have long recognized the O-somatic antigens as important surface antigens
of Gram-negative bacteria. Early investigations suggested that the O antigen of *Shigella
shigae* for example, was a 'complex composed of polysaccharide, phospholipin and con-
jugated protein residues [9]'. The specific polysaccharide 'haptens' were concluded to be
major components of the O antigens of many Gram-negative bacteria and much effort was
made to solubilize the complete antigenic complex [9]. The outstanding work of Westphal
and his colleagues provided the breakthrough which opened up the whole field of the
chemistry, immunochemistry and biology of the lipopolysaccharides of Gram-negative
bacteria, ultimately leading to our current extensive knowledge of the genetics and biosynthesis of these cell surface macromolecules. The hot-water-phenol (45% aqueous phenol) extraction method was developed by Westphal et al. [84] and permitted them to purify and characterize the lipopolysaccharides. Four decades of extensive chemical, genetic
and biochemical investigations have ensued and have led to an extraordinary body of
detail, a rich and voluminous literature and many valuable monographs devoted entirely
to lipopolysaccharides. Research on the chemistry and biology of lipid A is today still an
extremely active area of work. The depth of understanding the nature, biosynthesis and
biology of these envelope components is perhaps only rivalled by our knowledge of the
wall peptidoglycans. It was soon recognized after the isolation of the lipopolysaccharides
that they were of course an integral part of the cell envelopes and we now know that they
are anchored in the outer leaflet of the outer membranes, thus exposing the great variety
of hydrophilic polysaccharide chains on the surface of the Gram-negative cells and accounting for the multitude of antigenic determinants in these organisms.

4.2. Outer membrane proteins

With the development of suitable methods for the separation of the inner and outer membranes of Gram-negative bacteria, many studies have been focussed on the characterization of the outer membrane proteins. Although both structures are relatively rich in proteins, the outer membrane is characterized by possessing a fairly small number of major
proteins, four or five dominant proteins in some species. By contrast, the inner cytoplasmic membrane, perhaps not surprisingly, possesses a large variety of protein bands on
SDS-polyacrylamide gels, a feature shared with the cytoplasmic membranes of Gram-
positive bacteria [14,85]. The importance of the major outer membrane components (e.g.
matrix protein) became apparent when permeability studies were carried out with vesicles
prepared with fractions from the isolated outer membranes. Outer membrane proteins
were required along with phospholipid and lipopolysaccharide for active vesicle preparations and ultimately only one of the outer membrane proteins of *E. coli* was found to be
responsible for permeability properties of the vesicles. Because of its ability to form
pores, this membrane protein was designated as a 'porin' by Nakae and Nikaido [86] and
Nakae [87]. On the other hand, three outer membrane proteins were required for activity

in the *Salmonella typhimurium* vesicle system [87]. Thus, certain major outer membrane proteins became classified as 'porins,' with important permeability barrier properties for the outer membrane structure. Nikaido [28] and his colleagues have extensively investigated the properties of the porins and defined the molecular size exclusion limits for a variety of hydrophilic and hydrophobic substances. The functions of some of the other major outer membrane proteins not involved in transport or metabolite uptake have yet to be established.

4.3. Lipoprotein

One of the unusual components of the Gram-negative cell envelope of *E. coli* has been the lipoprotein discovered by Braun and Rehn [88] linked to the peptidoglycan structure. This lipoprotein was later found by Inouye [89] and Inouye et al. [90] to be in free form in the outer membrane structures. Braun [27] established some of the unusual features of this 58 amino acid residue protein including a linkage between the lysine residue through an amide bond to the ε-amino group of diaminopimelic acid in the peptidoglycan structure, in addition to a diglyceride linked as a thioester to cysteine. All of the evidence supported the conclusion that this covalently linked structure helped the maintenance of the outer membrane thereby aiding its barrier functions. Although the lipoprotein linked to peptidoglycan in *E. coli* was also found in other Gram-negative bacteria, it was not detected by Braun et al. [91] in *Proteus* and Pseudomonas species examined. The discovery of this unusual covalently linked lipoprotein in certain Gram-negative bacterial envelopes added yet another novel chemical entity to the complexity of these structures, but one which clearly has an important role in the stability and function of the envelope barrier.

In summary then, the Gram-negative cell envelope such as that of *E. coli* can be visualized as a complex structure of an asymmetric outer membrane possessing lipid A-anchored lipopolysaccharides in an outer leaflet of a bilayer membrane containing transmembrane porin-protein channels, stabilized by the covalent linkage of a lipoprotein with thioester-bonded diglyceride at N-terminal cysteine residues and a C-terminal lysine amide-linked to the ε-amino group of α,ε-diaminopimelic acid in the peptidoglycan. Although similar lipoproteins may exist in other Gram-negative envelopes, the molecular arrangement in those lacking this unusual component have yet to be elucidated. It will not be surprising if the level of complexity of the Gram-negative cell envelope increases with new molecular discoveries of the future. At least for the time being, we have a fairly clear impression of the principal elements of its construction and many specific details of the functional entities of this fascinating and complex structure.

5. Comparative biochemistry of cell envelopes of Gram-positive and Gram-negative bacteria and the Gram stain

The relationship between the Gram stain procedure and the structural, physical, chemical and biochemical differences between the two broad groups of bacteria has intrigued and fascinated microbiologists ever since the introduction of this technique by Christian Gram in 1884. As Bartholomew and Mittwer [80] pointed out in their excellent review, Gram recognized the difference in response of certain types of bacteria but 'he did not divide bacteria into the now well-known gram negative and gram positive types.' They did, however, suggest a plausible reason in that, 'his failure to recognize the taxonomic values of his stain was probably due to uncertainty resulting from his observation that while most of his pneumococci retained the gentian violet, some strains of pneumonia-producing bacteria were decolorized.' Bartholomew and Mittwer [80] concluded some 40 years ago that, 'Today Gram's staining procedure is generally recognized as a fundamental contribution to biological science. In bacteriology it is the first and a very valuable step in diagnosis and classification.' And this is certainly just as true today. With the developments of wall, envelope and membrane structure, chemistry and function, much is now known and many of the puzzles and uncertainties are readily explainable in clear molecular terms. Thus, for example, the 'nicking' of the cell wall peptidoglycan by an autolytic enzyme can break or damage the continuity of the wall of a Gram-positive organism and thereby render it Gram-negative, often as a 'ghost'.

Perhaps of greater importance for microbial biochemistry and physiology is that we now have a wealth of information of the major differences between Gram-positive and Gram-negative walls, envelopes and membranes and many of these, together with other parameters are summarized in Table I. As in any comparison it is realized that all is not 'black and white' and that exceptions exist and are bound to exist. Thus, 'presence' and 'absence' should be taken to mean for the majority of groups or strains. More often than not, a difference may become a gradation or spectrum. However, some differences are more 'absolute' than others and with one reported exception, that of *Listeria monocytogenes* wall studied by Wexler and Oppenheim [92], lipopolysaccharides appear to be uniquely confined to Gram-negative bacteria. Again with the teichoic acids, they appear to be restricted to certain groups of Gram-positive bacteria with one exception, that of *Butyrivibrio fibrisolvens*. It stains Gram-negative but its wall or envelope profile is of the Gram-positive type, although thinner than most as shown by Cheng and Costerton [93]. The lipoprotein covalently attached to the peptidoglycan layer [27] is present in some, but not all Gram-negative organisms, and it does not appear to have a counterpart in Gram-positive organisms so far studied.

Another interesting distinction between the envelopes of Gram-positive and Gram-negative organisms is the possession of two types of membrane (cytoplasmic) amphiphiles in the former group [70,71,94]. These may be of the lipoteichoic acid type, glycerophosphate polymers, or of the lipomannan variety [95,96] either succinylated [97] or without succinyl residues [98] as found in species of the *Micrococcus* genus. Both the lipoteichoic acids and the lipomannans terminate in diglyceride residues which provide a lipid membrane anchor thereby differing from the ribitol or glycerol teichoic acids covalently linked to wall peptidoglycan. Such negatively charged polymers could function in the sequestration of divalent cations and perhaps also basic metabolites. Evidence suggests that lipoteichoic acid chains can be expressed as cell surface antigens [71] and hence must be able to 'penetrate' through the peptidoglycan layer. Although Gram-negative bacteria appear to be devoid of the Gram-positive lipoteichoic acids and lipomannans, the Enterobacterial 'Common Antigen' amphiphile may be an analogous structure possessing some similar biological activities (see [94]). Some of the biological properties of these membrane amphiphiles are briefly noted in a subsequent section below.

14

TABLE I

Some characteristic differences in properties of Gram-positive and Gram- negative Eubacteria

	Gram-positive	Gram-negative
Peptidoglycan	Present	Present
	Thick	Thin
–% (wt.) of envelope/wall	ca. 40–95%	10–20%
	Generally	
– Diaminoacid	Lysine or DAP	DAP
– Teichoic acids	Present in many	Absent
(ribitol and/or glycerol)		
– Teichuronic acids	Present in some	Absent
– Lipopolysaccharides	Absent[a]	Present
– Lipoprotein	Absent	Present[a]
Outer membrane	Absent	Present
'S Layer'[b]	Present in some	Present in some
Membrane (cytoplasmic)		
Lipoteichoic acids	Present	Absent[a]
or lipomannans		
Susceptibility to antibacterial agents		
– Basic dyes	G+ > G–	
– Anionic surfactants	G+ > G–	
– Cationic surfactants	G+ > G–	
– Actinomycin D	G+ > G–	
– β-Lactams	G+ > G–	
– Azides	G+ < G–	
– Tellurites	G+ < G–	
– BPIP[c]	G+ < G–	
– Antibody/complement lysis	Resistant	Susceptible
''Internal' osmotic pressure	20 atm	<10 atm
Isoelectric points	pH 2–3	pH 5–6

Compiled from [4,21,25–27,31,33,45,70,71,73,74,80,94,99].
[a]See exceptions mentioned in text.
[b]See Sletyr and Messner [31], Glauert and Thornley [33] for crystalline surface layers (S layers).
[c]BPIP, bactericidal permeability increasing protein [102,103].

The generally greater susceptibility of Gram-positive bacteria to the growth inhibitory and killing effects of various antibiotics and antibacterial agents has been another intriguing phenomenon. Apart from intrinsic differences in target structures, this phenomenon can at least be understood in part by the protective role played by the outer membrane barrier of the Gram-negative bacteria [28]. The dramatic change in sensitivity to a variety of agents including actinomycin D, crystal violet, detergents, novobiocin and rifampicin following release of outer membrane lipopolysaccharide by EDTA treatment further attested to the barrier properties of this external membrane in certain Gram-negative bacteria investigated by Lieve [99], Muschel and Gustafson [100] and Nikaido [28]. Indeed, as Nikaido [28] pointed out 'deep-rough' mutants of Salmonella exhibited increased sensitivity approaching that of Gram-positive bacteria on exposure to a number of these antibacterial agents. In addition, the compartmentalization of β-lactamases in the periplasmic space could also contribute to the destruction of β-lactam antibiotics in transit to their target sites. Thus the organization of the Gram-negative envelope can contribute much to the resistance or sensitivity levels in Gram-negative bacteria through the 'permeability' restrictions of the outer membrane and the retention of hydrolytic enzymes in the peripheral zones (periplasm) of the cells. In marked contrast, the peptidoglycan network of the Gram-positive wall is a more porous structure and has a much higher molecular exclusion limit of the order of 10^5 molecular weight as determined by Scherrer and Gerhardt [101]. Although the thicker, robust wall of the Gram-positive organism undoubtedly provides the cell with a fairly rigid, protective 'shell,' it does not have the molecular exclusion properties attributable to the outer membrane of the Gram-negative cell envelope.

The difference in architecture and chemical constitution between Gram-positive and Gram-negative organisms is further emphasized by the responses observed in the Pfeiffer's phenomenon, lysis induced by the antibody-complement system. Dubos [21] noted that 'Whereas practically all species of Gram-negative bacteria are susceptible to the bactericidal effect of immune serum, the Gram-positive species are only little if at all affected by it.' In the light of what we now know of the enhanced sensitivity of 'deep-rough' mutants to a variety of antibacterial agents, it is particularly interesting to recall that Dubos [21] drew attention to the extreme susceptibility of the 'R' variants of Gram-negative organisms to lysis by complement and the greater resistance of the 'S forms' possessing the O polysaccharide antigens. Yet another interesting additional difference has been the discovery by Elsbach and Weiss [102] of the macrophage 'bactericidal permeability increasing protein' (BPIP) targeted to Gram-negative organisms and not Gram-positive species. The cationic protein involved has been shown by Elsbach and Weiss [103] to have considerable potential as a bactericidal agent for Gram-negative sepsis.

Armed with all this knowledge of the differences in envelope structure, chemistry and biochemistry of Gram-positive and Gram-negative bacteria, what has it told us about the reaction to Gram's staining procedure? The search for a component or components conferring Gram positivity has gone on for almost the entire existence of the Gram stain. Almost every class of cellular macromolecule has been implicated at one stage or another and has included polysaccharides, nucleoproteins, proteins, lipids, either singly or in combinations and finally 'positic acid', the polyglycerophosphate polymer of the staphylococcal wall isolated by Mitchell and Moyle [104]. None of these components, not even the possession of wall peptidoglycan, could account for Gram positivity or the positive reactions observed with certain yeasts and fungi (eukaryotic microorganisms). The search for a single Gram-positive substance thus proved to be fruitless and led Salton [30] to the conclusion that the relative thickness of the cell wall, be it peptidoglycan or polysaccharide complexes (as in yeasts and fungi) was a key factor in determining the inextractability of the crystal violet–iodine complex (and hence its retention within the cell) during the differentiation step with alcohol (or other solvent system). Moreover, it was suggested that the high lipid content of the Gram-negative envelope and its solubility in the differentiating solvent may also be a contributory factor to the negative reaction of these organisms [29]. Using more sophisticated methods of electron microscopy and an electron-dense analogue to replace I_2, Beveridge and his colleagues [105,106] were able to confirm the earlier conclusion by Salton [30] that the solvent-differentiating step resulted in the entrapment of the Gram complex within the cells of Gram-positive bacteria and its extraction, together with lipid, from the Gram-negative organisms. The conclusions explaining the nature of the Gram stain reaction are thus entirely compatible with what has been deduced, over the years, of the architecture, physical and chemical properties of the envelopes of the two groups of organisms investigated by Beveridge and Davies [106] and the necessity for an intact wall, unbreached by autolytic enzyme attack, in Gram-positive organisms. Such an explanation also accommodates the Gram stain responses of the Archaebacteria with their entirely different envelope components, including the 'pseudomurein' polymer replacing the peptidoglycan (murein) of Gram-positive species, together with the other 'unusual' surface structures described by Kandler and König [34]. The cell envelope profiles of the Gram-negative archaebacteria differ from those of the eubacteria in that they possess neither a specific sacculus polymer nor an outer membrane structure but instead a structured array of protein or glycoprotein subunits [34], or a complex layered structure surrounded by a fibrillar protein sheath as in *Methanospirillum* examined by Sprott et al. [107]. These discoveries further emphasize the variety of chemical entities of the surface structures conferring either Gram-positivity or Gram-negativity on the bacterial cell. Thus the anatomy of the surface profiles as seen in electron micrographs of thin sections and the mechanical integrity of the cell wall or sacculus structure appear to correlate well with a Gram-positive response to Christian Gram's (1884) staining procedure.

6. Perspectives on the functions and biological properties of cell envelopes and some of their constituent polymers

There now seems little doubt that the cell wall or sacculus provides the bacterial cell with a fairly rigid structure conferring the characteristic rod or coccal shape and protection of the more fragile underlying cytoplasmic membrane. The protective action of the wall against osmotic lysis was clearly established when Weibull [108] isolated intact protoplasts of *Bacillus megaterium* by removing the wall with lysozyme in an osmotically stabilizing medium, an observation abundantly confirmed subsequently with other Gram-positive bacteria and with spheroplasts of Gram-negative organisms. Such a structure could therefore be envisioned as protecting the cell from 'osmotic explosion' on encountering an osmotically unsuitable environment. It is of course important to realize that structures other than peptidoglycan and pseudomurein can also contribute to the overall stability of the cell envelope, as for example, those of the halophilic bacteria [34].

In addition to providing the cell with mechanical integrity, the roles played by the wall envelope in molecular sieving and barrier functions (outer membranes of Gram-negative bacteria) have already been mentioned. The surface layers and their components provide the immediate contacts of the cell between its external and internal environments and they are the sites of an array of surface antigens, bacteriophage and metabolite receptors and they play an important role in the interactions, economy and responses of the bacterial cell. Much of this forms part of our extensive knowledge of the biochemistry and physiology of microorganisms, so the ensuing remarks on the biological properties will thus be limited only to a selection of some of the surface envelope components. Envelopes are the 'carriers' of so many molecular entities of the bacterial surface with diverse functions, extensive enough in their array to preclude a review of their nature and properties in this introduction. Accordingly, some of the biological properties of the major components, the peptidoglycans, lipopolysaccharides and teichoic/lipoteichoic acids are listed in Table II and briefly discussed below.

The adjuvant properties residing in the surface of mycobacterial cells have long been recognized and Freund's adjuvant became as important to the serologist as the Gram stain did for the microbiologist. Through the pioneer work of Lederer [109] the chemistry of the biologically significant entities of the mycobacterial cell was elucidated and focussed attention on the peptidoglycolipids. This eventually led to the recognition of the dramatic adjuvant properties of cell wall muramyl peptides and the consequent synthesis of a variety of muramyl derivatives with biological potentials as adjuvants by Adam et al. [110] and Adam et al. [111]. In a somewhat analogous manner, another interesting biological property of somnogenic effects was traced by Krueger et al. [112] to muramyl compounds, apparently of bacterial cell wall origins. Somnogenic compounds of bacterial origin would obviously have to be small enough to cross the blood–brain barrier and it is

TABLE II

A selection of some of the biological properties of bacterial envelope-wall-membrane components or derived fractions

Component or fraction	Property/activity
Peptidoglycan	Antigenic
Peptidoglycolipid	Adjuvant
Muramyl peptides	Somnogenic
Teichoic acids	Antigenic; phage receptor; divalent cation/metabolite
Teichuronic acids	uptake; autolysis inhibition/regulation; antitumor
Lipopolysaccharide/lipid A	O-somatic antigen; endotoxin; lethal toxicity; pyrogenic; mitogenicity; cell binding; limulus lysate reaction; complement activation
Lipoteichoic acids	Immunogenic; mitogenic; cell adhesion; limulus lysate reaction; phage receptor; complement activation; autolysin inhibition/regulation
Lipoprotein	Cell binding; immunogenic; mitogenic
Lipomannan	Cell binding; immunogenic; mitogenic; limulus lysate reaction

From [70,71,94,110–114].

of interest to speculate that lysozyme capable of digesting the insoluble wall peptidogly-can polymer could have a role in the production of suitably sized substances with such properties. Both types of biological activity (adjuvant and somnogenic) thus open the way for synthetic chemists to construct new muramyl compounds for studies of structure-activ-ity relationships.

Lipopolysaccharides (LPS) of Gram-negative bacteria can also be classified as am-phiphiles and have for long attracted the attention of immunochemists and biochemists. It is therefore not surprising that a great deal has been learned about the biological proper-ties of these macromolecules as endotoxins, pyrogens and cell surface antigens. Although the antigenic specificity of the lipopolysaccharides resides in the O-polysaccharide chains, many of the biological properties of LPS have been traced to the 'lipid A' entity. Thus lipid A appears to be responsible for the lethal toxicity of endotoxins, pyrogenicity, adjuvant and antitumor activity, mitogenicity and other biological properties [94,113]. With the discovery that so many of the biological effects of LPS can be traced to the lipid A moiety, it is perhaps not surprising that much effort has been directed to the elucidation of its structure and the synthesis of analogues by Lüderitz et al. [113] and Kusumoto et al. [114]. It was of interest to note that although some of the synthetic compounds exhibited endotoxic activities, they were lower than those of the natural lipid A and were not of identical structure. In the light of its potent biological activities, there is certainty that the chemistry of lipid A and its analogues will be vigorously pursued in the future.

The possible functional role of teichoic and teichuronic acids in sequestering divalent cations and cationic metabolites has already been addressed above and in addition the structure of the pneumococcal teichoic acid has been shown to have important implica-tions for autolysis regulation by Höltje and Tomasz [115]. In addition to the wall teichoic acids, an intracellular form of the glycerolphosphate polymer was discovered by Kelemen and Baddiley [116], and ultimately identified as a membrane component by Hay et al. [117] and designated as a lipoteichoic acid. As Shockman [70] has pointed out, it is sometimes a problem to distinguish between wall teichoic acid of the glycerol type and the membrane lipoteichoic acid, also a glycerophosphate polyol polymer. The close rela-tionships of wall teichoic acids with the lipoteichoic acids of bacterial membranes may suggest further functions for these cell wall polymers. The recognition of the existence and structure of the lipoteichoic acids and lipomannans have stimulated much interest in the biological properties of these macromolecules, a selection of which has been included in Table II.

Apart from lethal toxicity and pyrogenicity, the lipoteichoic acids appear to share all of the other biological properties exhibited by lipopolysaccharides [94]. As a group, the membrane amphiphiles including LPS, lipoprotein of certain Gram-negative species, Enterobacterial 'common antigen' and lipomannan appear to be involved in binding or adhesion to eukaryotic cells and could thus play a role in host cell-bacterial cell interac-tions and pathogenesis (see [94]). In addition to so many of the shared interactions, the classical reaction of LPS in the Limulus lysate assay is also exhibited by lipoteichoic ac-ids and lipomannans of Micrococci. The array of similar biological properties exhibited by all of these bacterial membrane amphiphiles is therefore most striking [94].

Finally in this brief review of the discovery of some of the biological properties of membrane amphiphiles, the ability of lipoteichoic acids to inhibit endogenous autolysins should be mentioned. This important area of amphiphile 'function' has been critically addressed by Shockman [70]. He concluded that 'currently we can only state that both LTA (lipoteichoic acid) and certain lipids (e.g. bovine cardiolipin) can effectively inhibit the action of the autolytic muramidase of *S. faecium* in at least some test systems [70].' As pointed out by Shockman [70], the intimate biosynthetic relationship between LTA and certain lipids is a complicating factor in dissecting the physiological role of this amphiphile. Thus, the in vivo function of LTA in controlling and regulating autolysis will require further exploration. At least, the in vitro inhibitory effects are impressive.

Although the final analysis of the physiological roles of membrane amphiphiles is not yet at hand, the information to date on their in vitro properties suggests important and plausible functions for the bacterial cells possessing these rather unique macromolecules.

7. Concluding remarks – the present and the future

Instead of covering familiar ground of our current knowledge so expertly presented herein, this introductory chapter has attempted to provide a historical background to all, or at least many of the exciting discoveries that have preceded and set the stage for THE PRESENT (this monograph) and THE FUTURE (more to come!). To have been involved in the early development of a field of scientific inquiry such as this has been fascinating and an intellectually rewarding experience for many of us and we can but marvel at the tremendous advances and sophistication of THE PRESENT knowledge and inherent predictions for THE FUTURE presented in this book. It literally all started with the first observations of rod, coccal and spiral shapes of bacteria and not being able to see what held them together (Leeuwenhoek, 1675). This book tells us how and why we know.

As Pirie [1] so appropriately said, it was 'no longer necessary to apologize for the intrusion of the biochemists into what used to be a purely bacteriological field,' so no apology is needed for the many omissions obvious in this introduction. For a field which has encompassed the primary structures of peptidoglycans, their variety of diamino acid cross-bridging molecules, the biosynthetic lipid intermediates, the complex assembly processes, the adhesion sites, the heterogeneity and distinctions between wall and membrane teichoic acids, omissions are inevitable and much has to be left to the present and the future. The truly remarkable facts that have emerged are that the surface envelopes of prokaryotic cells differ so much structurally and biochemically from those of eukaryotic cells, thereby permitting selective antibiotic action – a piece of 'good fortune' for human beings. What is rarely realized is that we are more prokaryotic (about 10^{15} bacteria/human) than eukaryotic (circa 10^{13} cells/68 kg human being) in terms of total cell numbers (not to mention the surface area differences)! So we have come a long, long way from the early observations of Leeuwenhoek in 1675 to the primordial era of cell walls and their chemical and biochemical nature of the 1950s, and to the realization that we carry a large population of prokaryotic cells and may be vulnerable to the effects of some. The knowledge we have gained has not only given us a deeper understanding of the nature of the microbial world around us but also the way in which the differences between 'them and us' can often be manipulated to our advantage.

20

Acknowledgements

I deeply appreciate and thank the Editors, Dr. Jean-Marie Ghuysen and Dr. Regine Hakenbeck, for their enthusiasm and encouragement to contribute to this monograph, and I congratulate them for the superb task they have achieved. I wish to thank Wendy Coucill for her excellent typing of this contribution.

References

1. Pirie, N.W. (1949) in: A.A. Miles and N.W. Pirie (Eds.), The Nature of the Bacterial Surface, Blackwell, Oxford, pp. 1–8.
2. Epstein, L.A. and Chain, E. (1940) Br. J. Exp. Pathol. 21, 339–355.
3. Salton, M.R.J. (1952) Nature 170, 746–747.
4. Ghuysen, J.M. (1968) Bacteriol. Rev. 32, 425–464.
5. Dawson, I.M. (1949) in: A.A. Miles and N.W. Pirie (Eds.), The Nature of the Bacterial Surface, Blackwell, Oxford, pp. 119–121.
6. Salton, M.R.J. and Horne, R.W. (1951) Biochim. Biophys. Acta 7, 177–197.
7. Salton, M.R.J. (1952) Biochim. Biophys. Acta 8, 510–519.
8. Heidelberger, M., Kendall, F.E. and Scherp, H.W. (1936) J. Exp. Med. 64, 559–572.
9. Morgan, W.T.J. (1949) in: A.A. Miles and N.W. Pirie (Eds.), The Nature of the Bacterial Surface, Blackwell, Oxford, pp. 9–28.
10. Weibull, C. (1948) Biochim. Biophys. Acta 2, 351.
11. Weidel, W. (1951) Z. Naturforsch. 6b, 251.
12. Weibull, C. (1953) J. Bacteriol. 66, 696–702.
13. Salton, M.R.J. and Freer, J.H. (1965) Biochim. Biophys. Acta 107, 531–538.
14. Osborne, J.J. and Munson, R. (1974) Methods Enzymol. 31, 642–653.
15. Salton, M.R.J. (1971) CRC Crit. Rev. Microbiol. 1, 161–197.
16. Dobell, C. (1958) in: C. Dobell (Ed.), Antony van Leeuwenhoek and His 'Little Animals', Plate XXIV, Russell & Russell, New York, pp. 109–255.
17. Mudd, S., Polevitsky, K., Anderson, T.F. and Chambers, L.A. (1941) J. Bacteriol. 42, 251.
18. Vincenzi, L. (1887) Z. Physiol. Chem. 11, 181.
19. Fischer, A. (1894) Jahvb. Wissen. Bot. 27, 1–163.
20. Knaysi, G. (1950) Elements of Bacterial Cytology, 2nd edition, Comstock, Ithaca, NY, pp. 114–126.
21. Dubos, R.J. (1949) The Bacterial Cell, Harvard University Press, Cambridge, MA, pp. 36–37, 252–254.
22. Salton, M.R.J. and Williams, R.C. (1954) Biochim. Biophys. Acta 14, 455–458.
23. Ryter, A. and Kellenberger, E. (1958) J. Biophys. Biochem. Cytol. 4, 671–678.
24. Salton, M.R.J. (1953) Biochim. Biophys. Acta 10, 512–523.
25. Murray, R.G.E., Steed, P. and Elson, H.E. (1965) Can. J. Microbiol. 11, 547–560.
26. Weidel, W., Frank, H. and Martin, H.H. (1960) J. Gen. Microbiol. 22, 158–166.
27. Braun, V. (1975) Biochim. Biophys. Acta 415, 335–377.
28. Nikaido, H. (1979) in: M. Inouye (Ed.), Bacterial Outer Membranes, Wiley, New York, pp. 361–407.
29. Salton, M.R.J. (1964) The Bacterial Cell Wall, Elsevier, Amsterdam, pp. 1–293.
30. Salton, M.R.J. (1963) J. Gen. Microbiol. 30, 223–235.
31. Sletyr, U.B. and Messner, P. (1983) Annu. Rev. Microbiol. 37, 11–39.
32. Lancy, P. and Murray, R.G.E. (1978) Can. J. Microbiol. 24, 162–176.
33. Glauert, A.M. and Thornley, M.J. (1969) Annu. Rev. Microbiol. 23, 159–198.
34. Kandler, O. and König, H. (1985) in: C.R. Woese and R.S. Wolfe (Eds.), The Bacteria, Vol. 8, Archaebacteria, Academic Press, New York, pp. 413–457.
35. Park, J.T. and Johnson, M.J. (1949) J. Biol. Chem. 179, 585–592.
36. Park, J.T. (1952) J. Biol. Chem. 194, 877–884.
37. Park, J.T. and Strominger, J.L. (1957) Science 125, 99–101.

38. Strominger, J.L. (1970) The Harvey Lectures (69) Ser. 64, 179–213.
39. Work, E. (1949) Biochim. Biophys. Acta 3, 400–411.
40. Work, E. (1951) Biochem. J. 49, 17–23.
41. Holdsworth, E.S. (1952) Biochim. Biophys. Acta 9, 19–28.
42. Cummins, C.S. and Harris, H. (1954) Biochem. J. 57, 32P.
43. Meadow, P. and Work, E. (1959) Biochem. J. 72, 400–407.
44. Gilvarg, C. (1958) J. Biol. Chem. 233, 1501.
45. Work, E. (1957) Nature 179, 841–847.
46. Strange, R.E. and Powell, J.F. (1954) Biochem. J. 58, 80–85.
47. Strange, R.E. and Dark, F.A. (1956) Nature 177, 186–188.
48. Strange, R.E. (1956) Biochem. J. 64, 23P.
49. Strange, R.E. and Kent, L.H. (1959) Biochem. J. 71, 333.
50. Zilliken, F. (1959) Fed. Proc. 18, 966.
51. Lambert, R. and Zilliken, F. (1960) Chem. Ber. 93, 187.
52. Salton, M.R.J. (1956) Biochim. Biophys. Acta 22, 495–506.
53. Stevens, C., Halpern, P.E. and Gigger, R.P. (1951) J. Biol. Chem. 190, 705–710.
54. Snell, E.E., Radin, N.A. and Ikawa, M. (1955) J. Biol. Chem. 217, 803.
55. Ikawa, M. and Snell, E.E. (1956) Biochim. Biophys. Acta 19, 576–578.
56. Toennies, G., Bakay, B. and Shockman, G.D. (1959) J. Biol. Chem. 234, 1376.
57. Ikawa, M. and Snell, E.E. (1960) J. Biol. Chem. 235, 1376–1382.
58. Armstrong, J.J., Baddiley, J. and Buchanan, J.G. (1960) Biochem. J. 76, 610–621.
59. Hoare, D.S. and Work, E. (1957) Biochem. J. 65, 441–447.
60. Lechevalier, M.P. and Lechevalier, H.A. (1980) in: A. Dietz and D.W. Thayer (Eds.), Actinomycete Taxonomy, Special Publication 6, Soc. for Industrial Microbiol., Arlington, VA, pp. 227–291.
61. Schleifer, K.H. and Kandler, O. (1972) Bacteriol. Rev. 36, 407–477.
62. Work, E. and Dewey, D.L. (1953) J. Gen. Microbiol. 9, 394–406.
63. Baddiley, J. and Mathias, A.P. (1954) J. Chem. Soc. 2723–2731.
64. Baddiley, J., Buchanan, J.G., Carrs, B., Mathias, A.P. and Sanderson, A.R. (1956) Biochem. J. 64, 599–603.
65. Armstrong, J.J., Baddiley, J., Buchanan, J.G., Carrs, B. and Greenberg, G.R. (1958) J. Chem. Soc. 4344–4355.
66. Baddiley, J. (1961) in: M. Heidelberger, O.J. Plescia and R.A. Day (Eds.), Immunochemical Approaches to Problems in Microbiology, Rutgers University Press, NJ, pp. 91–99.
67. Baddiley, J. and Davison, A.L. (1961) J. Gen. Microbiol. 24, 295–299.
68. Archibald, A.R., Armstrong, J.J., Baddiley, J. and Hay, J.B. (1961) Nature 191, 570–572.
69. Heptinstall, S., Archibald, A.R. and Baddiley, J. (1970) Nature 225, 519–521.
70. Shockman, G.D. (1981) in: G.D. Shockman and A.J. Wicken (Eds.), Chemistry and Biological Activities of Bacterial Surface Amphiphiles, Academic Press, New York, pp. 21–40.
71. Knox, K.W. and Wicken, A.J. (1973) Bacteriol. Rev. 37, 215–257.
72. Archibald, A.R. and Coapes, H.E. (1972) J. Gen. Microbiol. 73, 581–585.
73. Janczura, E., Perkins, H.R. and Rogers, H.J. (1961) Biochem. J. 80, 82–93.
74. Rogers, H.J., Perkins, H.R. and Ward, J.B. (1980) Microbial Cell Walls and Membranes, Chapman and Hall, London, pp. 1–543.
75. Perkins, H.R. (1963) Biochem. J. 86, 475–483.
76. Hase, S. and Matsushima, Y. (1970) J. Biochem. 68, 723–728.
77. Rohr, T.E., Levy, G.N., Stark, N.J. and Anderson, J.S. (1977) J. Biol. Chem. 252, 3460–3465.
78. Ellwood, D.C. and Tempest, D.W. (1969) Biochem. J. 111, 1–5.
79. Gram, C. (1884) Fortschr. Med. 2, 185–189.
80. Bartholomew, J.W. and Mittwer, T. (1952) Bacteriol. Rev. 16, 1–29.
81. Birdsell, D.C. and Cota-Robles, E.H. (1967) J. Bacteriol. 93, 427–437.
82. Schaitman, C.A. (1970) J. Bacteriol. 104, 890–901.
83. Pollock, J.J., Nguyen-Disteche, M., Ghuysen, F.M., Coyette, J., Linder, R., Salton, M.R.J., Kim, K.S., Perkins, H.R. and Reynolds, P. (1974) Eur. J. Biochem. 41, 439–446.
84. Westphal, O., Lüderitz, O. and Bister, F. (1952) Z. Naturforsch. 76, 148–155.

85. Salton, M.R.J. (1976) Methods Membr. Biol. 6, 101–150.
86. Nakae, T. and Nikaido, H. (1975) J. Biol. Chem. 250, 7359–7365.
87. Nakae, T. (1976) J. Biol. Chem. 251, 2176–2178.
88. Braun, V. and Rehn, K. (1969) Eur. J. Biochem. 10, 426–438.
89. Inouye, M. (1971) J. Biol. Chem. 246, 4834–4838.
90. Inouye, M., Shaw, J. and Shen, C. (1972) J. Biol. Chem. 247, 8154–8159.
91. Braun, V., Rehn, K. and Wolff, H. (1970) Biochemistry 9, 5041–5049.
92. Wexler, H. and Oppenheim, J.D. (1979) Infect. Immun. 23, 845–857.
93. Cheng, K.J. and Costerton, J.W. (1977) J. Bacteriol. 129, 1506–1512.
94. Wicken, A.J. and Knox, K.W. (1981) in: G.D. Shockman and A.J. Wicken (Eds.), Chemistry and Biological Activities of Bacterial Surface Amphiphiles, Academic Press, New York, pp. 1–9.
95. Pless, D.D., Schmit, A.S. and Lennarz, W.J. (1975) J. Biol. Chem. 250, 1319–1327.
96. Powell, D.A., Duckworth, M. and Baddiley, J. (1975) Biochem. J. 151, 387–397.
97. Owen, P. and Salton, M.R.J. (1975) Biochem. Biophys. Res. Commun. 63, 875–880.
98. Lim, S. and Salton, M.R.J. (1985) FEMS Microbiol. Lett. 27, 287–291.
99. Leive, L. (1974) Ann. N.Y. Acad. Sci. 235, 109–129.
100. Muschel, L.H. and Gustafson, L. (1968) J. Bacteriol. 95, 2010–2013.
101. Scherrer, R. and Gerhardt, P. (1971) J. Bacteriol. 107, 718–735.
102. Elsbach, P. and Weiss, J. (1992) in: J.I. Gallin, I.M. Goldstein and R. Snyderman (Eds.), Inflammation, Basic Principles and Clinical Correlates, Raven Press, New York, pp. 603–636.
103. Elsbach, P. and Weiss, J. (1993) Curr. Opinion Immunol. (Innate Immunol.) 5, 103–107.
104. Mitchell, P. and Moyle, J. (1950) Nature 166, 218–220.
105. Davies, J.A., Anderson, G.K., Beveridge, T.J. and Clark, H.C. (1983) J. Bacteriol. 156, 837–845.
106. Beveridge, T.J. and Davies, J.A. (1983) J. Bacteriol. 156, 846–858.
107. Sprott, G.D., Colvin, J.R. and McKellar, R.C. (1979) Can. J. Microbiol. 25, 730–738.
108. Weibull, C. (1953) J. Bacteriol. 66, 688–695.
109. Lederer, E. (1971) Pure Appl. Chem. 25, 135–165.
110. Adam, A., Ciorbaru, R., Petit, J.F. and Lederer, E. (1972) Proc. Natl. Acad. Sci. USA 69, 851–854.
111. Adam, A., Devys, M., Souvannavong, V., Lefrancier, P., Choay, J. and Lederer, E. (1976) Biochem. Biophys. Res. Commun. 72, 339–346.
112. Krueger, J.M., Pappenheimer, J.R. and Karnovsky, M.L. (1982) Proc. Natl. Acad. Sci. USA 79, 6102–6106.
113. Lüderitz, O., Tanamoto, K.-I., Galanos, C., Westphal, O., Zahringer, H., Reitschal, E., Kusumoto, S. and Shiba, T. (1983) in: L. Anderson and F.M. Unger (Eds.), Bacterial Lipopolysaccharides, ACS Symposium Series 231, American Chemical Society, Washington, pp. 3–17.
114. Kusumoto, S., Inage, M., Chaki, H., Imoto, M., Shimamoto, T. and Shiba, T. (1983) in: L. Anderson and F.M. Unger (Eds.), Bacterial Lipopolysaccharides, ACS Symposium Series 231, American Chemical Society, Washington, pp. 237–254.
115. Höltje, J.-V. and Tomasz, A. (1975) Proc. Natl. Acad. Sci. USA 72, 1690–1694.
116. Kelemen, M.V. and Baddiley, J. (1961) Biochem. J. 80, 246–254.
117. Hay, J.B., Wicken, A.J. and Baddiley, J. (1963) Biochim. Biophys. Acta 71, 188–190.

J.-M Ghuysen and R. Hakenbeck (Eds.), *Bacterial Cell Wall*
© 1994 Elsevier Science B.V. All rights reserved

23

CHAPTER 2

Bacterial peptidoglycan: overview and evolving concepts

HARALD LABISCHINSKI and HEINRICH MAIDHOF

Robert Koch-Institute of the Federal Health Office, Nordufer 20, D-13353 Berlin 65, Germany

1. Introduction

The essential cell wall polymer of most eubacteria, peptidoglycan (synonym: murein), has attracted and fascinated scientists from many different disciplines since its distinct chemical composition and its unique role as an 'exoskeleton' were discovered more than 30 years ago (Chapter 1). Within a relatively short time, the basic chemical structure of this large, bag-like molecule (also called the sacculus, see Fig. 1) was elucidated, giving access to an increasingly refined knowledge of its spatial and morphological arrangement. Because the mode of action of the most important group of antibacterial drugs, the β-lactam antibiotics, and the mechanisms that the bacteria have developed to survive in the presence of these antibiotics, are related to the biosynthesis, three-dimensional structure and morphogenesis of the peptidoglycan, even more efforts have been undertaken to gain a detailed knowledge of the chemistry and structural features of the murein network. There are a number of reviews which give a vivid picture about the basic murein structure [see e.g. 1–8]. However, during the last 10 years, new insights into the complex dynamics, function and fine molecular structure of the peptidoglycan network have been reached and are still developing. The aim of this chapter is to discuss these recent, sometimes still controversial, and somewhat speculative new concepts. Whenever possible, *Escherichia coli* and *Staphylococcus aureus* will serve as 'representatives' for Gram-negative and Gram-positive bacteria, respectively.

2. Overall structure and architecture of peptidoglycan

2.1. Basic chemical structure

The peptidoglycan is a heteropolymer consisting of glycan strands crosslinked by peptides. Figure 2 depicts a simple scheme of the primary structure of the building block of the peptidoglycan. By using four chemical linkages (two at the reducing and non-reducing end groups of the disaccharide repeating unit, respectively, and two at the carboxyl terminal of the stem peptide and the amino-terminal of the diamino acid of the peptide moiety),

24

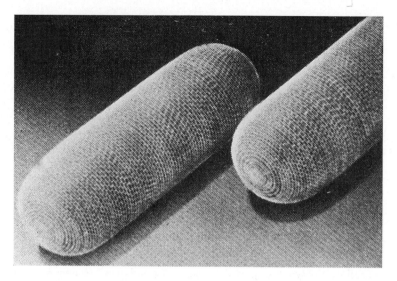

Fig. 1. Schematic representation of the structure of a *Escherichia coli* sacculus (from [9]). Two cells surrounded by a monolayer of peptidoglycan are shown. The drawing overestimates the actual state of order; the sugar chains, running parallel to the short axis of the sacculus, are much shorter than indicated. The short peptides (mainly dimers) which crosslink the sugar strands are oriented parallel to the long axis of the sacculus.

the giant network structure of the sacculus is formed (Fig. 1). In spite of this almost uniform construction principle, about 100 different chemical peptidoglycan types have been identified (for review, see [4,6,8]). They are characterized by a few modifications within the *N*-acetylglucosamyl-*N*-acetylmuramic acid sugar unit (e.g. acetylation of the muramyl-6-hydroxyl group, absence or modification of *N*-acetyl groups), and mostly by a large number of variations within the peptide moiety including the interpeptide bridges to crosslink peptides from different glycan strands in many Gram-positive bacteria. While the crosslinking in most bacteria extends from the omega amino group of the diamino acid in position 3 (group A peptidoglycan), group B bacteria use a diamino acid to crosslink the α-carboxyl group of the glutamic acid residue in position 2 of the stem peptide to the D-alanine residue in position 4 of another peptide. In all cases, peptide crosslinking by transpeptidation [9] involves removal of the D-alanine in position 5 (for more details see Chapters 3–6).

Although the chemical nature of the peptidoglycan building block allows infinitely long chains of the sugar and peptide moieties to be formed, the actual glycan chain lengths are quite low at least in the bacteria investigated as compared to the overall dimensions of the sacculus. In *Escherichia coli*, about 70% of the sugar chains have an average length of 9 disaccharide units, while the remaining material has an average length of 45 disaccharides, yielding a total average length of 21 disaccharide units [10]. Since the length of one disaccharide unit is approximately 1 nm, the majority of sugar chains being 5–10 units long, they span only 5–10 nm. Typically the *E. coli* rod-shaped sacculus is approximately 1500 nm long and 600 nm wide. Likewise, nearly 50% of the peptides are not involved in crosslinking, and the great majority of the remaining peptides just

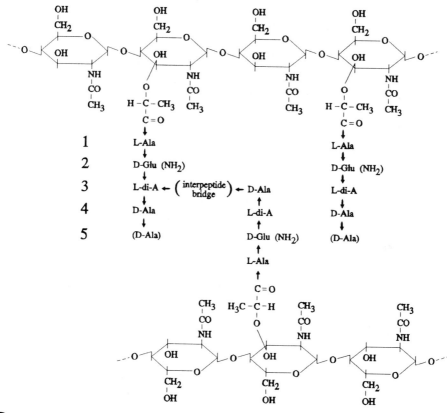

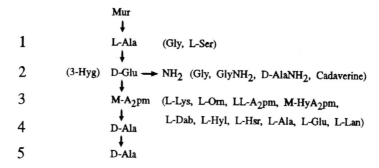

Fig. 2. (A) Primary structure of peptidoglycan. Two sugar strands crosslinked by a peptide-dimer are shown. diA, diamino acid. Substituents in brackets may be missing. (B) Variations of the primary structure of the muropeptide subunit. Amino acids in parentheses may replace the corresponding amino acids or substituents. Unusual abbreviations: A₂pm, 2,6-diaminopimelic acid; Dab, diaminobutyric acid; HyA₂pm, 2,6 -diamino-3-hydroxypimelic acid; Hyg, threo-β-hydroxyglutamic acid; L-Hyl, L-hydroxylysine; Hsr, homoserine; Lan, lanthionine. Adapted from [6].

forms dimers [11]. The peptide chain length is generally longer in Gram-positive bacteria, yet the most extensively crosslinked peptidoglycan, that of *Staphylococcus aureus*, has an average chain length of approximately 15 peptide units [1] and a maximal chain length of 30–40 peptide units [12,13].

Peptidoglycan hydrolases are present in all bacteria (Chapter 7). They modify the sugar and peptide chain length in the dynamic cell wall fabric, and, therefore, the afore-mentioned values reflect a kind of equilibrium between synthesis and modification of the sacculus. These chain lengths may differ substantially also in different topological parts of the sacculus (polar caps and cylindrical regions in cylindrically shaped bacteria, cross- and peripheral wall etc. [14,15]).

2.2. Conformational aspects

Because the murein polysaccharide, cellulose and chitin have similar chemical structures and perform similar functions, the first three-dimensional models proposed for the pepti-doglycan rested upon a flat and straight arrangement of the sugar chains, in that: (i) two consecutive sugar monomers were rotated against each other by 180°; (ii) the peptides protruded in the same direction with respect to the sugar chain axis; (iii) a paracrystalline, tight network was formed [16–18]. The use of physical chemistry methods such as X-ray diffraction, infrared spectroscopy and NMR studies [19–23], revealed instead a non-crys-talline type of order of the peptidoglycan network, and highlighted its enormous flexibil-ity and elasticity [24–27]. As suggested by molecular modelling studies [19,24–26], the bulky lactyl ether group at the muramic acid residue prevents formation of a chitin-like arrangement and forces the glycan strands to adopt a relatively regular helical conforma-tion (four- to five-fold helices with respect to the disaccharide units, Fig. 3), so that con-secutive peptides along a sugar chain point to different directions, roughly downwards, right, upwards and left [21,22,24,28].

The conformation of the peptide part is dictated by its alternating L,D sequence and the unusual γ-peptide bond between D-glutamic acid and the diamino acid residue. While its relaxed, stress-free conformation gives rise to a nearly ring-shaped arrangement in close contact to the sugar strand [24,25], the peptide can be drastically extended under stress by up to 400% in length [26,29]. In the situation experienced by the bacteria during normal growth, the available data point to a state intermediate between these two extremes and

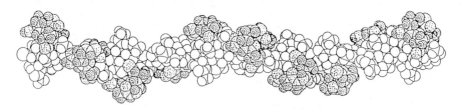

Fig. 3. Space-filling representation of the spatial arrangement of a peptidoglycan strand, constructed from ten disaccharide-pentapeptide units (actual length approx. 10 nm) The atoms of the sugar chain are drawn as white spheres, those belonging to the peptides are dotted. Consecutive peptides follow the helical glycan chain arrangement and are shown in their folded conformation (adapted from [24]).

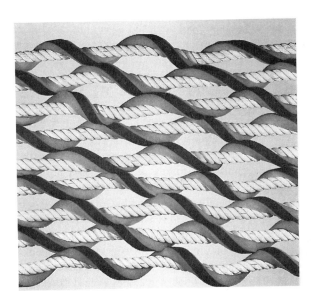

Fig. 4. Schematic representation of murein packing in a two-dimensional single layered architecture. The sugar chains of the peptidoglycan are replaced by more or less straight rods (light) around which the helical cross-linkage region (dark) is arranged. Crosslinking can occur only at those parts where the 'dark' regions of different peptidoglycan strands are in close contact. Therefore, well separated peptide dimers are characteristic for this packing type. Adapted from [30].

the bacteria can respond to changes in osmolarity by corresponding volume adaption [26,29].

The helical arrangement of the sugar chains and the resulting spatial arrangement of the peptide substituents drastically limit the attainable degree of crosslinking, since only those peptides which are close in space can be crosslinked to each other. In a single layered, two-dimensional network with approximately parallel arranged sugar strands as it probably occurs in the Gram-negative bacteria (see below), peptide dimers only are likely to be formed leading to a relatively low degree of crosslinking (approx. 25–30%) (see Fig. 4). In the Gram-positive walls, the multilayered arrangement of the peptidoglycan allows a higher degree of crosslinking to be achieved (up to 70%, depending on the packing type [28,30]). The extremely high degree of crosslinking observed in staphylococci (up to 90%) is possible only because their long and flexible pentaglycine interpeptide bridges are able to span distances between peptides otherwise much too far apart to be crosslinked. Hence, a simple physical conformational parameter, namely the helical structure of the sugar strands, is probably sufficient to explain the different degrees of crosslinking found amongst the bacterial species.

2.3. Remarks on cell wall morphology

In early electron micrographs of *E. coli* cells prepared by conventional embedding and thin-sectioning techniques, the cell envelope seemed to enclose a translucent compartment located between the outer and inner membranes. This was designated as the periplasmic space and contained a thin, 2–3 nm peptidoglycan layer (Fig. 5a, for details about outer membrane architecture and components, see Chapters 12–19). From these data, a monolayered arrangement of the Gram-negative murein was proposed. More recently, electron microscopy data using refined preparation methods as freeze-substitution and special staining techniques led to alternative models (Fig. 5b,c [30–33]). The periplasm now appears as a gel effectively spanning the whole space between the outer and inner membranes [30]. A peptidoglycan with a thickness of about 6–8 nm in exponentially growing cells and 10–11 nm in stationary phase cells was deduced, suggesting multilayered murein arrangements. The still controversial mono- versus multilayer murein architecture in Gram-negative organisms is discussed below using information derived from other experimental techniques. Electron microscopy data were also used for following cross wall formation and cell separation on a refined level in the division process [34,35].

Gram-positive walls appear as thick and relatively structureless organelles with no apparent localization of their main constituents, the murein and the teichoic acid [36]. However, electron microscopy of staphylococcal cell walls revealed a highly sophisticated architecture (Fig. 6a–c). As in all staphylococci, consecutive cell divisions are initiated at an angle of 90° in three dimensions, sometimes even before completion of the cell separation process of the first division plane (Fig. 6a). Cell division is achieved by the formation of a highly organized cross wall (Fig. 6b), which is initiated asymmetrically and eventually fuses in the center of the cell to form a complete cross wall. Based on distinct morphological features as revealed by electron microscopy, several autolytic systems have been suggested to be involved in cell division process. One of these, the splitting system, appears as a ring of periodically arranged tubules in the center of the cross wall [37–39]. Once the cross wall has been completed, cell separation is initiated by highly organized entities called murosomes which punch tiny holes into the peripheral wall along the division plane [40–42] (see Fig. 6b).

Teichoic acid-like material has been chemically associated with the splitting system [43–45]. During isolation and purification of the peptidoglycan, the morphological appearance of the splitting system remains detectable until the final purification step, i.e. removal of the teichoic acid. If the splitting system can no longer be detected, two separated pieces of cross wall are visible (Fig. 6c, own unpublished results).

3. Cell wall rigidity and elasticity: no contradiction

Osmotic pressure differences between the inside and the outside of the bacteria appear to be enormous. Gram-negative bacteria have to deal with a pressure of about 2–5 atm (about 200–500 kPa) while the cell wall of the Gram-positive bacteria has to withstand pressures up to 25 atm (approx. 2.5 MPa) [46–48]. Together with the observation that isolated sacculi of bacteria retain the shape of the intact microorganism and that bacteria devoid of an intact cell wall tend to form spheres, these data underline the rigidity and the shape defining properties of the peptidoglycan network. Its rigidity is attributable to the restricted flexibility of the sugar chains in which only very small rotations around the β1–4 linkage are permitted and abrupt bending of the chains is not possible. As a conse-

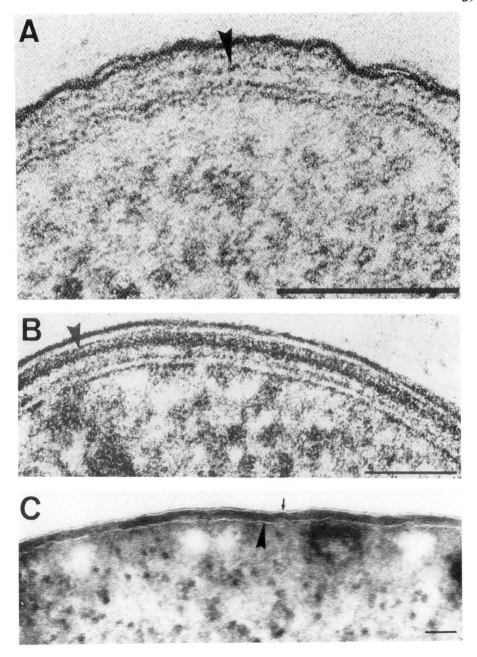

Fig. 5. Envelope profiles of *Escherichia coli*. (A) Prepared by conventional embedding (bar, 0.1 μm). The peptidoglycan appears as a central dark structure of 2–3 nm thickness (arrow head). From [32]. (B) Prepared by freeze-substitution (bar, 0.05 μm). The peptidoglycan forms part of the periplasmic gel (arrow head). From [32]. (C) Prepared by the modified Rambourg-method [112] (bar, 0.1 μm). The peptidoglycan forms part of an electron dense layer that fills up the space from the inner membrane (arrow head) to the outer membrane (small arrow). From [33].

quence, parallel arranged sugar chain arrays within a two-dimensional peptidoglycan plane do not undergo considerable changes in chain length upon stress. However, volume increase of the bacteria demonstrating elasticity of the murein can be induced by changes of the osmotic pressure. This can be explained by the flexibility of crosslinked peptides

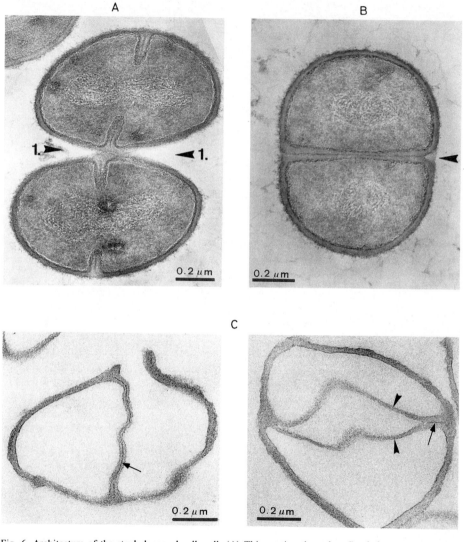

Fig. 6. Architecture of the staphylococcal cell wall. (A) Thin section through a *Staphylococcus aureus* cell demonstrating the alternation of division plane. 1. first division plane. The dark structure in the middle of the growing cross wall corresponds to the splitting system (from [37]). (B) The cell has just started murosome induced cutting of the peripheral wall to initiate cell separation (arrow head). From [37]. (C) Thin sections through isolated sacculi of *S. aureus*. The splitting system (arrows) is still detectable in most sacculi before the extraction of teichoic acid (left). If the splitting system is missing (in some of the sacculi or after removal of teichoic acid), the septal wall material appears as two, possibly independently preformed wall pieces (arrow heads).

where a fourfold length difference between folded and extended conformation has been determined [26,49–51].

Using 'bacterial threads' of up to 1 m in length, Mendelson and Thwaites demonstrated that despite its rigidity, the peptidoglycan has properties analogous to those of viscoelastic polymers. Threads containing up to 20 000 cellular filaments in parallel alignment were produced from a *Bacillus subtilis* mutant suppressed in cell separation, and were examined by using conventional textile engineering techniques [52–54]. They could be extended to up to 70% at high humidity and were able to bear an osmotic pressure of up to 24 atm. These values are probably lower limits since sliding of the individual filaments of the thread cannot be excluded. In fact, results obtained by Arthur Koch's and our own group using low angle laser light scattering and coulter counter volume measurements, respectively, showed that the sacculus can be expanded up to 300% in agreement with the flexibility of the crosslinked peptide [49–51].

Although probably not stretched to its elastic limit, even under normal growth conditions of the cell, the peptidoglycan has to bear a considerable stress. Important consequences for the bacterial growth have been compiled in the surface stress theory of Arthur Koch [55–57]. According to this, bacteria can utilize the turgor pressure to initiate cell elongation, to cause division, and to develop special shapes. Its main implications are: (i) all growth models should consider that newly formed peptidoglycan material has to be added in a relaxed state but will be transformed to a stressed state upon incorporation into the existing network; (ii) the activity and specificity of peptidoglycan interacting enzymes may vary drastically since stress-induced changes can influence the energy of a bond to be cleaved as well as the three-dimensional conformation of the wall substrate [55–58].

4. Complexity and variability of peptidoglycan

4.1. Models of the Gram-negative sacculus

The simple question of whether the thin murein sacculus of Gram-negative bacteria is comprised of one or more than one layer is still a matter of debate. Different answers are reflected in different growth models for the sacculus. Figure 7 schematically depicts some of them including topological variations such as multilayered polar caps and/or septa (model 4) or the existence of patches of multilayered material as proposed by Höltje and Glauner (model 3) [59] or the 'make-before-break' strategy (model 2) discussed in detail by Koch [60].

For the *E. coli* sacculus, there is now ample evidence that none of the 'extreme models' 1 or 5 (Fig. 7) describes the situation correctly. The single layer model is not compatible with the finding that *E. coli* can grow and divide with no obvious change in morphology with a drastically reduced murein content [61], and is difficult to reconcile with cell wall turnover [62–64] (see, however, [65]). Vice versa, thorough investigation of the amount of peptidoglycan in *E. coli* cell walls did not reveal enough cell wall material to cover even two complete layers [66]. Neutron and X-ray small angle scattering experiments performed on sacculi isolated from exponentially growing cells indicated that al-

32

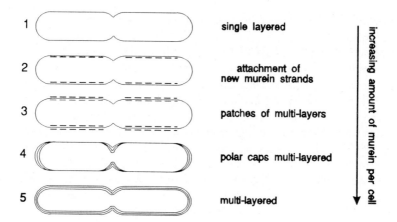

Fig. 7. Simplified sketches of models for the thickness profiles of the *Escherichia coli* sacculus. From [67].

most 75–80% of the sacculus surface corresponded to a 2.5 nm thick monolayer and that 20–25% represented 7.5 nm thick triple-layered regions [67,68].

A more precise location of the multilayered regions has not been possible. *E. coli* minicells have been used as model systems for polar caps. In isolated minicell sacculi, regions containing even more than three layers were detected, but most of the surface remained monolayered [66]. Electron micrographs of unstained *E. coli* sacculi recorded by mass dependent contrast did not detect any differences between polar cap and cylindrical wall areas (Nanninga, personal communication). The recent 'three for one model' of cell growth postulates triple-layered peptidoglycan in the ingrowing septal region [69] (Fig. 8). For the formation of the cylindrical wall portion, peptidoglycan synthesis is proposed to proceed in a three step fashion: (i) a prefabricated triplet of crosslinked glycan strands is hooked underneath a pre-existing 'template' strand by forming trimeric cross bridges; (ii) the template strand is then removed by trimer specific enzymes (a muramidase and an endopeptidase activity have to cooperate to perform this task); (iii) the new triplet can now be transported into the stress bearing layer. Septum formation would require direction of addition of new triplets exclusively to the site of future cell division and should be accompanied by an increased rate of murein synthesis in agreement with results obtained with synchronized cultures by Wientjes and Nanninga [70]. Inhibition of peptidoglycan synthesis could lead to a drastic reduction of triple-layered regions. Indeed, in sacculi isolated from mecillinam-treated, i.e. PBP2 inhibited cells, almost exclusively monolayered peptidoglycan was detected [66].

4.2. Complexity of muropeptide composition

For a long time, the peptidoglycan was thought to be a simple polymer consisting of a few basic subunits as detected on paper chromatograms of lysozyme digested *E. coli* sacculi [71]. However, analysis of muramidase digests of *E. coli* sacculi by reversed phase HPLC revealed a surprising complexity [11,64,72]. More than 80 different muropeptides were separated forming a fingerprint of the murein. In *E. coli,* this complexity is due to any combination of seven different stem peptides (varying in length from di- to pentapeptide with amino acid variations in position 4 and 5) within the monomer, dimer, trimer and even tetramer fraction. Furthermore, the muropeptides are crosslinked not only by the 'normal' mechanism described in Section 2.1 but also by an unusual L,D-peptide bond between two diaminopimelic acid residues. Finally, muropeptides positioned at the reducing end of a glycan chain are characterized by a 1,6-anhydro bond of the muramic acid [64].

The power of the HPLC method for identifying murein components specifically associated with different physiological conditions is enormous. The combination of the HPLC method with modern desorption mass spectrometric techniques, either on-line or off-line, has additionally paved the way for the detailed structural analysis of the separated muropeptides [73–76]. Gram-positive cell walls contain a much greater variety of muropeptide oligomers, aggravating HPLC analysis. Details on the peptidoglycan structure of pneumococci and staphylococci have been reported recently [77–81].

The reason for the complexity of muropeptides is not known. It might simply reflect the limited accuracy of the biosynthetic machinery in combination with that of other murein metabolizing enzymes. On the other hand, it may well represent a sophisticated device regulating spatial and temporal processes involved in bacterial morphogenesis [5,59].

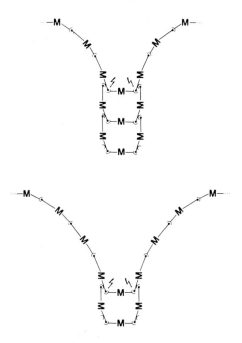

Fig. 8. Schematic view of the cell constriction (septum formation) according to the 'three for one' mechanism for Gram-negative bacteria. A projection along the glycan strands (M) running perpendicular to the plane of the drawing is shown. Donor peptide, —● ; acceptor peptide, —C ; cleavage site of enzymes specially recognizing trimeric crossbridges, ⋀⋀ . From [69].

4.3. Murein chemistry and cell shape

It has been known for a long time that the murein chemistry is dependent on growth phase (exponential or stationary growth phase, sporulation) [27,82–84] and depends on environmental factors such as glycine or D-amino acids content of the growth medium [85–87] or presence of antibiotics [3,35,88–97]. In this and the following section, we discuss murein chemistry in relation to cell morphology and the impact of cell wall inhibitors, e.g. β-lactam antibiotics. Cell wall alterations observed in vancomycin resistant strains are discussed in Chapter 26.

It has often been proposed that the cylindrical part of *E. coli* is chemically distinct from the polar caps. All attempts to reveal such differences failed when muropeptides derived after muramidase digestion were analyzed by HPLC [98–100]. In these studies, cells enriched in cylindrical portion (e.g. filaments) were compared to polar cap cells (minicells), or cells were harvested from synchronized cultures at different stages during the cell cycle for comparison of dividing versus elongating cells. On the other hand, amidase treatment of sacculi, which allows the determination of glycan chain length, revealed substantial differences using antibiotic induced spherical and filamentous cells, or *E. coli* *ftsZ* and *envA* mutants defective in cell separation before and after induction of septum formation [101]. Further support for 'non-uniform' chemistry of the murein layer was derived from determination of glycan chain length before and after the addition of penicillin to the growth medium [101], which induced lytic processes restricted to the presumptive division zone. All these results suggested somewhat shorter glycan chains in spherical versus cylindrical sacculus regions. However, no variation in glycan chain length could be detected during the division cycle in synchronized cultures (Nanninga, personal communication). It is therefore an open question if morphological changes due to the addition of certain compounds are comparable to those that occur in the division cycle of wild-type cells.

Other studies have followed murein synthesis in vivo by distinguishing between newly synthesized and pre-existing strands, based on the assumption that only new murein strands serve as donors for the crosslinking transpeptidation reaction. A pair-wise insertion of new strands at a constant rate was proposed [102]. Experiments by DeJonge et al. [98] on synchronized cultures suggested that during cell elongation peptidoglycan was inserted as single strands and during constriction a multistranded insertion occurred, and that the changed mode of insertion rather than the chemical structure of the peptidoglycan is responsible for the change in cell shape.

4.4. Murein structure in penicillin resistant bacteria

Despite the high complexity of the peptidoglycan composition, the HPLC profile of muropeptides isolated from the same species grown under identical conditions is relatively reproducible and stable. It was therefore a surprise to see major changes in the peptidoglycan structure of pneumococcal strains that differed in resistance to penicillin due to modified penicillin-binding proteins ([103]; see also Chapter 25). Resistant strains contained more hydrophobic, branched peptides suggesting an altered specificity of their PBPs that apparently did not affect growth behaviour.

In contrast to the situation with penicillin resistant pneumococci, peptidoglycan isolated from methicillin resistant *S. aureus* (MRSA) strains (see Chapters 6, 24, 25) that contain an additional PBP (PBP2') did not differ from that of susceptible *S. aureus* isolates, suggesting that the activity of PBP2' cannot be differentiated from those of the exsting PBPs, or that it is not functioning in cell wall biosynthesis, as long as these strains grow in the absence of β-lactam antibiotics [80,104]. When β-lactam antibiotics saturate the normal set of staphylococcal PBP, the cell wall still produced due to the presence of PBP2' was drastically hypocrosslinked. The expression of methicillin resistance does depend on several genes named *fem* [105–109], some of which have recently been shown to be involved in late steps of murein biosynthesis [79,81,104,110,111] Both, the *femA* and *femB* genes are involved in the synthesis of the pentaglycine interpeptide bridge. Peptides produced after inactivation of the *femA* gene contain only one glycine residue [79,81,104,110], whereas peptides of *femB* inactivated cells have three glycine residues attached [111; and unpublished results]. Their inactivation results in loss of resistance of MRSA strains and induction of hypersensitivity of sensitive staphylococci [104].

References

1. Rogers, H.J, Perkins, H.R and Ward, J.B. (1980) Microbial Cell Walls and Membranes, Chapman & Hall. London, Ch. 6.
2. Beveridge, T.J. (1988) Int. Rev. Cytol. 72, 229–297.
3. Shockman, G.D. and Barrett, J.F. (1983) Annu. Rev. Microbiol. 37, 501–527.
4. Schleifer, K.H. and Kandler, O. (1972) Bacteriol. Rev. 36, 407–477.
5. Höltje, J.-V. and Schwarz, U. (1985) in: N. Nanninga (Ed.), Molecular Cytology of *Escherichia coli*, Academic Press, London, pp. 77–119.
6. Seidl, P.H. and Schleifer, K.H. (1986) in: P.H. Seidl and K.H. Schleifer (Eds.), Biological Properties of Peptidoglycan, Walter de Gruyter, Berlin, pp.1–20.
7. Cooper, S. (1991) Microbiol. Rev. 55, 649–674.
8. Ghuysen, J.-M. (1968) Bacteriol. Rev. 32, 425–464.
9. Matsuhashi, M., Wachi, M. and Ishino, F. (1990) Res. Microbiol. 141, 89–103.
10. Harz, H., Burgdorf, K. and Höltje, J.-V. (1990) Anal. Biochem.190, 120–128.
11. Glauner, B., Höltje, J.-V. and Schwarz, U. (1988) J. Biol. Chem. 263, 10088–10095.
12. Snowden, M.A. and Perkins, H.R. (1990) Eur. J. Biochem. 191, 373–377.
13. Henze, U., Sidow, T., Wecke, J., Labischinski, H. and Berger-Bächi, B. (1993) J. Bacteriol. 175, 1612–1620.
14. Romeis, T., Kohlrausch, U., Burgdorf, K. and Höltje, J.V. (1991) Res. Microbiol. 142, 325–332.
15. Gmeiner, J. and Kroll, H.-P (1981) Eur. J. Biochem. 117, 171–177.
16. Kelemen, M.V. and Rogers, H.J. (1971) Proc. Natl. Acad. Sci. USA 68, 992–996.
17. Formanek, H., Formanek, S. and Wawra, H. (1974) Eur. J. Biochem. 46, 279–294.
18. Oldmixon, E.H., Glauser, S. and Higgins, M.L. (1974) Biopolymers 13, 2037–2060.
19. Burge, R.E., Fowler, A.G. and Reaveley, D.A. (1977) J. Mol. Biol. 117, 927–953.
20. Burge, R.E., Adams, R., Balyuzi, H.H.M. and Reaveley, D.A. (1977) J. Mol. Biol. 117, 955–974.
21. Labischinski, H., Barnickel, G., Bradaczek, H. and Giesbrecht, P. (1979) Eur. J. Biochem. 95, 147–155.
22. Naumann, D., Barnickel, G., Bradaczek, H., Labischinski, H. and Giesbrecht, P. (1982) Eur. J. Biochem. 125, 505–515.
23. Labischinski, H., Barnickel, G., Leps, B., Bradaczek, H. and Giesbrecht, P. (1980) Arch. Microbiol. 127, 195–201.
24. Barnickel G., Naumann, D., Bradaczek, H., Labischinski, H. and Giesbrecht, P. (1983) in: R.

36

Hakenbeck, J.-V. Höltje and H. Labischinski (Eds.), The Target of Penicillin, Walter de Gruyter, Berlin, pp. 61–66.

25. Barnickel, G., Labischinski, H., Bradaczek, H. and Giesbrecht, P. (1979) Eur. J. Biochem. 95, 157–165.

26. Labischinski, H., Barnickel, G. and Naumann, D. (1983) in: R. Hakenbeck, J.-V. Höltje and H. Labischinski (Eds.), The Target of Penicillin, Walter de Gruyter, Berlin, pp. 49–54.

27. Marquis, R.E. (1988) in: P. Actor, L. Daneo-Moore, M.L. Higgins, M.R. Salton, G.D. Shockman (Eds.), Antibiotic Inhibition of Bacterial Cell Surface Assembly and Function, American Society for Microbiology, Washington, DC, pp. 21–32.

28. Labischinski, H., Barnickel, G., Naumann, D. and Keller, P. (1985) Ann. Inst. Pasteur Microbiol. 136A, 45–50.

29. Koch, A.L. and Woeste, S. (1992) J. Bacteriol. 174, 4811–4819.

30. Labischinski, H. and Johannsen, L. (1986) in: P.H. Seidl and K.H. Schleifer (Eds.), Biological Properties of Peptidoglycan, Walter de Gruyter, Berlin, pp. 37–42.

31. Hobot, J.A., Carlemalm, E., Villiger, W. and Kellenberger, E. (1984) J. Bacteriol. 160, 143–152.

32. Graham, L.L., Harris, R., Villiger, W. and Beveridge, T.J. (1991) J. Bacteriol. 173, 1623–1633.

33. Leduc, M., Frehel, C., Siegel, E. and Van Heijenoort, J. (1989) J. Gen. Microbiol. 135, 1243–1254.

34. Nanninga, N. (1991) Mol. Microbiol. 5, 791–795.

35. Wientjes, F.B. and Nanninga, N. (1991) Res. Microbiol. 142, 333–344.

36. Rogers, H.J, Perkins, H.R and Ward, J.B. (1980) Microbial Cell Walls and Membranes, Chapman & Hall, London, Ch. 1. pp. 2ff.

37. Giesbrecht, P., Wecke, J. and Reinicke, B. (1976) Int. Rev. Cytol. 44, 225–318.

38. Giesbrecht, P. and Wecke, J. (1980) in: P. Brederoo and W. de Priester (Eds.), Proc. 7th European Congress on Electron Microscopy, Vol. 2, Seventh European Congress on Electron Microscopy Foundation, Leiden, pp. 446–453.

39. Amako, K. and Umeda, A. (1984) Microbiol. Immunol. 28, 1293–1301.

40. Giesbrecht, P. (1984) in: C. Nobela (Ed.), Microbial Cell Wall Synthesis and Autolysis, Elsevier, Amsterdam, pp. 177–186.

41. Giesbrecht, P., Kersten, T. and Wecke, J. (1992) J. Bacteriol. 174, 2241–2252.

42. Giesbrecht, P., Labischinski, H. and Wecke, J. (1985) Arch. Microbiol. 141, 315–324.

43. Morioka, H., Tachibana, M. and Suganuma, A. (1987) J. Bacteriol. 169, 1358–1362.

44. Suganuma, A. (1991) in: Jeljaszewicz, Ciborowski (Eds.), The staphylococci, Zbl. Bakt. Suppl. 21, Gustav Fischer Verlag, Stuttgart, pp. 27–35.

45. Umeda, A., Ikebuchi, T. and Amako, K. (1980) J. Bacteriol. 141, 838–844.

46. Mitchell, P. and Moyle, J. (1957) J. Gen. Microbiol. 16, 184–194.

47. Koch, A.L. and Pinette, M.F.S. (1987) J. Bacteriol. 169, 409–417.

48. Pinette, M.F.S and Koch, A.L. (1988) J. Bacteriol. 169, 4737–4742.

49. Koch, A.L. and Woeste, S. (1992) J. Bacteriol. 174, 4811–4819.

50. Braun, V., Gnirke, H., Henning, U. and Rehn, K. (1973) J. Bacteriol. 114, 1264–1270.

51. Koch, A.L. (1990) ASM News. 57(12), 633–637.

52. Thwaites, J.J. and Mendelson, N.H. (1985) Proc. Natl. Acad. Sci. USA. 82, 2163–2167.

53. Mendelson, N.H. and Thwaites, J.J. (1989) J. Bacteriol. 171, 1055–1062.

54. Thwaites, J.J. and Mendelson, N.H. (1989) Int. J. Biol. Macromol. 11, 201–206.

55. Koch, A.L. (1988) Adv. Microbiol. Rev. 52, 337–353.

56. Koch, A.L. (1988) Microbiol. Rev. 52, 337–353.

57. Koch, A.L. (1991) American. Sci. 78, 327–341.

58. Koch, A.L. (1991) FEMS Microbiol. Rev. 88, 15–26.

59. Höltje, J.-V. and Glauner, B. (1990) Res. Microbiol. 141, 75–89.

60. Koch, A.L. (1990) Res. Microbiol. 141, 529–541.

61. Prats, R. and De Pedro, M.A. (1989) J. Bacteriol. 171, 3740–3745.

62. Goodell, E.W. (1985) J. Bacteriol. 163, 305–310

63. Goodell, E.W. and Schwarz, U. (1985) J. Bacteriol.162, 391–397.

64. Höltje, J.-V. and Glauner, B. (1990) Res. Microbiol. 141, 75–89.

65. Park, T.J. (1993) J. Bacteriol. 175, 7–11.

66. Wientjes, F.B., Woldringh, C.L. and Nanninga, N. (1991) J. Bacteriol. 173, 7684–7691.

67. Labischinski, H., Hochberg, M., Sidow, T., Maidhof, H., Henze, U., Berger-Bächi, B. and Wecke, J. (1993) in: M.A. De Pedro, J.-V. Höltje and W. Löffelhardt (Eds.), Bacterial Growth and Lysis: Metabolism and Structure of the Bacterial Sacculus, Plenum Press, New York, pp. 9–21.
68. Labischinski, H., Goodell, E.W., Goodell, A. and Hochberg, M.L (1991) J. Bacteriol. 173, 751–756.
69. Höltje, J.-V. (1993) in: M.A. De Pedro, J.-V. Höltje and W. Löffelhardt (Eds.), Bacterial Growth and Lysis: Metabolism and Structure of the Bacterial Sacculus, Plenum Press, New York, pp. 419–426.
70. Wientjes, F.B. and Nanninga, N. J. (1989) J. Bacteriol. 171, 3412–3419.
71. Weidel, W. and Pelzer, H. (1964) Adv. Enzymol. 26, 193–232.
72. Glauner, B.(1988) Anal. Biochem. 172, 451–464.
73. Martin, S.A., Rosenthal. R.S. and Bieman, K. (1987) J. Biol. Chem. 262, 7514–7522.
74. Martin, S.A. (1988) in: P. Actor, L. Daneo-Moore, M.L. Higgins, M.R. Salton, G.D. Shockman (Eds.), Antibiotic Inhibition of Bacterial Cell Surface Assembly and Function, American Society for Microbiology, Washington, DC, pp. 129–145.
75. Pittenauer, E., Allmaier, G. and Schmid, E.R. (1993) in: M.A. De Pedro, J.-V. Höltje and W. Löffelhardt (Eds.), Bacterial Growth and Lysis: Metabolism and Structure of the Bacterial Sacculus, Plenum Press, New York, pp. 39–46.
76. Pittenauer, E., Rodriguez, M.C., de Pedro, M.A., Allmaier, G. and Schmid, E.R. (1993) in: M.A. De Pedro, J.-V. Höltje and W. Löffelhardt (Eds.), Bacterial Growth and Lysis: Metabolism and Structure of the Bacterial Sacculus, Plenum Press, New York, pp. 31–38.
77. Garcia-Bustos, J.F., Chait, B.T. and Tomasz, A. (1987) J. Biol. Chem. 262, 15400–15405.
78. Garcia-Bustos, J.F., Chait, B.T. and Tomasz, A. (1988) J. Bacteriol. 170, 2143–2147.
79. De Jonge, B.L.M, Chang, Y.-S, Gage, D. and Tomasz, A. (1992). J. Biol. Chem. 276, 11255–11259.
80. De Jonge, B.L.M, Chang, Y.-S, Gage, D. and Tomasz, A. (1992) J. Biol. Chem. 276, 11248–11254.
81. De Jonge, B.L.M, Sidow, T., Chang, Y.-S, Labischinski, H., Berger-Bächi, B., Gage, D. and Tomasz, A. (1993) J. Bacteriol 175, 2779–2782.
82. Pisabarro, A.G., de Pedro, M.A. and Vazquez, D. (1985) J. Bacteriol. 161, 238–242..
83. Glauner, B., Höltje, J.V. and Schwarz, U. (1988) J. Biol. Chem. 263, 10088–10095
84. Warth, A.D. and Strominger, J.L. (1972) Biochemistry 11, 1389–1396.
85. Schleifer, K.H., Hammes, W.P. and Kandler, O. (1976) Adv. Microbiol. Physiol. 13, 245–292
86. Ottolenghi. A.C., Caparrós, M. and de Pedro, M.A. (1993) J. Bacteriol, 175 1537–1542.
87. Caparrós, M., Torrecuadrada, J.L.M. and de Pedro, M.A. (1991) Res. Microbiol. 142, 345–350.
88. Kohlrausch, U. and Höltje, J.-V. (1991) J. Bacteriol. 173, 3425–3431.
89. Sidow, T., Johannsen, L. and Labischinski, H. (1990) Arch. Microbiol. 154, 73–81.
90. Martin, H.-H. and Gmeiner, J. (1979) Eur. J. Biochem. 95, 487–495.
91. Blundell, J.K. and Perkins, H.R. (1981) J. Bacteriol. 147, 633–641.
92. Johannsen, L., Labischinski, H., Reinicke, B. and Giesbrecht, P. (1983) FEMS Microbiol Lett. 16, 313–316.
93. Martin, H. and Gmeiner, J. (1979) Eur. J. Biochem. 95, 487–495.
94. Qoronfleh, M.W. and Wilkinson, B.J. (1986) Antimicrob. Agents Chemother. 29, 250–257
95. Sinha, R.K. and Neuhaus, F.C. (1991) Antimicrob. Agents Chemother. 35, 1753–1759.
96. Snowden, M.A. and Perkins, H.R. (1991) J. Gen. Microbiol. 137, 1661–1666.
97. Waxman, D.J. and Strominger, J.L (1983) Annu. Rev. Biochem. 52, 825–869
98. De Jonge, B.L.M., Wientjes, F.B., Jurida, I., Driehuis, F., Wouters, J.T.M. and Nanninga, N. (1989) J. Bacteriol. 171, 5783–5794.
99. Driehuis, F. and Wouters, J.T.M. (1987) J. Bacteriol. 169, 97–101.
100. Tuomanen, E. and Cozens, R. (1987) J. Bacteriol. 169, 5308–5310.
101. Romeis, T., Kohlrausch, U., Burgdorf, K. and Höltje, J.V. (1991) Res. Microbiol. 142, 325–332.
102. Burman, L.G. and Park, J.T. (1984) Proc. Natl. Acad. Sci. USA 81, 1844–1848.
103. Garcia-Bustos, J and Tomasz, A. (1990) Proc. Natl. Acad. Sci. USA 87, 5415–5419.
104. Labischinski, H. (1992) Med. Microbiol. Immunol. 181, 241–265.
105. Berger-Bächi, B. (1992) Chemother. J. 1, 58–64.
106. Berger-Bächi, B. (1983) J. Bacteriol. 154, 479–487.
107. Berger-Bächi, B., Strässle, A. and Kayser, F.H. (1986) Eur. J. Clin. Microbiol. 5, 697–701.
108. Berger-Bächi, B., Barberis-Maino, L., Strässle, A. and Kayser, F.H. (1989) Mol. Gen. Genet. 219,

38

263–269.

109. Berger-Bächi, B., Strässle, A., Gustafson, J.E. and Kayser, F.H. (1991) Antimicrob. Agents Chemother. 36, 1367–1373.

110. Maidhof, H., Reinicke, B., Blümel, P., Berger-Bächi, B. and Labischinski, H. (1991) J. Bacteriol. 173, 3507–3513.

111. Henze, U., Sidow, T., Wecke, J., Labischinski, H. and Berger-Bächi, B. (1993) J. Bacteriol. 175, 1612–1620.

112. Roland, J.C., Lembi, C.A. and Morre, D.J. (1972) Stain. Technol. 47, 195–200.

J.-M Ghuysen and R. Hakenbeck (Eds.), *Bacterial Cell Wall*

39

CHAPTER 3

Biosynthesis of the bacterial peptidoglycan unit

JEAN VAN HEIJENOORT

Unité de Recherche Associée 1131 du CNRS, Biochimie Moléculaire et Cellulaire, Université Paris-Sud,
Bât. 432, 91405 Orsay, France

1. Introduction

The biosynthesis of bacterial peptidoglycan is a complex two stage process. The first stage concerns the formation of the disaccharide peptide monomer unit and the second the polymerization reactions accompanied by the insertion of the newly made peptidoglycan material into the cell wall. This latter stage is reviewed in the following chapter.

The assembly of the peptidoglycan unit proceeds by a series of cytoplasmic and membrane reactions. Two essential features characterize this pathway. First, the high specificity of each step reflects the unusual structural characteristics of peptidoglycan [1,2], many of which are encountered in its monomer unit (alternating D and L residues in peptide linkages, γ-D-glutamyl linkage, presence of *N*-acetylmuramic acid and often of diaminopimelic acid). Secondly, in the final steps the monomer unit is transferred from the cytoplasm to the externally located sites of incorporation into growing peptidoglycan. This implies a passage through the hydrophobic environment of the membrane. Lipid intermediates are involved in this process. The various steps of the pathway have been identified in one bacteria or another, and a general scheme established that is valid for both Gram-positive and negative eubacteria [1]. For convenience, three groups of successive reactions can be considered (Fig. 1): the formation of UDP-*N*-acetylmuramic acid (UDP-MurNAc); the formation of the UDP-MurNAc-peptides; and the formation of the lipid intermediates.

UDP-MurNAc arises in two steps from UDP-*N*-acetylglucosamine (UDP-GlcNAc) which can be considered as the initial precursor of the pathway. In the first step, the addition of enolpyruvate to position 3 of the *N*-acetylglucosamine residue is catalysed by a transferase to yield UDP-GlcNAc-enolpyruvate. In the second step, the reduction of the enolpyruvyl moiety to D-lactate is catalysed by a reductase to yield UDP-MurNAc.

The UDP-MurNAc peptides are formed by the sequential addition in most cases of L-alanine, D-glutamic acid, a diamino acid (usually diaminopimelic acid or lysine) and D-alanyl-D-alanine on the D-lactyl group of UDP-MurNAc. Each step is catalysed by a specific synthetase using ATP.

The membrane steps involve first a translocase which catalyses the transfer of the phospho-MurNAc-pentapeptide moiety of UDP-MurNAc-pentapeptide to the membrane

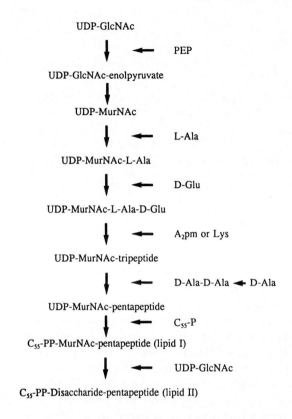

Fig. 1. General scheme of the biosynthesis of the peptidoglycan unit.

acceptor, undecaprenyl phosphate, yielding MurNAc(pentapeptide)-pyrophosphoryl-undecaprenol (lipid I). Thereafter, a transferase catalyses the addition of *N*-acetyl-glucosamine yielding GlcNAc-MurNAc(pentapeptide)-pyrophosphoryl-undecaprenol (lipid II), which, in the most simple cases, is the substrate for the polymerization reactions.

The functioning of this pathway is dependent on a number of metabolites (PEP, ATP, L-alanine, D-glutamic acid, diaminopimelic acid or L-lysine, D-alanyl-D-alanine, unde-caprenyl phosphate and UDP-GlcNAc) used as substrate at one step or another. Many are involved in other cell reactions. However, D-glutamic acid and D-alanyl-D-alanine appear as highly specific of the synthesis of peptidoglycan. In particular, the dipeptide is formed by dimerization of D-alanine.

Over the past 40 years, many features of the assembly of the peptidoglycan unit were investigated. Work up to the early 1980s has been critically reviewed [1,3]. Since this date, not only have biochemical researches continued, but many new genetic and physi-ological aspects have been developed. In the present review, emphasis is placed on the more recent results, and already reviewed data are summarized as concisely as possible.

2. Genetic analysis

2.1. Introduction

Genetic studies were undertaken to determine the position on the chromosome of the genes directly involved in the biosynthesis of peptidoglycan and to determine whether they are in any way related to other cell functions, and particularly with cell division. Peptidoglycan being an essential component for the survival of the bacterial cell, any mutation specifying an enzyme catalysing a step in the formation of its monomer unit will be characterized by a phenotype of osmotic fragility or cell lysis. Therefore, only conditional-lethal mutations can be considered. Temperature-sensitive mutants showing lysis at the restrictive temperature were isolated from *E. coli* [4–7]. Growth of several mutants was restored by addition of an osmotic stabilizer. Temperature-sensitive mutants of *S. aureus* with defective cell wall biosynthesis were differentiated from other ts mutants by their requirement for osmotic stabilization [8,9]. Biochemical analyses (accumulation of precursors and enzyme assays) showed that many mutations involved genes coding for precursor synthetases [4,6,8,9]. Similar deficiencies were described in L-phase variants of different Gram-positive bacteria [see 1 for references].

2.2. Genetic analysis of the E. coli system

In *E. coli*, the genes directly involved in the assembly of the peptidoglycan unit (Fig. 1) have now been identified, cloned and sequenced. Initially, a number of these genes were mapped with the help of temperature-sensitive mutants not forming peptidoglycan at 42°C [5–7] and were found to be in two clusters designated as mra and mrb. The mra cluster is located at 2 min on the *E. coli* chromosome and contains genes *murC, murE, murF* and *ddl* (Fig. 2) coding for the L-alanine, diaminopimelic acid (A_2pm) and D-alanyl-D-alanine adding enzymes, and the D-alanine:D-alanine ligase, respectively [6,11 and refs. therein]. More recently the subcloning of this 2-min region enabled the *murD* gene coding for the D-glutamic acid-adding enzyme to be identified and located [12]. Likewise, the genes *mraY* and *murG* coding for the translocase and transferase which catalyse the formation of lipid intermediates I and II were identified and located [13,14]. The 2-min region also contains genes involved in various steps of the cell division process: *pbpB, ftsW, ftsQ, ftsA, ftsZ* and *envA* (Fig. 2). The products of genes *pbpB* and *ftsW* play a role in the syn-

E. coli

-*fts36-pbpB-murE-murF-mraY-murD-ftsW-murG-murC-ddl-ftsQ-ftsA-ftsZ-envA-secA*-

B. subtilis

-*murE-mraY-murD-spoVE-murG-ORF2-dds-ORF4-ORF5-sbp-ftsA-ftsZ*-

Fig. 2. Comparison of the chromosomal organizations of the *E. coli* 2-min and *B. subtilis* 133° regions.

thesis of septum peptidoglycan (see Chapter 4). With the cloning and sequencing of all the genes from *pbpB* to *envA*, the organization of this region is now completely elucidated and its physical map established [14,15,16,18]. The genes are tightly packed and overlap in many cases. Homologies were found in the deduced amino acid sequences of the four UDP-MurNAc-peptide synthetases, products of *murC, murD, murE* and *murF* [15]. In particular, two domains were well conserved, one corresponding presumably to an ATP binding site. Although not rigorously proven by null allele experiments, the genes coding for the various precursor synthetases are considered unique. However, it has now been established that there are two distinct *ddl* genes (*ddlA* and *ddlB*) encoding ligases which catalyse the formation of D-alanyl-D-alanine [19]. Gene *ddlB* is located immediately downstream of *murC* [20], whereas *ddlA* is located in the 6-min region (K.E. Rudd, personal communication). There is 35% homology between the DdlA and DdlB enzymes. The *ddlA* gene was initially described in *S. typhimurium* [21].

The mrb cluster, approximately located at 90 min close to *argH* and *metB*, was initially found to probably contain the *murA* and *murB* genes coding for the transferase and the reductase, respectively, which catalyse the two-step synthesis of UDP-MurNAc [5,6, 22]. However, the two genes were not precisely mapped. Recently, the physical and genetic mapping of *murB* was accomplished after its subcloning [23,24]. It is located at 89.9 min immediately upstream of gene *birA* in a previously sequenced region of 15 kb (*btu-rpoBC*), in which all genes had been identified except one ORF. Subsequently, this ORF was identified unambiguously as the *murI* gene encoding the glutamic acid racemase [25–27]. Apparently no *murA* gene was located in the 15-kb fragment considered around 90 min. More recently, a gene encoding the transferase was cloned, sequenced, and mapped to 69.3 min on the *E. coli* chromosome [28]. It was designated as murZ to distinguish it from the *murA* gene. The different map positions of murA and murZ could suggest that *E. coli* may yet have two UDP-GlcNAc-enolpyruvyl transferases [28]. The gene for the transferase from *E. cloacae* was also recently cloned and sequenced [29]. There is a very high homology between the amino acid sequences of the two enzymes.

2.3. Genetic analysis of the B. subtilis system

A cluster homologous to the mra region of *E. coli* was found in *B. subtilis* [30–33]. Analysis of the primary structure of the product of cell division gene *ftsW* of the 2-min region of *E. coli* (Fig. 2) indicated high homology with the SpoVE protein, which functions in some way in the formation of the spore cortex of *B. subtilis*, and with the RodA protein which functions in the cell growth of *E. coli* [31,32,34]. When genes flanking *spoVE* located at 133° on the chromosome of *B. subtilis* were cloned and sequenced, the predicted amino acid sequences of their products appeared homologous to those of the *murE*, *mraY*, *murD* and *murG* gene products of *E. coli* [30,31,33]. The transcription of *murD*, *spoVE* and *murG* was analysed during growth and sporulation [31]. Insertional in vitro mutagenesis of the region revealed that *murD* and *murG* are essential for normal vegetative growth of *B. subtilis*, but *spoVE* is only required for sporulation [31]. Genes of *B. subtilis* involved in the growth and division of the cell have so far not been studied as extensively as have those in *E. coli*. However, recently homologues of the *E. coli* cell division genes *ftsA* and *ftsZ* were located close to and downstream of *spoVE* in *B. subtilis* [35 and refs. therein]. A possible alignment of the genes of the mra region of *E. coli* and of the *murE-ftsZ* region of *B. subtilis* is shown (Fig. 2). These results suggest the evolutionary relationship among genes involved in peptidoglycan synthesis, cell division and sporulation.

3. Biochemical studies

3.1. Introduction

The elucidation of the pathway leading to the complete peptidoglycan unit was established by isolating and characterizing the muramic acid containing precursors, and by developing a specific in vitro assay for the enzymatic activity catalysing each step [1,3]. Assays involving more than one step have also been developed. The most important one concerns the in vitro formation of polymerized peptidoglycan material from UDP-GlcNAc and UDP-MurNAc-pentapeptide as substrates, initially developed with particulate preparations from *S. aureus* [36,37]. This assay has thereafter been used in many instances with various cell-free systems (particulate preparations, crude cell walls, ether- or toluene- treated cells) and implies the transitory formation of the lipid intermediates (see [1,3] for references]. More recently, multi-enzyme assays for peptidoglycan synthesis in ether-treated *E. coli* cells were developed to include all enzymatic steps from UDP-GlcNac [38] or UDP-MurNAc-tripeptide [39]. Such assays are convenient for the screening of potential inhibitors.

3.2. The uridine nucleotide precursors

UDP-MurNAc precursors were originally found to accumulate in penicillin-treated cells of *S. aureus* [40] from which they were isolated and characterized [41]. Similar nucleotide precursors were shown to be present in a variety of both Gram-positive [42–45] and negative bacteria [46,47 and refs. therein]. These nucleotides accumulate not only in antibiotic-treated cells but also in resting cells [48,49], in glycine-treated cells [50] and in conditional-lethal mutants with altered enzymes of the pathway [4,6,8,9,11 and refs. therein; 132]. They can be quantitatively recovered from whole cells by various extraction techniques (boiling water, trichloroacetic acid, formic acid) [4,10,41,47,49,51,52]. Owing to their possible degradation by endogenous hydrolases, the use of lysed cells should be avoided [53]. The nucleotide precursors were initially purified from crude extracts by ion exchange chromatography, adsorption on and elution from charcoal, and paper chromatography or electrophoresis. It was later shown that they could be separated from the other low molecular weight cell metabolites by gel filtration on Sephadex G25 [44]. More recently, reverse phase HPLC techniques were introduced [54]. The association of the gel filtration and HPLC techniques has led to convenient methods for their quantitative analysis and preparation [52,55,56].

3.3. Membrane precursors

The detection, quantitative analysis and preparation of the lipid intermediates of peptidoglycan synthesis are difficult problems that have attracted too little attention to date.
The main reasons are undoubtedly their very low cell content, their amphipathy and also
presumably to some extent their lability. They were originally detected by paper or column chromatography after formation from UDP-MurNAc-pentapeptide and UDP-
GlcNAc in cell-free systems [57,58] or after their specific in vivo radiolabelling [59–61].
However, these techniques do not distinguish between lipid intermediates. More recently,
lipids I and II were quantitatively analysed in E. coli by techniques involving partial hydrolysis, chemical reduction and HPLC [14,61–63]. Furthermore, small amounts (ca.
30 nmol) of lipid I were secured after its accumulation in cell-free systems from S. aureus
[64] or B. megaterium [65]. Lipid II was secured in much larger amounts after its accumulation in cell-free systems from M. luteus [66] and E. coli [63,67]. Similarly, tripeptide
lipid II was recently accumulated and purified [63].

3.4. Modified precursors

In principle, the most simple structure for bacterial peptidoglycan is that of a heteropolymer in which the linear glycan strands have repeating disaccharide peptide units and
in which the peptide cross-bridges are directly established between two monomer units
(see Chapter 2). However, in many organisms, if not in all, the final peptidoglycan material has one or more additional structural feature: modifications of the hexosamine residues (O-acetylation, O-phosphorylation, N-glycolylation, 1,6-anhydro-MurNAc cycliza-
tion, etc.); modifications of an amino acid residue of the peptide moiety by amidation or
by addition of another amino acid; formation of more or less complex indirect cross-link-
ages [1,3]. These modifications are introduced at different steps of peptidoglycan synthesis. A few occur at the level of the nucleotide precursors [68–70]. Others occur at the
level of the lipid intermediates as clearly established in a few instances [1,3]. However, as
far as we are aware, no such modified lipid intermediate has yet been accumulated and
isolated. Finally, a number of modifications are concomitant with the polymerization reactions or take place thereafter (see Chapter 4) [1,3].

3.5. Precursor synthetases

The enzymes catalysing the formation of the various peptidoglycan precursors have been
investigated in one bacteria or another [1]. Most have at least been partially purified, but
only a few to homogeneity, owing presumably to their low cell content and instability. In
recent cases, the overproduction by genetic engineering has been extremely useful.

The UDP-GlcNAc-enolpyruvyl transferase activity was initially detected in cell extracts from S. aureus and E. cloacae by the transfer of enolpyruvate to UDP-GlcNAc
[71,72]. It was partially purified from Micrococcus luteus [73] and S. epidermidis [74],
and to homogeneity from E. cloacae [29,75] and E. coli [28]. The reaction it catalyses is
reversible [73] and moreover specifically and irreversibly inhibited by fosfomycin [76].
The UDP-GlcNAc-enolpyruvate reductase was partially purified from E. cloacae [77]

and *S. aureus* [78], and to homogeneity from *E. coli* [79,80]. In all cases, NADPH is used as cofactor.

The synthetases which catalyse the assembly of the peptide moiety of the peptidoglycan unit all appear as cytoplasmic. Their possible in vivo association has not yet been examined. The L-alanine-adding activities from *B. cereus* [81], *B. subtilis* [81], *E. coli* [82] and *S. aureus* [83] were investigated. In the four cases, the activity was markedly stimulated by reducing agents. The D-glutamic acid-adding enzyme was partially purified from *S. aureus* [84] and totally from *E. coli* [85]. The synthetases catalysing the addition of the third amino acid (usually A_2pm or L-Lys) were investigated in a number of instances [1]. The A_2pm or L-Lys-adding enzymes from *B. cereus* [86], *B. sphericus* [87], *C. xerosis* [49,88], *E. coli* [89,90] and *S. aureus* [49,91] were partially purified. The D-alanyl-D-alanine-adding activities from *B. subtilis* [92], *E. coli* [93,94], *S. aureus* [95] and *S. faecalis* [96] have been investigated. The enzyme was totally purified from *E. coli* [94].

These nucleotide peptide synthetases all catalyse the formation of a peptide bond which is coupled with the cleavage of ATP into ADP and inorganic phosphate. It is therefore reasonable to speculate that they all operate by an essentially similar mechanism. This would most likely entail carboxyl activation of a C-terminal amino acid residue of the nucleotide substrate to an acyl-phosphate intermediate followed by nucleophilic attack by the amino group of the condensing amino acid or dipeptide, with the elimination of phosphate and subsequent peptide bond formation. The reversibility of the reaction was established directly with the reaction products in a number of instances [84–86,90,92]. Another approach has been to examine the exchange reaction between the amino acid or dipeptide substrate and the nucleotide reaction product [84,86,90,92]. No reverse or exchange reaction was observed with the L-lysine-adding activity from *S. aureus* [91]. ADP-ATP or ATP-Pi exchange reactions were described for the D-alanyl-D-alanine-adding enzymes from *B. subtilis* and *E. coli* [91,93].

The translocase activity catalysing the formation of lipid I was investigated in *S. aureus* [64], *M. luteus* [97] and *E. coli* [98] and these results were reviewed in detail [1]. In the three cases, the transferase reaction is fully reversible. The activity was also found to catalyse an exchange reaction between UMP and UDP-MurNAc-pentapeptide. The lipid requirements for both the transferase and exchange reactions were investigated [64,97,98]. Moreover, the importance of the micro-environment was studied in detail in the case of the *S. aureus* [99 and refs. therein]. Recently, the translocase from *E. coli* has been identified as the product of gene *mraY* [13]. The base sequence indicated a high hydrophobicity of the MraY protein, which has regularly repeated hydrophobic and hydrophilic domains, strongly suggesting that the protein spaces the cytoplasmic membrane several times. The translocase reaction is inhibited by tunicamycin [1,3] and by mureidomycin A [100]. The transferase which catalyzes the formation of lipid II was investigated in *M. luteus* [101] and *S. aureus* [102]. It was purified from *B. megaterium* [65]. This latter activity was markedly stimulated by the presence of a crude lipid extract. Recently, the transferase of *E. coli*, identified as the product of the *murG* gene [14], was found to be mainly present in particulate fractions even upon high overproduction. However, the hydrophobicity analysis of its deduced amino acid sequence [103,104] failed to indicate any transmembrane region. It was also shown that the transferase was associated with the cytoplasmic face of the inner membrane [105].

The enzymatic mechanism by which D-glutamic acid is produced in bacteria has been investigated in a limited number of species. Two different routes for D-glutamic acid bio-synthesis were identified: (i) a transamination process catalysed by a D-amino acid tran-saminase using D-alanine and α-ketoglutarate, as demonstrated in *Bacillus* species [106,107]; or (ii) a direct conversion of L-glutamic acid to D-glutamic acid by a glutamate racemase, as observed in *E. coli* [26] and in *Lactobacillus* and *Pediococcus species* [108–110]. D-Alanyl-D-alanine is formed from L-alanine via D-alanine. A racemase con-verts L-alanine into D-alanine and a ligase catalyses the formation of the dipeptide [1,3]. D-Alanine:D-alanine ligases were purified from *E. coli* [19,111], *S. typhimurium* [112], *S. aureus* [95] and *S. faecalis* [113,114] and their properties investigated. This activity is inhibited by D-cycloserine [1,76] and the reversibility of the reaction was established [92,113]. The mechanism of the reaction was extensively studied with the ligase from *S. typhimurium* [115,116].

3.6. Specificity of the precursor synthetases

Studies of the specificity of the UDP-MurNAc-peptide synthetases were carried out with isolated activities or in some instances by examining the in vivo incorporation of unusual constituents into peptidoglycan. The specificity for the amino acid or dipeptide substrate can vary with the bacterial species and the position considered in the pentapeptide se-quence. At position 1, it was shown that the L-alanine adding enzyme can catalyse the addition of glycine or serine to UDP-MurNAc [81–83]. The growth of *S. aureus* in the presence of glycine will also lead to the replacement of L-alanine at this position by this amino acid [50 and refs. therein]. The specificity of the D-glutamic acid-adding enzyme for D-glutamic acid appears very strict, at least when the activity from *E. coli* [85,117] or *S. aureus* [84] was examined. The meso-A_2pm adding-enzyme from *E. coli* will not ac-cept L-lysine [91,118], but LL-A_2pm, cystathionine and lanthionine are substrates [88,119,120]. Conversely, the lysine adding-enzyme from *S. aureus* will not accept A_2pm as substrate [91,118]. Such a strict specificity was also encountered in *C. poinsettiae* [121]. The specificity of the D-alanyl-D-alanine-adding enzyme for the dipeptide has been extensively investigated [39,96]. Many dipeptides can substitute for D-alanyl-D-alanine. Recently, the enzyme from *E. coli* was shown to accept the D-alanyl-D-lactic acid dep-sipeptide as substrate with the same efficiency as D-alanyl-D-alanine [122].

Each UDP-MurNAc-peptide synthetase of a given strain is specific for its nucleotide substrate and will not accept any of the other nucleotide precursors of the pathway as substrate. However, it can accept certain structural modifications of its nucleotide sub-strate. For instance, the replacement of the uracil moiety by dihydrouracil [89,117] or fluorouracil [123 and refs. therein], or that of one of its amino acid residues by another [50,119–121] can still lead to good substrates. The phospho-MurNAc-L-Ala used as sub-strate by the D-glutamic acid-adding enzyme from *E. coli* represents the most important substrate modification yet encountered for these synthetases [117]. The fairly high speci-ficity of the UDP-MurNAc-peptide synthetases for their nucleotide substrate was further substantiated by the study of the inhibitory effects of structural analogues mimicking to various extents one part or another of these precursors [89,117]. Structural requirements for a good interaction were defined and it was concluded that the integrity of the whole structure of the precursor was more or less required for a proper recognition by the enzyme.

The specificities of the translocase and transferase which catalyse the formation of lipid intermediates I and II have been examined only to a limited extent. The translocase can accept various modifications of the peptide moiety of its nucleotide substrate: shorter or longer peptide chains, substitution of an amino acid by another one. These results were obtained with cell-free systems from Gram-positive bacteria (see [1] for references) and *E. coli* [63 and refs. therein] or by observing the incorporation of unusual amino acids into the pathway [50,119,120]. Regarding other parts of the nucleotide substrate, it was found that fluoro-UDP-MurNAc pentapeptide was not an effective substrate for the translocase [123]. Certain modifications of the peptide moiety of the peptidoglycan unit will also be accepted by the transferase since they are encountered in the final peptidoglycan material [50,119,120].

The specificity of the D-alanine:D-alanine ligase was studied with the enzyme from *S. faecalis* [124] which accepts only D-amino acid or glycine. The study of its donor and acceptor sites showed that the specificity is high for the D-amino acid in the N-terminal residue and low in the C-terminal residue of the dipeptide. Moreover, dipeptides glycyl-D-alanine and D-alanyl-glycine are presumably formed in vivo by the ligase when *S. aureus* cells are grown in the presence of a high concentration of glycine [50]. Another important modification of the dipeptide metabolism is the substitution of D-alanyl-D-alanine by D-alanyl-D-lactic acid depsipeptide upon induction of the resistance to vancomycin in *Enterococci* [125]. In this case, a new ligase is induced [126] and the normal D-alanyl-D-alanine synthesis is repressed [10].

In the pathway of a given bacterial species, the specificities of the various synthetases differ from one another. This will have a restrictive effect limiting possible structural variations in the final peptidoglycan unit and thus assure the preservation of its structure [1,3].

4. Physiological studies

4.1. Introduction

The possibility of quantitatively analysing precursor pool levels and the availability of accurate enzymatic in vitro assays for most of the steps of the pathway have opened the way to more systematic investigations of its in vivo functioning. The main approaches have dealt with the study of the variations of the precursor pool levels and of the specific activities of their synthetases under various conditions. Another useful approach has been the study of the variations of precursor pool levels upon treatment with specific antibiotics interfering at different steps of the pathway or by the use of specific mutants. Further interesting data were obtained by comparing of pool levels, K_m values and rates of peptidoglycan synthesis. To date, most of this work has concerned *E. coli*.

4.2. Precursor pool levels

The pool levels of the cytoplasmic precursors were determined under various conditions

[52,55,56,127,128]. In exponential-phase cells, minor differences were observed with growth rates [52] and between different *E. coli* strains [127]. Moreover, a similar distribution of the nucleotide precursors was encountered in *B. megaterium* [129]. Among the nucleotide precursors, UDP-GlcNAc and UDP-MurNAc-pentapeptide are at the highest cell concentrations, whereas UDP-GlcNAc-enolpyruvate is at a surprisingly low level (Table I). This latter nucleotide was only recently detected in vivo, isolated and quantified [55]. Previously, it had been obtained by in vitro synthesis from UDP-GlcNAc and characterized [72,130]. In stationary phase cells, a sharp depletion of certain nucleotide precursors was observed in cells grown in glucose overnight [128]. The pool levels of the lipid intermediates have been investigated only to a very limited extent. Recently, the cell contents of lipids I and II in *E. coli* were estimated at 700 and 2000 copies per cell, respectively [63].

When the pool levels of the various peptidoglycan precursors are compared to the average rate of synthesis of the macromolecule, which is ca. 1000 units per second per cell for *E. coli* growing on glucose [56], the turnover rate of the most abundant nucleotide (UDP-GlcNAc and UDP-MurNAc-pentapeptide) is a few minutes, whereas that of the less abundant precursors is only a few seconds. Consequently, the blocking of any step of the pathway will lead to rapid depletion of the pool of the downstream precursors, to a concomitant accumulation of upstream precursors and to the slowing down or arrest of peptidoglycan synthesis. A given step of the pathway can be inhibited by the use of conditional mutants [4,6,8,9] or in some cases by a specific antibiotic [1,76]. However, not all steps of the pathway are amenable to inhibition by antibiotics. This is surprisingly the case for the four UDP-MurNAc-peptide synthetases which nonetheless catalyse very specific reactions. The study of the accumulation of precursors has been a well established procedure for characterizing the target of antibiotics suspected of interfering with peptidoglycan metabolism [1,76].

TABLE I

Pool levels of peptidoglycan precursors in *E. coli* K-12 (exponential-phase cells grown in minimal glucose medium)

	Pool levels, concentration
UDP-GlcNAc	12.5×10^{-5}
UDP-GlcNAc-enolpyruvate	2.5×10^{-6}
UDP-MurNAc	4.6×10^{-5}
UDP-MurNAc-L-Ala	1.4×10^{-5}
UDP-MurNAc-dipeptide	1.1×10^{-5}
UDP-MurNAc-tripeptide	0.6×10^{-5}
UDP-MurNAc-pentapeptide	17.5×10^{-5}
D-Alanine	5×10^{4}
L-Alanine	5×10^{-3}
D-Glutamic acid	1×10^{-3}
Diaminopimelic acid	7.5×10^{-4}
D-Alanyl-D-alanine	2.5×10^{-4}
Lipid I	<700 copies per cell
Lipid II	1000–2000 copies per cell

4.3. Levels of synthetase activities

The specific activities of the synthetases of the uridine nucleotide precursors and of the D-alanyl-D-alanine ligase were determined under various growth conditions [56,128,131]. Little variation with the growth rate was observed. This result suggested that these activities may be constitutive. In fast-growing cells, certain of the specific activities determined in vitro appeared as corresponding approximately to the amounts of in vivo synthesized peptidoglycan if pseudo-physiological conditions were considered [52,55]. Otherwise, and especially in slower growing cells, these activities are in excess. The maintenance of a high capacity to synthesize peptidoglycan precursors could be an advantage for survival when cells are submitted to a rapid shift-up. This was further emphasized by the fact that overnight stationary phase cells retained at least 50% or more of these activities [56,128]. Possible variations of the translocase and transferase activities catalysing the formation of the lipid intermediates have not yet been examined.

4.4. In vivo functioning of the pathway: regulatory mechanisms

From our present knowledge of the precursor synthetases and pool levels a number of features of the in vivo functioning of this pathway have emerged, at least for exponential phase cells of *E. coli*. The rate of synthesis of the peptidoglycan will vary several fold between fast and slow growing cells [56]. This can be brought about by variations in the synthetase activities or in the rate of input of substrates or both. Since the enzymatic machinery is more or less constitutive, the control of the synthetase activities could involve cell effectors. A mechanism of this type was proposed for the regulation of the pathway by feed-back inhibition of its first step. In vitro experiments have shown that the UDP-GlcNAc-enolpyruvyl transferase activity from several bacterial species is ca. 50% inhibited by its reaction product at 10^{-4} M [55], by UDP-MurNAc at 10^{-3} M [55,74] and by UDP-MurNAc-tripeptide or UDP-MurNAc-pentapeptide at 5 mM and higher [55,75, 132]. However, the cell concentrations of these precursors in *E. coli* growing cells (Table I) are 20–200-fold lower than the IC_{50} values. Therefore, this mechanism could perhaps have a physiological significance only in the case of extremely high levels of precursor accumulation. This is further substantiated by the fact that high levels of UDP-MurNAc pentapeptide can accumulate to a high level in *S. aureus* upon inhibition of peptidoglycan polymerization by vancomycin [102]. Similarly, the 50% inhibition of the D-Glu-adding-enzyme by its product UDP-MurNAc-L-Ala-D-Glu or of the A_2pm-adding enzyme by its product UDP-MurNAc-tripeptide, both observed in vitro at 1 mM [52], is also probably of little physiological significance. When the synthesis of A_2pm or D-alanyl-D-alanine was inhibited by antibiotics [128,133] the high accumulation of UDP-MurNAc-dipeptide or UDP-MurNAc-tripeptide closely paralleled the deficiency of synthesized peptidoglycan and very limited or no accumulation of the upstream nucleotides was observed. This indicated the absence of a slowing down of any upstream step at least during the lapses of time considered.

If the rate of peptidoglycan synthesis is dependent on the rate of input of substrates into the pathway, a distinction must be made between those like PEP, ATP, L-alanine and A_2pm which are generally at a cell concentration saturating the corresponding synthetases and those like the nucleotide precursors (other than UDP-MurNAc-pentapeptide), D-glutamic acid and D-alanyl-D-alanine which are at cell concentrations lower or close to the K_m values of the corresponding synthetases [52,55,85,90,94]. The substrates of the first type also play a role in other metabolic pathways, whereas those of the second type are specific of the biosynthesis of peptidoglycan. If only the second type of substrates are considered, the question arises as to whether the side-pathways leading to D-glutamic acid and D-alanyl-D-alanine have a limiting effect. A control of the pool of D-alanyl-D-alanine by product inhibition of the D-alanine:D-alanine ligase activities has been investigated. This effect was observed with crude extracts and partially or totally purified activities from E. coli [19,128,131]. It was studied in detail with the ligase from S. faecalis [113]. The ligase specific activity does not vary greatly with growth conditions [128] and is thus generally in excess. It has been proposed that its tight regulation is necessary to prevent the L-alanine cell pool from being converted to D-alanyl-D-alanine via D-alanine [3]. This mechanism controls the excess input of dipeptide but does not seem to limit the formation of UDP-MurNAc-pentapeptide since the pool of UDP-MurNAc-tripeptide is at a fairly low level (Table I). Recent results [26] suggest that in E. coli, the formation of D-glutamic acid is also adjusted to the sole requirements of peptidoglycan synthesis by a mechanism which presumably avoids an excess conversion of L-glutamic acid into its D-form, but which does not, however, limit the formation of UDP-MurNAc-L-Ala-D-Glu, since the UDP-MurNAc-L-Ala pool is also at a low level (Table I).

The relatively high pool level of UDP-MurNAc-pentapeptide (Table I) is presumably a consequence of the reversibility of the reaction leading to lipid I (Fig. 1) in which the equilibrium is greatly in favor of the nucleotide precursor [64,97,98]. In E. coli growing cells, the ratio of UDP-MurNAc-pentapeptide to lipid I was estimated at 140:1 [14] or 300:1 [62]. This could explain why the pool of lipid I is so low. Furthermore, UDP-MurNAc-pentapeptide is apparently at a saturating concentration for the translocase catalysing the formation of lipid I since its cell concentration ($>10^{-4}$ M) [52] is quite higher than the K_m value of ca. 10^{-5} M [98]. The rate of formation of lipid I could therefore eventually be dependent on the pool level of undecaprenyl-phosphate. The question remains as to why the pool of lipid I has to be so low (Table I). It can be speculated that it is essential to limit the pools of the lipid intermediates in order to avoid a too low pool of undecaprenyl phosphate, which is perhaps only a few fold higher than that of lipids I or II [63] and which is used for the synthesis of other cell wall polymers. A second possibility is that higher pool levels of the lipid intermediates could be incompatible with the stability of the membrane. Finally, it should be considered that lipid intermediate pool levels (Table I) imply very fast rates of turnover of ca. 1 s. This raises the question as to whether these intermediates freely diffuse over the whole membrane or whether they are associated with the translocase and transferase which are membrane bound. In this latter case, there could in some way be a coupling between these two activities and also with the subsequent transport of lipid II across the membrane and it use in the polymerization reactions. Evidence for lipid intermediate-protein interactions which could indicate an association with the translocase in a peptidoglycan synthesizing complex has been brought with S. aureus [99 and refs. therein].

The fact that all the nucleotide precursors between UDP-GlcNAc and UDP-MurNAc-tripeptide are not at saturating concentrations for their respective synthetase suggests an unrestricted functioning of the pathway up to UDP-MurNAc-pentapeptide and therefore its dependence upon the rate of input of UDP-GlcNAc, which is controlled at some yet unidentified upstream step. Interestingly, the inhibition of protein synthesis was shown to lead to considerable increases in the pool levels of UDP-GlcNAc and UDP-MurNAc-pentapeptide [127]. This effect not only further stressed the unrestricted functioning of the intermediate steps, but also provided an explanation of how an inhibition of protein synthesis can insure the formation of enough UDP-MurNAc-pentapeptide to sustain peptidoglycan synthesis at a rate that will antagonize fosfomycin- or D-cycloserine-induced lysis [128,134]. The possible involvement of the *relA* gene product in the regulation of the system has also been examined [127,135].

5. Conclusion

The rapid progress made recently in genetic analysis of the biosynthesis of the peptidoglycan unit, initially in *E. coli* and more recently in *B. subtilis,* is noteworthy. The clustering of many of the genes in this pathway with genes involved in septation as well as the similarities between *E. coli* and *B. subtilis* organization are fascinating features. More detailed studies of the transcription of all these genes must now be developed. The possibility of overproducing any precursor synthetase by genetic engineering has already led to the purification to homogeneity of a few of them. In turn, the availability of large amounts of purified synthetases will lead to the development of detailed studies of the mechanism of the reactions they catalyse. This could open the way to the design of efficient inhibitors based on mechanistic considerations and perhaps subsequently to new antibacterial agents. The genetic engineering approach will also be of great help for studying the organization of the translocase and transferase in the membrane.

Undoubtedly, under certain physiological, antibacterial, or mutational constraints, the peptidoglycan of a given bacteria can undergo structural modifications. The extent of such variations is still poorly understood. Further studies of the specificities of the precursor synthetases could perhaps help to better define some of the limits of these structural variations. The determination of the precursor pool levels under various conditions has led, at least in *E. coli*, to a first synoptic view of the in vivo functioning of this pathway between UDP-GlcNAc and UDP-MurNAc pentapeptide. UDP-GlcNAc appears as a crucial point in the regulation of the pathway. Its formation and its dependency upon protein synthesis will have to be analysed in greater detail. The analysis of the pools of the different lipid intermediates also appears as essential to the understanding of the functioning of the membrane steps. This will imply the development of new efficient analytical methods and of detailed studies of the membrane organization of both the enzymes and the lipid intermediates.

References

1. Rogers, H.J., Perkins, H.R. and Ward, J.B. (1980) in: Microbial Cell Walls and Membranes, Chapman and Hall, London, pp. 239–297.

2. Schleifer, K.H. and Kandler, O. (1972) Bacteriol. Rev. 36, 407–477.
3. Ward, J.B. (1984) Pharmacol. Ther. 25, 327–369.
4. Lugtenberg, E.J.J., de Haas-Menger, L. and Ruyters, W.H.M. (1972) J. Bacteriol. 109, 326–335.
5. Matsuzawa, H. and Matsuhashi, M. (1969) Biochem. Biophys. Res. Commun. 36, 682–689.
6. Miyakawa, T., Matsuzawa, H., Matsuhashi, M. and Sugino, Y. (1972) J. Bacteriol. 112, 950–958.
7. Wijsman, H.J.W. (1972) Genet. Res. Comb. 20, 65–74.
8. Chatterjee, A.N. and Young, F.E. (1972) J. Bacteriol. 111, 220–230.
9. Good, C.M. and Tipper, D.J. (1972) J. Bacteriol. 111, 231–241.
10. Billot-Klein, D., Gutmann, L., Collatz, E. and van Heijenoort, J. (1992) Antimicrob. Agents Chemother. 36, 1487–1490.
11. Lugtenberg, E.J.J. and van Schjndel-van Dam, A. (1973) J. Bacteriol. 113, 96–104.
12. Mengin-Lecreulx, D., Parquet, C., Desviat, L.R., Pla, J., Flouret, B., Ayala, J.A. and van Heijenoort, J. (1989) J. Bacteriol. 171, 6126–6134.
13. Ikeda, M., Wachi, M., Jung, H.K., Ishino, F. and Matsuhashi, M. (1991) J. Bacteriol. 173, 1021–1026
14. Mengin-Lecreulx, D., Texier, L., Rousseau, M. and van Heijenoort, J. (1991) J. Bacteriol. 173, 4625–4636.
15. Ikeda, M., Wachi, M., Jung, H.K., Ishino, F. and Matsuhashi, M. (1990) J. Gen. Appl. Microbiol. 36, 179–187.
16. Kröger, M., Wahl, R. and Rice, P. (1991) Nucleic Acids Res. 19, 2023–2043.
17. Mengin-Lecreulx, D., Texier, L. and van Heijenoort, J. (1990) Nucleic Acids. Res. 18, 2810.
18. Michaud, C., Parquet, C., Flouret, B., Blanot, D. and van Heijenoort, J. (1990) 269, 277–280.
19. Zawadzke, L.E., Bugg, T.D.H. and Walsh, C.T. (1991) Biochemistry 30, 1673–1682.
20. Robinson, A.C., Kenan, D.J., Sweeney, J. and Donachie, W.D. (1986) J. Bacteriol. 167, 809–817.
21. Daub, E., Zawadzke, L.E., Botstein, D. and Walsh, C.T. (1988) Biochemistry 27, 3701–3708.
22. Wu, H.C. and Venkateswaran, P.S. (1974) Ann. N.Y. Acad. Sci. 235, 587–592.
23. Doublet, P., van Heijenoort, J. and Mengin-Lecreulx, D. (1993) in: Bacterial Growth and Lysis: Metabolism and Structure of Bacterial Sacculus, Plenum, New York, pp. 139–146.
24. Pucci, M.J., Discotto, L.F. and Dougherty, T.J. (1992) J. Bacteriol. 174, 1690–1693.
25. Doublet, P., van Heijenoort, J. and Mengin-Lecreulx, D. (1992) J. Bacteriol. 174, 5772–5779.
26. Doublet, P., van Heijenoort, J., Bohin, J.P. and Mengin-Lecreulx, D. (1993) J. Bacteriol. 175, 2970–2979.
27. Dougherty, T.J., Thanassi, J.A. and Pucci, M.J. (1993) J. Bacteriol. 175, 111–116.
28. Marquardt, J.L., Siegele, D.A., Kolter, R. and Walsh, C.T. (1992) J. Bacteriol. 174, 5748–5752.
29. Wanke, C., Falchetto, R. and Amrhein, N. (1992) FEBS Lett. 301, 271–276.
30. Daniel, R.A. and Errington, J. (1993) J. Gen. Microbiol. 139, 361–370.
31. Henriques, A.O., de Lencastre, H. and Piggot, P.J. (1992) Biochimie 74, 735–748.
32. Ikeda, M., Sato, T., Wachi, M., Jung, H.K., Ishino, F., Kobayashi, Y. and Matsuhashi, M. (1989) J. Bacteriol. 171, 6375–6378.
33. Miyao, A., Yoshimura, A., Sato, T., Yamamoto, T., Theeragool, G. and Kobayashi, Y. (1992) Gene 118, 147–148.
34. Joris, B., Dive, G., Henriques, A., Piggot, P.J. and Ghuysen, J.M. (1990) Mol. Microbiol. 4, 513–517.
35. Beall, B. and Lutkenhaus, J. (1989) J. Bacteriol. 171, 6821–6834.
36. Chatterjee, A.N. and Park, J.T. (1964) Proc. Natl. Acad. Sci. USA 51, 9–16.
37. Meadow, P.M., Anderson, J.S. and Strominger, J.L. (1964) Biochem. Biophys. Res. Commun. 14, 382–387.
38. Metz, R., Henning, S. and Hammes, W.P. (1983) Arch. Microbiol. 136, 297–299.
39. Pelzer, H. and Reuter, W. (1980) Antimicrob. Agents Chemother. 18, 887–892.
40. Park, J.T. and Johnson, M.J. (1949) J. Biol. Chem. 179, 585–592.
41. Park, J.T. (1952) J. Biol. Chem. 194, 877–904.
42. Chatterjee, A.N. and Perkins, H.R. (1966) Biochem. J. 100, 32 p.
43. Nakatani, T., Araki, Y. and Ito, E. (1968) Biochim. Biophys. Acta 156, 210–212.
44. Rosenthal, S. and Sharon, N. (1964) Biochim. Biophys. Acta 83, 378–380.
45. Takayama, K., David, H.L., Wang, L. and Goldman, D.S. (1970) Biochem. Biophys. Res. Commun. 39, 7–12.

46. Anwar, R.A., Roy, C. and Watson, R.W. (1963) Can. J. Biochem. Physiol. 41, 1065–1072.
47. Lilly, M.D., Clarke, P.H. and Meadow, P.M. (1963) J. Gen. Microbiol. 32, 103–116.
48. Ito, E. and Saito, M. (1963) Biochim. Biophys. Acta 78, 237–247.
49. Ito, E., Nathenson, S.G., Dietzler, D.N., Anderson, J.S. and Strominger, J.L. (1966) Methods Enzymol. VIII, 324–337.
50. Hammes, W., Schleifer, K.H. and Kandler, O. (1973) J. Bacteriol. 116, 1029– 1053.
51. Bochner, B.R. and Ames, B.N. (1982) J. Biol. Chem. 257, 9759–9769.
52. Mengin-Lecreulx, D., Flouret, B. and van Heijenoort, J. (1982) J. Bacteriol. 151, 1109–1117.
53. Schwarz, U. and Weidel, W. (1965) Z. Naturforsch. 20b, 147–153.
54. Flouret, B., Mengin-Lecreulx, D. and van Heijenoort, J. (1981) Anal. Biochem. 114, 59–63.
55. Mengin-Lecreulx, D., Flouret, B. and van Heijenoort, J. (1983) J. Bacteriol. 154, 1284–1290.
56. Mengin-Lecreulx, D. and van Heijenoort, J. (1985) J. Bacteriol. 163, 208–212.
57. Anderson, J.S., Matsuhashi, M., Haskin, M.A. and Strominger, J.L. (1965) Proc. Natl. Acad. Sci. USA 53, 881–889.
58. Izaki, K., Matsuhashi, M. and Strominger, J.L. (1966) Proc. Natl. Acad. Sci. USA 55, 656–663.
59. Braun, V. and Bosch, V. (1973) FEBS Lett. 34, 302–306.
60. Braun, V. (1975) Biochim. Biophys. Acta 415, 335–377.
61. Ramey, W.D. and Ishiguro, E.E. (1978) J. Bacteriol. 135, 71–77.
62. Kohlrausch, U., Wientjes, F.B. and Höltje, J.V. (1989) J. Gen. Microbiol. 135, 1499–1506.
63. van Heijenoort, Y., Gomez, M., Derrien, M., Ayala, J. and van Heijenoort, J. (1992) J. Bacteriol. 174, 3549–3557.
64. Pless, D.D. and Neuhaus, F.C. (1973) J. Biol. Chem. 248, 1568–1576.
65. Taku, A. and Fan, D.P. (1976) J. Biol. Chem. 251, 6154–6156.
66. Higashi, Y., Strominger, J.L. and Sweeley, C.C. (1967) Proc. Natl. Acad. Sci. USA 57, 1878–1884.
67. Umbreit, J.N. and Strominger, J.L. (1972) J. Bacteriol. 112, 1306–1309.
68. Gateau, O., Bordet, C. and Michel, G. (1976) Biochim. Biophys. Acta 421, 395–405.
69. Plapp, R. and Strominger, J.L. (1970) J. Biol. Chem. 245, 3667–3674.
70. Staudenbauer, W. and Strominger, J.L. (1972) J. Biol. Chem. 247, 5095–5102.
71. Gunetileke, K.G. and Anwar, R.A. (1966) J. Biol. Chem. 241, 5741–5743.
72. Strominger, J.L. (1958) Biochim. Biophys. Acta 30, 645–646.
73. Cassidy, P.J. and Kahan, F.M. (1973) Biochemistry 12, 1364–1374.
74. Wickus, G.G. and Strominger, J.L. (1973) J. Bacteriol. 113, 287–290.
75. Zemell, R.I. and Anwar, R.A. (1975) J. Biol. Chem. 250, 3185–3192.
76. Gale, E.F., Cundliffe, E., Reynolds, P.E., Richmond, M.H. and Waring, M.J. (1981) in: The Molecular Basis of Antibiotic Action, 2nd edition, Wiley, London, pp. 49–174.
77. Taku, A., Gunetileke, G. and Anwar, R.A. (1970) J. Biol. Chem. 245, 5012–5016.
78. Wickus, G.G., Rubenstein, P.A., Warth, A. and Strominger, J.L. (1973) J. Bacteriol. 113, 291–294.
79. Anwar, R.A. and Vlaovic, M. (1979) Can. J. Biochem. 57, 188–196.
80. Benson, T.E., Marquardt, J.L., Marquardt, A.C., Etzkorn, F.A. and Walsh, C.T. (1993) Biochemistry 32, 2024–2030.
81. Hishinuma, F., Izaki, K. and Takahashi, H. (1971) Agric. Biol. Chem. 35, 2050–2058.
82. Liger, D., Blanot, D. and van Heijenoort, J. (1991) FEMS Microbiol. Lett. 80, 111–116.
83. Mizuno, Y., Yaegashi, M. and Ito, E. (1973) J. Biochem. 74, 525–538.
84. Nathensen, S.G., Strominger, J.L. and Ito, E. (1964) J. Biol. Chem. 239, 1773–1776.
85. Pratviel-Sosa, F., Mengin-Lecreulx, D. and van Heijenoort, J. (1991) Eur. J. Biochem. 202, 1169–1176.
86. Mizuno, Y. and Ito, E. (1968) J. Biol. Chem. 243, 2665–2672.
87. Anwar, R.A. and Vlaovic, M. (1986) Biochem. Cell. Biol. 64, 297–303.
88. Dietzler, D., Threnn, R., Ito, E. and Strominger, J.L. (1972) An. Asoc. Quim. Argent. 60, 141–148.
89. Abo-Ghalia, M., Michaud, C., Blanot, D. and van Heijenoort, J. (1985) Eur. J. Biochem. 153, 81–87.
90. Michaud, C., Mengin-Lecreulx, D., van Heijenoort, J. and Blanot, D. (1990) Eur. J. Biochem. 194, 853–861.
91. Ito, E. and Strominger, J.L. (1964) J. Biol. Chem. 239, 210–214.
92. Egan, A., Lawrence, P. and Strominger, J.L. (1973) J. Biol. Chem. 248, 3122–3130.
93. Comb, D.G. (1962) J. Biol. Chem. 237, 1601–1604.

54

94. Duncan, K., van Heijenoort, J. and Walsh, C.T. (1990) Biochemistry 29, 2379–2386.
95. Ito, E. and Strominger, J.L. (1962) J. Biol. Chem. 237, 2696–2703.
96. Neuhaus, F.C. and Struve, W.G. (1965) Biochemistry 4, 120–131.
97. Umbreit, J.N. and Strominger, J.L. (1972) Proc. Natl. Acad. Sci. USA 69, 1972–1974.
98. Geis, A. and Plapp, R. (1978) Biochim. Biophys. Acta 527, 414–424.
99. Weppner, W.A. and Neuhaus, F.C. (1979) Biochim. Biophys. Acta 552, 418–427.
100. Isono, F. and Inukai, M. (1991) Antimicrobiol. Agents Chemother. 35, 234–236.
101. Anderson, J.S., Meadow, P.M., Haskin, M.A. and Strominger, J.L. (1966) Arch. Biochem. Biophys. 116, 487–515.
102. Park, J.T. and Chatterjee, A.N. (1966) Methods Enzymol. VIII, 466–472.
103. Ikeda, M., Wachi, M., Ishino, F. and Matsuhashi, M. (1990) Nucleic Acids Res. 18, 1058.
104. Mengin-Lecreulx, D. and van Heijenoort, J. (1990) Nucleic Acids Res. 18, 183.
105. Bupp, K. and van Heijenoort, J. (1993) J. Bacteriol. 175, 1841–1843.
106. Martinez-Carrob, M. and Jenkins, W.T. (1965) J. Biol. Chem. 240, 3538–3546.
107. Tanizawa, K., Asano, S., Masu, Y., Kurimitsu, S., Kagamiyama, H., Tanaka, H. and Soda, K. (1989) J. Biol. Chem. 264, 2450–2454.
108. Diven, W.D. (1969) Biochim. Biophys. Acta 191, 702–706.
109. Nakajima, N., Tanizawa, K., Tanaka, H. and Soda, K. (1986) Agric. Biol. Chem. 50, 2823–2830.
110. Nakajima, N., Tanizawa, K., Tanaka, H. and Soda, K. (1988) Agric. Biol. Chem. 52, 3099–3104.
111. Al-bar, A.M., O'Connor, C.D., Giles, I.G. and Akhtar, M. (1992) Biochem. J. 282, 747–752.
112. Knox, J.R., Lui, H., Walsh, C.T. and Zawadzke, L.E. (1989) J. Mol. Biol. 205, 461–463.
113. Carpenter, C.V. and Neuhaus, F.C. (1972) Biochemistry 11, 2594–2598.
114. Neuhaus, F.C. (1962) J. Biol. Chem. 237, 778–786.
115. Duncan, K. and Walsh, C.T. (1988) Biochemistry 27, 3709–3714.
116. Mullins, L.S., Zawadzke, L.E., Walsh, C.T. and Raushel, F.M. (1990) J. Biol. Chem. 265, 8993–8998.
117. Michaud, C., Blanot, D., Flouret, B. and van Heijenoort, J. (1987) Eur. J. Biochem. 166, 631–637.
118. Ito, E. and Strominger, J.L. (1973) J. Biol. Chem. 248, 3131–3136.
119. Mengin-Lecreulx, D., Michaud, C., Richaud, C., Blanot, D. and van Heijenoort, J. (1988) J. Bacteriol. 170, 2031–2039.
120. Richaud, C., Mengin-Lecreulx, D., Pochet, S., Johnson, E., Cohen, G. and Marlière, P. (1993) J. Biol. Chem. 268, in press.
121. Wyke, A.W. and Perkins, H.R. (1975) J. Gen. Microbiol. 88, 159–168.
122. Bugg, T.D.H., Wright, G.D., Dutka-Malen, S., Arthur, M., Courvalin, P. and Walsh, C.T. (1991) Biochemistry 30, 10408–10415.
123. Stickgold, R.A. and Neuhaus, F.C. (1967) J. Biol. Chem. 242, 1331–1337.
124. Neuhaus, F.C. (1962) J. Biol. Chem. 237, 3128–3135.
125. Wright, G.D. and Walsh, C.W. (1992) Acc. Chem. Res. 25, 468–473.
126. Courvalin, P. (1990) Antimicrob. Agents Chemother. 34, 2291–2296.
127. Mengin-Lecreulx, D., Siegel, E. and van Heijenoort, J. (1989) J. Bacteriol. 171, 3282–3287.
128. de Roubin, M.R., Mengin-Lecreulx, D. and van Heijenoort, J. (1992) J. Gen. Microbiol. 138, 1751–1757.
129. Mengin-Lecreulx, D., Allen, N.E., Hobbs, J.N. and van Heijenoort, J. (1990) FEMS Microb. Lett. 69, 245–248.
130. Gunetileke, K.G. and Anwar, R.A. (1968) J. Biol. Chem. 243, 5770–5778.
131. Lugtenberg, E.J.J. (1972) J. Bacteriol. 110, 26–34.
132. Venkateswaran, P.S., Lugtenberg, E.J.J. and Wu, H.C. (1973) Biochim. Biophys. Acta 293, 570–574.
133. Le Roux, P., Blanot, D., Mengin-Lecreulx, D. and van Heijenoort, J. (1989) in: Peptides 1988, Walter de Gruyter, Berlin, pp. 347–350.
134. Mengin-Lecreulx, D. and van Heijenoort, J. (1990) FEBS Microbiol. Lett. 66, 129–134.
135. Ishiguro, E.E. and Ramey, D.W. (1978) J. Bacteriol. 135, 766–774.

J.-M Ghuysen and R. Hakenbeck (Eds.), *Bacterial Cell Wall*
55

Utilization of lipid-linked precursors and the formation of peptidoglycan in the process of cell growth and division: membrane enzymes involved in the final steps of peptidoglycan synthesis and the mechanism of their regulation

MICHIO MATSUHASHI

Department of Biological Science and Technology, Tokai University, School of High Technology for Human Welfare, Nishino 317, Numazu-shi, Shizuoka-ken, 410-03 Japan

1. Introduction: the general scheme of peptidoglycan biosynthesis through the formation of undecaprenyl-pyrophosphate-linked intermediates

Peptidoglycans are basic constituents of bacterial cell walls. Their structure is common to most eubacteria. However, there are certain divergences among the peptidoglycans characterizing various species of bacteria. Moreover, they are also thought to differ according to the topology of the cell. The enzymes and enzymatic reactions involved in their biogenesis and the mechanism of their regulation are therefore diverse to a certain extent among various species and stages of their cell cycles.

The major pathway of peptidoglycan synthesis from the nucleotide precursors and genes or proteins involved is illustrated in Fig. 1 with *Escherichia coli,* a Gram-negative bacilli form bacterium which has been studied most extensively [1,2], as an example. The pathway consists of (1) formation of UDP-MurNAc-pentapeptide (MurNAc, N-acetylmuramate) through a sequence of cytoplasmic enzyme reactions, (2) transfer of MurNAc-pentapeptide to undecaprenyl-phosphate on the membrane and addition of GlcNAc to form undecaprenyl-pyrophosphate-linked disaccharide (GlcNAc-MurNAc)-pentapeptide (L-Ala-D-Glu-meso-A_2pm-D-Ala-D-Ala) (A_2pm, diaminopimelate), and (3) transfer of disaccharide-pentapeptide residues to form peptidoglycan. This final step of the biosynthesis consists of polymerization of the repeating unit disaccharide-pentapeptide (transglycosylation) and penicillin-sensitive formation of crosslinkages between the amino terminal of meso-A_2pm (D-center) and the carboxy terminal of the penultimate D-alanine residue of the pentapeptide side chains, with removal of one molecule of the terminal D-alanine residue of the pentapeptide for each crosslinkage formed (transpeptidation). In addition, reactions for preparation of the acceptor peptidoglycan and peptidoglycan maturation may occur. The degree of crosslinking of peptidoglycans in the cell wall is usually 20–30%, that is, about one A_2pm residue out of four is involved in the

PATHWAY OF PEPTIDOGLYCAN SYNTHESIS   GENE

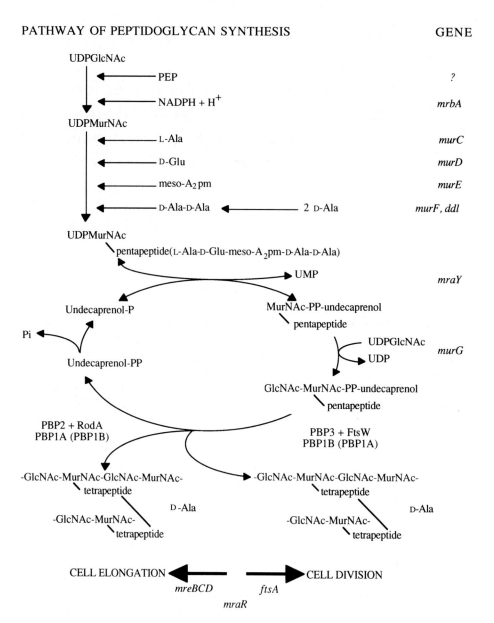

Fig. 1. The major pathway of peptidoglycan synthesis in *E. coli*, showing the genes and proteins involved. The release of the terminal phosphate residue of undecaprenyl pyrophosphate has not yet been thoroughly studied.

crosslinking. Differences in the enzymes involved in the synthesis are believed to cause topological differences in the fine structure of the peptidoglycan network [3].

In Gram-positive bacteria, the peptidoglycan forms a thick multilayer and is chemically more diverse than in Gram-negative bacteria. Its biosynthesis may subsequently diversify still further. The peptidoglycan of *Staphylococcus aureus* contains D-isoglutamine and L-lysine residues in its tetrapeptide side chain [4] and is almost 100% crosslinked, that is, every ε-amino group of L-lysine participates in crosslinkage. Moreover, the ε-amino group of L-lysine and the carboxyl group of D-alanine are crosslinked by a pentaglycine bridge [4]. During biosynthesis, the pentaglycine moiety is formed on the ε-amino group of L-lysine in the lipid-linked MurNAc- or disaccharide-pentapeptide [5–7], and then crosslinkage occurs between the amino group of glycine and the carboxy group of D-alanine. The de novo synthesis of peptidoglycan and its thickening are probably catalyzed by different enzymes [8]. This chapter includes a discussion of methicillin-resistant S. *aureus* because of its current clinical importance.

2. Early experiments on the enzymatic synthesis of peptidoglycans

Early experiments on peptidoglycan synthesis were conducted without sufficient attention to the genetic background of the biosynthesis [9–16]. Particulate membrane fractions had been obtained from exponentially growing normal cells, nucleotide-linked radiolabeled precursors were added, and the radiolabeled, lysozyme-digestible products assayed. Lipid-linked intermediates were isolated in these early experiments by Anderson et al. [10,11,17], and the lipid was identified as undecaprenyl-pyrophosphate [18]. The radiolabel of the nucleotide precursors was incorporated first into hydrophobic materials which migrated close to the front of the paper chromatogram with isobutyric acid/1 M ammonia (5:3) as solvent, and was then converted to the product peptidoglycan which remained at the origin of the paper chromatogram. The purified lipid-linked intermediates were also used as the substrate for direct assay of the transglycosylase activity [11,19–29].

The products formed with membrane preparations from cells of S. *aureus* or *Micrococcus luteus* (previously called M. *lysodeikticus)* were linear peptidoglycan chains lacking crosslinkages (transglycosylation) [5,6,9–11]. However, by using particulate membrane preparations of E. *coli* cells [14–16], the biosynthesis of crosslinked peptidoglycans, the mechanism of crosslinking, and the mode of action of penicillin that inhibits this reaction were elucidated. The membrane preparations of E. *coli* displayed activities of both transglycosylase and transpeptidase. The product peptidoglycan was crosslinked to a reasonable degree (slightly less than 20% as estimated by later research) and the crosslinking reaction was inhibited by β-lactam antibiotics, penicillins and cephalosporins [14,15]. The transpeptidase reaction caused concomitant release of D-alanine, which was also inhibited by β-lactam antibiotics. Although the release of one molecule of D-alanine per crosslinking reaction was expected, the value observed always exceeded the extent of crosslinking. Another reaction causing the release of D-alanine from UDP-MurNAc-pentapeptide, lipid-linked intermediates and the product peptidoglycan was identified [14,15]. This D-alanine carboxypeptidase reaction was thought to maturate the peptidoglycans by terminating the formation of crosslinkages. In fact, multiple D-alanine carboxypeptidase activities were present in E. *coli* [30,31].

Formation of the crosslinked peptidoglycan in S. *aureus* and M. *luteus* with concomitant release of D-alanine was also subsequently demonstrated by Mirelman et al. [32,33], using preparations of a cell wall–membrane complex [32,33]. In both of these reactions, the transglycosylation and transpeptidation in the cell membrane–wall system of these bacteria seemed to be sensitive to penicillin [32,33]. The results suggested that in *M. luteus* and *S. aureus,* formation of the crosslinkage between pre-existing peptidoglycan in the cell walls and a newly synthesized fragment of peptidoglycan is required for efficient chain extension. However, other experiments incorporating radiolabeled glycine or lysine into the peptidoglycan of *S. aureus* cells showed that the formation of uncrosslinked peptidoglycan in vivo was penicillin-insensitive [6].

3. Penicillin-binding proteins are enzymes of peptidoglycan synthesis

3.1. Detection of penicillin-binding proteins

Suginaka et al. [34] solubilized membranes of several bacteria and separated certain proteins, which they called penicillin-binding components, by column chromatography. They found multiple penicillin-binding components in each bacterial species. These penicillin-binding components were thought to be the enzymes catalyzing the peptidoglycan crosslinking reaction, but it was not understood why there were multiple penicillin-binding components. In 1975, Spratt and Pardee [35] utilized the new technique of protein separation by sodium dodecylsulfate polyacrylamide gel electrophoresis (SDS-PAGE) [36] and fluorography for the detection of membrane proteins binding penicillins (penicillin-binding proteins, PBPs). They separated six *E. coli* PBPs which they named PBP1 to PBP6 (PBP1 to 3 being the higher and PBP4 to 6 the lower molecular weight PBPs). The probable functions of each of the penicillin-binding proteins of *E. coli* were ascertained by isolating mutants of *E. coli* defective in some of the penicillin-binding proteins (PBP2 and PBP3) and by binding experiments on specific β-lactam compounds (cephaloridine for PBP1, amidinopenicillin mecillinam for PBP2, and cephalexin for PBP3) [35,37]. The highest molecular weight PBP1 (separated afterwards into PBP1A and a group of PBPs, PBP1B-α, -β and -γ) [38–40] was believed to function in cell elongation, PBP2 in determination of the cell shape, and PBP3 in the formation of septa [37].

3.2. PBP1Bs of E. coli are two-headed enzymes with activities of peptidoglycan transglycosylase and penicillin-sensitive transpeptidase

The correlation of penicillin-binding proteins and peptidoglycan synthetic enzyme activities was, however, first established with PBP1Bs by isolation of an *E. coli* mutant lacking in PBP1Bs [39]. Membrane preparations obtained from the cells of the PBP1B⁻ mutant displayed no peptidoglycan biosynthesizing activity. They lacked not only transpeptidase but also transglycosylase activity [39]. At that time, these two enzyme activities were thought to be the properties of two different enzymes, the explanation being that the two enzyme proteins are so tightly assembled in membranes that a mutation in one of the two proteins influences the activity of the other. PBP1Bs are rather stable, for instance, against heat and detergent treatment [41]. Surprisingly, the heat stability of the penicillin-binding activity of PBP1Bs and that of the transglycosylase activity of the membrane preparations were similar.

PBP1Bs were then isolated from membranes of *E. coli* cells over-producing these proteins and purified to single proteins. Finally, transglycosylase and transpeptidase activities were detected in purified preparations of PBP1Bs [19]. Subsequently, it became clear that both of these two activities reside in the same peptide chain [19–22]. Each component α, β and γ displayed the two enzyme activities [21,22].

The proposed mechanism of synthesis of the crosslinked peptidoglycan catalyzed by PBP1Bs is shown in Fig. 2 [22]. The peptidoglycan strand probably elongates through linkage to undecaprenyl pyrophosphate by transglycosylation and the growing peptidoglycan strands are crosslinked by transpeptidation.

The transglycosylase activity displayed simple kinetics during incubation time, with the degree of crosslinkage of the product formed remaining almost constant at 20–25% during the growth of the glycan chain [22,42]. Lysozyme treatment of the product resulted mainly in bis(disaccharide-peptide) and disaccharide-peptide [22]. The transpeptidase reaction is competitively inhibited by penicillin, a structural analogue of D-alanyl-D-

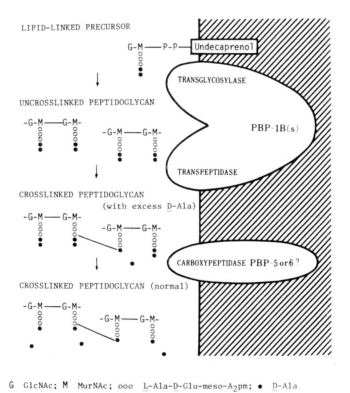

G GlcNAc; M MurNAc; ooo L-Ala-D-Glu-meso-A₂pm; ● D-Ala

Fig. 2. Transglycosylase and transpeptidase reactions of *E. coli* PBP1Bs and carboxypeptidase reaction of low-molecular-weight PBPs. PBP1B components α, β and γ are active two-headed enzymes. The enzymatic activities of PBP1B component δ have not been studied. PBP4 may also function in maturation of peptidoglycan. The reducing end of the oligosaccharides is thought to be bound to undecaprenyl pyrophosphate in the membrane. Reproduced from [22].

alanine [4,43]. On the other hand, penicillin and other β-lactam antibiotics do not inhibit but rather significantly enhance the transglycosylase activity at concentrations inhibiting the transpeptidase activity [15,22]. The reason for this enhancement of the transglycosylase activity of E. coli PBP1Bs by penicillin is unknown.

3.3. Other high molecular weight PBPs of E. coli also display dual enzyme activities

Membranes of E. coli express multiple transglycosylase-transpeptidase activities which are possessed by other higher molecular weight membrane PBPs, that is, 1A, 2 and 3, yet only the activities of PBP1Bs among the many PBPs were detected under the conventional assay conditions [14,15].

Subsequently, all the other higher molecular weight PBPs of E. coli were isolated and their transglycosylase and transpeptidase activities were detected either in purified PBP preparations (PBP1A [20,26,28,29] and PBP3 [27]) or in membrane preparations lacking PBP1Bs and containing disproportionately large amounts of the target PBP (PBP2 [44,45] and PBP3 [46,47]). The properties and proposed functions of E. coli PBPs are summarized in Table I.

Specific conditions were required for the enzyme assay of E. coli PBPs other than PBP1Bs. Lower pH (pH 6.0) and some divalent metal ions such as cobalt were stimulative for the activities of PBP1A [26,29]. The enzyme activities of PBP2 required a higher pH (pH 8.5) and were stimulated by high concentrations of Mg-EDTA (14–22 mM) (EDTA, ethylenediaminetetraacetate) [44,45] while those of PBP3 also required a higher pH (pH 8.0) and were stimulated by $CoCl_2$ (20 mM) and Mg-EDTA (5–15 mM) [27,46,47]. At first, for reasons not yet completely understood, utilization of the lipid-linked intermediate was not observed in membrane preparations of E. coli and purified E. coli PBP1Bs, as in the membranes of Gram-positive bacteria. This difficulty was overcome by pretreating the lipid-intermediate with a buffer and methanol [22]. Heijenoort and Heijenoort [23] used deoxycholate to effect utilization of the lipid intermediate. Purified PBP1A and PBP3 from E. coli did not require such pretreatment but required lipid-linked intermediates of different purities [29]. This may be one of the reasons why other investigators failed to demonstrate their activities [25].

3.4. Peptidoglycan synthetic enzyme activities of E. coli PBP1A

The kinetics of the transglycosylase activity of PBP1A displayed a typical sigmoidal pattern as a function of the time of incubation and the degree of crosslinkage increased with time of incubation [42]. After prolonged incubation, the product was hyper-crosslinked, and the degree of crosslinking reached a maximum of 39%. Lysozyme digestion of the hypercrosslinked product caused formation of tris(disaccharide-peptide), bis(disaccharide-peptide) and disaccharide-peptide [28]. The crosslinking reaction was hypersensitive to most β-lactam antibiotics [20,26,48], but surprisingly, transglycosylase activity in this reaction was also strongly inhibited by low concentrations of certain β-lactam antibiotics [48]. For example, 0.05 μg/ml of the carbapenem imipenem caused 50% inhibition of transglycosylase activity. This inhibition of transglycosylase activity by β-lactam antibiotics was, however, not complete even at concentrations as high as 1 μg/ml

TABLE I

Penicillin-binding proteins of *E. coli*

PBP	Relative molecular mass	Gene and mutant phenotype	Physiological function	Enzymatic activities
1A	93 500	mrcA(ponA) None[a]	Cell elongation	Peptidoglycan-transglycosylase-transpeptidase
1Bs	94 100[b]	mrcB(ponB) Cell lysis	Cell elongation (and cell division)	Peptidoglycan-transglycosylase-transpeptidase
2	70 867	pbpA(mrdA) Rounding of cell[c]	Cell shape determination or initiation of cell growth in association with RodA protein	Peptidoglycan-transglycosylase-transpeptidase[d]
3	63 850	ftsI(pbpB) Filamentous cell	Cell division in association with FtsW protein	Peptidoglycan-transglycosylase-transpeptidase[e]
4	49 568	dacB None[f]	Cell wall lysis and maturation	D-alanine carboxy-peptidase and DD-endopeptidase[g]
5	41 340	dacA None[f]	Cell wall maturation	D-Alanine carboxy-peptidase
6	–	– None[f]	Cell wall maturation	D-Alanine carboxy-peptidase

[a]Lethal when combined with lack of PBP1Bs

[b]Four components of PBP1B, α (M_r 94 100), β, γ (M_r 88 800) and δ.

[c]Deletion mutation is lethal.

[d]Coupled transglycosylase-transpeptidase activities measured in crude membrane preparations containing disproportionately large amounts of PBP2 and RodA.

[e]Requirement for FtsW measured in crude membrane preparations containing disproportionately large amounts of PBP3 and FtsW.

[f]Single mutations (also deletion mutation) are non-lethal. Double mutation in PBP4 and PBP5 also non-lethal.

[g]Endopeptidase activity that splits D-Ala-meso-A$_2$pm linkage of peptidoglycan is a type of carboxypeptidation.

of imipenem, whereas the transpeptidase activity was inhibited by 50% with 0.05 μg/ml and almost 100% with 1 μg/ml of imipenem. Further studies are required in order to explain the mechanism of inhibition of the transglycosylase activity of PBP1A.

The formation of uncrosslinked oligomers (dimer and up) of disaccharide-pentapeptide linked to lipid was detected [29] in experiments on transglycosylase-transpeptidase assay with purified *E. coli* PBP1A. These lipid-linked oligomers could be intermediates of peptidoglycan synthesis, and this hypothesis, if correct, suggests that the mechanism of elongation of the peptidoglycan chain is as follows: GlcNAc-MurNAc(-pentapeptide)-PP-undecaprenol + [GlcNAc-MurNAc(-pentapeptide)]$_n$-PP-undecaprenol $\rightarrow$ [GlcNAc-MurNAc(-pentapeptide)]$_n$-GlcNAc-MurNAc(-pentapeptide)-PP-undecaprenol + PP-undecaprenol, where PP-undecaprenol linked to the n-mer is removed. This mechanism of chain elongation is consistent with the finding that the extension of glycan chains in *Bacillus licheniformis* occurs by addition of the newly synthesized disaccharide-pentapeptide unit at the reducing end [49].

3.5. PBP2 and RodA of E. coli

The enzymatic activities of PBP2 require the presence of another membrane protein RodA [44,45], the product of the *rodA* (*mrdB*) gene [50–52]. The transglycosylase and the mecillinam-sensitive transpeptidase activities of PBP2 could be measured in membrane preparations containing large amounts of PBP2 and RodA, when the interfering activities of PBP1Bs had been eliminated from the membrane by using a PBP1Bs⁻ mutant [44,45]. Moreover, both transglycosylase and transpeptidase activities in the membranes became thermosensitive when membranes containing the thermosensitive RodA protein were used [45]. The β-lactam antibiotic mecillinam, which specifically binds to PBP2, inhibited the transpeptidase activity of the above membranes almost completely at the concentration of 1 μg/ml, and simultaneously enhanced the transglycosylate activity to 130–150% of the original. Purified PBP2 did not display these enzyme activities, probably due to the loss of the RodA protein from purified preparations. Experiments intended to reassemble the two proteins, PBP2 and RodA, by sonication of a mixture of cells containing PBP2 and cells containing RodA were unsuccessful [45].

3.6. PBP3 and FtsW of E. coli

Like PBP2, PBP3 also seemed to require another membrane protein for its enzymatic activities. This was suggested by the genetic information that the protein FtsW, the product of the *ftsW* gene, is required for cell division [53], and is structurally very similar to RodA [54]. Purified preparations of PBP3 obtained from membranes containing excess amounts of PBP3 and FtsW displayed transglycosylase and transpeptidase activities [27]. The transpeptidase activity was inhibited by low concentrations of β-lactam antibiotics that inhibit cell division and cause formation of filamentous cells. Both RodA and FtsW proteins are highly hydrophobic, and detection of these proteins in PBP preparations by gel electrophoresis is uncertain. The purified PBP3 preparations probably contained the FtsW protein bound tightly to PBP3. Hence, a reconstitution experiment was attempted. Unlike PBP2 and RodA, the reconstitution of PBP3 and FtsW seemed to work [46,47]. In control experiments, particulate membrane preparations obtained by sonication of cells overproducing PBP3 and that of cells overproducing FtsW displayed low activities of peptidoglycan synthesis from the UDP-linked precursors. However, after mixing the above two cell suspensions at a ratio of 1:1 and sonicating, the 'reconstituted' particulate membrane preparation displayed twice as high a level of peptidoglycan synthesis activity as in the control experiments [46,47]. More surprisingly, the quantity of lipid-linked intermediates was also twice as high as in the control experiments, suggesting a function of the FtsW protein in the utilization of lipid-linked intermediates [46,47]. One possible explanation for the accumulation of lipid-linked intermediates would be that FtsW, in association with PBP3 enhances the rate of membrane transport of lipid-linked disaccharide pentapeptide. This is a tempting speculation, although nothing definite, either biochemically or genetically, is known about the mechanism of membrane transport of the lipid-linked intermediates.

RodA and FtsW proteins possess highly similar amino acid sequences and hydrophobicity patterns [54]. These two proteins also exhibit homology with the SpoV protein of *Bacillus subtilis,* which functions in spore formation. The RodA, FtsW and SpoVE proteins probably function in association with each partner PBP in the elongation of the cell (RodA and PBP2), the formation of septa (FtsW and PBP3) and the formation of spores (SpoVE and unknown PBP) [54], as illustrated in Fig. 3.

3.7. Low molecular weight PBPs of E. coli

PBP4 possesses a highly penicillin-sensitive D-alanine carboxypeptidase activity that splits the terminal D-alanine molecule of the pentapeptide side chain and also the D-alanyl-meso-A_2pm linkage of the crosslinked peptidoglycan [55,56]. PBP5 and PBP6 catalyze normal D-alanine carboxypeptidase reactions moderately sensitive to penicillin [57–59]. The major functions of lower molecular weight PBPs could include both the maturation of peptidoglycans [14,15] and the preparation of the site of de novo peptidoglycan synthesis [3,60].

3.8. Future problems concerning enzymatic functions

Although experiments with purified PBPs provided an enormous amount of information about the mechanism of enzymatic synthesis of peptidoglycans, exact information about the functions and activities of each of these proteins is sometimes unavailable from in vitro experiments alone. Activities of peptidoglycan transglycosylases which are not PBPs have been detected in *E. coli* [61], *M. luteus* [9–11,62], *S. aureus* [9–11,62] and *Streptococcus pneumoniae* [63]. However, since no mutants lacking these enzyme activities have been isolated, their functions are unknown. Membrane preparations and isolated PBPs act with similar efficiencies upon several substrates for transglycosylation, namely, lipid-linked disaccharides with pentapeptide, tetrapeptide and tripeptide, and those containing A_2pm and L-lysine. The crosslinking reaction is much more specific; only a substrate with an appropriate pentapeptide containing the regular amino acid is utilized. Purified *E. coli* PBP1A, 1Bs and 3 have so far required lipid-linked disaccharide pentapeptide containing A_2pm as the sole substrate for formation of crosslinked products. A

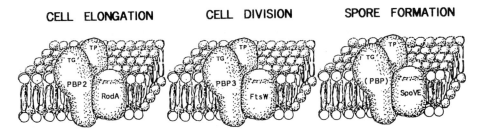

Fig. 3. Putative scheme of association of RodA (*E. coli*), FtsW (*E. coli*) and SpoVE (*B. subtilis*) proteins with their respective partner PBPs for functions in cell elongation, cell division and spore formation, respectively. Top, external surface of cytoplasmic membrane; bottom, internal surface.

more organized semi-vitro system using ether-treated permeabilized cells was reported to demonstrate that substrates containing tripeptide and pentapeptide are required for septum peptidoglycan synthesis [64]. A cell membrane-wall system of *Gaffkya homari* required substrates with tetrapeptide and pentapeptide for crosslinking [65]. Complex enzymes in the membranes probably function together for the synthesis of the fine structure of peptidoglycans. It is necessary to reconstruct precise complex enzyme systems with mechanisms more closely related to natural biosynthesis. This would require purification of all the related proteins and reconstruction of a complete biosynthetic system [46,47,66]. Another approach would be to crystallize the complex of associating proteins and study the crystallographic structure. This step may ultimately be necessary, but would be extremely time-consuming.

3.9. Structural analysis of PBPs

The structure of *E. coli* PBPs have been studied by peptide analysis [57,67–71], and their amino acid sequences have been deduced from the DNA sequences of the corresponding cloned structural genes (PBP1A [72], 1Bs [72], 2 [73], 3 [74], 4 [75] and 5 [76]). The two components α and γ are encoded on the same sequence and translated using alternative initiation sites [72,77]. The component β is formed by proteolysis of the component α [78] and likewise the smallest component δ is formed from the component γ by proteolysis [79]. In PBP1Bs, the transglycosylase domain is located at the N-terminal, and the transpeptidase domain at the C-terminal end [22]. All other higher molecular weight PBPs have amino acid sequences similar to PBP1Bs, and the two enzymatic domains have been deduced from those of PBP1Bs. The penicillin-binding serine residue is located close to the head of the transpeptidase domains of PBP1A [70], PBP1Bs [68,70], PBP2 [70,71] and PBP3 [69,70] of *E. coli*. In order to investigate their three-dimensional structure, some researchers have attempted to crystallize PBPs either in detergents (because of their strong hydrophobicity) or in buffers, after removing the membrane-spanning regions which interfere with solubilization of the protein. *E. coli* PBP1B-δ, the smallest component of PBP1Bs, was crystallized in a buffer containing β-octylglucoside, forming large hexagonal pillars with sharp edgelines [79], but the crystals were not adequate for crystallography. The truncated PBP5 has also been crystallized, and studied crystallographically [80].

4. The penicillin-binding protein of methicillin-resistant Staphylococcus aureus

The role of PBPs in the development of penicillin resistance has been investigated in several organisms [34,81]. This topic is further discussed in other chapters in this book. Here, only a study of the PBP of *S. aureus* is described. A PBP present in methicillin-resistant *S. aureus* (MRSA), first isolated in England in 1960 [82], is interesting both structurally and evolutionarily [83–85]. This PBP [86], which is called MRSA-PBP [83,85], PBP2' or PBP2a, is produced in large amounts in MRSA, often by induction with β-lactam compounds [87,88], and displays very low affinities to most β-lactam antibiotics. Therefore, this is the only PBP in the cell that is responsible for the crosslinking reaction of peptidoglycans in the presence of a high concentration of β-lactam antibiotic. The gene *mecA* which encodes MRSA-PBP was cloned in the author's laboratory [89], and shown to cause methicillin-resistance in a common *S. aureus* strain [90]. The deduced amino acid sequence of the *mecA* gene [83] included the putative C-terminal transpeptidase domain and the transglycosylase domain upstream. Major portions of the two domains were highly similar to the PBP2 and PBP3 of *E. coli*. A highly interesting feature of this protein or its coding DNA is that the N-terminal sequences are similar to those of the penicillinase of *S. aureus*, indicating that the molecule has two ancestors, with the short N-terminal portion originating from a penicillinase and the major C-terminal portion from a PBP [83]. The *mecA* gene probably evolved by fusion of a penicillinase gene and a gene coding for a PBP with low binding affinities to β-lactam antibiotics (Fig. 4). It is not known when the fusion of the two genes causing the evolution of MRSA-PBP took place, and from what strains of bacteria these two ancestral genes were derived. The *mecA* DNA did not hybridize with any DNA-fragment of *S. aureus* [84,85], so the regular PBP genes in the contemporary *S. aureus* strains are not the ancestors of the PBP-portion of the MRSA-PBP gene *mecA*. DNA sequences of several staphylococcal *mecA* genes have been compared and so far all of them have been similar. In particular, the region of the putative junction of the penicillinase-like sequence and PBP-like sequence was completely identical for different staphylococcal strains (J.H. Lee, unpublished experimental data). The result suggests that the evolution of MRSA-PBP by gene fusion was a unique event [84,85]. This conclusion has also been supported by other investigators [91].

For purposes of enzymological and crystallographical experiments, MRSA-PBP has been isolated from membranes of methicillin-resistant *S. aureus* and from *E. coli* that carried a plasmid containing *mecA* introduced in order to overproduce this protein. MRSA-PBP should also display dual activities of transglycosylase and transpeptidase with low sensitivities to β-lactam antibiotics. However, efforts to demonstrate these activities have so far not succeeded.

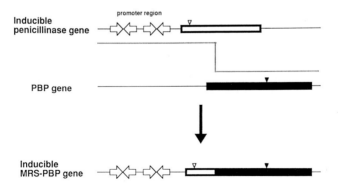

Fig. 4. Evolution of an inducible penicillin-binding protein of methicillin-resistant *S. aureus* by fusion of an inducible penicillinase gene and a gene coding for a penicillin-binding protein with low affinities to β-lactam antibiotics [2]. Large arrows indicate repeated palindrome sequences at the promoter region, to which the repressor is thought to bind. Small arrowheads indicate the penicillin-binding serine residue. Regulatory genes for the induction of penicillinase and MRSA-PBP are located head-to-head on the other side of the promoter regions.

5. Genes involved in peptidoglycan synthesis and its regulation

Genes involved in peptidoglycan synthesis are located on the *E. coli* chromosome mostly as a relatively small number of gene clusters together with other genes involved in cell growth and division (Fig. 5). The *mra* (murein a) [92] cluster region at 2 min on the *E. coli* chromosome map [93] is the largest one, encompassing nine structural genes for the peptidoglycan synthetic enzymes PBP3, MurE, MurF, MraY, MurD, MurG, MurC and Ddl and an associated protein FtsW, flanked by several genes for cell growth and division, such as the regulator genes *mraR, ftsA, ftsQ, ftsZ* and *envA* (Table II). The *mraR* gene is involved in cell division and probably also in cell growth [53,94].

  ftsQ, ftsA and *ftsZ* are genes involved in cell division and *envA* in the separation of cells after the division. Most of the genes in the *mra* area are closely adjacent to each other, and there are overlaps of frames in seven places (marked with * in Table II). Most of the *mra* area probably forms an operon, or operons. There are SOS box-like sequences closely upstream from *mraR,* although it is not known whether the region is under SOS control. It has also been observed that amplification of short chromosomal fragments encompassing the *mraR* gene and at least a part of the *ftsI* gene exerted a lethal effect on *ftsQ, ftsA* and *ftsZ* mutant cells at their original permissive temperatures [95]. The small

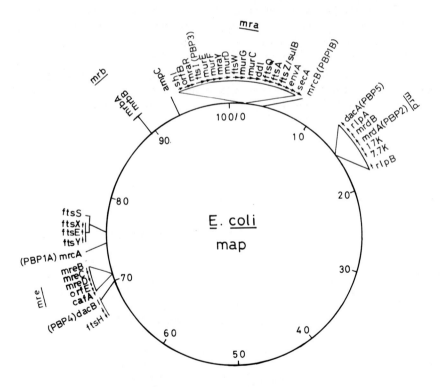

Fig. 5. *E. coli* genetic map showing clusters of genes involved in peptidoglycan synthesis and related genes. For genetic symbols, see the text and reference [93].

TABLE II

Alignment of genes of *mra* and flanking regions of *E. coli* chromosome

Gene	Codon number	Function (activity of product protein)	Ref.[a]
shl	334	Transcription regulator	(1)
orfB *b	248	Unknown	(2)
mraR	121	Control of cell growth and division	(3,4)
ftsl *	588	Septum peptidoglycan synthesis (peptidoglycan transglycosylase-transpeptidase)	(5)
murE *	495	UDP-MurNAc-tripeptide synthesis (meso-diaminopimelate adding enzyme)	(6,7)
murF *	452	UDP-MurNAc-pentapeptide synthesis (D-alanyl-D-alanine adding enzyme)	(6,8)
mraY	360	Lipid-linked MurNAc-pentapeptide synthesis (undecaprenyl-phosphate-UDP-MurNAc-pentapeptide phospho-MurNAc-pentapeptide transferase)	(9)
murD	438	UDP-MurNAc-dipeptide synthesis (D-glutamate adding enzyme)	(10)
ftsW *	414	Septum-peptidoglycan synthesis (PBP3-associated protein)	(3,11)
murG	355	Lipid-linked GlcNAc-MurNAc-pentapeptide synthesis (undecaprenyl-PP-MurNAc-pentapeptide-UDPGlcNAc GlcNAc transferase)	(12,13)
murC *	491	UDP-MurNAc-L-alanine synthesis (L-alanine adding enzyme)	(14,15)
ddl	306	D-alanyl-D-alanine synthesis	(14,16)
ftsQ *	276	Cell division	(16)
ftsA	420	Positive control of cell division	(17,18)
ftsZ	383	Cell division	(19)
envA	305	Cell separation	(20)

[a]References to functions or enzymatic activities: (1) Leclerc, G., Noël, S. and Drapeau, G.R. (1990) J. Bacteriol. 172, 4696–4700; (2) Gómez, M.J., Fluoret, B., van Heijenoort, J. and Ayala, J.A. (1990) Nucleic Acids Res. 18, 2813; (3) Ishino, F., Jung, H.K., Ikeda, M., Doi, M., Wachi, M. and Matsuhashi, M. (1989) J. Bacteriol. 171, 5523–5530; (4) Ueki, M., Wachi, M., Jung,. H.K., Ishino, F. and Matsuhashi, M. (1992) J. Bacteriol. 174, 7841–7843; (5) Ishino, F. and Matsuhashi, M. (1981) Biochem. Biophys. Res. Commun. 101, 905–911; (6) Lugtenberg, E.J.J. and van Schijndel-van Dam, A. (1972) J. Bacteriol. 110, 35–46; (7) Michaud, C., Parquet, C., Flouret, B., Blanot, D. and van Heijenoort, J. (1990) Biochem. J. 269, 277–280; (8) Parquet, C., Flouret, B., Mengin-Lecreulx, D. and van Heijenoort, J. (1989) Nucleic Acids Res. 17, 5379; (9) Ikeda, M., Wachi, M., Ishino, F. and Matsuhashi, M. (1991) J. Bacteriol. 173, 1021–1026; (10) Mengin-Lecreulx, D. and van Heijenoort, J. (1990) Nucleic Acids Res. 18, 183; (11) Ikeda, M., Sato, T., Wachi, M., Jung, H.K., Ishino, F., Kobayashi, Y. and Matsuhashi, M. (1989) J. Bacteriol. 171, 6375–6378; (12) Mengin-Lecreulx, D., Texier, L., Rousseau, M. and van Heijenoort, J. (1991) J. Bacteriol. 173, 4625–4636; (13) Ikeda, M., Wachi, M. and Matsuhashi, M. (1992) J. Gen. Appl. Microbiol. 38, 53–62; (14) Miyakawa, T., Matsuzawa, H., Matsuhashi, M. and Sugino, Y. (1972) J. Bacteriol. 112, 950–958; (15) Ikeda, M., Wachi, M., Jung, H.K., Ishino, F. and Matsuhashi, M. (1990) Nucleic Acids Res. 18, 4014; (16) Robinson, A.C., Kenan, D.J., Sweeney, J. and Donachie, W.D. (1986) J. Bacteriol. 167, 809–817; (17) Donachie, W.D., Begg, K.J., Lutkenhaus, J.F., Salmond, G.P.C., Martinez-Salas, E. and Vicente, M. (1979) J. Bacteriol. 140, 388–394; (18) Tormo, A. and Vicente, M. (1984) J. Bacteriol. 157, 779–784; (19) Lutkenhaus, J.F., Wolf-Watz, H. and Donachie, W.D. (1980) J. Bacteriol. 142, 615–620; (20) Normark, S. (1970) Genet. Res. 16, 63–78.
[b]Overlapping of frames.

area upstream from *ftsI* may well become a target of intense investigation on the regulatory mechanisms of cell division and growth.

The *mrd* (murein d) region at 14.5 min [51] encompasses genes for PBP2, PBP5 and RodA as well as genes encoding two lipoproteins [96]. The *mre* (murein e) region at 71 min encompasses regulator genes *mreB, mreC* and *mreD* [48,66,97], the functionally unknown gene *orfE* [98], and the possible cytoskeletal gene *cafA* [98,99]. The *mreB* gene negatively regulates cell division, with repression of PBP3 formation [100], while *ftsA* positively regulates this function [101,102]. However, these two proteins display some mutual similarity in their amino acid sequences [103,104], as well as similarities with the heat shock protein Hsp70 of *Xenopus* and DnaK of *E. coli* [1]. These proteins may constitute a FtsA-MreB superfamily. MreC and MreD are structurally different from MreB, but function similarly to MreB in the cell cycle, their defects or deletion commonly causing a rounded cell shape [66,97]. One peculiar functional feature of the three *mre* genes is that all their mutations cause simultaneous overproduction of PBP1Bs and PBP3 [66,105]. The simultaneous regulation of PBP1Bs and PBP3 by the *mre* genes suggests that these two kinds of PBPs function in a common process, i.e. cell division. It was previously reported that a suppressive thermoresistant mutation of an *mrcB* (PBP1B⁻) strain, which was tentatively called the *mreA* mutation [48], caused simultaneous over-production of PBP1A and 2 [39]. Its gene locus has not been precisely determined and the mechanism of its regulatory function has not been studied. However, the simultaneous regulation of PBP1A and PBP2 by *mreA* suggests a related function of these two PBPs in the determination of cell shape or initiation of cell elongation. On the other hand, the high affinity of PBP1Bs to the β-lactam antibiotic cephaloridine, which causes cell lysis, as well as the mutual compensatability of PBP1A and 1Bs [39,40], suggest that PBP1A and PBP1Bs perform certain functions in cell elongation. PBP1Bs probably function in both cell elongation and division, and PBP1A at least in cell elongation.

The structural gene of PBP4, *dacB* [55], belongs to another cluster, while the genes of PBP1A, *mrcA* (*ponA*) [40] and PBP1B, *mrcB* (*ponB*) [39,40] are located separately.

The genes involved in peptidoglycan synthesis and its possible regulatory pattern are shown in Fig. 1.

References

1. Matsuhashi, M., Wachi, M. and Ishino, M. (1990) Res. Microbiol. 141, 89–103.
2. Matsuhashi, M., Wachi, M., Ishino, F., Ikeda, M., Okada, Y., Jung, H.K., Ueki, M., Tomioka, S., Pankrushina, A.N. and Song., M.D. (1993) in: H. Kleinkauf (Ed.), 50 Years of Penicillin Application – History and Trends, Videopress, Prague, in press.
3. Höltje, J.-V. and Glauner, B. (1990) Res. Microbiol. 141, 75–89.
4. Strominger, J.L., Izaki, K., Matsuhashi, M. and Tipper., D.J. (1967) Fed. Proc. 26, 9–22.
5. Matsuhashi, M., Dietrich, C.P. and Strominger, J.L. (1965) Proc. Natl. Acad. Sci. USA 54, 587–594.
6. Matsuhashi, M., Dietrich, C.P. and Strominger., J.L. (1967) J. Biol. Chem. 242, 3191–3206.
7. Kamiryo, T. and Matsuhashi, M. (1972) J. Biol. Chem. 247, 6306–6311.
8. Curtis, N.A.C., Hayes, M.V., Wyke, A.W. and Ward., J.B. (1980) FEMS Microbiol. Lett. 9, 263–266.
9. Anderson, J.S., Meadow, P.M., Haskin, M.A. and Strominger., J.L. (1966) Arch. Biochem. Biophys. 116, 487–515.

10. Anderson, J.S., Matsuhashi, M., Haskin, M.A. and Strominger, J.L. (1965) Proc. Natl. Acad. Sci. USA 53, 881–889.
11. Anderson, J.S., Matsuhashi, M., Haskin, M.A. and Strominger, J.L. (1967) J. Biol. Chem. 242, 3180–3190.
12. Chatterjee, A.N. and Park., J.T. (1964) Proc. Natl. Acad. Sci. USA 51, 9–16.
13. Struve, W.G. and Neuhaus., F.C. (1965). Biochem. Biophys. Res. Commun. 18, 6–12.
14. Izaki, K., Matsuhashi, M. and Strominger., J.L. (1966) Proc. Natl. Acad. Sci. USA 55, 656–663.
15. Izaki, K. Matsuhashi, M. and Strominger., J.L. (1968) J. Biol. Chem. 243, 3180–3192.
16. Araki, Y., Shimada, A. and Ito, E. (1966) Biochem. Biophys. Res. Commun. 23, 518–525.
17. Anderson, J.S. and Strominger., J.L. (1965) Biochem. Biophys. Res. Commun. 21, 516–521.
18. Higashi, Y., Strominger, J.L. and Sweeley, C.C. (1967) Proc. Natl. Acad. Sci. USA 57, 1878–1884.
19. Nakagawa, J., Tamaki, S. and Matsuhashi, M. (1979) Agric. Biol. Chem. 43,1379–1380.
20. Matsuhashi, M., Ishino, F., Nakagawa, J., Mitsui, K., Nakajima-Iijima, S., Tamaki, S. and Hashizume, T. (1981) in: M.R.J. Salton and G.D. Shockman (Eds.), β-Lactam Antibiotics – Mode of Action, New Developments, and Future Prospects, Academic Press, New York, pp. 169–184.
21. Nakagawa, J. and Matsuhashi, M. (1982) Biochem. Biophys. Res. Commun. 105, 1546–1553.
22. Nakagawa, J., Tamaki, S., Tomioka, S. and Matsuhashi., M. (1984) J. Biol. Chem. 259, 13937–13946.
23. van Heijenoort, Y. and van Heijenoort., J. (1980) FEBS Lett. 110, 241–244.
24. Suzuki, H., van Heijenoort, Y., Tamura, T., Mizoguchi, J., Hirota, Y. and van Heijenoort., J. (1980) FEBS Lett. 110, 245–249.
25. Tamura, T., Suzuki, H., Nishimura, Y., Mizoguchi, J. and Hirota, Y. (1980) Proc. Natl. Acad. Sci. USA 77, 4499–4503.
26. Ishino, F., Mitsui, K., Tamaki, S. and Matsuhashi, M. (1980) Biochem. Biophys. Res. Commun. 97, 287–293.
27. Ishino, F. and Matsuhashi, M. (1981) Biochem. Biophys. Res. Commun. 101, 905–911.
28. Tomioka, S., Ishino, F., Tamaki, S. and Matsuhashi, M. (1982) Biochem. Biophys. Res. Commun. 106, 1175–1182.
29. Tomioka, S., Ishino, F. and Matsuhashi, M., unpublished results.
30. Nguyen-Distèche, M., Pollock, J.J., Ghuysen, J.M., Puig, J., Reynolds, P., Perkins, H.R., Coyette, J. and Salton, M.R.J. (1974) Eur. J. Biochem. 41, 457–463.
31. Tamura, T., Imae, Y. and Strominger, J.L. (1976) J. Biol. Chem. 251, 414–423.
32. Mirelman, D., Bracha, R. and Sharon, N. (1972) Proc. Natl. Acad. Sci. USA 69, 3355–3359.
33. Mirelman, D., and Sharon, N. (1972) Biochem. Biophys. Res. Commun. 46,1909–1917.
34. Suginaka, H., Blumberg, P.M. and Strominger, J.L. (1972) J. Biol. Chem. 247, 5279–5288.
35. Spratt, B.G. and Pardee, A.B. (1975) Nature 254, 516–517.
36. Laemmli, U.K. and Favre, M. (1973) J. Mol. Biol. 80, 575–599.
37. Spratt, B.G. (1975) Proc. Natl. Acad. Sci. USA 72, 2999–3003.
38. Spratt, B.G., Jobanputra, V. and Schwarz, U. (1977) FEBS Lett. 79, 374–378.
39. Tamaki, S., Nakajima, S. and Matsuhashi, M. (1977) Proc. Natl. Acad. Sci. USA 74, 5472–5476.
40. Suzuki, H., Nishimura, Y. and Hirota, Y. (1978) Proc. Natl. Acad. Sci. USA 75, 664–668.
41. Nakagawa, J., Matsuzawa, H. and Matsuhashi, M. (1979) J. Bacteriol. 138, 1029–1032.
42. Matsuhashi, M., Nakagawa, J., Tomioka, S., Ishino, F. and Tamaki, S. (1982) in: S. Mitsuhashi (Ed.), Drug Resistance in Bacteria, Genetics, Biochemistry, and Molecular Biology, Japan Scientific Societies Press, Tokyo and Thieme-Stratton, New York, pp. 297–310.
43. Tipper, D.J. and Strominger, J.L. (1965) Proc. Natl. Acad. Sci. USA 54, 1133–1141.
44. Ishino, F., Tamaki, S., Spratt, B. G. and Matsuhashi, M. (1982) Biochem. Biophys. Res. Commun. 109, 689–696.
45. Ishino, F., Park, W., Tomioka, S., Tamaki, S., Takase, I., Kunugita, K., Matsuzawa, H., Asoh, S., Ohta, T., Spratt, B.G. and Matsuhashi, M. (1986) J. Biol. Chem. 261, 7024–7031.
46. Matsuhashi, M., Pankrushina, A.N., Ikeda, M., Wachi, M. and Ishino, F. (1993) in: H. Kleinkauf (Ed.), 50 Years of Penicillin Application – History and Trends, Videopress, Prague, in press.
47. Matsuhashi, M., Pankrushina, A. N. and Ikeda, M. (1992) Bulletin of School of High-Technology for Human Welfare Tokai University 1, 183–200.
48. Matsuhashi, M., Ishino, F., Tamaki, S., Nakajima-Iijima, S., Tomioka, S., Nakagawa, J., Hirata, A.,

70

Spratt, B.G., Tsuruoka, T., Inouye, S. and Yamada, Y. (1982) in: H. Umezawa, A.L. Demain, T. Hata and C.R. Hutchinson (Eds.), Trends in Antibiotic Research, Japan Antibiotics Research Association, Tokyo, pp. 99–114.

49. Ward, J.B. and Perkins, H.P. (1973) Biochem. J. 135, 721–728.
50. Matsuzawa, H., Hayakawa, K., Sato, T. and Imahori, K. (1973) J. Bacteriol. 115, 436–442.
51. Tamaki, S., Matsuzawa, H. and Matsuhashi, M. (1980) J. Bacteriol. 141, 52–57.
52. Spratt, B.G., Boyd, A. and Stoker, N. (1980) J. Bacteriol. 143, 569–581.
53. Ishino, F., Jung, H.K., Ikeda, M., Doi, M., Wachi, M. and Matsuhashi, M. (1989) J. Bacteriol. 171, 5523–5530.
54. Ikeda, M., Sato, T., Wachi, M., Jung, H. K., Ishino, F., Kobayashi, Y. and Matsuhashi, M. (1989) J. Bacteriol. 171, 6375–6378.
55. Matsuhashi, M., Takagaki, Y., Maruyama, I. N., Tamaki, S., Nishimura, Y., Suzuki, H., Ogino, U. and Hirota, Y. (1977) Proc. Natl. Acad. Sci. USA 74, 2976–2979.
56. Iwaya, M. and Strominger, J. L. (1977) Proc. Natl. Acad. Sci. USA 74, 2980–2984.
57. Amanuma, H. and Strominger, J. L. (1980) J. Biol. Chem. 255,11173–11180.
58. Matsuhashi, M., Maruyama, I.N., Takagaki, Y., Tamaki, S., Nishimura, Y. and Hirota. Y. (1978) Proc. Natl. Acad. Sci. USA 75, 2631–2635.
59. Matsuhashi, M., Tamaki, S., Curtis, S.J. and Strominger, J.L. (1979) J. Bacteriol. 137, 644–647.
60. Mirelman, D., Yashouv-Gan, Y. and Schwartz, U. (1976) Biochemistry 15, 1781–1790.
61. Hara, H., and Suzuki, H. (1984) FEBS Lett. 168,155–160.
62. Park, W. and Matsuhashi, M. (1984) J. Bacteriol. 157, 538–544.
63. Park, W., Seto, H., Hakenbeck, R. and Matsuhashi, M. (1985) FEMS Microbiol. Lett. 27, 45–48.
64. Pisabarro, A.G., Prats, R., Vazquez, D. and Tébar, A.R. (1986) J. Bacteriol. 168, 199–206.
65. Hammes, W.P. and Kandler, O. (1976) Eur. J. Biochem. 70, 97–106.
66. Wachi, M., Doi, M., Tamaki, S., Park, W., Nakajima-Iijima, S. and Matsuhashi, M. (1987) J. Bacteriol. 169, 4935–4940.
67. Nicholas, R.A., Ishino, F., Park, W., Matsuhashi, M. and Strominger, J.L. (1985) J. Biol. Chem. 260, 6394–6397.
68. Nicholas, R.A., Strominger, J.L., Suzuki, H. and Hirota, Y. (1985) J. Bacteriol. 164, 456–460.
69. Nicholas, R.A., Suzuki, H., Hirota, Y. and Strominger, J. L. (1985) Biochemistry 24, 3448–3458.
70. Keck, W., Glauner, B., Schwarz, U., Broome-Smith, J.K. and Spratt, B.G. (1985) Proc. Natl. Acad. Sci. USA 82, 1999–2003.
71. Takasuga, A., Adachi, H., Ishino, F., Matsuhashi, M., Ohta, T. and Matsuzawa, H. (1988) J. Biochem. 104, 822–826.
72. Broome-Smith, J.K., Edelman, A., Yousif, S. and Spratt, B.G. (1985) Eur. J. Biochem. 147, 437–446.
73. Asoh, S., Matsuzawa, H., Ishino, F., Strominger, J.L., Matsuhashi, M. and Ohta, T. (1986) Eur. J. Biochem. 160, 231–238.
74. Nakamura M., Maruyama, I.N., Soma M., Kato, J., Suzuki, H. and Hirota, Y. (1983) Mol. Gen. Genet. 191, 1–9.
75. Mottl, H., Terpstra, P. and Keck, W. (1991) FEMS Microbiol Lett. 78, 213–220.
76. Broome-Smith, J., Edelman, A. and Spratt, B.G. (1983) in: R. Hakenbeck, J.-V. Höltje and H. Labischinski (Eds.), The Target of Penicillin, Walter de Gruyter, Berlin, pp. 403–408.
77. Kato, J., Suzuki, H. and Hirota, Y. (1984) Mol. Gen. Genet. 196, 449–457.
78. Suzuki, H., Kato, J., Sakagami, Y., Mori, M., Suzuki, A. and Hiorota, Y. (1987) J. Bacteriol. 169, 891–893.
79. Ishino, F., Wachi, M., Ueda, K., Ito, Y., Nicholas, R.A., Strominger, J.L., Senda, T., Ishikawa, K., Mitsui, Y. and Matsuhashi, M. (1988) in: P. Actor, L. Daneo-Moore, M.L. Higgins, M.R.J. Salton and G.D. Shockman (Eds.), Antibiotic Inhibition of Bacterial Cell Surface Assembly and Function, American Society for Microbiology, Washington, DC, pp. 285–291.
80. Ferreira, L.C.S., Schwarz, U., Keck, W., Charlier, P., Dideberg, O. and Ghuysen. J.M. (1988) Eur. J. Biochem. 171, 11–16.
81. Jackson, G.E.D. and Strominger, J.L. (1984) J. Biol. Chem. 259, 1483–1490.
82. Jevons, M.P. (1961) Br. Med. J. 1961,124–125.
83. Song, M.D., Wachi, M., Doi, M., Ishino, F. and Matsuhashi, M. (1987) FEBS Lett. 221, 167–171.

84. Song, M.D., Maesaki, S., Wachi, M., Takahashi, T., Doi, M., Ishino, F., Maeda, Y, Okonogi, K., Imada, A. and Matsuhashi, M. (1988) in: P. Actor, L. Daneo-Moore, M.L. Higgins, M.R.J. Salton and G.D. Shockman (Eds.), Antibiotic Inhibition of Bacterial Cell Surface Assembly and Function, American Society for Microbiology, Washington, DC, pp. 352–359.

85. Matsuhashi, M., Song, M.D., Wachi, M., Maesaki, S., Ubukata, K. and Konno, M. (1990) in: R.P. Novick (Ed.), Molecular Biology of the Staphylococci, VCH, New York, pp. 457–470.

86. Hayes, M.V., Curtis, N.A.C., Wyke, A.W. and Ward., J.B. (1981) FEMS Microbiol. Lett. 10, 119–122.

87. Ubukata, K., Yamashita, N. and Konno, M. (1985) Antimicrob. Agents Chemother. 27, 851–857.

88. Rossi, L., Tonin, E., Cheng, Y.R. and Fontana, R. (1985) Antimicrob. Agents Chemother. 27, 828–831.

89. Matsuhashi, M., Song, M.D., Ishino, F., Wachi, M., Doi, M., Inoue, M., Ubukata, K., Yamashita, N. and Konno, M. (1986) J. Bacteriol. 167, 975–980.

90. Ubukata, K., Nonoguchi, R., Matsuhashi, M. and Konno, M. (1989) J. Bacteriol. 171, 2882–2885.

91. Kreiswirth, B., Kornblum, J., Arbeit, R.D., Eisner, W., Maslow, J.N., McGeer, A., Low, D.E. and Novick, R.P. (1993) Science 259, 227–230.

92. Miyakawa, T., Matsuzawa, H., Matsuhashi, M. and Sugino, Y. (1972) J. Bacteriol. 112, 950–958.

93. Bachmann, B.J. (1990) Microbiol. Rev. 54, 130–197.

94. Ueki, M., Wachi, M., Jung, H.K., Ishino, F. and Matsuhashi, M. (1992) J. Bacteriol. 174, 7841–7843.

95. Jung, H.K., Ishino, F. and Matsuhashi, M. (1989) J. Bacteriol. 171, 6379–6382.

96. Takase, I, Ishino, F., Wachi, M., Kamata, H., Doi, M., Asoh, S., Matsuzawa, H., Ohta, T. and Matsuhashi, M. (1987) J. Bacteriol. 169, 5692–5699.

97. Wachi, M., Doi, M., Okada, Y. and Matsuhashi, M. (1989) J. Bacteriol. 171, 6511–6516.

98. Wachi, M., Doi, M., Ueda, T., Ueki, M., Tsuritani, K., Nagai, K. and Matsuhashi, M. (1991) Gene 106, 135–136.

99. Okada, Y., Wachi, M., Hirata, A., Suzuki, K., Nagai, K. and Matsuhashi, M. (1994) J. Bacteriol., in press.

100. Wachi, M., and Matsuhashi, M. (1989) J. Bacteriol. 171, 3123–3127.

101. Donachie, W.D., Begg, K.J., Lutkenhaus, J.F., Salmond, G.P.C., Martinez-Salas, E. and Vicente, M. (1979) J. Bacteriol. 140, 388–394.

102. Tormo, A., and Vicente, M. (1984) J. Bacteriol. 157,779–784.

103. Doi, M., Wachi, M., Ishino, F., Tomioka, S., Ito, M., Sakagami, Y., Suzuki, A. and Matsuhashi, M. (1988) J. Bacteriol. 170, 4619–4624.

104. Robinson, A.C., Kenan, D.J., Hatfull, G.F., Sullivan, N.F., Spiegelberg, R. and Donachie, W.D. (1984) J. Bacteriol. 160, 546–555.

105. Okada, Y., Wachi, M., Nagai, K. and Matsuhashi, M. (1992) J. Gen. Appl. Microbiol. 38, 157–163.

J.-M Ghuysen and R. Hakenbeck (Eds.), *Bacterial Cell Wall*
73

CHAPTER 5

Molecular biology of bacterial septation

JUAN A. AYALA[1], TERESA GARRIDO[2], MIGUEL A. DE PEDRO[1] and MIGUEL VICENTE[2]

[1]*Centro de Biología Molecular, CSIC. Universidad Autónoma de Madrid, Facultad de Ciencias, Cantoblanco, 28049 Madrid, Spain and* [2]*Centro de Investigaciones Biológicas, CSIC, C/ Velázquez, 144, 28006 Madrid, Spain*

1. Bacterial cell division

Bacterial division has been studied for many years (for a review, see [1]). Three conclusions are sufficient to introduce the topic of this chapter: (i) bacterial division is very well regulated in time and space [2]; (ii) division is a discontinuous event in topography and chronology; (iii) bacterial cells divide as a consequence of continuous growth. How does continuous growth trigger the discontinuities involved in cell division is the topic discussed in the following sections.

Many advances in the knowledge of division controls are derived from studies on unicellular organisms as yeasts [3]. Cell division in eukaryotic cells includes the partition of two structures, the nucleus and the cytoplasm. In bacteria, the nucleoid divides after replication by segregation, a simpler mechanism. This simplicity facilitates the study of cytokinesis as a separate event. Cytokinesis in *Escherichia coli* occurs by growth of a septum perpendicular to the long axis of the cell.

Morphological events during the cell division cycle in prokaryotes and eukaryotes are similar (Fig. 1). Growth induces a discontinuous event that leads to the initiation of DNA replication; this is one of the several transitions which regulate the exit from one cell cycle stage and the entrance into the next one. The molecular mechanisms which signal some transition points are well described for the cell cycles of bakers and fission yeasts [4]. They are not so well known for *E. coli*.

Conditional mutations in some *E. coli* genes interrupt division at different stages [5]. It has been postulated that these gene products form a structure active in cell division called septator [2]. The molecular biology of bacterial division is aimed at identifying: (i) the molecules involved in septation and their cellular localization; (ii) the biochemical activities of these molecules and their possible interactions; and (iii) the mechanisms that govern expression of the genes involved in division.

2. Growth, replication and division of Escherichia coli

E. coli DNA replication is initiated once per cycle at *ori*C after attainment of a mass that

PROKARYOTES

EUKARYOTES

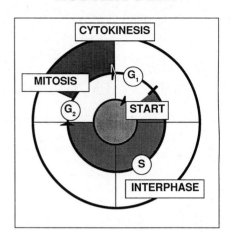

Fig. 1. Comparison of prokaryotic and eukaryotic cell division cycles. The naturalistic description of the division cycles of prokaryotic and eukaryotic cells share a basic pattern of chained and parallel events. Continuous growth during the I period, or G_1 phase (inner circle) triggers DNA replication when the cell attains a certain mass, the initiation mass (M_i) of bacteria, which is an event analogous to the start signal of eukaryotic cells. DNA replication (C or S phase) proceeds until completion of chromosome replication (mid-circle) and, in typical cycles, is a prerequisite for nucleoid segregation or nuclear partition (mitosis). Completion of replication and partition of the genetic material is required for cell division. Division is usually delayed from termination of DNA replication by a certain time, the D period or G_2 phase. All these morphologically salient events are represented in the outer circle of the cycles.

is multiple of a value named the initiation mass [6]. Replication takes nearly 40 min to be completed so that at cell doubling times shorter than 40 min, the cells contain more than one copy of the chromosome whose replication may have started in the grandmother cells. The first signs of septation are visualized after termination of chromosome replication and segregation of the nucleoid. Genetic defects in replication or segregation produce abnormal patterns of division or completely abolish it. Septation proceeds through the concerted action of several gene products. The product of *fts*Z acts at an early stage of septation and that of *fts*A at a terminal stage. Essential for the formation of a peptidoglycan crosswall is the action of PBP3, the penicillin-binding protein encoded by *pbp*B (= *fts*I). Mutations in the *fts* genes cause filamentation by allowing normal growth and replication while preventing septation. Septation takes nearly 20 min to be completed and is followed by cell separation which requires the action of the product of *env*A. After separation, the newly formed poles, which instants before formed part of an active division site, become inactivated by the action of the *min*C, D and E products.

Biosynthesis of the huge bag-like sacculus takes place in three cellular compartments: (i) the cytoplasm, where the soluble precursors are synthesized (see Chapter 3); (ii) the cytoplasmic membrane, where linking of the soluble precursor to the lipid carrier and translocation of the substrate to the outer face of the cytoplasmic membrane occur; and (iii) the periplasm, where the monomeric subunits are assembled by means of two types of enzymatic activities, a transglycosylase that joins the sugar moieties and a transpeptidase that crosslinks the peptide chains. In *E. coli*, the final stages of peptidoglycan biosynthesis are catalysed by a set of four enzymes, the high molecular weight (HMW) penicillin-binding proteins (PBPs) 1A, 1B, 2 and 3 (see Chapters 4 and 6).

Based on studies with specific β-lactam antibiotics and with mutants, different physiological functions have been ascribed to the four synthesising HMW-PBPs [7–9]. PBP1A/1B are responsible for major peptidoglycan synthesis, PBP2 for maintaining the rod shape and PBP3 for cell constriction.

3. The minC, D, E, dicB, F, and sulA genes code for natural inhibitors of septation

Much knowledge on bacterial septation derives from the study of cell-encoded molecules that inhibit the process. Placement of a septum can be inhibited by normal physiological events and by responses to injuries that compromise the survival of the cell. The *min* gene products prevent suicidal division at the newly formed pole of a newly born cell. Strains harbouring genetic lesions in the *min*B *loci* produce offspring containing minute spherical cells, the minicells. Minicells are formed by aberrant division at the cell poles at the expense of division events that should occur at the centre of the cell. The *min* strains show an abnormally wide distribution of lengths from minicells to cells several times the length of the normal cell. This is consistent with the idea that cells contain a limited amount of division potential per cell and per cell cycle.

The Min system (Fig. 2) is formed by three components. MinC is a low efficiency inhibitor able to act alone at any potential septation site. MinD converts MinC into an efficient inhibitor able to block any site of a growing cell, including the normal septation sites. Immunoelectron microscopy studies locate MinD in the cytoplasmic membrane. The activation of MinC depends on the ATPase activity of MinD; this has been deduced from the inability of one mutated MinD form in which a highly conserved residue at the ATP binding site has been altered, to activate the MinC mediated division inhibition [10]. MinD is also responsible for the topological specificity exhibited by the complex MinC–MinD when it is in the presence of MinE, the other component of the *min* system. It is postulated that MinE can distinguish between the newly formed septum primordia and the old division site moieties remaining at the poles. The presence of MinE at one given site prevents the inhibitory action of the MinC–MinD complex, perhaps by preventing activation of MinC by MinD. Under normal circumstances, MinE should exhibit a higher affinity for septum primordia which would escape inhibition, than for old septal moieties at the poles which would be inhibited preferentially. When overexpressed, MinE can aim both at the poles and at the centre of the cell preventing the activation of MinC by MinD at all the potential septation sites. This causes strains overexpressing MinE to behave as *min*C, D phenocopies [11].

The product of *dic*B is an inhibitor of septation. It is not normally expressed in the cell. The *dic*B gene belongs to a tightly repressed operon located in the DNA replication terminus [12]. The physiological significance of this inhibitor remains obscure. It acts on both septal primordia and old site moieties by activating the inhibitory action of MinC

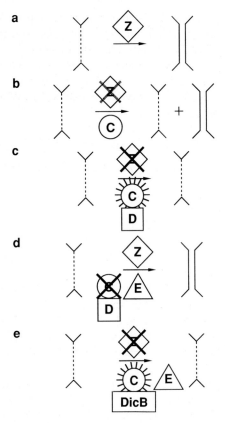

Fig. 2. The inhibition of septation mediated by the Min system. The action of FtsZ (Z) on a septal primordia (a). MinC (C) partially inhibits the action of FtsZ (b). Full activation of MinC by MinD (D) inhibits the action of FtsZ (c). The presence of MinE (E) at the central position of the cell blocks the inhibitory effect of MinC (d). MinC is activated by DicB independently of the presence of MinE and blocks septation (e).

independently of the presence of MinE. The *dic loci* may be remnants of a genetic element once inserted into the chromosome.

The behaviour of strains overexpressing the *fts*Z gene suggests that FtsZ is the target of the inhibitory action of MinC. Moderate amounts of FtsZ can overcome the effect of the MinC product, even when this is combined with any of its two activators. Direct proof of the interaction between MinC and FtsZ is likely to be found in the future.

The *dic*F gene codes for another inhibitor of division which, like *dic*B, is not normally expressed. The product of *dic*F is an RNA molecule complementary to the initial portion of the *fts*Z messenger. It is a natural antisense mRNA. It acts by preventing efficient translation of the essential FtsZ message [13].

Cells in which DNA has sustained damages due to external injuries such as UV light irradiation, elicit the SOS response which prevents the cells from entering into a fatal condition. The SOS response is mediated by the product of *lex*A, a repressor molecule that controls transcription of a regulon [14]. Part of the regulon is the *sul*A (= *sfi*A) gene whose product is a division inhibitor. The *sul*A gene has been detected in *E. coli* by the isolation of mutants which fail to induce filamentation when the SOS response is triggered [15]. The same behaviour is exhibited by the *sul*B (= *sfi*B) mutants which map at a different *locus* [16]. The SulA protein is short-lived. It blocks cell division after DNA has been damaged or replication abnormally stopped. The target of SulA is the FtsZ protein as an excess of FtsZ may overcome the inhibition by SulA [16,17]. Under normal circumstances, SulA is rapidly degraded by the Lon protease [18]. The *sul*B mutations map at the *fts*Z gene and code for a form of the FtsZ protein that is not sensitive to the SOS-induced division inhibition [16]. These results support the idea that FtsZ has a crucial role in the onset of septation.

4. Many division genes are clustered

Many of the genes that are required for division are grouped in a few clusters in the chromosome [19–21]. At min 14, the *mrd* cluster contains genes involved in the maintenance of cell shape [21]. Min 76.5 contains the *fts*S, Y, E, X genes involved in protein export. These genes share regulatory features with the heat-shock response [20]. The largest cluster, at min 2.5 [19], contains genes whose products are involved in the cytoplasmic and periplasmic stages of peptidoglycan biosynthesis. Among them *pbp*B which codes for PBP3, is close to the head of the cluster. In the same region three *fts* genes are found next to each other, the *fts*Q, A, and Z genes. They are followed by the *env*A and *sec*A genes involved in cell separation and protein export, respectively. Although this cluster has been called the *mra* region, it seems reasonable to name it the *dcw* (**d**ivision and **c**ell **w**all) cluster as it harbours a large number of genes involved in cell wall biosynthesis and cell division.

The presence of this grouping may have regulatory and evolutionary implications. A remarkably similar grouping is found in the Gram-positive *Bacillus subtilis*. In this microorganism, septation can follow two different developmental pathways, cell division or sporulation. Gene expression controls have to operate under both circumstances. Genes homologous to *fts*Z, *fts*A and *pbp*B have been found in *B. subtilis* [22]. Other ORFs in the region which are likely to be equivalents of the *E. coli* genes are being studied. Particularly interesting is the presence of two genes coding for the equivalent of PBP3 (see Chapter 8). One of them is involved in the synthesis of the septum in the vegetative cell and the second in the synthesis of spore peptidoglycan (Errington, personal communication). The protein encoded by *div*IB [23], a gene found in the *B. subtilis dcw* cluster, has a predicted secondary structure resembling that of FtsQ, i.e. a hydrophobic amino-proximal domain which may play also a role in the localization of the protein within the membrane [23]. DivIB and FtsQ have no significant homology at the nucleotide or amino acid level. The possibility that the two proteins play a similar role remains unsolved.

A morphogene *bol*A, at min 9.5, forms part of a small grouping with *tig*, the gene coding for trigger factor. The *bol*A gene has been identified in a chromosomal fragment able to complement the *wee* mutations present in strain OV25. The complementation was probably due to the presence of trigger factor in the fragment, because, after storage of the original clones, both trigger factor expression and complementing ability were lost due to spontaneous insertion of IS*10* (Garrido et al., unpublished).

5. Gearbox promoters yield constant amounts of transcripts per cell cycle

Both the *fts*Q and *bol*A contain one gearbox in their regulatory region. A gearbox is a promoter that yields a constant amount of transcript per cell and per cycle, indepéndently of the growth rate of the culture (Fig. 3b). Most of the metabolic genes are driven by

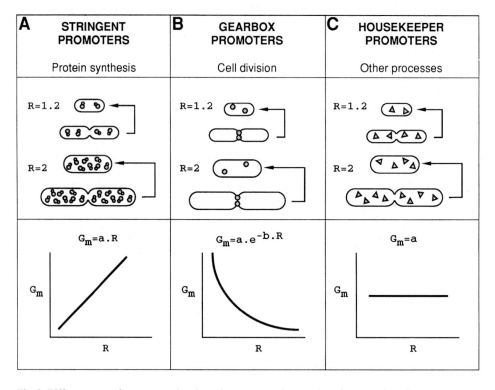

Fig. 3. Different types of gene expression dependence on growth rate. The stringent and gearbox promoters are responsible for direct and inverse growth-rate dependence, respectively. The housekeeper promoters include all the other growth-rate independent promoters. The middle row is a cartoon representing the cell sizes and the absolute amounts per cell that a gene product would attain when expressed by the three kinds of promoters at two different growth rates (R) (2.0 and 1.2 doublings per hour). The contents per cell mass of a gene product expressed by each type of promoter, (G_m), are plotted as a function of the growth rate in the bottom row. Housekeeper promoters (column C) produce constant amounts per cell mass independently of the growth rate. Stringent promoters (column A) are more active at higher growth rates so that, for example, the rRNA content per cell mass is proportional to the growth rate. Since cell mass (M), is an exponential function of growth rate ($M = M_0 \times e^{bR}$ with $M_0 =$ the cell mass at $R = 0$ and $b = 0.85$ for the *E. coli* K-12 strains used), any protein synthesized at a constant amount per cell would show relative contents to cell mass (P) being inversely dependent on the growth rate ($P = P_0 \times e^{-bR}$ with $P_0 =$ relative content at $R = 0$) (column B). For further details, see [2].

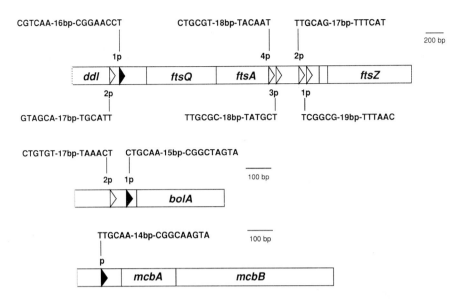

Fig. 4. Regulatory region of genes containing gearbox promoters. Open arrowheads indicate growth rate-independent promoters identified by mapping of transcript initiation points. Black arrowheads indicate the positions of the gearboxes. The proposed sequence of the −35 and −10 regions, as well as the spacing, are indicated for each promoter. Note that the scale is different for the *fts* operon. Data are from [23] for *mcb*A, [22] for /*bol*A and [24] for *fts*.

housekeeper promoters (Fig. 3c) ensuring a constant transcription rate relative to the total transcription rate of the cell. Some rRNA and ribosomal protein genes are controlled by stringent promoters in which the relative rate of transcription is higher at fast growth rates to fit the substantial increased demands on translation (Fig. 3a). In contrast, gearbox promoters originate a small part of the total transcripts found in a fast growing cell. As growth rate diminishes, the rate of transcription of housekeeper and stringent promoters per cell and cycle also decreases while that of gearboxes remains constant. Consequently, the relative rate of transcription from a gearbox is higher at slow growth rates and lower at fast growth rates (Fig. 3b).

The regulatory region of *bol*A (Fig. 4) contains two promoters [24]. One, *bolA1p*, is a gearbox that shares homology in the −10 region with the promoter of *mcb*A, a gene involved in the production of microcin B17 (a low molecular weight compound with antibiotic activity against *E. coli* produced at higher levels when the cells enter stationary phase) [25]. The detection of this homology has allowed the identification and study of the gearboxes. Transcription from the isolated *bolA1p* is not exactly constant per cell and per cycle at all growth rates. This is achieved only when it is placed downstream from *bolA2p*, a weak housekeeper promoter (Fig. 5).

The *fts*Q promoters are located inside the structural region of the upstream *ddl* gene (Fig. 4). One of them, *ftsQ1p*, also shows homology with the gearbox. This promoter, together with *ftsQ2p*, can direct transcription at constant levels per cell and per cell cycle independently of the growth rate of the culture [26].

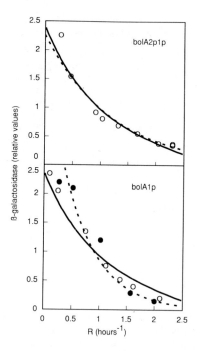

Fig. 5. Exact inverse growth-rate dependence of the gearbox promoters when preceded by a housekeeper pro-moter. The relative β-galactosidase activities produced by two transcriptional fusions [24] contained in lambda MAK400 (*bolA2p,1p::lacZ*, top panel), and lambda MAV103 (*bolA1p::lacZ*, bottom panel) are shown. Solid circles are the values obtained from exponential cultures grown in different media. Open circles are the relative values obtained from a culture grown in LB medium up to stationary phase, which are plotted versus instantaneous growth rate at each point. β-Galactosidase activities were made relative by regression to that cor-responding to $R = 1$. Continuous lines correspond to the theoretical curve for a gene product produced at con-stant amounts per cell. Discontinuous lines were obtained by regression of the plotted values using an inverse exponential dependence versus growth rate.

Gearboxes form a family of related signals. Differences in molecular details govern their expression. The three gearbox promoters mentioned above may not be transcribed by the same form of RNA polymerase. Transcription of *bolA1p* depends on the presence of the intact RpoS sigma factor (the product of *katF*) [27,28] while transcription of *mcbAp* is enhanced when *rpo*S (= *kat*F) is inactivated by insertion. Transcription directed by the combination of *ftsQ1p, 2p* is also affected by disruption of *rpo*S (Ballesteros et al., unpublished). In addition, RNA polymerase containing RpoS can direct transcription from housekeeper promoters although with low efficiency (Ishihama, personal communi-cation).

6. Expression of the dcw cluster involves a variety of regulatory mechanisms

The pattern of expression of the *dcw* cluster is more complex than that of non-essential operons. There is an abundance of potential promoters. Not all are proven to originate a transcript and some occur within the structural region of upstream genes. No transcription terminators are found from the head of the cluster until the end of *sec*A. In all likelihood, the whole cluster is not transcribed as a single molecule. Indeed, almost 90% of the *fts*Z transcripts, one of the distal genes, originate from the promoters found immediately upstream within *fts*A [29,29a], while 10% only originate from further upstream. In addition, post-transcriptional regulations are involved in the expression of the cluster.

6.1. Gene expression of the pbpB region

The *pbp*B region maps at the beginning of the *dcw* cluster. It contains *pbp*B, the gene that codes for penicillin-binding protein 3. This gene was first cloned from a plasmid in the Clark and Carbon bank carrying a chromosomal insert of strain W3110 [30]. Two years later the sequence of a 1764 bp open-reading-frame of the 2-min region that codes for a protein of 63 850 Da was published [31].

6.1.1. Structural genes and regulation of transcription

Eleven putative promoter regions have been identified upstream from *pbp*B, in a *Sma*I-*Mlu*I fragment (Fig. 6 and Table I) [32], by homology with the consensus –10 sequence of *E. coli* promoters. Analysis of transcription activity was first carried out by complementation of the thermosensitive phenotype with plasmids containing different inserts of the region and no vector-derived promoter. A complex regulatory region was predicted. It seems that the promoters in the 1830 bp $PvuII_{26}$-*Mlu*I fragment are inefficient for complementing when present in a low-copy-number vector but they can originate enough protein when cloned in a high-copy-number plasmid [33,34]. Expression of PBP3 from promoters in the 460 bp *Cla*I-*Mlu*I fragment cloned in a pBR322-derived plasmid can complement *pbp*B-thermosensitive mutants. Hence, promoters 5' upstream of $PvuII_{26}$ site must be also involved in expression of *pbp*B in the chromosome.

The levels of PBP3 as detected by a labelled antibiotic, appear to remain constant all along the cell cycle [35]. However, few details are known on the regulatory mechanisms that operate in transcription of *pbp*B. Recently, the nucleotide sequences from *leu*A to *mut*T of the *dcw* cluster have been assembled and the overlapping regions have been verified (unpublished data; EMBL accession number X55034). The sequence starts at the last nucleotide of the $HindIII_6$ site, it contains 28 277 nucleotides and it ends at $PvuII_{35}$ site (Fig. 6). Upstream of *pbp*B there is a 2400 bp region containing three open-reading frames which code for the proteins MraZ, MraW (formerly orfB, [36]) and MraR, and a non-coding region at the 3' end of gene *shl* (Fig. 6). Using *pho*A- and *lacZ*-gene fusions, it has been shown that *mra*R codes for a transmembrane protein [37; Ayala et al., unpublished). New mutations (*fts36* and *lts33*) showing filamentous and lytic thermosensitive phenotypes, have been described in *mra*R [38,39]. As the former phenotype is the result of inhibition of cell division and the latter may be a defect in cell growth, a possible role of this gene in regulation of both processes has been postulated. A polar effect of these mutations on the transcription of *pbp*B as it occurs in a *rod*A (*sui*) amber mutation, that affects expression of its adjacent genes *pbp*A and *dac*A [40], cannot be excluded.

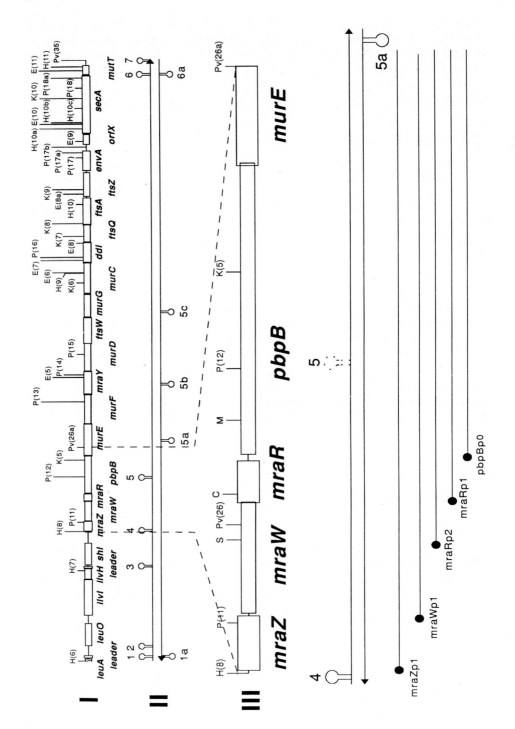

S1 mapping analysis of these new open-reading frames upstream of the gene *pbp*B reveals start-points for mRNA at positions 6126 for *mra*Z, 6540 for *mra*W, and 7219 and 7542 for *mra*R. Potential candidates for promoter sequences, ribosomal binding sites, and lengths of the 5' end non-coding region of the mRNA are shown in Table I. No specific mRNA for *pbp*B gene is detected by Northern blot analysis of total RNA extracted from wild type strains of *E. coli* [34], indicating that it must be present in extremely low amounts. Determination of a start point for *pbp*B mRNA has failed even by using the S1 mapping analysis that is ten times more sensitive than Northern blots. Stability of the *pbp*B mRNA is low [41,42]. Presumably it may be part of a long polycistronic transcript, as predicted from the absence of transcription termination sequences (see below). The low stability may also be the result of a strong coupling between transcription and translation.

Using the program Terminator of the Wisconsin package [43], a search for potential transcription termination signals in the sequence of the *pbp*B region has been done. Results, summarized in Fig. 6 and Table II, indicate that transcriptional terminators are not present in a long DNA stretch. Experimental evidence for the absence of transcription termination within the genes *mra*Z through *mur*F has been obtained by transcription of the 7.6 kbp *Hind*III$_8$-*Eco*RI$_5$ fragment driven by the *lac* promoter. IPTG-induced expression from the *lac* promoter results in an increase of the downstream MurE and MurF activities (see Chapter 3; J. van Heijenoort, personal communication).

Three genes (*mre*B, *mre*C, *mre*D) which map outside this region (at min 71), may exert some control on *pbp*B transcription [44]. These shape-determining genes exert a negative regulatory effect on cell division, and a positive effect on cell elongation, i.e. expression of additional copies of the *mre*B gene causes cell filamentation while defect results in round cells [45]. The step of cell division that is inhibited by MreB is unknown (see Chapter 4) although the facts that a *mre*B mutant overproduces PBP3 and a *mre*B,C,D deletion mutant overproduces both PBP1B and 3, suggest that MreB may exert a negative control on *pbp*B expression.

6.1.2. Translation of the pbpB messenger

The *pbp*B gene has a ribosomal binding site (RBS) which is located 12 nucleotide apart from the ATG start codon and shows low homology with the consensus sequence (Table I). The frequency of codon usage for PBP3 is that of a protein expressed at very low levels in *E. coli* [31]. As a result, and given the low amounts of specific *pbp*B mRNA, the levels of PBP3 are very low, about 50 molecules per wild type cell [46]. Expression of PBP3 at moderate levels (26-fold increase) can be achieved by placing the *pbp*B gene (with its own RBS) under the control of the lambda P_R promoter and the *cI857* repressor

---

Fig. 6. Schematic representation of the assembled sequences at the 2.0-min region of the *E. coli* chromosome. The names of the genes and the relevant restriction sites (H, *Hind*III; E, *Eco*RI; P, *Pst*I; K, *Kpn*I; Pv, *Pvu*II; S, *Sma*I; C, *Cla*I; M, *Mlu*I) are shown in lanes I and III. Sites are numbered starting from the first site for each restriction enzyme at the 0 min of the *E. coli* genetic map. Lane II shows the putative termination sequences as deduced with the use of the Wisconsin software package (see Table II). In lanes III, the fragment between the H$_8$ and Pv$_{26a}$ sites is enlarged and the transcription start points for the genes in this region are shown (see Table II). Putative transcription terminator 5 is drawn in discontinuous line because it is not functional in vitro (see text).

TABLE I

Promoters sequences and lengths of the known and predicted 5' ends of non-coding mRNA regions for proteins encoded in the *E. coli* genome 2.0-min region

Gene	mRNA start	nt to ATG	Predicted RBS	nt from RBS to ATG	Protein start
mraZ	6126	38	TAAGGGGTGA	10	6164
mraW	6540	84	GCAGGACTTG	11	6624
mraR	7219	343			
	7542	20	AGAGGACGAA	5	7562
pbpB	(7921)	(22)	TAAGGATAAA	12	7943
ftsQ	19275	410			
	19400	285			
	(19555)	(130)	TCTGGAACTG	11	19685
ftsA	(20107)	(405)			
	(20133)	(379)			
	(20460)	(52)	TCAGGCACAG	16	20512
			ACAGGCAGAA	7	
ftsZ	21169	666			
	21226	609			
	21576	259			
	21626	210	ATCGGAGAGA TAAGGAGGTT (consensus)	7	21835

Promoter	−35 region	Distance to −10 (bp)	−10 region	Distance to mRNA start
mraZ	TTGACA	18	TAAACT	13
mraW	TGGACA	19	TGGGAT	13
mraRp2	CTGCTG	16	TAAGTT	5
mraRp1	CTGAGA	16	TTCAGT	7
pbpBp0	TTGATC	13	TATCGT	<10
ftsQp2	GTAGCA	17	TGCATT	12
ftsQp1	CGTCAA	16	CGGAACCT	12
ftsQp0	TTGCAA	21	TATGCT	<130
ftsAp2	TTGCGC	16	TATGGT	<405
ftsAp1	TGGTAG	18	TCAGCG	<379
ftsAp0	TTGCGT	19	TAGGCT	<52
ftsZp4	CTGCGT	18	TACAAT	10
ftsZp3	TTGCGC	18	TATGCT	9
ftsZp2	TTGCAG	17	TTTCAT	5
ftsZp1	TCGGCG	19	TTTAAC	6

Numbers for mRNA and protein start point correspond to the assembled sequence from *Hind*III$_6$ to *Pvu*II$_{35}$. Numbers in parentheses indicate that the site for mRNA start has not been obtained by S1 mapping, but inferred from the sequence.

[47]. Replacement of the PBP3 RBS with the LacZ RBS produces a β−galactosidase-PBP3 fusion protein that is expressed at high levels (10% of total cell protein). As mRNA produced from the P$_R$ promoter is rather stable, these results indicate that the efficacy of the PBP3 RBS is very low.

6.2. Post-translational modifications and export of PBP3

PBP3 catalyses reactions that occur on the outer surface of the cytoplasmic membrane or within the periplasm. It must be located with much of its bulk in the periplasmic space and, consequently it must be exported. Exportable proteins are normally translated as pre-proteins with an N-terminal hydrophobic extension that serves for translocation through the cytoplasmic membrane. These N-terminal signal peptides are cleaved in secreted pro-teins by specific proteases, Spase I and Spase II [48,49], but in some membrane proteins they are not processed and serve as membrane anchors.

6.2.1. Lipoprotein nature and proteolytic processing of PBP3

PBP3 is synthesized as a preprotein [31] and is processed in vivo as a mature form having a higher electrophoretic mobility. The primary structure of PBP3 contains an N-terminal sequence that mimics the structural domains found in signal peptides: a basic region (M_1-E_{15}), a hydrophobic core $(H_{16}-P_{50})$, and putative sequences for processing by SpaseI $(A_{34}LA_{36})$ and SpaseII $(L_{27}CGC_{30})$ (Fig. 7). Hayashi et al. [50] have predicted that PBP3 may be modified by mechanisms comparable to those involved in lipoprotein export to the outer membrane, resulting in the substitution of Cys_{30} by an acylglyceride moiety. However, there is no proof that PBP3 itself is a lipoprotein for the following reasons: (i) only 15% of the total PBP3 is acylated under conditions in which the protein is highly overexpressed; (ii) no change in electrophoretic mobility is observed due to the modifica-tion (the whole population of PBP3 molecules migrates to the same position as does the processed mature form); (iii) processing of PBP3 is insensitive to globomycin, an antibi-otic which specifically inhibits the processing of pro-lipoproteins; (iv) site-directed mut-

TABLE II

Putative terminator sequences at the 2.0-min region of the *E. coli* chromosome

Number in Fig. 6 lane II	Sequence position (direction)[a]	Primary[b]	Secondary[c]
1	109–136 (cw)	6.72	23
1a	136–109 (ccw)	8.25	87
2	589–615 (cw)	3.51	84
3	4402–4424 (cw)	5.14	20
4	6029–6051 (cw)	6.11	25
5	8653–8674 (cw)	3.66	63
5a	10383–10358 (ccw)	3.66	67
5b	13013–12999 (ccw)	5.13	26
5c	16352–16331 (ccw)	3.99	79
6	27534–27559 (cw)	6.09	83
6a	27559–27534 (ccw)	3.80	60
7	28223–28240 (cw)	5.53	45

[a]The position (in bp) corresponds to the assembled sequence starting at *Hind*III$_6$ (see text). (cw): same direction of transcription as the *phpB* gene, (ccw): opposite direction of transcription to gene *phpB*.
[b,c]Values calculated by using the Terminator program of the Wisconsin package for primary and secondary structures of the putative terminator sequences. The thresholds are fixed to 5.0 or 3.0 (b) depending on the sec-ondary structure values, and to 20 or 60 (c) depending on the primary structure values.

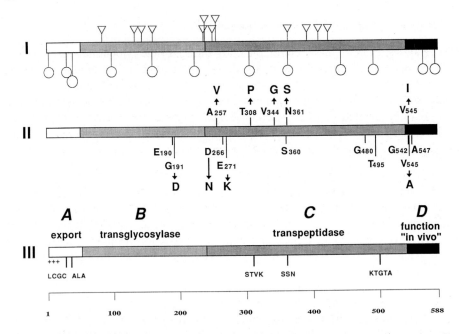

Fig. 7. Schematic representation of the gene fusions, mutations and functional domains of *E. coli* PBP3. Lane I shows the sites where fusion proteins with PhoA (upper) and Bla (bottom) have been obtained. Lane II displays the single point mutations that block or reduce the β-lactam binding without inhibiting the in vivo enzyme activity (upper), and those that block the function in vivo without a significant change in the β-lactam binding (bottom). Lane III shows the four functional domains described for the protein. Some relevant features are highlighted: +++, hydrophobic region; LCGC and ALA, putative processing sites for signal peptidases II and I, respectively. STVK, SSN and KTGAT, conserved sequences corresponding to the SXXK, SDN and KTG boxes found in DD-endopeptidases and β-lactamases [81], respectively.

agenesis of the sequence $L_{27}CGC_{30}$ [51] with changes of Leu_{27} to Phe, Gly_{29} to Ser and Cys_{30} to Arg which should abolish acylation, have no effects on maturation and function of PBP3. Note finally that PBP3 also undergoes proteolytic processing by removal of the undecapeptide I_{578}–S_{588} at the C-terminus of the protein [52].

6.2.2. Effect of secA, secY, ftsH and prc on PBP3 export

Experiments with *sec*A and *sec*Y secretion mutants indicate that the processing and function of PBP3 are dependent on the protein translocation machinery [53; unpublished results]. The N-terminal region serves as a signal for interactions with the translocation machinery. Indeed, those molecules whose 40 N-terminal residues are removed fail to translocate and accumulate in the cytoplasm [54]. Conversely, the first 36 N-terminal amino acids of PBP3 are able to translocate the β-lactamase moiety of a *pbp*B-*bla* gene fusion protein in the periplasm [55] (Fig. 7). PBP3 production is strongly reduced in a *fts*H mutant [56]. FtsH is a membrane-bound protein probably acting as a specific chaperonin for the export of PBP3 [57].

The gene *prc* responsible for the C-terminal processing of PBP3 has been cloned, mapped at 40.4 min and sequenced [53]. The *prc* gene product is involved in the protection of cells from thermal and osmotic stresses but it is dispensable for growth at low temperatures and in high osmolarity media. It is located on the periplasmic side of the cytoplasmic membrane [55] suggesting that PBP3 processing occurs probably outside the membrane. Surprisingly, maturation of PBP3 is not related to the growth defect of *prc* mutants. Processing does not occur under permissive conditions that allow the *prc* mutant cells to divide normally [58] suggesting that the unprocessed form of PBP3 rather than the processed form may be the functional protein engaged in cell division in vivo. C-terminal processing may be a first degradation step restricting the function of PBP3 even if the processed protein retains the ability to bind β-lactams. The point mutation Val$_{545}$Ala (Fig. 7) yields a mutant protein which retains β-lactam binding capacity but is unable to perform transpeptidation [42; unpublished results]. Moreover, cleavage of the 11 C-terminal residues is not essential for cell division, because non-processed forms, those deleted in the 12 C-terminal residues [58] or those containing an internal deletion at the C-terminal end that block processing [51], are able to complement a *pbp*B(Ts) mutation.

6.2.3. PBP3 turnover
The levels of mature PBP3 are constant throughout the cell cycle [35] and diminish very rapidly in the stationary growth phase [59]. No turnover number has been reported for PBP3. In the stationary phase, the protein is actively degraded, this degradation being dependent on the activity of *rel*A [60] and probably on specific periplasmic proteases [61]. After restart of growth, normal levels of PBP3 are quickly restored. Constant levels could result from a steady state between synthesis and degradation. Disappearance of the protein in stationary phase could be due to arrest of synthesis, rapid degradation or both. Why does the cell maintain a constant level of a protein needed only during one-third of the cell cycle? Possible explanations include: (i) expression is discontinuous with PBP3 being produced only during septation. The non-processed form would be rapidly used as active enzyme in septal peptidoglycan biosynthesis perhaps coupled with the appearance of a specific substrate (see below) and then processed into a mature stable form. This mature form would be the PBP3 detected at constant levels by β-lactam antibiotics. (ii) Alternatively, the cell could produce PBP3 continuously throughout the cell cycle and PBP3 could mature continuously, but at the moment of septum formation, the recently formed non-processed PBP3 could be activated by a specific substrate and/or by interaction with other components of the septator. These two models emphasize the complex situation of cell division in vivo although some of the features are debatable. For example, the affinity of the PBP3 mature form for some specific β-lactams (furazlocillin, cephalexin, azthreonan) correlates well with their differential ability to inhibit growth (as reflected by their respective ID$_{50}$ and MIC values). This may indicate that the affinity of the non-processed form for the β-lactam may be the same as that of the mature form. As PBP3 is maintained at a constant level along the cell cycle, being degraded only in non-growing cells, the question arises as to whether processing has any relevant, although dispensable, function.

6.3. Gene expression of the fts region

Transcription of the *fts*Q, A, Z genes is particularly complex. Two promoters inside the

88

ddl gene direct transcription of *fts*Q and A. As already mentioned, one, *ftsQ1p*, is a gearbox (Fig. 4). The other, *ftsQ2p*, is a housekeeper promoter. The SdiA protein [62] activates transcription from *ftsQ2p*. The activation is strong enough to increase the amount of FtsZ up to levels that can compensate for the inhibition of division caused by artificially induced levels of MinC–MinD. The existence of specific promoters for *fts*A [63–65] is not supported by mapping of the transcription initiation sites [26] nor by the identification of the transcripts [29]. Most of the *fts*Z transcripts originate at four promoters within the *fts*A gene (Fig. 4). Nearly 79% of the transcripts are initiated at the two proximal promoters, *ftsZ1p* and *2p*. Transcripts from these two promoters follow a cell cycle-dependent pattern of synthesis, while those from the distal *ftsZ3p* and *4p* merely increase when gene dosage doubles [29a]. Disruption of the natural transcription regulation by introducing an inducible *tac* promoter in the intergenic region between *fts*A and *fts*Z alters the normal timing of septation. These modified cells divide at sizes bigger than the wild type although the amounts of FtsZ per cell are similar [29a]. These results suggest that expression of *fts*Z involves additional effectors. One possible effector molecule that can increase the expression of *fts*Z is the RcsB protein [66]. Sequences homologous to DnaA boxes are found in the *fts*Q, A [67] genes but they do not influence the transcription of the ftsZ gene [29a]. Less specific transcriptional controls such as supercoiling, do not affect transcription of the *fts*Z gene (Garrido et al., unpublished).

Post-transcriptional regulation [68] may contribute to the different relative amounts of FtsQ and FtsA present in a normal cell. These two proteins are translated from a single messenger but the cell contents of FtsA, the downstream product, is nearly sixfold the amount of FtsQ. The amount of FtsA per cell as measured by immunoblotting with a monoclonal antibody, is proportional to the mass of the cell [69]. This observation would suggest that transcripts initiated from the gearbox upstream from *fts*Q do not produce a constant amount of FtsA per cell and cycle. A trivial explanation would be that the monoclonal antibody used for the titration of FtsA recognizes only one particular form of the total FtsA population. The same observation might result from post-transcriptional processing or indicate that most of the *fts*Q, A transcripts are not initiated at this promoter. Another possibility is that the transcription from *fts*Z promoters in the *fts*A structural region, perturbs the elongation of mRNA inside *fts*A in a growth dependent manner.

An additional feature of the expression of the *fts* region is the presence of translational frame shifting. Initiation and termination codons are very close or shared between *ddl-ftsQ* [70] and *ftsQ-fts*A [63]. The coupling may be required for the coordinated expression of these genes but its exact meaning is not known.

7. Influence of murein structure on septation

At the level of the cell wall, the *E. coli* division cycle comprises a phase of cell elongation followed by a phase of cell septation resulting in the formation of two new poles. The switch between the two phases may be triggered by cyclic changes in DD-carboxypeptidase levels [71], by changes in the amount or activity of PBP3 [72] and/or by reciprocal changes in RodA and PBP2 [40]. Such changes in protein levels or activities may be under the control of regulatory genes [8]. The rod shape of the cell may be maintained by a complex balance between cell division and cell elongation. A model for peptidoglycan biosynthesis, known as 'primer' model, has been described [73] in which PBP1B is responsible for the initiation of the synthesis of new crosslinked peptidoglycan at the new insertion sites, i.e. 'the primer'. PBP2 and PBP3 would be primer-dependent peptidoglycan synthetases specific for wall elongation and septation, respectively. This model fits most of the observations on the rate of incorporation of precursors, on the macromolecular peptidoglycan composition, and on the function of the biosynthetic enzymes. Figure 7 and Table III summarize some recent findings about the biosynthesis of septal peptidoglycan.

Elongation of the sacculus and septation share a common physical substrate: the pre-existing murein layer. Septa are initiated in the cylindrical part of the sacculus at defined positions and time [74–78]. No evidence supports the existence of specialized areas on the sacculus. No specific components of septal murein are known. Furthermore, the meaning of the differences detected in some structural parameters (crosslinkage, sugar-chain length) is still under debate [79–81] (see Chapters 2 and 4).

The major unsolved problem in studying the specific function of a specific PBP in the cell cycle is to measure and distinguish its in vivo enzymatic activity. β-Lactam antibiotics have a three-dimensional structure that closely resembles the donor site (X-D-Ala-D-Ala) of DD-transpeptidation and as a consequence they bind and inactivate the catalytic site of PBPs (for a review see [82] and Chapter 6). The only experimental approach used to study the enzymatic activities of the PBPs in vivo is to measure the effects of a specific β-lactam for a given PBP on the rate of peptidoglycan synthesis and on peptidoglycan composition. Transpeptidation is the rate-limiting step of the reaction cascade and is dependent on the product of transglycosylation. A transglycosylase activity has been proposed for all the HMW-PBPs [8] (see Chapter 4) but it has been consistently demonstrated in vitro for PBP1B, only at a 2–3% of the activity expected in vivo (Table III). Methods to analyse the polymerization process as a whole are not available. A new HPLC-based procedure [83] which separates the glycan strands according to their length provides no information on the coupling of the polymerization activities, because it needs an amidase treatment to separate the peptide moiety.

The initiation of septation is still an undefined event concerning the biochemical details of murein metabolism. The first sign of septation is the appearance of a reversible constriction at the future site of division [84]. The early steps of septation are apparently carried out by proteins other than PBP3 [81]. Involvement of murein hydrolases is likely but still uncertain [85] (see Chapter 7). The initial invagination defines an area where compartmentalization occurs by completion of the periseptal annuli [86]. However, it is important to notice that, up to that moment, murein in the preseptal region is a genuine cylindrical wall. Therefore septum-making enzymes should accept cylindrical murein as substrate at septum initiation. Hence, models describing septation should account for the effect, or the lack of effect, of treatments eliciting structural alterations in the murein layer. The parameters defining the structure, morphology and composition of the sacculus are susceptible to experimental manipulation.

Morphological alterations of the sacculus are potential division-disturbing situations. Mutants of E. coli with coccal shape are able of stable growth [21,87,88]. These strains

TABLE III

Some data related to peptidoglycan biosynthesis by the HMW-PBPs in *E. coli*

	Value	Reference
Surface of a cell with a $L/2R = 2$	$8.75 \pm 0.8 \mu$ m^2	73
Length of a disaccharide unit in a 3D helical structure	1.0 ± 0.05 nm	138
Average surface area per disaccharide unit	2.5 nm^2	73
Number of DAP molecules per sacculus	$3.5 \pm 0.6 \times 10^6$	73
	Range $2.7–5.6 \times 10^6$	1
Percentage of inhibition of ^{3}H-DAP incorporation by furazlocilin (amount of peptidoglycan made by PBP3)	$24 \pm 5\%$	105
	35%	73
Number of DAP molecules incorporated by PBP3, assumed to be the unique septal peptidoglycan synthetase:		
Minimal[a]	6×10^5	This work
Maximal[a]	19×10^5	This work
Number of molecules of PBP3 per cell	50	46
Number of monomers to be incorporated per second per molecule of PBP3:		
Minimal[b]	10 (40)	This work
Maximal[b]	31 (124)	This work
Total number of molecules of peptidoglycan-synthesizing HMW-PBPs per cell	300	46
Number of monomers to be incorporated per second per HMW-PBP molecule (in vivo)[c]	5	This work
Pool of undecaprenyl-P-P-MurNAc(penta)-GlcNAc [lipid II(penta)]	2000	47
Pool of undecaprenyl-P-P-MurNAc(tri)-GlcNAc [lipidII(tri)]	100	47
PBP1b incorporation (in vitro) (only transglycosylase)[d]	0.1 ± 0.03	47

[a]Calculated as 24% of the minimal (2.7×10^6) or 35% of the maximal (5.6×10^6) number of DAP molecules per cell.
[b]Calculated assuming that 50 molecules of PBP3 incorporate $6–19 \times 10^5$ monomers during 20 min (septation). Numbers in parentheses are the rate of incorporation expected for a strain possessing a reduced amount of PBP3, as in the case of Ax655 strain (25% of normal amount).
[c]Calculated assuming that all HMW-PBPs (300 molecules) incorporate 5.6×10^6 monomers during 60 min (cell cycle).
[d]Number of DAP molecules incorporated by one molecule of PBP1B molecule per second by polymerization in an in vitro standard assay.

grow and divide accordingly to regular patterns. Viability is as high as for the respective wild-types [89]. Growth in cell mass of coccal *E. coli* occurs mainly by changes in the cell diameter [89], a parameter that is constant in growing rods [1]. Septation of these cells reduces the diameter to that of the newly born parental cell [89]. These observations suggest that the septum forming machinery can work properly even in cells whose diameter is changing along the cell cycle. For a given mass, the diameter is bigger in coccal cells than in bacillar cells. Since division of coccal cells is also expected to be FtsZ dependent, the FtsZ ring structure (see below) must be sufficiently flexible to change its own diameter and, therefore, the number of monomers needed to build it up [90]. Nevertheless, septation is partially affected in spherical cells generated by mutations in the genes *pbp*A or *rod*A. These strains show a strong tendency to produce minicells under restrictive conditions [91]. This phenotype, rather than indicating a deficiency in the ability of the cells to make septa, suggests that errors are generated in the location of the division site.

The influence of the composition of pre-existing murein on septation can be considered under two aspects: the chemical composition of murein and the relative proportions of muropeptides (see Chapter 2). Changes in the chemical composition of *E. coli* murein can be induced in vivo by two different mechanisms: (i) feeding cells with incorporable analogues of the physiological precursors; (ii) exploiting the property of *E. coli* cells to catalyse a D-amino acid exchange reaction on macromolecular murein [92–95]. The last reaction results in the accumulation of muropeptides with a residue of the exchanged D-amino acid instead of D-alanine at the C-terminal position [96]. The specificity of the murein biosynthetic machinery in *E. coli* is rather high (see Chapter 3) and, consequently, the number of known incorporable analogues is small. The best studied is lanthionine [92,93]. This analogue of *meso* diaminopimelic acid is readily incorporated in the murein of diaminopimelic acid auxotrophic strains. Incorporation proceeds by the regular biosynthetic pathway and leads to a complete elimination of diaminopimelic acid from the murein. Initial investigations indicate that *E. coli* cells are capable of stable growth in lanthionine without noticeable morphological alterations [97]. These results indicate that biosynthetic complexes making lateral wall and septa have a similar specificity regarding the nature of the diamino acid. Murein from lanthionine-treated cells lacks the lanthionine-containing equivalent of one of the major physiological muropeptides. The missing muropeptide is probably the equivalent of a DAP-DAP crosslinked peptide in diaminopimelic acid-grown cells. Therefore, a critical involvement of DAP-DAP crosslinked muropeptides in septation is questionable.

Treatment with D-amino acids (methionine is the best studied) results in murein alterations at three levels: exchange of the terminal D-alanine in many tetrapeptide side chain-containing muropeptides, reduction in crosslinkage, decrease in the rate of murein synthesis [95] and reduction in the amount of murein per cell. Treated cells maintain normal morphology, viability and growth rate with 40% of the total muropeptides modified, with crosslinkage reduced by 30%, and with the rate of murein synthesis reduced by 50% [95]. This implies that septation is very flexible regarding the structure of the murein network. Particularly intriguing is the apparent dispensability of part of the murein in the sacculus. Rod morphology is maintained in cells grown in D-amino acids, then the reduction of murein in these cells reflects a decrease in the amount of murein per unit of surface area. This observation is supported by results obtained with diaminopimelic acid auxotrophic strains subjected to diaminopimelic acid limitation [98]. Stable growth and normal division are both possible with about one half the normal amount of murein. Hence, about one half of the murein is superfluous or, alternatively, the material left in the sacculus can be stretched up to a certain limit compatible with growth and septation. In growing cells, the sacculus is far from fully stretched (see Chapter 2). Such a mechanism could allow for a certain degree of tolerance during transient inhibitions of murein biosynthesis.

Another structural alteration of murein to be considered is the impairment of Braun's lipoprotein synthesis [99] (see Chapter 14). The murein-bound form of the lipoprotein is apparently responsible for the attachment of the outer membrane to the sacculus [99].

Lipoprotein deficient mutants are viable and divide normally. This reinforces the view that septation depends exclusively on the murein sacculus itself as the basic structural element [99]. The same idea is also supported by the phenotype of *env*A mutants. They are able to synthesize a complete septum but their outer membrane does not invaginate preventing separation of the daughter cells [100].

The observations discussed above emphasize that septation is an extraordinarily flexible process regarding the properties of its physical substrate, the murein layer. Models on peptidoglycan synthesis during cell division must account for the high built-in flexibility of the septum-making apparatus.

8. Function of PBP3 in septal peptidoglycan synthesis during the cell cycle

Three structural domains can be observed in the primary structure of PBP3 (Fig. 7): domain A is involved in the export of the protein (see above), domain B contains the putative transglycosylase activity and domain C is responsible for the transpeptidase activity (see Chapter 6). An additional functional domain, domain D, has been predicted from a number of observations: Several amino acid substitutions found during the analysis of mutated forms of PBP3 [101,102] occur at the C-terminal side of the KTG motif (Fig. 8) [82]. Deletions of the C-end of the protein obtained in vitro do not complement *pbp*B mutants but are able to bind β-lactams. Although the structural analysis of the class A β-lactamases and HMW-PBPs [103] assigns no function to this region, the frequent identification of mutants with amino acid substitutions located in domain D (Fig. 8) suggests that this domain is important for the more complex transpeptidation reaction catalyzed by the enzyme or for interactions with other components of the septator.

As already mentioned, a cell division-specific muropeptide has not been detected but, during septation, a shift from a mono-stranded to a multi-stranded mode of insertion occurs [81,104]. Pentapeptides are the preferred monomers for extension of the glycan chain, and the DD-Tetra-Tetra form is the major dimeric muropeptide (see Table IV). However, peptidoglycan synthesized de novo in a defilamentation system that mimics the activity of PBP3 in vivo, has a high degree of crosslinkage as well as a higher ratio of tripeptide-containing crosslinked subunits [72]. As tripeptide subunits cannot be used as donors for DD transpeptidation, the above results suggest that tripeptides are the acceptor substrates for PBP3. The change from mono- to multi-stranded insertion and the increase of tripeptide-containing subunits at the leading edge of septum formation might occur without a significant change in peptidoglycan composition of non-synchronously growing cells as follows. Polymerization of precursors into new chains, both tri- and pentapeptides, could be performed by the transglycosylase activity of PBP1B; coupling between the new and the old glycan strands could be performed by the Penta > Tetra DD-transpeptidase activity of PBP1B; and coupling between new strands may involve either the PBP1B activity or the Penta > Tri DD-transpeptidase activity of PBP3.

Involvement of PBP3 exclusively during septum formation is suggested by the close parallelism that exists between the specific blocking of the protein (by β-lactams or mutation), the inhibition of septation, and the significant decrease in murein synthesis during septation. [74,105,106]. However, impairment of PBP3 by mutation or by the PBP3-spe-

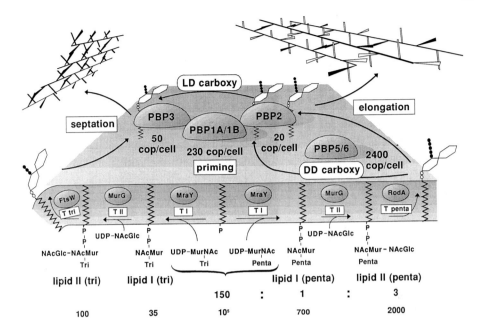

Fig. 8. Diagram of the enzymes and substrates that may be involved in the biosynthesis of murein during the cell cycle of *E. coli*. Soluble precursors are incorporated and made accessible to the membrane by a series of translocases TI, TII, Ttri and Tpenta. The *mra*Y [139] and *mur*G [140] gene products are responsible for the translocase I and II activities. The specific translocase activities for lipid II(tri) and lipid II(penta) are tentatively assigned to FtsW and RodA. Penicillin-binding proteins 1A or 1B (230 copies per cell) would be responsible for the initiation of new glycan chains (priming) using lipidII(penta) as donor. The rapid action of the DD-carboxypetidase activity (DD-carboxy) of PBP5 and 6 (2400 copies per cell) would produce the preferred substrate lipidII(tetra) for PBP2 (20 copies per cell) during elongation. Subsequent action of a LD-carboxypeptidase would also produce the lipidII(tri) precursor. Polymerization of both lipids II by PBP1B and the coupled activity of this PBP with the Penta > Tri DD-transpeptidase activity of PBP3 would produce the insertion of tripeptide-containing subunits at the leading edge of septum formation (septation) (see text). The ratio of soluble precursors is shown in the bottom lanes.

cific β-lactam furazlocillin does not affect DAP incorporation during initiation of constriction. However, it does so at later stages [74] indicating that a PBP3-independent system might be responsible for the change from lateral to septal peptidoglycan biosynthesis. The role of PBP3 as the septal-peptidoglycan biosynthetic enzyme derives from the fact that in vitro purified forms of the protein may perform both transpeptidase [107] and transglycosylase activities [108]. However, the transglycosylase activity has not always been detected with purified or membrane-bound PBP3 obtained from hyperproducing strains [47]. The *fts*W gene which codes for a protein with high homology to RodA, is required for the transglycosylase activity of PBP3. An increase in insoluble peptidoglycan and lipid precursor have been observed in an in vitro assay using cell envelopes of FtsW- and PBP3-overproducing strain [109] (see Chapter 4).

PBP3 seems to be the only HMW-PBP essential for cell growth. PBP1A and 1B compensate for each other [110,111] and PBP2 is not essential in the presence of the *lov* allele [112]. The first evidence that PBP3 is an essential protein rested upon the observation

TABLE IV

Potential types of dimers that could be synthesized by the different types of transpeptidases using the possible substrates as donor and acceptor[a]

Enzyme	Donor substrate	Acceptor substrate	Type of transpeptidation	Type of dimer[b]
D-D-transpeptidase	LipidII (penta)	LipidII (penta)	DD Penta > Penta	DD-Tetra-Penta
		(tetra)	DD Penta > Tetra	**DD-Tetra-Tetra**
		(tri)	DD Penta > Tri	**DD-Tetra-Tri**
L-D-transpeptidase	LipidII (penta)	LipidII (penta)	LD Penta > Penta	LD-Tri-Penta
		(tetra)	LD Penta > Tetra	**LD-Tri-Tetra**
		(tri)	LD Penta > Tri	LD-Tri-Tri
D-D-transpeptidase	LipidII (tetra)	LipidII (penta)	DD Tetra > Penta	Not possible
		(tetra)	DD Tetra > Tetra	Not possible
		(tri)	DD Tetra > Tri	Not possible
L-D-transpeptidase	LipidII (tetra)	LipidII (penta)	LD Tetra > Penta	LD-Tri-Penta
		(tetra)	LD Tetra > Tetra	**LD-Tri-Tetra**
		(tri)	LD Tetra > Tri	LD-Tri-Tri
D-D-transpeptidase	LipidII (tri)	LipidII (penta)	DD Tri > Penta	Not possible
		(tetra)	DD Tri > Tetra	Not possible
		(tri)	DD Tri > Tri	Not possible
L-D-transpeptidase	LipidII (tri)	LipidII (penta)	LD Tri > Penta	Not possible
		(tetra)	LD Tri > Tetra	Not possible
		(tri)	LD Tri > Tri	Not possible

[a]The most abundant species (>3%) found in peptidoglycan of exponentially growing cells of *Escherichia coli* are shown in bold type.
[b]In each dimer name, DD or LD indicates the type of crosslinking between donor (second part of the name) and acceptor (third part).

that only thermosensitive mutants can be isolated. Two independent results confirm this proposal, namely the analysis of the *rod*A (*sui*) mutation [113] which suppresses the thermosensitive filamentation of a *fts123* mutant strain and the *sfp* [114] mutation which suppresses filamentation of the *sep2158* (= *fts1655*) mutation [113]. The results indicate that PBP3, although present in low amounts in these strains grown at 42°C, is required for cell growth. An additional proof for the essential function of PBP3 is provided by a strain that contains a deletion/insertion mutation in *pbp*B and is unable to survive unless complemented by a wild type copy of the gene [115].

These results raise the question of how so few molecules of PBP3 can be sufficient for septation. Table III summarizes the data on the peptidoglycan biosynthetic activities of the HMW-PBPs. The minimal number of monomers incorporated per molecule of PBP3 in a wild type strain is 10/s. Spratt [46] has estimated that a normal cell contains about 50 molecules of PBP3. In a *pbp*B thermosensitive strain grown at the permitted temperature, only 30% of the normal level is found (about 20 molecules). Due to the low thermostability of the mutated PBP3, this number must still be smaller at 42°C. However, cells harbouring both a *pbp*B mutation and either the *sui* or the *sfp* mutation [40,114] are able to septate at the restrictive temperature in the presence of PBP1B or by overproducing PBP5, respectively.

9. PBP3, FtsA, FtsQ and FtsZ may exert their action at a place called septator

Regulation of gene expression provides one mechanism to temporally control cell division. Also mechanisms of spatial control should operate in the cell, as suggested by the behaviour of the *min* mutants. The primary signal that differentiates a site in the cell envelope, transforming it into a septum primordia, is unknown. FtsZ is an early component of the septation pathway but it does not have the capacity of initiating a septum. Overexpression of *fts*Z activates septation exclusively at the poles producing minicells, but no septa seem to be formed at other sites along the cell length as would be expected if FtsZ were sufficient to initiate differentiation of septal sites. Some theories, such as the nucleoid occlusion model [116], invoke purely physical reasons to explain the generation of septa at specific sites. Others, such as that based on periseptal annuli [117], invoke the action of proteins specific for septation. Thus, it seems likely that both kinds of mechanism, physical and biochemical, may contribute to septal differentiation. When the *bol*A gene is overexpressed together with *fts*Q, A and Z, aberrant septal primordia are placed at sites that do not correspond to the locations expected for the wild type. In some of these cells, an even number of septa can be observed producing, sometimes, a *guillotine* effect [118] on the unsegregated nucleoid (Fig. 9). This phenotype is not found in strains overexpressing *bol*A only. They have a round morphology probably due to increased levels of PBP6 [24]. Overexpression of *fts*Q, A and Z, or Z alone, produces minicells [119]. Hence, septation requires the concerted action of several genes and is dependent on the proper ratio of some gene products, such as FtsA and FtsZ [120,121].

The existence of a septator, the site at which proteins active in septation exert their action [2], is also reinforced by studies on the cellular localization of the Fts and PBP3 proteins. The cellular content of FtsQ is very low, around 25 molecules per cell. Cell fractionation shows that FtsQ is located in the cytoplasmic membrane and fusions of FtsQ with alkaline phosphatase that the protein is indeed located in the membrane with the C-terminal domain protruding into the periplasm [122]. FtsQ contains an aminoproximal hydrophobic domain essential for proper localization and biological function [123]. This domain contains a putative leucine zipper motif whose presence in an intact form may be required for the activity of the protein but not for its localization in the membrane (Sánchez et al., unpublished).

Although translated from the same transcripts as FtsQ, FtsA occurs in higher amounts, nearly 150 molecules per cell. Results of experiments performed on maxicells suggest a cytoplasmic membrane localization for FtsA [124]. This has been confirmed by cell frac-

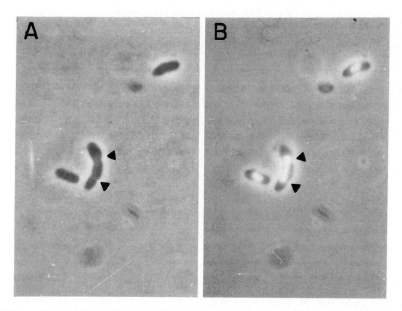

Fig. 9. The *guillotine* effect on the nucleoid caused by extra septation sites. Cells in which the *bol*A gene is overexpressed in the presence of high level expression of the *fts*Q, A and Z genes. Panel A (phase contrast image) and panel B (fluorescence of the DAPI-stained nucleoid) show the same individual cells. The strain used is *E. coli* VIP54 containing both pZAQ [118] and pMAK543 [134]. The pMAK543 plasmid contains *bol*A under the control of the *lac* promoter in a pBR-compatible replicon and pZAQ contains the *fts*Q,A,Z genes in a pBR322 type of plasmid. 1 mM IPTG has been added to a culture growing exponentially in LB broth and the culture subsequently kept below 5×10^7 cells/ml by dilution with pre-warmed medium containing IPTG. The photographs are those of a sample taken 150 min after addition of IPTG. For DAPI staining, the cells have been previously incubated for 10 min at 42°C with 100 μg/ml chloramphenicol to condense the nucleoid, spun down, washed in ice-cold azide saline and finally resuspended in azide-saline. DAPI staining was carried out as described [135]. The arrowheads in A point to two constrictions that are aberrantly initiated within the same cell. The arrowheads in B point to the stranglings observed in the unsegregated nucleoid, which closely match the position of the constrictions (these data were obtained by M.V. while visiting W.D. Donachie's laboratory).

tionation and trypsin accessibility in vesicles showing, in addition, that FtsA is facing towards the cytoplasm [125].

FtsZ is mostly localized in the cytoplasm [126]. It is present in over 5000 molecules per cell. As shown by immunoelectron microscopy, the cell population of FtsZ migrates towards the central part of the cell forming a FtsZ ring at the septum [90]. The mean half-life of the FtsZ molecules is more than one division cycle [127] but it is not known if the molecules that form the FtsZ ring remain active during more than one cycle when the ring is disassembled after division. Whatever the case, the level of FtsZ is likely to be replenished by cell cycle-dependent transcription as described above.

The existence of a septator is also supported by two results indicating the existence of interactions between PBP3 and both FtsA and FtsZ. The interaction between PBP3 and FtsA derives from genetic evidence. Binding of radiolabeled iodo-ampicillin to PBP3 is abolished when the only form of FtsA present in the cell is the inactivated FtsA3 protein [128]. Further interpretations of this result should be done cautiously since it has not been proven that the form of PBP3 harboured by D3, the strain used for the experiment, is identical to the wild type one.

Evidence for the interaction between PBP3 and FtsZ [129] rests upon the unlocalized distribution of an overproduced LacZ-PBP3 fusion in *fts*Z mutants. In contrast, this fusion protein is specifically localized at central and polar positions in wild type, *fts*Q, A and *pbp*B strains. Other interactions involving the three FtsQ, A, and Z proteins have been suggested [130] but they remain to be proven. Given the low abundance of PBP3 molecules in the cell, a cooperative septum-forming structure must be formed at the time of cell division. Considering that the non-processed form of PBP3 could be the active protein in septation (see above), one may hypothesize that only the specifically regulated, de novo synthesized PBP3 is used in septal peptidoglycan biosynthesis .

10. The biochemistry of cell division regulation

There have been considerable advances in the understanding of the biochemical modifications of proteins involved in eukaryotic cell division [3]. This contrasts with the paucity of similar data for bacterial division. This situation has recently begun to change with the finding of a GTPase activity for FtsZ [131–134], an ATPase activity for MinD [10] and the availability of a structural model for FtsA [135].

The GTPase activity of the FtsZ protein was discovered after finding that it contained a sequence motif of seven amino acids homologous to that found in tubulins. It would be interesting if the FtsZ molecules were located in different cell fractions depending on the presence or the absence of bound GTP. It is tempting to speculate that a correlation might exist between all the observed properties of FtsZ, namely the migration of the GTP-bound form of FtsZ to the septator where it triggers division, and the activation of other proteins, such as PBP3, by specific phosphorylation. This would not correlate well with the widely different number of molecules since FtsZ is at least in a 100-fold excess over PBP3. It is not known if all the molecules of FtsZ are active for division and, therefore, this imbalance in the number of molecules could be less spectacular if the populations of active PBP3 and FtsZ were compared. Another possibility is that FtsZ may form part of a contractile structure in the periseptal annuli, its phosphorylation being involved in the molecular mechanism causing contraction.

A valuable tool for analysis of these biochemical activities is provided by the methods of structural analysis, in particular for FtsA. The FtsA protein has been proposed to belong to a family of ATP-binding proteins that include both prokaryotic members like DnaK, a chaperonin, and eukaryotic members like actin, a contractile protein involved in cytokinesis [135]. Validation of the predicted biochemical activity of FtsA would support a mechanoprotein model of cell division. Although these hypotheses are very attractive, no data are available to reject or confirm them.

References

1. Cooper, S. (1991) Bacterial Growth and Division. Academic Press, Inc. San Diego, California.
2. Vicente, M., Kushner, S.R., Garrido, T. and Aldea, M. (1991) Mol. Microbiol. 5, 2085–2091.

3. Moreno, S., Hayles, J. and Nurse, P. (1989) Cell 58, 361–372.
4. Reed, S.I. (1992) Annu. Rev. Cell. Biol. 8, 529–561.
5. Donachie, W.D. and Robinson, A.C. (1987) in: F.C. Neidhardt, J.L. Ingraham, K. Brooks Low, B. Magasanik, M. Schaechter and H.E. Umbarger (Eds.), *Escherichia coli* and *Salmonella typhimurium.* Cellular and Molecular Biology, Vol. 2, American Society for Microbiology, Washington, DC, pp. 1578–1593.
6. Donachie, W.D. (1968) Nature (London) 219, 1077–1079.
7. Spratt, B.G. (1975) Proc. Natl. Acad. Sci. USA 72, 2999–3003.
8. Matsuhashi, M. Wachi, M. and Ishino, F. (1990) Res. Microbiol. 141, 89–103.
9. Waxman, D.J. and Strominger, J.L. (1983) Annu. Rev. Biochem. 52, 825–869.
10. de Boer, P.A.J., Crossley, R.E., Hand, A.R. and Rothfield, L.I. (1991) EMBO J. 10, 4371–4380.
11. de Boer, P.A.J., Crossley, R.E. and Rothfield, L.I. (1989) Cell 56, 641–649.
12. Faubladier, M., Cam, K. and Bouché, J.-P. (1990) J. Mol. Biol. 212, 461–471.
13. Tétart, F. and Bouché, J.-P. (1992) Mol. Microbiol. 6, 615–620.
14. Brent, R. and Ptashne, M. (1981) Proc. Natl. Acad. Sci. USA 78, 4202–4208.
15. George, J., Castellazi, M. and Buttin, G. (1975) Mol. Gen. Genet. 140, 309–332.
16. Lutkenhaus, J.F. (1983) J. Bacteriol. 154, 1339–1346.
17. Ward, J.E. and Lutkenhaus, J.F. (1984) J. Bacteriol. 157, 815–820.
18. Gottesman, S., Halpern, E. and Trisler, P. (1981) J. Bacteriol. 148, 265–273.
19. Bachmann, B.J. (1990) Microbiol. Rev. 54, 130–197.
20. Crickmore, N. and Salmond, G.P.C. (1992) J. Bacteriol. 174, 7202–7206.
21. Spratt, B.G., Boyd, A. and Stoker, N. (1980) J. Bacteriol. 143, 569–581.
22. Beall, B., Lowe, M. and Lutkenhaus, J. (1988) J. Bacteriol. 170, 4855–4864.
23. Harry, E.J., Steward, B.J. and Wake, R.G. (1993) Mol. Microbiol. 7, 611–621.
24. Aldea, M., Garrido, T., Hernández-Chico, C., Vicente, M. and Kushner, S.R. (1989) EMBO J. 8, 3923–3931.
25. Connell, N., Han, Z., Moreno, F. and Kolter, R. (1987) Mol. Microbiol. 1, 195–210.
26. Aldea, M., Garrido, T., Pla, J. and Vicente, M. (1990) EMBO J. 9, 3787–3794.
27. Lange, R. and Hengge-Aronis, R. (1991) J. Bacteriol. 173, 4474–4481.
28. Bohannon, D.E., Connell, N., Keener, J., Tormo, A., Espinosa-Urgel, M., Zambrano, M.M. and Kolter, R. (1991) J. Bacteriol. 173, 4482–4492.
29. Ghelardini, P., Lauri, P., Ruberti, I., Orlando, V. and Paolozzi, L. (1991) Res. Microbiol. 142, 259–268.
29a. Garrido, T., Sánchez, M., Palacios, P., Aldea, M. and Vicente, M. (1993) EMBO J. 12, 3957–3965.
30. Kraut, H. Keck, W. and Hirota, Y. (1981) in: Annual Report of National Institute of Genetics, National Institute of Genetics, Tokyo, pp. 24–29.
31. Nakamura, M. Maruyama, I.N. Soma, M. Kato, J. Suzuki, H. and Hirota. Y. (1983) Mol. Gen. Genet. 191, 1–9.
32. Kohara, Y. Akiyama, K. and Isono, K. (1987) Cell 50, 495–508.
33. Stoker, N.G., Fairweather, N.F. and Spratt, B.G. (1982) Gene 18, 335–341.
34. Gómez, M. (1991) Doctoral Dissertation, Universidad Autónoma de Madrid, Madrid.
35. Wientjes, F.B., Olijhoek, T.J.M., Schwarz, U. and Nanninga, N. (1983) J. Bacteriol. 153, 1287–1293.
36. Gómez, M.J., Fluoret, B., van Heijenoort, J. and Ayala, J.A. (1990) Nucleic Acids Res. 18, 2813.
37. Guzmán, L.M., Barondess, J.J. and Beckwith, J. (1992) J. Bacteriol. 174, 7716–7728.
38. Ishino, F. Jung, H.K., Ikeda, M., Doi, M., Wachi, M. and Matsuhashi, M. (1989) J. Bacteriol. 171, 5523–5530.
39. Ueki, M. Wachi, M. Jung, H.K., Ishino, F. and Matsuhashi, M. (1992) J. Bacteriol. 174, 7841–7843.
40. Begg, K. Takasuga, J.A., Edwards, D.H., Dewar, S.J., Spratt, B.G., Adachi, H., Ohta, T., Matsuzawa, H. and Donachie, W.D. (1990) J. Bacteriol. 172, 6697–6703.
41. Maruyama, I.N., Yamamoto, A., Maruyama, T. and Hirota, Y. (1983) in: R. Hakenbeck, J.-V. Höltje and H. Labischinski (Eds.), The Target of Penicillin, Walter de Gruyter, New York, pp. 393–402.
42. Spratt, B.G. and Cromie, K.D. (1988) Rev. Infect. Dis. 10, 699–711.
43. Devereux, J., Haeberli, P. and Smithies, O. (1984) Nucleic Acids Res. 12, 387–395.
44. Wachi, M. and Matsuhashi, M. (1989) J. Bacteriol. 171, 3123–3127.
45. Lutkenhaus, J.F. (1990) Trends Genet. 6, 22–25.

46. Spratt, B.G. (1977) Eur. J. Biochem. 72, 341–352.
47. van Heijenoort, Y. Gómez, M. Derrien, M. Ayala, J. A. and van Heijenoort, J. (1992) J. Bacteriol. 174, 3549–3557.
48. Dalbey, R.E.(1991) Mol. Microbiol. 5, 2855–2860
49. Tokunaga, M., Loranger, J.M. and Wu, H.C. (1984) J. Cell. Biochem. 24, 113–120.
50. Hayashi, S., Hara, H., Suzuki, H. and Hirota, Y. (1988) J. Bacteriol. 170, 5392–5395.
51. Gómez, M. Desviat, L.R. Merchante, R. and Ayala, J.A. (1992) in: M.A. de Pedro, J.-V. Höltje and W. Löffelhardt (Eds.), Bacterial Growth and Lysis: Metabolism and Structure of the Bacterial Sacculus, Plenum Press, London, pp. 309–318.
52. Nagasawa, H., Sakagami, Y., Suzuki, A., Suzuki, H., Hara, H. and Hirota, Y. (1989) J. Bacteriol. 171, 5890–5893.
53. Hara, H., Yamamoto, Y., Higashitani, A., Suzuki, H. and Nishimura, Y. (1991) J. Bacteriol. 173, 4799–4813.
54. Bartholomé-De Belder, J., Nguyen-Distèche, M., Houba-Herin, N., Ghuysen, J.M., Maruyama, I.N., Hara, H., Hirota, Y. and Inouye, M. (1988) Mol. Microbiol. 2, 519–525.
55. Bowler, L.D. and Spratt, B.G. (1989) Mol. Microbiol. 3, 1277–1286.
56. Ferreira, L.C.S., Keck, W., Betzner, A. and Schwarz, U. (1987) J. Bacteriol. 169, 5776–5781.
57. Ogura, T., Tomoyasu, T., Yuki, T., Morimura, S., Begg, K.J., Donachie, W.D., Mori, H., Niki, H. and Hiraga, S. (1991) Res. Microbiol. 142, 279–282.
58. Hara, H., Nishimura, Y., Kato, J.-I., Suzuki, H., Nagasawa, H., Suzuki, A. and Hirota, Y. (1989) J. Bacteriol. 171, 5882–5889.
59. De la Rosa, E.J., de Pedro, M.A. and Vázquez, D. (1982) FEMS Microbiol. Lett. 14, 91–94.
60. De la Rosa, E.J., de Pedro, M.A. and Vázquez, D. (1983) FEMS Microbiol. Lett. 19, 165–167.
61. Prats, R., Gómez, M.J., Pla, J., Blasco, B. and Ayala, J.A. (1989) J. Bacteriol. 171, 5194–5198.
62. Wang, X., de Boer, P.A.J. and Rothfield, L.I. (1991) EMBO J. 10, 3363–3372.
63. Robinson, A.C., Kenan, D.J., Hatfull, G.F., Sullivan, N.F., Spiegelberg, R. and Donachie, W.D. (1984) J. Bacteriol. 160, 546–555.
64. Yi, Q.-M., Rockenbach, S., Ward, J.E. and Lutkenhaus, J. (1985) J. Mol. Biol. 184, 399–412.
65. Dewar. S.J. and Donachie, W.D. (1990) J. Bacteriol. 172, 6611–6614.
66. Gervais, F.G., Phoenix, P. and Drapeau, G.R. (1992) J. Bacteriol. 174, 3964–3971.
67. Masters, M., Paterson, T., Popplewell, A.G., Owen-Hughes, T., Pringle, J.H. and Begg, K. (1989) Mol. Gen. Genet. 216, 475–483.
68. Mukherjee, A. and Donachie, W.D. (1990) J. Bacteriol. 172, 6106–6111.
69. Wang, H. and Gayda, R.C. (1992) Mol. Microbiol. 6, 2517–2524.
70. Robinson, A.C., Kenan, D.J., Sweeney, J. and Donachie, W.D. (1986) J. Bacteriol. 167, 809–817.
71. Markiewicz, Z., Broome-Smith, J.K., Schwarz, U. and Spratt, B.G. (1982) Nature 297, 702–704.
72. Pisabarro, A.G., Prats, R., Vázquez, D. and Rodríguez-Tébar, A. (1986) J. Bacteriol. 168, 199–206.
73. Wientjes, F.B. and Nanninga, N. (1991) Res. Microbiol. 142, 333–344.
74. Wientjes, F.B. and Nanninga, N. (1989) J. Bacteriol. 171, 3412–3419.
75. Woldringh, C.L., Huls, P., Pas, E., Brakenhoff, G.J. and Nanninga, N. (1987) J. Gen. Microbiol. 133, 575–586.
76. Glauner, B. and Höltje, J.-V. (1990) J. Biol. Chem. 265, 18988–18996.
77. Cooper, S. (1991) Microbiol. Rev. 55, 649–674.
78. Höltje, J.-V. and Schwarz, U. (1985) in: N. Nanninga (Ed.), Molecular Cytology of *Escherichia Coli*, Academic Press, New York, pp. 77–119.
79. De Jonge, B.L.M., Wientjes, F.B., Jurida, I., Driehuis, F., Wouters, J.T.M. and Nanninga, N. (1989) J. Bacteriol. 156, 136–140.
80. Romeis, T., Kohlrausch, U., Burgdorf, K. and Höltje, J.-V. (1991) Res. Microbiol. 142, 325–332.
81. Nanninga, N. (1991) Mol. Microbiol. 5, 791–795.
82. Ghuysen, J.M. (1991) Annu. Rev. Microbiol. 45, 37–67.
83. Höltje, J.-V. and Glauner, B. (1990) Res. Microbiol. 141, 75–89.
84. Begg, K.J. and Donachie, W.D. (1985) J. Bacteriol. 163, 615–622.
85. Höltje, J.-V. and Tuomanen, E. (1991) J. Gen. Microbiol. 137, 441–454.
86. Rothfield, L.I. and Cook, W.R. (1988) Microbiol. Sci. 5, 182–185.

100

87. Iwaya, M., Jones, C.W., Khorana, J. and Strominger, J.L. (1978) J. Bacteriol. 133, 196–202.
88. D'Ari, R., Jaffe, A., Bouloc, P. and Robin, A. (1988) J. Bacteriol. 170, 65–70.
89. Woldringh, C.L., Huls, P., Nanninga, N., Pas, E., Taschner, P.E.M. and Wientjes, F.B. (1988) in: P. Actor, L. Daneo-Moore, L.M. Higgins, M.R.J. Salton and G.D. Shockman (Eds.), Antibiotic Inhibition of Bacterial Cell Surface Assembly and Function, American Society for Microbiology, Washington, DC, pp. 66–78.
90. Bi, E. and Lutkenhaus, J. (1991) Nature (London) 354, 161–164.
91. Caparrós Rodríguez, M. and de Pedro, M.A. (1990) FEMS Microbiol. Lett. 72, 235–240.
92. Knüsel, F., Nüesch, J., Scherer, M. and Schmid, K. (1967) Path. Microbiol. 30, 871–879.
93. Chaloupka, J., Strnadová, M., Cáslavská, J. and Vere, K. (1974) Z. Allg. Mikrobiol. 14, 283–296.
94. Tsuruoka, T., Tamura, A., Miyata, A., Takei, T., Iwamatsu, K., Inouye, S. and Matsuhashi, M. (1984) J. Bacteriol. 160, 889–894.
95. Caparrós, M., Pisabarro, A.G. and de Pedro, M.A. (1992) J. Bacteriol. 174, 5549–5559.
96. Caparrós, M., Pittenauer, E., Schmidt, E.R., de Pedro, M.A. and Allmaier, G. (1993) FEBS Lett. 316, 181–185.
97. Caparrós, M., Quintela, J.C., Leguina, J.I. and de Pedro, M.A. (1993) in: M.A. de Pedro, J.-V. Höltje and W. Löffelhard (Eds.), Bacterial Growth and Lysis: Metabolism and Structure of the Bacterial Sacculus, Plenum Press, London, pp. 147–159.
98. Prats, R. and de Pedro, M.A. (1989) J. Bacteriol. 171, 3740–3745.
99. Wu. H.C. (1987) in: M. Inouye (Ed.), Bacterial Outer Membranes as Model Systems, Wiley, New York, pp. 37–71.
100. Normark, S. (1970) Genet. Res. 16, 63–78.
101. Hedge, P.J. and Spratt, B.G. (1985) Eur. J. Biochem. 151, 111–121.
102. Hedge, P.J. and Spratt, B.G. (1985) Nature 318, 478–480.
103. Englebert, S., Kharroubi, A.E., Piras, G., Joris, B., Coyette, J., Nguyen-Disteche, M. and Ghuysen, J.M. (1992) in: M.A. de Pedro, J.-V. Höltje and W. Löffelhard (Eds.), Bacterial Growth and Lysis: Metabolism and Structure of the Bacterial Sacculus, Plenum Press, London, pp. 319–333.
104. de Jonge, B.L.M., Wientjes, F.B., Jurida, I., Driehuis, F., Wouters, J.T.M. and Nanninga, N. (1989) J. Bacteriol. 171, 5783–5794.
105. Botta, G.A. and Park, J.T. (1981) J. Bacteriol. 145, 333–340.
106. Spratt, B.G. and Pardee, A.B. (1975) Nature 254, 516–517.
107. Adam, M., Damblon, C., Jamin, M., Zorzi, W., Dusart, V., Galleni, M., El Kharroubi, A., Piras, G., Spratt, B.G., Keck, W., Coyette, J., Ghuysen, J.M., Nguyen-Distèche, M. and Frèere, J.M. (1991) Biochem. J. 279, 601–604.
108. Ishino, F. and Matsuhashi, M. (1981) Biochem. Biophys. Res. Commun. 101, 905–911.
109. Matsuhashi, M., Wachi, M., Ishino, F., Ideda, M., Okada, Y., Jung, H., Ueki, K.M., Tomioka, S., Pankrushina, A.N. and Song, M.D. (1992) in: H. Kleinhauf and H. von Döhren (Eds.), Fifty Years of Penicillin Application – History and Trends, Video Press, Prague, in press.
110. Yousif, S.Y., Broome-Smith, J.K. and Spratt, B.G. (1985) J. Gen. Microbiol, 131, 2839–2845.
111 Kato, J.-I., Suzuki, H. and Hirota, Y. (1985) Mol. Gen. Genet. 200, 272–277.
112. Bouloc, P., Jaffé, A. and D'Ari, R. (1989) EMBO J. 8, 317–323.
113. Begg, K.J., Spratt, B.G. and Donachie, W.D. (1986) J. Bacteriol. 167, 1004–1008.
114. Del Portillo, F.G., de Pedro, M.A. and Ayala, J.A. (1991) FEMS Microbiol. Lett. 84, 7–14.
115. Hara, H. and Park, J.T. (1992) in: M.A. de Pedro, J.-V. Höltje and W. Löffelhard (Eds.), Bacterial Growth and Lysis: Metabolism and Structure of the Bacterial Sacculus, Plenum Press, London, pp. 303–308.
116. Woldringh, C.L., Mulder, E., Valkenburg, J.A.C., Wientjes, F.B., Zaritsky, A. and Nanninga, N. (1990) Res. Microbiol. 141, 39–49.
117. Rothfield, L., de Boer, P.A. and Cook, W.R. (1990) Res Microbiol. 141, 57–63.
118. Niki, H., Jaffé, A., Imamura, R., Ogura, T. and Hiraga, S. (1991) EMBO J. 10, 183–193.
119. Ward, J.E. and Lutkenhaus, J. (1985) Cell 42, 941–949.
120. Dai, K. and Lutkenhaus, J. (1992) J. Bacteriol. 174, 6145–6151.
121. Dewar, S.J., Begg, K. and Donachie, W.D. (1992) J. Bacteriol. 174, 6314–6316.
122. Carson M.J., Barondess, J. and Beckwith, J. (1991) J. Bacteriol. 173, 2187–2195.

123. Dopazo, A., Palacios, P., Sánchez, M., Pla, J. and Vicente, M. (1992) Mol. Microbiol. 6, 715–722.
124. Chon, Y. and Gayda, R. (1988) Biochem. Biophys. Res. Commun. 152, 1023–1030.
125. Pla, J., Dopazo, A. and Vicente, M. (1990) J. Bacteriol. 172, 5097–5102.
126. Pla, J., Sánchez, M., Palacios, P., Vicente, M. and Aldea, M. (1991) Mol. Microbiol. 5, 1681–1686.
127. Pla, J., Palacios, P., Sánchez, M., Garrido, T. and Vicente, M. (1993) in: M.A. de Pedro, J.-V. Höltje and W. Löffelhard (Eds.), Bacterial Growth and Lysis: Metabolism and Structure of the Bacterial Sacculus, Plenum Press, London, pp. 363–368.
128. Tormo, A., Ayala, J.A., de Pedro, M.A., Aldea, M. and Vicente, M. (1986) J. Bacteriol. 166, 985–992.
129. Ayala, J.A., Pla, J., Desviat, L.R. and de Pedro, M.A. (1988) J. Bacteriol. 170, 3333–3341.
130. Descoteaux, A. and Drapeau, G.R. (1987) J. Bacteriol. 169, 1938–1942.
131. de Boer, P.A.J., Crossley, R. and Rothfield, L. (1992) Nature (London) 359, 254–256.
132. Roy Chauduri, D. and Park, J.T. (1992) Nature (London) 359, 251–254.
133. Mukherjee, A., Dai, K. and Lutkenhaus, J. (1993) Proc. Natl. Acad. Sci. USA 90, 1053–1057.
134. Vinella, D., Bouloc, P. and D'Ari, R. (1993) Curr. Biol. 3, 65–66.
135. Bork, P., Sander, C. and Valencia, A. (1992) Proc. Natl. Acad. Sci. USA 89, 7290–7294.
136. Aldea, M., Hernández-Chico, C., de la Campa, A.G., Kushner, S.R. and Vicente, M. (1988) J. Bacteriol. 170, 5169-5176.
137. Hiraga, S., Niki, H., Ogura, T., Ichinose, C., Mori, H., Ezaki, B. and Jaffé, A. (1989) J. Bacteriol. 171, 1496–1505.
138. Barnickel, G., Naumann, D., Bradaczek, H., Labischinski, H. and Giesbrecht, P. (1983) in: R. Hakenbeck, J.-V. Höltje and H. Labischinski (Eds.), Walter de Gruyter, New York, pp. 61–66.
139. Ikeda, M., Wachi, M., Jung, H.K., Ishino, F. and Matsuhashi, M. (1991) J. Bacteriol. 173, 1021–1026.
140. Menguin-Lecreux, D., Texier, L., Rousseau, M. and van Heijenoort, J. (1991) J. Bacteriol. 173, 4625–4636.

J.-M Ghuysen and R. Hakenbeck (Eds.), *Bacterial Cell Wall*
103

CHAPTER 6

Biochemistry of the penicilloyl serine transferases

JEAN-MARIE GHUYSEN and GEORGES DIVE

Centre d'Ingénierie des Protéines, Université de Liège, Institut de Chimie, B6,
B-4000 Sart Tilman (Liège 1), Belgium

1. Introduction

The penicilloyl serine transferases catalyze cleavage of the cyclic amide bond of penicillin via formation of a serine S* ester-linked penicilloyl enzyme (Fig. 1). They bear a unique signature implying three conserved amino acid groupings according to the following motifs: S*XXK, SXN or YXN and K(H,R)T(S)G occurring at equivalent places along their amino acid sequences [1]. The enzymes, which are of known structure [2–6], consist of an all-α domain and an α/β domain whose core is a five-stranded β-sheet protected by additional helices on both faces (Fig. 2). In this structure, the three conserved motifs are brought close to each other and form part of the active site located between the all-α and α/β domains. The tetrad S*XXK is central to the cavity, at the amino end of helix $\alpha2$. The triad, SXN or YXN, is on a loop connecting two helices on one side of the cavity. The triad K(H)T(S)G is on the innermost $\beta3$ strand of the β-sheet on the other side of the cavity. Some penicilloyl serine transferases are involved in wall peptidoglycan metabolism and assembly. They are the targets of penicillin action. Others protect the bacteria against the deleterious effects of penicillin.

2. A putative ancestor of the penicilloyl serine transferases

With X = a diamino acid residue, wall peptidoglycan assembly from lipid-transported N-acetylmuramyl(L-Ala-γ-D-Glu-L-X-D-Ala-D-Ala)-N-acetylglucosamine disaccharide pentapeptide units is made by transglycosylation (glycan chain elongation) and transpeptidation (peptide crosslinking) reactions (see Chapters 2–5). Assuming that the primitive Eubacteria possessed unlinked transglycosylases and transpeptidases located on the outer face of the membrane, a putative ancestor of the interpeptide crosslinking enzyme might have been a DD-transpeptidase with a particular serine residue S* central in the catalysis (Fig. 3). Peptide crosslinking was made by transfer of the R-L-X-D-alanyl electrophilic group of a R-L-X-D-alanyl-D-alanine carbonyl donor peptide to the enzyme serine residue

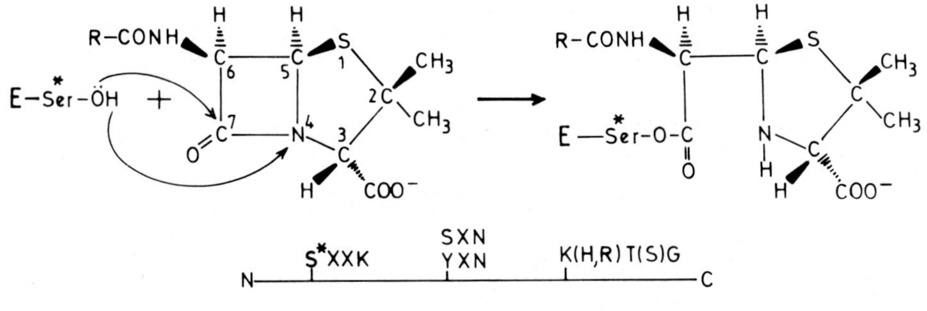

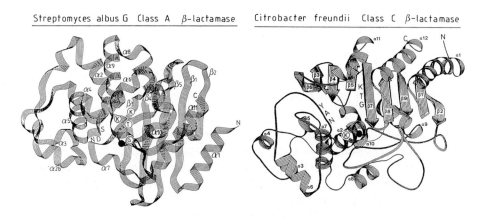

Fig. 1. The penicilloyl serine transferases superfamily. Formation of serine ester-linked acyl (penicilloyl) enzyme and active-site defining motifs. S*, essential serine residue.

S* with release of the leaving group D-alanine and formation of a serine ester-linked acyl (R-L-X-D-alanyl) enzyme, and from this acyl enzyme, to the ω-amino group of the L-X residue of another peptide acting as acceptor. Efficacy of the transpeptidation rested upon the hydrolytic inertness of the acyl enzyme intermediate. Water was not an acceptor of the transfer reaction.

D-Alanyl-D-alanine and penicillin, a cyclic tripeptide secondary metabolite, were isosteric. Acylation of the enzyme serine residue S* by penicillin generated a leaving group which remained part of the acyl (penicilloyl) enzyme and could not diffuse away from the active site (Fig. 3). As a result of this occupancy, the penicilloyl enzyme was both hydrolytically and aminolytically inert and the inactivated DD-transpeptidase behaved as a penicillin-binding protein (PBP).

Fig. 2. Three-dimensional structures of β-lactamases. α, α-helix; β, β-strand. The essential serine (filled circle at the amino terminus of α2) and the other active-site defining motifs are indicated. Adapted from [4] (*C. freundii* enzyme) and [5] (*Streptomyces albus* G enzyme).

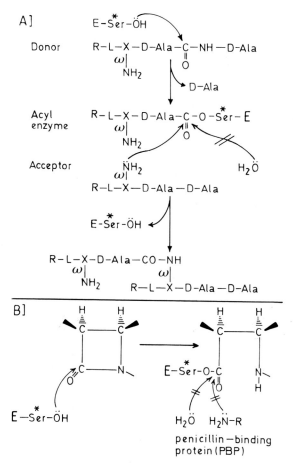

Fig. 3. Putative ancestor of the peptidoglycan crosslinking enzyme. (A) Reaction with R-L-X-D-alanyl-D-alanine-terminated peptides; (B) reaction with penicillin.

3. Evolution features

The present-day penicilloyl serine transferases comprise the *Streptomyces* K15 DD-transpeptidase/PBP, the monofunctional DD-peptidases/PBPs, the β-lactamases, the penicillin sensory PBPs and the bi(multi)functional high M_r PBPs.

3.1. The Streptomyces K15 DD-transpeptidase/PBP

A mechanistic model of the putative peptidase ancestor is provided by the DD-transpeptidase/PBP of *Streptomyces* K15 [7–9]. Upon cleavage of the signal peptide of the precursor, the exported 262 amino acid residue protein remains in interaction with the outer face of the membrane (from which it can be detached by high salt concentrations). The enzyme consists of one catalytic module. The tetrad S*35TTK and the triad K217TG are close to the amino and carboxy ends of the polypeptide chain, respectively, and the triad S96GC (a variation of the SXN motif) is located 61 amino acid residues downstream from S*35. The enzyme has been overexpressed and crystallized in a form suitable for X-ray analysis.

The acyl transfer reactions that the *Streptomyces* K15 enzyme catalyzes on diacetyl(Ac_2)-L-lysyl-D-alanyl-D-alanine (an analogue of the R-L-X-D-alanyl-D-alanine-terminated peptidoglycan precursors), proceed through formation of a serine S*35 ester-linked acyl (Ac_2-L-Lys-D-alanyl) enzyme. The reactions are similar to those shown in Fig. 3. However, hydrolysis of the tripeptide is negligible and the enzyme is seemingly silent because the released D-alanine residue competes successfully with 55.5 M H_2O, performs aminolysis of the acyl enzyme and regenerates the original carbonyl donor substrate. Glycyl-glycine at millimolar concentrations overcomes the acceptor activities of D-alanine and water. In the presence of Gly-Gly, Ac_2-L-Lys-D-Ala-D-Ala is quantitatively converted into the transpeptidated product Ac_2-L-Lys-D-Ala-Gly-Gly with a catalytic efficiency (see Section 4) of 65 M^{-1} s^{-1}.

D-Lactate, an alcohol analogue of D-alanine, has no acceptor activity. Hydrolysis of the depsipeptide Ac_2-L-Lys-D-Ala-D-lactate proceeds with a catalytic efficiency of 650 M^{-1} s^{-1}. In the presence of Gly-Gly, hydrolysis and aminolysis occur on a competitive basis.

The *Streptomyces* K15 enzyme is inactivated by β-lactams in the form of a PBP. The second-order rate constant of acylation of S*35 is 150 M^{-1} s^{-1} for benzylpenicillin and 850 M^{-1} s^{-1} for cefoxitin.

3.2. Monofunctional DD-peptidases/PBPs, β-lactamases and penicillin sensors

By reference to the *Streptomyces* K15 DD-transpeptidase/PBP, features of evolution re-sulted in increased preference for water as attacking nucleophile of the serine ester-linked acyl (R-L-X-D-alanyl) enzyme with conservation of penicilloyl enzyme inertness. The emerging DD-peptidases/PBPs performed concomitant hydrolysis and transpeptidation of R-L-X-D-alanyl-D-alanine peptides on a competitive basis. Membrane anchoring was achieved by fusion of the carboxy end of the extracellular catalytic module to a 50–100 amino acid residue extension, the end of which contained a signal-like segment that served as membrane anchor [10,11]. All the bacteria possess one or several membrane-bound DD-peptidases/PBPs. Cloning vectors are available which allow *Escherichia coli* PBP5 to be overexpressed in the form of an anchor-free, catalytically active water-soluble derivative [12,13]. Few DD-peptidases/PBPs are secretory proteins. The *Streptomyces* R61 [14] and *Actinomadura* R39 [15] DD-peptidases/PBPs have been studied extensively (see Section 4.2). The *Streptomyces* R61 enzyme precursor has a cleavable N-terminal peptide signal and a cleavable C-terminal extension [14]. C-terminal processing involves cleavage of the tyrosyl-glycine peptide bond of the sequence KPTTGE which has homol-ogy with the 'wall associating' LPXTGX motif (see Chapter 9). Similar to the reactions that these extracellular enzymes catalyze in vitro, the membrane-bound DD-pepti-dases/PBPs probably help control the extent of peptide crosslinking, are involved in peptidoglycan-remodelling during the cell life cycle and play important roles in cell mor-phogenesis [16-18] and sporulation (see Chapter 8).

Loss of peptidase activity and acquisition of catalyzed hydrolysis of penicilloyl enzyme gave rise to the defensive β-lactamases. The β-lactamases are secretory proteins (which accumulate in the periplasm of the Gram-negative bacteria). They consist of one catalytic module. The S*XXK motif is 30–40 amino acid residues downstream from the amino end of the mature protein and the KT(S)G motif is 50–60 amino acid residues upstream from the carboxy end of the protein.

Loss of peptidase activity without acquisition of catalyzed hydrolysis of penicilloyl enzyme gave rise to PBPs whose only function was penicillin binding. Fusion of the amino end of the catalytic module of these PBPs to the carboxy end of transmembrane proteins, gave rise to penicillin-sensory transducers involved in β-lactams-induced derepression of β-lactamase synthesis in *Bacillus licheniformis* and *Staphylococcus aureus* and derepression of low-affinity PBP2' synthesis in *S. aureus* (see Chapter 24). Cloning vectors are available which allow the *B. licheniformis* BlaR penicillin sensor to be exported as an active PBP in the periplasm of *E. coli*, independently of the transducer [19].

Features of evolution also resulted in the emergence of distinct classes of DD-peptidases/PBPs and β-lactamases. Enzymes belonging to a given class are related in their primary structures by similarity scores which are at least five standard deviations (but may be as high as fifty standard deviations) above that expected for a run of twenty randomized pairs of proteins having the same amino acid compositions as the pairs of enzymes under comparison.

Using the ABL amino acid numbering [20], the active-site defining motifs of the β-lactamases of class A are S*70XXK, S130DN and K234T(S)G. Those of *S. aureus*, *B. licheniformis*, *Streptomyces albus* G and *E. coli* (the TEM enzyme) and the β-lactamase of class C of *Citrobacter freundii* are of known 3-D structure [2–6]. The structure of the *Streptomyces* R61 DD-peptidase/PBP is partially resolved [21]. On the basis of these data, the distribution patterns of the secondary structure elements and active-site defining motifs along the amino acid sequences [22] reveal a common basic polypeptide scaffolding and highlights features of divergence (Fig. 4). Variations in the configuration of the active sites are due to the presence of an SXN motif in the β-lactamases of class A and an YXN motif in the β-lactamase of class C and *Streptomyces* R61 DD-peptidase/PBP. By reference to the β-lactamases of class A, the *C. freundii* β-lactamase and *Streptomyces* R61 DD-peptidase/PBP have additional surface loops, small helices and β-strands occurring away from the active site and connecting conserved structures.

Amino acid alignments and hydrophobic cluster analysis suggest that the DD-peptidases/PBPs, β-lactamases and penicillin sensors have comparable 3-D structures. The DD-peptidases/PBPs, except the *Streptomyces* R61 enzyme (see above), are of the SXN type [9,14,15,23–26]. The *Actinomadura* R39 DD-peptidase/PBP [15] and *E. coli* PBP4 [26], however, possess a large additional module 170–180 amino acid residues long (and of unknown function) inserted within the all-α domain. The β-lactamases of class D and the penicillin sensors have similarity scores above 30 deviation units [27]. They are of the YXN type. By reference to the β-lactamases of class A, they lack several α-helices and may have a partially uncovered five-stranded β-sheet and a more accessible active site [28]. The metallo (zinc) β-lactamases of class B are not included in this review.

108

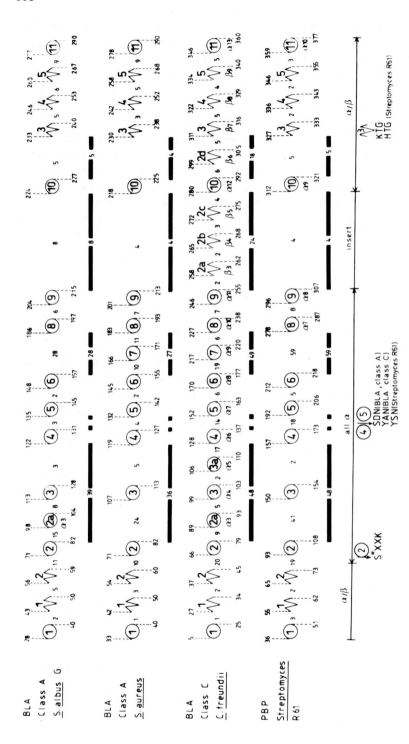

Fig. 4. Occurrence of secondary structures and active-site defining motifs along the amino acid sequences of the penicilloyl serine transferases of known three-dimensional structure. BLA, β-lactamase; open circles, α-helices; jagged lines, β-strands. The ABL amino acid numbering is used for the class A β-lactamases. The amino acid numbering used for the *C. freundii* β-lactamase of class C and the *Streptomyces* R61 DD-peptidase/PBP refers to the mature proteins. The secondary structures are numbered according to the ABL consensus scheme. The original numbering is also indicated when different from the ABL consensus. Heavy lines are regions where insertions occur. Adapted from [22].

3.3. Bi(multi)functional high M_r PBPs

The high M_r PBPs illustrate the principle according to which enzyme molecules are often constructed from distinct catalytic/binding modules that are linked together in a single polypeptide chain.

3.3.1. Modular design
Figure 5 gives a schematic representation of the molecular organization of the high M_r PBPs of known primary structure: i.e. PBP1A [29], PBP1B [29], PBP2 [30] and PBP3 [31] of *E. coli*, PBP1a [32], PBP2x [33] and PBP2B [34] of *Streptococcus pneumoniae*, PBP5 [35] and PBP3^r [36] of *Enterococcus hirae*, PBP2' of *Staphylococcus aureus* [37] and PBP2 of *Neisseria meningitidis* and *N. gonorrhoeae* [38]. The catalytic, penicillin-binding module is now part of a multi-module protein. The proposed modular design rests upon pair-wise comparison of the amino acid sequences and identification of modules having similar patterns of distribution of hydrophobic clusters and bearing conserved amino acid groupings (boxes 1–8 in Fig. 5) [39].
  The simplest pattern is that of PBP2 and PBP3 of *E. coli* and PBP2 of *N. meningitidis* and *N. gonorrhoeae*. A hydrophobic region close to the amino terminus anchors the bulk of the protein on the outer face of the plasma membrane. The membrane anchor is linked to an extracellular 200 amino acid residue N-terminal module, itself linked to a 250 amino acid residue, penicillin-binding module.
  Increased complexity results from acquisition of additional modules. An 'AE' amino extension is inserted between the membrane anchor and the N-terminal module in PBP1B of *E. coli*, PBP2' of *S. aureus* and PBP3^r and PBP5 of *E. hirae*. An 'IN' internal extension is inserted between the N-terminal module and the penicillin-binding module in PBP1A of *E. coli*. A 'CE' carboxy extension is fused to the carboxy end of the penicillin-binding module in PBP1A and PBP1B of *E. coli* and PBP1a and PBP2x of *S. pneumoniae*. Inserts also occur within the N-terminal module of PBP2B of *S. pneumoniae* and the penicillin-binding module of PBP1A of *E. coli*. All these inserts, 100 amino acid residues long or more, are large enough to possess their own particular folding, perform a separate function and confer multifunctionality to the PBPs. Finally, 'mosaic' PBPs (not shown in Fig. 5) can emerge by inter-species recombinational events that replace parts of a PBP gene with the corresponding parts from homologous PBPs-encoding genes of closely related species (see Chapter 25).
  The high M_r PBPs bear obvious fingerprints of divergence. The penicillin-binding modules possess the active-site defining motifs S*XXK (box 6), SXN or analogue (box 7) and KT(S)G (box 8). However, by reference to the β-lactamases of class A and the catalytic modules of the DD-peptidases/PBPs (of the SXN type), their amino acid sequences have diverged so far that other traces of similarity in hydrophobic cluster and secondary structure potential patterns have almost completely disappeared.
  PBP2 and PBP3 of *E. coli*, PBP2 of *N. meningitidis*, PBP2 of *N. gonorrhoeae*, PBP2x of *S. pneumoniae*, PBP2' of *S. aureus* and PBP3^r and PBP5 of *E. hirae* are of class B. Boxes 1–4 of their N-terminal modules possess several conserved amino acid residues (R, H, D, E, N, S, T, Q) that may play a role in catalysis. Box 4 is located at the carboxy end of the N-terminal module and is followed by box 5, itself located at the amino end of the

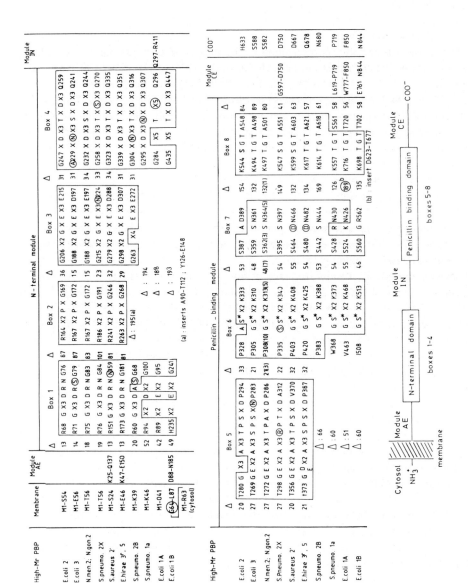

Fig. 5. Modular design of the bi(multi)functional high M_r PBPs.

penicillin-binding module. PBP1A and PBP1B of *E. coli* and PBP1a of *S. pneumoniae* are of class A. They lack boxes 2, 3, 5 and boxes 1 and 4 have low similarity with the equivalent boxes of the high M_r PBPs of class B. Finally, PBP2B of *S. pneumoniae* lacks boxes 2 and 5 analogous to the high M_r PBPs of class B and it possesses boxes 1, 3 and 4 analogous to the high M_r PBPs of class A.

3.3.2. Unsolved questions
The main contributions to wall peptidoglycan assembly are provided by the high M_r PBPs. How do they perform their functions is still a major unsolved question.

On the basis of the reactions that they catalyze in vitro on lipid-linked disaccharide-pentapeptide precursors, the *E. coli* PBP1A and PBP1B are bifunctional transglycosylases (performing glycan chain elongation)/transpeptidases (performing peptide crosslinking) (see Chapter 4). However, the reactions proceed with a low turnover number indicating that the in vitro conditions poorly mimic the in vivo situation. In addition, PBP1A and PBP1B and all the high M_r PBPs lack acyl transfer activity on Ac_2-L-Lys -D-Ala-D -Ala.

Penicilloylation of S*510 of the penicillin-binding (transpeptidase) module of PBP1B inhibits peptide crosslinking and, under certain conditions, greatly increases formation of uncrosslinked peptidoglycan, suggesting that the uncoupling of the two reactions stimulates the transglycosylase. Moenomycin, which inhibits the transglycosylation reaction, also inhibits or prevents peptide crosslinking, suggesting that transpeptidation requires prior or concomitant glycan chain elongation [40]. PBP1B might have only one type of pentapeptide (donor)/tetrapeptide (acceptor) transpeptidase activity [41].

E. coli PBP2 and RodA (an intrinsic membrane protein) form a protein complex which is linked to the ribosomes via the *lov* gene and to the cyclic AMP (cAMP)-receptor protein complex [42,43]. In addition, ppGpp regulates the transcription of a gene whose product may be part of a chain of interacting elements which coordinate ribosomal activity with that of the PBP [44]. Penicilloylation of S*330 of PBP2 causes dysfunctioning of the complex, loss of the rod-shape of the cells and, eventually, cell death. However, PBP2 is dispensable when ppGpp synthesis is induced [44].

E. coli PBP3, FtsW (which is very similar to RodA), FtsA, FtsQ and FtsZ form a large protein complex known as divisome [45] or septator [46], which encompasses the cytosol, the membrane and the periplasm (see Chapter 5). Penicilloylation of S*307 of PBP3 causes dysfunctioning of the protein complex, inhibits cell septation and causes cell death. Maturation of PBP3 involves removal of the C-terminal undecapeptide I578–S588 [47]. Deletions and/or mutations affecting the penicillin-binding domain and the C-terminal region downstream from the KTG motif give rise to two types of protein mutants. Some have a reduced penicillin-binding capacity but exhibit genetic complementation activity. Others lack the in vivo activity but bind penicillin (see Chapter 5). All attempts to isolate the penicillin-binding module of PBP3 independently of the N-terminal module in a stable and catalytically active form, have failed (unpublished results).

Whether the ~100 amino acid residue 'AE' insert present in the low affinity PBP2' of *S. aureus* and PBP3^r and PBP5 of *E. hirae* [36] is involved in PBP-mediated penicillin resistance is another open question.

There is a close interplay between the constituent modules of the high M_r PBPs and between the PBPs and other cellular components but the nature of the interactions is unknown. By analogy with the penicillin-sensory transducers which provide examples of how modules with different functions can combine and give rise to a new function, i.e. control of gene expression, high M_r PBPs are components of multiprotein complexes in which they may function as signalling devices of one kind or another.

Resolution of the 3-D structures of the high M_r PBPs is a research topic that should receive high priority. Membrane anchor-free, water-soluble derivatives of E. coli PBP3 [48,49], S. aureus PBP2' [50] and S. pneumoniae 2x [51] can be produced by using appropriate secretion vectors. Water-soluble fragments of the E. hirae PBP3^r and PBP5 can be obtained by controlled proteolytic digestion of isolated membranes [52]. All these derivatives are catalytically active in terms of penicillin binding. Some of them have been crystallized.

4. Kinetics of enzyme-catalyzed reactions

Mechanistically and kinetically, the penicilloyl serine transferases are related to the usual serine peptidases of the trypsin and subtilisin families. They obey the same general equations.

With the definition E = penicilloyl serine transferase, D = carbonyl donor substrate, HY = acceptor co-substrate, E·D = Michaelis complex, P1 = leaving group of the enzyme acylation step, P2 = second reaction product , [Z] = concentration of compound Z and assuming [D] >> [E], the catalyzed three-step reaction is

$$E + D \underset{k_{-1}}{\overset{k_{+1}}{\rightleftharpoons}} E \cdot D \underset{-P1}{\overset{k_{+2}}{\longrightarrow}} acyl\ enzyme \underset{+HY}{\overset{k_{+3}}{\longrightarrow}} E + P2 \qquad (1)$$

For $k_{+2} \ll k_{-1}$, $(k_{-1} + k_{+2}))/k_{+1}$ simplifies to k_{-1}/k_{+1} which is the dissociation constant K of E·D, i.e. [E]·[D]/[E].

With the definition k_a = pseudo-first-order rate constant of enzyme acylation at a given [D] value, k_{cat}/K_m = catalytic efficiency at low [D] and k_{+2}/K = second order-rate constant of enzyme acylation, the following equations apply [53]:

$$k_a = \frac{k_{+2}}{1 + (K/[D])} \quad \text{or} \quad \frac{1}{k_a} = \frac{1}{k_{+2}} + \frac{K}{k_{+2}[D]} \qquad (2)$$

which for [D] $\ll K$ simplifies to

$$k_a = \frac{k_{+2}[D]}{K} \qquad (3)$$

$$k_{cat} = \frac{k_{+2}k_{+3}}{k_{+2} + k_{+3}} \qquad (4)$$

$$K_m = \frac{K k_{+3}}{k_{+2} + k_{+3}} \tag{5}$$

$$\frac{k_{cat}}{K_m} = \frac{k_{+2}}{K} \tag{6}$$

Central to the catalytic mechanism is the formation of the acyl enzyme. When at a given [D], the system reaches the steady state, the concentrations of free enzyme E, E·D and acyl enzyme remain stable. The ratio of total enzyme [E]$_0$ to [acyl enzyme] is

$$\frac{[E]_0}{[\text{acyl enzyme}]_{ss}} = 1 + \frac{k_{+3}}{k_{+2}} + \frac{K k_{+3}}{k_{+2}[D]} \tag{7}$$

On the basis of eq. (7), Fig. 6 gives the percentage of total enzyme in the form of acyl enzyme at the steady state of the reaction for varying carbonyl donor concentrations and varying k_{+2}/k_{+3} ratio values. Under saturating conditions ([E]$_0$ >> K), for $k_{+2} = k_{+3}$ and at the steady-state, the reaction proceeds with 50% of total enzyme in the form of acyl enzyme.

Three types of carbonyl donors (β-lactams, esters/thioesters and peptides) are examined successively.

4.1. Reaction with endocyclic β-lactam amide carbonyl donors

The scissile amide bond in penicillin is endocyclic and, therefore, the 'leaving group' of the enzyme acylation step remains part of the acyl enzyme. With HY = H$_2$O, reaction (1) becomes

$$E + D \underset{k_{-1}}{\overset{k_{+1}}{\rightleftharpoons}} E \cdot D \xrightarrow{k_{+2}} \text{acyl enzyme} \xrightarrow[+H_2O]{k_{+3}} E + P \text{ (penicilloate)} \tag{8}$$

The β-lactamases effectively hydrolyze the β-lactams into biologically inactive metabolites (high k_{+2} and k_{+3} values). The E. coli TEM β-lactamase-catalyzed hydrolysis of benzylpenicillin proceeds with $k_{+1} \approx 120\ \mu M^{-1}\ s^{-1}$, $k_{-1} \approx 12\ 000\ s^{-1}$, $k_{+2} = 3000\ s^{-1}$, $k_{+3} = 1500\ s^{-1}$, $k_{cat} \approx 1000\ s^{-1}$, k_{cat}/K_m or $k_{+2}/K \approx 25\ \mu M^{-1}\ s^{-1}$, $K_m \approx 45\ \mu M$ and $K_s \approx 100\ \mu M$ [54]. The enzyme catalytic centre turns over 1000-fold per second. Under saturating concentrations of penicillin, 76% of the total enzyme occurs as acyl enzyme at the steady state of the reaction.

In their interaction with penicillin, the PBPs remain immobilized at the abortive level of the acyl enzyme, at least for a long time. Not only is k_{+3} very much less than k_{+2} but k_{+3} has a very small absolute value. The higher the k_{+2}/K and the smaller the k_{+3} are, the lower the β-lactam concentration required to inactivate 100% of the enzyme in the form of acyl enzyme at the steady state of the reaction. The values of K, k_{+2} and k_{+3} or at least k_{+2}/K and k_{+3} which govern the inactivating potency of a β-lactam for a given PBP are experimentally available data [53].

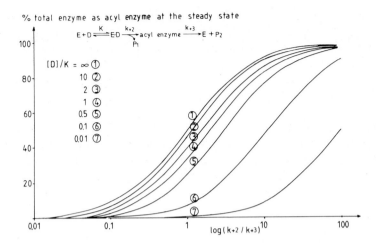

Fig. 6. Accumulation of acyl enzyme at the steady state of the reaction as a function of varying concentrations of carbonyl donor and varying values k_{+2} and k_{+3}.

The value of k_{+3} is related to the half-life of the acyl enzyme by half-life $= -\ln 0.5/k_{+3}$. Knowing k_{+3}, k_a can be computed from

$$\frac{[\text{acyl enzyme}]}{[\text{E}]_0} = \frac{k_a}{k_{+3} + k_a}\{1 - \exp[-(k_{+3} + k_a)t]\} \tag{9}$$

which for $k_a \gg k_{+3}$ simplifies to

$$k_a t = -\ln\left(1 - \frac{[\text{acyl enzyme}]}{[\text{E}]_0}\right) \tag{10}$$

Use of eq. (10) is justified if the time of incubation of the β-lactam with the enzyme is much shorter than the half-life of the acyl enzyme and if the β-lactam, at the concentration used, causes complete acylation of the enzyme at the steady state of the reaction. Knowing k_a, eq. (3) applies if [D] is very much less than K and eq. (2) applies if [D] is not very much less than K. The kinetic constants for the interaction of benzylpenicillin with the *Streptomyces* R61 DD-peptidase/PBP are $K = 13$ mM, $K_{+2} = 180$ s^{-1}, $K_{+2}/K = 1300$ M^{-1} s^{-1} and $k_{+3} = 1.4 \times 10^{-4}$ s^{-1} [55]. The enzyme catalytic centre turns over once every 120 min ($k_{\text{cat}} = k_{+3}$). Under saturating concentrations of penicillin and at the steady state of the reaction, 99. 9% of total enzyme occurs as acyl enzyme. The half-life of the acyl enzyme is about 1 h.

The distinction between β-lactamases and PBPs in terms of lability versus stability of the acyl (penicilloyl) enzyme is only quantitative. Some β-lactamases may be partially or completely inactivated by β-lactam substrates. They show hysteretic kinetics, the observed substrate-induced inactivation being attributed to the enzyme conformational motility [56]. Other β-lactamases also give rise to long-lived acyl enzymes and these acyl enzymes may undergo various types of intramolecular rearrangements [57]. The common feature of β-lactamase inactivation is that of a branched pathway in which β-elimination leads to a hydrolytically inert acyl enzyme and, sometimes, further modification of some residues of the active site.

The inertness of the acyl (penicilloyl) enzyme formed by reaction of penicillin with the PBPs is not absolute. Slow breakdown generates penicilloate. Concomitant slow intramolecular rearrangement of the penicilloyl moiety causes rupture of the C5–C6 bond and release of the leaving group (eventually as N-formylpenicillamine) [58]. As a result of the vacancy thus created in the active site, the newly formed acyl(phenylacetylglycyl) enzyme is both hydrolytically and aminolytically labile [59].

4.2. Reaction with acyclic ester/thioester carbonyl donors

The ester and thioester shown in Fig. 7 are acyclic analogues of penicillin. Compounds of this type are hydrolyzed with varying efficacies by all the penicilloyl serine transferases tested: β-lactamases, *Streptomyces* K15 DD-transpeptidase/PBP, exocellular DD-peptidases/ PBPs and water-soluble derivatives of high M_r PBPs [60,61]. The interest of these penicillin analogues is that the leaving group of the enzyme acylation step vacates the active site. As a corollary, the enzyme activity can undergo partitioning between alternate acceptors, H_2O and amino compounds.

Simple bi-substrate models do not explain the observed variations of aminolysis over hydrolysis caused by varying concentrations of the ester/thioester carbonyl donors and amino acceptors. The scheme as best representing the mechanism of the *Streptomyces* R61 DD-peptidase/PBP-catalyzed reactions on the co-substrates C_6H_5–CONH–CH_2–COS–CH_2–COO^- and D-alanine is shown in Fig. 7 [62]. An amino acceptor molecule binds to the acyl enzyme made at the expense of one carbonyl donor molecule and a second carbonyl donor molecule binds to the ternary complex acyl enzyme-acceptor to yield a quaternary complex acyl enzyme-acceptor-donor which is productive in hydrolysis.

The simplest model representing the mechanism of the *E. cloacae* P_{99} β-lactamase-catalyzed reactions on the co-substrates *m*-[(phenylacetylglycyl)oxy] benzoic acid and D-phenylalanine is also shown in Fig. 7 [63]. It suggests the presence of two different 1:1 enzyme/donor complexes only one of which leads to hydrolysis and aminolysis (donor in site 1) and a 1:2 enzyme/donor complex which leads only to hydrolysis (donor in both sites 1 and 2).

Comparable complex kinetics of concomitant hydrolysis and aminolysis are observed with the water-soluble derivatives of PBP2x and PBP2B of *S. pneumoniae* and PBP3 of *E. coli*. Under certain conditions, these high M_r PBPs perform preferential aminolysis over hydrolysis of thioester substrates [61[.

4.3. Reaction with acyclic amide carbonyl donors

Ac$_2$-L-Lys-D-Ala-D-Ala is an acyclic analogue of penicillin in the D-alanyl-D-alanine moiety and an analogue of the peptidoglycan precursor in both the Ac$_2$-L-lysyl and D-alanyl-D-alanine moieties. It is a substrate of the monofunctional DD -peptidases/PBPs.

116

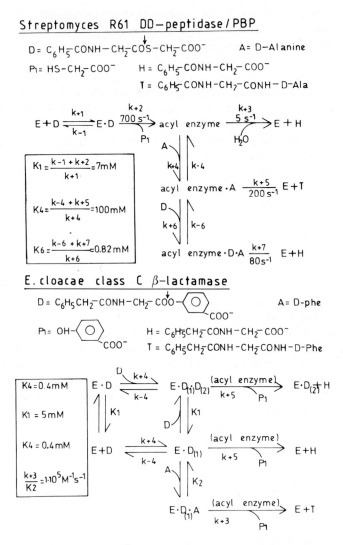

Fig. 7. Catalyzed hydrolysis and aminolysis of thioester and depsipeptide carbonyl donors by the *Streptomyces* R61 DD-peptidase/PBP and *E. cloacae* β-lactamase. D, carbonyl donor; A, amino acceptor; P1, leaving group; H, hydrolysis product; T, aminolysis product. Adapted from [62] for the *Streptomyces* R61 enzyme and [63] for the *E. cloacae* enzyme.

The *Streptomyces* K15 DD-transpeptidase/PBP has been discussed in Section 3.1. The *Streptomyces* R61 and *Actinomadura* R39 DD-peptidases/PBPs hydrolyze the tripeptide with k_{cat}/K_m values of 4000 M^{-1} s^{-1} and 53 000 M^{-1} s^{-1}, respectively. They catalyze transfer of the electrophilic group Ac$_2$L-Lys-D-Ala to amino acceptors structurally related to the wall peptidoglycan [55,64]. They perform dimerization of properly structured peptides (Fig. 8) [65,66]. Kinetic models of the type shown in Fig. 7 with ester/thioester substrates, best explain the observed variations of transpeptidation over hydrolysis as a function of the relative and absolute concentrations of the co-substrates. Under certain conditions, the rate of transpeptidation exceeds the rate of hydrolysis.

From the foregoing, it follows that the penicilloyl serine transferases have similar extended binding sites and much of the same catalytic machinery except in the respect that the DD-peptidases/PBPs only utilize the tripeptide Ac_2-L-Lys-D-Ala-D-Ala and other peptidoglycan precursor analogues as carbonyl donors. Isosterism between penicillin and Ac_2-L-Lys-D-Ala-D-Ala is only partial. Differences in the electronic properties of the two carbonyl donors may explain the specific activity of the DD-peptidases/PBPs.

Productive binding of lactams belonging to different chemical families in terms of PBP inactivating potency (and β-lactamase substrate activity) requires a relatively strict spatial disposition of the triad formed by the carbon atom of the carboxylate, the carbonyl of the scissile amide bond and the oxygen atom of the exocyclic carbonyl group (or COH in thienamycin) (Fig. 9) [67]. One requirement is an almost co-planarity of the three functional groups. In one of the most stable extended conformations of Ac_2-L-Lys-D-Ala-D-Ala, the carboxylate, the carbonyl of the D-alanyl-D-alanine scissile peptide bond and the carbonyl of the L-lysyl-D-alanine peptide bond have a spatial disposition comparable to that of the equivalent triad in penicillin. These functional groups generate a typical electronic property and this electronic distribution gives rise to an electrostatic potential of definite shape and volume (see Section 5.1).

Comparison of the electrostatic potential isocontours (at -10 kcal/mol) (Fig. 9) shows that the N^a- and N^c-acetyl-L-lysine amide bonds of the tripeptide generate two extended negative electrostatic wells which are not present in the penicillin map. In all likelihood, corresponding subsites allowing the tripeptide to undergo correct positioning with respect to the enzyme functional groups exist in the DD-peptidases/PBPs but not in the β-lactamases. The reason why the high M_r PBPs lack peptidase activity on Ac_2-L-Lys-D-Ala-D-Ala is not understood (see Section 3.3).

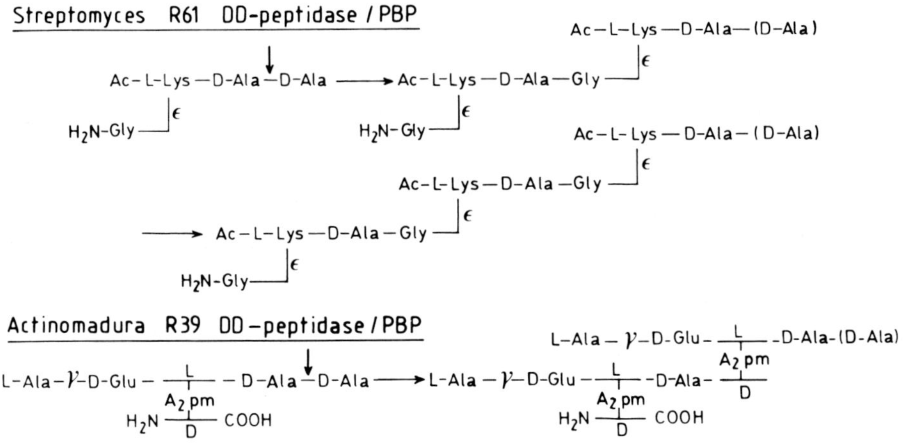

Fig. 8. DD-peptidase-catalyzed peptide dimerization. Vertical arrow, scissile peptide bond. Adapted from [64–66].

118

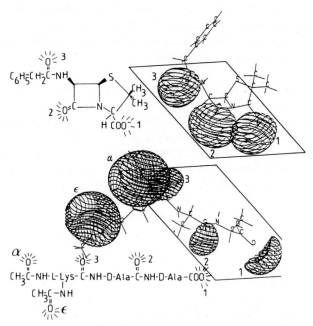

Fig. 9. Electrostatic potential isocontours at −10 kcal/mol of benzylpenicillin and the most stable extended conformer of Ac$_2$-L-Lys- D-Ala-D-Ala. The views are shown along the plane of the scissile amide bond. The α- and ε-side chains of L-lysine are above and below the reference plane, respectively.

5. Mechanisms of enzyme-catalyzed reactions

Understanding how the penicilloyl serine transferases function requires knowledge of the entire energy profiles of the catalyzed reactions giving the energies and geometries of substrates, intermediates and transition states bound to the active site and identifying those interactions important for transition states stabilization. Experimental techniques, kinetic data, NMR, X-ray crystallography, site-directed mutagenesis and theoretical methods are needed.

5.1. Empirical force fields methods and molecular orbital calculations

Theoretical methods that use empirical force fields expressing the energy of a molecular system given by simple functions of interatomic distances and other internal coordinates, are applicable to systems containing up to 5000 atoms. Since parametrization is set up so that properties of stable molecules at the equilibrium are reproduced, these methods are not suitable for computations on unstable intermediates and transition states of vanishing small life-time. Static molecular mechanics methods used for geometry optimization and structure refinement and molecular dynamics simulations able to give useful insight into conformational changes provide starting structures for quantum chemical calculations.

Quantum chemical descriptions of the electronic distributions of the molecules can be achieved only by ab initio molecular orbital (MO) calculations. Each molecular orbital is treated as a combination of a basis set of functions representing atomic orbitals and all the integrals are then evaluated analytically. Ab initio methods are applicable to systems of up to a few tens of atoms (using program packages such as Gaussian 92) [68]. For larger systems of up to 100–200 heavy atoms, semi-empirical MO calculations are used in which the more intractable integrals are replaced by simplified functions containing parameters chosen to reproduce experimental results for a large set of molecules. When applied to molecules related to those in the reference set, self-consistent field MO calculations (containing the AM1 [69], MNDO [70], PM3 [71] semi-empirical Hamiltonians parametrized for many elements) give results of comparable accuracy to those obtained by ab initio calculations with small basis sets (STO-3G [72], Minil [73]).

The atomic rearrangements and electronic redistributions which occur in an enzyme–ligand interacting system are best described by potential energy hypersurfaces. The surface generated by a system containing N atoms, has $3N - 6$ dimensions. Surfaces up to 300 dimensions can be computed with acceptable accuracy. Hence, semi-empirical quantum calculations are restricted to reduced models of up to 100 heavy atoms. Amino acid residues assumed to be catalytically important and represented by small molecules (a histidine by an imidazole ring, an aspartate by a formate, etc.) in the same orientation as in the X-ray structure serve as the starting structure for optimization. Such models allow the effects of local structural features to be studied and the likelihood of varying catalytic mechanisms to be estimated by finding the transition states and measuring the associated activation barriers. A transition state is a point on the potential energy hypersurface where the energy is maximum along the reaction coordinate and minimum in all the other directions (as an example, see Fig. 12 in Section 5.4). Algorithms find a true transition state as a point having one negative hessian eigenvalue [74].

Effects of the surrounding protein and solvent atoms can be studied by combining the quantum mechanical representation of the interactive part of the enzyme with a simpler empirical valence force fields description of the rest of proteins [75]. Approaches of this type allow the electronic properties of the protein to be computed and expressed in the form of 3-D electrostatic potential maps. An electrostatic potential map is the best fingerprint of a molecule [76–78]. It takes into account its volume, conformation and electronic distribution. At each point on the map, the electrostatic potential expresses the energy of interaction with a unitary positive charge. Figure 9 (Section 4.3) shows the electrostatic potential maps of two small molecules, benzylpenicillin and the tripeptide Ac_2-L-Lys-D-Ala-D-Ala.

Interactions of greatest importance in enzyme catalysis are those which govern the initial enzyme-ligand binding and those which lower the activation energy barrier. The electrostatic interactions are the main driving forces for the formation of non-covalent complexes and, therefore, the major component of the interaction energy is the electrostatic energy [79]. It can be calculated by integrating numerically [80] and/or analytically [81] the electronic density of the ligand and the electrostatic potential of the enzyme.

Free energy perturbation methods used in combination with molecular dynamics allow differences in binding free energies for varying ligands and the effects of site-directed mutagenesis to be evaluated [82].

120

5.2. β-Lactamases catalytic machinery

The double proton shuttle which takes place during enzyme-catalyzed hydrolysis of penicillin is schematically represented in Fig. 10. Enzyme acylation is achieved by transfer of the proton of the γOH of the essential serine residue S* to the nitrogen atom of the β-lactam amide bond with concomitant nucleophilic attack of the adjacent carbonyl carbon atom by the activated S*Oγ atom. Enzyme deacylation involves transfer of the proton from a water molecule to the S*Oγ atom of the acyl enzyme with concomitant attack of the adjacent carbonyl carbon atom by OH and re-entry of a water molecule. The reactions are facilitated by polarization of the carbonyl groups of the scissile amide and ester bonds by backbone NH groups. The mechanism is concerted. The acyl enzyme is the only chemical intermediate which can be trapped. The tetrahedral intermediates 1 and 2 vanish as they are formed. The 'molecular mechanics' sees the catalyzed reaction as a series of discrete steps occurring along the pathway from substrate to product. The 'quantum chemistry' view is that of a continuum with transition states along the reaction coordinate, which is the combination of all the degrees of freedom involved in the molecular rearrangement.

5.2.1. β-Lactamases of class A
Using the ABL numbering, the four β-lactamases of known 3-D structure [2–6] possess a conserved pentapeptide E166XELN170 in addition to the active-site defining motifs

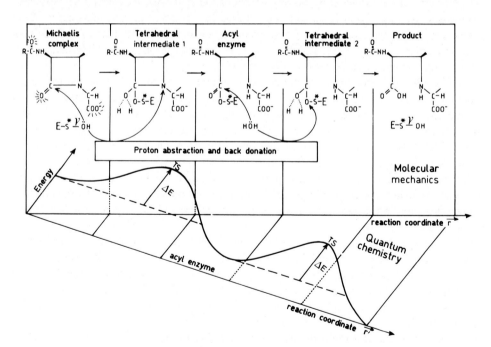

Fig. 10. β-Lactamase-catalyzed hydrolysis of penicillin. Energy profiles following the reaction coordinates $\vec{r}$ and $\vec{r}$ ' in molecular mechanics and quantum chemistry frameworks, respectively.

S*70XXK, S130DN and K234T(S)G. The *Streptomyces albus* G enzyme serves to illustrate the underlying catalytic mechanism [5].

The 3-D structure of a protein as derived from X-ray crystallography, is correct within the limits of experimental error. It must be refined. The geometry of that portion of the β-lactamase containing all the amino acid residues within a 15 Å radius around the α carbon of S*70 and forming the three innermost shells of the active site is optimized by energy minimization within the molecular mechanics AMBER V3 framework. The atomic positions of water molecules generated by a Monte Carlo bath are refined together with the protein coordinates. Figure 11A gives a schematic view of the immediate boundary of the optimized active site. A dense hydrogen bonding network interconnects the water molecules W1 and W2 and the side chains of nine of the fifteen amino acid residues present, namely S*70 and K73 (of the S*TTK motif on helix α2), S130 and N132 (of the SDN motif connecting helices α4 and α5), K234, T235 and A237 (of the KTGA motif on strand β3) and E166 and N170 (of the EPELN motif on a loop connecting helices α6 and α8 at the entry of the cavity).

When acting on a good β-lactam substrate, the β-lactamase catalytic centre turns over many times per second. Based on the 'molecular mechanics' view of the reaction (Fig. 10), modelling the five discrete reaction steps may shed light on the likely mechanism of proton shuttle. The geometries of the ligand (benzylpenicillin), acyl enzyme, product (penicilloate) and tetrahedral intermediates 1 and 2 (treated as stable entities and in which the –O–S*–E moiety is replaced by O–CH$_3$) are optimized by energy minimization within the MO semi-empirical AM1 framework. The optimized molecules are docked separately in the optimized enzyme active site by geometrical adjustments and the geometries of the most logical complexes are energy-minimized within the molecular mechanics AMBER V3 framework [5].

The picture which emerges from this modelling exercise is that formation of the non-covalent Michaelis complex is driven by hydrogen bonding interactions between the three functional groups of the benzylpenicillin molecule (Fig. 9, Section 4.3) and several amino acid side chains of the enzyme active site (Fig. 11B). The carboxylate at the end of the ligand interacts with the γ-OH groups of S130 and T235; the exocyclic CONH amide bond at the other end of the ligand interacts with the side chain amino group of N132 and the backbone carbonyl of A237; and the carbonyl of the scissile β-lactam amide bond interacts with the backbone NH groups of S*70 and A237 in a way reminiscent of the oxyanion-hole hydrogen bonds of the serine peptidases of the trypsin and subtilisin families.

As a result of these interactions, the bound penicillin molecule is in an optimal position (Fig. 11B) to allow the proton of the γOH of S*70 to be abstracted by the Oε1 group of E166 via the water molecule W1, the activated OγS*70 to perform attack of the β-lactam carbonyl carbon on the well exposed α-face of the molecule and the abstracted proton to be donated back to the nitrogen atom through the hydrogen bonding subnetwork W2, K73 and S130 acting as final proton donor. As a result of these intramolecular rearrangements (not shown), the acyl (penicilloyl) enzyme is now in an optimal position to allow the proton of the water molecule W1 to be abstracted by the Oε1 group of E166, the activated OH$^-$ to perform attack of the carbonyl carbon of the acyl enzyme and the abstracted proton to be donated back to the S*70 Oγ atom, thus achieving penicillin hydrolysis and enzyme regeneration.

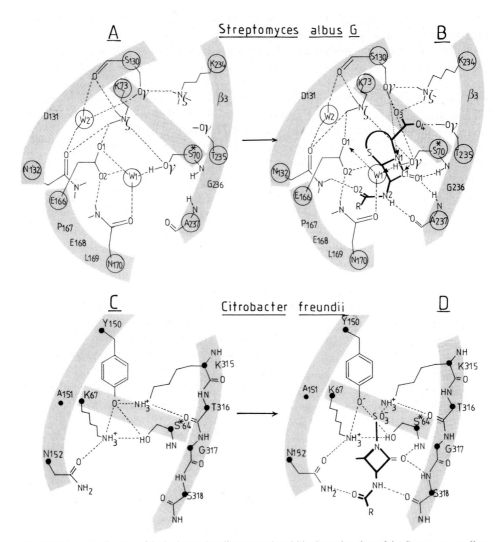

Fig. 11. Schematic diagram of the hydrogen bonding networks within the active sites of the *Streptomyces albus* G β-lactamase of class A and the *Citrobacter freundii* β-lactamase of class C. A and C, ligand-free cavities; B, Michaelis complex formed with benzylpenicillin; D, Michaelis complex formed with aztreonam.

According to this mechanism, the acylation and deacylation steps are 'mirror' images involving the same general base catalyst, namely E166. The effects caused by the selective mutation of E166 into A, N or D in several β-lactamases of class A give rise to conflicting interpretations.

As reported for the *B. licheniformis* β-lactamase [83], the E166A mutation causes a million-fold reduction of the enzyme catalytic activity and stoichiometric accumulation of the acyl enzyme. As derived from burst kinetics and stopped-flow studies, the acylation step (with nitrocefin as substrate) remains rapid suggesting that the major role of E166 is on that water molecule poised to attack the acyl enzyme carbonyl. The crystal structures of the native and mutated enzymes are very similar except that the hydrolytic water molecule seen in the native enzyme shifts markedly in the enzyme mutant.

As reported for the *E. coli* TEM β-lactamase [6], the E166N mutation also causes accumulation of the acyl enzyme. Based on independent determinations of the crystal structures of the wild-type enzyme, the enzyme mutant and its acyl (benzylpenicilloyl) enzyme, K73 would act as general base in the acylation step, abstracting the proton from the γOH of S*70, while the deacylation step only would be E166-dependent. This conclusion, however, suffers from a lack of analysis of the kinetic properties of the enzymes. The wild type enzyme hydrolyzes benzylpenicillin with similar values of k_{+2} (3000 s^{-1}) and k_{+3} (1500 s^{-1}) (see Section 4.1). Not knowing the k_{+2} and k_{+3} values for the enzyme mutant [84], conversion of the β-lactamase into a PBP is not proof that the E166N mutation decreases only the k_{+3} value. Assuming for the enzyme mutant, a k_{+2} value of 0.1 s^{-1}, which is four orders of magnitude smaller than that of the wild-type enzyme, and a k_{+3} value of 0.001 s^{-1}, 99% of the enzyme would accumulate as acyl enzyme under saturating conditions and at the steady state of the reaction.

As reported for the *B. cereus* β-lactamase [85], the E166D mutation (which is much more conservative than the E166A and E166N mutations) causes a 2000-fold decrease of the values of both k_{+2} and k_{+3} for benzylpenicillin hydrolysis while the mutation K73R causes only a 100-fold decrease of the k_{+2} value. Hence, the acylation step appears to be much more E166-dependent than K73-dependent. Computer modelling of the corresponding *Streptomyces albus* G β-lactamase mutants [5] suggests that as a result of the E166D mutation, W1 no longer connects the γOH of S*70 to the dicarboxylic acid, thus severely hampering proton abstraction. In the K73R mutant, W1 is oriented toward the Oε2 atom of E166 (and not to the Oε1 atom as in the wild-type enzyme) thus likely decreasing the efficiency of proton abstraction.

5.2.2. The C. freundii β-lactamase of class C

Using the numbering of the mature protein, the active-site defining motifs are S*64VSK, Y150AN and K315TG. At variance with the β-lactamases of class A, the *C. freundii* β-lactamase lacks the pentapeptide EXELN and it possesses a tyrosine residue Y150 instead of the serine residue S130.

The refined structure of the *C. freundii* β-lactamase [4] reveals a remarkably dense hydrogen-bonding network involving S*64, K67, Y150, N152 and K315 (Fig. 11C). Superimposition experiments highlight the similarity of the active site with that of the β-lactamases of class A. The β3 strands are similarly situated with respect to the γ-OH of the essential serine residues. The YAN motif of the *C. freundii* β-lactamase is about 4–5 Å farther away from β3 than the corresponding SDN motif in the β-lactamases of class A but as a consequence of this displacement, the phenolic hydroxyl group of the tyrosine residue of the YAN motif and the γ-OH group of the serine residue of the SDN motif are similarly disposed with respect to the γ-OH of the essential serine residue.

Positioning of a β-lactam ligand in the active site of the *C. freundii* β-lactamase has been derived from the refined X-ray structure of the stable acyl enzyme formed with the monobactam inhibitor aztreonam [4]. The aztreonam molecule bound to the class C β-lactamase has the same general orientation as the benzylpenicillin molecule bound to the class A β-lactamases (Fig. 11D). In particular, the carbonyl group of the scissile amide bond of aztreonam forms hydrogen bonds to the backbone NH groups of S*64 and S318, the equivalent of the oxyanion hole of the class A β-lactamases.

Superimposition of the *C. freundii* β-lactamase active site onto trypsin using the essential serine and the oxyanion hole as reference points, leads to the conclusion that the phenolic oxygen of Y150 is <0.5 Å from the Nε1 position of the essential histidine in trypsin, suggesting that Y150, as its anion, would act as a general base during β-lactam-catalyzed hydrolysis in a way similar to that of H57 in trypsin [4]. Modelling of the Michaelis complex, nucleophilic attack, tetrahedral intermediate formation and collapse to give the acyl enzyme, support the view that Y150 is in an optimal position to act as both proton acceptor and proton donor [4].

5.3. PBPs catalytic machinery

The *Streptomyces* R61 DD-peptidase/PBP is of partially known 3-D structure [21]. It possesses a YSN motif, it lacks the pentapeptide EXELN and has a HTG triad instead of the usual KT(S)G motif. Based on superimposition experiments, the active site appears to be comparable to that of the *C. freundii* β-lactamase of class C. The β3 strands, helices α2 (with the essential serine residue) and YSN/YAN motifs have similar spatial dispositions in the two enzymes. Mutants of the *Streptomyces* R61 enzyme have been produced by chemical methods and site-directed mutagenesis [86–90]. The significance of the observed effects must await establishment of the refined structure of the wild type enzyme. Interestingly, however, replacement of Y159 of the YSN motif by the corresponding serine residue of the class A β-lactamase gives rise to an enzyme mutant which retains a large proportion of the thioesterase activity (on Ac$_2$-L-Lys-D-Ala-D-thiolactate) and penicillin-binding capacity but has much reduced peptidase activities [90].

All the monofunctional DD-peptidases/PBPs (except the *Streptomyces* R61 enzyme) and penicillin-binding modules of the bi(multi?)functional high M_r PBPs possess an SXN motif and several dicarboxylic acid residues occurring at places roughly equivalent to the conserved motif E166XELN of the β-lactamase of class A [1]. Whether these dicarboxylic acid residues are catalytically important or not is not known. The roles of D447 in the *E. coli* PBP2 [91] and of E396, D409 and E411 in the *E. coli* PBP3 [92] have been investigated. The D447E PBP2 mutant retains both penicillin-binding and genetic complementation activities. The D447N PBP2 mutant binds penicillin but lacks complementation activity. The D447A protein mutant is inert in both respects. The E396A, D409A and E411A PBP3 mutants bind penicillin and have complementation activity.

5.4. Transition states

It is not yet possible to express the evolution features described in Section 3 and the kinetic properties described in Section 4 in atomic terms. But we are learning much.

Due to the high density of their hydrogen bonding networks, the active sites of the penicilloyl serine transferases are structures of high motility. Local modifications causing disappearance or weakening of any hydrogen bond may propagate their effects far from the mutated amino acid residues and alter the entire hydrogen bonding configuration of the cavity. A cavity damaged by mutations with respect to a given ligand, may regain functionality upon binding of another ligand either by readapting a configuration comparable to that of the wild-type enzyme or by utilizing an alternate route of proton shuttle. Quantum chemistry calculations only may allow the functioning of the charge relay system to be elucidated, the amino acid residues acting as general base catalysts to be identified and the mechanism through which these amino acid residues have their proton affinity sufficiently decreased for activation of the nucleophile to be understood. Decreased proton affinity of amino acid residues such as lysine (K73 in the TEM β-lactamase) or ty-

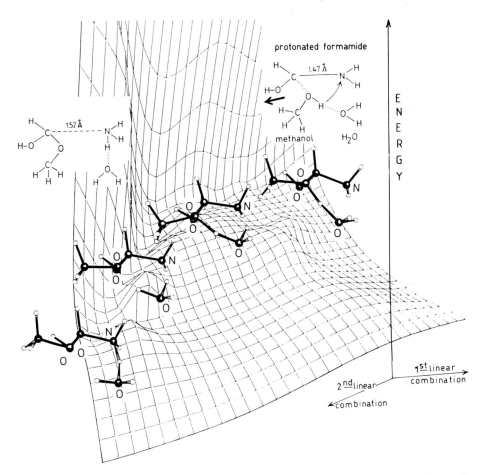

Fig. 12. Energy hypersurface of water-assisted methanolysis of protonated formamide. The reaction coordinate from the optimized protonated formamide/water/methanol molecular system to product is from right to left. The 42 dimension hypersurface ($N = 16$ atoms) calculated by using the STO-3G basis set is projected into a three-dimensional space (with more than 94% of total information conserved). Proton donation to the nitrogen atom is via H_2O.

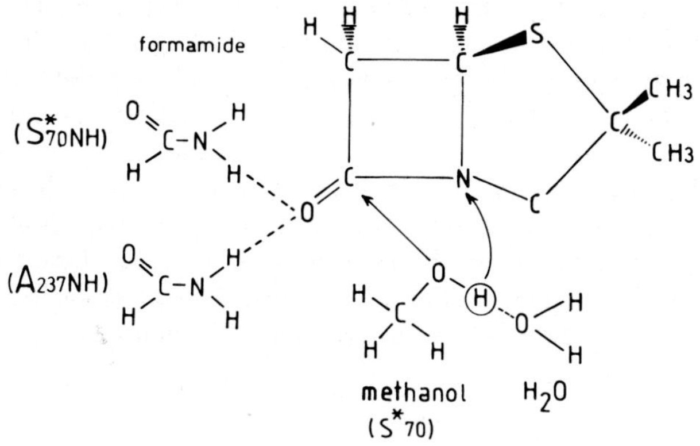

Fig. 13. Water-assisted methanolysis of a bicyclic penam nucleus. S*70 and A237 refer to amino acid residues of the active site of the β-lactamases of class A (see Fig. 12A). For details, see text. Proton donation to the nitrogen atom is via H_2O.

rosine (Y150 in the *C. freundii* β-lactamase) to which a role as general base catalyst is tentatively assigned, is one of the many unsolved problems.

A molecular system containing the active-site amino acid residues, the structurally ordered water molecules and a bound benzylpenicillin molecule generate an energy hypersurface close to one thousand dimensions. Useful information can be provided by smaller models. Calculation of the 42 dimension energy hypersurface of the water-assisted methanolysis of protonated formamide (Fig. 12) allows the transition state to be identified and the associated activation barrier to be estimated. Similarly, rupture of the β-lactam amide bond can be studied on models consisting of benzylpenicillin, methanol (mimicking the active site serine) and two formamides (mimicking the oxyanion hole) in the same spatial disposition as in the refined Michaelis complex (Fig. 13). From this starting conformation, the water-assisted nucleophilic attack of methanol on the α-face of penicillin is a 120 dimension hypersurface accessible to ab initio STO-3G minimal basis calculation (unpublished data). The energy barrier of the transition state along the reaction coordinate to penicilloyl methyl ester is less than 15 kcal/mol. Computation of similar models in which lactam or lactone compounds belonging to varying chemical families replace penicillin, suggest that activation barriers for rupture of the putative scissile bond exceeding 25 kcal/mol cannot be overcome.

8. Concluding remarks

Techniques able to calculate energy profiles for enzymic reactions have a wide range of applications. In the field of the penicilloyl serine transferases, they could allow effective inhibitors to be designed as transition state analogues and/or mechanistic perturbators of the normal reaction pathway. Such molecules should be useful lead compounds in the design of new antibacterial drugs. Future advances depend on the availability of crystal structures for an increasing number of enzymes belonging to the various groups and classes described, and on the refinement and increased power of the computational techniques.

Acknowledgements

Original work from this laboratory was supported in part by the Belgian programme on Interuniversity Poles of attraction initiated by the Belgian State, Prime Minister's Office, Science Policy Programming (PAI no. 19), the Fonds de la Recherche Scientifique Médicale (contract no. 3.4531.92), and a Convention tripartite between the Région wallonne, SmithKline Beecham, U.K. and the University of Liège. G.D. is chercheur qualifié du Fonds National de la Recherche Scientifique (FNRS, Brussels).

References

1. Ghuysen, J.M. (1991) Annu. Rev. Microbiol. 45, 37–67.
2. Herzberg, O. and Moult, J. (1987) Science 236, 694–701.
3. Moews, P.C., Knox, J.R., Dideberg, O., Charlier, P. and Frère, J.M. (1990) Proteins Struct. Funct. Genet. 7, 156–171.
4. Oefner, C., D'Arcy, A., Daly, J.J., Gubernator, K., Charnas, R.L., Heinze, I., Hubschwerlen, C. and Winkler, F.K. (1990) Nature 243, 284–288.
5. Lamotte-Brasseur, J., Dive, G., Dideberg, O., Charlier, P., Frère, J.M. and Ghuysen, J.M. (1991) Biochem. J. 279, 213–221.
6. Strynadka, N.C.J., Adachi, H., Jensen, S.E., Johns, K., Sielecki, A., Betzel, C., Sutoh, K. and James, M.N.G. (1992) Nature 359, 700–705.
7. Nguyen-Distèche, M., Leyh-Bouille, M., Pirlot, S., Frère, J.M. and Ghuysen, J.M. (1986) Biochem. J. 235, 167–176.
8. Leyh-Bouille, M., Nguyen-Distèche, M., Pirlot, S., Veithen, A., Bourguignon, C. and Ghuysen, J.M. (1986) Biochem. J. 235, 177–182.
9. Palomeque-Messia, P., Englebert, S., Leyh-Bouille, M., Nguyen-Distèche, M., Duez, C., Houba, S., Dideberg, O., Van Beeumen, J. and Ghuysen, J.M. (1991) Biochem. J. 279, 223–230.
10. Pratt, J.M., Jackson, M.E. and Holland, I.B. (1986) EMBO J. 5, 2399–2405.
11. Jackson, M.E. and Pratt, J.M. (1988) Mol. Microbiol. 2, 563–568.
12. Ferreira, L.C.S., Schwarz, U., Keck, W., Charlier, P., Dideberg, O. and Ghuysen, J.M. (1988) Eur. J. Biochem. 171, 11–16.
13. Van Der Linden, M.P.G., De Haan, L. and Keck, W. (1993) Biochem. J. 289, 593–598.
14. Duez, C., Fraipont-Piron, C., Joris, B., Dusart, J., Urdea, M.S., Martial, J.A., Frère, J.M. and Ghuysen, J.M. (1987) Eur. J. Biochem. 162, 509–518.
15. Granier, B., Duez, C., Lepage, S., Englebert, S., Dusart, J., Dideberg, O., Van Beeumen, J., Frère, J.M. and Ghuysen, J.M. (1992) Biochem. J. 282, 781–788.
16. Schuster, C., Dobrinski, B. and Hakenbeck, R. (1990) J. Bacteriol. 172, 6499–6505.
17. Severin, A., Schuster, C., Hakenbeck, R. and Tomasz, A. (1992) J. Bacteriol. 174, 5152–5155.
18. Stocker, N.G., Broome-Smith, J.K., Edelman, A. and Spratt, B.G. (1983) J. Bacteriol. 155, 847–859.
19. Joris, B., Ledent, P., Kobayashi, T., Lampen, J.O. and Ghuysen, J. M. (1990) FEMS Microbiol. Lett. 70, 107–114.
20. Ambler, R.P., Coulson, A.F.W., Frère, J.M., Ghuysen, J.M., Joris, B., Forsman, M., Levesque, R.C., Tiraby, G. and Waley, S.G. (1991) Biochem. J. 276, 269–272.
21. Kelly, J.A., Knox, J.R., Zhao, H., Frère, J.M. and Ghuysen, J.M. (1989) J. Mol. Biol. 209, 281–295.
22. Joris, B., Ledent, P., Dideberg, O., Fonzé, E., Lamotte-Brasseur, J., Kelly, J.A., Ghuysen, J.M. and Frère, J.M. (1991) Antimicrob. Agents Chemother. 35, 2294–2301.

23. Broome-Smith, J.K., Ioannidis, I., Edelman, A. and Spratt, B.G. (1988) Nucleic Acids Res. 16, 1617.
24. Todd, J.A., Roberts, A.N., Johnston, K., Piggot, P., Winter, G. and Ellar, D.J. (1986) J. Bacteriol. 167, 257–264.
25. Mottl, H., Terpstra, P. and Keck, W. (1991) FEMS Microbiol. Lett. 78, 213–220.
26. Mottl, H., Nieland, P., Dekort, G., Wuringa, J.J. and Keck, W. (1992) J. Bacteriol. 174, 3261–3269.
27. Zhu, Y.F., Curran, J.J., Joris, B., Ghuysen, J.M. and Lampen, J.O. (1990) J. Bacteriol. 172, 1137–1141.
28. Zhu, Y., Englebert, S., Joris, B., Ghuysen, J. M., Kobayashi, T. and Lampen, J.O. (1992) J. Bacteriol. 174, 6171–6178.
29. Broome-Smith, J.K., Edelman, A., Yousif, S. and Spratt, B.G. (1985) Eur. J. Biochem. 147, 437–446.
30. Asoh, S., Matsuzawa, H., Ishino, F., Strominger, J.L., Matsuhashi, M. and Ohta, T. (1986) Eur. J. Biochem. 160, 231–238.
31. Nakamura, M., Maruyama, I.N., Soma, M., Kato, J.I., Suzuki, H. and Hirota, Y. (1983) Mol. Gen. Genet. 191, 1–9.
32. Martin, C., Briese, T. and Hakenbeck, R. (1992) J. Bacteriol. 174, 4517–4523.
33. Laible, G., Hakenbeck, R., Sicard, M.A., Joris, B. and Ghuysen, J.M. (1989) Mol. Microbiol. 3, 1337–1348.
34. Dowson, C.G., Hutchinson, A. and Spratt, B.G. (1989) Nucleic Acids Res. 17, 7518.
35. El Kharroubi, A., Jacques, P., Piras, G., Coyette, J., Van Beeumen, J. and Ghuysen, J.M. (1991) Biochem. J. 280, 463–469.
36. Piras, G., Raze, D., El Kharroubi, A., Hastir, D., Englebert, S., Coyette, J. and Ghuysen, J.M. (1993) J. Bacteriol. 175, 2844–2852.
37. Song, M.D., Wachi, M., Doi, M., Ishino, F. and Matsuhashi, M. (1987) FEBS Lett. 221, 167–171.
38. Zhang, Q.Y. and Spratt, B.G. (1989) Nucleic Acids Res. 17, 5383.
39. Englebert, S., Piras, G., El Kharroubi, A., Joris, B., Coyette, J., Nguyen-Distèche, M. and Ghuysen, J.M. (1993) in: M.A. de Pedro, J.-V. Höltje and W. Löffelhardt (Eds.), Bacterial Growth and Lysis: Metabolism and Structure of the Bacterial Sacculus, FEMS Symposium Proceedings, Lluc, Mallorca, Plenum Press, New York, pp. 319–333.
40. Van Heijenoort, Y., Leduc, M., Singer, H. and Van Heijenoort, J. (1987) J. Gen. Microbiol. 133, 667–674.
41. den Blaauven, T., Aarsman, M. and Nanninga, N. (1990) J. Bacteriol. 172, 63–70.
42. Bouloc, P., Jaffé, A. and D'Ari, R. (1989) EMBO J. 8, 317–323.
43. Ogura, T., Bouloc, P., Niki, H., D'Ari, R., Hiraga, S. and Joffé, A. (1989) J. Bacteriol. 171, 3025–3030.
44. Vinella, D., D'Ari, R. and Bouloc, P. (1992) EMBO J. 11, 1493–1501.
45. Nanninga, N. (1991) Mol. Microbiol. 5, 791–795.
46. Vicente, M., Palacios, P., Dopazo, A., Garrido, T., Pla, J. and Aldea, M. (1991) Res. Microbiol. 142, 253–257.
47. Nagasawa, H., Sakagami, Y., Suzuki, A., Suzuki, H., Hara, H. and Hirota, Y. (1989) J. Bacteriol. 171, 5890–5893.
48. Bartholomé-De Belder, J., Nguyen-Distèche, M., Houba-Herin, N., Ghuysen, J.M., Naruyama, I.N., Hara, H., Hirota, Y. and Inouye, M. (1988) Mol. Microbiol. 2, 519–525.
49. Fraipont, C., Adam, M., Nguyen-Distèche, M., Keck, W., Van Beeumen, J., Ayala, J.A., Granier, B., Hara, H. and Ghuysen, J.M. (1993) Biochem. J., in press.
50. Wu, C.Y.E., Hoskins, J., Blaszczak, L.C., Preston, D.A. and Skatrüd, P.L. (1992) Antimicrob. Agents Chemother. 36, 533–539.
51. Laible, G., Keck, W., Lurz, R., Mottl, H., Frère, J.M., Jamin, M. and Hakenbeck, R. (1992) Eur. J. Biochem. 207, 943–949.
52. El Kharroubi, A., Jacques, P., Piras, G., Coyette, J. and Ghuysen, J.M. (1988) in: P. Actor P., L. Daneo-Moore, M.L. Higgins, M.R.J. Salton and G.D. Shockman (Eds.), Antibiotic Inhibition of Bacterial Cell Surface Assembly and Function, American Society for Microbiology, Washington, DC, pp. 367–376.
53. Ghuysen, J.M., Frère, J.M., Leyh- Bouille, M., Nguyen-Distèche, M. and Coyette, J. (1986) Biochem. J. 235, 159–165.
54. Christensen, H., Martin, M.T. and Waley, S.G. (1990) Biochem. J. 266, 853–861.
55. Frère, J.M. and Joris, B. (1985) CRC Crit. Rev. Microbiol. 11, 299–396.
56. Samuni, A. and Citri, N. (1979) Mol. Pharmacol. 16, 250–255.

129

57. Cartwright, S.J. and Waley, S.G. (1983) Medicinal Res. Rev. 4, 341–382.
58. Frère, J.M., Ghuysen, J.M., Degelaen, J., Loffet, A. and Perkins, H.R. (1975) Nature 258, 168–170.
59. Marquet, A., Frère, J.M., Ghuysen, J.M. and Loffet, A. (1979) Biochem. J. 177, 909–916.
60. Adam, M., Damblon, C., Plaitin, B., Christiaens, L. and Frère, J.M. (1990) Biochem. J. 270, 525–529.
61. Adam, M., Damblon, C., Jamin, M., Zorzi, W., Dusart, V., Galleni, M., El Kharroubi, A., Piras, G., Spratt, B.G., Keck, W., Coyette, J., Ghuysen, J.M., Nguyen-Distèche, M. and Frère, J.M. (1991) Biochem. J. 279, 601–604.
62. Jamin, M., Wilkin, J.M. and Frère, J.M. (1993) Biochemistry 32, 7278–7285.
63. Pazhanisamy, S. and Pratt, R.F. (1989) Biochemistry 28, 6875–6882.
64. Ghuysen, J.M. (1977) in: W.E. Brown (Ed.), E.R. Squibb Lectures on Chemistry of Microbial Products, University of Tokyo Press, Tokyo, 162 pp.
65. Ghuysen, J.M., Leyh-Bouille, M., Campbell, J.N., Moreno, R., Frère, J.M., Duez, C., Nieto, M. and Perkins, H.R. (1973) Biochemistry 12, 1243–1251.
66. Frère, J.M., Ghuysen, J.M., Zeiger, A.R. and Perkins, H.R. (1976) FEBS Lett. 63, 112–116.
67. Lamotte-Brasseur, J., Dive, G. and Ghuysen, J.M. (1984) Eur. J. Med. Chem. 19, 319–330.
68. Gaussian 92: M.J. Frisch, G.W. Trucks, M. Head-Gordon, P.M.W. Gill, M.W. Wong, J.B. Foresman, B.G. Johnson, H.B. Schlegel, M.A. Robb, E.S. Replogle, R. Comperts, J.L. Andres, K. Raghavachari, J.S. Binkley, C. Gonzalez, R.L. Martin, D.J. Fox, D.J. Defrees, J. Baker, J.J.P. Stewart and J.A. Pople, Gaussian Inc., Pittsburgh PA, 1992.
69. Dewar, M.J.S., Zoebisch, E.G. and Healy, E.F. (1985) J. Am. Chem. Soc. 107, 3902–3909.
70. Dewar, M.J.S. and Thiel, W. (1977) J. Am. Chem. Soc. 99, 4899–4905.
71. Stewart, J.J.P. (1989) J. Comp. Chem. 10, 209–216.
72. Hehre, W.J., Stewart, R.F. and Pople, J.A. (1969) J. Chem. Phys. 51, 2657–2664.
73. Tatewaki, H. and Huzinaga, S. (1980) J. Comp. Chem. 1, 205–217.
74. Culot, P., Dive, G., Nguyen, V.H. and Ghuysen, J.M. (1992) Theor. Chim. Acta 82, 189–205.
75. Yang, A.-S., Gunner, M.R., Sampogna, R., Sharp, K. and Honig, B. (1993) Proteins 15, 252–265.
76. Lamotte-Brasseur, J., Dive, G., Dehareng, D. and Ghuysen, J. M. (1990) J. Theor. Biol. 145, 215–220.
77. Dehareng, D., Dive, G. and Ghuysen, J.M. (1991) Theor. Chim. Acta 79, 141-152.
78. Dive, G. and Dehareng, D. (1993) Int. J. Quant. Chem. 46, 127–136.
79. Dive, G., Dehareng, D. and Ghuysen, J.M. (1993) Theor. Chim. Acta 85, 409–421.
80. Dehareng, D., Dive, G., Lamotte-Brasseur, J. and Ghuysen, J.M. (1989) Theor. Chim. Acta 76, 85–94.
81. Dehareng, D., Dive, G. and Ghuysen, J.M. (1993) Int. J. Quant. Chem. 46, 701–734.
82. Rao, S.N., Singh, V.C., Bash, P.A. and Kollman, P.A. (1987) Nature 328, 551–554.
83. Knox, J.R., Moews, P. C., Escolar, W.A. and Fink, A.L. (1993) Protein Eng. 6, 11–18.
84. Adachi, H., Ohta, T. and Matsuzawa, H. (1991) J. Biol. Chem. 266, 3186–3191.
85. Gibson, R.M., Christensen, H. and Waley, S.G. (1990) Biochem. J. 272, 613–619.
86. Bourguignon-Bellefroid, C., Wilkin, J.M., Joris, B., Aplin, R.T., Houssier, C., Prendergast, F.C., Van Beeumen, J., Ghuysen, J.M. and Frère, J.M. (1992) Biochem. J. 282, 361–367.
87. Hadonou, A. M., Jamin, M., Adam, M., Joris, B., Dusart, J., Ghuysen, J.M. and Frère, J.M. (1992) Biochem. J. 282, 495–500.
88. Bourguignon-Bellefroid, C., Joris, B., Van Beeumen, J., Ghuysen, J.M. and Frère, J.M. (1992) Biochem. J. 283, 123–128.
89. Hadonou, A.M., Wilkin, J.M., Varetto, L., Joris, B., Lamotte-Brasseur, J., Klein, D., Brown, G., Ghuysen, J.M. and Frère, J.M. (1992) Eur. J. Biochem. 207, 97–102.
90. Wilkin, J.M., Jamin, M., Damblon, C., Zhao, G.H., Joris, B., Duez, C. and Frère, J.M. (1993) Biochem. J. 291, 537–544.
91. Adachi, H., Ishiguro, M., Imajoh, S., Ohta, T. and Matsuzawa, H. (1992) Biochemistry 31, 430–437.
92. Goffin, C., Ayala, J.A., Nguyen-Distèche, M. and Ghuysen, J.M. (1993) FEMS Microbiol. Lett., in press.

J.-M Ghuysen and R. Hakenbeck (Eds.), *Bacterial Cell Wall*

131

CHAPTER 7

Microbial peptidoglycan (murein) hydrolases

G.D. SHOCKMAN[1] and J.-V. HÖLTJE[2]

[1]*Department of Microbiology and Immunology, Temple University School of Medicine, Philadelphia,
PA 19140, USA and* [2]*Max-Planck-Institut für Entwicklungsbiologie, Abteilung Biochemie,
Spemannstrasse 35, 72076 Tubingen, Germany*

1. Introduction

Environmental conditions that lead to the dissolution of bacterial cells (bacteriolysis) have long been known. It was soon discovered that bacteriolysis was catalyzed by 'factors' in extracts of tissues, plants and microorganisms and in some instances by microbial culture filtrates. At one time it was thought that bacteriolysis was a complex phenomenon accompanied by a variety of enzymatic processes [1]. However, soon after the observations of Fleming [2], who found that a powerful substance that he called 'lysozyme' was capable of rapidly lysing thick suspensions of certain bacteria, especially an organism he called *Micrococcus lysodeikticus* (now known as *Micrococcus luteus*), it became clear that this enzyme was widely distributed in nature. Lysozyme was found to be present in high concentration in hen egg white [2], from which it was crystallized [3], enabling extensive studies of its action on bacterial cells [4,5].

Studies of the action of lysozyme and of other bacteriolytic enzymes [6,7] revealed that this bacteriolytic action was the result of hydrolysis of specific bonds in the protective and shape-maintaining bacterial exoskeleton, the peptidoglycan (murein) sacculus of the cell wall [8,9; see also Chapter 2].

In addition, and concurrently, the experiments of Weibull [10] clearly demonstrated that clarification of dense suspensions of *Bacillus megaterium* by lysozyme could be prevented by the presence of a sufficient, iso-osmotic concentration of solutes, such as 0.3 M sucrose, that are unable to permeate into the bacteria. Instead the rod-shaped bacilli were transformed into spherical, osmotically fragile bodies. Electron microscopy showed that these spherical bodies were no longer surrounded by the characteristic, approximately 20-nm thick wall but were contained in a thin, typical membrane structure.

It soon became clear that, at least in some instances, it was not necessary to add lysozyme or other enzymes or extracts to observe bacteriolysis or, under appropriate iso-osmotic environments, formation of osmotically fragile protoplasts. Cultures of some organisms, such as *Streptococcus pneumoniae*, various species of *Bacilli* or *Enterococcus hirae* ATCC9790 (formerly called *Streptococcus faecalis*, *Streptococcus faecium* and *Enterococcus faecium*) would dissolve (autolyze) when incubated under conditions of 'unbalanced growth' resulting in an inhibition of further cell wall peptidoglycan synthesis or, for that matter, in an appropriate buffer. The observations that bacteriolysis could be prevented, with the formation of spherical, osmotically fragile bodies, by the provisions of an iso-osmotic environment again demonstrated that autolysis was the result of hydrolysis of a sufficient number of bonds in the protective cell wall peptidoglycan to cause its dissolution and removal from the bacterial surface. Thus, it was concluded that these bacteria are equipped with their own, endogenous, peptidoglycan hydrolases that eventually can autolyze the cell.

2. Enzyme specificities

Autolysins, then, are defined as endogenous enzymes that hydrolyze specific bonds in the bacterial cell wall (peptidoglycan) resulting in damage to the integrity and protective properties of the two- or three-dimensional structure of the peptidoglycan [11]. This definition does not include all the enzymes that are capable of hydrolyzing bonds in the peptidoglycan. Included in the term peptidoglycan hydrolases but excluded from this definition of autolysins would be those enzymes that hydrolyze bonds that are not relevant for the mechanical stability of the peptidoglycan sacculus, such as DD-carboxypeptidases that remove terminal D-alanine residues, or enzymes that can hydrolyze bonds in peptidoglycan subunits but are not capable of hydrolyzing bonds in intact, insoluble peptidoglycan.

A broad variety of enzyme specificities have been detected, including peptidases and glycosidases (Fig. 1). Two types of glycosidases are known. β-N-acetylmuramidases (lysozymes) and β-N-acetylglucosaminidases, which hydrolyze the β-1,4-glycosidic bond between MurNAc and GlcNAc and between GlcNAc and MurNAc, respectively. A peculiar muramidase-like enzyme has been isolated from *Escherichia coli* [12,13] and phage lambda [14,15]. These latter enzymes not only split the β-1,4-glycosidic bond but concomitantly catalyze a transfer of the glycosyl bond onto the hydroxyl group of the carbon

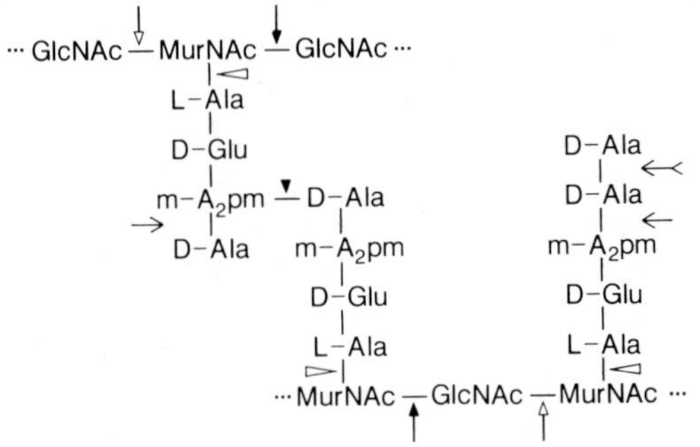

Fig. 1. Structure of the murein of *E. coli*. N-Acetylglucosamine (GlcNAc) and N-acetylmuramic acid (MurNAc) are interlinked in an alternating sequence by β-1,4-glycosidic bonds forming polysaccharide chains which are crosslinked by peptides consisting of L- and D-alanine, D-glutamic acid and *m*-diaminopimelic acid (m-A$_2$pm) as indicated. Murein hydrolases cleave specific bonds: →, lytic transglycosylase; –▷, β–N-acetylglucosaminidase; ▷, N-acetylmuramyl-L-alanine-amidase; ▶, D,D-endopeptidase; →, L,D-carboxypeptidase; ⟩→, D,D-carboxypeptidase. (From ref. 12.)

6 of the same muramic acid, thereby forming 1,6-anhydromuramic acid. These enzymes are called lytic transglycosylases. They must not be confused with the synthetic transglycosylases such as some of the well known high molecular weight, penicillin-binding proteins (PBPs; see Chapter 6).

Amidases (*N*-acetylmuramoyl-L-alanine amidases) that specifically cleave the amide bond between the lactyl group of muramic acid and the α-amino group of L-alanine, the first amino acid of the stem peptide, are found in many bacteria and effectively separate the peptide components from the amino sugar chains of the peptidoglycan [16].

The peptide moieties of peptidoglycan are attacked by exo- as well as by endo-peptidases. DD-carboxypeptidases convert pentapeptide into tetrapeptide moieties by cleaving the terminal D-alanyl-D-alanine bond of crosslinked and uncrosslinked stem peptides. DD-endopeptidases have been described and isolated which cleave the crucial DD-peptide bond that crosslinks the stem peptides to yield the characteristic peptidoglycan network. Both DD-carboxypeptidases and some of the DD-endopeptidases that are inhibited by β-lactam antibiotics are addressed in detail in another chapter of this volume (see Chapter 6). The penultimate peptide bond which is an L–D bond between the amino acid in position 3 (almost always either L-lysine or *m*-diaminopimelic acid) and the D-alanine in position 4 of the stem peptide is cleaved by LD-carboxypeptidases. Only a few other peptidases have been characterized so far. However, other peptidase specificities must exist, because of the existence of certain peptidoglycan turnover products, such as GlcNAc-MurNAc L-Ala-D-Glu [17].

3. Enzyme assay systems

Fleming's classical assay for peptidoglycan hydrolases is a rather unspecific one that simply follows the dissolution of cells of *M. luteus*, usually by measuring a decrease in turbidity. Similar and somewhat more specific assays follow the decrease in turbidity of insoluble cell wall or peptidoglycan substrates. Use of such substrates has led to measurements of released amino- or carboxy-terminal and/or reducing groups [18]. In addition, the specificity of the bond hydrolyzed can be determined by identification of the released terminus. For example, the action of an amidase results in an increase in N-terminal L-alanine whereas the action of a muramidase or glucosaminidase results in an increase in reducing MurNAc or GlcNAc, respectively. More sophisticated and sensitive assays include the use of specifically radioactively labeled peptidoglycan, allowing the determination of radioactivity in the supernatant after separation of the insoluble peptidoglycan by centrifugation [19,20].

Recently an elegant procedure has been reported by which the action of a whole set of peptidoglycan hydrolases in crude cell extracts can be visualized after SDS-PAGE [21]. Suspensions of bacterial cells or walls in the gels are used. Activity is visualized as a clear band in the turbid gel.

As noted by Foster [22], detection of bands of lysis after SDS-PAGE on acrylamide gels containing wall or other substrates is unlikely to include all of the autolysins of an organism such as *B. subtilis*. The methodology requires that the activity (1) survives SDS-PAGE, (2) can be renatured to an enzymatically active form after SDS-PAGE, (3) is active in monomeric form or composed of identical subunits that can reassociate, (4) hydrolyzes a sufficient number of bonds to cause visible dissolution of the substrate, (5) hydrolyzes and dissolves the substrate incorporated into the gel (the substrate specificity problem), and (6) does not require cofactors or activators such as processing by the action of a protease. Proteolysis can present an additional problem in the generation of more than one proteolytically processed polypeptide that may or may not retain enzymatic activity capable of dissolving the selected substrate. Clearly the method will not detect non-autolytic peptidoglycan hydrolase activities.

By means of HPLC-based separation of peptidoglycan subunits [23], specific enzyme assays for various peptidoglycan hydrolases including carboxypeptidases, endopeptidases [24] and amidases [16] have been established.

4. Distribution of peptidoglycan hydrolases

As discussed below, the presence of peptidoglycan hydrolases in bacteria has been predicted on the basis of the unique feature of the structure of peptidoglycan that envelopes the cell with a gigantic, bag-shaped, covalently crosslinked network. To us, enlargement of this peptidoglycan sacculus seems conceivable only by an intricate interplay of synthesizing and hydrolyzing enzymes to allow new peptidoglycan subunits to be inserted into the pre-existing sacculus [9,25,26]. Indeed, peptidoglycan hydrolases turn out to be widely distributed among bacteria [8,27]. Until now, a mutant has not been isolated that completely lacks peptidoglycan hydrolase activity; in part, probably due to the presence of more than one peptidoglycan hydrolase in the species examined thus far. On the other hand, final proof for the assumption that peptidoglycan hydrolases are indispensable to the bacterial cell does not exist either. A list of some of the bacteria possessing peptidoglycan hydrolases is presented in Table I [12,28–68]. This table, which is based on that in Rogers et al. [27], gives an idea of the breadth of both Gram-positive and Gram-negative genera and species in which bacterial autolysins have been found, although one or more enzymes could easily be a bacteriophage- or plasmid-coded enzyme and not all of the listed activities have been well characterized by current criteria. It is of interest to note that many of the organisms listed in Table I possess peptidoglycan hydrolases of more than one specificity. In addition, as shown in Table II, which lists some recently well characterized microbial peptidoglycan hydrolase activities and some of their interesting properties [13,33,35–38,41,43–45,61–66,69–109], several bacterial species, including *E. coli*, *E. hirae* and *B. subtilis*, possess more than one enzyme that hydrolyzes the same bond. As discussed below, the presence of multiple peptidoglycan hydrolases complicates determinations of putative function(s) of these activities.

Interestingly, peptidoglycan hydrolases have also been found in other microorganisms and even in higher organisms, with hen egg white lysozyme the best known. Lysozymes have been widely found in animal tissues, tears, milk, cervical mucus and urine and are also present in plants. In addition, an amidase has been found in human and other mammalian sera [110,111]. The presence of peptidoglycan hydrolases in organisms other than bacteria may be considered a natural antibacterial defense mechanism.

TABLE I

Bacteria that have been reported to possess peptidoglycan hydrolase activities[a]

Organism	Muramidases					Ref.
	A	B	C	D	E	
Arthrobacter crystallopoietes	+					28
Bacillus cereus			+	+		29
Bacillus licheniformis			+	+		30
Bacillus sphaericus				+	+	31
Bacillus stearothermophilus				+		32
Bacillus subtilis			+	+		33
Bacillus thuringiensis	+			+	+	34
Brucella abortus				+		35
Clostridium acetobutylicum	+			+		36–38
Clostridium botulinum			+	+		39
Clostridium welchii	+		+			40, 41
Enterococcus hirae	+					42–45
Escherichia coli		+	+	+	+	12
Lactobacillus acidophilus	+					46
Listeria monocytogenes	(+)		(+)	+		47
Micrococcus luteus			+	+		48
Mycobacterium smegmatis				+	+	49
Myxobacter sp.				+	+	50, 51
Neisseria gonorrhoeae		+	+	+	+	52–55
Proteus vulgaris				+		35
Pseudomonas aeruginosa					+	56
Salmonella typhimurium				+		57
Staphylococcus spp.			+	+	+	58, 59
Streptococcus hygrocopicus			+			60
Streptococcus pneumoniae			+	+		61–66
Streptococcus pyogenes	+					67
Streptomyces spp.	+		+	+	+	68

[a]Modified from ref. 27.
[b]A, lysozyme (EC 3.2.1.17); B, lytic transglycosylase (EC 3.2.1.–); C, endo-β-N-acetyl-glucosaminidase (EC 3.2.1.30); D, N-acetylmuramyl-L-alanine-amidase (EC 3.5.1.28); E, peptidoglycan-DD-endopeptidase (EC 3.4.99).
+, Evidence of the activity obtained; (+), a glycanase activity detected. Bond specificity not determined.

In addition, the action of peptidoglycan hydrolases could result in the production of (or hydrolysis of) peptidoglycan fragments that possess bioactive properties such as immunostimulating [112–114] or sleep-inducing [115,116] activities. The production of peptidoglycan hydrolases by fungi or bacterial species in natural habitats such as soils or aqueous environments may limit the microbiota to bacterial species that are, for one reason or another, resistant to the particular peptidoglycan hydrolase activity or activities present.

136

TABLE II
Reasonably well characterized peptidoglycan hydrolases

Organism and enzyme	Cell and/or wall lytic	Substrate	Bond hydrolyzed	M_r (kDa)	Signal peptide	Comments	Ref.
Streptomyces albus G DD-Carboxypeptidase	+	Muramyl pentapeptide; wall peptidoglycans	C-terminal D-Ala-D-Ala; C-terminal D-Ala-D-diamino acid (α to COO⁻)	18	42 a.a.	Sequenced, crystal structure determined	69–74
B. sphaericus Endopeptidases							
I	–	x-γ-D-Glu-(L)-meso A₂pm; x-γ-D-Glu-(L)-(meso)-A₂pm-D-Ala	γ-D-Glu-(L)mesoA₂pm peptidyl dipeptide hydrolase	44.7	–	2 N-terminal repeats similar to *E. hirae* repeats; C-terminal catalytic domain	75
II	–	L-Ala-γ-D-Glu-L-diamino acid-X	γ-D-Glu-L-meso A₂pm; requires free N-terminal L-Ala	30	–		76
B. subtilis 168 Amidases							
CwlA (Foster) M59232 (GenBank)	+	Homologous walls and spore cortex; walls of other Gram-positive spp. (e.g., *M. luteus*)	MurNAc-L-Ala (DNP L-Ala)	29.9	A₃₉–A₄₀	Identical nucleotide sequence. Proteolytically processed at C-terminal end from 30 kDa to 21–23 kDa. Both forms active. C-terminal has homology with N-terminal of *S. albus* G peptidase. Central region is highly homologous with a.a. sequence of *Bacillus* sp. autolysin and probably contains enzyme-active site. Two short repeats at C-terminus (233°)	77
CwlA M37710 (GenBank)		Homologous SDS walls					78,79

LytC	+	Homologous walls	52.6–49.9		MurNAc-L-Ala (DNP L-Ala)	3 N-terminal to central (57 a.a.) repeats; enzyme active site in C-terminal domain (307°)	80,81
CwlB M61747 (GenBank)				24 a.a. A_{24}–D_{25}			82
Modifier protein CwbA LytB D10388 (DDBJ)	−	Homologous walls	74.1–75	A_{25}–A_{26}	None	Binds to and modifies activity of CwlB (LytC) from random to processive. N-terminal region highly homologous with N-terminal region of LytC (CwlB). C-terminal region has homology with SpoIID of *B. subtilis* and *B. amyloliquefaciens* (307°)	81,83
Glucosaminidase	+	Walls of *B. subtilis, B. licheniformis, M. luteus*. Teichoic or teichuronic acids not required	90		GlcNAc-MurNAc endo-*N*-acetyl	Activated by 0.1–0.2 M HCl	33,41
Bacillus sp. Amidase (?)	+	Homologous SDS-treated cells. *M. luteus* cells (autoclaved or SDS treated)	27.6	A_{37}–I_{38}	Not determined	Enzyme active site probably in N-terminal domain	84
B. licheniformis FD0120 Amidase CwlM X62116 (EMBL)	+	Walls of *B. licheniformis, B. subtilis, M. luteus*	27.5	—	MurNAc-L-Ala	N-terminal has homology with C-terminal catalytic domain of CwlB; two C-terminal repeats. Stimulated by modifier protein	85

TABLE II (*continued*)

Organism and enzyme	Cell and/ or wall lytic	Substrate	Bond hydrolyzed	M_r (kDa)	Signal peptide	Comments	Ref.
Strep. pneumoniae Amidases							
LytA	+	Pneumococcal walls with choline	MurNAc-L-Ala (N-terminal L-ala)	36.5	–	C-terminal repeats	61
φ HB-3, Hbl-3	+	Pneumococcal walls with choline	MurNAc-L-Ala	36.5		Extensive homology with *lytA* gene	62
Muramidases							
φ Cp-1, Cpl-1	+	Pneumococcal walls with choline	MurNAc-GlcNAc	39.1		6 C-terminal repeats with homology to LytA repeats; Asp 10, Glu 37	63, 64
φ Cp-9, Cpl-9	+	Pneumococcal walls with choline	MurNAc-GlcNAc	39.1		6 C-terminal repeats with homology to LytA repeats; Asp 10, Glu 37	65
φ Cp-7, Cpl-7	+	Pneumococcal walls; no choline	MurNAc-GlcNAc	38.5		3 C-terminal repeats different from above repeats; does not require choline; Asp 10, Glu 37	65
Glucosaminidase	+	Pneumococcal walls with choline	GlcNAc-MurNAc	64		Not sequenced	66
Staph. simulans NRRL B-2628							
Lysostaphin	+	Cells and walls of *Staph.*	Gly-Gly in cross-bridges	51.7– 26.9	ca. 38 a.a.	Preproenzyme with 7–14, 13 a.a. tandem repeats at N-terminal. Plasmid coded. Extracellular	86–88
Staph. aureus NCTC8325 (PS47) Amidase							
LytA M59424 (GenBank)	+	*S. aureus* walls *M. luteus* cells	MurNAc-L-Ala (DNP-Ala)	53.8	–	A prophage gene; C-terminal sequence homology with lysostaphin	89–91

Enzyme		Substrate/specificity	Bond cleaved	MW		Notes	Ref.
Staph. aureus A8 Glucosaminidase SaG	+	*M. luteus* cells	GlcNAc-MurNAc	ca. 80		Cloned; not sequenced	92
Lactococcus lactis SK112 Amidase (?) φ US3, LytA	+	Homologous cells; strains of *L. lactis*	Not determined	29	?	Homology to pneumo amidase	93
Lactobacillus delbrueckii subsp. *bulgaricus* Muramidase (?) φ mv1 *lysA*, LysA M35235 (GenBank)	+	Cells of *L. bulgaricus*, *L. helveticus*, *S. salivarius*	Probably MurNac-GlcNAc based on homology	21.1	–	Homology with CH muramidase; Asp 8, Glu 35	94
Chalaropsis sp. Muramidase CH	+	*S. aureus* walls	OAcMurNAc-OAcGlnAc MurNAc-GlcNAc	22.4		Sequenced; not cloned; Asp 6, Glu 33	95
Strep. globosporus ATCC 21553 Muramidase (mutanolysin) M1 (ACM)	+	Broad range of bacterial walls and cells	MurNAc-GlcNAc OAc MurNAc-OAcGlcNAc	23.6	77 a.a.	Homology with Chalaropsis CH muramidase Asp 9, Glu 36	96
Clostridium acetobutylicum ATCC 824 Muramidase LyC M68865 (GenBank)	+	*C. acetobutylicum* walls; *C. acetobutylicum* peptidoglycan; not inhibited by choline	$MurNH_2$-$GlcNH_2$; Not active on re-N-acetylated peptidoglycan	34.9 (sequence) 41	–	Homology with CH muramidase, Asp 5, Glu 32; 5 C-terminal repeats; C-terminal region has homology with N-terminal of *S. albus* G peptidase	36,37

TABLE II (*continued*)

Organism and enzyme	Cell and/ or wall lytic	Substrate	Bond hydrolyzed	M_r (kDa)	Signal peptide	Comments	Ref.
C. acetobutylicum WCIB 8052 (CECT 806) Amidase	+	*S. pneumoniae* walls	Probably MurNAc-L-Ala; choline dependent	115	−	Also in *C. saccharop-erbutylacetonicum* (more active)	38
Enterococcus hirae ATCC 9790 Muramidases M-1	+	*E. hirae* cells or walls; re-N-acetylated peptidoglycan	MurNAc-GlcNAc	137–87		Not cloned or sequenced. Latent form (130 kDa) proteinase activated. Glucoenzyme, nucleotidlylated with ca. 12 monomeric, 5-mercaptouridyllic acids. Processively hydrolyzes soluble glycan chains	43
M-2 M77639 (GenBank)	+	*M. luteus* walls and peptidoglycan; *E. hirae* peptidoglycan; N-Ac-peptidoglycan; *L. acidophilus* walls or peptidoglycan; *S. aureus*, tar1 N-acetylated peptidoglycan; *S. aureus* peptidoglycan	MurNAc-GlcNAc	70.7	$A_{49}-D_{50}$	6 45 a.a. repeats at C-terminal end; S&T (ca. 51%) rich regions at C-terminal	44,45, 97,98
Strep. faecalis Muramidase (?)	+	*M. luteus* cells; *S. faecalis* walls	Not determined; probably MurNAc-GlcNAc	74	+	Homology with *E. hirae* muramidase; 4 direct 68 a.a. repeats at C-terminus	99

Enzyme	Lytic	Substrate	Bond cleaved			Comments	References
Escherichia coli							
Lytic transglycosylases							
Slt 70	+	*E. coli* walls	MurNAc-GlcNAc yields 1,6 anhydroMurNAc	70	27 a.a.	Cloned sequenced 3-D structure determined, 99.7 min	13, 100, 101
Slt 35	+	*E. coli* walls	MurNAc-GlcNAc yields 1,6 anhydroMurNAc	35	?	26 N-terminal a.a. sequenced with similarity to Slt 70	102, 103
Mlt 38	+	*E. coli* walls (MurNAc-GlcNAc)n	MurNAc-GlcNAc yields 1,6 anhydroMurNAc	38		Not sequenced. Membrane bound	104
Endopeptidases							
PBP 4 (DacB)	?	*E. coli* walls, muropeptides	meso-A$_2$pm(D)-D-Ala; D-Ala-D-Ala	49	20 a.a.	Cloned, sequenced, membrane bound, penicillin sensitive, 68.6 min; also has carboxypeptidase activity	105, 106, 253a
MepA	?	*E. coli* walls, muropeptides	meso-A$_2$pm(D)-D-Ala	30	19 a.a.	Cloned, sequenced, soluble protein, penicillin insensitive, 50.4 min	107, 108, 254
Amidase							
AmiA	-	Muropeptides	MurNAc-L-Ala	39		Not cloned or sequenced, purified, 51 min	35, 258
Glucosaminidase	-	Muropeptides	GlcNAc-MurNAc	36		Not cloned or sequenced, purified, 25–27 min	109

5. Physiological functions in bacteria

Early on in the relatively brief history of bacterial cell wall science, the presence of auto-lytic peptidoglycan hydrolase activities in cultures of rapidly growing and dividing bacteria led to the idea that such potentially suicidal enzymes may play one or more roles in cell wall assembly and bacterial growth [9,25,26]. In fact, in several bacterial species the capacity of bacterial cells to autolyze is maximal or near maximal during the exponential growth phase [25,117–119]. Such impressions have since been reinforced by data which positively correlated autolytic capacity or autolysin activity with growth rate [120,121], and by studies of changes in peptidoglycan hydrolase activities during the division cycle of synchronized cultures [120,122,123]. Although data supporting several of the proposed roles have been obtained, the presence of multiple peptidoglycan hydrolase activities have in most cases complicated interpretations. Direct proof of such roles awaits comprehensive studies of well characterized mutants.

A number of potential roles for peptidoglycan hydrolase activities in bacterial surface growth and division have been proposed, with indirect experimental evidence in support of some of them [27,124–127]. These roles include (1) hydrolytic action to provide new acceptor sites for the addition of peptidoglycan precursors in the bag-shaped macromolecule surrounding the bacterium, (2) insertion of new subunits into the stress-bearing peptidoglycan layer by a 'make-before-break', inside-to-outside growth mechanism, (3) hydrolysis of bonds in selected areas of the cell wall surface causing modifications of the shape (remodeling) of previously assembled wall, (4) cell division, i.e. compartmentalization into two new cell units separated by both membrane and wall, (5) participation in the final stages of cell separation, (6) peptidoglycan turnover and recycling, (7) transformation of competent cells, (8) spore formation and germination.

5.1. Role in the provision of new acceptor sites

Peptidoglycan hydrolases of only two specificities would provide new acceptor sites consistent with current knowledge of peptidoglycan biosynthesis (see Chapters 3 and 4). These would be N-acetylmuramidases (including the lytic transglycosylases) and appropriate endopeptidases [11,12,27,124–127]. Whereas transglycosidation reactions such as that carried out by hen egg white lysozyme [128] would not create additional MurNAc reducing ends and GlcNAc non-reducing ends, such an activity could result in the relocation of such potential acceptor sites. The available data concerning the average glycan chain length and extent of peptide crosslinking of walls of a number of bacterial species (summarized in [27,129,130] are not consistent with a deficit of available peptide or glycan acceptor sites. Also inconsistent with an overall lack of acceptor sites is the continued assembly of wall in the form of thickened wall rather than wall surface expansion that is not associated with the sole autolytic N-acetylmuramidase activity of $E. hirae$ [131].

5.2. Enlargement of the peptidoglycan sacculus

In order to expand the sacculus, new peptidoglycan subunits have to be inserted into the pre-existing network. We are unable to visualize the occurrence of this process in the absence of the rupture of covalent bonds to allow enlargement of the cell envelope. For rod-shaped Gram-positive bacteria, it has been shown that the peptidoglycan grows according to an inside-to-outside growth mechanism [131–135]. New peptidoglycan is first attached underneath the pre-existing peptidoglycan layer. This is followed by specific cleavage of covalent bonds in the stress-bearing layer by peptidoglycan hydrolases. As a result of this, the new material is automatically pulled into the layer under stress. Thus peptidoglycan hydrolases may indeed be considered pacemaker enzymes for the enlargement of the peptidoglycan sacculus. Even for the thin murein layer of Gram-negative organisms a kind of inside-to-outside growth mechanism following the safe strategy of 'make before break' [136] has been proposed [137].

5.3. Remodeling functions of peptidoglycan hydrolases

To our knowledge, this idea was first proposed by Rogers [127]. There is no question that bacterial cells change in shape and in surface area to volume ratios throughout their cell division cycle [125]. Also, such dimensions vary with growth rate of the bacteria. This is not only true for rod-shaped species, but also for cocci. Peptidoglycan hydrolases of any specificity could serve this function. The number of bonds hydrolyzed at specific topological sites and at specific times during the division cycle could be relatively small, making detection of hydrolysis products difficult. Recent analyses of the composition of the peptidoglycans of several different bacteria, including *E. coli, S. aureus, S. pneumoniae* and *Gaffkya homari*, by HPLC of muramidase-hydrolyzed peptidoglycan fragments [23,24,138–141] have revealed the presence of a large number of components, some in quantitatively minor amounts. Whereas some of these components probably result from the biosynthetic action of previously undescribed transferases, at least some probably result from the action of hydrolases. The possible role(s) of these newly discovered peptidoglycan structural units in cell shape, changes in shape and availability for biosynthetic (e.g. acceptor) activities, remains to be determined.

5.4. In cell division and cell separation

There are technical difficulties in distinguishing between the end of cell division and the beginning of cell separation. The latter is usually defined as the point in time at which the number of distinct and countable particles increases. This event usually takes the form of a doubling in number, but in the case of separation of chains of bacteria, it can be a much larger increase in cell numbers. Intuitively, it seems highly likely that hydrolysis of some bonds in the peptidoglycan (rather than a velcro-like peeling apart) should be involved in the separation of the two new cell poles. Correlations between deficiencies in autolytic activity and failure of cells to separate of several coccal and rod-shaped species have been observed [142–149]. In some instances, autolytic deficiencies were due to only partially characterized mutations, resulting in reduced autolytic activity or to environmentally induced phenotypic changes in wall substrate-suppressed lytic activity [144]. In at least some instances, cell separation could be induced by the addition of extracts containing autolytic activity or by the addition of hen egg white lysozyme. Thus, curing of, for example, a presumed amidase defect by the addition of a muramidase clearly demonstrates a lack of hydrolytic specificity for the cell separation function [150].

Separation of fully divided bacterial cells is not essential to further growth, division and survival of the bacteria. It does inhibit the dissemination of daughter cells and thus could affect distribution, for example, towards nutrients or away from inhibitors, or the dissemination of organisms in an infection. Thus, cell separation could influence pathogenesis [151]. In Gram-negative bacteria, cell separation overlaps with the formation of the septum. Cleavage of the septum along its middle line starts before the septum has been completed. This gives rise to the formation of the typical V-shaped constrictions rather than septation as is the case for most Gram-positive rods. Mutants of *E. coli* defective in cell separation (*env*A) form chains of cells [152].

5.5. Peptidoglycan turnover

Cell wall turnover is usually defined as the loss of previously assembled, insoluble wall from the bacterial cells. In the case of Gram-positive bacteria, the presence of the wall on the outside of the membrane permeability barrier and the absence of an outer membrane exterior to the peptidoglycan means that turnover can be measured by the release of labeled peptidoglycan fragments from the cells to the supernatant culture medium. These methods appear to be valid, providing the released fragments are not re-utilized for wall assembly. For Gram-negative species, such as *E. coli*, measurement of loss of peptidoglycan is complicated by the presence and properties of the outer membrane. Relatively recently, it was discovered that substantial amounts of peptidoglycan fragments are released from the peptidoglycan sacculus of *E. coli* [153] and may be re-utilized for peptidoglycan assembly [154–156].

Loss of wall and wall peptidoglycan has been observed during growth of a number of bacterial species including *B. subtilis, B. megaterium, Lactobacillus acidophilus. S. aureus, Listeria monocytogenes, E. coli* and *Neisseria gonorrhoeae* (see [129,157] for recent reviews). On the other hand, peptidoglycan turnover has not been observed during growth of *E. hirae* [158], even under conditions of stress, in a variety of strains of *S. mutans* and *S. sanguis* [159], or in a Lys⁻DAP⁻ mutant of *B. megaterium* [160]. Thus, peptidoglycan turnover does not appear to be essential for wall surface growth in these bacteria. In *B. subtilis*, thickened wall made after inhibition of protein synthesis does not appear to turn over [161].

Clearly peptidoglycan hydrolases must be involved in the loss of peptidoglycan fragments from assembled, peptide-crosslinked wall. In *B. subtilis*, an amidase appears to be the primary enzyme activity involved in wall turnover as determined from analysis of the peptidoglycan fragments present in the culture supernate [162]. These investigators were unable to detect amidase in the growth medium and thus postulated that enzyme bound to the cell surface catalyzed removal of wall material. In *L. acidophilus,* the only detectable peptidoglycan hydrolase activity, muramidase [46], must be responsible for the loss of peptidoglycan from the cells. In *E. coli*, lytic transglycosylases and endopeptidases are responsible for turnover, whereas amidases, glucosaminidases and perhaps carboxypeptidases are involved in further hydrolysis of the turnover products to smaller units that can be re-utilized for peptidoglycan assembly after uptake back into the cytoplasm via active transport systems [155].

5.6. Autolysins involved in transformation

Indirect evidence suggests a role for autolytic activity in the ability of three different bacterial species, *B. subtilis, S. gordonii* (formerly *S. sanguis)* and *Pneumococci*, to be transformed. Correlations have been observed between increased competence for transformation and the rate of autolysis of intact cells and of isolated walls of *B. subtilis* [163] and of *S. gordonii* [164, 165]. However, in at least two instances, autolysis-deficient (but not - negative) mutants remained fully transformable. Both a mutant of the pneumococcus deleted in the *lyt*A gene [166] and *lyt* mutants of *B. subtilis* [167] appear to retain full transformability. The recent observations that each of these organisms possesses more than one autolytic peptidoglycan hydrolase (see [22,168] and Sections 9.2 and 9.3) are a complicating factor, requiring additional experiments with well defined single, and perhaps multiple, mutants in peptidoglycan hydrolase activities.

5.7. Peptidoglycan hydrolases in sporulation and spore germination

It seems highly likely that peptidoglycan hydrolases play one or more important roles in (1) the liberation of mature spores from the mother cell and (2) the hydrolysis of the unique cortical peptidoglycan upon spore germination. A variety of peptidoglycan hydrolases of varying specificities, including amidases, peptidases and hexosaminidases, have been detected in various bacilli and in some cases appear to increase in activity during sporulation or germination (summarized in [27]; see also Chapter 8). More recently, using the production of lytic bands after SDS-PAGE of extracts on gels containing a suspension of *B. subtilis* 168 walls, Foster [22] detected at least seven different polypeptide bands of lytic activity, more than one of which increased in intensity during sporulation of *B. subtilis*. Sorting out the functions of each of these activities and their potential roles in the development of spores will require application of modern genetic and biochemical studies.

6. Subcellular localization

It seems justified to describe peptidoglycan hydrolases as unusual enzymes for an unusual substrate. The substrate for this group of enzymes is a unique, insoluble, high molecular weight structure of the size of the bacterium. The shape-maintaining peptidoglycan sacculus located outside of the cell's permeability barrier consists of structurally and functionally different substructures, such as hemispherical polar caps and a cylindrical middle part in the case of rod-shaped bacteria, raised wall bands, peripheral wall and the leading edge of the invagination in streptococci. The latter structure is just one example of a strictly localized growth zone [169] which contrasts with multiple growth sites in *E. coli* [170,171]. There are the stress-bearing layers of peptidoglycan but also the outermost layers which in *B. subtilis* may be undergoing dissolution [135]. In *E. coli,* nascent pepti-

146

doglycan differs from mature peptidoglycan [172,173]. The enzymes may be specifically directed to certain areas of the macromolecular sacculus by specific recognition of defined structures. Binding of many of the hydrolases to peptidoglycan has been found to be quite strong. In some cases salt concentrations of up to 5–8 M LiCl [174,175], 0.01 N NaOH [119,176] or 4–8 M guanidine–HCl [177,178] were necessary to release the enzymes from the peptidoglycan. In at least two instances, the number of these high-affinity binding sites appears to be limited [177–179]. The soluble lytic transglycosylase of *E. coli* could still be found bound to the murein sacculus after boiling in 4% sodium dodecylsulfate [179].

The specific subcellular distribution of two individual enzymes on the peptidoglycan sacculus was visualized by immunoelectron microscopy. The amidase of *B. subtilis* was found to be bound at the site of cell division, that is, in the septa of dividing cells [180]. Similarly the amidase of *S. pneumoniae* was also found to be located mainly in the septal region [181]. The soluble lytic transglycosylase of *E. coli* was shown to be bound exclusively to the outer surface of the peptidoglycan sacculus [179].

Necessarily, the peptidoglycan hydrolases have to be transported across the cytoplasmic membrane in order to reach their peptidoglycan substrate. Several possibilities exist for the final localization of the exported enzymes. Either they remain attached or anchored to the outer surface of the cytoplasmic membrane, or they bind directly to the peptidoglycan. In addition, in Gram-negative bacteria, the enzymes can remain in a soluble form in the periplasm or can be bound to the outer membrane. Examples for all these cases exist. A leader peptide is present [101] to direct the soluble lytic transglycosylase (Slt70) of *E. coli* to the periplasmic space, where a limited amount binds to the peptidoglycan [179,182]. By contrast, the amidase of *S. pneumoniae* is synthesized without an N-terminal signal sequence [181]. Therefore, the mechanism by which this enzyme is translocated across the membrane is not clear. After export, the enzyme seems to be retained at the outer surface of the membrane by a specific interaction with the lipoteichoic acid [183]. A membrane bound lytic transglycosylase has been claimed to reside and be active in the outer membrane of *E. coli* [184].

The lysis system of phage lambda consists of a highly sophisticated mechanism that includes not only a gene *R* for a lytic transglycosylase, known as lambda endolysin [15], but in addition a gene *Rz* that seems to be needed for efficient lysis in the presence of Mg^{2+} [185,186], and furthermore of a gene *S* which codes for a pore-forming protein to facilitate the specific translocation of the soluble enzyme across the inner membrane of *E. coli* [187]. Analogous export systems for bacterial peptidoglycan hydrolases cannot be ruled out.

7. Regulation and control of peptidoglycan hydrolases

As expected, for such potentially suicidal activities, peptidoglycan hydrolase activities must be exquisitely regulated and integrated with other biosynthetic and perhaps other degradative processes. Regulation of peptidoglycan hydrolase activities may occur at several levels.

First, such enzymes, that are, like all proteins, synthesized on cytoplasmic ribosomes, must be transported through the cytoplasmic membrane to the exterior wall of Gram positives, or to the periplasm of Gram negatives, to have access to the substrate. Thus, control of general access to the insoluble substrate is an important factor. For example, Hartmann et al. [188] proposed a 'barrier' that prevented access of peptidoglycan hydrolases to the peptidoglycan of *E. coli*. Indeed, the autolytic transglycosylase, Slt70, can be up to 30-fold overproduced in *E. coli* without causing lysis of the cell [100]. Such access could even be limited to specific topologically located areas on the surface, such as at or near nascent cross walls in the case of *E. hirae* [189,190] or, in the case of *E. coli*, to a periplasmic ring structure, formed by the periseptal annuli [191]. Similarly, intact, growing cells of *E. hirae* are resistant to the action of its muramidases [192], even though muramidase-2 is known to be present in an active form in the culture medium [177,178]. In this instance, an exterior molecule could protect the peptidoglycan from hydrolysis. Muramidase-1 of *E. hirae* is present in a latent, zymogen form, that requires proteolytic activation [193–195]. Finally, recently obtained data suggest that the state of protein folding (or refolding) could be an important factor in governing the activity of muramidase-2 of *E. hirae* [177,178].

The ability of such enzymes to bind to and maintain contact with the insoluble substrate, particularly outside of the cellular permeability barrier, is also important. As mentioned in Section 6, several peptidoglycan hydrolases have been shown to bind to their substrates, with very high affinity, and, in at least two instances, bind to a limited number of binding sites [178,179]. Muramidase-2 of *E. hirae* [97], the autolysin of *S. faecalis* [99], as well as the amidase and bacteriophage-coded amidases and glycosidase of pneumococci [64] have been shown to possess long amino acid repeat sequences that could be involved with substrate binding. Wall polymers other than the peptidoglycan substrate, such as a polysaccharide, teichoic acid or teichuronic acid, in some instances appear to be important for autolysin binding and can, therefore, affect substrate specificity. For example, the choline-containing wall teichoic acid of the pneumococcus appears to be important in the binding of the pneumococcal amidase to the wall [196].

Substrate specificity is also important. Therefore, changes in the chemistry of the peptidoglycan may control the action of the peptidoglycan hydrolases. O- and N-acetylation of amino sugars of the peptidoglycan itself are two examples. O-acetylation decreases the susceptibility of peptidoglycan to hydrolysis by hen egg white lysozyme but not to the *Chalaropsis* muramidase (see [197] for a recent summary). The presence, absence, and/or exact chemical structure of non-peptidoglycan wall polymers such as teichoic acids, that interact with either enzyme or substrate, perhaps at specific topological sites, and that modify [198,199] or inhibit hydrolase activities, such as cardiolipin [200,201], lipoteichoic acids [202], and the Forssman antigen of the pneumococcus [203], are known. Besides these latter structures, which are only present in Gram-positive bacteria, an additional candidate for an endogenous inhibitory compound has been described that could be involved in the regulation of peptidoglycan hydrolases in *E. coli*. Coenzyme A-*S*-*S*-glutathione (CoAS-SG), isolated from *E. coli* in a search for regulatory substances, was found to inhibit the endogenous autolytic enzymes [204]. Although proof of an in vivo function of this compound was not obtained, mutants defective in the synthesis of CoAS-SG lysed spontaneously when resuspended in 0.01 M phosphate (pH 6.8) at 37°C.

The control of the peptidoglycan hydrolases is connected with other central metabolic processes of the cell; for example, with the stringent response [205–207] and with heat shock control [208]. An important consequence of this is that penicillin and other peptidoglycan synthesis inhibitors operate as bacteriostatic rather than bacteriolytic agents when stringency has been induced, for example, by amino acid starvation. Inhibition of bacteriolysis of many bacterial strains cultured at rather low pH values has been observed, giving rise to the phenomenon of phenotypic tolerance [209,210]. The primary inhibition of penicillin-sensitive enzymes, which results in specific morphological alterations, is fully expressed under these conditions. However, secondary bacteriolysis does not become apparent. The energized state of the membrane of *B. subtilis* appears to be important in regulating its amidase activity [211]. Chaotropic and other agents such as ethanol [212], osmotic shifts up and down [213], and treatments with EDTA, sucrose or NaCl [188] have all been found to induce spontaneous bacteriolysis. Furthermore, as mentioned in Section 5, correlations have been made between the capacity of bacterial cells to autolyze and growth rate [120], the cell division cycle in synchronized cultures [120,122,123], and phase of growth [121,131]. For example, when cultures of *E. hirae* ATCC 9790 enter the stationary phase due to exhaustion of the supply of valine or threonine (required amino acids that are components of protein but not of peptidoglycan), the capacity of the bacteria to autolyze decreases rapidly, although their content of autolytic enzyme activity (probably muramidase-1, see Section 9.1) remains virtually unchanged [131].

Recently, the antibiotic bulgecin, a glucosamine derivative, was shown to specifically inhibit the major autolytic enzyme of *E. coli*, the soluble lytic transglycosylase (Slt70) [214]. Inhibition of this enzyme was determined to be non-competitive. Since a deletion of the *slt* gene showed no obvious phenotype, it was not surprising to find that bulgecin inhibition of the enzyme did not change the growth characteristics of the wild-type *E. coli*. However, when combined with furazlocillin, which alone causes filamentation, bulges were formed prior to rapid bacteriolysis and the MIC for mecillinam was significantly lowered. These data indicate a severe disturbance of the cell's peptidoglycan metabolism when Slt is inhibited or the gene deleted.

8. Bacteriolysis by endogenous autolysins

In general, lysis of bacterial cells is observed after inhibition of further synthesis of peptidoglycan either nutritionally or by the addition of an appropriate antibiotic [215,216]. This lytic response is irrespective of the site of action of the antibiotic. Inhibition of an early step, such as the formation of the UDP-MurNAc-pentapeptide by cycloserine, inhibition of the final crosslinking reaction by a β-lactam antibiotic, as well as a block at any other intermediate step of the biosynthetic pathway usually leads to bacteriolysis. The molecular details are not known. Thus, we have to face the fact that more than 50 years after Fleming's discovery of penicillin, its characteristic bacteriolytic effect is still not understood.

The reason for this situation is the lack of knowledge of the intracellular control mechanisms for autolytic enzymes [12,125]. Although ignored for a long time, bacteriolysis in the presence of inhibitors of peptidoglycan synthesis is the result of an uncontrolled action of the endogenous peptidoglycan hydrolases. In general, inhibition of peptidoglycan synthesis triggers a sequence of events that merges into a common pathway finally causing bacteriolysis by hydrolysis of a sufficient number of bonds in the peptidoglycan sacculus.

Two hypotheses have been proposed to explain the bacteriolytic effect of penicillin and related inhibitors: an enzymic balance model and a peptidoglycan hydrolase trigger mechanism. Weidel and Pelzer [9] assumed that growth of the sacculus requires that the two enzyme systems, hydrolases and synthetases, must cooperate in a well balanced fashion. Inhibition of the peptidoglycan synthesis system would result in a loss of this critical balance, with the peptidoglycan hydrolases getting the upper hand. According to this model, the peptidoglycan hydrolases involved in antibiotic-induced lysis would also be involved in normal growth of the peptidoglycan sacculus.

However, the observation of the phenomenon of tolerance due to a defect in the peptidoglycan hydrolase system is thought to be inconsistent with this proposal [216,217]. Since tolerant mutants grow normally [218,219], one has to conclude that the affected peptidoglycan hydrolase(s) is (are) not absolutely necessary for growth. In the case of *S. pneumoniae,* it became evident that the amidase enzyme apparently responsible for antibiotic-induced lysis was indeed not needed for normal growth and division. It may be only involved in the final splitting of the septum to allow separation of the daughter cells [144]. However, it must be noted that pneumococci that lack or have an inactivated amidase still possess a second autolytic peptidoglycan hydrolase, a glucosaminidase [166,168]. Thus, it remains possible that for normal wall surface growth and division, the residual glucosaminidase activity can serve the required function even though it hydrolyzes a different bond in the peptidoglycan.

Loss of this amidase activity simply was accompanied by growth of *S. pneumoniae* in chains of unseparated but septated cells [144]. Earlier reports on this system put forward the idea that inhibition of peptidoglycan synthesis, for example by penicillin, triggered activity of the enzyme normally involved in splitting of the septum, at a premature, false time point in the cell division cycle [216,220].

Detailed ultrastructural studies of penicillin-induced lysis of *S. aureus* by Giesbrecht and co-workers [221,222] suggest another mechanism for the induction of bacteriolysis. In the presence of low concentrations of penicillin, murein synthesis was shown to continue, but the new wall appeared to accumulate at previously initiated nascent crosswalls. New crosswalls failed to be formed, but splitting of the peripheral wall was obtained at the presumed next division site. Cellular lysis occurs due to this 'morphogenetic defect'.

A similar conclusion was drawn from experiments analyzing the changes in the composition of the peptidoglycan sacculus of *E. coli* during antibiotic-induced bacteriolysis [223]. The results indicated that lysis by cycloserine, moenomycin and penicillin is likely to be by a zipper-like action of an exo-muramidase (e.g. a lytic transglycosylase) that is normally responsible for the precise cutting of the septum. Thus, it seems that inhibition of peptidoglycan synthesis which affects proper synthesis of a septum (thereby causing a morphogenetic defect) cannot be regulated by the cell, with the result that the action of the septum splitting system can no longer be cancelled. Electron microscopy [224] showed that lysis occurred predominantly at sites of future cell divisions and not randomly as one would expect for the case of a general weakening of the peptidoglycan sacculus.

A mechanism that triggers silent peptidoglycan hydrolases may be involved in the lysis process induced by the lysis proteins of certain bacteriophages such as $\Phi\chi174$ and MS2. These low molecular weight, hydrophobic proteins coded for by the phage do not possess peptidoglycan hydrolase activity. Since lysis of the host cells depends on the presence of endogenous bacterial peptidoglycan hydrolases and consequently does not occur under tolerant conditions, it seems that these proteins act by triggering host enzymes [225–227]. The molecular details are not known. Interestingly, however, it has been shown that MS2 lysis protein not only induces the formation of adhesion sites between inner and outer membrane but also gets inserted into these sites [228]. Obviously, besides inhibition of peptidoglycan synthesis, other mechanisms exist by which the endogenous bacterial peptidoglycan hydrolases can get out of control.

9. Well characterized peptidoglycan hydrolase systems

Peptidoglycan hydrolase activities, primarily autolysin activities, have been examined in a variety of both Gram-positive and Gram-negative bacterial species. Based on the putative relationship of these degradative activities to wall peptidoglycan assembly, the, at least superficially, simplest, and perhaps most primitive, systems are those present in Gram-positive species that are normally unable to construct cylindrically shaped wall and that can divide only in a single plane [229]. These models are thus the group broadly defined as streptococci. In this respect, the streptococcal species studied in most detail are the autolytic systems of *E. hirae* and *S. pneumoniae*. In this classification, next in physical complexity come cocci which possess information permitting them to divide in more than one plane: the *Micrococci* and the *Staphylococci*, with *S. aureus* most extensively studied. Third in order of increasing complexity are Gram-positive species that can form a cylindrical surface and are therefore normally rod-shaped. In this class, the organism for which the most information is available is *B. subtilis*, with additional data available from studies of other *Bacillus* species including *Bacillus licheniformis* and *Bacillus cereus*, and from some studies of *Clostridia, Lactobacilli,* and *Listeria*.

Studies of peptidoglycan hydrolase activities in Gram-negative bacteria have the added complication of dealing with a more complex surface structure, but at least in the case of the best studied species, *E. coli*, have the advantage of availability of a relatively enormous amount of genetic information and the easy availability of excellent genetic techniques. Limited, but in some cases important, studies of other Gram-negative species such as *Klebsiella* and of *Neisseria* which are nearly coccal shaped and divide in at least two planes, add to our information. The availability of data accumulated from studies of morphologically, and apparently biochemically, different systems enables comparisons of the systems used to attain similar results and helps to sort out common principles even when sometimes important details differ.

9.1. The peptidoglycan hydrolases of E. hirae ATCC 9790

A broad variety of data are completely consistent with the presence of only one specificity of peptidoglycan hydrolase in *E. hirae* (summarized in [125,126]). The only detectable bond cleaved is the β-1,4 bond between MurNAc and GlcNAc. The presence of only muramidase activity is consistent with the belief that this coccal-shaped organism that divides in only one plane is a relatively simple model of wall surface growth and degradation. However, the observations [44] that *E. hirae* possesses two separate and biochemically distinct peptidoglycan hydrolases as well as a growing assortment of data which show that, in contrast to hen egg white lysozyme, both muramidases are high molecular weight complex proteins, indicates that this model is far from simple.

9.1.1. Some properties of muramidase-1
Muramidase-1 is synthesized as a large molecular weight (130 kDa) zymogen that appears to be first translocated to the wall [195], where it is tightly bound and then proteolytically cleaved to an 87 kDa active form [43]. It is a *gluco*enzyme that is multiply substituted with oligosaccharides containing only glucose units [43]. It also possesses about 12 phosphodiester-linked monomeric 5-mercaptouridine monophosphate residues [230]. There is some evidence that nucleotidylation plays a role in enzyme activity, although the exact role remains to be elucidated (Dolinger and Shockman, unpublished data).

Muramidase-1 processively hydrolyzes β-1,4 bonds between MurNAc and GlcNAc in soluble peptidoglycan chains [231–233]. Processive hydrolysis starts from the non-reducing ends of the disaccharide-peptide glycan chains, resulting in virtually only one product, monomeric disaccharide-peptide units [232,233].

9.1.2. Muramidase-2
Muramidase-2 possesses very little ability to *dissolve* intact walls or cells of *E. hirae* despite the fact that it can hydrolyze muramidase-sensitive bonds in the wall peptidoglycan of *E. hirae*, and dissolves walls of *M. luteus* [44]. Muramidase-2 does not seem to be a glycoprotein, and differs from muramidase-1 in its abilities to bind to and hydrolyze various wall substrates [42,234]. Although it is remotely possible that muramidase-2 is a non-processed (non-glycosylated and/or non-nucleotidylated) form of muramidase-1, substantial evidence exists that each is a product of separate genes [235].

Muramidase-2 is also a complex protein. Two polypeptide bands, with apparent molecular weights of about 125 and 70 kDa on SDS-PAGE, possess peptidoglycan hydrolase activity *and* have the ability to bind radioactively labeled penicillin with low affinity [45]. These two polypeptides appeared to have the same electrophoretic mobility on SDS-PAGE as PBP1 and PBP5 present in membrane preparations of *E. hirae*. However, immunochemical and molecular data clearly demonstrate that PBP5 and the 70 kDa form of muramidase-2 are separate and distinct proteins [235].

9.1.3. Cloning and sequence analysis of the gene for muramidase-2
Recently [97], a gene that codes for the extracellular form of muramidase-2 was cloned and sequenced. A 4.5-kb insert cloned in λgt11 hybridized with six DNA bands at about

4.5, 5.0, 6.1, 7.1, 9.1 and 12.5 kb of *Eco*RI-digested chromosomal DNA of *E. hirae*. The nucleotide sequences of fragments of the 4.5-kb band subcloned into plasmids were determined. The deduced amino acid sequence of the open reading frame that codes for a 666-amino-acid-long protein showed (1) a 49-amino-acid-long putative signal peptide, with a potential protease cleavage site between A49 and D50 (the virtual identity of the determined N-terminal amino acid sequence of mature extracellular muramidase-2 with the deduced sequence, clearly identified this cleavage site); and (2) the presence of six 45-amino-acid-long, highly homologous, amino acid repeat units, separated by intervening sequences that are highly enriched for serine and threonine (ca. 51% S plus T), at the C-terminal end of the protein. Although these repeats have little homology with repeat units found in a variety of different proteins of Gram-positive bacterial species, including those in the pneumococcal amidase gene [64] and the muramidase genes of several pneumococcal phages [65], substantial homology was observed with four contiguous 68-amino acid-long repeat units at the C-terminus of an autolysin of unspecified specificity of *S. faecalis* [99]. The amino acid sequence of the *S. faecalis* protein showed other areas of substantial homology with muramidase-2 including the putative signal peptide and much of the rest of the two polypeptides, except for a 104-amino acid insertion between amino acids 64 and 65 of the *E. hirae* muramidase-2, and for the serine- plus threonine-enriched sequences between the C-terminal repeats of muramidase-2 [98].

Analogous to the interpretations of García and co-workers [64,65], it seems likely that the six 45-amino-acid-long repeats of muramidase-2 may be involved with the high-affinity binding [177,178] of this enzyme to the wall peptidoglycan of *E. hirae*.

Further analysis of the derived amino acid sequence of muramidase-2 [98] showed the presence of an SXXK-active site motif and of several amino acid groupings such as SGN and KSG, that are characteristic motifs of serine β-lactamases and PBPs [236]. Although the spacing of the motifs in muramidase-2 are not inconsistent with those found in other penicillin-interactive proteins, the shape and distribution of hydrophobic and hydrophilic clusters along the amino acid sequences of muramidase-2 differ from those found in a broad assortment of serine β-lactamases and in a number of PBPs [236]. Thus, it is possible that the mechanism of penicillin interaction may differ from that of the typical serine-active site penicillin-interactive protein [236].

9.1.4. Possible functions of the two muramidases of E. hirae in surface growth and division

The presence of two peptidoglycan hydrolases in *E. hirae* that hydrolyze the same bond complicates earlier interpretations and hypotheses concerning the possible functions of autolysins in surface growth and division. On the other hand, the presence of two separate and distinct enzymes that can hydrolyze the same bond, contributes to our lack of success in isolating autolytic-*negative* mutants of *E. hirae* despite considerable and varied efforts over many years. It is also of interest to note the dearth of autolytic-*negative* strains of other bacterial species.

While the presence of multiple enzymes that catalyze the same type of reaction suggests that the cells are carrying 'insurance' for a presumably essential function, it is equally clear that only a low level of peptidoglycan hydrolase activity is essential to the survival and growth of the bacterial cell. Evidence for this is the existence of species and strains, such as some Group A streptococci, that fail to autolyze under conditions that result in the rapid autolysis of other strains [67]. Recent successes in cloning, sequencing and expressing peptidoglycan hydrolases of several bacterial species (see Table II) should lead to solving the questions of functions and essentiality of these various peptidoglycan hydrolases in the not too distant future.

9.2. The autolytic system of S. pneumoniae

S. pneumoniae has long been known to possess an amidase activity [237,238], which was purified and characterized [239]. Although analyses of the products of the action of partially purified preparations of the autolytic enzyme activity of S. pneumoniae on cell wall preparations of the pneumococcus, that results in the release of amino sugars from the walls, suggested the presence and action of one or more glycosidases [240], substantial evidence for the presence of a second peptidoglycan hydrolase activity was obtained only relatively recently. In studies of a mutant of S. pneumoniae deleted in the lytA gene coding for the amidase, Sánchez-Puelles et al. [166,168] demonstrated the presence of a second peptidoglycan hydrolase, which was purified and shown to be a 64 kDa endo N-acetyl-glucosaminidase [66]. To our knowledge, this enzyme has not yet been cloned and sequenced.

9.2.1. Choline-dependent amidases
The pneumococcal amidase has a number of interesting and unusual properties. The enzyme appears to recognize and interact with the choline-containing wall teichoic acid [241]. The enzyme appears to be synthesized in the 'E' form which seems to be present only in the cytoplasmic fraction [183], has little to no peptidoglycan hydrolase activity, and in the absence of a wall choline-containing teichoic acid, is not transported through the membrane, nor is it converted to the active C form. Conversion to the C form appears to be due to the formation of a choline-amidase complex and an allosteric change in the enzyme. On the other hand, high choline concentrations (e.g. 2%) result in inhibition of cellular lysis and chain formation.

The gene for the pneumococcal amidase, lytA, has been cloned in E. coli, sequenced [61,242] and expressed as the E form [61]. The gene codes for a protein of 36.5 kDa. A mutant strain, M31, was isolated that possessed a completely deleted lytA gene. This mutant grew at a normal rate, could be transformed, failed to lyse in the stationary growth phase, displayed a tolerant response to the action of β-lactam antibiotics, and grew in short chains of 6–8 cells. The insertion of a plasmid containing the lytA gene resulted in amidase production at a rate about five times faster than in the parental strain and restored the ability to autolyze in stationary phase, to lyse in the presence of penicillin and to form normal diplococci [243]. Similarly, insertional inactivation of the lytA gene failed to affect its phenotype, except for resistance to autolysis [244].

9.2.2. Modular organization of enzymes
Extensive studies of several lytic peptidoglycan hydrolases of several pneumococcal bacteriophages, including the construction of chimeric proteins, resulted in the concept of

modular organization of these lytic enzymes. The N-terminal domain appears to contain the catalytic site, which in one class catalyzes an amidase activity (e.g. LytA, Hbl-3) and in the other class catalyzes a muramidase activity (e.g. Cpl-1, Cpl-9). The N-terminal domains of the bacteriophage muramidases not only had homology with each other but also contained aspartic and glutamic residues (D9, E36) at sites that are homologous to those observed in other microbial lysozymes including the *Chalaropsis* CH (D6, E33) [95] and *Streptomyces globosporus* M1 (D9, E36) [96] lysozymes.

The C-terminal domains, which contain three or six amino acid repeats, are thought to be important in the recognition of choline residues in the pneumococcal wall. The choline-recognizing enzymes, including the LytA amidase and the Cpl-1 and Cpl-9 muramidases, possess six 20-amino-acid-long tandem repeats, while Cpl-7 muramidase, whose lytic action is independent of the presence of choline in the wall, possesses 2.8 48-amino-acid-long tandem repeats. The two classes of repeats lack homology with each other.

9.3. The peptidoglycan hydrolases of B. subtilis and other Bacilli

Various strains of *B. subtilis*, *B. licheniformis* and other *Bacillus* species have been observed to possess enzyme activities that can hydrolyze bonds in their peptidoglycan [22,129]. These activities include those that (1) appear to be specific for spore cortex peptidoglycan and may be involved with spore germination, (2) increase in amount post-exponentially during sporulation and may play a role in sporulation, (3) hydrolyze bonds in the peptidoglycans but, because of their substrate specificities, are unlikely to be involved with peptidoglycan or cellular dissolution, such as the exo-*N*-acetylglucosaminidase [245,246] and exo-*N*-acetylmuramidase [247] of a strain of *B. subtilis*, and finally (4) are truly autolytic peptidoglycan hydrolases.

9.3.1. The amidase of B. subtilis
The predominant autolytic peptidoglycan hydrolase activity of several *Bacillus* species appears to be an amidase [129]. This is the enzyme specificity that seems to be responsible for loss of cell wall peptidoglycan by turnover of walls of bacilli [157].

A number of years ago Herbold and Glaser [198,199] purified and characterized an amidase of *B. subtilis* ATCC 6051. This enzyme has a molecular weight of about 50 kDa. Activity of this amidase is modulated by an approximately 80-kDa modifier protein which stoichiometrically combines with the enzyme and changes the pattern of wall hydrolysis from random to processive hydrolysis of amide bonds on individual glycan strands. Wall glyceroteichoic acid appeared to be required for the binding of the enzyme to walls and for the functional interaction of the enzyme and modifier protein, since the enzyme failed to bind to, and hydrolyze, walls of *B. subtilis* W23 that possesses a wall ribitol teichoic acid instead of the wall glyceroteichoic acid found in walls of strain 6051. However, the amidase binds strongly to walls of strain 6051 containing teichuronic acid in the place of wall glyceroteichoic acid [248].

Subsequently Lindsay and Glaser [249] purified and partially characterized an amidase from a different strain of *B. subtilis*, strain W23, that possesses a ribitol teichoic acid in its wall, in contrast to the glyceroteichoic acid in walls of the ATCC 6051 strain. Although the W23 amidase was similar in molecular size to the 6051 amidase, the two enzymes differed. The W23 amidase failed to interact with the modifier protein, and differed from the 6051 amidase in both pH optimum and substrate specificity. Furthermore, various wall teichoic acids affected the enzymes differently, and the two enzymes failed to immunochemically cross-react [248]. Thus, similar to the peptidoglycan hydrolases of other bacterial species, the high-molecular-weight amidases of *B. subtilis* strains possess complex substrate specificity, substrate binding and protein interaction profiles [198,199, 248,249].

9.3.2. Cloning and sequencing of autolysins of Bacilli
Recently, several laboratories have cloned in *E. coli* and sequenced the genes for autolysins of *Bacilli*. These include the genes for a 28-kDa autolysin of undetermined specificity of a *Bacillus* species [84], a 30-kDa amidase of *B. subtilis* 168 [77–79], a 50- to 53-kDa amidase of *B. subtilis* 168 [80–82], an amidase of *B. licheniformis*, FD0120 [85], as well as the modifier protein of the 50- to 53-kDa amidase of *B. subtilis* 168 [81,85]. Cloning and sequencing of the gene for the *N*-acetylglucosaminidase (*Lyt*D; map position 310°) has not yet been reported.

9.3.3. The 28-kDa autolysin of a Bacillus sp. and the 30-kDa amidase (CwlA) of B. subtilis 168
As mentioned above, two laboratories [77,78] have cloned in *E. coli* and sequenced a gene, called *cwl*A (map position 233°), that codes for an amidase activity that has the identical nucleotide (and amino acid) sequence. The amino acid sequence of the N-terminal and central regions showed a high degree of homology with the 28-kDa *Bacillus* sp. autolysin. There appears to be a signal peptide that is cleaved after A39 [78]. Furthermore, in *E. coli*, the enzyme seems to be proteolytically processed, probably at the C-terminus of the protein, to a 21- to 23-kDa polypeptide [77,78]. Both the 30-kDa and 21- to 23-kDa forms were enzymatically active. Thus, the N-terminal end of the protein probably carries the catalytic domain. Interestingly, the C-terminal amino acid sequence of CwlA possesses two short repeated sequences and displayed extensive homology with the N-terminal sequence of the *Streptomyces albus* G wall-lytic carboxypeptidase [250]. The cloned CwlA amidase has a broad substrate specificity, as it dissolved walls of a variety of Gram-positive species, including walls of several *Bacillus* spp., walls of *M. luteus* and *Lactobacillus arabinosus*, as well as the spore cortex of *B. megaterium*. It did not hydrolyze walls of *E. coli*.

9.3.4. The 53-kDa amidase (CwlB or LytC) of B. subtilis 168
Again, two laboratories have cloned in *E. coli* and sequenced the gene for this enzyme, called *cwl*B by Kuroda and Sekiguchi [82] and *lyt*C (map position 307°) by Margot and Karamata [80] and Lazarevic et al. [81]. The gene codes for a 52.6-kDa polypeptide that carries a putative signal sequence that is cleaved after A24 to yield the 49.9-kDa amidase protein. This amidase apparently failed to exhibit amino acid homology with other known amidases or with other proteins in the data bank, except for a similarity to the CwlM amidase of *B. licheniformis*.

9.3.5. The 27.5-kDa amidase (CwlM) of B. licheniformis [85]

There appear to be similarities between this amidase and the 53-kDa CwlB amidase of *B. subtilis* 168. The N-terminal region I4 to L154 of CwlM has striking homology with the C-terminal region I323 to L474 of CwlB (37% identity). Since a truncated form of CwlM, CwlMt, which lacks 64 amino acids at its C-terminal end, retains amidase activity, the catalytic domain of CwlM probably resides in its N-terminal region. The C-terminal domain of CwlM contains two 31-amino-acid-long repeats. In contrast, CwlB amidase contains three repeated amino acid sequences in its non-catalytic N-terminal domain. It is of some interest to note that the CwlMt, which lacks 64 amino acids at its C-terminal end, including the two 31-amino-acid-long repeats, retained both catalytic and wall-binding activities to walls of both *B. subtilis* and *M. luteus*, indicating that, in this instance, the non-catalytic C-terminal domain of CwlM containing the repeated sequence did not appear to play a role in binding to the wall substrate.

9.3.6. The modifier protein of the 52.6-kDa amidase (CwbA or LytB) of B. subtilis 168 [81,83]

Upstream from the *cwl*B (or *lyt*C; map position 307°) gene is the gene for the 76.7-kDa protein that not only stimulates the activity of the 52.6-kDa amidase but also changes its mode of hydrolysis from random to processive [198,199], called CwbA by Kuroda et al. [83] and LytB by Lazarevic et al. [81]. The amino acid sequence of the N-terminal portion of CwbA (or LytB) has homology to the N-terminal sequence of CwlB and its entire sequence showed significant amino acid homology to the *spo*IID gene product of both *B. subtilis* and *Bacillus amyloliquefaciens* [251,252]. The genes for both the 76.7-kDa modifier and the 52.6-kDa amidase appear to be part of a regulatory unit that is divergently transcribed and has been called a 'divergon' by Lazarevic et al. [81]. The transcribed genes include *lyt*ABC, and *lyt*R, which is transcribed in the opposite direction. The LytA (or LppX) product appears to be a 9.4-kDa highly acidic peptide that is probably a lipoprotein [81,83]. LytR is a 35-kDa polypeptide which acts as an attenuator of the expression of both the *lyt*ABC and *lyt*R operons [81].

9.3.7. Possible functions of the amidases of Bacilli

It seems clear that amidase activity is responsible for cell wall turnover in *Bacilli*. However, insertionally inactivated mutants in the *lyt*B and *lyt*C genes produced only a 60 min delay in the onset of loss of wall by turnover [80]. After this time, the slopes of the curves were about the same as for the parent strain. Although rates of lysis of intact cells and/or isolated native walls containing endogenous wall-bound autolysins of *lyt*B, *lyt*C and *lyt*C insertionally inactivated mutants were much slower than those of walls of the parent strain, wall autolysis continued for an extended time interval (e.g. 24 h). Growth rate, average chain length, cell shape, and flagellar motility did not seem to be affected by the absence of the 52.6-kDa amidase and/or the modifier protein. However, swarming of cells of amidase-deficient mutants was considerably reduced. Strains lacking the 52.6-kDa amidase and/or the modifier, sporulated normally, their spores germinated normally, and cells yielded comparable numbers of transformants. However, additional autolysins, including at least one other amidase and a glucosaminidase, remained present [80]. Recently using zymograms after SDS-PAGE on gels containing suspensions of purified cell walls from vegetative cells of *B. subtilis* 168 or *B. megaterium* KM spore cortex, Foster [22] observed in SDS extracts of vegetative bacteria the presence of four polypeptide bands that dissolved walls of *B. subtilis*. Two polypeptides, A3 and A4, at 34 and 30 kDa, respectively, plus another polypeptide, A5 at 23 kDa, were able to dissolve the spore cortex peptidoglycan of *B. megaterium* KM. Band A4 showed a large increase in activity 6 h after induction of sporulation (just before mother cell lysis and release of the mature spores), band A5 appeared transiently at 5 h postinduction, and a quantitatively minor band at 41 kDa, A6, that appeared to be cortex specific, was observed in dormant spores. Studies of insertionally inactivated mutants were used to identify bands A1 and A2 as the previously biochemically characterized 90-kDa glucosaminidase and 50-kDa amidase, respectively [33,198,199]. Since band A4 was not lost in a mutant lacking the 30-kDa amidase, Foster was unable to identify this band as the product of a cloned gene that codes for a 30-kDa amidase [77]. Thus, the presence of more than one peptidoglycan hydrolase appears to complicate the precise determinations of possible biological significance and function(s).

9.4. The murein hydrolases of E. coli

With more than nine different murein (peptidoglycan) hydrolases, *E. coli* has one of the more complex systems of peptidoglycan hydrolases [12,130] (Table II). The actual autolytic system, that is, the group of enzymes that could autolyze the cell by cleaving bonds in the high-molecular-weight sacculus, consists of enzymes of two specificities, endopeptidase and muramidase activity. A unique characteristic feature of the muramidase of *E. coli* is its peculiar reaction mechanism. Unlike lysozyme these enzymes do not simply hydrolyze the β-1,4-glycosidic bond between MurNAc and GlcNAc but, concomitantly with cleavage of the bond, form a 1,6-anhydromuramic acid ring structure [13]. The significance of this reaction is not known. It may be a means of conserving the bond energy for further rearrangement reactions.

9.4.1. Lytic transglycosylases
Three lytic transglycosylases have been isolated from *E. coli* [13,103,104]. One of these, the soluble lytic transglycosylase (Slt70), has been studied in detail. The *slt*70 gene (99.7 min) has been cloned [100] and the sequence has been determined [101]. The enzyme protein is synthesized with a leader peptide and translocated across the membrane and processed to a protein with a molecular mass of 70 kDa. The products of *sec*A, *sec*Y and *sec*B have been shown to be involved [182]. No obvious homology with other muramidases, including lambda endolysin, a lytic transglycosylase, could be detected. Recently, the crystal structure of Slt70 was obtained at a resolution of 2.8 Å [253]. It is the first autolytic and the second bacterial peptidoglycan hydrolase for which the complete three-dimensional structure has been determined. (The first is the *S. albus* G DD-carboxypeptidase [70,71].) The 360 N-terminal amino acids represent a horseshoe-like structure possessing exclusively α helices. The C-terminal domain (aa384–aa618) is a globular domain and possibly contains the catalytic site. Homology in the structure but not in the amino acid sequence exists between the carboxyl domain and T4 lysozyme (Thunmissen and Dijkstra, personal communication). From the mutant carrying a deletion

in the *slt*70 gene, a second lytic transglycosylase has been purified and determined to be a protein of 35 kDa [103]. Thus, this enzyme has been called Slt35. Furthermore, recently a membrane-bound lytic transglycosylase (Mlt38) with a molecular weight of about 38 kDa has been purified [104]. An interesting property of this enzyme is its capacity to degrade isolated glycan strands, stripped of their peptides by treatment with an amidase.

9.4.2. Endopeptidases

Two endopeptidase activities are present in *E. coli*, a penicillin-sensitive enzyme that is identical with PBP4, the *dacB* (68.6 min) gene product [106] and a penicillin-insensitive one, the *mepA* (50.4 min) gene product [107,253a,254]. Both enzymes have been purified and characterized. They fail to show significant sequence homology with each other and the conserved Ser-Xaa-Xaa-Lys motif of PBPs is not present in the penicillin-insensitive endopeptidase. Overproduction of either or both endopeptidases did not result in a visible destabilization of the peptidoglycan sacculus. Although PBP4 has been considered to be a membrane protein, it does not appear to be anchored in the membrane. For example, in a strain overproducing the enzyme, about 80% of the total amount was detected in the soluble fraction in an active form. The protein is translated with a cleavable signal peptide of 20 amino acids [108].

9.4.3. Additional murein hydrolases

Besides two DD-carboxypeptidases, PBP5 and PBP6 (see Chapter 6) [255], LD-carboxypeptidase activities have also been identified [123,256]. A nocardicin A-sensitive LD-carboxypeptidase (MW 32 kDa) has been purified [257]. Such an enzyme activity might be involved in the regulation of cell division, since its activity was found to increase in synchronized cultures during cell division [123].

Two enzyme specificities present in the periplasmic space of the cell envelope of *E. coli* participate in the efficient recycling of murein turnover products, an amidase [35,259] and a β-N-acetylglucosaminidase [109]. Since mainly the murein monomers, the 1,6-anhydrodisaccharidetetra- and -tripeptides, have been found to be released into the periplasm as a result of turnover [153], amidase activity is needed to set free the peptide moieties which are accepted by the oligopeptide transport system (*opp*) of the cell to be taken into the cytoplasm for re-utilization [155]. Recently, indirect evidence for a second permease system has been obtained [156]. Therefore, efficient recycling may obscure true rates of turnover [156]. The disaccharides are likely to be cleaved by the endo-β-N-acetylglucosaminidase known to be present in the periplasm [109]. The monosaccharides may then be transported into the cell by sugar transport systems.

9.5. Autolytic peptidoglycan hydrolases of other bacterial species

As listed in Table I, a broad variety of bacterial species have been shown to produce peptidoglycan hydrolases. Since the original list was compiled in 1980 [27], experimental evidence that a number of additional bacterial species produce peptidoglycan hydrolases has been obtained (Tables I and II). For example, evidence that *N. gonorrhoeae* [52,53] *Pseudomonas aeruginosa* [56], other strains of *Clostridia* [36–38], and *Streptococci* [60,67] produce peptidoglycan hydrolases has been obtained. The question then arises: do all bacteria produce such enzymes? This question is important in terms of the potential role such activities have been postulated to play in cell wall assembly and cell division [26,124–126]. Since some of the enzyme activities described are extracellular and others are coded for by genes of bacteriophages or plasmids, their role in the economy of the cell is questionable. On the other hand, significant is the demonstration of McDowell and Lemanski [67] that *S. pyogenes*, an organism that fails to demonstrate overt autolysis even when challenged with cell wall antibiotics, possesses an apparent hexosamidase activity.

Recently, the genes for several peptidoglycan hydrolases, including several putative autolysins, have been cloned and sequenced (Table II). Further studies and comparisons of these genes and proteins should be revealing.

10. Concluding remarks

Until now, convincing proof showing that peptidoglycan hydrolases are indispensable for growth of bacteria containing peptidoglycan in their walls fails to exist. However, it is difficult to think of a mechanism by which the peptidoglycan network could be enlarged and divided without the help of hydrolytic enzymes. Thus, theoretically, peptidoglycan hydrolases must be considered pacemaker enzymes for cell wall growth and therefore for bacterial growth in general. If we accept this concept, then indeed this class of enzymes represents unique bacterial proteins. On the one hand, they enable the cell to grow, but on the other hand, they can also kill the cell by means of autolysis. Because of both features, these Janus-faced enzymes are of great interest for chemotherapy.

Although the molecular details are still not understood, it seems clear that the bacteriolytic action of cell wall synthesis inhibitors such as penicillins is due to the uncontrolled action of endogenous peptidoglycan hydrolases [215,216]. In addition to inhibition of peptidoglycan synthesis, other ways of inducing uncontrolled action of these enzymes are feasible. The mode of action of certain low molecular weight phage-coded lysis proteins are one such example [225–227].

As pacemaker enzymes for bacterial growth, peptidoglycan hydrolases should be an ideal target for antibiotics. The problem, however, is the multiplicity of hydrolases found in bacteria. This fact has not only inhibited the isolation of mutants completely lacking peptidoglycan hydrolytic activity but also is an obstacle in attempts to demonstrate that a complete block of these hydrolases results in bacteriostasis as would be expected for true pacemaker enzymes.

An interesting situation is caused by the inhibition of only one peptidoglycan hydrolase with the other hydrolases left unaffected. Inhibition of the major autolytic enzyme present in *E. coli*, the Slt70, by bulgecin resulted in increased lysis sensitivity [214]. Probably because of the presence of additional autolytic enzymes, the inhibition of just one pacemaker enzyme together with some PBPs yielded a synergistic, bacteriolytic effect.

Peptidoglycan hydrolases, which appear to be involved in the growth of the shape-maintaining structure of the bacterial cell, necessarily have to be under strict topological and temporal control. Thus, by studying the cellular control mechanisms for peptidoglycan hydrolases, a fundamental problem in biology is being addressed, namely, how a specific shape is realized, maintained and modified in a well controlled fashion in a bio-

160

logical system. One clue for an understanding of these mechanisms may turn out to be the specific subcellular distribution of the enzymes. Since some of them are tightly bound to peptidoglycan, it is possible that the sacculus serves as a matrix to guarantee synthesis of a peptidoglycan network identical in shape to the pre-existing structure. In analogy to DNA replication, it is quite likely that multi-enzyme complexes are involved in these processes. In particular, for a coordination of the insertion of new peptidoglycan subunits with the cleavage of bonds to provoke enlargement of the sacculus, holoenzyme-like complexes are reasonable possibilities. Therefore, progress in understanding growth and division of the bacterial cell wall, which is the molecular basis for the morphogenesis of the bacterium during its cell cycle, calls for intense investigations of the interplay of peptidoglycan-synthesizing and -hydrolyzing enzymes.

Acknowledgments

We thank those investigators who provided us with preprints of papers before publication. We also thank Drs. L. Daneo-Moore, P. Piggot and R. Kariyama for their critical comments on the manuscript, and Mr. G. Harvey for editorial assistance. The research in the laboratory of G.D.S. was supported by research grant AI05044 from the National Institute of Allergy and Infectious Diseases, US Public Health Services.

References

1. Dubos, R. (1945) The Bacterial Cell. Harvard University Press, Cambridge, MA.
2. Fleming, A. (1922) Proc. R. Soc. (London; Ser. B) 93, 306–317.
3. Alderton, G., Ward, W.H. and Fevold, H.L. (1945) J. Biol. Chem. 157, 43–58.
4. Salton, M.R.J. (1957) Bacteriol. Rev. 21, 82–100.
5. Salton, M.R.J. (1956) in: Bacterial Anatomy – Sixth Symposium of the Society for General Microbiology, Cambridge, University Press, Cambridge, UK, pp. 81–110.
6. McCarty, M. (1952) J. Exp. Med. 96, 569–580.
7. Ghuysen, J.-M. (1957) Arch. Int. Phys. Biochem. 45, 173–305.
8. Ghuysen, J.-M. (1968) Bacteriol. Rev. 32, 425–464.
9. Weidel, W. and Pelzer, H. (1964) Adv. Enzymol. 26, 193–232.
10. Weibull, C. (1953) J. Bacteriol. 66, 688–695.
11. Shockman, G.D. and Barrett, J.F. (1983) Annu. Rev. Microbiol. 37, 501–527.
12. Höltje, J.-V. and Tuomanen, E.I. (1991) J. Gen. Microbiol. 137, 441–454.
13. Höltje, J.-V., Mirelman, D., Sharon, N. and Schwarz, U. (1975) J. Bacteriol. 124, 1067–1076.
14. Taylor, A., Das, B.C. and van Heijenoort, J. (1975) Eur. J. Biochem. 53, 47–54.
15. Bienkowska-Szewczyk, K., Lipinska, B. and Taylor, A. (1981) Mol. Gen. Genet. 184, 111–114.
16. Harz, H., Burgdorf, K. and Höltje, J.-V. (1990) Anal. Biochem. 190, 120–128.
17. Gmeiner, J. and Kroll, H.-P. (1981) FEBS Lett. 129, 142–144.
18. Ghuysen, J.-M., Tipper, D.J. and Strominger, J.L. (1966) Methods Enzymol. 8, 685–699.
19. Shockman, G.D., Pooley, H.M. and Thompson, J.S. (1967) J. Bacteriol. 94, 1525–1530.
20. Hartmann, R., Höltje, J.-V. and Schwarz, U. (1972) Nature (London) 235, 426–429.
21. Leclerc, D. and Asselin, A. (1989) Can. J. Microbiol. 35, 749–753.
22. Foster, S.J. (1992) J. Bacteriol. 174, 464–470.
23. Glauner, B. (1988) Anal. Biochem. 172, 451–464.

24. Glauner, B., Höltje, J.-V. and Schwarz, U. (1988) J. Biol. Chem. 263, 10088–10095.
25. Shockman, G.D., Kolb, J.J. and Toennies, G. (1958) J. Biol. Chem. 230, 961–977.
26. Shockman, G.D. (1965) Bacteriol. Rev. 29, 345–358.
27. Rogers, H.J., Perkins, H.R. and Ward, J.B. (1980) Microbial Cell Walls and Membranes. Chapman and Hall, London.
28. Krulwich, T.A. and Ensign, J.C. (1968) J. Bacteriol. 96, 857–859.
29. Hughes, R.C. (1971) Biochem. J. 121, 791–802.
30. Hughes, R.C. (1970) Biochem. J. 119, 849–860.
31. Guinand, M., Michel, G. and Tipper, D.J. (1974) J. Bacteriol. 120, 173–184.
32. Grant, W.D. and Wicken, A.J. (1970) Biochem. J. 118, 859–868.
33. Rogers, H.J., Taylor, C., Rayter, S. and Ward, J.B. (1984) J. Gen. Microbiol. 130, 2395–2402.
34. Kingau, S.L. and Ensign, J.C. (1968) J. Bacteriol. 96, 629–638.
35. Van Heijenoort, J., Parquet, C., Flouret, B. and Van Heijenoort, Y. (1975) Eur. J. Biochem. 58, 611–619.
36. Croux, C. and García, J.L. (1991) Gene 104, 25–31.
37. Croux, C., Canard, B., Goma, G. and Soucaille, P. (1992) Appl. Environ. Microbiol. 58, 1075–1081.
38. García, J.L., García, E., Sánchez-Puelles, J.M. and López, R. (1988) FEMS Microbiol. Lett. 52, 133–138.
39. Takumi, K., Kawata, T. and Hisatsune, K. (1971) Jpn. J. Microbiol. 15, 131–141.
40. Martin, H.H. and Kemper, S. (1970) J. Bacteriol. 102, 347–352.
41. Williamson, R. and Ward, J.B. (1979) J. Gen. Microbiol. 114, 349–354.
42. Shockman, G.D., Dolinger, D.L. and Daneo-Moore, L. (1988) in: P. Actor, L. Daneo-Moore, M.R.J. Salton and G.D. Shockman (Eds.), Antibiotic Inhibition of Bacterial Cell Surface Assembly and Function, American Society for Microbiology, Washington, DC., pp. 195–210.
43. Kawamura, T. and Shockman, G.D. (1983) J. Biol. Chem. 258, 9514–9521.
44. Kawamura, T. and Shockman, G.D. (1983) FEMS Microbiol. Lett. 19, 65–69.
45. Dolinger, D.L., Daneo-Moore, L. and Shockman, G.D. (1989) J. Bacteriol. 171, 4355–4361.
46. Coyette, J. and Ghuysen, J.-M. (1970) Biochemistry 9, 2952–2955.
47. Tinelli, R. and Andr-Romain, M. (1965) C. R. Acad. Sci. Paris, 261, 4265–4267.
48. Silcock, R., Weston, A. and Perkins, H.R. (1978) Proc. Soc. Gen. Microbiol. 5, 63–64.
49. Kilburn, J.O. and Best, G.K. (1977) J. Bacteriol. 129, 750–755.
50. Ensign, J.C. and Wolfe, R.S. (1965) J. Bacteriol. 90, 395–402.
51. Hungerer, K.D., Fleck, J. and Tipper, D.J. (1969) Biochemistry 8, 3567–3573.
52. Rosenthal, R.S. (1979) Infect. Immun. 24, 869–878.
53. Chapman, S.J. and Perkins, H.R. (1983) J. Gen. Microbiol. 129, 877–883.
54. Sinha, R.K. and Rosenthal, R.S. (1980) Infect. Immun. 29, 914–925.
55. Gubish, E.R., Jr., Chen, K.C.S. and Buchanan, T.M. (1982) J. Bacteriol. 151, 172–176.
56. Brito, N., Falcon, M.A., Carnicero, A., Gutierrez-Navarro, A.M. and Mansito, T.B. (1989) Res. Microbiol. 140, 125–137.
57. Pizarro, R.A., Fernandez, R.O. and Orce, L.V. (1988) Arch. Intern. Physiol. Biochem. 96, 171–177.
58. Tipper, D.J. (1969) J. Bacteriol. 97, 837–847.
59. Wadstrom, T. and Vesterberg, O. (1971) Acta Path. Microbiol. Scand. 79, 248–264.
60. Irhuma, A., Gallagher, J., Hackett, T.J. and McHale, A.P. (1991) Biochim. Biophys. Acta 1074, 1–5.
61. García, E., García, J.L., Ronda, C., García, P., and López, R. (1985) Mol. Gen. Genet. 201, 225–230.
62. Romero, A., López, R. and García, P. (1990) J. Bacteriol. 172, 5064–5070.
63. Sánchez, J.M. and García, J.L. (1990) Eur. J. Biochem. 187, 409–416.
64. García, E., García, J.L., García, P., Arrarás, A., Sánchez-Puelles, J.M. and López, R. (1988) Proc. Natl. Acad. Sci. USA 85, 914–918.
65. García, P., García, J.L., García, E., Sánchez-Puelles, J.M. and López, R. (1990) Gene 86, 81–88.
66. García, P., García, J.L., García, E. and López, R. (1989) Biochem. Biophys. Res. Commun. 158, 251–256.
67. McDowell, T.D. and Lemanski, C.L. (1988) J. Bacteriol. 170, 1783–1788.
68. Munoz, E., Ghuysen, J.-M., Leyh-Bouille, M., Petit, J.-F. and Tinelli, R. (1966) Biochemistry 5, 3091–3098.

162

69. Duez, C., Lakaye, B., Houba, S., Dusart, J. and Ghuysen, J.-M. (1990) FEMS Microbiol. Lett. 71, 215–220.
70. Dideberg, O., Charlier, P., Dive, G., Joris, B., Frre, J.M. and Ghuysen, J.M. (1982) Nature 299, 469–470.
71. Dideberg, O., Charlier, P., Dupont, L., Vermeire, M., Frre, J.-M. and Ghuysen, J.-M. (1989) FEBS Lett. 117, 212–214.
72. Wry, P. (1987) Ph.D. thesis, University of Liege.
73. Ghuysen, J.-M., Dierickx, L., Coyette, J., Leyh-Bouille, M., Guinand, M. and Campbell, J.M. (1969) Biochemistry 8, 213–222.
74. Ghuysen, J.-M., Leyh-Bouille, M., Bonaly, R., Nieto, M., Perkins, H.R., Schleifer, K.H. and Kandler, O. (1970) Biochemistry 9, 2955–2961.
75. Hourdou, M.-L., Drez, C., Joris, B., Vacheron, M.-J., Guinand, M., Michel, G. and Ghuysen, J.-M. (1992) FEMS Microbiol. Lett. 91, 165–170.
76. Hourdou, M.-L., Guinaud, M., Vacheron, M.-J., Michel, G., Denoroy, L., Duez, C., Englebert, S., Joris, B., Weber, G. and Ghuysen, J.-M. (1993) Biochem. J. 292, 563–570.
77. Foster, S.J. (1991) J. Gen. Microbiol. 137, 1987–1998.
78. Kuroda, A. and Sekiguchi, J. (1990) J. Gen. Microbiol. 136, 2209–2216.
79. Kuroda, A., Imazeki, M. and Sekiguchi, J. (1991) FEMS Microbiol. Lett. 81, 9–14.
80. Margot, P. and Karamata, D. (1992) Mol. Gen. Genet. 232, 359–366.
81. Lazarevic, V., Margot, P., Soldo, B. and Karamata, D. (1992) J. Gen. Microbiol. 138, 1949–1961.
82. Kuroda, A. and Sekiguchi, J. (1991) J. Bacteriol. 173, 7304–7312.
83. Kuroda, A., Rashid, M.H. and Sekiguchi, J. (1992) J. Gen. Microbiol. 138, 1062–1076.
84. Potvin, C., Leclerc, D., Tremblay, G., Asselin, A. and Bellemare, G. (1988) Mol. Gen. Genet. 214, 241–248.
85. Kuroda, A., Sugimoto, Y., Funahashi, T. and Sekiguchi, J. (1992) Mol. Gen. Genet. 234, 129–137.
86. Recsei, P.A., Gruss, A.D. and Novick, R.P. (1987) Proc. Natl. Acad. Sci. USA 84, 1127–1131.
87. Heinrich, P., Rosenstein, R. Böhmer, M., Sonner, P. and Götz, F. (1987) Mol. Gen. Genet. 209, 563–569.
88. Heath, L.S., Heath, H.E. and Sloan, G.L. (1984) FEMS Microbiol. Lett. 44, 129–133.
89. Jayaswal, R.K., Lee, Y.-I. and Wilkinson, B.J. (1990) J. Bacteriol. 172, 5783–5788.
90. Wang, X., Wilkinson, B.J. and Jayaswal, R.K. (1991) Gene 102, 105–109.
91. Wang, X., Mani, N., Pattee, P.A., Wilkinson, B.J. and Jayaswal, R.K. (1992) J. Bacteriol. 174, 6303–6306.
92. Biavasco, F., Pruzzo, C. and Thomas, C. (1988) FEMS Microbiol. Lett. 49, 137–142.
93. Platteeuw, C. and de Vos, W.M. (1992) Gene 118, 115–120.
94. Boizet, B., Lahbib-Mansais, Y., Dupont, L., Ritzenthaler, P. and Mata, M. (1990) Gene 94, 61–67.
95. Felch, J.W., Inagami, T. and Hash, J.H. (1975) J. Biol. Chem. 250, 3713–3720.
96. Lichenstein, H.S., Hastings, A.E., Langley, K.E., Mendiaz, E.A., Rohde, M.F., Elmore, R. and Zukowski, M.M. (1990) Gene 88, 81–86.
97. Chu, C.-P., Kariyama, R., Daneo-Moore, L. and Shockman, G.D. (1992) J. Bacteriol. 174, 1619–1625.
98. Joris, B., Englebert, S., Chu, C.-P., Kariyama, R., Daneo-Moore, L., Shockman, G.D. and Ghuysen, J.-M. (1992) FEMS Microbiol. Lett. 91, 257–264.
99. Beliveau, C., Potvin, C., Trudel, J., Asselin, A. and Bellemare, G. (1991) J. Bacteriol. 173, 5619–5623.
100. Betzner, A.S. and Keck, W. (1989) Mol. Gen. Genet. 219, 489–491.
101. Engel, A., Kazemier, B. and Keck, W. (1991) J. Bacteriol. 173, 6773–6782.
102. Mett, H., Keck, W., Funk, A. and Schwarz, U. (1980) J. Bacteriol. 144, 45–52.
103. Engel, H., Smink, A.J., van Wijngaarden, L. and Keck, W. (1992) J. Bacteriol. 174, 6394–6403.
104. Ursinus, A. and Höltje, J.-V. (1993) unpublished.
105. Mottl, H., Terpstra, P. and Keck, W. (1991) FEMS Microbiol. Lett. 78, 213–220.
106. Korat, B., Mottl, H. and Keck, W. (1991) Mol. Microbiol. 5, 675–684.
107. Keck, W., van Leeuwen, A.M., Huber, M. and Goodell, E.W. (1990) Mol. Microbiol. 4, 209–219.
108. Tomioka, S. and Matsuhashi, M. (1978) Biochem. Biophys. Res. Commun. 84, 978–984.
109. Yem, D.W. and Wu, H.C. (1976) J. Bacteriol. 125, 324–331.

110. Valinger, Z., Ladešić, B. and Tomašić, J. (1982) Biochim. Biophys. Acta 701, 63–71.
111. Mollner, S. and Braun, V. (1984) Arch. Microbiol. 140, 171–177.
112. Kotani, S., Watanabe, Y., Shimono, T., Narita, T., Kato, K., Stewart-Tull, D.E.S., Kinoshita, F., Yokogawa, K., Kawata, S., Shiba, T., Kusumoto, S. and Tarumi, Y. (1975) Z. Immun.-Forsch. Bd. 149, 302–319.
113. Adam, A., Petit, J.-F., Lefrancier, P. and Lederer, E. (1981) Mol. Cell. Biochem. 41, 27–47.
114. Kamisango, K., Saiki, I., Tanio, Y., Okumura, H., Araki, Y., Sekikawa, I., Azuma, I. and Yamamura, Y. (1982) J. Biochem. 92, 23–33.
115. Krueger, J.M., Pappenheimer, J.R. and Karnovsky, M.L. (1982) Proc. Natl. Acad. Sci. USA 79, 6102–6106.
116. Rosenthal, R.S. and Krueger, J.M. (1987) Antonie van Leeuwenhock 53, 523–532.
117. Mitchell, P. and Moyle, J. (1957) J. Gen. Microbiol. 16, 184–194.
118. Young, F.E. (1966) J. Biol. Chem. 241, 3462–3467.
119. Coyette, J. and Shockman, G.D. (1973) J. Bacteriol. 114, 34–41.
120. Hinks, R.P., Daneo-Moore, L. and Shockman, G.D. (1978) J. Bacteriol. 133, 822–829.
121. Tuomanen, E., Cozens, R., Tosch, W., Zak, O. and Tomasz, A. (1986) J. Gen. Microbiol. 137, 1297–1304.
122. Hakenbeck, R. and Messer, W. (1977) J. Bacteriol. 129, 1239–1244.
123. Beck, B.D. and Park, J.T. (1976) J. Bacteriol. 126, 1250–1260.
124. Ghuysen, J.-M. and Shockman, G.D. (1973) in: L. Leive (Ed.), Bacterial Membranes and Walls, Marcel Dekker, New York, pp. 37–130.
125. Daneo-Moore, L. and Shockman, G.D. (1977) in: G. Poste and G.L. Nicolson (Eds.), The Synthesis, Assembly and Turnover of Cell Surface Components, Elsevier/North-Holland, Amsterdam, pp. 597–715.
126. Shockman, G.D. (1992) FEMS Microbiol. Lett. 100, 261–268.
127. Rogers, H.J. (1970) Bacteriol. Rev. 34, 194–214.
128. Sharon, N. and Seifter, S. (1964) J. Biol. Chem. 239, PC2398–PC2399.
129. Doyle, R.J. and Koch, A.L. (1987) CRC Crit. Rev. Microbiol. 15, 169–222.
130. Höltje, J.-V. and Schwarz, U. (1985) in: N. Nanninga (Ed.), Molecular Cytology of *Escherichia coli*, Academic Press, London, pp. 77–119.
131. Pooley, H.M. and Shockman, G.D. (1970) J. Bacteriol. 103, 457–466.
132. Pooley, H.M. (1976) J. Bacteriol. 125, 1127–1138.
133. Pooley, H.M. (1976) J. Bacteriol. 125, 1139–1147.
134. Archibald, A.R. and Coapes, H.E. (1976) J. Bacteriol. 125, 1195–1206.
135. Koch, A.L. and Doyle, R.J. (1985) J. Theor. Biol. 117, 132–157.
136. Koch, A.L. (1990) Res. Microbiol. 141, 529–541.
137. Höltje, J.-V. (1993) in: M.A. de Pedro, J.-V. Höltje and W. Löffelhardt (Eds.), Bacterial Growth and Lysis: Metabolism and Structure of the Bacterial Sacculus, FEMS Symposium, Plenum, New York, pp. 419–426.
138. Glauner, B. and Schwarz, U. (1983) in: R. Hakenbeck, J.-V. Höltje and H. Labischinski (Eds.), The Target of Penicillin, Walter de Gruyter, Berlin, pp. 29–34.
139. de Jonge, B.L.M., Chang, Y.-S., Gage, D. and Tomasz, A. (1992) J. Biol. Chem. 267, 11248–11254.
140. García-Bustos, J.F., Chait, B.T. and Tomasz, A. (1987) J. Biol. Chem. 262, 15400–15405.
141. Sinha, R.K. and Neuhaus, F.C. (1991) Antimicrob. Agents Chemother. 35, 1753–1759.
142. Pooley, H.M., Shockman, G.D., Higgins, M.L. and Porres-Juan, J. (1972) J. Bacteriol. 109, 423–431.
143. Shungu, D.L., Cornett, J.B. and Shockman, G.D. (1979) J. Bacteriol. 138, 598–608.
144. Tomasz, A. (1968) Proc. Natl. Acad. Sci. USA 59, 86–93.
145. Forsberg, C.W. and Rogers, H.J. (1971) Nature (London) 229, 272–273.
146. Fan, D.P. and Beckman, M.M. (1973) J. Bacteriol. 114, 798–803.
147. Chatterjee, A.N., Mirelman, D., Singer, H.J. and Park, J.T. (1969) J. Bacteriol. 100, 846–853.
148. Yamada, M., Hirose, A. and Matsuhashi, M. (1975) J. Bacteriol. 123, 678–686.
149. Wolf-Watz, H. and Normark, S. (1976) J. Bacteriol. 128, 580–586.
150. Fan, D.P. (1970) J. Bacteriol. 103, 494–499.

164

151. Berry, A.M., Lock, R.A., Hansmann, D. and Patton, J.C. (1989) Infect. Immun. 57, 2324–2330.
152. Normark, S., Norlander, L., Grundström, T., Bloom, G.D., Boquet, P. and Frelat, G. (1976) J. Bacteriol. 128, 401–412.
153. Goodell, E.W. and Schwarz, U. (1985) J. Bacteriol. 162, 391–397.
154. Goodell, E.W. (1985) J. Bacteriol. 163, 305–310.
155. Goodell, E.W. and Higgins, C.F. (1987) J. Bacteriol. 169, 3861–3865.
156. Park, J.T. (1993) J. Bacteriol. 175, 7–11.
157. Doyle, R.J., Chaloupka, J. and Vinter, V. (1988) Microbiol. Rev. 52, 554–567.
158. Boothby, D., Daneo-Moore, L., Higgins, M.L., Coyette, J. and Shockman, G.D. (1973) J. Biol. Chem. 248, 2161–2169.
159. Mychajlonka, M., McDowell, T.D. and Shockman, G.D. (1980) Infect. Immun. 28, 65–73.
160. Pitel, D.W. and Gilvarg, C. (1970) J. Biol. Chem. 245, 6711–6717.
161. Glaser, L. and Lindsay, B. (1977) J. Bacteriol. 130, 610–619.
162. Fiedler, F. and Glaser, L. (1973) Biochim. Biophys. Acta 300, 467–485.
163. Young, F.E. and Spizizen, J. (1963) J. Biol. Chem. 238, 3126–3130.
164. Ranhand, J.M. (1973) J. Bacteriol. 115, 607–614.
165. Ranhand, J.M., Leonard, C.G. and Cole, R.M. (1971) J. Bacteriol. 106, 257–268.
166. Sánchez-Puelles, J.M., Ronda, C., García, J.-L., López, R. and García, E. (1986) Eur. J. Biochem. 158, 289–293.
167. Fein, J.E. and Rogers, H.J. (1976) J. Bacteriol. 127, 1427–1442.
168. Sánchez-Puelles, J.M., Ronda, C., García, E., Méndez, E., García, J.L. and López, R. (1986) FEMS Microbiol. Lett. 35, 163–166.
169. Higgins, M.L. and Shockman, G.D. (1970) J. Bacteriol. 101, 643–648.
170. Burman, L.G., Reichler, J. and Park, J.T. (1983) J. Bacteriol. 156, 386–392.
171. Woldringh, C.L., Huls, P., Pas, E., Brakenhoff, G.J. and Nanninga, N. (1978) J. Gen. Microbiol. 133, 575–586.
172. de Pedro, M.A. and Schwarz, U. (1981) Proc. Natl. Acad. Sci. USA 78, 5856–5860.
173. Glauner, B. and Höltje, J.-V. (1990) J. Biol. Chem. 265, 18988–18996.
174. Pooley, H.M., Porres-Juan, J.M. and Shockman, G.D. (1970) Biochem. Biophys. Res. Commun. 38, 1134–1140.
175. Fan, D.P. (1970) J. Bacteriol. 103, 488–493.
176. Cornett, J.B., Johnson, C.A. and Shockman, G.D. (1979) J. Bacteriol. 138, 699–704.
177. Kariyama, R. and Shockman, G.D. (1993) in: M.A. de Pedro, J.-V. Höltje and W. Löffelhardt (Eds.), Bacterial Growth and Lysis: Metabolism and Structure of the Bacterial Sacculus, FEMS Symposium, Plenum, New York, pp. 229–234.
178. Kariyama, R. and Shockman, G.D. (1992) J. Bacteriol. 174, 3236–3241.
179. Walderich, B. and Höltje, J.-V. (1991) J. Bacteriol. 173, 5668–5676.
180. Hobot, J.A. and Rogers, H.J. (1991) J. Bacteriol. 173, 961–967.
181. Díaz, E., García, E., Ascaso, C., Mndez, E., López, R. and García, J.L. (1989) J. Biol. Chem. 264, 1238–1244.
182. Romeis, T. and Höltje, J.-V. (1993) in: M.A. de Pedro, J.-V. Höltje and W. Löffelhardt (Eds.), Bacterial Growth and Lysis: Metabolism and Structure of the Bacterial Sacculus, FEMS Symposium, Plenum, New York, pp. 235–240.
183. Briese, T. and Hakenbeck, R. (1985) Eur. J. Biochem. 146, 417–427.
184. Höltje, J.-V. and Keck, W. (1988) in: P. Actor, L. Daneo-Moore, M.R.J. Salton and G.D. Shockman (Eds.), Antibiotic Inhibition of Bacterial Cell Surface Assembly and Function, American Society for Microbiology, Washington, DC, pp. 181–188.
185. Young, R., Way, J., Yin, S. and Syvanen, M. (1979) J. Mol. Biol. 132, 307–322.
186. Hanych, B., Kedzierska, S., Walderich, B. and Taylor, A. (1993) in: M.A. de Pedro, J.-V. Höltje and W. Löffelhardt (Eds.), Bacterial Growth and Lysis: Metabolism and Structure of the Bacterial Sacculus, FEMS Symposium, Plenum, New York.
187. Zagotta, M.T. and Wilson, D.B. (1990) J. Bacteriol. 172, 912–921.
188. Hartmann, R., Bock-Hennig, S.B. and Schwarz, U. (1974) Eur. J. Biochem. 41, 203–208.
189. Higgins, M.L., Pooley, H.M. and Shockman, G.D. (1970) J. Bacteriol. 103, 504–512.

190. Joseph, R. and Shockman, G.D. (1974) J. Bacteriol. 118, 735–746.
191. Foley, M., Brass, J.M., Birmingham, J., Cook, W.R., Garland, P.B., Higgins, C.F. and Rothfield, L.I. (1989) Mol. Microbiol. 3, 1329–1336.
192. Joseph, R. and Shockman, G.D. (1976) J. Bacteriol. 127, 1482–1493.
193. Conover, M.J., Thompson, J.S. and Shockman, G.D. (1966) Biochem. Biophys. Res. Commun. 23, 713–719.
194. Shockman, G.D., Thompson, J.S. and Conover, M.J. (1967) Biochemistry 6, 1054–1065.
195. Pooley, H.M. and Shockman, G.D. (1969) J. Bacteriol. 100, 617–624.
196. Guidicelli, S. and Tomasz, A. (1984) J. Bacteriol. 158, 1188–1190.
197. Clarke, A.J. and Dupont, C. (1992) Can. J. Microbiol. 38, 85–91.
198. Herbold, D.R. and Glaser, L. (1975) J. Biol. Chem. 250, 1676–1682.
199. Herbold, D.R. and Glaser, L. (1975) J. Biol. Chem. 250, 7231–7238.
200. Cleveland, R.F., Wicken, A.J., Daneo-Moore, L. and Shockman, G.D. (1976) J. Bacteriol. 126, 192–197.
201. Cleveland, R.F., Daneo-Moore, L., Wicken, A.J. and Shockman, G.D. (1976) J. Bacteriol. 127, 1582–1584.
202. Cleveland, R.F., Höltje, J.-V., Wicken, A.J., Tomasz, A., Daneo-Moore, L. and Shockman, G.D. (1975) Biochem. Biophys. Res. Commun. 67, 1128–1135.
203. Höltje, J.-V. and Tomasz, A. (1975) Proc. Natl. Acad. Sci. USA 72, 1690–1694.
204. Höltje, J.-V. and Schwarz, U. (1983) in: R. Hakenbeck, J.-V. Höltje and H. Labischinski (Eds.), The Target of Penicillin, Walter de Gruyter, Berlin, pp. 185–190.
205. Kusser, W. and Ishiguro, E.E. (1985) J. Bacteriol. 164, 861–865.
206. Tuomanen, E. and Tomasz, A. (1986) J. Bacteriol. 167, 1077–1080.
207. Betzner, A.S., Ferreira, L.C.S., Höltje, J.-V. and Keck, W. (1990) FEMS Microbiol. Lett. 67, 161–164.
208. Young, K.D., Anderson, R.J. and Hafner, R.J. (1989) J. Bacteriol. 171, 4334–4341.
209. Goodell, E.W., López, R. and Tomasz, A. (1976) Proc. Natl. Acad. Sci. USA 73, 3293–3297.
210. López, R., Ronda-Lain, C., Tapia, A., Waks, S.B. and Tomasz, A. (1976) Antimicrob. Agents Chemother. 10, 697–706.
211. Jolliffe, L.K., Doyle, R.J. and Streips, U.N. (1981) Cell 25, 753–763.
212. Ingram, L.O. (1981) J. Bacteriol. 146, 331–336.
213. Leduc, M. and Heijenoort, J.V. (1980) J. Bacteriol. 142, 52–59.
214. Templin, M.F., Edwards, D.H. and Höltje, J.-V. (1992) J. Biol. Chem. 267, 20039–20043.
215. Shockman, G.D., Daneo-Moore, L., Cornett, J.B. and Mychajlonka, M. (1979) Rev. Inf. Dis. 1, 787–796.
216. Tomasz, A. (1979) Annu. Rev. Microbiol. 33, 113–137.
217. Tomasz, A., Albino, A. and Zanati, E. (1970) Nature 227, 138–140.
218. Kitano, K. and Tomasz, A. (1979) J. Bacteriol. 140, 955–963.
219. Harkness, R.E. and Ishiguro, E.E. (1983) J. Bacteriol. 155, 15–21.
220. Tomasz, A. and Waks, S. (1975) Proc. Natl. Acad. Sci. USA 72, 4162–4166.
221. Reinicke, B., Blümel, P., Labischinski, H. and Giesbrecht, P. (1985) Arch. Microbiol. 141, 309–314.
222. Giesbrecht, P., Labischinski, H. and Wecke, J. (1985) Arch. Microbiol. 141, 315–324.
223. Kohlrausch, U. and Höltje, J.-V. (1991) J. Bacteriol. 173, 3425–3431.
224. Schwarz, U., Asmus, A. and Frank, H. (1969) J. Mol. Biol. 41, 419–429.
225. Lubitz, W., Harkness, R.E. and Ishiguro, E.E. (1984) J. Bacteriol. 159, 385–387.
226. Höltje, J.-V. and Duin, J.V. (1984) in: C. Nombela (Ed.), Microbial Cell Wall Synthesis and Autolysis, Elsevier, Amsterdam, pp. 195–199.
227. Walderich, B., Ursinus-Wössner, A., van Duin, J. and Höltje, J.-V. (1988) J. Bacteriol. 170, 5027–5033.
228. Walderich, B. and Höltje, J.-V. (1989) J. Bacteriol. 171, 3331–3336.
229. Higgins, M.L. and Shockman, G.D. (1971) CRC Crit. Rev. Microbiol. 1, 29–72.
230. Dolinger, D.L., Schramm, V.L. and Shockman, G.D. (1988) Proc. Natl. Acad. Sci. USA 85, 6667–6671.
231. Barrett, J.F. and Shockman, G.D. (1984) J. Bacteriol. 159, 511–519.
232. Barrett, J.F., Schramm, V.L. and Shockman, G.D. (1984) J. Bacteriol. 159, 520–526.

166

233. Barrett, J.F., Dolinger, D.L., Schramm, V.L. and Shockman, G.D. (1984) J. Biol. Chem. 259, 11818–11827.
234. Shockman, G.D., Kawamura, T., Barrett, J.F. and Dolinger, D. (1983) in: R. Hakenbeck, J.-V. Höltje and H. Labischinski (Eds.), The Target of Penicillin, Walter de Gruyter, Berlin, pp. 165–172.
235. Shockman, G.D., Chu, C.-P., Kariyama, R., Tepper, L.K. and Daneo-Moore, L. (1993) in: M.A. de Pedro, W. Löffelhardt and J.-V. Höltje (Eds.), Bacterial Growth and Lysis: Metabolism and Structure of the Bacterial Sacculus, FEMS Symposium, Plenum, New York, pp. 213–227.
236. Ghuysen, J.-M. (1991) Annu. Rev. Microbiol. 45, 37–67.
237. Mosser, J.L. and Tomasz, A. (1970) J. Biol. Chem. 245, 287–298.
238. Howard, L.V. and Gooder, H. (1974) J. Bacteriol. 117, 796–804.
239. Höltje, J.-V. and Tomasz, A. (1976) J. Biol. Chem. 251, 4199–4207.
240. Lee, C.J. and Liu, T.Y. (1977) Int. J. Biochem. 8, 573–580.
241. Höltje, J.-V. and Tomasz, A. (1975) J. Biol. Chem. 250, 6072–6076.
242. García, P., García, J.L., García, E., and López, R. (1986) Gene 43, 265–272.
243. Ronda, C., García, J.L., García, E., Sánchez-Puelles, J.M., and López, R. (1987) Eur. J. Biochem. 164, 621–624.
244. Tomasz, A., Moreillon, P. and Pozzi, G. (1988) J. Bacteriol. 170, 5931–5934.
245. Ortiz, J.M., Gillespie, J.B. and Berkeley, R.C.W. (1972) Biochim. Biophys. Acta 289, 174–186.
246. Berkeley, R.C.W., Brewer, S.J., Ortiz, J.M and Gillespie, J.B. (1973) Biochim. Biophys. Acta 309, 157–168.
247. Del Rio, L.A., Berkeley, R.C.W., Brewer, S.J. and Roberts, S.F. (1973) FEBS Lett. 37, 7–9.
248. Munson, R.S. and Glaser, L. (1981) in: V. Ginsburg and P. Robbins (Eds.), Biology of Carbohydrates, Vol. 1, Wiley, New York, pp. 91–123.
249. Lindsay, B. and Glaser, L. (1976) J. Bacteriol. 127, 803–811.
250. Ghuysen, J.M., Dierickx, L., Leyh-Bouille, M., Strominger, J.L., Bricas, E. and Nicot, C. (1965) Biochemistry 4, 2237–2244.
251. Lopez-Diaz, I., Clarke, S. and Mandelstam, J. (1986) J. Gen. Microbiol. 132, 341–345.
252. Turner, S.M. and Mandelstam, J. (1986) J. Gen. Microbiol. 132, 3025–3035.
253. Rozeboom, H.J., Dijkstra, B.W., Engel, H. and Keck, W. (1990) J. Mol. Biol. 212, 557–559.
253a. Yamada, M., Yasuda, S., Suzuki, S. and Hirota, H. (1981) Plasmid 6, 86–98.
254. Keck, W. and Schwarz, U. (1979) J. Bacteriol. 139, 770–774.
255. Amanuma, H. and Strominger J.L. (1980) J. Biol. Chem. 255, 11173–11180.
256. Metz, R., Henning, S. and Hammes, W.P. (1986) Arch. Microbiol. 144, 175–180.
257. Ursinus, A., Steinhaus, H. and Höltje, J.-V. (1992) J. Bacteriol. 174, 441–446.
258. Tomioka, S., Nikaido, T., Miyakawa, T. and Matsuhashi, M. (1983) J. Bacteriol. 156, 463–465.

J.-M Ghuysen and R. Hakenbeck (Eds.), *Bacterial Cell Wall*
© 1994 Elsevier Science B.V. All rights reserved

CHAPTER 8

Cell wall changes during bacterial endospore formation

C.E. BUCHANAN[1], A.O. HENRIQUES[2] and P.J. PIGGOT[3]

[1]*Department of Biological Sciences, Southern Methodist University, Dallas, TX 75275 0376, USA,*
[2]*Centro de Tecnologia Química e Biológica, Apartado 127, 2780 Oeiras, Portugal and* [3]*Department of Microbiology and Immunology, Temple University School of Medicine, Philadelphia, PA 19140, USA*

1. Description of sporulation

1.1. Overview

Bacterial endospore formation is a primitive cellular differentiation that is induced by nutrient depletion. An early stage is an asymmetrically located division which yields two distinct cells that have radically different developmental fates. The smaller cell (the prespore) is engulfed by the larger cell (the mother cell) and develops into the mature spore; the mother cell ultimately lyses. The mature spore is characterized by its resistance to a variety of environmental stresses. There are several periods when wall metabolism during the developmental cycle is substantially different from vegetative wall metabolism; some of these involve new wall synthesis whereas others are degradative.

1.2. Stages of sporulation

Early microscopy studies enabled investigators to identify a series of intermediates in the morphological transition from vegetative cell to spore; these stages are now conventionally designated with Roman numerals [1] (Fig. 1), with the vegetative bacterium designated stage 0 (zero). The overall process is very similar for all species of Bacillaceae that have been studied [2]. Soon after the start of sporulation, the nucleoid rearranges to form an axial filament (stage I); the significance of this rearrangement remains unclear. Formation of the asymmetrically located spore septum is defined as stage II of sporulation. Accompanying septation, nucleoids segregate to both cell types, with the nucleoid in the prespore being much more condensed than that in the mother cell [3]. The sporulation septum contains much less visible wall material than is found in a vegetative division septum or in the cylindrical wall of the bacillus. Loss of that wall material converts the septum from a rigid to a flexible structure and permits complete engulfment of the prespore by the mother cell [4]. This event is associated with a gross change in the septal membrane as indicated by the reduction in the number of electron-dense layers visualized by electron microscopy, a change that may indicate membrane fusion [5].

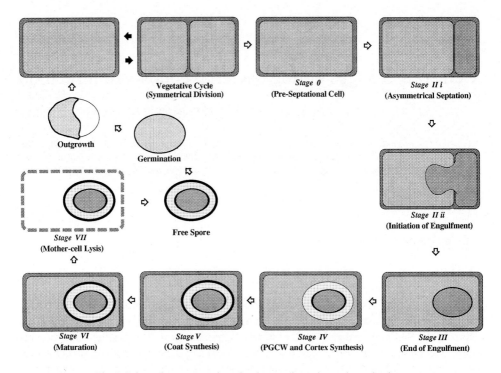

Fig. 1. Schematic representation of endospore formation and germination.

The engulfed prespore is commonly called the forespore. Soon after the forespore has been formed (stage III), it becomes surrounded by two types of peptidoglycan. There is an inner layer, the primordial germ-cell wall (PGCW), and an outer layer, the cortex. The PGCW ultimately evolves to become the wall of the germinated spore, whereas the cortex is degraded upon spore germination. In so far as they can be distinguished, the enzymes for cortex synthesis are considered to be associated with the mother cell, while the enzymes for PGCW synthesis are considered to be associated with the forespore [6,7]. Formation of these two peptidoglycans is defined as stage IV of sporulation. By this stage, the forespore is increasing in density and is becoming distinguishable by light microscopy as a refractile body within the mother cell.

Subsequent stages of sporulation include deposition of spore coat proteins around the forespore (stage V), 'maturation' of the forespore (stage VI), and release of the mature spore upon lysis of the mother cell (stage VII). Between stage III and stage VII, the volume of the forespore/spore cytoplasm decreases to less than half its original value, with the mature spore being dehydrated [8].

Germination of a dormant spore to give an actively-metabolizing, heat-sensitive cell can occur within a few minutes. Degradation of the cortex occurs in this interval [9]. The germinated spore then grows to become a vegetative cell in about 2 h by a process known as outgrowth.

2. Cell wall structures and metabolism

2.1. Peptidoglycan synthesis during sporulation

There are three periods of peptidoglycan synthesis during sporulation. These correspond to (1) synthesis of a small amount associated with spore septum formation, (2) synthesis of the PGCW which is quite similar to vegetative wall, and (3) synthesis of the cortex. The timing of the different periods of synthesis was established biochemically by measuring increased incorporation of cell wall-specific components such as diaminopimelic acid and by determining which stages of sporulation were sensitive to inhibition by penicillin. The location of peptidoglycan in the developing spore was visualized by electron microscopy, and subsequently was confirmed by fractionation and biochemical analysis of sporulating cells.

2.2. Sporulation division septum

After the last vegetative cell division, there is an interval of 1–2 h during which little or no peptidoglycan synthesis occurs [10,11].Then, if conditions are favorable for sporulation, a septum is formed at one pole of the cell. Although asymmetric septation is comparable to vegetative septation in many respects (for an excellent review, see [12]), and is at least as sensitive as vegetative division is to inhibition by penicillin [13], it involves only a very small amount of peptidoglycan synthesis. Electron micrographs reveal a thin dense layer of wall in the completed septum. It is formed as in the vegetative cells by centripetal growth from the sides of the cell coincident with invagination of the cell membrane [1,2,14–16]. The wall is not destined to become exocellular and so is not necessary for osmotic protection. The small amount of peptidoglycan in the wall may be required simply to ensure growth of the septum in the correct direction [4].

The peptidoglycan contained within the sporulation septum is thought to be chemically similar to vegetative peptidoglycan, primarily because of two observations. First, septation at stage II is not preceded by an overall increase of cell wall-synthetic enzymes or the appearance of novel enzymes, which suggests that the residual amounts of the vegetative enzymes are adequate. Second, cells at stage II of sporulation will resume vegetative growth if they are transferred to a nutritionally favorable medium. When this occurs, the completed sporulation septum acquires more crosswall material and becomes the wall that separates the two resulting vegetative cells [4,17,18].

The spore septum is short-lived. Soon after its completion (designated stage IIi by Illing and Errington [16]), the crosswall begins to 'dissolve'. The dissolution begins in the center of the septal disk and is marked by bulging of the prespore into the mother cell as the septum loses its rigidity. This has been designated stage IIii [16]. During stage IIiii, hydrolysis of the septal wall continues outwards to the sides of the cell, and the membranous septum, which is now completely flexible, begins to grow and engulf the prespore (Fig. 1).

There are several types of *spo* mutants that are blocked at different points within stage II, and these give some insight into the process of spore septum formation. The *spoIIE* and *spoIIA(P-)* mutants form the sporulation septum, but do not degrade the peptidogly-can within it; the *spoIIE* mutants ultimately make a thick spore crosswall containing as much peptidoglycan as a vegetative septum [16,19]. Further analysis of these loci may yield the key to understanding what normally limits peptidoglycan synthesis at this time in development. The *spoIIA* locus contains three genes, one of which encodes an RNA po-lymerase sigma factor, σ^F, whose activation may be crucial for the compartmentalization of gene expression [20]. Less is known about the product of the *spoIIE* locus except that it appears to be a membrane protein (Youngman, personal communication cited in [20]).

The *spoIID* mutation causes the accumulation of cells that have only a partially hydro-lyzed crosswall and thus are unable to proceed through the engulfment stage of sporula-tion (stage IIii [19,21]). The protein sequence derived from the *spoIID* gene is similar to that of the C-terminal portion of the LytB (CwbA) protein, which modifies *N*-acetylmu-ramoyl-L-alanine amidase activity in vegetative cells [22–24]. The similarity is consistent with the notion that SpoIID protein has a direct role in septal wall autolysis and is re-quired to complete hydrolysis of the peptidoglycan so that engulfment can proceed.

The question that remains is what initiates the hydrolysis of peptidoglycan in the spore septum. It should be noted that this event is unlike the splitting of the vegetative septum that occurs during separation of daughter cells. The septal wall in the sporulating cell is not split into two parts, but rather it is completely dissolved. Moreover, loss of peptidoglycan during stage II begins in the center of the septum and proceeds outwards, whereas autolysis of the vegetative septum begins on the periphery of the septum, which is in contact with the external medium, and proceeds inwards. Thus, a sporulation-specific autolytic enzyme, or a sporulation-specific autolysin activator, may be required for this process.

2.3. Germ cell wall

Immediately after engulfment, the forespore lacks peptidoglycan, and appears to be un-stable as mutants blocked at this stage tend to lyse (*spoIIIA* [25], *spoIIIE* [26], *spoIIIJ* [27]). The structure of the membrane surrounding the spore at this stage is not well re-solved; several micrographs suggest that it may have the character of a single membrane, although this remains to be established [5]. After peptidoglycan synthesis has com-menced, the forespore cytoplasm is seen to be separated from the mother-cell cytoplasm by an inner forespore membrane (ifm) and an outer forespore membrane (ofm) whose 'outside' surfaces face one another, creating an extracellular space between them [28,29].

During stage IV of sporulation, two different types of wall are laid down between the ifm and the ofm. They can be distinguished by electron microscopy, which reveals a rela-tively thin, darker layer of material apposed to the outer surface of the ifm (the PGCW) and a thicker region of less dense cortex [1,2,14,15]. Vinter [10] pulse-labelled sporulat-ing cells of *Bacillus cereus* with ^{14}C-diaminopimelic acid (DAP) and determined that the DAP-containing peptidoglycan made first during this stage did not have the same fate as that made later. The structure that was synthesized first was more stable during germina-tion of the mature spores and was retained by the cells during the outgrowth phase. It was suggested that this stable structure serves as the cell wall for the young vegetative cells (the PGCW). The DAP-containing structure that is degraded during germination corresponds to the cortex layer of the spore (see Section 2.4). Vinter's conclusions were confirmed by Pearce and Fitz-James [30], and later extended by Cleveland and Gilvarg [31] who observed that the peptidoglycan of the PGCW of *Bacillus megaterium* was more highly crosslinked than the cortical peptidoglycan, thereby providing a possible mechanism for PGCW resistance to the autolytic enzymes that are active on the cortex during germination.

The chemical structure of the PGCW closely resembles that of vegetative wall. For example, if the vegetative cell wall is sensitive to lysozyme, the PGCW from spores of that species has a similar sensitivity [7]. Also, the amino acid composition of the peptidoglycan peptides is invariably the same for PGCW as for vegetative cell wall, even when the cortical peptidoglycan has a different amino acid composition. These similarities are not surprising, as the fate of the PGCW is to serve as the primer or basement layer for vegetative wall synthesis during outgrowth of the spore. Nevertheless, the PGCW is not identical to the vegetative wall. Not only is it thinner, but also it apparently lacks teichoic acid [32–34] and in *Bacillus sphaericus* the vegetative T layer protein is absent [7].

While it is a logical prediction that the PGCW is synthesized by enzymes in the forespore and ifm, it has not been easy to verify this for most species of *Bacillus*. This is because most, if not all, of the enzymes needed for PGCW synthesis are also required for synthesis of the cortex, so they will also be present in the mother cell compartment and ofm. *B. sphaericus* is a useful exception to this generalization, because its cortex contains DAP while its vegetative and germ cell walls contain lysine. Thus, Linnett and Tipper [7,11] were able to distinguish the locations of germ wall peptidoglycan synthesis and cortical peptidoglycan synthesis in this species. They observed that synthesis of all the enzymes needed for vegetative wall peptidoglycan precursors ceased at about the start of sporulation (t_0). De novo synthesis of these enzymes resumed after completion of the septum (in the system used, this was at t_2, that is, 2 h after the start of sporulation) and continued throughout the engulfment phase. Although the pre-existing vegetative enzymes were stable and thus were still present, their specific activity had decreased as a consequence of continued synthesis of other proteins. Assays from samples taken between t_2 and t_4 determined that the vegetative enzyme Lys ligase was present in both the mother cell and the prespore/forespore, but its specific activity was significantly greater in the prespore/forespore. This suggests that the de novo synthesis was being directed from the forespore genome. At about t_4 when synthesis of Lys ligase stopped, synthesis of cortex-specific DAP ligase commenced exclusively in the mother cell. Meanwhile, enzymes that were needed for both types of peptidoglycan had similar specific activities in both forespore and mother cell.

The cell wall biosynthetic enzymes are generally quite stable, and it is likely that some of the enzymes synthesized in the forespore are retained by the mature spore in an inactive form. Later they may be activated and used during the early stages of germination.

2.4. Cortex

Cortex formation is the major event to occur during the morphological stage IV (Fig. 1). Cortex is laid down between the germ cell wall and the ofm. The cortex is composed of peptidoglycan with features that distinguish it from the vegetative form. It is a very highly conserved structure among the various species of *Bacillus*; that is, despite the differences that may exist among their vegetative peptidoglycan structures, all bacilli appear to have the same chemotype of cortex [35] (Fig. 2). In essence, it is composed of peptidoglycan of chemotype I [37] that contains *meso*-DAP and direct interpeptide linkages between D-Ala of one peptide and the D-amino group of *meso*-DAP on another peptide. This basic structure is also typical of the vegetative peptidoglycan of *Bacillus subtilis* and closely related species. In contrast, the vegetative peptidoglycan of *B. sphaericus* is chemotype II, which contains L-Lys instead of DAP and has isoasparaginyl interpeptide bridges [38].

It is likely that the UDP-*N*-acetylmuramyl (MurNAc) pentapeptide precursor of the cortex is synthesized in the mother cell cytoplasm by vegetative enzymes. This is supported by the fact that the only new enzyme required for its synthesis by *B. sphaericus* is DAP ligase, and this enzyme is only detectable in the mother cell compartment [11]. Synthesis of DAP ligase is coordinated with an increase in the cytoplasmic activities of the vegetative cell wall biosynthetic enzymes, except for Lys ligase. Evidently the same D-Ala-D-Ala ligase recognizes both the vegetative and DAP-containing precursors in this species.

The cortical peptidoglycan is subject to sporulation-specific modifications either during or after its assembly on the surface of the ofm. In the mature cortex of *B. subtilis* and other species of *Bacillus* almost half of the muramic acid residues are present as the lactam rather than being N-acetylated, and thus they cannot be linked to a peptide. The *N*-acetylglucosamine-muramoylactam units alternate regularly with the *N*-acetylglucosamine-MurNAc disaccharides, which suggests that this unique structure may serve some purpose (Fig. 2) [36,39]. Although removal of the peptide from muramic acid may be done with the cortex-associated enzyme MurNAc-L-alanine amidase [40], the enzymes responsible for the subsequent synthesis of the muramoylactam structure have not been identified, nor do we know how this modification is so tightly regulated.

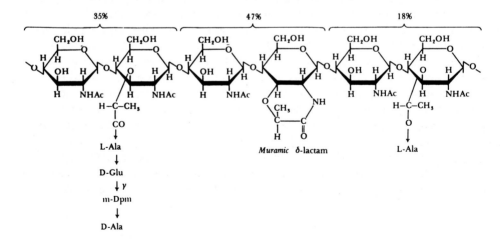

Fig. 2. Structure of peptidoglycan found in the cortex of *B. subtilis* spores [36]. Approximately 19% of the diaminopimelic acid residues (Dpm) are linked by their ε-amino group to the carboxyl group of D-alanine on a nearby peptide chain. Reprinted with permission from ref. [35].

No more than 35% of the muramic acid residues in the cortex of *B. subtilis* spores are substituted with a peptide, and about 18% are linked only to L-alanine (Fig. 2) [36]. In contrast, all of the muramic acid residues in vegetative peptidoglycan are linked to a peptide [41]. It is clear that having fewer muramylpeptides per disaccharide reduces the number of crosslinks that are possible in the cortex compared with the vegetative structure. The degree of crosslinkage in the cortex is apparently even less than would be predicted from this structure. Warth and Strominger [36,40] measured the amount of DAP in vegetative and spore walls that had a free amino group available for dinitrophenylation. In vegetative cells, 35% of the DAP residues were blocked by crosslinking, but only 19% in the cortex [36,41]. The peptide crosslinking in vegetative *B. subtilis* (35%) is somewhat lower than has been reported for other species of *Bacillus* [31,41], whereas the value for crosslinking in its cortex (19%) is quite similar to that reported for *B. sphaericus* (20%) and *B. megaterium* (17%) [31,42]. Marquis and Bender have challenged these results and have reported high crosslinking in the cortex [43]; however recently Popham and Setlow [44] have been unable to repeat Marquis and Bender's results and have obtained data similar to Warth and Strominger's [36] (see also Section 2.5).

There are other structural differences between vegetative and cortical wall. Tripeptides are very common in vegetative wall peptidoglycan [41], but there is a greater proportion of tetrapeptides than tripeptides in cortical peptidoglycan of both *B. subtilis* [36] and *B. sphaericus* [42]. Teichoic acid is not detected in the cortex [36]. Finally, the D-carboxyl group of *meso*-DAP is amidated in vegetative wall peptidoglycan of *B. subtilis*, but not in the spore peptidoglycan [39]. This feature, when combined with the presence of L-alanine residues that have a free carboxyl group and the greater proportion of peptide chains not involved in crosslinking, accounts for the fact that the cortical peptidoglycan is more negatively charged than vegetative peptidoglycan [6]. It has been suggested that a polyanionic structure may be essential for the function of cortex (see Section 2.5).

There are still many unresolved questions about cortical peptidoglycan synthesis. For example, it is not known how the muramic lactam and the MurNAc residues substituted only with L-alanine are made. It is assumed that they are both derived from MurNAc-pentapeptide, but the enzymes needed to modify this precursor have not all been identified. Tipper and Gauthier [6] proposed two routes by which lactams could be formed, but one path required an unidentified transacylase and the other required a novel transpeptidase.

The removal of four of the amino acids from the pentapeptide chain to leave just L-alanine is believed to occur by a stepwise process [45]. First D-Ala-D-Ala-carboxypeptidase removes the terminal D-alanine, and then D-Glu-(L)-*meso*-DAP endopeptidase I releases the dipeptide (L)-*meso*-DAP-D-Ala. Both of these activities have been detected in sporulating cells of *B. sphaericus* and *B. subtilis* [40,42]. Moreover, they both appear to be membrane bound, which would be consistent with hydrolytic modification of the lipid-linked peptidoglycan precursors. Further evidence for this pathway is that an increasing amount of the dipeptide product is produced during sporulation [42]. However, the other product (MurNAc-L-Ala-D-Glu) has not been detected. This suggests that there is very rapid removal of the D-glutamate residue by an L,D-carboxypeptidase. Although there is an L,D-carboxypeptidase that increases during sporulation of *B. sphaericus*, it only hydrolyses the L-Lys-D-Ala linkage in vegetative cell walls, and it is most active in the forespore where it is probably needed to generate tripeptides in the germ cell wall [45]. It does not hydrolyze the L-Ala-D-Glu linkage of the cortex. Thus, another enzyme remains to be identified.

Among the sporulation enzymes whose specific roles in cell wall metabolism are uncertain are an L,D-dipeptidase and D-Glu-L-diamino acid endopeptidase II. The dipeptidase appears late in sporulation of *B. sphaericus* and is active on either L-Lys-D-Ala or (L)-*meso*-DAP-D-Ala [45]. It is located exclusively in the forespore cytoplasm. It has been suggested that this is a salvage enzyme, but that role does not explain why the enzyme is only present in the forespore, what directs the dipeptides to the forespore rather than the mother cell, or what use the forespore could have for the free DAP. In contrast, the other enzyme, D-Glu-L-diamino acid endopeptidase II, is located exclusively in the cytoplasm of the mother cell and it first appears early in sporulation [46–48]. Unlike endopeptidase I described above, endopeptidase II is less specific for the diamino acid in its peptide substrate, but it requires a peptide with an unsubstituted N-terminal L-Ala. Thus, its action can only follow that of MurNAc-L-alanine amidase. Since amidase is a wall-bound enzyme [40–50], the peptides generated by its action must return to the interior of the mother cell for digestion by endopeptidase II. It is not clear if this particular endopeptidase serves a special role in the sporulating cell or if its appearance is simply associated with stationary-phase starvation conditions.

2.5. Relationship of cortex content and structure to heat resistance

The essential role of cortex in the heat resistance of spores is illustrated by the phenotypes of three abnormal spore variants: spores that developed in the presence of penicillin, spores deficient in cortex, and spores lacking cortex [51]. Penicillin has profound effects on the developing spore. If it is added to the culture when cortex is being synthesized, DAP continues to be incorporated but there is less cortex made and it appears distorted. In addition, the spores do not become fully refractile, and they tend to lyse when liberated from the mother cell [2,52,53]. Those spores that survive the penicillin treatment and do not lyse are heat sensitive [53]. The heat sensitivity and tendency to lyse are also typical of cortexless mutants [30,54].

Wickus et al. [55] observed that the increase in muramic lactam content in sporulating samples of either *B. cereus* or *B. megaterium* paralleled the increase in number of heat-resistant spores formed. Imae and Strominger [54,56] extended this finding to *B. sphaericus*. They also manipulated the *meso*-DAP concentration in the medium of a DAP-requiring mutant of *B. sphaericus* to generate mutant spores with varying amounts of cortex. They concluded that maximum heat resistance required over 90% of the maximum cortex content, whereas no more than 20% of the normal amount of cortex is required for spores to become refractile and to accumulate dipicolinic acid.

There are two features of the cortex structure that are believed to be relevant to its involvement in heat resistance. It has a high net negative charge and it is loosely crosslinked. The relatively small amount of crosslinking should permit the cortex to expand or contract, depending on the concentration and type of cations present [35]. This flexibility is considered to be crucial to the heat resistance of spores, although the actual mechanism remains controversial [35,57–59].

The view of the cortex as a loosely crosslinked, negatively charged structure has been challenged by Marquis and Bender [43], who have suggested that the cell walls used in the earlier studies were badly damaged during sonication or shaking with glass beads. In contrast, their samples of cortical peptidoglycan prepared by chemical extraction of decoated spores appeared to be tightly crosslinked (>90%), because very few free amino groups could be detected. This type of structure would be consistent with the observation that intact cortex has very low conductivity, and therefore does not have as many charged groups as previously believed [60]. It was postulated that a compact, uncharged, and rigid cortex would help maintain the dehydrated state of the core by resisting the osmotic swelling of the spore protoplast [43]. However, a recent attempt to confirm the tight crosslinking in chemically extracted spores did not do so, but rather gave results that were consistent with a more loosely crosslinked structure [44] as previously proposed by Warth and Strominger [36,41].

2.6. Roles for the vegetative penicillin-binding proteins (PBPs) in spore wall metabolism

There are six PBPs (numbered 1, 2A, 2B, 3, 4, 5) that can be routinely observed in membranes from vegetative cells of B. subtilis [61,62]. The PBPs are designated by number, in descending order of their molecular weights, and thus PBPs with the same number from different species are not necessarily similar to one another. Despite intensive biochemical analyses, there is still very little known about their catalytic activities [63,64].

The amounts of PBP1, 2A, 4 and 5 (as indicated by penicillin binding) decrease substantially soon after the end of exponential growth in both sporulating and non-sporulating cells [61,65]. It is tempting to suggest that the residual amount of these proteins is sufficient for whatever role they might have in spore septum formation, PGCW, or cortex synthesis, but the fact that these events still occur in null mutants of PBP1, 4, or 5 argues against any essential role for those PBPs during sporulation [66,67].

It has been proposed that PBP2A may be required specifically for synthesis of the sidewalls of the cell during vegetative growth [68], and hence it may be the functional equivalent of the shape-maintaining PBP2 of Escherichia coli. Two observations are consistent with this proposal. First, the membrane-bound amount of PBP2A rapidly drops to less than 20% of its vegetative level (as assayed by penicillin binding) when cells enter stationary phase where length extension is greatly reduced [61,62,65]. Second, de novo synthesis of PBP2A resumes early in germination at a time when one would predict that the biosynthetic enzymes for cell elongation are being made [68]. Unfortunately no useful mutants of PBP2A exist and the structural gene has not yet been cloned.

There are several reasons to suggest that PBP2B is probably required for both vegetative and spore-specific septum formation. First, there is enhanced synthesis of PBP2B during stage II of sporulation but not in non-sporulating stationary-phase cells [61]. Second, PBP2B is not detectable in the mature spore [69], but its synthesis is restored shortly before the first vegetative cell division following germination (i.e. substantially later than synthesis of PBP2A) [68]. Thus, PBP2B from B. subtilis may play a role in cell division that is similar to that of PBP3 from E. coli [70]. The genes for both of these PBPs have been cloned and sequenced [71] (Yanouri and Buchanan, unpublished results). Their derived amino acid sequences are 28% identical over a length of 630 residues. The

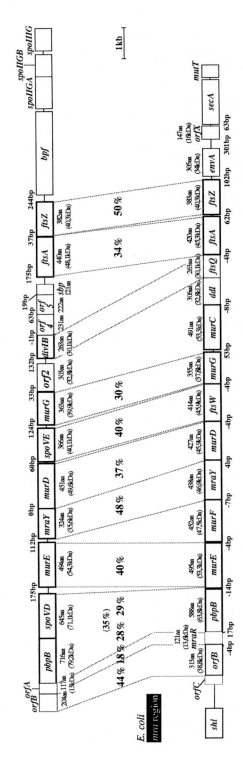

Fig. 3. Comparison of the 133° region of the *B. subtilis* genetic map with the *mra* region of the *E. coli* genetic map. Similar pairs of genes are joined by dotted lines and similarities in predicted amino acid sequences are indicated as % identity. The gaps between adjacent coding regions are indicated in base pairs (bp).

two genes are located in homologous regions of the chromosomes (Fig. 3). For these reasons the gene for PBP2B has been given the same name as its *E. coli* counterpart *pbpB*.

There is enhanced synthesis of PBP3 during stage III, which is the period when the enzymes needed for cortical peptidoglycan synthesis are made [62]. In addition, this PBP is rapidly restored to its vegetative level during the outgrowth phase of germination [68]. The timing of its expression suggests that PBP3 might fulfill a function that is common to both vegetative and spore wall synthesis. No useful mutants of this PBP exist and its gene has not yet been identified

2.7. Sporulation-specific PBPs

Two PBPs are synthesized exclusively during sporulation [61,62]. PBP4* appears earlier than PBP5* and also reaches its maximum amount earlier (in [61,62] they are named 4a and 5a). Although PBP4* is never detected in vegetative cells or in non-sporulating mutants, it is not clear if it is actually required for sporulation. This uncertainty stems from the fact that there is strain-specific variation in the amount of PBP4* produced during sporulation, some sporulating strains apparently do not produce any at all, and that inactivation of its structural gene has no obvious phenotypic effect [61] (Popham, personal communication). The gene for PBP4* has recently been isolated and sequenced (Popham, personal communication). Interestingly, its sequence is most homologous to a newly discovered D-aminopeptidase from an unrelated soil organism [72]. What role such an enzyme would serve in spore wall metabolism is unknown.

The appearance of sporulation-specific PBP5* occurs at roughly the same time as the enhanced synthesis of vegetative PBP3. However, in contrast to PBP3, which is detectable in both the ifm and ofm of the mature spore, PBP5* appears to be located exclusively in the ofm [69]. This is consistent with its likely role in cortex synthesis and with the observation that the PBP is synthesized only in the mother cell. The *dacB* gene that encodes PBP5* has been cloned and sequenced. The transcription control of *dacB* by σ^E is consistent with the specific location of PBP5* in the mother cell [73] (Buchanan and Hancock, unpublished results).

The amino acid sequence of PBP5* is 30–35% identical to that of known D,D-carboxypeptidases from a variety of organisms [73], and a partially purified sample of the protein had a weak D,D-carboxypeptidase activity [74]. If this is indeed its function in vivo, it raises the question of why the sporulating cell should need a second D,D-carboxypeptidase when there is still an abundance of the more active vegetative PBP5 present. Apparent redundancy of D,D-carboxypeptidases is also found in *E. coli* [75–78]. A null mutant of PBP5* grew normally, but formed spores that were very heat-sensitive [79], which suggests that PBP5* is needed for normal cortex synthesis. Thus, there is a requirement specifically for PBP5* during sporulation, which PBP5 cannot fulfill. PBP5* is apparently anchored in the ofm by a carboxy-terminal amphiphilic helix, which is similar to the membrane anchor of other low molecular weight PBPs [73]. Its amino-terminal, active-site domain extends out from the membrane into the cortical space between the ofm and ifm, where it can act upon nascent cortical peptidoglycan. This cortical space is clearly not comparable to the extracellular surface where the vegetative PBPs normally are found. The pH, ion concentration, and even substrate availability may be quite different in the two locations [59]. These differences, which are ill-defined at present, may ultimately explain the requirement for a new carboxypeptidase enzyme for cortex synthesis, but they leave open the question of how vegetative PBP3 can apparently be active in both environments.

The product of *dacF* is believed to be a third D,D-carboxypeptidase, because its derived amino acid sequence is 34% identical to that of PBP5*, and over 30% identical to that of vegetative PBP5 as well as the D,D-carboxypeptidases from *E. coli* [73,80]. The DacF protein has not yet been detected in cell extracts by the penicillin-binding assay. The *dacF* gene is expressed as part of a long transcript that also includes the *spoIIA* operon [80] and is expressed specifically in the prespore/forespore [80] (Schuch and Piggot, unpublished observations). *dacF* is not part of an earlier period of *spoIIA* transcription which occurs before formation of the spore septum (see Section 3.2). It is not yet known what the function of the DacF protein is likely to be, as null mutants of *dacF* form normal heat-resistant spores that are also capable of normal germination [80]. Possibly there is redundancy of function with DacA.

Immediately downstream from *pbpB* is located the *spoVD* gene, which is expressed only during sporulation. The derived amino acid sequence for SpoVD is 35% identical to the sequence for PBP2B over a length of 675 residues, and it is 29% identical to that for PBP3 of *E. coli* (Daniel et al., unpublished results). It has recently been demonstrated that the product of the *spoVD* gene cloned in *E. coli* binds to penicillin (Buchanan, unpublished observations), but the SpoVD protein has not yet been identified in membranes of *B. subtilis* by the penicillin-binding assay. *spoVD* mutants are defective in cortex [19], so it is likely that this protein is indeed a penicillin-sensitive cell wall enzyme specifically required during sporulation.

Located in the same cluster as *spoVD* is the *spoVE* locus (Fig. 3). The protein encoded by *spoVE* is similar to that encoded by *ftsW* which occupies the corresponding map position in *E. coli* (Fig. 3), and to the protein encoded by *rodA* of *E. coli* [81,82]. The three proteins have very similar hydropathy profiles with the potential for the formation of at least nine transmembrane segments [81,82], and the likely membrane association was verified in the case of the RodA protein [83–85]. They are not themselves PBPs, but are thought to interact with PBPs. In *E. coli*, the RodA protein is known to interact with PBP2 and both are associated with elongation [86]; it is suggested that FtsW interacts with PBP3 and that the pair are associated with septation [87,88]. It is tempting to speculate that there is a similar interaction between SpoVE and SpoVD. Mutations in *spoVE* and in *spoVD* result in a defect in cortex synthesis [19]. Moreover, both genes are expressed in the mother cell.

2.8. Autolytic activity during spore formation and germination

From the initiation of sporulation to the germination of the spores, there are a series of changes in autolytic activity. Moreover, several novel lytic activities appear during the developmental cycle, of which the cortex-lysing enzyme isolated from spores by Strange and Dark [89] is an early example. The changes in autolysis often appear to be crucial to further development, although this cannot yet be said for the changes that have been observed at the start of sporulation [90]. Shockman and Höltje (Chapter 7) give an authoritative review of the autolysins present during exponential growth of sporeforming (and other) bacteria.

The role of autolysins during the transition from Stage II to Stage III is considered in Section 2.2. The gross reorganization of the septal membrane which results from wall autolysis is speculated to be a major factor in the compartmentalization of gene expression, as well as in the subsequent interaction between the two cell types [5]. The role of lytic enzymes in cortex synthesis is discussed in Section 2.4. Lytic enzymes are also involved in mother cell lysis, and in cortex breakdown during germination. Foster [91] has recently re-examined autolysins from sporulating *B. subtilis*. Several distinct activities against cortex and/or vegetative cell wall were detected; at present it is not clear how these relate to the enzymes of Guinand et al. [41] (see Section 2.4).

Activation of a cortex autolysin has long been considered to be an important early germination event [41,92,93]. A variety of cortex-lytic enzymes have been extracted from germinating spores, and from dormant spores (reviewed in [9]). The toughness of dormant spores has frustrated attempts to establish which, if any, of these enzyme activities is the prime cause of cortex breakdown, and to establish how early cortex autolysis is in the sequence of germination events [9].

Spores do not contain wall teichoic acid [32–34]. Teichoicase activity is detected during sporulation [94–96] and might be responsible for the lack of wall teichoic acid in spores. It is not known if the teichoicase is regulated in concert with any of the peptidoglycan-lytic enzymes, but it seems reasonable to mention it in the context of the latter activities.

3. Genes associated with peptidoglycan metabolism during sporulation

3.1. Morphogenes and morphogene clusters required for sporulation

A genetic map of loci involved, directly or indirectly, in wall metabolism during sporulation is shown in Fig. 4, and the loci are described in Table I. The *spo* loci, the *ger* loci, *dacB*, and *dacF* are only transcribed during sporulation. Many of the *spo* loci do not code for proteins directly involved in wall metabolism but rather for transcription regulators of wall-metabolism genes. The list also includes loci for wall metabolism during vegetative growth. These loci are, or are thought to be, required also for spore formation.

A large group of 'morphogenes' are clustered in the region centered at 133°, and were identified on the basis of similarities of their respective products to proteins encoded by genes located within a similar cluster in the *mra* region of the *E. coli* genetic map [127]. The two regions are compared in Fig. 3. Interestingly, the *B. subtilis* 133° cluster does not contain the equivalents of the *E. coli ddl*, *murC* and *murF* genes, but does contain two sporulation-specific loci, *spoVD* and *spoVE*.

3.2. Regulation of gene expression and the coupling of transcription activation with the course of morphogenesis

The later stages of spore formation rely on the cooperation between the two cell types that are formed as a result of the sporulation division. The cell types, the prespore/forespore and the mother cell, differ in developmental fates as well as in size [19,128]. There is now a large body of evidence that distinct, but not totally independent, programs of gene expression are activated in each of them soon after septation [129–133]. Temporally regulated and compartmentalized gene expression during sporulation is in part a consequence of the existence of distinct RNA polymerase sigma factors which are activated in one or the other of the cell types [20]. The first sigma factors to show cell specificity are σ^F and σ^E. The sporulation genes that encode them, *spoIIAC* and *spoIIGB*, respectively, are transcribed and translated prior to septation [107,134,135]. However, the activities of σ^F and σ^E have only been detected after septation, and have been found to be segregated to the prespore and to the mother cell, respectively (reviewed in [20]). σ^E is required for the mother cell-specific transcription of the peptidoglycan-metabolism genes *dacB* and *spoIID* [73,120,132], and possibly for *spoVD* and *spoVE* [112] (Daniel et al., unpublished results). σ^F is required for the prespore/forespore-specific transcription of *dacF* (Schuch and Piggot, unpublished observations).

The immediate product of *spoIIGB* is not σ^E, but rather an inactive precursor pro-σ^E. The formation of the spore septum somehow triggers processing of pro-σ^E to its mature form [122,136,137]. Mutations in *spoIIA*, and *spoIIE* that block sporulation at stage IIi

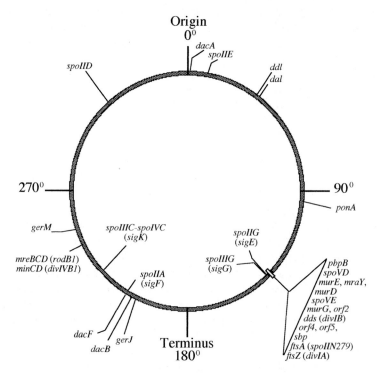

Fig. 4. Location of wall-associated genetic loci on the genetic map of *B. subtilis* 168 [97]. Alternative names for loci are indicated in parentheses. The origin and terminus of chromosome replication are also indicated.

TABLE I

Genetic loci associated with wall metabolism during development of *B. subtilis*

Locus	Map position (°)	Comments	Refs.
dacA	0	PBP5 (45.5 kDa); vegetative D-Ala carboxypeptidase	[61,74,79,98–101]
dacB	208	PBP5* (40.1 kDa); cortex synthesis (D-Ala carboxypeptidase)	[61,62,65,69,73,79,102, 103]
dacF	211	DacF protein similar to PBP 5 and PBP5*, but has not been detected by penicillin-binding assay. Gene co-transcribed with *spoIIA* late in sporulation	[80]
dal	38	Alanine racemase	[104]
ddl	36	D-Ala-D-Ala synthetase	[104]
dds	134	Deficient in division and sporulation; also called *div1B*	[105,106]
ftsA	134	Septum formation during growth and sporulation	[105,107,108]
ftsZ	134	Septum formation during growth and sporulation	[105,107,108]
gerJ	206	Defective germination; mutants do not complete cortex hydrolysis during germination; spore resistance properties develop later than normal	[109]
gerM	251	Defective germination; mutants do not complete cortex hydrolysis during germination. Possibly encodes a lipoprotein	[109]
minC,D	242	Mutations cause minicell production	[110,111]
mraY	133	Encodes the UDP-N-AcMur (pentapep-tide): undecaprenyl-phosphate phospho-N-AcMur (pentapeptide) transferase (involved in the first step of the lipid cycle reactions).	(Daniel et al., unpublished)
mreB,C,D	242	Three ORFs similar to the shape-determining *mre* genes of *E. coli*; the *divIVB1* mutation maps in this region	[110,111]
murD	134	Product similar to the D-glutamate adding enzymes of *E. coli*.	[112] (Daniel et al., unpublisjed)
murE	133	Product similar to the diaminopimelic acid adding enzymes of *E. coli*.	(Daniel et al., unpublished)
murG	134	Encodes the UDP-N-AcGlc:NAcMur (pentapeptide) pyrophosphoryl-undecaprenol N-AcGlc transferase (involved in the formation of lipid intermediate II from lipid intermedi-ate I and UDP-GlcNAc).	[112,113]

TABLE I (*continued*)

Locus	Map position (°)	Comments	Refs.
(pbpA)[a]		PBP2A (78.8 kDa); cell elongation.	[61,65,68]
pbpB	133	PBP2B (77.6 kDa); septum formation during growth and sporulation; PBP2B is similar in sequence to PBP3 of E. coli	[61,65,68] (Buchanan, unpublished observations)
(pbpC)	NM	PBP3 (70.7 kDa); cortex and vegetative wall synthesis	[61,68,69]
(pbpD)	NM	PBP4 (67.0 kDa); vegetative role as backup enzyme?	[61,66,68]
pbpE	296	PBP4* (51.4 kDa); unknown sporulation function	[61,65] (Popham, personal communication)
ponA	93	Unidentified protein with 36% identity to PBP1A of E. coli	[114]
(pon)		PBP1 (100.5 kDa) vegetative transglycosylase/transpeptidase; gene is not the same as ponA	[63,67,115,116] (Popham, personal communication)
spoIIA	211	Tricistronic operon; spoIIAC (also called sigF) encodes σ^F; spoIIAA and spoIIAB encode proteins that regulate σ^F activity	[117–119]
spoIID	316	Product has similar sequence to modifier of amidase	[22,120]
spoIIE	10	Mutants form thick septa	[19,121]
spoIIG	135	Dicistronic operon; spoIIGB (also called sigE) encodes pro-σ^E; spoIIGA thought to encode the protease that activates pro-σ^E	[122–125]
spoVB	239	Mutants defective in cortex formation	[19,126]
spoVD	133	Mutan's have altered cortex; protein product similar in sequence to PBP2B	[19,125] (Daniel et al., unpublished)
spoVE	134	Mutants defective in cortex formation; protein product similar in sequence to proteins FtsW and RodA of E. coli	[81,82,112]

[a]Parentheses indicate that the locus has not yet been identified.

prevent pro-σ^E processing [16,138]. Thus, septation per se is insufficient to trigger processing. Higgins and Piggot [5] have suggested that the gross septal membrane reorganization (membrane fusion?) that occurs subsequent to septal wall autolysis may be the trigger for processing. It is at this period of membrane reorganization that prespores lose their ability to return directly to vegetative growth upon the addition of nutrients, and become committed to develop into mature spores.

A group of genes expressed in the mother cell after engulfment is under the control of another sigma factor, σ^K. The *sigK* gene, encoding σ^K, is generated by a mother cell-specific rearrangement that joins in phase the *spoIVCB* and the *spoIIIC* loci [139–141]. *sigK* and the gene required for its formation require σ^E for their transcription [142]. The morphology of *sigK* mutants [19] indicates that σ^K is not needed for the formation of PGCW, but is required for cortex and coat synthesis. σ^K is formed from an inactive precursor, pro-σ^K, whose activation requires the forespore-specific σ^G [20]. Since σ^G itself does not become active until after completion of engulfment [143–146], it follows that σ^K-dependent synthesis of cortex and coat is also delayed until then. Candidates for a direct involvement in cortical peptidoglycan synthesis include the *spoVB*, *spoVD*, *spoVE*, *gerJ* and *gerM* loci, as mutations in these loci impair cortex formation. There are experimental grounds for thinking that σ^K may be required for transcription of these various loci [73,109,126] (Henriques and Piggot, unpublished results; Daniel et al., unpublished results). However, because of the considerable overlap in promoter specificity between σ^E and σ^K [142,147–149], it is not always clear if one or other or both of these sigma factors functions in vivo, so that this suggested mechanism for coupling cortex synthesis to the completion of engulfment is by no means established.

Acknowledgments

We would like to thank J. Errington, D. Karamata, Y. Kobayashi, D. Popham and T. Sato for the communication of information prior to publication, and A. I. Estrela for helpful discussions. A.O. Henriques was the recipient of a fellowship from Junta Nacional de Investigação Científica e Tecnolgica (J.N.I.C.T.). The work in P.J.P.'s laboratory was supported by grant GM43577, and in C.E.B.'s laboratory by grant GM43564, both from the National Institutes of Health.

References

1. Ryter, A. (1965) Ann. Inst. Pasteur, Paris 108, 40–60.
2. Fitz-James, P. and Young, E. (1969) in: G.W. Gould and A. Hurst (Eds.), The Bacterial Spore, Academic Press, New York, pp. 39–72.
3. Setlow, B., Magill, N., Febbroriello, P., Nakhimovsky, L., Koppel, D.E. and Setlow, P. (1991) J. Bacteriol. 173, 6270-6278
4. Freese, E. (1972) Curr. Top. Dev. Biol. 7, 85–124.
5. Higgins, M.L. and Piggot, P.J. (1992) Mol. Microbiol. 6, 2565–2571.
6. Tipper, D.J. and Gauthier, J.J. (1972) in: H.O. Halvorson, R. Hanson and L.L. Campbell (Eds.), Spores V, American Society for Microbiology, Washington, DC, pp. 3–12.
7. Tipper, D.J. and Linnett, P.E. (1976) J. Bacteriol. 126, 213–221.
8. Murrell, W.G. (1981) in: H.S. Levinson, A.L. Sonenshein and D.J. Tipper (Eds.), Sporulation and Germination, American Society for Microbiology, Washington, DC, pp. 64–77.
9. Foster, S.J. and Johnstone, K. (1989) in: I. Smith, R.A. Slepecky and P. Setlow (Eds.), Regulation of Procaryotic Development, American Society for Microbiology, Washington, DC, pp. 89–108.
10. Vinter, V. (1965) Folia Microbiol. 10, 280–287.
11. Linnett, P.E. and Tipper, D.J. (1974) J. Bacteriol. 120, 342–354.
12. Hitchins, A.D. (1978) Can. J. Microbiol. 24, 1103–1134.
13. Hitchins, A.D. and Slepecky, R.A. (1969) J. Bacteriol. 97, 1513–1515.
14. Walker, P.D. (1969) J. Appl. Bacteriol. 32, 463–467.
15. Holt, S.C., Gauthier, J.J. and Tipper, D.J. (1975) J. Bacteriol. 122, 1322–1338.
16. Illing, N. and Errington, J. (1991) J. Bacteriol. 173, 3159–3169.
17. Fitz-James, P.C. (1963) Colloq. Int. Centre Natl. Rech. Sci. (Paris) 124, 529–544.

184

18. Magill, N.G. and Setlow, P. (1992) J. Bacteriol. 174, 8148–8151.
19. Piggot, P.J. and Coote, J.G. (1976) Bacteriol. Rev. 40, 908–962.
20. Losick, R. and Stragier, P. (1992) Nature 355, 601–604.
21. Coote, J.G. (1972) J. Gen. Microbiol. 71, 1–15.
22. Lopez-Diaz, I., Clarke, S. and Mandelstam, J. (1986) J. Gen. Microbiol. 132, 341–354.
23. Kuroda, A., Rashid, M.H. and Sekiguchi, J. (1992) J. Gen. Microbiol. 138, 1067–1076.
24. Lazarevic, V., Margot, P., Doldo, B. and Karamata, D. (1992) J. Gen. Microbiol. 138, 1949–1961.
25. Waites, W.M., Kay, D., Dawes, I.W., Wood, D.A., Warren, S.C. and Mandelstam, J. (1970) Biochem. J. 118, 667–676.
26. Stragier, P. (1989) in: I. Smith, R.A. Slepecky and P. Setlow (Eds.), Regulation of Procaryotic Development, American Society for Microbiology, Washington, DC, pp. 243–254.
27. Errington, J., Appleby, L., Daniel, R.A., Goodfellow, H., Partridge, S.R. and Yudkin, M.D. (1992) J. Gen. Microbiol. 138, 2609–2618.
28. Wilkinson, B.J., Deans, J.A. and Ellar, D.J. (1975) Biochem. J. 152, 561–569.
29. Ellar, D.J. (1978) Symp. Soc. Gen. Microbiol. 28, 295–325.
30. Pearce, S.M. and Fitz-James, P.C. (1971) J. Bacteriol. 105, 339–348.
31. Cleveland, E.F. and Gilvarg, C. (1975) in: P. Gerhardt, R.N. Costilow and H.L. Sadoff (Eds.), Spores VI, American Society for Microbiology, Washington, DC, pp. 458–464.
32. Boylen, C.W. and Ensign, J.C. (1968) J. Bacteriol. 96, 421–427.
33. Chin, T., Younger, J. and Glaser, L. (1968) J. Bacteriol. 95, 2044–2050.
34. Johnstone, K., Simion, F.A. and Ellar, D.J. (1982) Biochem. J. 202, 459–467.
35. Warth, A.D. (1978) Adv. Microb. Physiol. 17, 1–45.
36. Warth, A.D. and Strominger, J.L. (1972) Biochemistry 11, 1389–1396.
37. Ghuysen, J.-M. (1968) Bacteriol. Rev. 32, 425–464.
38. Hungerer, K.D. and Tipper, D.J. (1969) Biochemistry 8, 3577–3587.
39. Warth, A.D. and Strominger, J.L. (1969) Proc. Natl. Acad. Sci. USA 64, 528–535.
40. Warth, A.D. and Strominger, J.L. (1971) Biochem. 10, 4349–4358.
41. Guinand, M., Michel, G. and Balassa, G. (1976) Biochem. Biophys. Res. Commun. 68, 1287–1293.
42. Guinand, M., Michel, G. and Tipper, D.J. (1974) J. Bacteriol. 120, 173–184.
43. Marquis, R.E. and Bender, G.R. (1990) Can. J. Microbiol. 36, 426–429.
44. Popham, D.L. and Setlow, P. (1993) J. Bacteriol., 75, 2767–2769.
45. Guinand, M., Vacheron, M.J., Michel, G. and Tipper, D.J. (1979) J. Bacteriol. 138, 126–132.
46. Hourdou, M.-L., Duez, C., Joris, B., Vacheron, M.-J., Guinand, M., Michel, G. and Ghuysen, J.-M. (1992) FEMS Microbiol. Lett. 91, 165–170.
47. Vacheron, M.-J., Guinand, M., Francon, A. and Michel, G. (1979) Eur. J. Biochem. 100, 189–196.
48. Bourgogne, T., Vacheron, M.-J., Guinand, M. and Michel, G. (1992) Int. J. Biochem. 24, 471–476.
49. Hobot, J.A. and Rogers, H.J. (1991) J. Bacteriol. 173, 961–967.
50. Herbold, D.R. and Glaser, L. (1975) J. Biol. Chem. 250, 1676–1683.
51. Murrell, W.G. (1969) in: G.W. Gould and A. Hurst (Eds.), The Bacterial Spore, Academic Press, New York, pp. 215–273.
52. Vinter, V. (1963) Experientia 19, 307–308.
53. Vinter, V. (1964) Folia Microbiol. 8, 58–71.
54. Imae, Y. and Strominger, J.L. (1976) J. Bacteriol. 126, 914–918.
55. Wickus, G.G., Warth, A.D. and Strominger, J.L. (1972) J. Bacteriol. 111, 625–627.
56. Imae, Y. and Strominger, J.L. (1976) J. Bacteriol. 126, 907–913.
57. Lewis, J.C., Snell, N.S. and Burr, H.K. (1960) Science 132, 544–545.
58. Gould, G.W. and Dring, G.J. (1975) Nature (London) 258, 401–405.
59. Gerhardt, P. and Marquis, R.E. (1989) in: I. Smith, R.A. Slepecky and P. Setlow (Eds.), Regulation of Procaryotic Development, American Society for Microbiology, Washington, DC, pp. 43–63.
60. Carstensen, E.L. and Marquis, R.E. (1975) in: P. Gerhardt, R.N. Costilow and H.L. Sadoff (Eds.), Spores VI, American Society for Microbiology, Washington, DC, pp. 563–571.
61. Sowell, M.O. and Buchanan, C.E. (1983) J. Bacteriol. 153, 1331–1337.
62. Todd, J.A., Bone, E.J., Piggot, P.J. and Ellar, D.J. (1983) FEMS Microbiol. Lett. 18, 197–202.
63. Waxman, D.J., Lindgren, D.M. and Strominger, J.L. (1981) J. Bacteriol. 148, 950–955.

64. Kleppe, G. and Strominger, J.L. (1979) J. Biol. Chem. 254, 4856–4862.
65. Buchanan, C.E. and Sowell, M.O. (1983) J. Bacteriol. 156, 545–551.
66. Buchanan, C.E. (1987) J. Bacteriol. 169, 5301–5303.
67. Buchanan, C.E. (1988) in: P. Actor, L. Daneo-Moore, M.L. Higgins, M.R.J. Salton and G.D. Shockman (Eds.), Antibiotic Inhibition of Bacterial Cell Surface Assembly and Function, American Society for Microbiology, Washington, DC, pp. 332–342.
68. Neyman, S.L. and Buchanan, C.E. (1985) J. Bacteriol. 161, 164–168.
69. Buchanan, C.E. and Neyman, S.L. (1986) J. Bacteriol. 165, 498–503.
70. Spratt, B.G. (1977) J. Bacteriol. 131, 293–305.
71. Nakamura, M., Maruyama, I.N., Soma, M., Kato, J.-I., Suzuki, H. and Hirota, Y. (1983) Mol. Gen. Genet. 191, 1–9.
72. Asano, Y., Kato, Y., Yamada, A. and Kondo, K. (1992) Biochemistry 31, 2316–2328.
73. Buchanan, C.E. and Ling, M.-L. (1992) J. Bacteriol. 174, 1717–1725.
74. Todd, J.A., Roberts, A.N., Johnstone, K., Piggot, P.J., Winter, G. and Ellar, D.J. (1986) J. Bacteriol. 167, 257–264.
75. Broome-Smith, J.K. and Spratt, B.G. (1982) J. Bacteriol. 152, 904–906.
76. Broome-Smith, J.K. (1985) J. Gen. Microbiol. 131, 2115–2118.
77. Matsuhashi, M., Maruyama, I.N., Takagaki, Y., Tamaki, S., Nishimura, Y. and Hirota, Y. (1978) Proc. Natl. Acad. Sci. USA 75, 2631–2635.
78. Spratt, B.G. (1980) J. Bacteriol. 144, 1190–1192.
79. Buchanan, C.E. and Gustafson, A. (1992) J. Bacteriol. 174, 5430–5435.
80. Wu, J.-J., Schuch, R. and Piggot, P. (1992) J. Bacteriol. 174, 4885–4892.
81. Ikeda, M., Sato, T., Wachi, M. Jung, H.K., Ishino, F., Kobayashi, Y. and Matsuhashi, M. (1989) J. Bacteriol. 171, 6375–6378.
82. Joris, B., Dive, G., Henriques, A., Piggot, P.J. and Ghuysen, J.M. (1990) Mol. Microbiol. 4, 513–517.
83. Stoker, N.G., Pratt, J.M. and Spratt, B.G. (1983) J. Bacteriol. 155, 854–859.
84. Ishino, F., Park, W., Tomioka, S., Tamaki, S., Takase, I., Kunugita, K., Matsuzawa, H., Asoh, S., Ohta, T., Spratt, B.G. and Matsuhashi, M. (1986) J. Biol. Chem. 261, 7024–7031.
85. Matsuzawa, H., Asoh, S., Kunai, K., Muraiso, K., Takasuga, A. and Ohta, T. (1989) J. Bacteriol. 171, 558–560.
86. Tamaki, S., Matsuzawa, H. and Matsuhashi, M. (1980) J. Bacteriol. 141, 52–57.
87. Ishino, F., Jung, H.K., Ikeda, M., Doi, M., Wachi, M. and Matsuhashi, M. (1989) J. Bacteriol. 171, 5523–5530.
88. Ishino, F. and Matsuhashi, M. (1981) Biochem. Biophys. Res. Commun. 101, 905–911.
89. Strange, R.E. and Dark, F.A. (1957) J. Gen. Microbiol. 16, 236–249.
90. Uratani, B., Lopez, J.M. and Freese, E. (1983) J. Bacteriol. 154, 261–268.
91. Foster, S.J. (1992) 11th Int. Spores Conf., Marine Biological Laboratories, Woods Hole, MA, Abstract 41.
92. Powell, J.F. and Strange, R.E. (1956) Biochem. J. 63, 661–668.
93. Brown, W.C. (1977) in: D. Schlessinger (Ed.), Microbiology – 1977, American Society for Microbiology, Washington, DC, pp. 75–84.
94. Wise, E.M., Glickman, R.S. and Teimer, E. (1972) Proc. Natl. Acad. Sci. USA 69, 233–237.
95. Grant, W.D. (1979) FEMS Microbiol. Lett. 6, 301–304.
96. Kusser, W. and Fielder, F. (1983) J. Bacteriol. 155, 302–310.
97. Anagnostopoulos, C., Piggot, P.J. and Hoch, J.A. (1993) in: A.L. Sonenshein, R. Losick and J.A. Hoch (Eds.), Bacillus subtilis and Other Gram-positive Bacteria: Physiology, Biochemistry and Molecular Biology, American Society for Microbiology, Washington, DC, pp. 425–461.
98. Blumberg, P.M. and Strominger, J.L. (1971) Proc. Natl. Acad. Sci. USA 68, 2814–2817.
99. Blumberg, P.M. and Strominger, J.L. (1972) J. Biol. Chem. 247, 8107–8113.
100. Buchanan, C.E. and Gustafson, A. (1991) J. Bacteriol. 173:1807–1809.
101. Sharpe, A., Blumberg, P.M. and Strominger, J.L. (1974) J. Bacteriol. 117, 926–927.
102. Todd, J.A., Bone, E.J. and Ellar, D.J. (1985) Biochem. J. 230, 825–828.
103. Warburg, R.J., Buchanan, C.E., Parent, K. and Halvorson, H.O. (1986) J. Gen. Microbiol. 132, 2309–2319.

186

104. Buxton, R.S. and Ward, J.B. (1980) J. Gen. Microbiol. 120, 283–293.
105. Beall, B. and Lutkenhaus, J. (1989) J. Bacteriol. 171, 6821–6834.
106. Harry, E.J. and Wake, R.G. (1989) J. Bacteriol. 171, 6835–6839.
107. Gholamhoseinian, A., Shen, Z., Wu, J.-J. and Piggot, P.J. (1992) J. Bacteriol. 174, 4647–4656.
108. Gonzy-Tréboul, G., Karamazyn-Campelli, C. and Stragier, P. (1992) J. Mol. Biol. 224, 967–979.
109. Moir, A. and Smith, D. (1990) Annu. Rev. Microbiol. 44, 531–553.
110. Levin, P.A., Margolis, P.S., Setlow, P., Losick, R. and Dun, D. (1992) J. Bacteriol. 174, 6717–6728.
111. Varley, A.W. and Stewart, G.C. (1992) J. Bacteriol. 174, 6729–6742.
112. Henriques, A.O., de Lencastre, H. and Piggot, P.J. (1992) Biochimie 74, 735–748.
113. Ikeda, M., Wachi, M., Jung, H.K., Ishino, F. and Matsuhashi, M. (1990) J. Gen. Appl. Microbiol. 36, 179–187.
114. Hansson, M. and Hederstedt, L. (1992) J. Bacteriol. 174, 8081–8093.
115. Jackson, G.E.D. and Strominger, J.L. (1984) J. Biol. Chem. 259, 1483–1490.
116. Kleppe, G., Yu, W. and Strominger, J.L. (1982) Antimicrob. Agents Chemother. 21, 979–983.
117. Sun, D., Stragier, P. and Setlow, P. (1989) Genes Dev. 3, 141–149.
118. Fort, P. and Piggot, P.J. (1984) J. Gen. Microbiol. 130, 2147–2153.
119. Schmidt, R., Margolis, P., Duncan, L., Coppolecchia, R., Moran, Jr., C.P. and Losick, R. (1990) Proc. Natl. Acad. Sci. USA 87, 9221–9225.
120. Rong, S., Rosenkrantz, M. and Sonenshein, A.L. (1986) J. Bacteriol. 165, 771–779.
121. Guzmán, P., Westpheling, J. and Youngman, P. (1988) J. Bacteriol. 170, 1598–1609.
122. Stragier, P., Bonamy, C. and Karmazyn-Campelli, C. (1988) Cell 52, 697–704.
123. Trempy, J.E., Bonamy, C., Szulmajster, J. and Haldenwang, W.G. (1985) Proc. Natl. Acad. Sci. USA 82, 4189–4192.
124. Masuda, E.S., Anaguchi, H., Sato, T., Takeuchi, M. and Kobayashi, Y. (1990). Nucleic Acids Res. 18, 657.
125. Errington, J. (1993) Microbiol. Rev. 57, 1–33.
126. Popham, D.L. and Stragier, P. (1991) J. Bacteriol. 173, 7942–7949.
127. Miyakawa, J., Matsuzawa, M., Matsuhashi, M. and Sugino, Y. (1972) J. Bacteriol. 112, 950–958.
128. Losick, R., Youngman, P. and Piggot, P.J. (1986) Annu. Rev. Genet. 20, 625–669.
129. Lencastre, H. de and Piggot, P.J. (1979) J. Gen. Microbiol. 114, 377–389.
130. Setlow, P. (1989) in: I. Smith, R.A. Slepecky and P. Setlow (Eds.), Regulation of Prokaryotic Development, American Society for Microbiology, Washington, DC, pp. 211–221.
131. Losick, R. and Kroos, L. (1989) in: I. Smith, R.A. Slepecky and P. Setlow (Eds.), Regulation of Prokaryotic Development, American Society for Microbiology, Washington, DC, pp. 223–241.
132. Driks, A. and Losick, R. (1991) Proc. Natl. Acad. Sci. USA 88, 9934–9938.
133. Margolis, P., Driks, A. and Losick, R. (1991) Science 254, 562–565.
134. Kenney, T.J., Kirchman, P.A. and Moran, Jr., C.P. (1988) J. Bacteriol. 170, 3058–3064.
135. Wu, J.-J., Howard, M.G. and Piggot, P.J. (1989) J. Bacteriol. 171, 692–698.
136. LaBell, T.L., Trempy, J.E. and Haldenwang, W.G. (1987) Proc. Natl. Acad. Sci. USA 84, 1784–1788.
137. Beall, B. and Lutkenhaus, J. (1991) Genes Dev. 5, 447–455.
138. Jonas, R.M. and Haldenwang, W.G. (1989) J. Bacteriol. 171, 5226–5228.
139. Stragier, P., Kunkel, B., Kroos, L. and Losick, R. (1989) Science 243, 507–512.
140. Sato, T., Samori, Y. and Kobayashi, Y. (1990) J. Bacteriol. 172, 1092–1098.
141. Kunkel, B., Losick, R. and Stragier, P. (1990) Genes Develop. 4, 525–535.
142. Kroos, L., Kunkel, B. and Losick, R. (1989) Science 243, 526–529.
143. Karmazyn-Campelli, C., Bonamy, C., Savelli, B. and Stragier, P. (1989) Genes Dev. 3, 150–157.
144. Sun, D., Cabrera-Martinez, R.M. and Setlow, P. (1991) J. Bacteriol. 173, 2977–2984.
145. Nicholson, W.L., Sun, D., Setlow, B. and Setlow, P. (1989) J. Bacteriol. 171, 2708–2718.
146. Karmazyn-Campelli, C. and Stragier, P. (1992) 11th Int. Spores Conf., Marine Biological Laboratories, Woods Hole, MA, Abstract 152.
147. Zheng, L. and Losick, R. (1990) J. Mol. Biol. 212, 645–660.
148. Foulger, D. and Errington, J. (1991) Mol. Microbiol. 5, 1363–1373.
149. Zheng, L., Halberg, R., Roels, S., Ichikawa, H., Kroos, L. and Losick, R. (1992) J. Mol. Biol. 226, 1037–1050.

J.-M Ghuysen and R. Hakenbeck (Eds.), *Bacterial Cell Wall*
© 1994 Elsevier Science B.V. All rights reserved

187

CHAPTER 9

Teichoic acid synthesis in *Bacillus subtilis*: genetic organization and biological roles

HAROLD M. POOLEY and DIMITRI KARAMATA

Institut de génétique et de biologie microbiennes, Rue César-Roux 19,
CH-1005 Lausanne, Switzerland

1. Introduction

The structural characterization of anionic cell wall polymers followed the discovery of their widespread occurrence among Gram-positive organisms [1,2]. Further studies revealed the biosynthetic pathways of the main chain of many representative examples of these secondary wall polymers [1,3–6]. The isolation of *Bacillus subtilis* 168 mutants deficient in the synthesis of the major wall teichoic acid [7,8], characterized by specific morphological disturbances, was not followed immediately by a thorough genetic analysis partly because of a lack of consensus regarding the nature of the biochemical function affected [7,9–11]. More recently, a major cluster of teichoic acid genes has been identified and, through DNA sequencing and biochemical characterization of mutants, their organization in *B. subtilis* 168 is emerging. In focusing on these developments, the present contribution seeks to draw out the more general implications in the hope that these will be relevant for other Gram-positive organisms.

Although the question of the biological role of the wall anionic polymers had been raised, that of the necessity of their synthesis was not addressed for a long time. Demonstration of this [12,13] is a key finding at the fundamental level. In turn, it raises the possibility of their being an antibiotic target, underlining the need to identify and to understand their role. The structure, biosynthesis and functions of anionic wall polymers have been the subject of recent reviews [5,6,14].

2. Synthesis of the Bacillus subtilis 168 cell wall teichoic acids

2.1. Poly(glucosylated glycerol phosphate)

2.1.1. Chemical structure
The glucosylated poly(glycerol phosphate) (poly(GroP)), with the equivalent ribitol containing poly(ribitol phosphate) (poly(RboP)), is the most widely distributed and fre-

quently encountered wall anionic polymer [1,4]. Structurally, it consists of two parts (Fig. 1): the main chain and the so-called linkage unit (LU) [15]. The former contains 50–60 glycerol 1,3 phosphate repeating units (J.H. Pollack and F. Neuhaus, personal communication), with the 2-hydroxyl groups bearing substituents, among the commonest of which are α or β C-1 linked glucose and ester-linked D-alanine. This chain is covalently linked by a phosphodiester to the muramic 6 C residue, through the LU [16–20]. The commonest form, present with minor variations in several members of the genera *Bacilli*, *Staphylococci* and *Lactobacilli* [21,22], consists of the disaccharide *N*-acetyl mannosamine (β1-4), linked to *N*-acetyl glucosamine (1–6), in turn linked to a muramic acid residue of peptidoglycan (PG). The non reducing terminus of this disaccharide is joined to the main polyol phosphate (see above) via one or several, frequently three, glycerol phosphate residues. Several variant forms of LU for anionic polymers in a variety of organisms have since been identified [15].

2.1.2. Biosynthesis and genetics
On the basis of their structural organization, four aspects of the biosynthesis of this polymer can be recognized:

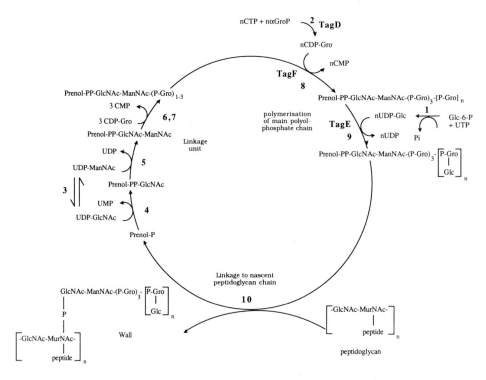

Fig. 1. Biosynthetic pathway for poly(glucosylated glycerolphosphate), the major wall teichoic acid of *B. subtilis* 168. This figure, slightly amended [39], is reproduced with permission of the *Journal of Bacteriology*, American Society for Microbiology.

(i)   the cytoplasmic synthesis of nucleotide sugar diphosphate linked soluble precursors (including steps 1, 2 and 3, Fig. 1),

(ii)   the polymerization of the LU through a pyrophosphate linkage to the membrane located polyprenol, by successive addition of N-acetylglucosamine (GlcNAc), N-acetylmannosamine (ManNAc) and, probably, three GroP units from nucleotide linked precursors (Fig. 1, steps 4–7);

(iii)   the polymerization, onto the preformed polyprenol linked LU, of the main polymer backbone by sequential addition to the terminal polyol phosphate residue of polyol phosphate units from the soluble CDP-glycerol (CDPGro) (Fig. 1, step 8). Hexose linkage to the free hydroxyl group (Fig. 1, step 9) goes hand in hand with the polymerization;

(iv)   the ligation of the completed teichoic acid molecule to a simultaneously synthesized PG chain occurs through formation of a phosphodiester link to muramic-6-hydroxyl together with liberation of polyprenol phosphate (Fig. 1, step 10).

2.1.2.1. The soluble precursors. Those required for the synthesis of the main polymer chain include CDPGro (or CDP-ribitol for poly[RboP]), and uridine diphosphoglucose (UDPG), while for LU synthesis UDP N-acetyl glucosamine (UDPGlcNAc), UDP N-acetyl mannosamine (UDPManNAc) as well as CDPGro are required.

(a)   CDPGro, a specific precursor for wall teichoic acid (WTA) synthesis, is formed by Gro-3-phosphate cytidylyl transferase, readily detectable in the cytosol fraction of disrupted *B. subtilis* 168 and W23 cells [23,24]. *tagD* was identified as the structural gene for this enzyme by localization of mutation *tagD1* [25], associated with an absence of a CDPGro pool at the non-permissive temperature [13].

(b)   UDPG: mutations in the *gtaB* locus provoke an absence of glucose in the isolated cell walls of *B. subtilis* 168 [26]. The absence of UDPG pyrophosphorylase activity (UDPGPPase), invariably associated with mutations in the *gtaB* locus [27], as well as recent DNA sequence studies of the *gtaB* region [28], revealing a high overall homology of the deduced amino acid sequence with several analogous prokaryotic enzymes, establish that *gtaB* is the structural gene for the UDPGPPase. No mutants blocked in formation of precursors, UDPManNAc (step 3, Fig. 1) and UDPGlcNAc, have been reported, and the structural genes of the enzymes responsible have not been identified (Table I).

(c)   UDPGlcNAc: this precursor, principally involved in the synthesis of the peptidoglycan polysaccharide chain, also forms part of the disaccharide unit linking this polymer to the main polyol phosphate chain.

(d)   UDPManNAc: UDP-N-acetyl-D-glucosamine 2-epimerase, responsible for the formation of UDPManNAc from UDPGlcNAc, was identified in cell extracts of *Bacillus cereus* [29], *B. subtilis* [20] and *E. coli* [30].

2.1.2.2. The linkage unit formation. The first step involves the transfer of GlcNAc from UDPGlcNAc to polyprenol (Fig. 1, step 4). In several Gram-positive organisms, this step was shown, in vitro, to be extremely sensitive to tunicamycin [31,32]. Subsequent steps in LU synthesis and in main chain polymerization are insensitive to this antibiotic [31,32].

TABLE I

Genes and enzymes involved in synthesis of wall teichoic acids in *B. subtilis* 168, poly(glucosylated glycerol phosphate) and poly(glucose *N*-acetylgalactosamine phosphate)

Biosynthetic step	Enzyme[a]	Cellular localization	Genetic organisation				Poly(glcgalNAcP)				
			Poly(glucosylated GroP)								
			Chromosome location of relevant gene(s)[b]	Gene	Operon	Ref.	Enzyme	Chromosome location of relevant gene(s)	Gene	Operon	Ref.
1. Soluble precursors	1[d]	Cytosol	308°	*gtaB*	*gtaB*	1	1[d]	308°	*gtaB*	*gtaB*	1
	2	Cytosol	308°	*tagD*	*gtaDEF*	2	2[e]	335°	*gne*	Unknown	6
	3[f]	Cytosol	Unknown	Unknown	Unknown		?[h]	Unknown	Unknown	Unknown	
2. Linkage unit	4–6[f]	Membrane	Unknown	Unknown	Unknown		?[h]	Unknown	Unknown	Unknown	
	7[f,g]	Membrane	Unknown	Unknown	Unknown		?[h]	Unknown	Unknown	Unknown	
3. Main chain formation	8	Membrane	308°	*tagF*	*tagDEF*	3	8	308°	*gga*	Unknown	7
	9	Membrane	308°	*tagE(gtaA)*	Unknown	4	9	308°	*gga*	Unknown	7
4. Teichoic peptidoglycan linking reaction	10[f]	Membrane	308°?	*tagABC* ?	*tagABC*	5	10[h]	Unknown	Unknown	Unknown	

[a]Numbers refer to steps shown in Fig. 1.

[b]Refers to map position on chromosome beginning at origin (0°) in a clockwise direction ending at origin again (360°)

[c]References: 1, Soldo et al.[28]; 2, Pooley et al. [13]; 3, Pooley et al. [39]; 4, Young [26]; 5, Fig.2; 6, Young [26]; 7, Estrela et al. [46] and this chapter.

[d]UDPG is used for the synthesis of both poly(glcGroP) and poly(glcgalNAcP). Mutational [27] or insertional [28] inactivation of *gtaB* blocks incorporation of glucose into both polymers, consistent with *gtaB* encoding a common enzyme.

[e]UDPGalNAc is a precursor unique to poly(glcgalNAcP).

[f]Possibly, common enzyme responsible for this step of poly(GroP) and of poly(glcgalNAcP) biosynthesis (see text).

[g]It is not known whether two separate enzymes are needed for this step. If the linkage unit contains three molecules of glycerol phosphate, conceivably, one activity could be sufficient. For strain W23, the linkage unit consists of two glycerol phosphate units.

[h]The existence of a linkage unit or of covalent linkage to PG has received little attention. The possibility of linkage via the same LU as poly(groP) remains open (see text).

Based on incorporation from CDP-[³H]Rbo by membrane preparations [33], mutant 52A5 of *S. aureus* was reported to be deficient in the activity of the first enzyme in LU formation. However, while ribitol was absent from the mutant, cell walls of mutant and wild type contained comparable amounts of glycerol (0.05–0.1 μmol/mg) and of muramic phosphate [34]. These findings are inconsistent with a block in the first step of LU formation.

The proposed biosynthetic pathway of the LU and the main chain of poly(ribitol phosphate) of *S. aureus* received strong support [35] through the isolation and identification, of the specific intermediates and their stepwise formation through the activity of enzymes present in membranes (including steps 4–7 in Fig. 1).

To date, there is no information on the LU genes in *B. subtilis*. Sequencing and analysis of a 21 kb segment localized near 308° (Fig. 2) of the *B. subtilis* chromosome [25,26,36,37; V. Lazarevic and D. Karamata, personal communication; P.-Ph. Freymond and D. Karamata, personal communication], containing nearly all genes involved in teichoic acid synthesis identified so far, leaves no room for genes encoding LU enzymes. Although, functions have not yet been unambiguously assigned to certain sequenced ORFs of the *tagABC* operon, the latter are likely to be involved in the ultimate steps of poly(GroP) incorporation into the wall (see below).

Among about 20 *B. subtilis* 168 mutants thermosensitive in teichoic acid synthesis, not one appears to be affected in the synthesis of the LU (Table I). This is surprising since, firstly, the number of relevant genes should represent half or more of the total of those involved in poly(GroP) synthesis (Fig. 1), and, secondly, their deficiency could lead to a conditional lethal phenotype. Indeed, tunicamycin (see below), known to inhibit the first step in LU formation, produces morphological changes and growth inhibition comparable to those which accompany mutations in *tag* genes [38]. A possible existence of two sets of LU genes might account for the absence of mutants deficient in LU synthesis.

2.1.2.3. The main chain. Nearly 30 years ago, poly(glycerol phosphate) polymerase was demonstrated in vitro with membrane preparations of *B. subtilis* and *B. licheniformis* [3]. It appeared probable that the gene encoding this enzyme would be found among those forming the *tagABCDEF* divergon [25]. Indeed, nearly all *tag* mutations identified so far were mapped to *tagB*, *tagD* or *tagF(rodC)*. Enzyme assays revealed polymerase deficiency to be invariably and exclusively associated with mutant alleles of *tagF* [39]. That *tagF*, which encodes a 88 063 Da product [36], was the structural gene was deduced from

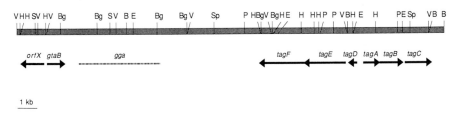

Fig. 2. Schematic map of genes in the 308° region of the *B. subtilis* 168 chromosome. This region is flanked on both sides by genes encoding lytic enzymes [37]. Reference restriction sites are indicated: B, *Bam*HI; Bg, *Bgl*II; E, *Eco*RI; H, *Hind*III; P, *Pst*I; S, *Sac*I; Sp, *Sph*I; V, *Eco*RV.

the thermosensitivity of this polymerase for mutants in which activity was detectable in vitro.

Glucosylation of poly(glycerol phosphate). Glycosyl transfer from UDPG to the C-2 position of the 1,3-phosphodiester linked glycerol units (step 9, Fig. 1) was readily demonstrated in isolated membrane preparations [3,40].

Mutations associated with a specific deficiency in this activity were all mapped to the *gtaA* locus [26]. This gene, sequenced under the name of *rodD*, encodes a 78 285 Da protein [36]. As the second gene in the *tagDEF* operon, it has been designated *tagE* [25]. The absence of a conditional lethal phenotype for *tagE(gtaA)* mutations implies that the non-glucosylated poly(GroP) is able to fulfil all functions essential for growth.

The identification of products encoded by all three genes of the *tagDEF* operon reveals a genetically coordinated expression of functions directly involved in successive steps of the main chain polymerization.

2.1.2.4. Attachment to newly synthesized peptidoglycan. The last step in the biosynthesis of WTA involves transfer of the complete molecule to a nascent PG chain through the formation of a phosphodiester linkage.

In *B. subtilis* 168, the attachment of WTA or teichuronic acid (TU) to simultaneously made PG chains was shown to occur in vivo [41]. So far, the *B. subtilis* 168 genes involved in the attachment step remain unidentified. Candidates for this function include *tagA* and *tagB*. The unique mutation mapped to these ORFs is *tagB1(tag-1, rodA)*. Following shift to the restrictive temperature, analysis of isolated cell walls revealed that phosphate incorporation is strongly and very comparably inhibited for strains bearing markers *tagB1, tagF1* or *tagD1* [42]. Nevertheless, when compared with strains bearing the latter mutations, incorporation of 2-[^{3}H]glycerol into whole cells bearing *tagB1* was significantly higher (unpublished observations), which is consistent with continuing TagF activity. A block at the PG linking step could account for the extremely low level of phosphate in isolated cell walls.

2.2. Poly(glucosyl N-acetylgalactosamine 1-phosphate)

2.2.1. Genetic and biochemical analysis of mutants

B. subtilis 168 walls contain a second, so-called minor, teichoic acid, poly(glucosyl N-acetylgalactosamine 1-phosphate) [43,44]. Galactosamine amounts to about 20 % of all wall hexosamine present in cells grown in rich medium at 30°C [27]. Genetic and biochemical analysis of 25 mutants specifically resistant to bacteriophage *Φ3T*, of which this polymer forms part of the receptor, led to the identification of genes concerned with the synthesis of this polymer. About one-quarter of the mutations were mapped to a *gne* locus around 335° [45,46], the remainder, named *gga*, were all localized to a second region very close to the *tagDEF* operon [25,46]. In contrast to results obtained with poly(GroP), mutations associated with absence of poly(glucosyl N-acetyl galactosamine 1-phosphate) (poly(glcgalNAcP)) did not have a conditional lethal phenotype. Mutations in other genes, *gtaC* (phosphoglucomutase deficient), *gtaE* (phosphoglucomutase and UDPGPP-ase deficient) and *gtaB* [27], cause a block in the synthesis of this polymer, in addition to that of the glucosylation of the major WTA.

2.2.2. Biosynthesis

Genes involved in the biosynthesis of the soluble sugar nucleotide linked precursors, UDPG and UDP *N*-acetyl galactosamine (UDPGalNAc), have been identified. As noted above, the former is synthesized by the product of the *gtaB* gene. UDPGalNAc is obtained by epimerization of UDPGlcNAc. The *gne* marker is presumed to be the structural gene for this enzyme, since a specific absence of UDPGlcNAc 4-epimerase activity was observed in cell extracts of six strains carrying mutations, of which all were mapped to this locus [46].

Physical mapping and insertional mutagenesis revealed an approximately 4 kb region associated with a $\Phi3T$ resistant phenotype, and localized between the *tagABCDEF* [47] and *gtaBorfX* [28] divergons. All *gga* mutations were localized in this region [46]. Preliminary observations suggest that this 3–4 kb region is unlikely to contain more than two ORFs (P.-Ph. Freymond and D. Karamata, personal communication).

Cell free membrane, as well as membrane and wall preparations, of strains deficient in UDPG synthesis are able to synthesize the poly(glcgalNAcP) following addition of UDPG and UDPGalNAc only. The possible involvement of a LU was apparently not examined [48].

In contrast to the monosaccharidic poly(GroP), a linearly coupled disaccharide phosphate is the repeating unit of this *gga* polymer. Thus, two alternative biosynthetic pathways appear possible. The first would involve the polymer synthesis, in the same direction and manner as that of the monomeric alditol phosphate, by addition to the end distal to that of the lipid carrier. This would mean the alternate addition, from the sugar nucleotides, of GalNAc phosphate and then of glucose linked directly to the GalNAc 3-hydroxyl group. Two enzymes would be required. The polymerization of galactosylglycerol phosphate, a disaccharide teichoic acid of *B. coagulans*, occurs via this pathway [49]. The second mechanism, likely to require three enzymes, entails the assembly of the disaccharide repeating unit on a lipid carrier, followed by successive addition from sugar nucleotides. This mechanism is characteristic of the assembly of disaccharide repeating units of PG or TU of several organisms, as well as, apparently, the disaccharide polymer of *B. licheniformis*, the glucosyl glycerol phosphate [6,50]. Synthesis of these polymers proceeds in the direction opposite to that of monosaccharidic teichoic acids, i.e. via the transfer of the carrier linked repeating unit to the growing polymer chain bound to a second lipid carrier molecule, by insertion between the lipid carrier and the polymer. Thus, the first (oldest) repeating unit added is the most distant from the lipid carrier.

It has been suggested [15] that the lipid carrier distal addition of monosaccharide phosphate units may not apply for teichoic acids consisting of repeating units of tri- or oligo-saccharide phosphates, which could be synthesized in a lipid carrier proximal direction, the polymer being directly attached to peptidoglycan without any LU. Should the presence of two genes involved in the polymerization of poly(glcgalNAcP) be confirmed, the lipid carrier distal addition of monomers would appear plausible (see above) and, in turn, imply the presence of a LU.

From the foregoing, it is apparent that the small number of *gga* and *gne* genes identified falls short of those required for synthesis of precursors, LU, polymerization and ligation to PG. However, sequencing studies reveal no further candidate ORFs in the 21 kb region (Fig. 2), containing *tag* genes, and mutations mapping elsewhere than in the two regions described here above were not identified. It is possible that defects in the unidentified genes have a different phenotype. Conceivably, any reactions common to the synthesis of several teichoic acid polymers could be achieved by a unique set of enzymes. Such functions include LU synthesis and, possibly, the final, PG linkage reaction (Table I). In view of the essentiality of poly(GroP) synthesis, inactivation of genes encoding any such enzymes are expected to be lethal. The existence of shared functions would imply an interdependence of the synthesis of the two WTA of *B. subtilis*.

2.2.3. Biological functions

The role of this second anionic polymer remains undefined, but it can be considered as a substitute for the major anionic polymer of *B. subtilis* 168 [46]. If this were true, the fact that cells grown at 45°C do not make the GalNAc containing polymer [7] could be partly responsible for the conditional lethal phenotype at this temperature of *tag* mutations.

3. Roles of poly(glycerol phosphate) synthesis

3.1. Introduction

Attempts to understand the biological roles of WTA followed their discovery and the realization of their widespread distribution among the Gram-positive bacteria. Their presence in the cell wall was judged to be important, for the negative surface charge [51], for the binding of divalent cations [51,52], for specific affinities for cell autolysins [53] and bacteriophages [26,54]. Their role as a phosphate reserve [55], and as a component of the cell permeability barrier [56] were also proposed.

3.2. Synthesis of poly(glycerol phosphate) in B. subtilis 168 is essential for growth

Until recently, the question of the necessity of WTA synthesis for cell growth had, surprisingly, not been addressed directly. With hindsight, this may be explained by the faculty of certain bacilli to produce a substitute polymer, TU [2,57], thereby showing that WTA was not, strictly speaking, indispensable. Nevertheless, TU is produced only under phosphate starvation; in phosphate replete media, TU is not made, even when WTA synthesis is blocked mutationally [58,59]. Isolation of mutants thermosensitive for teichoic acid synthesis unable to sustain growth in non-permissive conditions strongly argues for the essentiality of WTA synthesis. That strain 168 cells make three distinct anionic wall polymers, i.e. poly(GroP), poly(glcgalNAcP) and TU, partly capable of substituting for one another [46,57], only reinforces the conclusion of the necessity for a wall polyanion synthesis for cell growth.

Decisive evidence for this conclusion came from insertional mutagenesis [12], and the identification of specific activities of teichoic acid enzymes encoded by genes which have suffered conditional lethal mutations [13,39,42]. The need to identify the specific functions fulfilled by these molecules has become even more evident.

3.3. Hybrid strains containing heterologous cell wall teichoic acid

Hybrid *B. subtilis* 168 recombinants in which poly(GroP) genes were replaced by poly(RboP) ones, originating from strain W23, were obtained by genetic exchange in the 308° region of the chromosome [60], through substitution and deletion, respectively, of what were subsequently shown to be *tar* and *tag* teichoic acid gene cassettes [25,36,61]. The surprisingly and, apparently, little altered phenotype of such hybrids provided new support for the suggestion [1] that, rather than specific chemical components, it was the anionic nature and overall structure that were important functionally.

3.4. The pH gradient across the cell wall

Together with the membrane anchored lipoteichoic acids (LTA), the role of polyanions in maintaining an appropriate ionic environment for membrane synthetic activities has long been recognized [52]. Their presence may also be critical for the generation of a pH gradient across the cell wall [62–64], by sequestering protons released at the membrane level by oxidative or fermentative processes. The low pH thus generated could, for instance, inhibit peptidoglycan hydrolases from degrading nascent PG [63]. Generating a pH gradient in a high medium pH should be more difficult. An increased pH in the vicinity of the membrane could lead to dysfunction that could block growth. Changes in cell wall composition shown by alkaliphilic bacteria [65] support this view. For a majority of alkaliphilic organisms of the genus *Bacillus*, transfer from pH 7 to pH 10 medium is accompanied by a substantial increase in polyanionic wall components [65]. This response should facilitate retention of protons near the membrane.

3.5. Teichoic acid and surface extension: cylinder elongation in rod-shaped organisms

Following extended incubation of thermosensitive *tag* mutants at the restrictive temperature, grossly deformed cell shapes [8,66] develop as the result of complex metabolic interactions. The time course of changes in the rate of synthesis of each of the major macromolecular fractions was not clearly or completely established [9,42]. Recently, it was found that the very first consequences of blocking WTA synthesis were: a halving in the rate of PG synthesis and, apparently, an immediate block in cell length increase [67]. This implies that cylinder expansion is dependent on the synthesis of WTA, which, in turn, is consistent with an essential role for anionic polymers for cell growth. Cell division, which underwent no interruption during the same period, can, apparently, be achieved through PG synthesis alone. Analysis of randomly isolated partial phenotypic revertants of conditional lethal *tag* gene mutants in *B. subtilis* [11] revealed that their cell wall phosphate content was invariably increased, which is consistent with the above conclusions. No secondary mutations of a suppressor type were identified, able to undergo more or less normal cell elongation in the absence of WTA synthesis.

3.6. Wall teichoic acid synthesis as a target for antibiotics

The conclusion of the necessity for growth provides strong grounds for exploring the possibility of wall anionic polymer synthesis being an antibiotic target [68]. Based on in vitro evidence [31], there is a case for tunicamycin being a model antiteichoic acid [68]. Tunicamycin at the minimum inhibitory concentration strongly inhibits 2-[^{3}H]glycerol incorporation into poly(GroP), while barely affecting incorporation of N-acetylglucosamine into PG (unpublished). Although, because of toxicity, the clinical use of tunicamycin is excluded, knowing that inhibition of teichoic acid synthesis stops growth opens a new perspective in the search for novel antibiotics.

4. Concluding remarks

Whereas there has been progress towards genetically defining teichoic acid synthesis in *B. subtilis* 168, several identified genes remain functionally uncharacterized, while, at present, no genes of the phylogenetically widespread peptidoglycan LU have been identified.

Establishing the necessity of the polyanionic wall synthesis for growth in *B. subtilis* 168 underlines the continuing lack of knowledge of the key function(s) of this polymer. Evidence for a role, in this rod-shaped organism, in the elongation of the cylindrical part of the surface, has provided no understanding of just how teichoic acid could participate in this process.

Acknowledgements

We are grateful for the contributions over the past decade of the following colleagues: F.-X. Abellan, M. Briehl, A.-I. Estrela, Ph. Margot, M. Monod, D. Paschoud and M. Young. The key contributions of Catherine Mauël to the genetic organization of poly-(GroP) synthesis and of its essential function are gratefully acknowledged. For communicating unpublished material, we would like to thank the following: P.Ph. Freymond, V. Lazarevic, F. Neuhaus, J.H. Pollack and B. Soldo.

References

1. Baddiley, J. (1972) Essays Biochem. 8, 35–77.
2. Janczura, E., Perkins, H.R. and Rogers, H.J. (1961) Biochem. J. 80, 82–93.
3. Burger, M.M. and Glaser, L. (1964) J. Biol. Chem. 239, 3168–3177.
4. Archibald, A.R. (1974) Adv. Microb. Physiol. 11, 53–95.
5. Hancock, I.C. and Baddiley, J. (1985) in: A.N. Martonosi (Ed.), The Enzymes of Biological Macromolecules, Vol. 2, Plenum Press, New York, pp. 279–307.
6. Archibald, A.R., Hancock, I.C. and Harwood, C.R. (1993) in: A.L. Sonenshein, J.A. Hoch and R. Losick (Eds.), *Bacillus subtilis* and other Gram-Positive Bacteria, American Society for Microbiology, Washington, DC, pp. 381–410.
7. Boylan, R.J., Mendelson, N.H., Brooks, D. and Young, F.E. (1972) J. Bacteriol. 110, 281–290.
8. Rogers, H.J., McConnell, M. and Burdett, I.D.J. (1970) J. Gen. Microbiol. 61, 155–171.
9. Rogers, H.J., Thurman, P.F., Taylor, C. and Reeve, J.N. (1974) J. Gen. Microbiol. 85, 335–350.
10. Rogers, H.J. and Taylor, C. (1978) J. Bacteriol. 135, 1032–1042.
11. Shiflett, M.A., Brooks, D. and Young, F.E. (1977) J. Bacteriol. 132, 681–690.
12. Mauël, C., Young, M., Margot, P. and Karamata, D. (1989) Mol. Gen. Genet. 215, 388–394.

13. Pooley, H.M., Abellan, F.X. and Karamata, D. (1991) J. Gen. Microbiol. 137, 921–928.
14. Ward, J.B. (1981) Microbiol. Rev. 45, 211–243.
15. Araki, Y. and Ito, E. (1989) Crit. Rev. Microbiol. 17, 121–135.
16. Coley, J., Archibald, A.R. and Baddiley, J. (1976) FEBS Lett. 61, 240.
17. Bracha, R. and Glaser, L. (1976) Biochem. Biophys. Res. Commun. 72, 1091–1098.
18. Bracha, R. and Glaser, L. (1976) J. Bacteriol. 125, 872–879.
19. Sasaki, Y., Araki, Y. and Ito, E. (1980) Biochem. Biophys. Res. Commun. 96, 529–534.
20. Harrington, C.R. and Baddiley, J. (1985) Eur. J. Biochem. 153, 639–645.
21. Kojima, N., Araki, Y. and Ito, E. (1985) J. Bacteriol. 161, 299.
22. Kojima, N., Araki, Y. and Ito, E. (1985) Eur. J. Biochem. 148, 29.
23. Glaser, L. and Loewy, A. (1979) J. Bacteriol. 137, 327–331.
24. Hancock, I.C. (1983) Arch. Microbiol. 134, 222–226.
25. Mauël, C., Young, M. and Karamata, D. (1991) J. Gen. Microbiol. 137, 929–941.
26. Young, F.E. (1967) Proc. Natl Acad. Sci. USA 58, 2377–2384.
27. Pooley, H.M., Paschoud, D. and Karamata, D. (1987) J. Gen. Microbiol. 133, 3481–3493.
28. Soldo, B., Lazarevic, V., Margot, Ph. and Karamata, D. (1993) J. Gen. Microbiol. 139, in press.
29. Kawamura, T., Kimura, M., Yamamori, S. and Ito, E. (1978) J. Biol. Chem. 253, 3595–3601.
30. Kawamura, T., Ichihara, N., Ishimoto, N. and Ito, E. (1975) Biochem. Biophys. Res. Commun. 66, 1506–1512.
31. Ward, J.B., Wyke, A.W. and Curtis, C.A.M. (1980) Biochem. Soc. Trans. 8, 164–166.
32. Wyke, A.W. and Ward, J.B. (1977) J. Bacteriol. 130, 1055–1063.
33. Bracha, R., Davidson, R. and Mirelman, D. (1978) J. Bacteriol. 134, 412–417.
34. Mirelman, D., Shaw, D.R.D. and Park, J.T. (1971) J. Bacteriol. 107, 239–244.
35. Yokoyama, K., La Mar, G.N., Araki, Y. and Ito, E. (1986) Eur. J. Biochem. 161, 479–489.
36. Honeyman, A.L. and Stewart, G.C. (1989) Mol. Microbiol. 3, 1257–1268.
37. Lazarevic, V., Margot, Ph., Soldo, B. and Karamata, D. (1992) J. Gen. Microbiol. 138, 1949–1961.
38. Takatsuki, A., Shimizu, K.-I. and Tamura, G. (1972) J. Antibiotics 25, 75–85.
39. Pooley, H.M., Abellan, F.X. and Karamata, D. (1992) J. Bacteriol. 174, 646–649.
40. Brooks, D., Mays, L.L., Hatefi, Y. and Young, F.E. (1971) J. Bacteriol. 107, 223–229.
41. Mauck, J. and Glaser, L. (1972) J. Biol. Chem. 247, 1180–1187.
42. Briehl, M., Pooley, H.M. and Karamata, D. (1989) J. Gen. Microbiol. 135, 1325–1334.
43. Duckworth, M., Archibald, A.R. and Baddiley, J. (1972) Biochem. J. 130, 691–696.
44. Shibaev, V.N., Duckworth, M., Archibald, A.R. and Baddiley, J. (1973) Biochem. J. 135, 383–384.
45. Estrela, A.I., de Lencastre, H. and Archer, L.J. (1986) J. Gen. Microbiol. 132, 411–415.
46. Estrela, A.I., Pooley, H.M., de Lencastre, H. and Karamata, D. (1991) J. Gen. Microbiol. 137, 943–950.
47. Mauël, C. and Karamata, D. (1989) 5th Conf. on Bacilli, Asilomar, USA, Abstract.
48. Hayes, M.V., Ward, J.B. and Rogers, H.J. (1977) Proc. Soc. Gen. Microbiol. 4, 85–86.
49. Yokoyama, K., Araki, Y. and Ito, E. (1987) Eur. J. Biochem. 165, 47–53.
50. Hancock, I.C. and Baddiley, J. (1972) Biochem. J. 127, 27–37.
51. Heptinstall, S., Archibald, A.R. and Baddiley, J. (1970) Nature 225, 519–521.
52. Hughes, A.H., Hancock, I.C. and Baddiley, J. (1973) Biochem. J. 132, 83–93.
53. Höltje, J.V. and Tomasz, A. (1975) J. Biol. Chem. 250, 6072–6076.
54. Archibald, A.R. (1980) in: L.L. Randall and L. Philipson (Eds.), Receptors and Recognition, Series B., Vol. 7, Virus Receptors, Chapman and Hall, London, pp. 5–26.
55. Grant, W.D. (1979) J. Bacteriol. 137, 35–43.
56. Bertram, K.C., Hancock, I.C. and Baddiley, J. (1981) J. Bacteriol. 148, 406–412.
57. Ellwood, D.C. and Tempest, D.W. (1972) Adv. Microbiol. Physiol. 7, 83–117.
58. Rosenberger, R.F. (1976) Biochim. Biophys. Acta 428, 516–524.
59. Prayitno, N. (1992) Ph. D. Thesis, University of Newcastle-upon-Tyne, UK.
60. Karamata, D., Pooley, H.M. and Monod, M. (1987) Mol. Gen. Genet. 207, 73–81.
61. Young, M., Mauël, C., Margot, P. and Karamata, D. (1989) Mol. Microbiol 3, 1805–1812.
62. Urrutia, M., Kemper, M., Doyle, R.J. and Beveridge, T.J. (1992) Appl. Environ. Microbiol. 58, 3837–3844.
63. Kemper, M. and Doyle, R.J. (1993) in: M.A. de Pedro, J.-V. Höltje and W. Löffelhardt (Eds.), Bacterial

Growth and Lysis: Metabolism and Structure of the Bacterial Sacculus, Plenum Press, New York, pp. 245-252.

64. Kemper, M., Urrutia, M., Beveridge, T.J., Koch, A.L. and Doyle, R.J. (1993) J. Bacteriol. 175, 5690–5696.

65. Aono, R. and Horikoshi, K. (1983) J. Gen. Microbiol. 129, 1083–1087.

66. Cole, R.M., Popkin, T.S., Boylan, R.J. and Mendelson, N.H. (1970) J. Bacteriol. 103, 793–810.

67. Pooley, H.M, Abellan, F.-X. and Karamata, D. (1993) in: M.A. de Pedro, J.-V. Höltje and W. Löffelhardt (Eds.), Bacterial Growth and Lysis: Metabolism and Structure of the Bacterial Sacculus, Plenum Press, New York, pp. 385–392.

68. Pooley, H.M. and Karamata, D. (1988) in: P. Actor, L. Daneo-Moore, M.L. Higgins, M.R.J. Salton and G.D. Shockman (Eds.), Antibiotic Inhibition of Bacterial Cell Surface Assembly and Function, American Society for Microbiology, Washington, DC, pp. 591–594.

J.-M Ghuysen and R. Hakenbeck (Eds.), *Bacterial Cell Wall*

199

CHAPTER 10

Lipoteichoic acids and lipoglycans

WERNER FISCHER

Institut für Biochemie der Medizinischen Fakultät, Universität Erlangen-Nürnberg, Fahrstrasse 17,
D-91054 Erlangen, Germany

1. Introduction

Glyceroglycolipids, lipoteichoic acids and lipoglycans are characteristic components of the cytoplasmic membrane of Gram-positive bacteria, whereas teichoic acids are components of the peptidoglycan layer. In analogy to teichoic acids [1], lipoteichoic acids may be defined as macroamphiphiles that contain alditolphosphates as integral parts of the hydrophilic chain. Lipoglycans possess a linear or branched homo- or heteropolysaccharide as hydrophilic moiety and may carry monomeric glycerophosphate (GroP) branches. The term lipoglycan is proposed in order to differentiate these macroamphiphiles from the structurally, physicochemically and functionally different class of lipopolysaccharides (LPS) of Gram-negative bacteria [2,3]. Lipoteichoic acids and lipoglycans do not occur together in the same organism and are thought to replace each other functionally. Lipoteichoic acids are prevailing in that subgroup of Gram-positive eubacteria which contains DNA of a guanine and cytosine (G + C) content of less than 50%, whereas lipoglycans occur preferentially in the other subgroup with DNA of a (G + C) content greater than 55% [4]. Several reviews on lipoteichoic acids and lipoglycans have appeared [4–7] covering methodological, metabolic, and functional aspects, with complete references.

2. Structure and occurrence

2.1. Poly(glycerophosphate) lipoteichoic acids

This classical type is most widespread and has been found in bacilli, enterococci, lactobacilli, lactococci, leuconostoc, listeria, staphylococci and certain streptococci (for literature, see [4]). As shown in Fig. 1(I), the hydrophilic moiety is a single unbranched 1,3-linked poly(glycerophosphate) chain which is covalently linked by a phosphodiester bond to O-6 of the non-reducing hexosyl terminus of a glyceroglycolipid [8–10]. The latter serves through the hydrocarbon chains of its fatty acids as a hydrophobic membrane anchor and is in most cases derived from a glycolipid found among the free lipids of the cytoplasmic membrane [4]. Because the glycolipid structures may vary in a genus- or

I

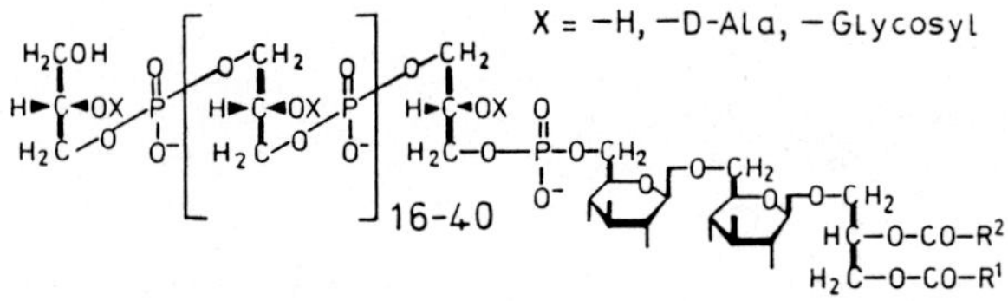
X = −H, −D-Ala, −Glycosyl
16-40

II

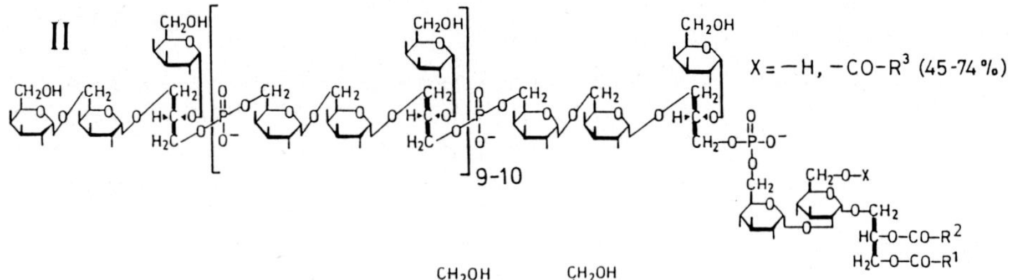
X = −H, −CO−R³ (45-74 %)
9-10

III

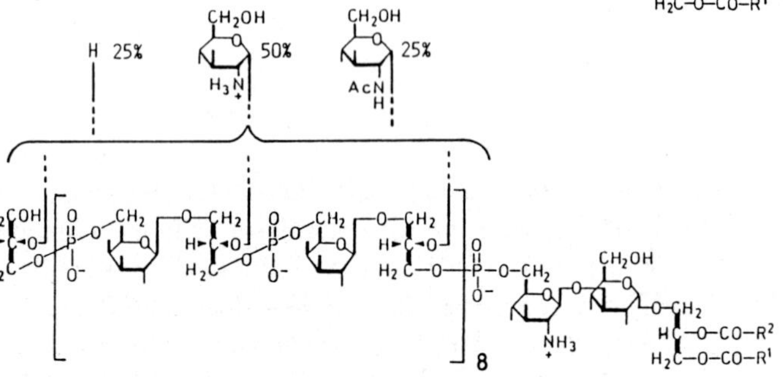
H 25%
50%
25%
8

IV

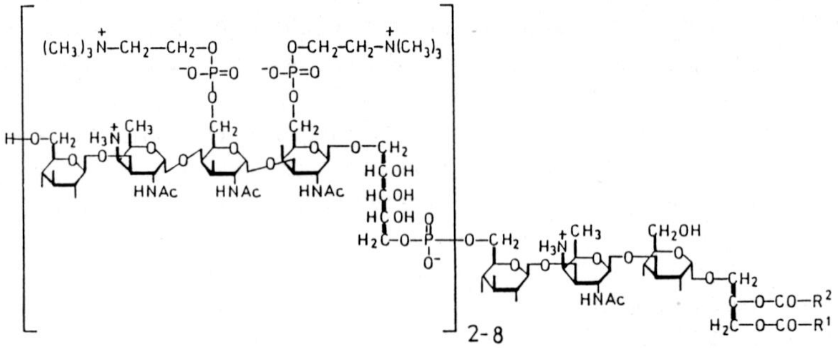
2-8

V

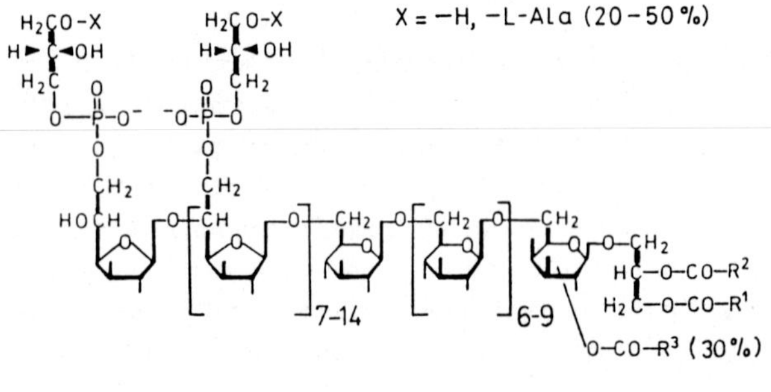
X = −H, −L-Ala (20−50 %)
7-14
6-9
O−CO−R³ (30 %)

TABLE I

Lipid anchors of poly(glycerophosphate) lipoteichoic acids (for references, see [6])

Glycolipid	Derivative	Occurrence
Glc(α1–2)Glc(α1–3)acyl$_2$Gro	–	*Streptococci, Leuconostoc*
Glc(α1–2)Glc(α1–3)acyl$_2$Gro	Glc(α1–2)Glc(α1–3)acyl$_2$Gro 6 └acyl	*Lactococci*
Glc(α1–2)Glc(α1–3)acyl$_2$Gro	Glc(α1–2)Glc(α1–3)acyl$_2$Gro 6 └Ptd	*Enterococci*
Glc(β1–6)Glc(β1–3)acyl$_2$Gro	–	*Bacilli, Staphylococci*
Gal(α1–2)Glc(α1–3)acyl$_2$Gro	Gal(α1–2)Glc(α1–3)acyl$_2$Gro 6 └acyl	*Streptococci, Lactobacilli*
Gal(α1–2)Glc(α1–3)acyl$_2$Gro	Gal(α1–2)Glc(α1–3)acyl$_2$Gro 6 └Ptd	*Listeria*
Glc(β1–6)Gal(α1–2)Glc(α1–3) acyl$_2$Gro	Glc(β1–6)Gal(α1–2)Glc(α1–3)acyl$_2$Gro 6 └acyl	*Lactobacilli*
	acyl$_2$Gro	*Bacilli*

species-specific manner, the lipid anchors of lipoteichoic acids vary accordingly (Table I). In a number of bacteria, the lipoteichoic acid consists of two species, one linked to the glyceroglycolipid, the other to an acyl or phosphatidyl derivative (Table I). Di-O-acylglycerol (acyl$_2$Gro) in place of a glycolipid was found in the lipoteichoic acids of certain bacilli.

The poly(glycerophosphate) chain is unbranched and is 16–40 glycerophosphate residues in length on average [4,6]. Position 2 of the glycerophosphate residues is substituted in part either with D-alanine ester or with D-alanine ester and glycosyl residues. Lipoteichoic acids lacking any substituent or carrying glycosyl substituents only are exceptional. The glycosyl substituents are common sugars, such as D-glucose, D-galactose and *N*-acetyl-D-glucosamine, and are in many cases attached to the glycerol (Gro) residues as monomeric branches. Mono- to tetra-α-D-glucopyranosyl (Glc) residues with (1–2) interglycosidic linkages have been found as chain substituents in the lipoteichoic acids of enterococci [9,11] and *Leuconostoc* [12]. A similar α-glucooligosaccharide series with (1–6) interglycosidic linkages occurs in the lipoteichoic acid of *Streptococcus sanguis* [13].

Fig. 1. Structures of lipoteichoic acids (I–IV) and a lipoglycan (V). For occurrence and references, see text.

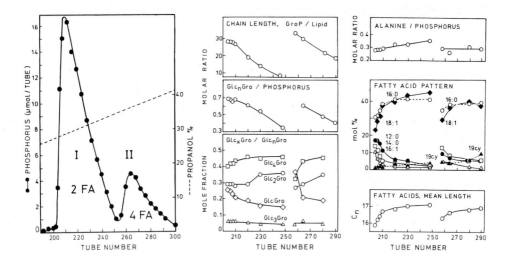

Fig. 2. Polydispersity of the lipoteichoic acid from *Enterococcus hirae*. The polydispersity concerns the number of fatty acids (left panel), the chain length (GroP/lipid), the extent of glycosylation (Glc$_n$Gro/phosphorus), the glycosylation pattern (Glc$_x$Gro/Glc$_n$Gro), the extent of alanylation (alanine/phosphorus) and the fatty acid pattern. Abbreviations: C$_n$, mean length of fatty acids; FA, fatty acids; Glc$_n$Gro, total glycosylated glycerol; Glc$_x$Gro, individual glycosylated glycerols, where x (1–4) is the number of glucosyl residues per glycerol. Examples for abbreviations of fatty acids: 12:0, dodecanoic acid; 16:1, hexadecenoic acid; 19cy, *cis*-11,12-methylenoctadecanoic acid. Taken from [21].

Detailed analyses of several lipoteichoic acids showed that all chains are partially substituted [12,14], and D-alanine ester and glycosyl substituents occur together on the same rather than on separate chains [12,15]. The distribution of glycosyl substituents along the chain is basically random but may be overlaid by a tendency to a regular distribution or an accumulation [16,17]. The alanyl residues occupy the free positions between the glycosyl substituents and are therefore distributed in a similar pattern.

A more recent finding is the polydispersity of lipoteichoic acids [12,18,19]. It concerns the chain length, the extent of glycosylation, and the composition of the fatty acids [12,20,21]. As an example, the composition of the lipoteichoic acid of *Enterococcus hirae* is depicted in Fig. 2. Molecular species were separated by hydrophobic interaction chromatography primarily according to the number of fatty acids and then within each of the resulting peaks according to the length of the hydrophilic chain.

2.2. Poly(glycosylalditolphosphate) lipoteichoic acids

This type of lipoteichoic acid is less widespread: structures II–IV (Fig. 1) have each been found in one bacterial species only. The lipoteichoic acid of *Lactococcus garvieae* (II) contains (α1–6)-linked digalactosyl residues between the glycerophosphate moieties, and position 2 of the latter is substituted with another α-D-galactopyranosyl (Gal) residue [22,23]. Because of the absence of D-alanine ester substituents, the chain is a pure polyanion. In the lipoteichoic acid of *Clostridium innocuum* (III) β-galactopyranosyl residues are intercalated between the glycerophosphate moieties. Approximately 25% of the latter are substituted with *N*-acetyl-D-glucosaminyl residues, 50% with positively charged D-glucosaminyl residues, and a positively charged glucosaminyl residue is also present in the glycolipid anchor (W. Fischer, unpublished). In the lipoteichoic acid of *Streptococcus pneumoniae* (IV), glycerophosphate is replaced by ribitolphosphate; between the ribitolphosphates a tetrasaccharide unit is intercalated which carries two phosphocholine residues on the *N*-acetyl-galactosaminyl residues [24]. The rare amino sugar 2-acetamido-4-amino-2,4,6-trideoxy-D-galactose confers a positive charge to the repeating unit and is also present in the glycolipid anchor. Polydispersity has been demonstrated for II [21] and IV [24]. Whereas the lipid anchors of II are derived from membrane glycolipids, those of III and IV are not because both *C. innocuum* and *S. pneumoniae* contain $Glc(\alpha 1-3)$-acyl$_2$Gro and $Gal(\alpha 1-2)Glc(\alpha 1-3)$acyl$_2$Gro as membrane glycolipids [25,26; W. Fischer, unpublished]. There is a unique situation in *S. pneumoniae* as the teichoic acid and lipoteichoic acid have identical chain structures [27]. In other Gram-positive bacteria, lipoteichoic acids and teichoic acids are unrelated entities, and even in those cases where both possess poly(glycerophosphate) chains, the glycerophosphate residues are enantiomers and biosynthetically derived from phosphatidylglycerol and CDP-glycerol, respectively (for references, see [6]).

Streptococcus oralis which is closely related to *Streptococcus pneumoniae* possesses a choline- and ribitolphosphate-containing teichoic acid [28] and possibly lipoteichoic acid [29]. The occurrence of a lipid-substituted poly(ribitolphosphate) with galactose and alanine substitution was suggested in the nutritionally variant streptococcus serotype I [30].

2.3. Lipoglycans

On the basis of the base composition of DNA, Gram-positive bacteria form two subdivisions with guanine + cytosine content higher and lower than 50%, respectively [31]. According to the present state of knowledge, lipoteichoic acids predominate in the low-guanine + cytosine subdivision and seem to be replaced by lipoglycans in the high-guanine + cytosine subdivision [4].

In lipoglycans, the hydrophilic moiety is a polysaccharide structure which in contrast to LPS is not made up by repeating units. The lipoglycan of *Bifidobacterium bifidum* (Fig. 1, V) contains a linear (1–5)-β-D-galactofuranan attached to a linear (1–6)-β-D-glucopyranan and this is glycosidically linked to D-Gal$p(\beta 1-3)$acyl$_2$Gro and an acylated derivative [32]. Both glycolipids occur among free membrane lipids [33]. Glycerophosphate residues are no longer an integral part of the chain, but attached as monomeric branches to O-6 of the galactofuranosyl residues. Part of the glycerophosphate is substituted with alanine ester which is, however, the L-form in contrast to the D-form in lipoteichoic acids [32]. The same lipoglycan structures have been found in other bifidobacteria [34,35].

Succinylated lipomannans were isolated from *Micrococcus luteus*, *Micrococcus flavus*, *Micrococcus sodonensis* [36,37] and a non-succinylated form from *Micrococcus agilis* [38]. The hydrophobic moiety contains 50–70 (1–2)-, (1–3)-, and (1–6)-linked α-D-mannopyranosyl (Man) residues and two 2,4-substituted branch points; the lipid moiety is diacylglycerol, possibly as part of Man(α1–3)Man(α1–3)acyl$_2$Gro, the major membrane glycolipid [39].

Mycobacterium leprae and *Mycobacterium tuberculosis* contain lipomannan and lipoarabinomannan which are substituted with ester-linked lactate and succinate and anchored to the membrane via phosphatidylinositol [40–43]. Species with one or two additional fatty acids and polydispersity in chain size have been detected [43]. The lipid-bound polysaccharide core is a (1–6)-α-D-mannopyranan with short (1–2)-α-D-mannopyranosyl side chains [40,42]. The outer part of the lipoarabinomannan consists of an O-3-branched (1–5)-α-D-arabinofuranan with β-D-Ara$f$(1–2)-α-D-Ara$f$ residues as nonreducing antigenic termini [44]. In virulent strains of *M. tuberculosis,* these termini are capped with (α1–2)-linked mono-, di-, and trimannosyl residues [45] which, as a kind of bacterial mimicry, imitate common sequences on mammalian cell surface glycoproteins [46]. Lipoarabinomannan, lipomannan, and the presumably biosynthetically related mannosylphosphatidylinositols [47–49] contain hexadecanoate and tuberculostearate as fatty acids, which differs from the fatty acid composition of mycobacterial membrane lipids. A lipoarabinomannan with an as yet unidentified lipid anchor was identified in *Mycobacterium paratuberculosis* [50].

Another lipomannan, thought to be membrane-anchored via phosphatidylinositol, was found in *Propionibacterium freudenreichii* [51].

3. Cellular location

Poly(glycerophosphate) lipoteichoic acids and lipoglycans are associated with the outer leaflet of the cytoplasmic membrane as was shown for numerous bacteria with various immunoelectron microscopy procedures (for references, see [4,52]). Lipid macroamphiphiles, detected towards the surface of the cell, may have lost the contact to the cytoplasmic membrane during secretion into the surroundings (see below). As shown by analysis, lipoteichoic acid amounts to 6 and 10 mol% of lipid amphiphiles in the membrane of *Staphylococcus aureus* and *Lactococcus lactis*, respectively, [53,54], which suggests that lipoteichoic acid constitutes every eighth or fifth lipid molecule in the outer leaflet of the membrane.

4. Biosynthesis

4.1. Formation of the poly(glycerophosphate) chain

The synthesis of poly(glycerophosphate) lipoteichoic acids occurs in linkage to the definite glycolipid anchor [53,55] and is accomplished by sequential addition of individual glycerophosphate residues distal to the lipid anchor [56,57]. The glycerophosphate donor is phosphatidyl glycerol [58–60], not CDP-glycerol which serves as the donor in the synthesis of poly(glycerophosphate) teichoic acids [61].

Two glycerophosphate transferases may be involved, one to recognize the glycolipid, the other to polymerize the chain:

$$\text{Hex}_2\text{acyl}_2\text{Gro} + \text{PtdGro} \rightarrow \text{Gro}P\text{-Hex}_2\text{acyl}_2\text{Gro} + \text{acyl}_2\text{Gro}$$

$$\text{Gro}P\text{-Hex}_2\text{acyl}_2\text{Gro} + n\ \text{PtdGro} \rightarrow (\text{Gro}P)_{n+1}\text{Hex}_2\text{acyl}_2\text{Gro} + n\ \text{acyl}_2\text{Gro}$$

Glycerophosphoglycolipids [4,6], the products of the first reaction [53], are usually detected in variable amounts among free membrane lipids [54]. Concomitantly found acylHex$_2$acyl$_2$Gro and PtdHex$_2$acyl$_2$Gro and their glycerophosphate derivatives may represent the precursors of those lipoteichoic acid species which carry three and four fatty acids on their lipid anchors (Table I). Polydispersity in chain length suggests that from a certain length, chain growth may stop at any point [12,20,21]. How the recently observed variations in the fatty acid composition, dependent on the length of the hydrophilic chain, are accomplished (Fig. 2), has not been clarified.

Owing to the use of phosphatidyl glycerol as glycerophosphate donor, large amounts of diacylglycerol are formed during lipoteichoic acid synthesis. For *S. aureus,* it has been calculated that in one bacterial doubling, the pool of phosphatidyl glycerol (50 mol% of total membrane lipids) turns over three times for lipoteichoic acid synthesis, and the diacylglycerol formed concomitantly is six times the amount present in the diacylglycerol pool [53]. The excess diacylglycerol is not degraded: approximately 15% is used for membrane glycolipid synthesis, and the major fraction is recycled via phosphatidic acid to phosphatidylglycerol (Fig. 3). Recycling of diacylglycerol in the course of lipoteichoic acid synthesis has also been observed with other bacteria [57,63–65].

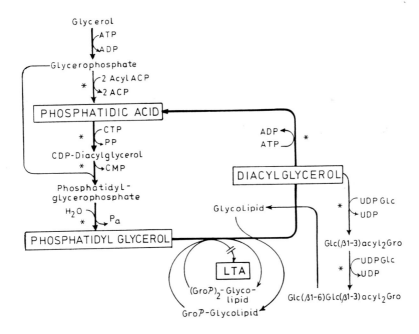

Fig. 3. Lipoteichoic acid biosynthesis and its relationship to membrane lipid metabolism in *S. aureus* [53]. Reactions on the cytosolic site of the membrane are marked by asterisks. For lipoteichoic acid synthesis to occur on the outer leaflet of the membrane, transmembrane movement of phosphatidylglycerol, glycolipid and diacylglycerol is required. The reactions of lipid metabolism have been reviewed by Pieringer [62].

4.2. Substitution of the polymer chain

Chain growth and addition of substituents to the chain are not necessarily dependent on each other [55] but may occur concomitantly in vivo [66,67]. On completed chains, glycosylation and elongation of oligosaccharide residues may continue [94]. Glycosylation requires hexosyl-1-phosphate derivatives of undecaprenol which are formed from nucleoside diphosphate (NDP) hexoses [68,69]:

NDP-hexose + phospho-undecaprenol ⇌ hexosyl-1-phospho-undecaprenol + NDP

hexosyl-1-phospho-undecaprenol + LTA ⇌ hexosyl-LTA + phospho-undecaprenol

The incorporation of D-alanine ester into lipoteichoic acid has been studied in *Lactobacillus casei* [70]. The cytosolic phase requires two proteins: D-alanine-activating enzyme and D-alanyl carrier protein (Dcp), previously named D-alanine:membrane acceptor ligase [71; M.P. Heaton and F.C. Neuhaus, personal communication].

D-alanine + ATP ⇌ D-alanyl-AMP + PP

D-alanyl-AMP + Dcp ⇌ D-alanyl-Dcp + AMP

Both reactions are catalyzed by the D-alanine-activating enzyme. The enzyme system which transfers the D-alanyl residue from D-alanyl carrier protein to lipoteichoic acid remains to be identified.

In growing *S. aureus*, alanine ester is rapidly lost from completed lipoteichoic acid by alanyl transfer to teichoic acid and spontaneous hydrolysis of the labile ester bond [72]. The loss is compensated for by incorporation of new alanine ester at a rate adjusted to the velocity of loss [73].

4.3. Location of lipoteichoic acid synthesis

The site of lipoteichoic acid synthesis on the membrane has not been identified. However, the use of glycolipid, phosphatidylglycerol and hexosyl-1-phosphoundecaprenol makes synthesis feasible to occur on the outer leaflet, the location of the completed polymer. A prerequisite was the movement of the lipid reactants through the membrane in a way similar to that of newly synthesized phosphatidylethanolamine in the membrane of *Bacillus megaterium* [74]. In contrast to lipoteichoic acid synthesis, the membrane-associated synthesis of peptidoglycan and teichoic acid use nucleotide- rather than lipid-activated precursors and are thought to be accomplished by membrane-spanning multienzyme complexes [75–77]. The replacement of glycerophosphate by ribitolphosphate in the lipoteichoic acid of *S. pneumoniae* (Fig. 1, V), suggests the requirement of CDP-ribitol for biosynthesis and therefore a route similar to that of teichoic acids. Moreover, owing to the identical chain structure of the lipoteichoic acid and teichoic acid, common biosynthetic intermediates might be envisaged.

4.4. Synthesis of lipoglycans

In *M. luteus*, Man(α1–3)Man(α1–3)acyl$_2$Gro, the putative anchor of lipomannan, is synthesized from GDP-mannose and diacylglycerol on the inner leaflet of the membrane [39]. The polymer synthesis, however, requires D-mannosyl-1-phospho undecaprenol as glycosyl donor [78,79] and might therefore be located on the outer leaflet. Noteworthy in context with this location is the observation that in *B. bifidum,* the glycerophosphate residues of the lipoglycan are derived from phosphatidylglycerol [80].

5. Metabolic fate of lipoteichoic acids

In pulse-chase experiments using [2-^{3}H]glycerol as the label, after the chase, there was no or negligible loss of lipoteichoic acid and lipids from growing cells of *Bacillus subtilis* [64] and *S. aureus* [53,81], whereas in experiments with *E. faecalis* [82] and *Bacillus stearothermophilus* [63], most of the radiolabel disappeared rapidly from cellular lipids and lipoteichoic acid. Part of the label lost from *B. stearothermophilus* was recovered in the culture medium as apparently unchanged lipid amphiphiles. Secretion of lipoteichoic acids in acylated or deacylated form has been described for various bacteria (for literature, see [83,84]). Under the action of penicillin and other inhibitors of cell wall synthesis, secretion of lipoteichoic acid may be enhanced considerably [85–87] and accompanied by secretion of lipids and proteins [19,88,89]. It has therefore been proposed that during normal growth, lipoteichoic acid may be secreted in the monomeric form, but in the presence of cell wall antibiotics as part of membrane-derived vesicles [19].

6. Physicochemical properties

The critical micellar concentrations of various lipoteichoic acids is in the range of 5×10^{-6} M [90] and 2.8×10^{-7} M to 6.9×10^{-7} M [91] which is three to four orders of magnitude higher than the value for dipalmitoylglycerophosphocholine (4×10^{-10} M) [92]. The lower hydrophobicity of lipoteichoic acids is caused by their large hydrophilic chains [4] and, accordingly, an inverse relationship between hydrophobicity and the size of the hydrophilic chain could be demonstrated [20,21,24,43]. Lipid macroamphiphiles are therefore less firmly anchored in the cytoplasmic membrane than lipids, and long-chain species more prone to extrusion than their short-chain homologues. Additional fatty acids increase the hydrophobicity [21].

X-Ray scattering analysis of various lipoteichoic acids (Fig. 1, I, II) and a lipoglycan (Fig. 1, V) showed that all of them adopt very similar micellar structures in aqueous dispersion, independent on their different chain structures and chain substituents over a wide range of temperature and at various salt concentrations [93]. Nearly independent on the extent of alanine substitution, an average micelle of *S. aureus* lipoteichoic acid contains 150 molecules arranged in a spherical assembly with a diameter of about 22 nm. The hydrophilic region occupies an outer shell of about 8.5 nm thickness, which indicates a highly coiled conformation of the hydrophilic chains because an extended chain of 25 glycerophosphate residues is 17.5 nm in length. Lipoteichoic acids and lipoglycans may also form coiled structures in vivo, presumably located between the peptidoglycan layer and the surface of the cytoplasmic membrane. The closer proximity to the membrane surface helps our understanding of how chain elongation [56], elongation of oligosaccharide substituents [94] and re-esterification of the chain with alanine (see above) may be accomplished. In order to serve as a donor in the alanylation of wall-linked teichoic acid [73], lipoteichoic acid may alternate in vivo between a coiled and a more or less extended conformation.

The micellar structure of lipoteichoic acids and lipoglycans contrasts with the bilayer structure which LPS, phospholipids, and glycerophospholipids adopt in aqueous dispersions [95–98]. In contrast to LPS, lipoteichoic acids seem therefore to be constructed for the insertion into bilayer membranes rather than for the formation of such structures. Accordingly, in studies with Langmuir Blodgett techniques, pure lipoteichoic acid forms monofilms of low collapse pressure. However, mixtures which contain dipalmitoylphosphatidylglycerol and limited amounts of lipoteichoic acid do form stable monolayers with collapse pressures similar to that of pure phosphatidylglycerol [99]. The mean area per molecule remains constant at lipoteichoic acid concentrations up to 20 mol%, but gradually decreases at increasing concentrations, possibly by squeezing out lipoteichoic acid micelles. Similar results were obtained with mixtures of lipoteichoic acid and natural lamellar phase lipids of S. aureus [100]. In X-ray powder diffraction experiments, these mixtures showed a gradual loss of lamellar order at increasing lipoteichoic acid concentrations [100]. It is noteworthy in this context that the lipoteichoic acid concentration in the outer layer of the membrane of S. aureus lies at 10–20 mol% in the range of stable monolayers.

7. Functional aspects

About 15 years ago, when almost only poly(glycerophosphate) lipoteichoic acids were known, essentially three possible functions were ascribed to them: carrier function in teichoic acid synthesis, control of the magnesium ion concentration near the membrane, and regulation of the activity of cellular autolytic enzymes [101]. The carrier activity in teichoic acid synthesis proved to be an artifact when native alanyl-substituted lipoteichoic acids became available, and it seems at present that alanine ester together with glycosyl substituents prevent the interference of lipoteichoic acid with the assembling of teichoic acid on its definitive linkage unit [6].

Together with wall-linked teichoic acids and teichuronic acids, membrane-associated lipoteichoic acid forms a polyionic network between the membrane and the surface of the cell. On the basis of numerous experiments, this network has been proposed to act as a reservoir of divalent cations and particularly to supply membrane-associated enzymes with magnesium ions [4,6]. In this context, the succinylated lipomannan of micrococci has been considered a substitute of lipoteichoic acid, and it is noteworthy that mycobacterial lipoglycans also contain succinyl residues (see above). As discussed previously, some conflicting results make a re-evaluation of the ion-scavenging role of lipoteichoic acids and lipoglycans desirable [4,6].

Autolytic enzymes which hydrolyze particular linkages of the peptidoglycan layer are thought to play a strictly controlled role in cell growth and cell separation. The first lipo-teichoic acid shown to be inhibitory to autologous autolysin was that of *S. pneumoniae* [102]. This autolysin, an *N*-acetylmuramyl-L-alanine-amidase, requires peptidoglycan-linked teichoic acid for activation and binding to its substrate, whereby the phos-phocholine residues of the teichoic acid are essential and cannot be replaced by phos-phoethanolamine [103,104]. In addition to the inhibitory effect on autolysin in vitro, lipo-teichoic acid added to growing pneumococcal cultures, inhibited cell separation and effected resistance to penicillin-induced and stationary-phase lysis [102,105]. Because lipoteichoic acid proved non-inhibitory in the monomeric form [106] and, as shown re-cently, possesses the same phosphocholine-containing chain structure as teichoic acid [24,27], the aforementioned effects seem to have been caused by trapping the enzyme on the surface of lipoteichoic acid micelles [106]. Membrane-associated lipoteichoic acid might therefore act in vivo as a topological barrier. However, temporary alterations of the autolysin-fixation potential, as required for a regulatory role, have still to be demon-strated.

Poly(glycerophosphate) lipoteichoic acids proved inhibitory to autolytic enzymes from organisms containing this type of lipoteichoic acid [107–109]. The acylated and deacylated form [107] and the alanine-free and alanylated acylated form of lipoteichoic acid were found to represent pairs of inhibitory and non-inhibitory species [110]. Moreover, dependent on the extent of alanylation, all intermediates between fully active and fully inactive forms can be observed [110]. It is still an open question whether this regulatory potential is used in the cell (for discussion, see [4]). On the other hand, mutant strains of *L. casei*, deficient in D-alanine ester content of lipoteichoic acid, show aberrant cell morphology and defective cell separation [111]. Moreover, the induction of autolysis of *Staphylococcus simulans* by cationic group A lantibiotics is apparently caused by re-lease of the cationic autolysins from the negatively charged lipoteichoic acid and teichoic acid [112].

As shown with *S. aureus*, lipoteichoic acid may function as D-alanyl-donor for the alanylation of teichoic acid [72]. It is not clear whether this is the only way to introduce alanyl ester into teichoic acid. Although many teichoic acids contain this substituent [113,114], the enzymatic reactions for its incorporation into teichoic acid have not been studied.

As noted above, lipoteichoic acid may influence the physical properties of the bilayer membrane. Changes of the membrane properties are likely to occur, at least between growing and stationary phase cells. A role of macroamphiphiles in this process would not be dependent on a certain hydrophilic structure and could therefore be exerted by the various lipoteichoic acids and lipoglycans equally well.

Whether the killing of enterococci and other Gram-positive bacteria by the cyclic lipopeptide antibiotic daptomycin [115] is caused by the demonstrated halt of lipoteichoic acid synthesis [116,117] requires further study.

8. Potential role in pathogenicity

Activities of lipoteichoic acid, which in the case of pathogenic bacteria may play a role in the host invader interplay, are summarized in Fig. 4. As noted above, lipoteichoic acids and lipoglycans can be secreted during bacterial growth, and secretion may be enhanced by penicillin and other cell wall antibiotics. During secretion, lipidated polymers are thought to become exposed at the cell surface and serve to mediate bacterial adhesion to various cells of the host organism (for references, see [7,118, 119]). The detection of protein adhesins led to a two-step model in which lipoteichoic acid is responsible for primary weak hydrophobic binding which may be followed by stronger binding, mediated by bacterial surface proteins [120]. After entering the host organism, lipoteichoic acids and lipoglycans interact with a number of humoral and cellular components of the host. Lipoteichoic acids have been shown to bind to fibronectin [121] and serum proteins, e.g. albumin [122], and a 28 kDa factor, previously thought to be specific for LPS [123]. There is a transient retention of acylated lipoteichoic acids in heart, liver and kidney, whereas deacylated lipoteichoic acid is rapidly excreted in the urine [124].

Lipoteichoic acids and lipoglycans are potent immunogens, in particular as components of inactivated bacterial cells (for literature, see [4,5]). Immunodeterminants of lipoteichoic acid are the poly(glycerophosphate), glycosyl substituents and alanine ester. Hypersensitive reactions against lipoteichoic acids have been observed, whereby cell-mediated type V hypersensitivity seems not to be involved [125–128].

Like other macroamphiphiles, lipoteichoic acids possess the ability to bind to mammalian cells which requires the hydrophobic moiety and is a saturable process [129–131]. Whether specific receptors are involved is still an open question. Lipoteichoic acids, attached to the cell surface, may bind antilipoteichoic acid antibodies and render the cell susceptible to lysis by complement. Complement-independent cytotoxic reactions of lipoteichoic acids on various mammalian tissue culture cells have also been reported (for references, see [4]). In addition to antibody-dependent complement activation, poly(glycerophosphate) lipoteichoic acids have per se the capacity to activate the direct

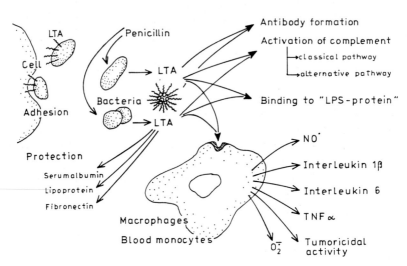

Fig. 4. Biological activities of lipoteichoic acid in the mammalian organism (for literature, see [4,5] and text).

complement pathway, whereby unshielded negative charges and micellar organization enhance this capacity [132]. Pneumococcal lipoteichoic acid, possibly due to its particular structure, activates the alternative complement pathway and, being inactive in micellar dispersion, requires insertion into the cell membrane for activity [133].

In contrast to LPS of Gram-negative bacteria, lipoteichoic acids are not pyrogenic and lack lethal toxicity [5]. A reason for this difference may be the simple lipid structure of lipoteichoic acids and lipoglycans because the endotoxicity of LPS resides in the lipid A moiety and is optimally expressed by a defined disaccharide structure, D-GlcN$p(\beta$1-6)-D-GlcNp, carrying two phosphate groups and six fatty acids including 3-acyloxyacyl groups in a defined location [3]. A key role in the expression of endotoxicity in vivo is played by LPS-reactive mononuclear cells [134] which produce and secrete as mediators hormone-like proteins such as tumor necrosis factor α (TNF) and interleukin 1 (Il-1) along with interleukin 6 (Il-6) [135]. In recent studies, certain lipoteichoic acids displayed a similar capacity of stimulating human blood monocytes and rat macrophages to secrete Il-1β, Il-6, TNFα [11,136,137], and tumoricidal activity [136,138]. Also induced was the formation of nitric oxide NO·and superoxide radical O_2^- in rat macrophages [136] and blood monocytes [139,140], respectively. Induction of nitric oxide synthase has been further observed in vascular smooth muscle cells [141]. The concentration of lipoteichoic acids required for these effects seems to be several orders of magnitude higher than for LPS [3,136]. This may normally prevent overproduction of monokines in vivo, avoid septic shock, and possibly promote beneficial effects. For example, parenterally administered lipoteichoic acid has been reported to lead to immunomodulation in early *Listeria monocytogenes* infection [142], antitumor activity on fibrosarcoma [143] and prevention of pulmonary tumor colonization [144].

Activities, possibly relevant for the host invader interplay, have also been reported for mycobacterial lipoarabinomannan: it is excreted in copious quantities, suppresses T-lymphocyte activation, inhibits γ-interferon activation of macrophages, induces the release of TFNα, and leads to a generalized inhibition of antigen presentation by antigen presenting cells (for literature, see [41,42,44]).

Acknowledgement

The continuous support of the work done in the author's laboratory by the Deutsche Forschungsgemeinschaft is gratefully acknowledged.

References

1. Baddiley, J. (1972) Essays Biochem. 8, 35–77.
2. Brade, H., Brade, L. and Rietschel, E.Th. (1988) Zbl. Bakteriol. Hyg. 268, 151–179.
3. Rietschel, E.Th., Kirikae, T., Feist, W., Loppnow, H., Zabel, P., Brade, L., Ulmer, A.J., Brade, H., Seydel, U., Zähringer, U., Schlaak, M., Flad, H.D. and Schade, U. (1991) in: H. Sies, L. Flohé and G. Zimmer (Eds.), Molecular Aspects of Inflammation, 42. Colloquium Mosbach 1991, Springer-Verlag, Berlin, pp. 207–231.
4. Fischer, W. (1990) in: D. Hanahan (Ed.), Handbook of Lipid Research, Vol. 6, Plenum Press, New York, pp. 123–233.

212

5. Wicken, A.J. and Knox, K.W. (1980) Biochim. Biophys. Acta 604, 1–26.
6. Fischer, W. (1988) in: A.H. Rose and D.W. Tempest (Eds.), Advances in Microbial Physiology, Vol. 29, Academic Press, London, pp. 233–302.
7. Sutcliff, I.C. and Shaw, N. (1991) J. Bacteriol. 173, 7065–7069.
8. Wicken, A.J. and Knox, K.W. (1970) J. Gen. Microbiol. 60, 293–301.
9. Toon, P., Brown, P.E. and Baddiley, J. (1972) Biochem. J. 127, 399–409.
10. Fischer, W., Mannsfeld, T. and Hagen, G. (1990) Biochem. Cell Biol. 68, 33–43.
11. Bhakdi, S., Klonisch, T., Nuber, P. and Fischer, W. (1991) Infect. Immun. 59, 4614–4620.
12. Leopold, K. and Fischer, W. (1991) Eur. J. Biochem. 196, 475–482.
13. Kochanowski, B., Fischer, W., Iida-Tanaka, N. and Ishizuka, I. (1993) Eur. J. Biochem., 214, 747–755.
14. Fischer, W. and Rösel, P. (1980) FEBS Lett. 119, 224–226.
15. Kochanowski, B., Leopold, K. and Fischer, W. (1993) Eur. J. Biochem. 214, 757–761.
16. Schurek, J. and Fischer, W. (1989) Eur. J. Biochem. 186, 649–655.
17. Leopold, K. (1991) Doctoral Thesis, University Erlangen-Nürnberg, Germany.
18. Maurer, J.J. and Mattingly, S.J. (1991) J. Bacteriol. 173, 487–494.
19. Pollack, J.H., Ntamere, A.S. and Neuhaus, F.C. (1992) J. Gen. Microbiol. 138, 849–859.
20. Leopold, K. and Fischer, W. (1992) Anal. Biochem. 201, 350–355.
21. Fischer, W. (1993) Anal. Biochem. 208, 49–56.
22. Koch, H.U. and Fischer, W. (1978) Biochemistry 17, 5275–5281.
23. Schleifer, K.H., Kraus, J., Dvorak, C., Kilpper-Bälz, R., Collins, M.D. and Fischer, W. (1985) System. Appl. Microbiol. 6, 183–195.
24. Behr, T., Fischer, W., Peter-Katalinić, J and Egge, H. (1992) Eur. J. Biochem. 207, 1063–1075.
25. Kaufman, B., Kundig, F.D., Distler, J. and Roseman, S. (1965) Biochem. Biophys. Res. Commun. 18, 312–318.
26. Brundish, D.E., Shaw, N. and Baddiley, J. (1967) Biochem. J. 104, 205–211.
27. Fischer, W., Behr, T., Hartmann, R., Peter-Katalinić, J. and Egge, H. (1993) Eur. J. Biochem. 215, 851–857.
28. Kilpper-Bälz, R., Wenzig, P. and Schleifer, K. H. (1985) Int. J. Systemat. Bacteriol. 35, 482–488.
29. Horne, D.S. and Tomasz, A. (1993) J. Bacteriol. 175, 1717–1722.
30. George, M. and van de Rijn, I. (1988) J. Immunol. 140, 2008–2015.
31. Woese, C.R. (1987) Microbial. Rev. 51, 221–271.
32. Fischer, W. (1987) Eur. J. Biochem. 165, 639–646.
33. Veerkamp, J.H. (1972) Biochim. Biophys. Acta 273, 359–367.
34. Op den Camp, H.J.M., Peeters, P.A.M., Oosterhof, A. and Veerkamp, J.H. (1985) J. Gen. Microbiol. 131, 661–668.
35. Iwasaki, H., Araki, Y., Ito, E., Nagaoka, M. and Yokokura, T. (1990) J. Bacteriol. 172, 845–852.
36. Pless, D.D., Schmit, A.S. and Lennarz, W.J. (1975) J. Biol. Chem. 250, 1319–1327.
37. Powell, D.A., Duckworth, M. and Baddiley, J. (1975) Biochem. J. 151, 387–397.
38. Lim, S. and Salton, R. J. (1985) FEMS Microbiol. Lett. 27, 287–291.
39. Lennarz, W.J. and Talamo, B. (1966) J. Biol. Chem. 241, 2707–2719.
40. Hunter, S.W., Gaylord, H. and Brennan, P.J. (1986) J. Biol. Chem. 261, 12345–12351.
41. Hunter, S.W. and Brennan, P.J. (1990) J. Biol. Chem. 265, 9272–9279.
42. Chatterjee, D., Hunter, S.W., McNeil, M. and Brennan, P.J. (1992) J. Biol. Chem. 267, 6228–6233.
43. Leopold, K. and Fischer, W. (1993) Anal. Biochem. 208, 57–64.
44. Chatterjee, D., Bozic, C.M., McNeil, M. and Brennan, J.P. (1991) J. Biol. Chem. 266, 9652–9660.
45. Chatterjee, D., Lowell, K., Rivoire, B., McNeil, M.R. and Brennan, P.J. (1992) J. Biol. Chem. 267, 6234–6239.
46. Kornfeld, R. and Kornfeld, S. (1981) in: W.J. Lennarz (Ed.), The Biochemistry of Glycoproteins and Proteoglycan, Plenum Press, New York, pp. 1–34.
47. Ballou, C.E., Vilkas, E. and Lederer, E. (1963) J. Biol. Chem. 238, 69–76.
48. Lee, Y.C. and Ballou, C.E. (1965) Biochemistry 4, 1395–1404.
49. Pangborn, M.C. and McKinney, J.A. (1966) J. Lipid Res. 7, 627–633.
50. Sugden, E.A., Bakhshish, S.S., Bundle, D.R. and Duncan, R. (1987) Infect. Immun. 55, 762–770.

51. Sutcliff, I.C. and Shaw, N. (1989) FEMS Microbiol. Lett. 59, 249–252.
52. Wicken, A.J. and Knox, K.W. (1975) Science 187, 1161–1167.
53. Koch, H.U., Haas, R. and Fischer, W. (1984) Eur. J. Biochem. 138, 357–363.
54. Fischer, W. (1981) in: G.D. Shockman and A.J. Wicken (Eds.), Chemistry and Biological Activities of Bacterial Surface Amphiphiles, Academic Press, New York, pp. 209–228.
55. Ganfield, M.-C.W. and Pieringer, R.A. (1980) J. Biol. Chem. 255, 5164–5169.
56. Cabacungan, E. and Pieringer, R.A. (1981) J. Bacteriol. 147, 75–79.
57. Taron, D.J., Childs, W.-C. III and Neuhaus, F.C. (1983) J. Bacteriol. 154, 1110–1116.
58. Glaser, L. and Lindsay, B. (1974) Biochem. Biophys. Res. Commun. 59, 1131–1136.
59. Emdur, L.I. and Chiu, T.H. (1975) FEBS Lett. 55, 216–219.
60. Pieringer, R.A., Ganfield, M.-C.W., Gustow, E. and Cabacungan, E. (1981) in: G.D. Shockman and A.J. Wicken (Eds.), Chemistry and Biological Activities of Bacterial Surface Amphiphiles, Academic Press, New York, pp. 167–179.
61. Hancock, I.C. and Baddiley, J. (1985) in: A.N. Martonosi (Ed.), The Enzymes of Biological Membranes, Vol. 2, Plenum Press, New York, pp. 279–307.
62. Pieringer, R.A. (1983) in: P.D. Boyer (Ed.), The Enzymes, Vol. XVI, Academic Press, London, pp. 255–306.
63. Card, G.L. and Finn, D.J. (1983) J. Bacteriol. 154, 294–303.
64. Koga, Y., Nishihara, M. and Morii, H. (1984) Biochim. Biophys. Acta 793, 86–94.
65. Lombardi, F.J., Chen, S.L. and Fulco, A.J. (1980) J. Bacteriol. 141, 626–634.
66. Brautigan, V.M., Childs, W.C. III and Neuhaus, F.C. (1981) J. Bacteriol. 146, 239–250.
67. Fischer, W. (1982) Biochim. Biophys. Acta 711, 372–375.
68. Mancuso, D.J. and Chiu, T.-H. (1982) J. Bacteriol. 152, 616–625.
69. Yokoyama, K., Araki, Y. and Ito, E. (1988) Eur. J. Biochem. 173, 453–458.
70. Neuhaus, F.C., Linzer, R. and Reusch, Jr., V.M. (1974) Ann. N.Y. Acad. Sci. 235, 501–516.
71. Heaton, M.P. and Neuhaus, F.C. (1992) J. Bacteriol. 174, 4707–4717.
72. Haas, R., Koch, H.U. and Fischer, W. (1984) FEMS Microbiol. Lett. 21, 27–31.
73. Koch, H.U., Döker, R. and Fischer, W. (1985) J. Bacteriol. 164, 1211–1217.
74. Rothman, J.E. and Kennedy, E.P. (1977) Proc. Natl. Acad. Sci. USA, 74, 1821–1825.
75. Harrington, C.R. and Baddiley, J. (1983) J. Bacteriol. 155, 776–792.
76. Harrington, C.R. and Baddiley, J. (1985) Eur. J. Biochem. 153, 639–645.
77. Baddiley, J. (1985) Biochem. Soc. Trans. 13, 992–994.
78. Scher, M. and Lennarz, W.J. (1969) J. Biol. Chem. 244, 2777–2789.
79. Scher, M., Lennarz, W.J. and Sweeley, C.C. (1968) Proc. Natl. Acad. Sci. USA 59, 1313–1320.
80. Op den Camp, H.J.M., Oosterhof, A. and Veerkamp, J. H. (1985) Biochem. J. 228, 683–688.
81. Raychaudhuri, D. and Chatterjee, A.N. (1985) J. Bacteriol. 164, 1337–1349.
82. Carson, D.D., Pieringer, R.A. and Daneo-Moore, L. (1981) J. Bacteriol. 146, 590–604.
83. Shockman, G.D. (1981) in: G.D. Shockman and A.J. Wicken (Eds.), Chemistry and Biological Activities of Bacterial Surface Amphiphiles, Academic Press, New York, pp. 21–40.
84. Knox, K.W. and Wicken, A.J. (1981) in: G.D. Shockman and A.J. Wicken (Eds.), Chemistry and Biological Activities of Bacterial Surface Amphiphiles, Academic Press, New York, pp. 229–237.
85. Alkan, M.L. and Beachy, E.M. (1978) J. Clin. Invest. 61, 671–677.
86. Horne, D. and Tomasz, A. (1979) J. Bacteriol. 137, 1180–1184.
87. Brisette, J.L., Shockman, G.D. and Pieringer, R.A. (1982) J. Bacteriol. 151, 838–844.
88. Hakenbeck, R., Martin, C. and Morelli, G. (1983) J. Bacteriol. 155, 1372–1381.
89. Ntamere, A.S. and Neuhaus, F.C. (1987) Abstracts of the Annual Meeting of the American Society for Microbiology 87, p. 226.
90. Courtney, H.S., Simpson, W.A. and Beachy, E.H. (1986) Infect. Immun. 51, 414–418.
91. Wicken, A.J., Evans, J.D. and Knox, K.W. (1986) J. Bacteriol. 166, 72–77.
92. Smith, R. and Tanford, C. (1972) J. Mol. Biol. 67, 75–83.
93. Labischinski, H., Naumann, D. and Fischer, W. (1991) Eur. J. Biochem. 202, 1269–1274.
94. Cabacungan, E. and Pieringer, R.A. (1985) FEMS Microbiol. Lett. 26, 49–52.
95. Naumann, D., Schultz, C., Born, J., Labischinski, H., Brandenburg, K., von Busse, G., Brade, H. and Seydel, U. (1987) Eur. J. Biochem. 164, 159–169.

96. Naumann, D., Schultz, C., Sabisch, A., Kastowsky, M. and Labischinski, H. (1989) J. Mol. Struct. 214, 213–246.
97. Cullis, P.R. and Hope, M.J. (1985) in: D.E. Vance and J.E. Vance (Eds.), Biochemistry of Lipids and Membranes, Benjamin/Cummins, Menlo Park, CA, pp. 25–72.
98. Shipley, G.G. (1973) in: D. Chapman and D.F.H. Wallach (Eds.), Biological Membranes, Vol. 2, Academic Press, London, pp. 1–89.
99. Gutberlet, Th., Markwitz, S., Labischinski, H. and Bradaczek, H. (1991) Macromol. Chem., Macromol. Symp. 46, 283–287.
100. Gutberlet, Th. and Bradaczek, H. (1992) 4th European Conference of Organized Thin Films, Bangor, 1992.
101. Lambert, P.A., Hancock, I.C. and Baddiley, J. (1977) Biochim. Biophys. Acta 472, 1–12.
102. Höltje, J.-V. and Tomasz, A. (1975) Proc. Natl. Acad. Sci. USA 72, 1690–1694.
103. Höltje, J.-V. and Tomasz, A. (1975) J. Biol. Chem. 250, 6072–6076.
104. Guidicelli, S. and Tomasz, A. (1984) J. Bacteriol. 158, 1188–1190.
105. Höltje, J.-V. and Tomasz, A. (1975) J. Bacteriol. 124, 1023–1027.
106. Briese, T. and Hakenbeck, R. (1985) Eur. J. Biochem. 146, 417–427.
107. Cleveland, R.F., Höltje, J.-V., Wicken, A.J., Tomasz, A., Daneo-Moore, L. and Shockman, G.D. (1975) Biochem. Biophys. Res. Commun. 67, 1128–1135.
108. Cleveland, R.F., Wicken, A.J., Daneo-Moore, L. and Shockman, G.D. (1976) J. Bacteriol. 126, 192–197.
109. Cleveland, R.F., Daneo-Moore, L., Wicken, A.J. and Shockman, G.D. (1976) J. Bacteriol. 127, 1582–1584.
110. Fischer, W., Rösel, P. and Koch, H.U. (1981) J. Bacteriol. 146, 467–475.
111. Ntamere, A.S., Taron, J.D. and Neuhaus, F.C. (1987) J. Bacteriol. 169, 1702–1711.
112. Bierbaum, G. and Sahl, H.G. (1991) in: G. Jung and H.G. Sahl (Eds.), Nisin and Novel Lantibiotics ESCOM, Leyden, pp. 386–396.
113. Archibald, A.R. and Baddiley, J. (1966) in: M.L. Wolfrom and R.S. Tipson (Eds.), Advances in Carbohydrate Chemistry, Academic Press, New York, pp. 323–375.
114. Knox, K.W. and Wicken, A.J. (1973) Bacteriol. Rev. 37, 215–257.
115. Eliopoulos, G.M., Thauvin, C., Gerson, B. and Moellering, Jr., R.C. (1985) Antimicrob. Agents Chemother. 27, 357–362.
116. Canepari, P., Boaretti, M., del Mar Lleò, M. and Satta, G. (1990) Antimicrob. Agents Chemother. 34, 1220–1226.
117. Boaretti, M., Canepari, P., del Mar Lleò, M. and Satta, G. (1993) J. Antimicrob. Chemother. 31, 227–235.
118. Beachy, E.H. and Courtney, H.S. (1987) Rev. Infect. Dis. 9, 475–481.
119. Courtney, H.S., Hasty, D.L. and Ofek, I. (1990) in: R.J. Doyle and M. Rosenberg (Eds.), Microbial Cell Surface Hydrophobicity, American Society for Microbiology, Washington, DC, pp. 361–386.
120. Hasty, D.L., Beachy, E.H., Courtney, H.S. and Simpson, W.A. (1989) in: S.E. Carsons (Ed.), Fibronectin in Health and Disease, CRC Press, Boca Raton, FL, pp. 89–112.
121. Courtney, H., Stanislawski, L., Ofek, I., Simpson, W.A., Hasty, D.L. and Beachy, E.H. (1988) Rev. Infect. Dis. 10, S360–S362.
122. Simpson, W.A., Ofek, I. and Beachy, E.H. (1980) Infect. Immun. 29, 119–122.
123. Brade, L., Brade, H. and Fischer, W. (1990) Microbial Pathogenesis 9, 355–362.
124. Hyzy, J., Sciotti, V., Albini, B. and Stinson, M. (1992) Microbial Pathogenesis 13, 123–132.
125. Fiedel, B.A. and Jackson, R.W. (1976) Infect. Immun. 13, 1585–1590.
126. Fiedel, B.A. and Jackson, R.W. (1979) Med. Microbiol. Immunol. 167, 251–260.
127. Jackson, D.E., Howlett, C.R., Wicken, A.J. and Jackson, G.D.F. (1981) Int. Arch. Allergy Appl. Immunol. 65, 304–312.
128. Aasjord, P., Nyland, H. and Mørk, S. (1980) Acta Pathol. Microbiol. Scand. (C) 88, 287–291.
129. Beachy, E.H., Dale, J.B., Simpson, W.A., Evans, J.D., Knox, K.W., Ofek, I. and Wicken, A.J. (1979) Infect. Immun. 23, 618–625.
130. Beachy, E.H., Simpson, W.A., Ofek, I., Hasty, D.L., Dale, J.B. and Whitnak, E. (1983) Rev. Infect. Dis. 5, S670–S677.

131. Simpson, W.A., Ofek, I., Sarasohn, C., Morrison, J.C. and Beachy, E.H. (1980) J. Infect. Dis. 141, 457–462.
132. Loos, M., Clas, F. and Fischer, W. (1986) Infect. Immun. 53, 595–599.
133. Hummell, D.S., Swift, A.J., Tomasz, A. and Winkelstein, J.A. (1985) Infect. Immun. 47, 384–387.
134. Freudenberg, M.A., Keppler, D. and Galanos, C. (1986) Infect. Immun. 51, 891–895.
135. Vogel, S.N. and Hogan, M.M. (1990) in: J.J. Oppenheim and E.M. Shevack (Eds.), Immunopharmacology- the Role of Cells and Cytokines in Immunity and Inflammation, Oxford University Press, New York, pp. 238–258.
136. Keller, R., Fischer, W., Keist, R. and Bassetti, S. (1992) Infect. Immun. 60, 3664–3672.
137. Riesenfeld-Orn, I., Wolpe, S., Garcia-Bustos, J.F., Hoffmann, M.K. and Tuomanen, E. (1989) Infect. Immun. 57, 1890–1893.
138. Hamada, S., Yamamoto, T., Koga, T., McGhee, J.R., Michalek, S.M. and Yamamoto, S. (1985) Zbl. Bakteriol. Parasitenkd. Infektionskr. Hyg. Abt. A 259, 228–243.
139. Oshima, Y., Beuth, J., Yassin, A., Ko, H.L. and Pulverer, G. (1988) Med. Microbiol. Immunol. 177, 115–121.
140. Levy, R., Kotb, M., Naganker, O., Majumdar, G., Alkan, M., Ofek, I. and Beachy, E.H. (1990) Infect. Immun. 58, 566–568.
141. Auguet, M., Lonchampt, M.O., Delaflotte, S., Goulin-Schulz, J., Chabrier, P. E. and Braquet, P. (1992) FEBS Lett. 297, 183–185.
142. Oshima, Y., Beuth, J., Ko, H.L., Roszkowski, K., Hauck, D. and Pulverer, G. (1988) Zbl. Bakteriol. Hyg. A 269, 251–256.
143. Usami, H., Yamamoto, A., Yamashita, W., Sugawara, Y., Hamada, S., Yamamoto, T., Kato, K., Kokeguchi, S., Ohokuni, H. and Kotani, S. (1988) Br. J. Cancer 57, 70–73.
144. Oshima, Y., Ko, H.L., Beuth, J. and Pulverer, G. (1988) Zbl. Bakteriol. Hyg. A 270, 213–218.

J.-M Ghuysen and R. Hakenbeck (Eds.), *Bacterial Cell Wall*
217

CHAPTER 11

Cell-wall-associated proteins in Gram-positive bacteria

MICHAEL A. KEHOE

*Department of Microbiology, University of Newcastle upon Tyne, Medical School,
Newcastle upon Tyne, NE2 4HH, UK*

1. Introduction

The cell walls of all Gram-positive bacteria contain proteins that play pivotal roles in cell growth and division, such as enzymes involved in the synthesis, modification and turn-over of peptidoglycan. The reader is referred to Chapters 2–7, 25 and 26 for detailed accounts of these processes and proteins. In addition to these 'housekeeping' proteins, many Gram-positive bacteria express wall-associated proteins that interact in various ways with the extracellular environment and this chapter focuses on this latter category of wall-associated proteins.

The first Gram-positive cell-wall-associated proteins to be described in the literature were the M antigens of group A streptococci (*Streptococcus pyogenes*), which were extracted with hot-acid and subjected to serological studies as early as the 1920s [1,2]. However, such extraction procedures resulted in highly fragmented, crude antigen preparations and this limited detailed biochemical or structural studies for many decades. With the exception of 'housekeeping' proteins, cell-wall-associated proteins in other Gram-positive species received little attention until the discovery of *Staphylococcus aureus* protein A in 1966 [3]. Lysostaphin, which had been discovered a year earlier, proved to be an effective means for releasing undegraded protein A from *S. aureus* cell walls [4], facilitating the first detailed biochemical and structural studies on a Gram-positive cell-wall-associated protein. Subsequently, it was found that pepsin releases a large fragment of the *S. pyogenes* M protein, called pepM antigen, which could be purified to apparent homogeneity, permitting detailed structural studies on M proteins in the late 1970s [5,6]. However, such studies were limited to only part of the M protein molecule and studies on other wall-associated proteins continued to be hindered by difficulties in extracting and purifying intact proteins. Progress remained slow for many years, but a real breakthrough was provided by the application of recombinant DNA techniques to the analysis of wall-associated protein genes from the mid-1980s onwards. This has produced a very rapid, exponential expansion in our knowledge of Gram-positive cell-wall-associated proteins and this chapter concentrates on those proteins that have been studied at a molecular level in recent years.

Recent molecular studies on Gram-positive cell-wall-associated proteins have been stimulated by interest in a small number of species of medical or commercial importance, rather than interests in Gram-positive cell walls per se. Most recent studies have focused on pathogenic bacteria, particularly staphylococcal and streptococcal species that remain in a predominantly extracellular location upon infecting human or animal hosts. Not surprisingly, these pathogens express wall-associated proteins that interact with various factors present in the extracellular matrix of mammalian tissues or circulating in body fluids. These include mammalian proteins such as collagen, complement components, antibodies, fibrinogen, fibronectin, kininogen, laminin, α2-macroglobulin, plasmin, prothrombin and salivary glycoproteins. A wide range of Gram-positive bacterial cell-wall-associated proteins that interact with such mammalian proteins have been characterized in considerable detail in recent years. Thus, the major part of this chapter concerns wall-associated proteins in pathogenic Gram-positive species, although a limited number of proteins from non-pathogenic species that have been studied at a molecular level in recent years are also discussed.

2. Protein secretion: novel Gram-positive components

The factors that determine whether a secreted protein is destined to be released from the cell or to form durable cell-wall associations are very closely linked with secretion of the protein through the cytoplasmic membrane. With two exceptions (see Sections 10 and 14), wall-associated proteins in Gram-positive bacteria are synthesized initially as precursors possessing short (up to 42 residues) N-terminal extensions that are removed by a membrane-associated signal-peptidase during protein translocation across the cytoplasmic membrane. Gram-positive N-terminal secretion-signal sequences are structurally very similar to, but tend to be longer than, their counterparts in other species and the reader is referred elsewhere [7,8] for a detailed account of these signals. Secretion pathways in Gram-positive bacteria have been studied in far less detail than in *Escherichia coli*, but homologues of several *E. coli* secretion components have been identified in *Bacillus subtilis*, suggesting that major secretion pathways are very similar in Gram-positive and Gram-negative species [7]. Further, most recombinant Gram-positive exoproteins that are expressed in *E. coli* are secreted to the periplasm. Despite these similarities, recent studies on *B. subtilis* mutants have identified a novel Gram-positive secretion component designated PrsA [9,10], which is of particular interest here in that it raises the possibility that extracytoplasmic chaperons determine the fate of at least some secreted proteins. PrsA-defective mutants display a 50-fold reduction in the secretion of a highly expressed recombinant α-amylase and a 18–62% reduction in the secretion of native *B. subtilis* exoproteins [9]. Mature α-amylase (i.e. lacking its N-terminal signal sequence) accumulates in membrane fractions of PrsA-defective mutants, indicating that both the association of the α-amylase precursor with the membrane and the processing of its N-terminal signal sequence occurs independently of PrsA [9]. Sequence analysis [10] suggests that PrsA is a hydrophilic lipoprotein, anchored to the external surface of the cytoplasmic membrane (Section 3.1). Together, these data suggest that PrsA plays a role at a late stage in the secretion process. A clue to its function is provided by recent studies showing that the folding of *B. subtilis* levansucrase is influenced by Fe^{3+} and that in the absence of Fe^{3+} mature levansucrase accumulates in the cell envelope [11]. By analogy, it has been suggested that PrsA acts as an extracytoplasmic chaperon, guiding the folding of exoproteins emerging from the membrane into conformations that are compatible with their efficient release from the cell [10].

Although the mechanism of action of PrsA remains to be confirmed, interestingly, it shares considerable homology with a protein called PrtM, which is required for the maturation of cell-wall-associated serine proteases produced by *Lactococcus lactis* [12,13]. Like many extracellular proteases, the plasmid-encoded *L. lactis* Wg2 and SK11 proteases are synthesized initially as preproproteins, with a large propeptide sequence between a typical N-terminal signal sequence and the N-terminal end of the fully mature protease [14,15]. These propeptides must be removed after translocation of the proprotease across the membrane in order to activate the protease. Propeptide processing of *Bacillus* subtilisins is predominantly autocatalytic [16]. However, unlike subtilisins, propeptide processing of proWg2 or proSK11 requires an accessory factor, PtrM, which is encoded by sequences adjacent to the Wg2 or SK11 genes on the plasmid [12,13]. Like the homologous PrsA, PtrM is predicted from sequence data to be an extracytoplasmic lipoprotein, but there is no evidence that it possesses protease activity [12,13]. It has been suggested that PtrM may act as a chaperon to hold proWg2 and proSK11 emerging from the membrane in conformations compatible with the autocatalytic removal of their propeptides [10]. The possibility that similar wall-associated chaperon proteins might play a role in directing certain secreted proteins into cell-wall-association pathways would be consistent with some of the studies outlined in Section 3.2, although it must be emphasized that at present there is no direct evidence that this is the case.

3. Localization of proteins in the cell wall

After translocation across the cytoplasmic membrane, the factors that influence the subsequent localization of secreted proteins are quite different in Gram-positive and Gram-negative bacteria. In Gram-negative species, the outer membrane forms an effective permeability barrier that retains most secreted proteins in the periplasm. To be released from the cell, a protein usually needs to possess specific signals that can be recognized by accessory secretion pathways, such as the HlyB-HlyD-TolC pathway in the case of *E. coli* α-haemolysin (Chapter 20). In contrast, in Gram-positive bacteria specific signals are often required to retain a protein in the cell wall, rather than to direct its release. Some secreted proteins diffuse rapidly through the thick Gram-positive cell wall, although limits in wall porosity or non-covalent affinities for anionic wall polymers can retard the release of many extracellular proteins [17]. However, the majority of proteins that form durable wall associations possess either distinctive N-terminal signals (lipoproteins) or, more commonly, distinctive C-terminal wall-associating signals, although a number of wall-associated proteins which are discussed in later sections possess neither of these types of signals.

3.1. N-terminal signals in lipoproteins

A number of wall-associated proteins in Gram-positive bacteria are anchored to the external surface of the cytoplasmic membrane via a covalently attached lipid moiety. These include wall-associated penicillinases in *Bacillus licheniformis, Bacillus cereus* and *S. aureus*, and the wall-associated adhesins SarA and SsaB in *Streptococcus gordonii* and *Streptococcus sanguis,* respectively [7,18–20, and Section 12]. In addition, sequence predictions strongly suggest that the PrsA and PtrM proteins described above [10,12,13], as well as the AmiA and MalX proteins of *Streptococcus pneumoniae* [21] and the MelE protein of *Streptococcus mutans* [22] are similarly anchored to the cytoplasmic membrane. Most studies on secreted bacterial lipoproteins have focused on Gram-negative species [18], but the mechanisms by which secreted lipoproteins acquire a lipid anchor are very similar in Gram-positive and Gram-negative bacteria. Like *E. coli* lipoproteins, there is direct evidence that the *B. licheniformis* penicillinase is synthesized as a glyceride-modified prepenicillinase [23]. Further, its secretion and processing are inhibited by globomycin, a cyclic peptide antibiotic that specifically inhibits the processing of lipoprotein precursors in *E. coli* [18,23]. Both Gram-negative and Gram-positive lipoproteins possess similar, distinctive, N-terminal signal-sequences, which tend to be shorter than those of other secreted proteins [18,24]. These unique signal sequences contain a tetrapeptide consensus at the cleavage site, consisting of Leu-X-Y-Cys, where X and Y are predominantly small neutral residues and signal-peptidase cleavage occurs between Y and Cys (Fig. 1a). This sequence directs either co- or post-translational modification, involving the transfer of glycerol from phosphatidylglycerol to the +1 Cys, followed by the transfer of fatty acids from phospholipid to the glyceryl-prelipoprotein, to produce a diglyceride-prelipoprotein [18]. In *E. coli*, processing of diglyceride-prelipoproteins requires a distinct signal peptidase (Spase II) which is specifically inhibited by globomycin [18]. Although a homologue of Spase II has yet to be identified in Gram-positive bacteria, the specific inhibition of *B. licheniformis* prepenicillinase processing by globomycin [23] suggests strongly that one exists.

3.2. C-terminal wall-associating sequences

The C-terminal ends of a large number of otherwise quite distinct Gram-positive wall-associated proteins share common structural features that are required to localize these proteins in the cell wall. This suggests that there has been a strong selective pressure for retaining or acquiring these structures during evolution. These C-terminal structures include a number of distinct features (Fig. 1b). At the extreme C-terminus there is a stretch of 15–22 hydrophobic residues, followed by a short tail of predominantly charged amino acids. Immediately upstream from this hydrophobic/charged-tail domain, there is a highly conserved Leu-Pro-X-Thr-Gly-X (LPXTGX) motif [25], which is usually preceded by a sequence containing a high proportion of regularly spaced proline (Pro) residues. Both the length and the primary sequence of this Pro-rich region can vary considerably between unrelated wall-associated proteins and in some cases, for example protein A (Section 8.1), the Pro-rich region can be located a considerable distance upstream from the LPXTGX motif, with the intervening sequences containing relatively few prolines. Early fractiona-

A. N-TERMINAL SIGNAL-SEQUENCES IN WALL-ASSOCIATED LIPOPROTEINS

B. licheniformis penicillinase	MKLWFSTLKLKKAAAVLLFSCVA LAG	C ANNQ......
S. aureus penicillinase	MKKLIFLIVIALV LSA	C NSNS......
S. sanguis adhesin SsaB	MKKLGFLSLLLLAVCT LFA	C SNQK.....
S. gordonii adhesin SarA	MKKGKILALAGVALLATGV LAA	C SNST.....
L. lactis PrtM	MKKKMRLKVLLASTATALLL LSG	C QSNQ......

Diglyceride

B. C-TERMINAL WALL-ASSOCIATING SIGNALS

PROTEIN		LPXTGX Motif		Hydrophobic region	Charged tail
M5	LAKLRAGKASDSQTPDTKPGNKAVPGKGQAPQAGTKPNQNKAPMKETKRQ	LPSTGE	TAN	PFFTAAALTVMATAGVAAVV	KRKEEN
Protein G	ASIPLVPLTPATPIAK DDTKK EDAKK PEAKK DDAKK AET	LPTTGE	GSN	PFFTAAALAVMAGAGALAVA	SRKKED
Inl A	TPPTTNNGNTTPPSANIPGSDTSNTSTGNSASTTSTMNAYDPYNSKEAS	LPTTGD	SDN	ALYLLGLLAVGTAMALY	KKARASK
FnBP B	PGKPIPPAKEEPKKPSKPVEQGKVVTPVIEINEKVKAVVPTKKAQSKKSE	LPETGG	EESTNN	GLMFGGLFSILGLALL	RRNKKNHKA
Protein A	NDIAKANGTTADKIAADNKLADKNMIKPGQELVVDKKQPANHADANKAQA	LPETGE	EN	PLIGTTVFGGLSLALGAALLAG	RRREL

Fig. 1. Cell-wall-associating signals. Only some examples of the large number of proteins possessing the types of wall-associating signals described in Section 3 are shown. The sequences shown are from the following references: B. licheniformis and S. aureus penicillinases [18], SsaB [67], SarA [19], PrtM [12,13], M5 [120], Protein G [42], Inl A [265], FnBPB [213] and protein A [182]. In part B, only the C-terminal ends of the Pro-rich regions in protein G (Section 8.1), Inl A (internalin; Section 14) and FnBP B (Section 9) are shown and the Pro-rich Xr region of protein A (Section 8.1) is not shown. The boxes upstream from the LPXTGX motif in protein G highlight repeated sequences.

tion studies showed that this Pro-rich region in *S. aureus* protein A is associated with the cell wall [26,27] and more recent studies have confirmed that this is also the case in group A streptococcal M proteins [28].

It can be predicted that the C-terminal hydrophobic residues form a membrane-spanning α-helical domain, with the downstream charged-tail acting as a stop-transfer signal [25,29]. This prediction is supported by the observation that at pH 5.5 (but see below), group A streptococcal M proteins remain tightly associated with protoplast membranes after removal of the cell wall [30]. Thus, the extreme C-terminal hydrophobic/charged-tail sequences are often described in the literature as a membrane 'anchor' and it has been widely assumed that proteins are held firmly on the cell surface by being anchored in the membrane via these C-terminal sequences. It must be emphasized that this assumption is not correct. As early as 1969, Forsgren [31] reported that protein A is released from *S. aureus* protoplasts and in 1972 Sjoquist et al. [32] described experiments suggesting strongly that protein A is covalently crosslinked to *S. aureus* wall peptidoglycan. These early studies have been confirmed by more recent observations, where protein A fractionated with cell-wall material and little or none remained associated with protoplast membrane fractions [33]. The apparent failure of the C-terminal hydrophobic/charged-tail sequences to act as a durable membrane anchor is not unique to *S. aureus* protein A. The C-terminal ends of streptococcal M proteins are highly susceptible to proteolytic cleavage [34] and a membrane-associated protease, called membrane-anchor cleaving enzyme (MACE) releases M protein from streptococcal protoplasts at pHs between 6.0 and 8.0 [30], despite the fact that this protease is inactive at or below pH 5.5 (see above). Similarly, *Streptococcus mutans* adhesin P1 (Section 12) possesses a typical C-terminal hydrophobic/charged-tail sequence [35], but it is released from *S. mutans* protoplasts by an endogenous enzymatic activity designated surface protein-releasing enzyme (SPRE) [36]. It could be argued that the release of these proteins from protoplasts is an artefact, resulting from digestion of the cell wall in vitro. However, sequencing of a *S. mutans* gene that encodes an extracellular fructanase has revealed that this enzyme also possesses both an LPXTGX motif and C-terminal hydrophobic/charged-tail sequences [37], yet it is released readily by intact cells. Similarly, studies on the localization of protein A deletion mutants and alkaline phosphatase-protein A fusion proteins expressed in *S. aureus* have confirmed that both the LPXTGX motif and the C-terminal hydrophobic/charged tail sequences are required to localize protein A in the *S. aureus* cell wall, but suggest that the hydrophobic/charged-tail sequences play a transient role in a more complex wall-associating pathway [33]. They may act to retard the release of the C-terminal end of the protein from the membrane while other steps in this pathway associate the protein with the cell wall, but some experiments have suggested that they may possess a more specific signalling function. Deleting the extreme C-terminal residue (Leu) from protein A has no effect on its normal incorporation into the cell wall, but replacing this C-terminal Leu by Cys results in complete release of the mutant protein, including the hydrophobic/charged-tail sequence, from intact cells [33]. Thus, one should not be misled by the term membrane 'anchor' into thinking that proteins are held firmly on the cell-surface via their C-terminal hydrophobic/charged-tail sequences.

Having being translocated across the membrane and directed into a 'wall-associating' pathway, the precise mechanisms by which proteins are then associated with the cell wall are not clearly understood. Based on sequence homologies with the C-terminal attachment site for glucosyl-phosphatidylinositol (GPI) anchors in a number of eucaryotic membrane-anchored proteins, it has been suggested that the LPXTGX motif might serve a similar function in bacterial proteins [30]. To date, however, GPI anchors have not been identified on bacterial surface proteins. Thus, the precise function of the LPXTGX motif remains unclear, although studies on mutations in this motif of protein A indicate that it is an important signal [33]. The strong conservation of LPXTGX motifs and of a hydrophobic/charged-tail domain suggests that different Gram-positive species possess very similar pathways to coordinate protein secretion with subsequent steps required to anchor proteins in the cell wall. These subsequent steps are likely to vary, depending on the particular species or protein. For example, *S. aureus* protein A appears to be covalently crosslinked to the cell wall [32], whereas streptococcal M proteins may not be covalently linked to other wall polymers [28]. An alkaline phosphatase fusion protein possessing only the protein A LPXTGX motif and hydrophobic charged-tail sequences is firmly anchored in the *S. aureus* cell wall [33], whereas a fructanase possessing a similar LPXTGX motif and hydrophobic/charged-tail sequence is released very readily from *S. mutans* cells [37]. As mentioned above, the primary sequences upstream from the LPXTGX motif vary considerably between unrelated wall-associated proteins and this might have a significant influence on wall-associating the mechanisms. In the literature, these upstream sequences have usually been described as a cell-wall 'spanning' region, primarily because it was assumed that they remain attached to the extreme C-terminal sequences anchored in the membrane but, as described above, current evidence suggests that this is not the case. Although sequences upstream from the LPXTGX motif are clearly associated with the cell wall [26–28], at present there is little direct evidence that they actually span the thick Gram-positive cell wall

The studies outlined above suggest the following working model. Like other extracellular proteins, a typical N-terminal signal-sequence initially directs wall-associated proteins into a secretion pathway. Then, unlike other extracellular proteins, additional C-terminal signals retard the release of the C-terminal end of the protein from the membrane and direct the protein into a wall-associating pathway. This pathway includes steps required to form durable wall associations, but the precise mechanisms may vary. In some cases, such as protein A, it may involve the formation of covalent crosslinks with other wall polymers. In other cases, it may simply involve protein release from the membrane being retarded while sufficient crosslinks are formed in newly incorporated peptidoglycan around the protein's Pro-rich region to hold the protein in the wall. For some proteins, durable wall association might require more specific non-covalent interactions with other wall or membrane-associated polymers. Irrespective of the precise wall-associating mechanism, having served a transient anchoring and/or signalling function, the C-terminal hydrophobic/charged-tail sequences are likely to be detached by a membrane-associated protease. At present, the advantages of utilizing an apparently complex pathway, rather than simply anchoring a protein firmly in the membrane via a C-terminal hydrophobic/charged-tail sequence are not clear. It might be necessary to avoid excessive protein bridging between the wall and membrane interfering with cell-wall growth and/or to avoid the membrane being saturated with hydrophobic/charged-tail anchors that might interfere with other membrane functions. Although our understanding is still limited, it is clear that

224

proteolytic cleavage of the C-terminal hydrophobic/charged-tail sequences must be carefully coordinated with other steps that anchor the protein in the wall. Variations in the efficiency of this coordination could explain why some wall-associated proteins are released from cell walls much more readily than others and why such release can vary considerable between strains and growth conditions.

4. Common structural themes

In addition to possessing common N-terminal and C-terminal signals involved in protein secretion and localization (Sections 2 and 3), many distinct Gram-positive cell-wall-associated proteins possess other common structural features.

4.1. Repeats

A particularly striking structural feature found in the vast majority of wall-associated proteins for which sequence data are available is the existence of tandemly repeated sequences. Some proteins possess a single type of tandemly repeated sequence whereas others may possess several distinct types of tandem repeats and in some cases a very large proportion of the molecule can consist of repeated sequences. Where a protein contains a number of distinct types of repeated sequences these are normally distinguished by letters (A repeats, B repeats, etc.), whereas individual copies of the same repeated sequence are normally distinguished by numbers (A1, A2, etc.). Tandem repeats can vary in length from as few as two to over 100 residues. Within any particular set of tandem repeats, the internal copies are often very highly conserved whereas the external copies can be more degenerate. This reflects the fact that internal copies are likely to have been generated more recently and thus would have had less time than the external older copies to accumulate divergent point mutations .

The mechanisms by which repeats are produced initially in wall-associated proteins are not clearly understood, although one could speculate that they may have resulted from slippages during replication or unequal intergenic recombination events. However, once two or more copies of a repeat exist, it is easy to visualize how replication slippage or homologous recombination events can duplicate or delete copies. Such events have been observed to occur at a frequency of about 1×10^{-3} in the case of tandem repeats in group A streptococcal M proteins, resulting in the same M protein expressed by different isolates varying in size [38,39]. The advantages of possessing apparently 'unstable' tandem repeats are not clear at present. It is possible that the ability to reversibly shorten or lengthen a cell-surface protein might facilitate an organism's ability to adapt to changing environmental conditions. Tandem repeats in many proteins have been associated with binding domains for other proteins or polysaccharides [40] and an ability to reversibly alter the number of such binding domains might also be important in adapting to changes in the environment. Further, at the DNA level, recombination events between tandem repeats provide a means for accelerating sequence diversity. If the number of nucleotides in a tandem repeat is not a precise multiple of three, homologous recombination between repeats would generate frame-shift mutations. In most cases, however, the number of nucleotides in repeated sequences of genes encoding wall-associated proteins is divisible by three and consequently one sees corresponding repeats in the protein. However, even in this case recombination between imperfect, divergent copies of a repeat can produce new sequence combinations and there is evidence that such recombination events can contribute to antigenic variation in the case of streptococcal M proteins [41].

Recombination or replication slippages involving repeats can sometimes give rise to structures that might appear at first sight to be quite different from the parent protein. A good example of this are the repetitive structures of the two protein G (Section 8.1) sequences depicted in Fig. 2, which correspond to protein G genes from distinct strains [42,43]. In the upper sequence, the N-terminal half of the molecule contains two copies of a 37-residue repeat (A1 and A2), which are separated by an unrepeated 38-residue 'spacer'. Similarly, in the C-terminal half of the molecule there are two copies of a 55-residue repeat (B1 and B2), separated by a 15-residue unrepeated spacer. However, recombination or replication slippages between A1/A2, and between B1/B2, can give rise to the lower structure. Note that in addition to generating a third copy of each of the original repeats, these recombination events have also duplicated the originally unrepeated spacer sequences, generating two additional types of repeats. Thus, some protein G sequences have been described as having two types of repeats designated A1-A2...B1-B2, whereas others have been described as having four types of repeats designated A1-B1-A2-B2-A3 ...C1-D1-C2-D2-C3, where repeat B in the former is equivalent to repeat C in the latter (Fig. 2). An alternative way of describing the latter structure would be to consider the A1-B1 sequence as a single repeat (say Z), in which case the N-terminal half of the molecule would have two complete copies of the tandem repeat Z, followed by a third partial copy (Fig. 2). Tandem repeats in many wall-associated proteins have been described in exactly this fashion, that is several complete copies followed by a further partial copy, and protein G provides a good example of how such structures might arise.

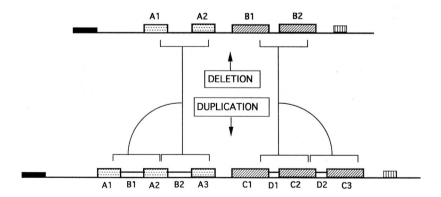

Fig. 2. Repetitive sequences in protein G. The structure of protein G (Section 8.1) from strain GX7809 [42] is depicted in the upper part of the diagram and the lower half depicts protein G from strain GX7805 [43]. The narrow black boxes on the extreme left correspond to the N-terminal signal sequences, and other boxes correspond to repeated sequences.

4.2. Protein fibrils, sequence variation and coiled-coil dimers

Under the electron microscope, a number of Gram-positive wall-associated proteins have been visualized as fibrils projecting outwards from the cell surface (Fig. 3). These fibrils are usually <200 nm long, although some can be as long as 400 nm [44]. Individual fibrils

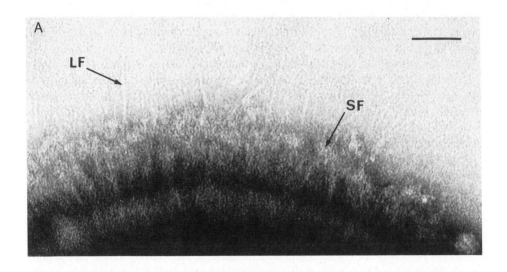

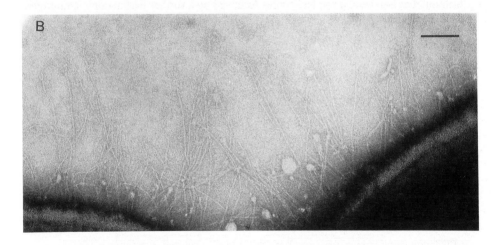

Fig. 3. Wall-associated protein fibrils and fimbriae. (A) *Streptococcus salivarius* strain HB (168 000×) expressing a dense fringe of short (ca. 90 nm) peritrichous fibrils (SF), and sparse longer (ca. 168 nm) fibrils (LF) that have aggregated into thicker filaments of varying widths (B) *S. salivarius* strain CHR (130 000×) expressing peritrichous fimbriae, 0.5–1.0 μm long and 3–4 nm wide. The cells were stained with 1% methylene tungstate and the bars represent 100 nm. This figure was generously provided by P.S. Handley and is reproduced with permission (© 1990 Harwood Academic Publishers GmbH) from [44].

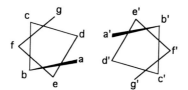

Fig. 4. Helical wheel representation of a coiled-coil dimer.

are very thin and their widths cannot be measured easily in electron micrographs [44]. They often form a peritrichous 'hairy fuzz' over the entire cell surface, but they may also aggregate into tufts.

In pathogenic species, the regions of fibrillar proteins (usually the N-terminal halves) that project outwards from the cell-surface are potential targets for protective host antibodies. This can impose a selective pressure for antigenic variation, giving rise to multiple serotypes of certain wall-associated proteins. The ability to undergo sequence variations in order to avoid recognition by host antibodies must be balanced by functional constraints on the types of structural changes that may be permitted. One means of accommodating very extensive sequence variation while maintaining a common overall structure is to produce a fibrillar protein possessing an α-helical coiled-coil secondary structure. A good example of this is group A streptococcal M proteins, which are expressed as extended (50–60 nm) α-helical coiled-coil dimers with highly variable N-terminal sequences projecting outwards from the cell surface [29]. This structure is maintained by a seven-residue periodicity which extends throughout most of the M protein molecule, with the exception of the extreme C-terminal sequences and a short non-helical region of variable length (usually 10–20 residues) at the N-terminus of the mature protein [45,46]. The heptapeptide periodicity accommodates very considerable variation in the primary sequence, since it requires only that certain types of residues occur periodically [47]. This is depicted in Fig. 4, where the residues in an α-helical heptapeptide repeat are designated by the letters a b c d e f g. An α-helical coiled-coil dimer is most likely if the residues at positions a and d are predominantly hydrophobic and those at positions b, c, e, f, g are predominantly helix-promoting hydrophilic residues. The residues at positions a and d are located on the same side of the α-helix, such that they interact with residues at positions d' and a', respectively, on the second α-helical polypeptide. The side-chains of these residues interlink to produce close contacts along the axis of each α-helical chain, producing a coiled-coil dimer with the axis of each polypeptide tilted at an angle of about 10° with respect to the other. This structure is stabilized by short range van der Waals attractive forces. If residues a and d are hydrophobic, exclusion of the aqueous solvent produces an energetically very favourable structure, which may be stabilized further by, for example, salt bridges between similarly charged residues at positions e and g [47]. Thus, in M proteins, non-polar residues predominate at positions a and d and charged residues occur frequently at positions e and g [29,48]. However, the precise characteristics of the seven residue periodicity varies to some extent in different M proteins, and even in different segments of the same M protein. There can, for example, be a higher proportion of polar residues at positions a and d than is found in other fibrous proteins such as mammalian tropomyosin. Nevertheless, physical studies have confirmed that M proteins possess an α-helical coiled-coil structure [45], although occasional discontinuities in the heptapeptide periodicity confer extended M protein fibrils with some degree of flexibility [29,46].

5. Complex cell wall-associated protein structures

5.1. Fimbriae

Some Gram-positive bacteria express fimbriae, which are morphologically quite distinct from protein fibrils (Fig. 3). Fimbriae can extend for several micrometres (500–4000 nm) beyond the cell wall and have constant, measurable widths of 3.0–10.0 nm along their entire length. Although they are not as common as their counterparts in Gram-negative organisms, they have been detected in an increasing number of Gram-positive species in recent years, including a variety of *Corynebacterium, Staphylococcus, Streptococcus* and *Actinomyces* species [44,49–52; P.S. Handley, personal communication].

Molecular studies on Gram-positive fimbriae have been sparse, or in most cases non-existent. Although first described over two decades ago [49], only one *Corynebacterium* fimbrial gene, which has yet to be sequenced, has been cloned [53]. Most studies have focused on the oral pathogens, *Streptococcus sanguis, Actinomyces viscosus* and *A. naeslundii*. Only some strains of *S. sanguis* (e.g. FW213) express fimbriae, which mediate adhesion to salivary glycoprotein agglutinins (SAG) adsorbed to tooth enamel [54,55]. Most *A. viscosus* strains express two distinct types of fimbriae: type 1 mediate adhesion to SAG and type 2 bind to polysaccharides on certain other oral bacteria, mediating bacterial co-aggregation [56–58]. Following direct adhesion of *S. sanguis* and *A. viscosus* to SAG-coated teeth, such co-aggregration reactions (as well as cell-cell bridging by SAGs and bacterial dextrans or levans) contributes to the development of a more complex oral flora. Most strains of *A. naeslundii* lack type 1 fimbriae, but express type 2 fimbriae similar to those of *A. viscosus* [56,59].

Studies on the subunit structures of Gram-positive fimbriae have been complicated by the fact that the subunits are very tightly associated and appear as ladders on SDS-polyacrylamide gels [56]. However, the expression of cloned genes in *E. coli* has allowed some progress to be made. Major subunits with M_r values of 65 000 and 59 000 have been identified in the case of *Actinomyces* type 1 and type 2 fimbriae, respectively [60–62]. These do not self-assemble when expressed in *E. coli*, suggesting that their strong association in native fimbriae may require interactions with additional fimbrial subunits or processing by accessory proteins involved in fimbrial biogenesis. However, such additional subunits or accessory proteins have yet to be identified. The cloned *A. viscocus* type 1 and *A. naeslundii* type 2 fimbrial genes have been sequenced and the deduced proteins contain typical N-terminal signal sequences and share considerable homology [62,63], but they share no significant homologies with Gram-negative fimbrial antigens. Perhaps surprisingly for fimbrial subunits, their sequences include typical C-terminal wall-associating signals of the type described in Section 3.2. It is possible that these sequences play a role in coordinating subunit secretion with fimbrial biogenesis and that they are then proteolytically detached. However, at present we know essentially nothing about the mechanisms involved in Gram-positive fimbrial biogenesis.

Studies of cloned genes in *E. coli* have also identified a M_r 30 000 protein (designated FimA) that reacts with antisera to purified *S. sanguis* FW213 fimbriae [64]. The deduced FimA sequence includes a N-terminal signal peptide, but is quite distinct from the larger *Actinomyces* fimbrial antigens and it does not possess obvious C-terminal wall-associating signals [65]. It is, however, highly homologous to antigen SsaB, a non-fimbrial adhesin produced by a distinct strain of *S. sanguis* [66,67; Section 12]. More recent studies identified additional FW213 fimbrial antigens, with M_r values of 53 000, 125 000 and 160 000, and suggested that the M_r 53 000 antigen may be the major structural subunit [68]. Thus, FimA might be analogous to the adhesive minor fimbrial subunits found in many Gram-negative fimbriae [69]. Further, FimA is encoded by a polycistronic message which includes open reading frames for additional polypeptides that might be analogous to accessory fimbrial biogenesis proteins in Gram-negative species [68,69]. However, given the very different architecture of their cell envelopes, the mechanisms involved in fimbrial biogenesis in Gram-positive species is likely to be quite different than those in Gram-negative organisms.

5.2. Flagella

Despite the differences in cell envelopes, Gram-positive flagella are morphologically very similar to their counterparts in Gram-negative species [70]. Flagella in *E. coli* and *Salmonella typhimurium* have been studied in considerable detail and over 35 distinct flagellar genes have been identified [71], but there have been few comparable studies on Gram-positive flagella. However, a flagella locus (*flaA*) from *B. subtilis* was recently sequenced and predicted protein sequences corresponding to 9 out of 12 open reading frames were found to share significant homologies with flagellar components of *E. coli* or *S. typhimurium* [72]. Thus, the reader is referred elsewhere [71] for reviews on the intensively studied flagella systems in *E. coli* and *S. typhimurium*, which are likely to be good models for their poorly studied counterparts in Gram-positive organisms.

5.3. S-Layers

Many bacteria secrete proteins that are assembled in a variety of geometric formats to produce crystalline surface layers (S-layers), which may coat the entire cell [73,74]. S-Layers are commonly associated with archaebacteria, but they have also been detected in over 100 distinct species of Gram-positive and Gram-negative eubacteria [75]. They usually consist of a single type of protein or glycoprotein subunit, with M_r values ranging between 40 000 and 200 000, but some contain two or more distinct subunits [73,74]. In Gram-positive eubacteria, they are attached by non-covalent interactions to the cell wall and when dissociating agents are removed, purified S-layer proteins reassemble spontaneously into two-dimensional lattices with the same geometric formats as those on the intact cell and they can reattach to cells and other surfaces [74,76]. Thus, the biogenesis of S-layers appears to be a predominantly self-assembly process. Molecular studies have been limited and primary sequence data have been reported for relatively few S-layer proteins, including those produced by four diverse Gram-positive species, namely *Acetogenium kivui* [77], *Bacillus brevis* [78,79], *Deinococcus radiodurans* [80] and *Lactobacillus brevis* [81]. These sequences include typical N-terminal signal peptides, share little or no primary sequence homologies, and lack the kinds of wall-associating signals described in Section 3. However, they all contain relatively high levels of hydroxylated residues (Thr, Ser or Tyr), suggesting that inter- and intra-molecular H-bonding may be an important features of S-layer structure.

An average S-layer contains 5×10^5 subunit monomers [74]. Thus, their biogenesis reflects a considerable expenditure of energy by the cell, suggesting that they are functionally very important structures. However, our understanding of these functions is limited. Given the diversity of bacterial species that produce S-layers, some may have adapted to serve diverse functions in different species. All S-layers contain regularly arrayed pores that can act as molecular sieves with homogeneous, well-defined exclusion limits for macromolecules and one common function may be to act as permeability barriers to protect underlying cell walls against harsh environmental conditions [73,74]. This may explain why many bacteria cease to produce an S-layer when grown under favourable laboratory conditions [73]. S-Layers are produced by a variety of pathogenic bacteria and could play an important role in virulence, although this has been demonstrated directly only in the case of the A-layer of *Aeromonas salmonicida* (a Gram-negative pathogen) [82]. However, since S-layers are often difficult to detect when organisms are grown in vitro [73], their role in virulence might well be underestimated. Further studies on S-layers might reveal novel virulence mechanisms and might also lead to the industrial exploitation of their molecular sieving properties and their highly efficient subunit expression systems.

5.4. Cellulosomes

In the early 1980s, it was discovered fortuitously that the cellulolytic Gram-positive bacterium *Clostridium thermocellum* adheres in a specific manner to its cellulose substrate, but not to other insoluble polysaccharides [83]. Subsequent studies found that the adhesin resides in 18-nm particles consisting of at least 14 distinct polypeptides, 13 of which possess endoglucanase, β-glucosidase or xylanase activity [83,84]. These cellulolytic particles have been found in both wall-associated and extracellular forms, and are now called the 'cellulosomes'. In the cellulosome, the enzymatic subunits associate with a non-catalytic M_r 210 000 scaffolding-protein, designated S_L or S1 [83,84]. Extensive molecular studies have shown that the enzymatic subunits contain similar 23 or 24-residue tandem repeats, which bind to one of seven repeated 147-residue 'cellulase-docking' domains in the non-catalytic S_L [85,86]. An intact 147-residue S_L repeat is the minimum structure to which a cellulase can bind [86]. Thus, it is possible that each S_L subunit binds no more than seven enzymatic subunits, but that different copies of S_L bind different combinations of the 13 enzymatic polypeptides found in cellulosomes. Size considerations suggest than an individual 18-nm cellulosomal particle is unlikely to contain more than a few copies of S_L and might even consist of just one S_L with seven associated cellulases.

Electron microscopic studies have visualized polycellulosomal protuberances decorating the surface of *C. thermocellum* [87]. Recent studies suggest that these structures bind to cellulose via a 167-residue unrepeated cellulose-binding domain (CBD) in S_L [88]. Concentrating cellulases in cellulosomes aligns multiple cellulases with the substrate, and this may allow the cell to achieve a more efficient degradation of this recalcitrant substrate [89]. The mechanism by which these enzymatic complexes are attached to the bacterial cell wall has yet to be determined. None of the sequenced cellulosomal subunits (including S_L) possess recognizable Gram-positive cell-wall-associating signals [86,88,90]. Further studies will be required to define the precise architecture of the cellulosomes and the mechanism of their attachment to the cell wall.

6. Staphylococcus aureus clumping-factor and coagulase

Binding of fibrinogen to *S. aureus* causes cells to clump in plasma and a number of studies have attributed this clumping to cell-bound coagulase [91,92]. However, McDevitt et al. [93] have demonstrated that while cell-bound coagulase can bind fibrinogen (see below), this is not responsible for the clumping reaction. These workers have recently cloned and sequenced a distinct clumping-factor gene (*clf*) (D. McDevitt and T.J. Foster, personal communication). The deduced CLF protein sequence possesses an N-terminal signal peptide and the extreme C-terminal end includes a typical wall-associating LPXTGX motif, followed by a hydrophobic/charged tail domain. However, the Pro-rich sequence that precedes the LPXTGX motif in most wall-associated proteins (Section 3.2) has been interrupted in CLF by a remarkable repetitive sequence, where the dipeptide Asp-Ser is repeated almost perfectly 150 times (D. McDevitt and T.J. Foster, personal communication). The binding site for *S. aureus* whole cells has been localized to a 15-residue sequence at the C-terminus of the fibrinogen γ-chain [94,95] which includes a number of positively charged residues that might interact with the negatively charged CLF Asp-Ser repeat. At present, however, there is no direct evidence that fibrinogen binds to the CLF Asp-Ser repeat region and it is quite possible that this sequence has a structural role. It is interesting to note that linker or hinge sequences between distinct domains in a variety of proteins have been found to possess similar dipeptide repeats [96], although those described to date are much shorter and would be predicted to be more flexible than the CLF Asp-Ser region, which could form a rigid extended structure. It is possible that this structure serves as a bridge extending from a wall anchor to project the N-terminal half of CLF well beyond the wall. The N-terminal half of CLF is not repetitive, but it has been noted to share significant homology with the N-terminal halves of *S. aureus* fibronectin-binding proteins (Section 9), suggesting a common evolutionary origin (D. McDevitt and T.J. Foster, personal communication).

As mentioned above, *S. aureus* coagulase also binds fibrinogen and although it is normally described as an extracellular protein it also exists in a cell-bound, wall-associated form [91,92]. The name coagulase is a misnomer, since it acts stoichiometrically and has no enzymatic activity. It coagulates plasma by binding to and activating prothrombin, which triggers the polymerization of fibrin [97,98]. Sequencing of cloned *coa* genes revealed that coagulase does not possess typical wall-associating signals, but the C-terminal ends of the deduced protein sequences contain 27-residue tandem repeats and share some structural similarities with the Pro-rich Xr-repeat region of protein A (Section 8.1), suggesting that these sequences might play a role in wall association [99,100]. The absence of more typical wall-associating signals might explain why coagulase is released very readily from the cell. The C-terminal halves of distinct serotypes of coagulase are highly conserved, but their N-terminal halves display considerable primary sequence diversity [99–102]. Despite this diversity, prothrombin-binding has been localized to the N-terminal half of coagulase [102], whereas fibrinogen binds to the more conserved C-terminal half of the molecule [93]. Thus, prothrombin and fibrinogen bind to distinct sites on coagulase, although neither binding-site has been mapped precisely. Textbooks usually describe coagulase as a virulence factor, and in vivo studies employing chemically induced mutants have indicated that coagulase-defective mutants display reduced virulence in animal models [103–106]. However, using genetically well-defined mutants constructed by allele-replacement mutagenesis [107], Phonimdaeng et al. [100] observed no reduction in the virulence of isogenic Coa$^+$ and Coa$^-$ derivatives of *S. aureus* strain 8325-4 in either mouse subcutaneous infection or mastitis models. Thus, the potential contribution of coagulase to virulence remains a matter of speculation.

7. Streptococcal M proteins

Early studies on group A streptococcal M proteins indicated that they are both important virulence factors and highly protective antigens. For this reason, they have been studied in considerably more detail than most other Gram-positive wall-associated proteins. The reader is referred to an earlier review [29] for a detailed account of studies up to the end of 1988 and to more recent reviews for a discussion on the immunological properties of M proteins [108,109]. Here, we focus primarily on more recent advances in our understanding of M protein structure, genetics and interactions with host proteins. However, to facilitate the subsequent discussion, it will be useful to begin with a brief account of traditional serotyping systems that have had a considerable, and to some extent misleading, influence on how M proteins have been defined.

M proteins were originally defined in the 1920s as type-specific protein antigens in crude hot-acid extracts of *S. pyogenes* [2]. Antisera were raised to individual clinical isolates and absorbed extensively with heterologous strains to remove non-strain-specific antibodies in order to produce a range of M serotyping-sera [2]. Essentially the same laborious and crude procedures are still used widely today to divide group A streptococcal strains into distinct M types (designated M1, M2, M3, etc.) and approximately 100 M types of *S. pyogenes* have been described to date. M serotyping has been supplemented by two other typing reactions. Some strains of group A streptococci express a serum opacity-reaction factor (designated OF or SOR), which is a rather poorly defined product that causes opalescence in serum, and typing systems have associated expression of OF with the expression of specific serotypes of M proteins [113,114]. In addition, treatment of *S. pyogenes* cells with trypsin exposes an antigen called T protein and 25 distinct serotypes of T antigen have been defined by antibody-absorbtion procedures similar to those described above. A single M type can express either one or more than one serotype of T antigen, but the T-serotyping pattern was found to be consistent within an M type [115]. One T antigen gene (*tee6*) has recently been sequenced and the deduced T6 protein sequence contains typical C-terminal wall-associating signals, but is otherwise quite distinct from M antigens [116]. Hybridization studies with *tee6* gene probes have suggested that T antigens may correspond to a variety of quite distinct wall-associated proteins [117]. However, since T proteins have no known role in virulence, most have not been studied in any detail and they will not be discussed further in this chapter. Here, it is sufficient to note the apparent parallels in typing studies between the M type, T type and OF phenotype.

Rebecca Lancefield's early studies indicated that M proteins are important antiphagocytic virulence factors and that immunity to *S. pyogenes* infections is due primarily to opsonic, type-specific anti-M antibodies [110]. Shortly after the discovery of M proteins, clumping of group A streptococci in plasma was first noted [111] and in 1965 Kantor [112] demonstrated that fibrinogen is precipitated by acid-extracted M antigen. These early observations were supported by subsequent studies, but it is important to note that neither in vivo nor in vitro phagocytosis and opsonization studies have been performed on many of the strains defined by serotyping as distinct M types. Nevertheless, it has been widely assumed that all virulent group A streptococci express a single, type-specific, fibrinogen-binding, anti-phagocytic M protein. This assumption is misleading. It is now known that M protein genes (*emm*) are members of a larger *emm*-like gene family, that many *S. pyogenes* strains express more than one M-like protein, and that only some M-like proteins bind fibrinogen and possess antiphagocytic properties. For reasons of clarity, a discussion of the *emm*-like gene family and its implications for the definition and nomenclature of M and M-like proteins is deferred until Section 7.3. Here, it is sufficient to note that 'type-specific M antigen' and 'antiphagocytic M protein' are not necessarily synonymous. Since the term 'M protein' has been considered for many decades to imply an antiphagocytic function, here this designation is restricted to those type-specific M antigens that have been clearly demonstrated to possess antiphagocytic properties The designation 'M antigen' is used in a more general sense, in that it refers to both antiphagocytic M proteins and type-specific M-like proteins that have been defined by serotyping rather than functional criteria and which, therefore, may or may not have antiphagocytic properties. Similarly, the term '*emm*-like' gene is used rather tham *emm* gene, except where there is direct evidence that the gene product is an antiphagocytic M protein.

7.1. Structure, structural relationships and antigenic variation

To date the complete sequences of cloned *emm* or *emm*-like genes corresponding to the type-specific M antigens of serotypes 2, 5, 6, 12, 24, 49 and 57 *S. pyogenes* have been reported [48,118–123], as well partial sequences of *emm*-like genes corresponding to serotypes 1, 3, 18 and 19 [124–126]. As summarized in Fig. 5, the sequences encoding the N-terminal signal peptides and sequences corresponding to the C-terminal halves of M antigens are very highly conserved between serotypes and they all possess typical C-terminal wall-associating signals. In contrast to these highly conserved regions, sequences corresponding to the N-terminal halves of mature (i.e. minus the signal peptide) M antigens that project outwards from the cell wall are highly variable. Despite this variation, M antigens are predicted to share a common structure, consisting of extended α-helical coiled-

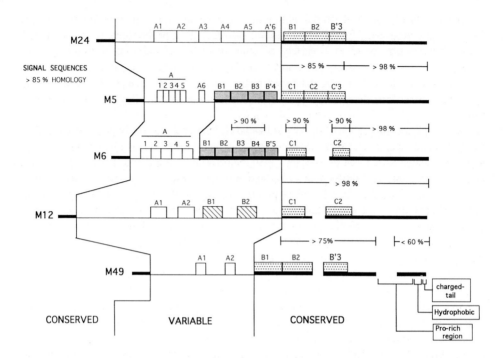

Fig. 5. Structural relationships between group A streptococcal M antigens. The narrow black boxes on the extreme left represent the N-terminal signal sequences and other boxes represent repeated sequences. It should be noted that the B repeats in M24 and M49 are similar to the C repeats in the other antigens depicted and consequently they are sometimes described as conserved 'C' repeats in the literature. Approximate levels of sequence homologies are indicated by the % between various M antigens. The sequences represented by the diagram are from the following references: M24 [121], M5 [120], M6 [118], M12 [119] and M49 [122].

coil dimers (Section 4.2), that appear under the electron microscope as hairy fibrils projecting from the streptococcal cell-surface [127]. A significant proportion of all sequenced M antigens consists of tandem repeats, although the number of distinct repeats varies between M types (Fig. 5). As mentioned in Section 4, tandem repeats are a common feature of Gram-positive wall-associated proteins and are often associated with binding domains for other proteins. By analyzing defined deletion mutants we have recently localized the binding site for fibrinogen in serotype 5 M protein to within the B repeats (J. Shields and M.A. Kehoe, unpublished data), which is consistent with ultrastructural studies indicating that fibrinogen binds to the central regions of M protein fibrils [128]. Although DNA isolated from about 20% of other M types hybridize to M5 B-repeat probes [129], including for example M6 (Fig. 5), the corresponding regions of other M types are quite distinct. Thus, fibrinogen appears to bind to a highly variable region of M proteins and it seems likely that distinct primary sequences can form a structurally similar fibrinogen-binding domain. The M protein binding-site on fibrinogen has been localized to a region distinct from the *S. aureus* CLF binding-site (Section 6), but has not been mapped precisely [130].

Although the C-terminal halves of M antigens are highly conserved, it has been found that certain monoclonal antibodies to the conserved C-repeats of M6 (Fig. 5) can distinguish OF$^+$ and OF$^-$ M types of *S. pyogenes*, suggesting that there are two evolutionary divergent lines of M antigens [131,132]. Thus, M antigens have been divided into two distinct groups, designated class I and class II, which closely parallel OF$^+$ and OF$^-$ M types [131,132]. This division was supported by available published sequences, which show that there are significant differences between the conserved C-terminal regions of M antigen genes cloned from OF$^+$ and OF$^-$ M types, and by subsequent PCR studies [125,133]. However, some discrepancies were noted [125,131]. In an attempt to determine if these classes reflect genuine evolutionary relationships, we recently sequenced conserved C-repeat and signal-peptide regions of over 40 distinct *emm*-like genes, divided about equally between OF$^+$ and OF$^-$ serotypes. Although at present we cannot be certain that all of these genes express a type-specific M antigen (see Section 7.3), these studies have confirmed that limited sequence variation in the conserved C-repeat regions does divide these *emm*-like genes into two divergent groups that generally parallel the above classes. However, they also provide evidence for the horizontal transfer of sub-segments of *emm* and *emm*-like genes between OF$^+$ and OF$^-$ M types, suggesting that the classification of M antigens into two distinct groups solely on the basis of PCR or antibody probes that recognize specific sites could be misleading (A. Whatmore, V. Kapur, J. Musser and M.A. Kehoe, unpublished data).

In published type-specific *emm*-like sequences, there is a striking contrast between the very high levels of sequence variation in regions corresponding to the N-terminal halves of mature M antigens and the very strong conservation of flanking signal-peptide and C-terminal regions (Fig. 5). Further, *emm* or *emm*-like gene probes corresponding to the variable N-terminal regions of a variety of distinct M antigens react only with DNA from the homologous M type, suggesting that these regions are very distinct in different serotypes [129,134a,b]. This suggested that variation in these regions might involve antigenic shifts, with intergenic recombination events introducing quite distinct sequences between the conserved regions of *emm* genes [120,124]. We have recently determined and compared sequences in the 5' variable regions of *emm*-like genes from over 80 distinct M types (A. Whatmore and M.A. Kehoe, unpublished data). Again it should be noted that at present we cannot be certain that all of these genes express a type-specific M antigen (see Section 7.3). Nevertheless, these comparisons have provided new evidence concerning possible mechanisms of antigenic variation, as well as highlighting some of the limitations of M serotyping. It is now clear that the variable regions of certain 'type-specific' *emm*-like genes share considerably more homology than previously recognized and the divergence of corresponding antigens could be accounted for by antigenic drift, with protective antibody selecting for the accumulation of base-substitutions and/or short deletions/insertions. Indeed, in some cases there is less variation in the 5' sequences of 'type-specific' *emm*-like genes from distinct M serotypes than there is between the corresponding sequences in different isolates of the same M type, highlighting the fact that M serotyping may provide a misleading impression of relationships between M antigens. Nevertheless, the variable regions of many *emm*-like genes are still very divergent and it appears that intergenic recombination events also contribute very significantly to antigenic variation. The studies outlined in Section 7.3 suggest that other *emm*-like genes in adjacent regions of the chromosome might contribute to such intergenic recombination events, but it is also clear that *emm*-like genes and subsegments of these genes are exchanged between group A streptococcal strains (A. Whatmore and M.A. Kehoe, unpublished data). Further, a recent report by Simpson et al. [135], provides evidence that suggests an *emm12* gene has been transferred between group A and G streptococci, suggesting that *emm*-like genes can be transferred not only between individual strains of group A streptococci but also between streptococci that are currently classified as distinct species.

7.2. Phase-switching and regulation

A long recognized characteristic of group A streptococci is that reversible loss of M antigen expression can occur when organisms are cultivated in vitro [110]. In 1980, it was reported that a phage isolated from an M12 strain mediated reversion of a type 76 M⁻ variant to the M76⁺ phenotype [136], but this phage was subsequently found to have no effect on M⁻ variants of other strains, including its parent M12 strain. In later studies, however, Simpson and Cleary [137] observed reversible phase-switching of a serotype 12 strain between M⁺ and M⁻ phenotypes, and this switching was associated with a change in colony opacity from Op⁺ to Op⁻. Although phase-switching of the M and Op phenotypes was normally simultaneous, it also occurred independently, suggesting that colony opacity was determined by a distinct factor that was coregulated with the M antigen [137]. More recently, these workers have reported that expression of a third determinant, namely the streptococcal C5a peptidase (SCP; Section 11) is also controlled by this phase-switch [138]. Although hybridization studies revealed no gross sequence differences between reversible M⁺ and M⁻ variants, studies on irreversible phase-locked M⁻ variants of M12 strains, and a Tn*916* insertion mutant of an M6 strain, have identified a locus several hundred bp upstream of these *emm* genes that activates transcription of both the *emm* and *scp* genes, as well as controlling the colony Op phenotype in M12 strains [119,137–139]. The factor determining colony opacity in the M12 strains has not been defined, but the *scp* gene has been cloned, sequenced and is located about 4 kb downstream from the *emm12* gene [140,141]. The *emm12* and *scp* genes encode distinct monocistronic mRNAs [138]. Thus a locus, designated VirR (**vir**ulence **r**egulator) by one group and Mry (**M** protein **R**NA **y**ield) by another, encodes a *trans*-acting product that activates expression of a number of distinct *S. pyogenes* genes [138,139]. More recent studies have identified a *mry* (*virR*) gene that encodes a 62-kDa protein, which shares some homology with effector proteins of two component sensor/effector regulators in other species [142]. Further, it has also been reported that expression of M protein is influenced by at least one environmental factor, namely carbon dioxide tension [143] and this would be consistent with Mry (VirR) playing a role in an environmental sensor/effector regulator system. Although Mry (VirR) controls at least three distinct genes (*emm, scp* and colony Op determinant in M12 strains), it has no effect on expression of streptokinase or hyaluronic acid capsule [139].

7.3. M-like proteins and gene families

Many group A streptococci express wall-associated proteins that bind the Fc regions of various human immunoglobulins (Section 8) and the sequences of these IgA or IgG Fc-binding proteins have been found to share significant homologies with the type-specific M antigens. Although sequences in the central and 5' regions of these genes can be quite variable, homologies between sequences encoding the N-terminal signal peptides and the C-terminal halves of M antigens and M-like Ig-binding proteins indicate that they are essentially members of a single gene family [144–146]. Over the past year, published descriptions of an increasing number of *emm*-like genes have revealed a complexity that challenges traditional definitions of 'M proteins', but an unfortunate lack of agreed definitions and nomenclature has been a source of some confusion for those not working in the field. Therefore, these studies are summarized below, with reference to Fig. 6.

Recently, Haanes et al. [147] reported hybridization studies indicating that the *emm*-like genes in all strains of group A streptococci they examined are located in the same position in the chromosome and are flanked by the *virR* (*mry*) and *scp* genes described in the previous section. However, while OF⁻ M types appeared to possess a single *emm*-like gene between *virR* and *scp*, in OF⁺ strains a triplet of *emm*-like genes has been found in this position (Fig. 6) [147]. Different laboratories have determined the sequences (or partial sequences) of multiple *emm*-like genes in M49 [122,147], M4 [145,148–151] and M2

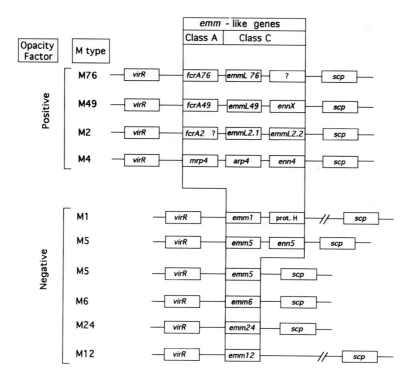

Fig. 6. Arrangement of *emm*-like genes in various strains of *S. pyogenes*. All of the genes within the central box are members of the *emm*-like gene family and are flanked by the regulatory gene *virR* (also called *mry*) and *scp*, which encodes a C5a peptidase. Individual *emm*-like genes are described in the text. The boxes in the M76 and M2 triplets containing a ? represent putative *emm*-like genes that have not been identified directly but that are likely to exist (see text).

[123] streptococci and available data suggest that each of these genes is monocistronic. It must be emphasized that published sequence data are limited to individual strains from relatively few M types. Nevertheless, a comparison of these sequences suggests that corresponding *emm*-like genes are located at a similar position in the *emm*-like gene triplet in different serotypes (Fig. 6), with the *emm*-like genes discussed in Section 7.1 corresponding to the central gene in these triplets. However, the proteins encoded by corresponding *emm*-like genes in different serotypes can have distinct biological properties and consequently they have been given quite distinct designations by different laboratories.

The central gene in the M49 triplet was described as an M protein gene (*emm49*) [122], because it corresponds to a previously described pepM49 antigen that had been shown to absorb opsonic, bactericidal antibodies from anti-M49 sera [152]. While the protective M49 antigen may well possess antiphagocytic properties, it is not clear if this has been tested directly and therefore, in order to be consistent, in Fig. 6 the designation *emm*L49 (for *emm*-like) is used. In contrast to the protective M49 antigen, in M4 streptococci the central gene in the *emm*-like triplet expresses an IgA Fc-binding protein (Arp4) and antibodies to the Arp4 protein are not opsonic [145,149]. The corresponding *emm*-like gene in M2 streptococci has been designated *emm*L2.1 and encodes a product that is recognized by M2 typing sera [123]. In each of these strains there is a highly homologous (>80% homology) *emm*-like gene located ca. 207 base-pairs (bp) downstream from these central genes. This has been designated *ennX* (M49), *enn4* (M4) or *emm*L2.2 (M2) [123,147,148]. In both M49 and M4 strains, this downstream *emm*-like gene appears to be transcriptionally silent [148,153], but in the M2 strain it expresses an IgA Fc-binding protein, although its expression is about 30-fold lower than that of the central (*emm*L2.1) *emm*-like gene [123].

The upstream *emm*-like gene in the M4 strain express IgG Fc-binding protein that has been designated Mrp4 (**M** related protein) [150]. Mrp4 is very similar to a previously described IgG Fc-binding protein from M76 streptococci, called FcrA76 (**Fc** receptor **A**), and the *fcrA76* gene is also located upstream from other *emm*-like genes [144], although sequence data for these other M76 *emm*-like genes have not been reported. In the M49 strain, there is a similar upstream *emm*-like gene which was designated *fcrA49* due to its homology with *fcrA76* [147], but the Ig-binding properties of the putative gene product have not been reported. The corresponding region in the M2 strain was not sequenced, but the observation that this strain binds IgG in addition to IgA [123], together with hybridization studies [147] strongly suggest that the M2 strain also contains a similar *emm*-like gene. A comparison of the complete sequences of Mrp4 and FcrA76 (only part of *fcrA49* was sequenced) showed that these proteins possess a single set of tandem repeats (designated A) that are very distinct from repeats observed to date in other M and M-like proteins, including the usually highly conserved repeats (often, but not always, designated C repeats; Fig. 5) in the C-terminal halves of these molecules [147,150]. Thus, some workers divide M-like proteins into two distinct groups (class A and C) on the basis of whether they possess Mrp4 A-like repeats or conserved C-like repeats found in other proteins [150]. According to this scheme, the class I and II M antigens described by others (Section 7.1) would represent subdivisions of the class C M-like proteins.

Fibrinogen-binding has been considered to be a characteristic property of M proteins, but Mrp4 and the closely related FcrA76 proteins have been found to bind both fibrinogen and IgG [150]. In addition, similar Ig- and fibrinogen-binding proteins have been found in M22, M28 and M60 strains [151]. This blurs the previously assumed distinction between fibrinogen-binding 'M proteins' and M-like Ig Fc-binding proteins. Further, hybridization studies indicate that all OF⁺ M types examined possess multiple *emm*-like genes similar to those described in detail above [147]. Although some of these *emm*-like genes may be transcriptionally silent (e.g. *ennX, enn4*), studies to date suggest that most OF⁺ strains are likely to express at least two, and in some cases three, distinct M-like proteins, each of which have N-terminal sequences that may vary between serotypes. This raises the question: what is an M antigen and how does one define an M protein? As noted above, sequence comparisons suggest that the *emm*-like genes discussed in Section 7.1 correspond to the central genes in *emm*-like triplets and where data are available (e.g. M2, M49), M-typing sera appear to recognize the product of this central gene. At present, however, a direct correlation between the M-typing reaction and a specific *emm*-like gene product has not been established for the vast majority of M types and many of the conventional typing-sera could well contain antibodies against more than one M-like protein. Even when typing-sera identify a single highly expressed M-like protein (e.g. EmmL2.1) when cells are grown in vitro, in the absence of direct evidence it is dangerous to assume that this protein is dominant under all in vivo growth conditions, or that it is an antiphagocytic protein, or that the typing-sera are opsonic. Indeed, although Mrp4 possesses many characteristics traditionally associated with M proteins (including a variable N-terminal sequence, a predicted coiled-coil structure and fibrinogen-binding), anti-Mrp4 sera are not opsonic and hence the designation Mrp rather than M protein [149]. It is clear that it is time to reconsider dogma and definitions that have emerged from traditional serotyping studies. For example, for the reasons outlined earlier, it may be useful to confine the term M protein to those M-like proteins that have been demonstrated directly to possess antiphagocytic properties. However, until clearer definitions are agreed, one must be careful not to be misled by the fact that different laboratories may use different designations for variants of very similar *emm*-like genes.

From the published literature, the situation in OF⁻ M types seems more straightforward in that they are reported to possess a single *emm*-like gene and in many cases there is direct evidence for the antiphagocytic and opsoninogenic properties of these *emm* gene products. The apparently distinct architecture of the *virR–emm–scp* regions (called the *vir* 'regulon') in OF⁺ and OF⁻, strains suggested that they represent evolutionary divergent lines [147], which are reflected in the class I and II M protein divisions discussed in Section 7.1. Two exceptions have been described, namely M80 (OF⁻) in which hybridization studies have identified an OF⁺ *vir* regulon architecture, and M1 (also OF⁻), which expresses an M-like IgG Fc-binding protein, called protein H [146,147]. However, we have recently found that such exceptions are not uncommon. We have identified at least two distinct *emm*-like genes in a wide variety of OF⁻ M types and in the case of one M5 strain have shown that an *emm*-like gene located 210 bp downstream from *emm5* expresses an IgG-binding protein when cloned in *E. coli*, although its expression by *S. pyogenes* has not been confirmed (A. Whatmore and M.A. Kehoe, unpublished data). Interestingly, there is not a corresponding downstream *emm*-like gene in other M5 strains which we have examined, indicating that significant variation in the architecture of the *vir* regulon can occur in strains expressing the same serotype of M protein. This, together with our studies of other strains, has provided strong evidence for the horizontal transfer of *emm*-like genes between strains (A. Whatmore and M.A. Kehoe, unpublished data). These observations, together with those outlined in Section 7.1, indicate that evolution of group A streptococcal *emm*-like genes is a very dynamic process, involving horizontal transfer of either entire genes and/or gene segments, as well as other events. Given this mixing and matching of genes and gene segments, one must be careful not to be misled by traditional typing methods or other subdivisions that tend to oversimplify complex evolutionary relationships.

7.4. Role of M proteins in pathogenesis

The antiphagocytic properties of types 5 and 6 M proteins have been demonstrated directly, by showing that expression of recombinant M proteins in *Streptococcus sanguis* (rM5) or an M⁻ strain of *S. pyogenes* (rM6) confers these cells with resistance to phagocytosis [154,155]. For a range of other M types, the antiphagocytic properties of M proteins have been inferred from the ability of M⁺ cells to resist phagocytic killing and/or the ability of type-specific, opsonic anti-M sera to overcome this resistance [110]. Detailed studies have focused on a limited number of serotypes, and in these cases it has been demonstrated that M-negative (M⁻) streptococci are effectively opsonized by the alternative complement pathway, whereas in the absence of type-specific opsonic antibody, M-positive (M⁺) cells resist phagocytosis [156–158]. The antiphagocytic properties of these M proteins have been attributed by different groups to their ability to bind either plasma fibrinogen [128,159,160] or complement factor H [161,162]. Fibrinogen binding to M proteins has been suggested to mask receptors for the complement C3b opsonin, whereas factor H downregulates production of C3b, by facilitating the inactivation of C3b and the decay of the C3bBb complex that amplifies C3b production through a positive feedback loop.

Different laboratories have reported apparently conflicting results concerning the extent to which fibrinogen-binding contributes to the antiphagocytic properties of M proteins. It is clear that expression of M protein on the cell surface reduces deposition of C3b to a significant extent even in the absence of fibrinogen. Jacks-Weis et al. found that C3b deposition on M6⁺ and M49⁺ cells in serum (i.e. absence of fibrinogen) was patchy, resulting in ineffective opsonization and almost complete resistance to phagocytosis [158]. In contrast, Whitnack and Beachey [160] found more uniform C3b deposition on M24⁺ cells in the absence of fibrinogen, resulting in effective opsonization and only partial resistance to phagocytosis, whereas in the presence of fibrinogen, the cells were completely resistant to phagocytosis. These differences might be explained by differences in the amounts of M protein expressed by different strains in these experiments.

It has been reported that M proteins can also bind the complement regulator, factor H, and it was suggested that this explains their antiphagocytic properties [161]. However, the affinity of M proteins for fibrinogen is at least 10-fold higher than their affinity for factor H [128,161] and the concentration of fibrinogen in plasma is higher than the concentration of factor H. Thus, in vivo, fibrinogen binding to M protein would be expected to greatly inhibit the binding of factor H to M protein. Indeed, Horstmann et al. [162] reported recently that physiological concentrations of fibrinogen reduce factor H binding to

$M6^+$ streptococci in vitro by 5–10-fold. However, it was also reported that factor H has a low binding-affinity for fibrinogen, that it binds to M^+ cells in the presence of fibrinogen, and that unlike factor H, fibrinogen does not alter the rate of decay of cell-bound C3bBb complexes [162]. The anti-opsonic properties of M proteins under in vivo conditions (i.e. in the presence of fibrinogen) were attributed to this low-affinity factor H binding and it was concluded that the contribution of fibrinogen-binding was small or neutral. It is difficult, however, to account for the difference in the resistance of M^+ and M^- cells to opsonization in plasma (i.e. in the presence of fibrinogen) solely in terms of low affinity factor H binding to M^+ cells. In the studies outlined above, the amount of factor H that bound to $M6^+$ streptococci in the presence of fibrinogen was no higher than the amount binding to M^- cells in its absence, and only two-fold higher than the amount binding to M^- cells in the presence of fibrinogen [162]. Since similar amounts of factor H binding to M^- cells does not inhibit opsonization, the potential ability of bound factor H to inhibit opsonization of M^+ cells must depend on other mechanisms ensuring that less C3b is deposited on M^+ than on M^- cells. In the above studies, it was found that binding of C3 to $M6^+$ cells in 50% serum was about 3-fold lower than binding to M^- cells, and in 50% plasma this difference was increased to 6-fold. Given that C3b acts through a positive feedback loop to amplify its own production, there is likely to be a very fine balance between the inhibitory activity of factor H and the ability of C3 deposition to overcome this inhibition. It seems no more speculative to suggest the two-fold reduction in C3 deposition caused by fibrinogen binding to M proteins helps to tip this balance in favour of the bacterial cell, than it is to suggest that low-affinity factor H binding is significant. In addition to reducing the deposition of C3b, fibrinogen binding to M proteins may contribute to resistance to phagocytosis by other mechanisms, for example, by masking conserved or partially conserved epitopes in the centre and C-terminal halves of M proteins [163]. During evolution, the N-terminal regions of M proteins containing fibrinogen-binding domains (Section 7.1) have undergone considerable primary sequence variation, suggesting that there has been a strong selective pressure to retain the ability to bind fibrinogen, despite the fact that this actually reduces factor H binding to M^+ cells. However, the relative contributions of fibrinogen or factor H-binding to the antiphagocytic properties of M proteins may well vary, depending on the particular strain, variations in physiological environments and on the immunological status of the host. Further, although M proteins are clearly very major antiphagocytic virulence factors in many strains, it must be remembered that this has not been demonstrated directly for many M types and that group A streptococci can also express other antiphagocytic factors including, for example, a hyaluronic acid capsule [164]. The relative contributions of M proteins and these other factors to virulence may also vary between different strains, physiological environments and immunological conditions. Thus, it is an oversimplification to attribute the antiphagocytic properties of all virulent group A streptococci solely to M proteins or to their interactions with one particular host factor.

In addition to their antiphagocytic properties, it has been suggested that M protein might play a role in adhesion of group A streptococci to host surfaces, either by binding directly to the surface of epithelial cells [165] or by acting as an anchor that binds LTA such that its lipid moiety is oriented outwards and binds to fibronectin on the surface of host epithelial cells [166]. However, using allele-replacement mutagenesis, Caparon et al.

242

[167,168] demonstrated recently that M protein is not required for adhesion of M6 streptococci to human buccal or tonsillar epithelial cells in vitro, although it does promote microcolony formation subsequent to adhesion. It has been argued, however, that the adhesion of group A streptococci to host surfaces may be a multifactorial process and it remains possible that M proteins may contribute to the overall adhesiveness of certain strains or to adhesion to certain types of host surfaces [169]. It has also been suggested that M antigens may contribute to the pathogenesis of post-streptococcal autoimmune diseases (acute glomerulonephritis and acute rheumatic fever) by eliciting harmful immunological reactions. The reader is referred to other recent reviews for a discussion of these properties [108,109].

8. Immunoglobulin Fc-binding proteins

In 1966, Forsgren and Sjoquist [6] first demonstrated that *S. aureus* protein A binds gamma-immunoglobulins via their Fc domains. For many years protein A was thought to be a unique bacterial Ig Fc-binding protein, but in the mid-1970s, it was found that IgG Fc-binding also occurs in groups A, C and G streptococci [170,171a], and in 1980 it was confirmed that 'non-immune' IgA-binding to group A streptococci [171b] also occurs via the IgA Fc domain [172]. During the past 10 years an increasing number of Gram-positive wall-associated proteins displaying a wide variety of Ig-binding affinities have been described [173]. These proteins are frequently referred to as Ig receptors. However, as pointed out by Bjorck and Akerstrom [174], the word *receptor* implies that a biological response is evoked upon ligand-binding, but there is no evidence that this is the case for bacterial Ig-binding proteins. Thus, here we avoid the term receptor, although in some cases it is implied by specific protein or gene designations. The roles of Ig-binding proteins in pathogenesis have not been clearly defined, although allele-replacement mutations in the *S. aureus* protein A gene have been found to modulate virulence to a limited extent in certain animal model systems [175]. The reader is referred elsewhere [173] for discussions on the potential mechanisms by which Ig-binding proteins might contribute to virulence.

8.1. IgG Fc-binding proteins

Bacterial IgG Fc-binding proteins were originally classified into five types (designated Types I–V) on the basis of differences in their binding affinities for human IgG subclasses and IgG from various animal species [176]. At the time this system was devised, it was assumed that an individual species expresses a single Type of IgG Fc-binding protein. This appears to be the case for *S. aureus* protein A (designated Type I), which accounts for all of the IgG Fc-binding activity associated with whole cells, is strongly conserved among different strains of *S. aureus* [173], and shares little homology with other IgG Fc-binding proteins (see below). The Type III IgG Fc-binding proteins, expressed by most strains of *S. equisimilis*, *S. dysgalactiae* (closely related group C species) and human isolates of group G streptococci, also represent a distinct, single type of IgG-binding protein, called protein G [177]. Recombination events between repeated sequences may alter both the size and binding properties of protein G in different strains (Section 4, Fig. 2) However, if one aligns protein G sequences leaving gaps to account for duplication/deletion events, one can see that their primary sequences are almost identical and are quite distinct from other IgG Fc-binding proteins [42,43,178,179].

Although protein A (Type I) and protein G (Type III) represent clearly distinct types of IgG Fc-binding proteins, it is now obvious that there is considerable diversity among the group A streptococcal IgG-binding proteins, which were originally defined as a single Type II group that bind all four subclasses of human IgG [176], but subsequently subdivided into Types IIo, II'o, IIa, IIb and IIc on the basis of variation in the binding properties of cell extracts [180]. These subdivisions, however, could be equally misleading and the variation in Ig-binding patterns is consistent with what we now know about the diversity and multiple copy number of *emm*-like genes in *S. pyogenes* (Section 7.3), where the M-like protein family includes variable IgG Fc-binding proteins that possess a variety of IgG Fc-binding specificities. Therefore, one must be careful not to be misled by the fact that the literature often refers to an individual variant as *the* Type II IgG Fc-binding protein or by the fact that one specific variant was designated protein H (human IgG Fc receptor) [181] before this diversity was appreciated. The position has been complicated further by the discovery of Mrp proteins that bind both IgG and fibrinogen (Section 7.3) and protein Sir (streptococcal Ig receptor) that binds both IgG and IgA [151]. It is now apparent that antigenic and host protein-binding properties of individual variants of IgG Fc-binding proteins in group A streptococci can be as distinct as those of protein A (Type I) and protein G (Type III). The classification of Type IV IgG Fc-binding proteins in bovine isolates group G streptococci and Types V and VI IgG Fc-binding proteins in *S. zooepidemicus* (group C) are based solely on antigenic relationships and Ig-binding affinities [173] and in the absence of sequence data, it is difficult to judge the validity of these divisions.

The sequences for protein A, protein G and a number of group A streptococcal *emm*-like IgG-binding protein genes have been reported [42,43,144–146,150,178,182a]. The predicted structures and relationships between these proteins are summarized in Fig. 7. Like most other wall-associated proteins, they all possess N-terminal signal sequences, tandemly repeated sequences, C-terminal LPXTGX motifs and hydrophobic/charged tail domains (Section 4.2), but the upstream wall-associated regions can be quite divergent. Thus, in protein G there are five copies of a pentapeptide repeat between a Pro/Gly-rich region and the LPXTGX motif (Fig. 1b). In protein A, there is an extensive repetitive region (12 × 8-residue repeats) designated Xr, which is rich in Pro/Gly, followed by region Xc which consists of an unrepeated 55-residue sequence containing only 6 Pro or Gly residues [182a]. The remainder of the protein A molecule consists of five copies of a 56/58-residue tandem repeats, each of which contains an IgG Fc-binding domain [28]. For historical reasons, each of these repeats is designated by a distinct letter (Fig. 7), which differs from the convention subsequently adopted for all other wall-associated proteins, where different letters are used to designate different types of repeats (Section 4.1). The N-terminal halves of sequenced M-like IgG Fc-binding proteins from group A streptococci (protein H, FcrA76 and Mrp4) consist of variable, unrepeated sequences and their C-terminal halves contain a single set of tandem repeats, which have been discussed in Section 7.3. The IgG Fc-binding sites in these M-like proteins have yet to be defined

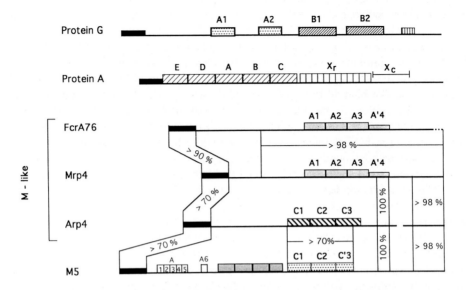

Fig. 7. Structures of immunoglobulin Fc-binding proteins. The black boxes on the extreme left represent N-terminal signal sequences and other boxes represent repeated sequences. The % indicated between M-like Ig-binding proteins represent approximate levels of sequence homologies and the (non-Ig Fc-binding) M5 protein is included for comparison. The sequence of the extreme C-terminal end of FcrA76, represented by the dashed line, was not reported but is likely to be almost identical to the corresponding regions of the other M-like proteins. The sequences represented by the diagram are described in the following references: protein G [42], protein A [182], FcrA76 [144], Mrp4 [150], Arp4 [145] and M5 [120].

precisely, but it has been reported that IgG binds to the C-terminal repeats in FcrA76 [182b] and it has been proposed that the IgG-binding site in protein H is more likely to be located in unrepeated, N-terminal sequences [146].

8.2. IgA Fc-binding proteins

The genes for two distinct types of IgA Fc-binding proteins have been sequenced. One type is represented by the M-like proteins Arp4 and EmmL2.2 from M4 and M2 group A streptococci, respectively [123,145]. Although these are clearly members of the same M-like protein family, differences between their conserved regions and in the relative locations of *arp4* and *emmL2.2* genes in the *emm*-like gene clusters of their respective hosts (Fig. 6) place them in distinct subdivisions within the M-like protein family (Section 7.3). In Arp4, the IgA Fc-binding site has been localized to unique, unrepeated sequences in the N-terminal half of the molecule [145,183,184]. A IgA Fc-binding protein, called Arp60, has been isolated from a *S. pyogenes* M60 strain [185]. Arp60 is very similar to Arp4, but like other M-like proteins, the sequences in the N-terminal halves of Arp4 and Arp60 are variable [183]. In addition to binding IgA Fc with high affinity, Arp4 and Arp60 display weak binding for IgG [183–185].

The second, type of IgA Fc-binding protein for which complete sequence data are available is a component of the 'C antigen complex' of group B streptococci, which is expressed by all capsular serotype Ib and many serotypes Ia and II strains [186]. The C antigen complex has been reported to include four distinct proteins [186,187], but only the α and β antigens, which have been shown to be protective, have been studied in any detail. Genes for both the α and β antigens have been cloned and sequenced [188–192]. The α antigen does not appear to bind Ig or other host proteins and its function has not been defined. The β antigen, which is an IgA Fc-binding protein, has been designated protein Bac (**B** streptococci IgA receptor, **C** complex protein) by some workers. Unlike most wall-associated proteins, the 123 786 Da protein Bac (β-antigen) does not contain extensive tandem repeats, and although it contains a typical C-terminal LPXTGX motif and hydrophobic/charged tail domain, there are only 5 prolines in the 84 residues preceding the LPXTGX motif. However, upstream from this there is a very Pro-rich 90-residue sequence with an unusual tripeptide XPZ periodicity, where every third residue is Pro, the residue at position X is variable but either hydrophobic or poorly hydrophilic and position Z alternates between a positively and negatively charged residue [191]. It has been reported that there are two distinct IgA Fc-binding sites in the β-antigen, although these do not correspond to obvious repeats [190]. However, the use of polyclonal human IgA in these experiments raises the possibility that some of the observed binding might have been due to antigen-specific antibody and the number of distinct Fc-binding sites remains to be confirmed. In addition to the IgG- and IgA-binding proteins described above, a *S. pyogenes* protein designated Sir (streptococcal **I**g **r**eceptor), which displays high-affinity binding for both IgG and IgA Fc-regions has been described [151]. Its sequence has not been published, but it has been found to be a class C (Section 7.3) M-like protein (G. Lindahl, personal communication).

8.3. Non-immune Fab-binding proteins

Non-immune binding of Fab regions of various immunoglobulin subclasses (IgG, IgA, IgM and IgE) to protein A and of IgG Fab to protein G has been reported [193]. This Fab-binding is sometimes called alternative non-immune Ig-binding, to distinguish it from Ig Fc-binding. Inhibition and binding experiments employing various Ig subfragments and different antibody isotypes have indicated that Fab-binding to protein A involves a site in the variable region of the Ig heavy chain, and is specific for the $V_H III$, explaining the ability of protein A to bind Fab regions of different immunoglobulin subclasses [194,195]. Fab-binding to protein G involves only heavy chain constant region sequences and has been defined precisely by crystallographic studies (see following section). Neither protein A nor protein G appear to interact with Ig light chains. However, a novel Ig-binding protein which specifically binds Ig light chains has been isolated from *Peptococcus magnus* and has been designated protein L (**L** chain) [196a,b]. The protein L gene has recently been cloned and sequenced [197] and the predicted protein L sequence shares many of the characteristic features of other wall-associated proteins, including a highly repetitive structure and a C-terminal hydrophobic/charged tail domain, preceded by the sequence LPKAGS, which is very similar to the more typical LPXTGX motif. The upstream Pro-rich region consists of 51 residues with a XPX tripeptide periodicity (analogous but not identical to that in protein Bac; above), followed by 6 hexapeptide tandem repeats [197]. The molecule also includes 5 copies of a 72–76 residue tandem repeat which contain the Ig light chain-binding domains, and two further distinct repeats whose function has not been defined. A distinctive structural feature of protein L is that the predicted N-terminal signal sequence (which is not similar to those of lipoproteins) is unusually short (18 residues) for a Gram-positive protein.

8.4. Defined Ig-binding sites in proteins A and G

Crystallographic studies have defined the IgG Fc-binding interaction between IgG and an isolated tandem repeat from protein A [198]. This binding involves predominantly hydrophobic contacts between residues in two α-helices in the protein A repeat and residues at the interface of the C_H2 and C_H3 domains in the Fc region of IgG. The IgG Fc-binding site in protein G has been localized to the C-terminal ends of the 55-residue C-terminal repeats (designated B in the GX7809 sequence and C in the GX7805 sequence; Fig. 2) [199]. There is no primary sequence homology between the well-defined IgG Fc-binding sites on protein A and protein G, nor between these sites and IgG Fc-binding regions of protein H. Nevertheless, inhibition experiments with synthetic peptides indicate that all three proteins bind to the same region of IgG Fc and it has been suggested that this reflects convergent evolution towards a common Fc-binding structure [199]. However, NMR studies on an isolated 55-residue C-terminal repeat from protein G have shown that the structure of the protein G Fc-binding region is quite distinct from that in protein A [200–202]. Further, recent studies on the crystal structure of Fab bound to an isolated 55-residue C-terminal repeat from protein G, have shown that this binding involves interactions between a β-strand in protein G aligned antiparallel to a β-strand in Fab, effectively extending a β-sheet structure in protein G into the Fab molecule [203]. This is a particularly interesting arrangement, since hydrogen bonding between atoms in the polypeptide backbone contributes significantly to the stability of the β-sheet, such that the binding domains could tolerate considerable variation in their primary sequences.

8.5. Binding affinities for other host proteins

In addition to non-immune binding of immunoglobulins, protein A, protein G and protein L have been reported to bind the plasma protease-inhibitors kininogen and α_2-macroglobulin (α_2M) [204]. The interaction between protein G and α_2M has been studied in most detail. Inhibition experiments have shown that IgG and α_2M compete for the same binding sites on protein G, and that the affinity of protein G for IgG is about twice that of its affinity for α_2M [204]. Further, since the molar concentration of IgG in plasma is about twenty times higher than that of α_2M, group A streptococci are likely to bind IgG much more readily than α_2M. Nevertheless, at physiological pH, binding of both IgG and α_2M to intact streptococci in vitro has been demonstrated [204]. Protein G also binds human serum albumin via domains in the N-terminal half of the molecule which are quite distinct from the C-terminal IgG-binding domains [174]. The potential significance of these binding reactions in vivo remains a matter for speculation.

9. Fibronectin-binding proteins

Fibronectin is an adhesive glycoprotein that is an important constituent of the extracellular matrix in human and animal tissues and it is also deposited on implanted medical catheters and other prosthetic devices. A variety of bacterial pathogens can bind to fibronectin on host surfaces or implanted prostheses [205–207], including *Staphylococcus aureus*, coagulase-negative staphylococci and groups A, C and G streptococci. Genes encoding wall-associated fibronectin-binding proteins (FnBP) have been cloned from *S. aureus* [208,209], *S. pyogenes* [210,211] and *S. dysgalactiae* [212].

Two distinct FnBPs, designated FnBP A and FnBP B, have been cloned from a single strain (8325-4) of *S. aureus* and both have been sequenced [209,213]. They contain typical C-terminal wall-associating signals, a number of distinct types of tandemly repeated sequences and share considerable homologies (Fig. 8). These genes are located close to each other (separated by 682 bp) in the *S. aureus* strain 8325-4 chromosome [209] and are likely to have arisen by gene duplication and divergence. Some *S. aureus* strains appear to possess only a single *fnb* gene, but the number of strains that have been examined to date is limited (T.J. Foster, personal communication). The fibronectin-binding site on FnBP A has been localized to the 38-residue D repeats (Fig. 8) and individual D1, D2 or D3 peptides inhibit binding of fibronectin to *S. aureus* cells, as well as binding of *S. aureus* to fibronectin-coated substrates [208,213,214]. By chemically modifying various amino acid side chains and examining subpeptides of D3, the sequence EEDT was identified as an important constituent of the fibronectin-binding site [215]. However, a synthetic peptide with the sequence FEEDTL failed to inhibit fibronectin-binding, indicating that adjacent residues also play a role in presenting the EEDT motif in the correct conformation [215].

Studies on truncated FnBP B proteins have shown that in addition to the D repeats, FnBP B contains an additional fibronectin-binding site which is not present in FnBP A

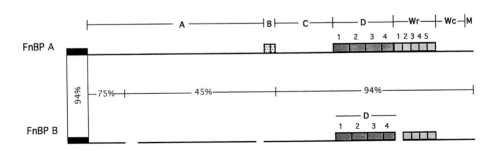

Fig. 8. Structures and relationships between *S. aureus* fibronectin-binding proteins. The black boxes on the extreme left represent N-terminal signal peptides and other boxes represent repeated sequences. Various regions of the FnBPs (including unrepeated regions) are designated by letters , as described in the original papers [209,213]. Approximate levels of sequence homologies are indicated by the % between FnBP A and B [209].

and therefore is likely to be located in the variable N-terminal half of FnBP B (region A, Fig. 8), which lacks an EEDT motif [215]. However, these sequences contain a number of sites displaying a similar charge distribution to the EEDT motif [215], suggesting that the fibronectin-binding domain consists of a structural motif that can tolerate some primary sequence variation. A partial sequence corresponding to the C-terminal region of a fibronectin-binding protein from *S. pyogenes* has been reported and, interestingly, it also contains short (37-residue) tandem repeats that bind fibronectin [216]. These repeats have a similar distribution of charged and uncharged residues as in the D repeats of *S. aureus* FnBP A and B [216] and earlier experiments have suggested that *S. pyogenes* and *S. aureus* share a common binding site on fibronectin [217]. This might reflect convergence to a common fibronectin-binding domain, but the possibility that these domains in the *S. pyogenes* and *S. aureus* proteins have diverged from a common ancestor cannot be excluded. The extreme C-terminal sequence of the partially sequenced *S. pyogenes* FnBP possess typical wall-associating signals [216].

Evidence that *S. aureus* FnBP plays a role in virulence has been provided by studies on Tn*918* insertion mutants that displayed both a reduced ability to bind fibronectin and a 100-fold reduction in virulence in a rat traumatized heart valve endocarditis model [218]. Although at the time these studies were performed, the precise nature of the mutants had not been defined, recent studies have indicated that the Tn*918* insertion is located either in or very close to a single *fnb* gene in this clinical isolate (T.J. Foster, personal communication). Using allele-replacement mutagenesis, Hanski and Capron [211] have demonstrated that inactivation of an FnBP gene (called *prtF*) in M6 *S. pyogenes*, abolishes the ability to adhere to respiratory epithelial cells, at least in vitro. Thus, the PrtF protein is likely to be an important virulence factor, although this needs to be confirmed by in vivo studies since, in addition to cell-matrix fibronectin, there is also a soluble form that circulates in body fluids and soluble fibronectin has been observed to inhibit binding of *S. pyogenes* to eucaryotic cells in vitro [205].

Hybridization studies have detected sequences homologous to *prtF* in a number of distinct clinical isolates of *S. pyogenes* [219], but Talay et al. [210] failed to detect hybridization between *fnb* genes cloned from two distinct clinical isolates, indicating that there is some diversity among *S. pyogenes fnb* genes. To date, there is no evidence that there is more than one *fnb* gene in a single *S. pyogenes* strain. However, two distinct *fnb* genes have been cloned from a single strain of *S. dysgalactiae*, but even although both cloned gene products bound fibronectin only one of these was detected on the surface of the parent organism [212]. All of the FnBPs characterized to date bind to a 29 kDa N-terminal fragment of fibronectin, suggesting that they share similar binding-domains, although their precise binding sites on fibronectin have not been defined.

10. Collagen- and plasmin-binding proteins and GAPDH

Some strains of *S. aureus* express a surface protein that adheres to collagen, which may contribute to their ability to colonize collagen-rich tissues such as bone or cartilage, and to the pathogenesis of osteomyelitis. A *S. aureus* collagen adhesin gene (*cna*) has recently been cloned and sequenced [220]. The predicted protein sequence (1185 residues) shares no homology with other sequences in databases and has yet to be studied in detail. However, it possesses features typical of a wall-associated protein, including three large tandem repeats in the C-terminal half of the molecule and typical C-terminal wall-associating signals [220].

The sequence of a *S. pyogenes* gene (*plr*) encoding a high-affinity plasmin-binding protein that is released from intact *S. pyogenes* cells by mutanolysin (a muralytic enzyme) has been described recently [221]. Surprisingly, the predicted primary sequence was found to share considerable homology with bacterial glyceraldehyde-3-phosphate dehydrogenases (GAPDH), which are normally cytoplasmic proteins, and preliminary data indicated that the recombinant Plr protein also possess GAPDH activity [221]. Similarly, another laboratory has recently reported that a major M_r 39 000 *S. pyogenes* surface-protein is a GAPDH [222]. This GAPDH was purified and found to bind to a variety of mammalian proteins, including fibronectin, lysozyme, myosin and actin, and anti-GAPDH sera reacted with a similar size polypeptide in wall-extracts of all streptococcal strains tested [222]. However, the purified GAPDH had only a weak binding affinity for plasmin, suggesting that it is not identical to the high affinity plasmin-binding Prl protein described above [221,222]. Thus, there might be a family of closely related, wall-associated GAPDHs in group A streptococci, possessing a variety of affinities for various host proteins. Although they have been reported to be major wall proteins [222], at present the functional significance of wall-associated GAPDHs in *S. pyogenes* is not clear. Their enzymatic activities are not consistent with our current understanding of the biochemistry of *S. pyogenes* cell-wall polymers. GAPDH would produce glyceric acid diphosphate from glyceraldehyde 3-phosphate and NAD^+. Extracellular polysaccharides produced by one Gram-negative species (*Pseudomonas elodea*) are known to contain glyceric acid substituents [266], but glyceric acid has not been found in *S. pyogenes* cell walls, at least when organisms are grown in vitro. It is not clear if the potential to produce glyceric acid diphosphate during infection (from substrates supplied by the host) has any significance for virulence, but the high affinities of *S. pyogenes* wall-associated GAPDHs for certain host proteins suggest that they might contribute to virulence by virtue of their binding properties. It is intriguing to find such binding activities associated with cell-wall GAPDHs, rather than a more conventional wall-associated binding protein.

The predicted sequence of the cloned *plr* gene product [221] shows that it is quite distinct from most other wall-associated proteins described in this chapter, in that it possesses neither tandem repeats nor obvious wall-associating signals of the type described in Section 4. Further, the predicted protein sequence shows that Plr lacks an N-terminal secretion-signal sequence and this has been confirmed by N-terminal amino acid sequencing mutanolysin-released Plr [221]. Nevertheless, both Plr and the homologous purified GAPDH from a different strain, appear to be strongly associated with the cell-wall fractions [221,222]. Thus, these proteins might be secreted and associated with the wall by novel mechanisms that are not, at present, understood. It is interesting to note that at least one other Gram-positive cell-wall-associated protein, namely *Listeria monocytogenes* internalin (Section 14), also lacks an N-terminal signal peptide.

11. Complement C5a peptidase

Group A streptococci express a wall-associated complement C5a peptidase (SCP), which has been characterized in some detail [140,141]. The *scp* gene has already been described as part of the *S. pyogenes vir* regulon in Sections 7.2 and 7.3. Sequencing of a cloned *scp* gene has shown that this protease possesses a typical N-terminal signal peptide and C-terminal wall-associating sequences, and that it shares some homology with the catalytic domain of *Bacillus* subtilisin serine proteases [141]. It is, however, highly specific for C5a [223], which is produced upon activation of complement and is an important inflammatory mediator and chemotaxin. Mouse intraperitoneal virulence assays on *scp*::Tn*916* insertion mutants [224] suggest that cleavage of the C5a chemotaxin delays the accumulation of polymorphonuclear leukocytes at the site of infection, although the mutation did not reduce the LD_{50} of streptococci in this mouse model. Wall-associated C5a peptidases are also expressed by human strains of group G streptococci [225] and some strains of group B streptococci [226] and hybridization studies suggest that these are very similar to the group A streptococcal enzyme [226,227].

12. Wall-associated proteins in oral streptococci

Interests in dental health have stimulated studies of surface proteins in oral pathogens and in addition to the fimbriae discussed in Section 5.1, a variety of non-fimbrial wall-associated proteins have been characterized. *Streptococcus salivarius* expresses either surface fimbriae (about 50% of strains) or fibrils (Fig. 3), suggesting that they might not be able to express both types of filaments simultaneously [44]. The fimbriae have been characterized to only a limited extent, but two distinct *S. salivarius* fibrils, designated antigen B (AgB) and AgC have been studied in more detail [228–230]. AgB mediates co-aggregation with *Veillonella parvula* and is called the veillonella-binding protein (VBP), whereas AgC mediates adhesion to oral surfaces and is called host adhesion factor (HAF) [228]. Neither the AgB nor AgC genes have been cloned, but the antigens have been purified and characterized at a biochemical level. Both AgB and AgC consist of a single, monomeric, but very high molecular weight polypeptides and AgC appears to be a glycoprotein [228]. AgB has a M_r of 380 000 and two forms of AgC have been described: an M_r 488 000 form (AgC_{in}) accumulates in the cytoplasm of certain non-adhesive mutants, whereas AgC isolated from the cell walls of wild-type strains (AgC_{cw}) has a M_r between 250 000 and 320 000 depending on extraction conditions [230]. The difference in the size of AgC_{in} and AgC_{cw} might be due to degradation during extraction procedures, but the possibility that it reflects specific processing events during secretion and localization in the wall has not been ruled out. Interestingly, the accumulation of AgC (but not AgB) in the cytoplasm of two distinct non-adhesive mutants [230] raises the possibility that additional accessory factors may be required for its localization in the wall. Ultrastructural and immunogold-labelling studies have shown that both AgB and AgC consist of flexible, very thin (ca. 2 nm) rods with swollen globular ends and that they form distinct 91 nm (AgB) and 73 nm (AgC) long fibrils on the cell surface [229]. Fibrillar *S. salivarius* strains also express 63 nm and 178 nm fibrils which have not been characterized at a bio-chemical level.

A non-fimbrial adhesin in *S. sanguis* strain 12, designated SsaB, has already been de-scribed as a lipoprotein (Section 3.1). The *ssaB* gene has been cloned and sequenced [66,67] and the deduced SsaB sequence shares 87% homology with the *S. sanguis* FW213 fimbrial subunit FimA (Section 5.1), and like FimA it adheres to saliva-coated hydroxyapatite [66]. Anti-SsaB sera cross-react with both FimA and a M_r 38 000 *S. gor-donii* PK488 adhesin that mediates co-aggregration with *A. naeslundi* [231]. Further, anti-PK488 adhesin sera has been shown to inhibit co-aggregation reactions mediated by a variety of other oral streptococci [231]. Thus, FimA, SsaB and the PK488 adhesin may be variants of a family of closely related adhesins distributed among oral streptococci, al-though they appear to have evolved different mechanisms for associating with the cell wall, since unlike the closely related FimA, SsaB is a lipoprotein.

Other strains of *S. sanguis* and *S. gordonii* have been found to express quite distinct adhesins. The cloning of a determinant expressing part of a $M_r > 84\ 000$ adhesin (expressed as an M_r 200 000 *lacZ* gene-fusion) from *S. sanguis* G9B [232], as well as the cloning and partial sequencing of a M_r 76 000 lipoprotein adhesin, designated SarA (Section 3.1) from *S. gordonii* strain Challis [19,234] has been described. A further large molecular weight (M_r 205 000) adhesin, called SSP-5, has been cloned from *S. sanguis* strain M5 [234]. The SSP-5 protein binds salivary agglutinin (SAG) and when the cloned *ssp5* gene was introduced into a normally non-aggregrating species, *Enterococcus fae-calis*, it was expressed on the cell-surface and caused *E. faecalis* to aggregate in the pres-ence of SAG [235]. SSP-5 is closely related to similar large molecular weight adhesins in a number of other species of oral streptococci, including protein SpaA from *S. sobrinus* [236] and the *S. mutans* P1 adhesin [237] (also called Pac [238,239], antigen B [237,240,241], antigen I/II [242] and IF [243]). Sequencing of cloned genes has shown that P1, SpaA and SSP-5 share considerable homologies [35,239,244–246]. The P1 ad-hesin possesses typical C-terminal cell-wall associating sequences. However, although SpaA contains a C-terminal Pro-rich region and an LPXTGX motif, it lacks an obvious hydrophobic/charged-tail domain [245] and the C-terminal end of SSP-5 has only a very short Pro-rich sequence and lacks both an LPXTGX motif and a hydrophobic/charged-tail domain [246]. Thus, despite the close relationships between P1, SpaA and SSP-5, it seems likely that they employ different pathways for associating with the cell wall. The signal sequences of neither SpaA nor SSP-5 possess characteristic features of a lipopro-tein signal and the mechanisms by which they associate with the cell wall have yet to be determined. All three of these proteins possess two distinct types of tandem repeats, one (called the HR1 or A-region) close to the N-terminal ends and a second (called the PR1 or P-region) in the C-terminal halves of the proteins. Interestingly, the N-terminal HR1 re-peats share some homology with group A streptococcal M proteins, although other re-gions of SSP-5 are quite distinct from M proteins [246].

In addition to the above adhesins, the gene for a wall-associated *S. mutans* antigen, designated *wapA* (**w**all-**a**ssociated **p**rotein A) has been cloned and sequenced [247]. The C-terminal end of the predicted protein sequence contains a LPXTGX motif and hydro-phobic/ charged-tail domain, but instead of a typical upstream Pro-rich region there is a 60 residue wall-associating region containing a very high (65%) proportion of threonine and serine. WapA is a potential candidate in an anti-caries vaccine, but to date there is no information available on its function.

13. Pneumococcal surface protein A (PspA)

The polysaccharide capsule of *Streptococcus pneumoniae* has long been recognized as an important virulence factor, but in addition to a capsule, pneumococci express a highly immunogenic cell-surface protein designated PspA. Anti-PspA antibodies protect mice upon challenge with virulent pneumococci [248,249] and *pspA*-defective insertion mutants display reduced virulence in mice [250]. Like streptococcal M proteins, PspA exhibits both size variation (M_r 67 000–99 000) and antigenic variation, but unlike M proteins, distinct serotypes are cross-protective indicating conservation of accessible protective epitopes [251–253]. The *pspA* gene from one pneumococcal strain has recently been cloned and sequenced and studies on pneumococcal mutants expressing defined truncated forms of PspA have shown that the C-terminal region is required to anchor PspA to the cell-surface [254,255]. However, although the deduced PspA primary sequence includes a typical N-terminal secretion signal sequence, it lacks C-terminal wall-associating signals of the type described in Section 3.2. The entire C-terminal half of the molecule consists of 10 copies of a 20-residue tandem repeat, followed by a short 12-residue tail [254]. Interestingly, these repeats share considerable homology with repeats in the C-terminus of pneumococcal autolysin (LytA) (Fig. 9), whereas other regions of PspA and LytA are quite distinct [254,256]. There is evidence that LytA interacts with choline residues on pneumococcal teichoic acids [257] and it has been suggested that PspA may be associated with the cell wall by similar interactions involving its LytA-homologous C-terminal repeats [254,255]. *S. pneumoniae* is the only species described to date in which teichoic acids possess phosphodiester-linked choline residues and these linkages are chemically much more stable that the corresponding labile ester-linkages of D-alanine residues in teichoic acids of other species [258]. It is possible that this difference in teichoic acid structure between pneumococci and other Gram-positive species permit wall-associated proteins to be anchored firmly in the wall by interactions with teichoic acids in pneumococci, whereas this mechanism might be much less effective in other species. Upstream

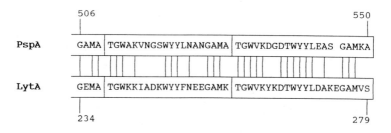

Fig. 9. Relationship between C-terminal repeats in PspA and LytA. Only a part of the homologous repeat regions is shown in the diagram to illustrate the extensive homology between these regions of PspA [255] and LytA.[156]. A more complete comparison is provided in ref. 255.

from the repeated sequences, in the centre of PspA, there is a sequence containing two regions (separated by 20 amino acids) where >40% of the residues are Pro and it has been suggested that this region may 'span' the cell wall [254]. If this is the case, then the entire repetitive region of PspA, accounting for almost half the molecule, is likely to be embedded in the wall. Upstream from the Pro-rich region, the N-terminal half of PspA (excluding 9 extreme N-terminal residues) possesses a heptapeptide periodicity that predicts a fibrillar, dimeric, α-helical coiled-coil structure (Section 5.2). The preponderance of hydrophobic residues at positions 1 and 4 and hydrophilic residues at other positions in the PspA heptapeptide repeats is significantly higher than in streptococcal M proteins [254], indicating that the predicted coiled-coil structure would be energetically very favourable (Section 5.2.). Although the predicted N-terminal coiled-coil region of PspA shares structural (and limited primary sequence) homology with group A streptococcal M proteins, and like M proteins, PspA appears to play a role in virulence, the function of this surface protein has still to be defined.

14. Listeria monocytogenes internalin

Invasion of non-professional phagocytes is a key step in the pathogenesis of listeriosis [259–264]. Recent studies on invasion-defective Tn*1545* insertion mutants of *L. monocytogenes* identified an invasion locus containing two ORFs, designated *inlA* and *inlB* (for 'internalization'), which are separated by 85 bp [265]. Subclones containing *inlA* alone were shown to confer an invasive phenotype on the normally non-invasive species *L. innocua*, and immunogold-labelling demonstrated that the *inlA* gene product is expressed on the cell surface. Thus, the 744-residue *inlA* product was called 'internalin'. An *inlB* product was not detected and it is not yet clear if this ORF is expressed.

The predicted sequence of internalin contains typical C-terminal Gram-positive wall-associating signals [265]. This is the first example of such C-terminal signals in a surface protein from a rod-shaped bacterium and indicates that the wall-associating mechanisms discussed in Section 3.2 are not confined to Gram-positive cocci. A further structural similarity with many other Gram-positive surface proteins is that about two-thirds of the internalin molecule consists of repeated sequences and one of two distinct types of repeated regions in internalin displays a periodicity in the distribution of non-polar Leu and Ile residues, suggestive of an α-helical coiled-coil structure [265]. Although internalin seems structurally similar to many other wall-associated proteins, it also possesses a number of unusual features. There is an unusually high (21%) content of Ser and Thr residues, distributed regularly through the protein. The significance of this is not clear, but it is interesting to note that a similar high content of Ser and Thr residues has been observed in the predicted sequence of group A streptococcal T6 protein [116]. A particularly striking feature is that the N-terminus of the deduced internalin sequence does not possess the characteristic features of a signal peptide, although the protein is clearly expressed on the cell surface [265]. This suggests that the translocation of internalin across the cytoplasmic membrane involves novel mechanisms (also see Section 10). Further studies will be required to determine if this is the case and to define functional domains in internalin.

15. Concluding remarks

The studies summarized in this chapter reflect a very rapid expansion in our knowledge of Gram-positive wall-associated proteins in recent years, and in many cases have implications well beyond interests in specific proteins or species. For example, the novel protein G Ig-binding domain described in Section 8.4, involving the extension of β-sheet structure between protein G and Ig Fab, has wider implications for protein–protein interactions in general and the dynamic evolution of the *emm*-like gene family described in Section 7.3, emphasizes the limitations of specialized classification or typing systems that are still widely used in microbiology. It is important to recognize, however, that our current knowledge of Gram-positive cell-wall-associated proteins is limited predominantly to similar types of proteins produced by a relatively small number of pathogenic species, many of which are closely related. There have been remarkably few detailed studies on wall-associated proteins in non-pathogenic species and it remains to be determined if common structural themes and patterns found in wall-associated proteins expressed by pathogenic bacteria extend to surface proteins of species that interact with very different types of environments. Even in the pathogenic species that have received most attention, certain types of wall-associated proteins have been studied to a very limited extent. For example, studies on Gram-positive bacterial fimbriae have been so sparse compared to their intensively studied counterparts in Gram-negative species that there is still a common misconception that Gram-positive bacteria do not produce fimbriae. Further, studies to date have focused primarily on proteins that are expressed well when organisms are grown in batch culture in vitro. It must be remembered that the repertoire of wall-associated proteins expressed by organisms growing in natural environments could be significantly different from that observed when organisms are grown under artificial conditions in the laboratory and more attention deserves to be paid to the effect of growth conditions on both the expression of wall-associated proteins and their interactions with the cell wall.

Acknowledgements

I am indebted to T.J. Foster, P.S. Handley and G. Lindahl for providing data prior to publication, and to T.J. Foster, G. Lindahl and A. Whatmore for providing constructive comments on the manuscript

References

1. Lancefield, R.C. (1928) J. Exp. Med. 47, 91–103.
2. Lancefield, R.C. (1928) J. Exp. Med. 47, 481–491.
3. Forsgren, A. and Sjoquist, J. (1966) J. Immunol. 97, 822–827.
4. Sjoquist, J., Meloun, B. and Hjelm, H. (1972) Eur. J. Biochem. 29, 572–578.
5. Beachey, E.H., Campbell, G.L. and Ofek, I. (1974) Infect. Immun. 9, 891–896.
6. Beachey, E.H., Stollerman, G.H., Chiang, E.Y., Chiang, T.M., Seyer, J.M. and Kang, A.H. (1977) J. Exp. Med. 145, 1469–1483.
7. Sarvas, M. (1986) Curr. Top. Microbiol. Immunol. 125, 103–125.

8. von Heijne, G. and Abrahmsen, L. (1989) FEBS Lett. 244, 439–446.
9. Kontinen, V.P. and Sarvas, M. (1988) J. Gen. Microbiol. 134, 2333–2344.
10. Kontinen, V.P., Saris, P. and Sarvas, M. (1991) Mol. Microbiol. 5, 1273–1283.
11. Chambert, R., Benyahia, F. and Petit-Glaton, M.-F. (1990) Biochem. J. 265, 375–382.
12. Haandrikman, A., Kok, J., Laan, H., Soemitro, S., Ledeboer, A.M., Konings, W.N. and Venema, G. (1989) J. Bacteriol. 171, 2789–2794.
13. Vos, P., van Asseldonk, M., van Jeveren, F., Siezen, R., Simons, G. and de Vos, W.M. (1989) J. Bacteriol. 171, 2795–2802.
14. Kok, J., Leenhouts, K.J., Haandrikman, A.J., Ledeboer, A.M. and Venema, G. (1988) Appl. Environ. Microbiol. 54, 231–238.
15. Vos, P., Simons, G., Siezen, R.J. and de Vos W.M. (1989) J. Biol. Chem. 264, 13579–13585.
16. Power, S.D., Adams, R.M. and Wells, J.A. (1986) Proc. Natl. Acad. Sci. USA 83,3096–3100.
17. Gould, A.R., May B.K. and Elliot, W.H. (1975) J. Bacteriol. 122, 34–40.
18. Wu, H.C. and Tokunaga, M. (1986) Curr. Top. Microbiol. Immunol. 125, 127–157.
19. Jenkinson, H.F. (1992) Infect. Immun. 60, 1225–1228.
20. Ganeshkumar, N., Arora, N. and Kohlenbrander, P.E. (1993) J. Bacteriol. 175, 572–574.
21. Gilson, E., Alloing, G., Schmidt, T., Claverys, J.P., Dudler, R. and Hofnung, M. (1988) EMBO J. 7, 3971–3974.
22. Russell, R.R.B., Aduse-Opoku, J., Tao, L. and Feretti, J.J. (1991) in: G.M. Dunny, P.P. Cleary and L.L. McKay (Eds.), Genetics and Molecular Biology of Streptococci, Lactococci and Enterococci, American Society for Microbiology, Washington, DC, pp. 244–247.
23. Hayashi, S. and Wu, H.C. (1983) J. Bacteriol. 156, 773–777.
24. Klein, P., Somorjai, R.L. and Lau, P.C.K. (1988) Protein Eng. 2, 15–20.
25. Fischetti, V.A., Pancholi, V. and Schneewind, O. (1990) Mol. Microbiol. 4, 1603–1605.
26. Sjodahl, J. (1977) Eur. J. Biochem. 73, 343–351.
27. Guss, B., Uhlen, M., Nilsson, B., Lindberg, M., Sjoquist, J. and Sjodahl, J. (1984) Eur. J. Biochem. 138, 413–420.
28. Pancholi, V. and Fischetti, V.A. (1989) J. Bacteriol. 170, 2618–2624.
29. Fischetti, V.A. (1989) Clin. Microbiol. Rev. 2, 285–314.
30. Pancholi, V. and Fischetti, V.A. (1989) J. Exp. Med. 170, 2119–2133.
31. Forsgren, A. (1969) Acta. Path. Microbiol. Scand. 75, 481–490.
32. Sjoquist, J., Movitz, J., Johansson, I.-B., and Hjelm, H. (1972) Eur. J. Biochem. 30, 190–194.
33. Schneewind, O., Model, P. and Fischetti, V.A. (1992) Cell 70, 267–281.
34. Kehoe, M. A., Poirier, T.P., Beachey, E.H. and Timmis, K.N. (1985) Infect. Immun. 48, 190–197.
35. Kelly, C.P., Evans, P., Bergmeier, L., Lee, S.F., Progulske-Fox, A., Bleiweis, A.S. and Lehner, T. (1989) FEBS Lett. 258, 127–132.
36. Lee, S.F. (1992) Infect. Immun. 60, 4032–4039.
37. Burne, R.A. and Penders, J.E.C. (1992) Infect. Immun. 60, 4621–4632.
38. Fischetti, V.A., Jarymowycz, M., Jones, K.F. and Scott, J.R. (1986) J. Exp. Med. 164, 971–980.
39. Hollingshead, S.K., Fischetti, V.A. and Scott, J.R. (1987) Mol. Gen. Genet. 207, 196–203.
40. Wren, B.W. (1991) Mol. Microbiol. 5, 797–803.
41. Jones, K.F., Hollingshead, S.K., Scott, J.R. and Fischetti, V.A. (1988) Proc. Natl. Acad. Sci. USA 85, 8271–8275.
42. Fahnestock, S.R., Alexander, P., Nagle, J. and Filpula, D. (1986) J. Bacteriol. 167, 870–880.
43. Filpula, D., Alexander, P. and Fahnestock, S. (1987) Nucleic Acids Res. 15, 7210.
44. Handley, P.S. (1990) Biofouling 2, 239–264.
45. Phillips, G.N., Flicker, P.F., Cohen, C., Manjula, B.N. and Fischetti, V.A. (1981) Proc. Natl. Acad. Sci. USA 78, 4689–4693.
46. Manjula, B.N. and Fischetti, V.A. (1980) J. Exp. Med. 151, 695–708.
47. Schulz, G.E. and Schirmer, R.H. (1979) Principles of Protein Structure, Springer-Verlag, New York.
48. Manjula, B.N., Khandke, K.M., Fairwell, T., Relf, W.A. and Sriprakash, K.S. (1991) J. Protein Chem. 10, 369–384.
49. Yanagawa, R., Otsuki, K. and Tokui, T. (1968) Jpn. J. Vet. Res. 16, 31–38.
50. Yanagawa, R. and Honda, E. (1976) Infect. Immun. 13, 1293–1295.

51. Honda, E. and Yanagawa, R. (1978) Am. J. Vet. Res. 39, 155–158.
52. Schmidt, H., Naumann, G. and Putzke, H.-P. (1988) Zbl. Bakteriol. Hyg. 268, 228–237.
53. Saito, T., Koga T., Ono E., Yanagawa, R., Ito, T., Kida, H. and Shimizu, Y. (1990) Jpn. J. Vet. Res. 52, 11–18.
54. Willcox, M.D.P., Wyatt, J.E. and Handley, P.S. (1989) J. Appl. Bacteriol. 66, 291–299.
55. Fachon-Kalweit, S., Elder, B.L. and Fives-Taylor, P. (1985) Infect. Immun. 48, 617–624.
56. Cisar, J.O. (1986) in: D. Mirelman (Ed.), Microbial Lectins and Agglutinins, Wiley, New York, pp. 183–196.
57. Gibbons, R.J., Hay, D.I., Cisar, J.O. and Clark, W.B. (1988) Infect. Immun. 56, 2990–2993.
58. McIntire, F.C., Crosby, L.K., Vatter, A.E., Cisar, O.J., McNeil, M.R., Bush, C.A., Tjoa, S.S. and Fennessey, P.V. (1988) J. Bacteriol. 170, 2229–2235.
59. Brennan, M.J., Cisar, J.O. and Sandberg, A.L. (1986) Infect. Immun. 52, 840–845.
60. Yeung, M.K., Chassy, B.M. and Cisar, J.O. (1987) J. Bacteriol. 169, 1678–1683.
61. Donkersloot, J.A., Cisar, J.O., Wax, M.E., Harr, R.J. and Chassy, B.M. (1985) J. Bacteriol. 162, 1075–1078.
62. Yeung, M.K. and Cisar, J.O. (1988) J. Bacteriol. 170, 3803–3809.
63. Yeung, M.K. and Cisar, J.O. (1990) J. Bacteriol. 172, 2462–2468.
64. Fives-Taylor, P.M., Macrina, F.L., Pritchard, T.J. and Peene, S.S. (1987) Infect. Immun. 55, 123–128.
65. Fenno, J.C., LeBlanc, D.J. and Fives-Taylor, P. (1989) Infect. Immun. 57, 3527–3533.
66. Ganeshkumar, N., Song, M. and McBride, B. (1988) Infect. Immun. 56, 1150–1157.
67. Ganeshkumar, N., Hannam, P.M., Kolenbrander, P.E. and McBride, B.C. (1991) Infect. Immun. 59, 1093–1099.
68. Fives-Taylor, P., Fenno, J.C., Holden, E., Linehan, L., Oligino, L. and Volansky, M. (1991) in: G.M. Dunny, P.P. Cleary, P.P. and L.L. McKay (Eds.), Genetics and Molecular Biology of Streptococci, Lactococci and Enterococci, American Society for Microbiology, Washington, DC, pp. 240–243.
69. de Graaf, F.K. (1990) Curr. Top. Microbiol. Immunol. 151, 29–53.
70. DePamphilis, M.L. and Adler, J. (1971) J. Bacteriol. 105, 384–395.
71. Macnab, R.M. (1987) in: F.C. Neidhardt, J.L. Ingraham, K.B. Low, B. Magasanik, M. Schaechter, E.H. Umbarger (Eds.), Escherichia coli and Salmonella typhimurium: Cellular and Molecular Biology, Vol. 1, American Society for Microbiology, Washington, DC, pp. 70–83.
72. Albertini, A.M., Caramori, T., Crabb, W.D., Scoffone, F. and Galizzi, A. (1991) J. Bacteriol. 173, 3573–3579.
73. Beveridge, T.J. and Graham, L.L. (1991) Microbiol. Rev. 55, 684–705.
74. Sleytr, U.B. and Messner, P. (1988) J. Bacteriol. 170, 2891–2897.
75. Sleytr, U.B. and Messner, P. (1988) Crystalline Bacterial Surface Layers, Springer-Verlag, Berlin.
76. Lortal, S., van Heijenoort, J., Gruber, K. and Sleytr, U.B. (1992) J. Gen. Microbiol. 138, 611–618.
77. Peters, j., Peters, M., Lottspeich, F. and Baumeister, W. (1989) J. Bacteriol. 171, 6307–6315.
78. Tsuboi, A., Uchihi, R., Tabata, R., Takahashi, Y., Hashiba, H., Sasaki, T., Yamagata, H., Tsukagoshi, N. and Udaka, S. (1986) J. Bacteriol. 168, 365–373.
79. Ebisu, S., Tsuboi, A., Takagi, H., Naruse, Y., Yamagata, H., Tsukagoshi, N. and Udaka, S. (1990) J. Bacteriol. 172, 1312–1320.
80. Peters, J., Peters, M., Lottspeich, F., Schafer, W. and Baumeister, W. (1987) J. Bacteriol. 169, 5216–5223.
81. Vidgren, G., Palva, I., Pakkanen, R., Lounatmaa, K. and Palva, A. (1992) J. Bacteriol. 174, 7419–7427.
82. Garduno, R.A., Lee, E.J.Y. and Kay, W.W. (1992) Infect. Immun. 60, 4373–4382.
83. Lamed, R. and Bayer, E.A. (1988) Adv. Appl. Microbiol. 33, 1–46.
84. Gilbert, H.J. and Hazelwood, G.P. (1993) J. Gen. Microbiol. 139, 187–194.
85. Tokatlidis, K., Salamitou, S., Beguin, P., Dhurjati, P. and Aubert, J.P. (1991) FEBS Lett. 291, 185–188.
86. Fujino, T., Beguin, P. and Aubert, J.P. (1992) FEMS Microbiol. Letts. 94, 165–170.
87. Bayer, E.A. and Lamed, R. (1986) J. Bacteriol. 167, 828–836.
88. Poole, D.M., Morag, E., Lamed, R., Bayer, E.A., Hazlewood, G.P. and Gilbert, H.J. (1992) FEMS Microbiol. Letts. 99, 181–186.
89. Mayer, F., Coughlan, M.P., Mari, Y. and Ljungdahl, L.G. (1987) Appl. Environ. Microbiol. 53, 785–792.

90. Henrissat, B., Claeyssens, M., Tomme, P., Lemesle, L. and Mornon, J.P. (1989) Gene 81, 83–95.
91. Boden, M.K. and Flock, J.-L. (1989) Infect. Immun. 57, 2358–2363.
92. Jeljaszewicz, J., Switalski, L.M. and Adlam, C. (1983) in: C.S.F. Easmon and C. Adlam (Eds.), Staphylococci and Staphylococcal Infections, Vol. 2, Academic Press, London, pp. 525–557.
93. McDevitt, D., Vaudaux, P. and Foster, T.J. (1992) Infect. Immun. 60, 1514–1523.
94. Hawiger, J., Timmons, S., Strong, D.D., Cottrel, B.A., Riley, M. and Doolittle, R.F. (1982) Biochemistry 21, 1407–1413.
95. Strong, D.D., Laudano, A.P., Hawiger, J. and Doolittle, R.F. (1982) Biochemistry 21, 1414–1420.
96. Gilkes, N.R., Henrissat, B., Kilburn, D.G., Miller, R.C. and Warren, R.A.J. (1991) Microbiol. Rev. 55, 303–315.
97. Hemker, H.C., Bas, B.M. and Muller, A.D. (1975) Biochim. Biophys. Acta 379, 180–188.
98. Kawabata, S., Morita, T., Iwanaga, S. and Igarashi, H. (1985) J. Biochem. 98, 1603–1614.
99. Kaida, S., Miyata, T., Yoshizawa, Y., Kawabata, S., Morita, T., Igarashi, H. and Iwanaga, S. (1987) J. Biochem. 102, 1177–1186.
100. Phonimdaeng, P., O'Reilly, M., Nowlan, P., Bramley, A.J. and Foster, T.J. (1990) Mol. Microbiol. 4, 393–404.
101. Kawabata, S., Miyata, T., Morita, T., Miyata, T., Iwanaga, S. and Igarashi, H. (1986) J. Biol. Chem. 261, 527–531.
102. Kawabata, S., Morita, T., Miyata, T., Iwanaga, S. and Igarashi, H. (1986) J. Biol. Chem. 261, 1427–1433.
103. Hasegawa, N. and San Clemente, C.L. (1978) Infect. Immun. 22, 473–479.
104. Masuda, S. (1983) Microbiol. Immunol. 27, 801–805.
105. Haraldsson, I. and Jonsson, P. (1984) J. Comp. Pathol. 94, 183–196.
106. Jonsson, P., Lindberg, M., Haraldsson, I. and Wadstrom, T. (1985) Infect. Immun. 49, 765–769.
107. Foster, T.J. (1992) in: C.E. Hormaeche, C.W. Penn and C.J. Smyth (Eds.), Molecular Biology of Bacterial Infection: SGM Symp. 49, Cambridge University Press, Cambridge, pp. 173–191.
108. Kehoe, M.A. (1991) Vaccine 9, 797–806.
109. Robinson, J.H. and Kehoe, M.A. (1992) Immunol. Today 13, 362–367.
110. Lancefield, R.C. (1962) J. Immunol. 89, 307–313.
111. Tillet, W.S. and Garner, R.L. (1934) Bull. Johns Hopkins Hosp. 54, 145–156.
112. Kantor, F.S. (1965) J. Exp. Med. 121, 849–859.
113. Top, F.H. and Wannamaker, L.W. (1968) J. Exp. Med. 127, 1013–1034.
114. Widdowson, J.P., Maxted, W.R. and Grant, D.L. (1970) J. Gen. Microbiol. 61, 343–353.
115. Rotta, J. (1978) in: T. Bergan and J.R. Norris (Eds.), Methods in Microbiology, Vol. 12, Academic Press, London, pp. 177–198.
116. Schneewind, O., Jones, K.F. and Fischetti, V.A. (1990) J. Bacteriol. 172, 3310–3317.
117. Jones, K.F., Schneewind, O., Koomey, J.M. and Fischetti, V.A. (1991) Mol. Microbiol. 5, 2847–2852.
118. Hollingshead, S.K., Fischetti, V.A. and Scott, J.R. (1986) J. Biol. Chem. 261, 1677–1686.
119. Robbins, J.C., Spanier, J.G., Jones, S.J., Simpson, W.J. and Cleary, P.P. (1987) J. Bacteriol. 169, 5633–5640.
120. Miller, L., Gray, L., Beachey, E.H. and Kehoe, M. (1988) J. Biol. Chem. 263, 5668–5673.
121. Mouw, A., Beachey, E.H. and Burdett, V. (1988) J. Bacteriol. 170, 676–684.
122. Haanes, E.J. and Cleary, P.P. (1989) J. Bacteriol. 171, 6397–6408.
123. Bessen, D.E. and Fischetti, V.A. (1992) Infect. Immun. 60, 124–135.
124. Haanes-Fritz, E., Kraus, W., Burdett, V., Dale, J.B., Beachey, E.H. and Cleary, P.P. (1988) Nucleic Acids Res. 16, 4667–4677.
125. Podbielski, A., Melzer, B. and Lutticken, R. (1991) Med. Microbiol. Immunol. 180, 213–227.
126. Podbielski, A., Baird, R. and Kaufhold, A. (1992) Med. Microbiol. Immunol. 181, 209–213.
127. Swanson, J., Hsu, K.C. and Gotschlich, E.C. (1969) J. Exp. Med. 130, 1063–1091.
128. Ryc, M. Beachey, E.H. and Whitnack, E. Infect. Immun. (1989) 57, 2397–2404.
129. Miller, L., Burdett, V., Poirier, T.P., Gray, L.D., Beachey, E.H. and Kehoe, M.A. (1988) Infect. Immun. 56, 2198–2204.
130. Whitnack, E. and Beachey, E.H. (1985) J. Exp. Med. 162, 1983–1997.
131. Bessen, D., Jones, K.F. and Fischetti, V.A. (1989) J. Exp. Med. 169, 269–283.

258

132. Bessen, D.E. and Fischetti, V.A. (1990) J. Exp. Med. 172, 1757–1764.
133. Relf, W.A. and Sriprakash, K.S. (1990) FEMS Microbiol. Lett. 71, 345–350.
134. (a) Scott, J.R., Hollingshead, S.K. and Fischetti, V.A. (1986) Infect. Immun. 52, 609–612. (b) Kaufhord, A., Podbielski, A., Johnson, D.R., Kaplan, E.L. and Lutticken, R. (1992) J. Clin. Microbiol. 30, 2391–2397.
135. Simpson, W.J., Musser, J.M. and Cleary, P.P. (1992) Infect. Immun. 60, 1890–1893.
136. Spanier, J.G. and Cleary, P.P. (1980) J. Exp. Med. 152, 1393–1406.
137. Simpson, W.J. and Cleary, P.P. (1987) Infect. Immun. 55, 2448–2455.
138. Simpson, W.J., Lapenta, D., Chen, C. and Cleary, P.P. (1990) J. Bacteriol. 172, 696–700.
139. Caparon, M.G. and Scott, J.R. (1987) Proc. Natl. Acad. Sci. USA 84, 8677–8681.
140. Chen, C.C. and Cleary, P.P. (1989) Infect. Immun. 57, 1740–1745.
141. Chen, C.C. and Cleary, P.P. (1990) J. Biol. Chem. 265, 3161–3167.
142. Perez-Casal, J., Caparon, M.G. and Scott, J.R. (1991) J. Bacteriol. 173, 2617–2614.
143. Caparon, M.G., Geist, R.T., Perez-Casal, J. and Scott, J.R. (1992) J. Bacteriol. 174, 5693–5701.
144. Heath, D.G. and Cleary, P.P. (1989) Proc. Natl. Acad. Sci. USA 86, 4741–4745.
145. Frithz, E., Heden, L.-O. and Lindahl, G. (1989) Mol. Microbiol. 3, 1111–1119.
146. Gomi, H., Hozumi, T., Hattori, S., Tagawa, C., Kishimoto, F. and Bjorck, L. (1990) J. Immunol. 144, 4046–4052.
147. Haanes, E.J., Heath, D.G. and Cleary P.P. (1992) J. Bacteriol. 174, 4967–4976.
148. Jeppson, H., Frithz, E. and Heden, L.-O. (1992) FEMS Microbiol. Lett. 92, 139–146.
149. Lindahl, G. (1989) Mol. Gen. Genet. 216, 372–379.
150. O'Toole, P., Stenberg, L., Rissler, M. and Lindahl, G. (1992) Proc. Natl. Acad. Sci. USA 89, 8661–8665.
151. Stenberg, L., O'Toole, P. and Lindahl, G. (1992) Mol. Microbiol. 6, 1185–1194.
152. Khandke, K.M., Fairwell, T. and Manjula, B.N. (1987) J. Exp. Med. 166, 151–162.
153. Cleary, P.P., LaPenta, D., Heath, D., Haanes, E.J. and Chen, C. (1991) in: G.M. Dunny, P.P. Cleary and L.L. McKay (Eds.), Genetics and Molecular Biology of Streptococci, Lactococci and Enterococci, American Society for Microbiology, Washington, DC, pp. 147–151.
154. Poirier, T.P., Kehoe, M.A., Whitnack, E., Dockter, M.E. and Beachey, E.H. (1989) Infect. Immun. 57, 29–35.
155. Scott, J.R., Guenthner, P.C., Malone, L.M. and Fischetti, V.A. (1986) J. Exp. Med. 164, 1641–1651.
156. Bisno, A.L. (1979) Infect. Immun. 26, 1172–1176.
157. Peterson, P.K., Schmeling, D., Cleary, P.P., Wilkinson, B.J., Kim, Y. and Quie, G. (1979) J. Infect. Dis. 139, 575–587.
158. Jacks-Weis, J., Kim, Y. and Cleary, P.P. (1982) J. Immunol. 128, 1897–1902.
159. Whitnack, E. and Beachey, E.H. (1985) J. Bacteriol. 164, 350–358.
160. Whitnack, E. and Beachey, E.H. (1982) J. Clin. Invest. 69, 1042–1045.
161. Horstmann, R.D., Sievertsen, H.J., Knobloch, J. and Fischetti, V.A. (1988) Proc. Natl. Acad. Sci. USA 85, 1657–1661.
162. Horstmann, R.D., Sievertsen, H.J., Leippe, M. and Fischetti, V.A. (1992) Infect. Immun. 60, 5036–5041.
163. Whitnack, E., Dale, J.B. and Beachey, E.H. (1984) J. Exp. Med. 159, 1201–1212.
164. Wessels, M.R., Moses, A.E., Goldberg, J.B. and DiCesare, T.J. (1991) Proc. Natl. Acad. Sci. USA. 88, 8317–8321.
165. Tylewska, S.K., Fischetti, V.A. and Gibbons, R.J. (1988) Curr. Microbiol. 16, 209–216.
166. Ofek, I., Simpson, W.A. and Beachey, E.H. (1982) J. Bacteriol. 149, 426–433.
167. Caparon, M.G., Stephens, D.S., Olsen, A. and Scott, J.R. (1991) Infect. Immun. 59, 1811–1817.
168. Hanski, E. and Caparon, M.G. (1992) Proc. Natl. Acad. Sci. USA 89, 6172–6176.
169. Hasty, D.L., Ofek, I., Courtney, H.S. and Doyle, R.J. (1992) Infect. Immun. 60, 2147–2152.
170. Kronvall G. (1973) J. Immunol. 111, 1401–1406.
171. (a) Christensen, P. Johansson, B.G. and Kronvall, G. (1976) Acta. Pathol. Microbiol. Scand. Sect. C 84, 73–76. (b) Christensen, P. and Oxelius, V.A. (1975) Acta. Pathol. Scand. 83, 184–188.
172. Schalen, C. (1980) Acta. Pathol. Microbiol. Scand. C 88, 271–274.
173. Boyle, M.D.P. (1990) Bacterial Immunoglobulin-Binding Proteins, Vol. 1, Academic Press, London.

174. Bjorck, L. and Akerstrom B. (1990) in: M.D.P. Boyle (Ed.), Bacterial Immunoglobulin-Binding Proteins, Vol. 1, Academic Press, London, pp. 113–126.
175. Patel, A.H., Nowlan, P., Weavers, E.D. and Foster, T.J. (1987) Infect. Immun. 55, 3103–3110.
176. Myhre, E.B. and Kronvall, G. (1982) in: S.E. Holm and P. Christensen (Eds.), Basic Concepts of Streptococci and Streptococcal Diseases, Reedbooks, Surrey, pp. 209–210.
177. Bjorck, L. and Kronvall, G. (1984) J. Immunol. 133, 969–974.
178. Olsson, A., Eliasson, M., Guss, B., Nilsson, B., Hellman, U., Lindberg, M. and Uhlen, M. (1987) Eur. J. Biochem. 168, 319–324.
179. Sjobring, U., Bjorck, L. and Kastern, W. (1991) J. Biol. Chem. 266, 399–405.
180. Raeder, R., Otten, R.A., Chamberlin, L. and Boyle, M.D.P. (1992) J. Clin. Microbiol. 30, 3074–3081.
181. Akesson, P., Cooney, J., Kishimoto, F. and Bjorck, L. (1990) Mol. Immunol. 27, 523–531.
182. (a) Uhlen, M., Guss, B., Nilsson, B., Gatebeck, S., Philipson, L. and Lindberg, M. (1984) J. Biol. Chem. 259, 1695–1702 (correction p. 13628). (b) Heath, D.G., Boyle, M.D.P. and Cleary, P.P. (1990) Mol. Microbiol. 4, 2071–2079.
183. Lindahl, G., Akerstrom, B., Stenberg, L., Frithz, E. and Heden, L.-O. (1991) in: G.M. Dunny, P.P. Cleary and L.L. McKay (Eds.), Genetics and Molecular Biology of Streptococci, Lactococci and Enterococci, American Society for Microbiology, Washington, DC, pp. 155–159.
184. Akerstrom, B., Lindqvist, A. and Lindahl, G. (1991) Mol. Immunol. 28, 349–257.
185. Lindahl, G. and Akerstrom, B. (1989) Mol. Microbiol. 3, 239–247.
186. Johnson, D.R. and Ferrieri, P. (1984) J. Clin. Microbiol. 19, 506–510.
187. Brady, L.J., Daphtary, U.D., Ayoub, E.M. and Boyle, M.D.P. (1988) J. Infect. Dis. 158, 965–972.
188. Cleat, P.H. and Timmis, K.N. (1987) Infect. Immun. 55, 1151–1155.
189. Michel, J.L., Madoff, L.C., Kling, D.E., Kasper, D.L. and Ausubel, F.M. (1991) Infect. Immun. 59, 2023–2028.
190. Jeristrom, P.G., Chhatwal, G.S. and Timmis, K.N. (1991) Mol. Microbiol. 5, 843–849.
191. Heden, L.-O., Frithz, E. and Lindahl, G. (1991) Eur. J. Immunol. 21, 1481–1490.
192. Michel, J.L., Madoff, L.C., Olson, K., Kling, D.E., Kasper, D.L. and Ausubel, F.B. (1992) Proc. Natl. Acad. Sci. USA 89, 10060–10064.
193. Myhre, E.B. (1990) in: M.D.P. Boyle (Ed.), Bacterial Immunoglobulin-Binding Proteins, Vol. 1, Academic Press, London, pp. 243–256.
194. Seppala, I., Kaartinen, M., Ibrahim, S. and Makela, O. (1990) J. Immunol. 145,2989–1993.
195. Sasso, E.H., Silverman, G.J. and Mannik, M. (1991) J. Immunol. 147, 1877–1883.
196. (a) Myhre, E.B. and Erntell, M. (1985) Mol. Immunol. 22, 879–885. (b) Borck, L. (1988) J. Immunol. 140, 1194–1197.
197. Kastern, W., Sjobring, U. and Bjorck, L. (1992) J. Biol. Chem. 267, 12820–12825.
198. Deisenhofer, J. (1981) Biochemistry 20, 2361–2370.
199. Frick, I.-M., Wikstrom, M., Forsen, S., Drakenberg, T., Gomi, H., Sjobring, U. and Bjork, L. (1992) Proc. Natl. Acad. Sci. USA 89, 8532–8536.
200. Gronenborn, A.M., Filpula, D.R., Essig, N.Z., Achari, A., Whitlow, M., Wingfield, P.T. and Clore, G.M. (1991) Science 253, 657–661.
201. Lian, L.-Y., Yang, J.C., Derrick, J.P., Sutcliffe, M.J., Roberts, G.C.K., Murphy, J.P., Goward, C.R. and Atkinson, T. (1991) Biochemistry 30, 5335–5340.
202. Lian, L.-Y., Derrick, J.P., Sutcliffe, M.J., Yang, J.C. and Roberts, G.C.K. (1992) J. Mol. Biol. 228, 1219–1234.
203. Derrick, J.P. and Wigley, D.B. (1992) Nature 359, 752–754.
204. Sjobring, U., Trojnar, J., Grubb, A., Akerstrom, B. and Bjorck, L. (1989) J. Immunol. 143, 2948–2954.
205. Hook, M., Switalski, L.M., Wadstrom, T. and Lindberg, M. (1989) in: D.F. Mosher (Ed.), Fibronectin, Academic Press, San Diego, CA, pp. 295–308.
206. Vaudaux, P.E., Waldvogel, F.A., Morgenthaler, J.J. and Nydegger, U.E. (1984) Infect. Immun. 45, 768–774.
207. Myhre, E.B. and Kuusela, P. (1983) Infect. Immun. 40, 29–34.
208. Flock, J.-I., Froman, G., Jonsson, K., Guss, B., Signas, C., Nilsson, B., Raucci, G., Hook, M., Wadstrom, T. and Lindberg, M. (1987) EMBO J. 6, 2351–2357.
209. Jonsson, K., Signas, C., Muller, H.-P. and Lindberg, M. (1991) Eur. J. Biochem. 202, 1041–1048.

260

210. Talay, S.R., Ehrenfeld, E., Chhatwal, G.S. and Timmis, K.N. (1991) Mol. Microbiol. 5, 1727–1734.
211. Hanski, E. and Caparon, M. (1992) Proc. Natl. Acad. Sci. USA 89, 6172–6176.
212. Lindgren, P.-E., Speziale, P., McGavin, M., Monstein, H.-J., Hook, M., Visai, L., Kostiainen, T., Bozzini, S. and Lindberg, M. (1992) J. Biol. Chem. 267, 1924–1931.
213. Signas, C., Raucci, G., Jonsson, K., Lindgren, P.-E., Anantharamaiah, G.M., Hook, M. and Lindberg, M. (1989) Proc. Natl. Acad. Sci. USA 86, 699–703.
214. Raja, R.H., Raucci, G. and Hook, M. (1990) Infect. Immun. 58, 2593–2598.
215. McGavin, M.J., Raucci, G., Gurusiddappa, S. and Hook, M. (1991) J. Biol. Chem. 266, 8343–8347.
216. Talay, S.R., Valentin-Weigand, P., Jerlstrom, P.G., Timmis, K.N. and Chhatwal, G.S. (1992) Infect. Immun. 60, 3837–3844.
217. Speziale, P., Hook, M., Switalski, L.M. and Wadstrom, T. (1984) J. Bacteriol. 157, 420–427.
218. Kuypers, J.M. and Proctor, R.A. (1989) Infect. Immun. 57, 2306–2312.
219. Hanski, E., Horwitz, P.A. and Caparon, M. (1992) Infect. Immun. 60. 5119–5125.
220. Patti, J.M., Jonsson, H., Guss, B., Switalski, L.M., Wiberg, K., Lindberg, M. and Hook, M. (1992) J. Biol. Chem. 267, 4766–4772.
221. Lottenberg, R., Broder, C.C., Boyle, M.D.P., Kain, S.J., Schroder, B.L. and Curtiss III, R. (1992) J. Bacteriol. 174, 5204–5210.
222. Pancholi, V. and Fischetti, V.A. (1992) J. Exp. Med. 176, 415–426.
223. Wexler, D.E., Chenoweth, D.E. and Cleary, P.P. (1985) Proc. Natl. Acad. Sci. USA 82, 8144–8148.
224. O'Connor, S.P. and Cleary P.P. (1987) J. Infect. Dis. 156, 495–504.
225. Cleary, P.P. Peterson, J., Chen. C. and Nelson, C. (1991) Infect. Immun. 59, 2305–2310.
226. Hill, H.R., Bohnsack, J.F., Morris, E.Z., Augustine, N.H., Parker, C.J., Cleary, P.P. and Wu, J.T.(1988) J. Immunol. 141, 3551–3556.
227. Cleary, P.P., Handley, J., Suvorov, A.N., Podbielski, A. and Ferrieri, P. (1992) Infect. Immun. 60, 4239–4244.
228. Weerkamp, A.H. and Jacobs, T. (1982) Infect. Immun. 38, 233–242.
229. Weerkamp, A.H., Handley, P.S., Baars, A. and Slot, J.W. (1986) J. Bacteriol. 165, 746–755.
230. Weerkamp, A.H., van dei Mei, H.C. and Liem, R.S.B. (1986) J. Bacteriol. 165, 756–762.
231. Kolenbrander, P.E. and Andersen, R. (1990) Infect. Immun. 58, 3064–3072.
232. Rosan, B., Baker, C.T., Nelson, G.M., Berman, R., Lamont, R.J. and Demuth, D.R. (1989) J. Gen. Microbiol. 135, 531–538.
233. Jenkinson, H.F. (1991) in: G.M. Dunny, P.P. Cleary and L.L. McKay (Eds.), Genetics and Molecular biology of Streptococci, Lactococci and Enterococci, American Society for Microbiology, Washington, pp. 284–288.
234. Demuth, D.R., Davis, C.A., Corner, A.M., Lamount, R.J., Leboy, P.S. and Malamud, D. (1988) Infect. Immun. 56, 2484–2490.
235. Demuth, D.R., Berthold, P., Leboy, P.S., Golub, E.E., Davis, C.A. and Malamud, D. (1989) Infect. Immun. 57, 1470–1475.
236. Goldschmidt, R.M. and Curtiss III, R. (1990) Infect. Immun. 58, 2276–2282.
237. Douglas, C.W. and Russell, R.R. (1984) FEMS Microbiol. Letts. 25, 211–214.
238. Okahashi, N., Sasakawa, C., Yoshikawa, M., Hamada, S. and Koga, T. (1989) Mol. Microbiol. 3, 221–228.
239. Okahashi, N., Sasakawa, C., Yoshikawa, M., Hamada, S. and Koga, T. (1989) Mol. Microbiol. 3, 673–678.
240. Russell, R.R.B. (1979) J. Gen. Microbiol. 114, 109–115.
241. Russell, R.R.B. (1980) J. Gen. Microbiol. 118, 383–388.
242. Russell, M.W. and Lehner, T. (1978) Arch. Oral Biol. 23, 7–15.
243. Hughs, M., MacHardy, S.M., Sheppard, A.J. and Woods, N.C. (1980) Infect. Immun. 27, 576–588.
244. Takahashi, I., Okahashi, N., Sasakawa, C., Yoshikawa, M., Hamada, S. and Koga, T. (1989) FEBS Lett. 249, 383–388.
245. LaPolla, R.J., Haron, J.A., Kelly, C.G., Taylor, W.R., Bohart, C., Hendricks, M., Pyati, R., Graff, R.T., Ma, J.K.-C. and Lehner, T. (1991) Infect. Immun. 59, 2677–2685.
246. Demuth, D.R., Golub, E.E. and Malamud, D. (1990) J. Biol. Chem. 265, 7120–7126.
247. Ferretti, J.J., Russell, R.R.B. and Dao, M.L. (1989) Mol. Microbiol. 3, 469–478.

248. McDaniel, L.S., Scott, G., Kearney, J.F. and Briles, D.E. (1984) J. Exp. Med. 160, 386–397.
249. McDaniel, L.S., Sheffield, J.S., DeLucchi, P. and Briles, D.E. (1991) Infect. Immun. 59, 222–228.
250. McDaniel, L.S., Yother, J., Vijayakumar, M., McGarry, L., Guild, W.R. and Briles, D.E. (1987) J. Exp. Med. 165, 381–394.
251. McDaniel, L.S., Scott, G., Widenhofer, K., Carroll, J. and Briles, D.E. (1986) Microb. Pathogenesis 1, 519–531
252. Waltman, W.D., McDaniel, L.S., Gray, B.M. and Briles, D.E. (1990) Microb. Pathog. 8, 61–69.
253. Crain, M.J., Waltman II, W.D., Turner, J.S., Yother, J., Talkington, D.F., McDaniel, L.S., Gray, B.M. and Briles, D.E. (1990) Infect. Immun. 58, 3293–3299.
254. Yother, J. and Briles, D.E. (1992) J. Bacteriol. 174, 601–609.
255. Yother, J., Handsom, G.L. and Briles, D.E. (1992) J. Bacteriol. 174, 610–618.
256. Garcia, P, Garcia, J.L., Garcia, E. and lopez, R. (1986) Gene, 43, 265–272.
257. Briese, T. and Hakenbeck, R. (1985) Eur. J. Biochem. 146, 417–427.
258. Archibald, A.R., Hancock, I.C. and Harwood, C.R. (1993) in: A.L. Sonenshein, J. Hoch and R. Losick (Eds.), *Bacillus subtilis* and Other Gram-positive Bacteria: Biochemistry, Physiology and Molecular Genetics, American Society for Microbiology, Washington, DC, pp. 381–410.
259. Gaillard, J.-L., Berche, P., Mounier, J., Richard, S. and Sansonetti, P. (1987) Infect. Immun. 55, 2822–2829.
260. Kuhn, M., Kathariou, S. and Goebel, W. (1988) Infect. Immun 56, 79–82.
261. Portnoy, D., Jacks, P.S. and Hinrichs, D. (1988) J. Exp. Med. 167, 1459–1471.
262. Cossart, P., Vicente, M.F., Mengaud, J., Baquero, F., Perez-Diaz, J.C. and Berche, P. (1989) Infect. Immun. 57, 3629–3636.
263. Tilney, L.G. and Portnoy, D.A. (1989) J. Cell Biol. 109, 1597–1608.
264. Mounier, J., Ryter, A., Coquis-Rondon, M. and Sansonetti, P.J. (1990) Infect. Immun. 58, 1048–1058.
265. Gaillard, J.-L., Berche, P., Frehel, C., Gouin, E. and Cossart, P. (1991) Cell, 65, 1127–1141.
266. Kuo, M.S., Mort, A.J. and Dell, A. (1986) Carbohydr. Res. 124, 123–133.

J.-M Ghuysen and R. Hakenbeck (Eds.), *Bacterial Cell Wall*
263

CHAPTER 12

Molecular organization and structural role of outer membrane macromolecules

ROBERT E.W. HANCOCK, D. NEDRA KARUNARATNE and
CHRISTINE BERNEGGER-EGLI

*Department of Microbiology, University of British Columbia, 300 – 6174 University Blvd.,
Vancouver, B.C., V6T 1Z3, Canada*

1. Introduction

The outer membrane has been the subject of intensive research over the past two decades. During this time, our image of this layer has matured from one of a rather simple capsule-like girdle, the lipopolysaccharide layer, to that of a sophisticated, unique and multifunctional membrane. This evolution arose from the research of pioneers like Leive, Nikaido and Nakae who recognized the importance of the outer membrane as a semi-permeable barrier [1,2]. A representative molecular model of a section of the outer membrane based on the data presented by several researchers is shown in Fig. 1. The reader is referred to several recent reviews [3–6] and to other chapters in this book for specific discussions of outer membrane constituents and functions. In this review, we attempt to present an overview of how the individual constituents of outer membranes are integrated into a complex multifunctional unit. Discussion is based on the well studied *Escherichia coli* and *Pseudomonas aeruginosa* outer membranes, with exceptions presented where appropriate.

2. Lipopolysaccharides

2.1. General principles

Lipopolysaccharide (LPS) is a major constituent of the bacterial cell envelope accounting for 3–8% of the dry weight of the cell [7]. It is an amphipathic molecule consisting of a hydrophilic portion represented by the O-antigenic polysaccharide and core oligosaccharide linked to the glycolipidic Lipid A residue. The molecular weight of individual LPS molecules can vary from about 8000 to 54 000 according to the lack or presence of variable numbers of the repeating saccharide units that comprise the O-antigenic polysaccharide. However, certain bacterial species, including *Neisseria* sp., *Haemophilus influenzae* and *Bordetella pertussis* do not produce long O-polysaccharides. Instead the carbohydrate

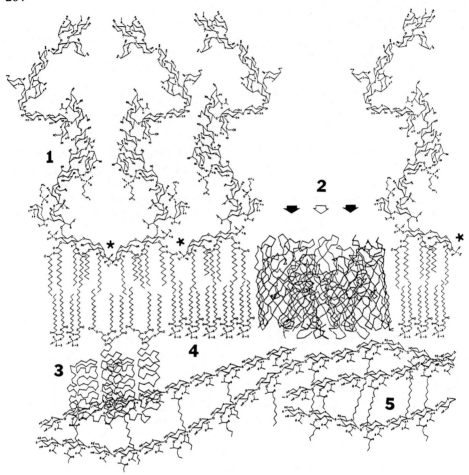

Fig. 1. Side view of a chemical model of part of the *E. coli* outer membrane. LPS (1), matrix porin OmpF (2), lipoprotein (3), phospholipids (4), peptidoglycan (5) and the proposed outer membrane stabilizing binding sites for divalent cations (*) are shown. The structure of LPS shows two O-polysaccharide units, however LPS can contain up to 40 of these pentasaccharides as indicated in Fig. 2. The structure of OmpF (kindly provided by S. Cowan) shows a section of the trimeric porin, having two channels in the front (solid arrows) and one in the back (open arrow). The lipid part of the two lipoproteins (that may be part of a trimeric arrangement) are inserted into the inner leaflet of the outer membrane. Their carboxy termini are linked (covalently or non covalently) to the peptidoglycan layer, which consists of crossbridged *N*-acetylmuramic acid-*N*-acetyl glucosamine-tetrapeptide units. For clarity only, the amino acid backbones of the crossbridging peptide chains of peptidoglycan, of OmpF, and of lipoprotein are shown. Phosphatidylethanolamine is the major lipid component in *E. coli* outer membrane, but other phospholipids such as phosphatidylglycerol and cardiolipin are also found.

portion attached to the Lipid A consists of about ten monosaccharides and hence these molecules are termed lipooligosaccharides (LOS; also termed R-type or rough LPS) [8].

The distribution of LPS in cells has been probed using immunoelectron microscopy, freeze fracture studies and enzyme accessibility studies [9,10]. LPS is exclusively localized in the outer leaflet of the outer membrane. Although the most predominant type of

LPS molecule in most Gram-negative bacteria is LPS that is unsubstituted with O-poly-saccharide (i.e. rough LPS), the protruding O-polysaccharide chains of the remaining smooth LPS molecules form a capsule-like coating over the bacterium [11]. For example, in *P. aeruginosa*, where smooth LPS species comprise less than 10% of the total LPS molecules, a polysaccharide matrix extending 40 nm from the cell surface has been observed.

LPS is anchored in the outer membrane in part, by the fatty acyl chains of its Lipid A portion [12], and thus it contributes substantially to the formation of the outer monolayer of the outer membrane bilayer. In addition, interactions with divalent cations [13,14] and with proteins [4,15–17] are important in stabilizing LPS in the outer membrane. Indeed all major outer membrane proteins studied have been found to interact with LPS. It is generally accepted that LPS comprises by far most of the lipidic material in the outer monolayer of the outer membrane of wild type bacteria [18]. However, in mutants with altered LPS composition, the picture is not as clear with some authors suggesting similar levels of LPS and some suggesting lesser amounts and the presence of lipidic patches [16,19]. In addition, such mutants can demonstrate alterations in the protein constituents of the outer membrane [17], suggesting that the LPS can in some way influence the overall outer membrane composition. It is known that such mutants have an increase in outer membrane fluidity with decreasing polysaccharide chain length [13].

2.2. Chemistry

As outlined above, changes in the structure of LPS lead to alterations in the structure and function of the outer membrane. Therefore, it is necessary to know the chemical structure of LPS to better understand the derivation of outer membrane functions. The LPS has been discussed in several reviews [7,20]. It is made of three general regions (Fig. 2): (a) O side chain polysaccharides, which are immunodominant, (b) a core oligosaccharide

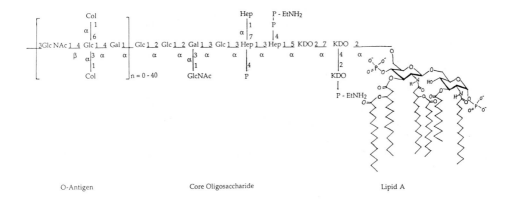

Fig. 2. The structure of LPS from *E. coli*. The three regions Lipid A, core and O-polysaccharide are shown here. The hexose region of the outer core is the R3 core structure found in *E. coli* 0111, whose O antigen contains the very labile sugar colitose. KDO, 3-deoxy-D-manno-2-octulosonic acid; Glc, glucose; Gal, galactose; GlcNAc, *N*-acetylglucosamine; Col, colitose; Hep, heptose; EtNH2, ethanol amine; P, phosphate.

usually containing heptose, 2-keto-3-deoxyoctulosonate (KDO), phosphate and hexose, and (c) the hydrophobic, biologically active endotoxin, Lipid A. Many publications on the chemical structures of these different regions of LPS, have appeared in recent years. From these studies, it is evident that the O side chains are highly variable in structure, composition and polymerization. The core oligosaccharide structure is conserved to a very high degree, only changing from species to species. Lipid A is even more highly conserved but can vary to some extent in different genera [21]. In addition to this heterogeneity, it is now clear that cells may contain more than one type of LPS [22]. For example, *Pseudomonas aeruginosa* cells are known to produce A band and B band LPS which are quite different [22]. Also *Bordetella pertussis* [23] and *Klebsiella* [24] cells can produce two major lipopolysaccharides LPS1 and LPS2 which differ in the side chain region. Structural microheterogeneity in the LPS of *Salmonella* and *E. coli*, and more recently in other species has been observed using SDS polyacrylamide gel electrophoresis to resolve the heterogeneous LPS fractions [25,26]. The capping frequency, or the extent of side chain length, introduces a heterogeneity in O side chain length visualized by a ladder pattern on polyacrylamide gel electrophoresis. Therefore, molecules of varying chain lengths from short chain LOS to long chain LPS may be seen in isolated LPS preparations [22].

O specific chains of LPS are made of repeating oligosaccharide units. The chemical structure of the side chains of several Gram-negative bacteria have been documented [27]. The sugars present in the repeating units may be a single sugar type with differences in linkage sequence resulting in a homopolymer, for example, a mannan in *E. coli* 09 [28] and a rhamnan in some *Pseudomonas* species [29]. In most cases, however, the repeating oligosaccharide contains units of 3–5 different sugars in specific linkages giving rise to a heteropolysaccharide. The O side chains of the *Enterobacteriaceae* have been extensively characterized. Classification into chemotypes according to the sugar composition, and into serotypes according to serological cross-reactivity has been performed [7,20]. A marked difference between O specific chains of pseudomonads and the enteric bacteria is the high content of amino sugars found in the former [30].

Chemical analysis of the core oligosaccharide has been greatly accelerated by the availability of mutants defective in LPS biosynthesis. In the case of *Salmonella*, mutants defective at each stage of biosynthesis of the core oligosaccharide have been used to study the core structure [7,25]. A similar study of the core structure of *E. coli* K12 using *rfa* gene deletions has been reported [31].

The inner core region contains 2–3 residues of the unique octulose, KDO, through which linkage to Lipid A occurs. The KDO residues are linked usually to two L-glycero-D-manno heptose residues. The outer core region consists of hexoses linked to the heptose in the inner core. Glucose, galactose, rhamnose and galactosamine are some of the common outer core hexoses found in the Enterobacteriaceae. Phosphate, pyrophosphate, and phosphoryl ethanolamine substituents may be attached to these sugars in varying degrees. Branching of the sugar core may diversify the structure of the outer core region. The inner core region of the Enterobacteriaceae seems fairly consistent, with variations being limited to the extent of phosphate, pyrophosphate and phosphoryl ethanolamine substitutions [25]. The presence of alanine amide linked to galactosamine in *Pseudomonas* [30] and the presence of galacturonic acid in *Proteus* sp. and *Morganella morganii* is indicative of the considerable variation in outer core structure between species [32].

A beta 1–6 linked diglucosamine disaccharide constitutes the backbone of most Enterobacteriaceae Lipid As. Fatty acids are attached as O- and N-acyl substituents to the glucosamine residues. Phosphates are usually attached to 4' and 1 position and may serve as linkage points for phosphoethanolamine, D-glucosamine, 4-amino-4-deoxy-L arabinose, ethanolamine or phosphate [21]. Variation of Lipid A structure from the regular backbone has been found in certain phototrophic bacteria as well as some non-phototrophic bacteria [33]. The fatty acids attached to the Lipid A disaccharide differ from species to species. The amide linked fatty acids are usually 3-hydroxy alkanoic acids. The number of carbon atoms in the fatty acids could vary from C_{10} to C_{21} [21,33]). In the *Salmonella* Lipid A, 3-hydroxy myristic acid is amide linked as well as ester linked [21].

2.3. Biophysics

Three types of lipids are present in the envelope: phospholipid, LPS and lipoprotein. The phospholipid is distributed approximately equally between the inner and outer membranes, although the ratio of phospholipid/protein in the inner membrane is more than twice that of the outer membrane [34]. The distribution of phospholipid in the outer membrane is mostly in the inner leaflet, with LPS replacing the phospholipid in the outer leaflet of the bilayer [35].

Formation of a lipid bilayer is required for membrane fluidity. Fluid membrane bilayers are important for normal cell functions, e.g. transport across the membrane and excretion. The fluidity of the membrane undergoes major changes in state with temperature. The composition of the membrane constituents (ratios of protein/phospholipid/LPS), the nature of the lipid group (length of the fatty acyl chain, unsaturation) and association of lipids with membrane proteins (lipid–protein interactions) affect the temperature of the phase transition [36]. Usually a lower transition temperature arising from the melting of lipid, and a second transition due to protein is observed with membrane bilayers. Transitions have been monitored by X-ray diffraction, deuterium nuclear magnetic resonance spectroscopy, fluorescent probes, spin probes and scanning calorimetry [36,37]. A single transition was observed in wild type live *E. coli* cells. However, in whole cells and envelopes containing both inner and outer membranes, two reversible transitions have been observed. The first transition is characteristic of live cells, the second appears only after exposure to high temperature, prolonged storage, sonication or lysozyme-EDTA treatment. LPS from *E. coli* undergoes a broad thermal transition with a mid-point at 22°C well within the range of the first phase transition [38]. Probing the LPS domains in the outer membranes of *E. coli* by electron spin resonance spectroscopy confirmed these data [37]. Since the beginning of the phase transition indicates the melting of the 'frozen' membrane, there is no growth observed below this temperature. The end of the transition occurs at the temperature when the membrane is almost fluid and is usually around or above the physiological growth temperature. Thus, we may conclude that outer membranes have a fluid hydrophobic core.

The size of the LPS molecule is dependent on the length of its O antigen. The protruding O antigens which will have the greatest interaction with the external environment of the cell are thus involved in the physiological properties of the outer membrane. On the basis of freeze fracture electron microscopy on *E. coli* K12, LPS was shown to occur in three different structures in the outer membrane: in a lamellar orientation, as hemi-micelles complexed with proteins, and as hemi-micelles introduced by divalent cations and/or polyamines [39]. Ferritin-labelled antibodies to the O antigen were reacted with the ribbon-like structures formed by purified LPS and subjected to electron microscopy. The electron-dense ferritin lay external to the polysaccharide ribbon and pictures indicated that LPS could extend outwards up to 30 nm [40]. A study using intact cells of *Pseudomonas* labelled with anti O-specific monoclonal antibody and a protein A–dextran–gold conjugate showed that the gold particles were located 30–40 nm beyond the outer membrane [11].

The electrostatic charge of the cell surface is a net charge resulting from the combined charges of the molecules comprising the cell surface and their counterions. At neutral pH, the net charge of several bacterial strains was found to be negative [41]. The largest contribution to charge is from the enterobacterial polysaccharide capsule in encapsulated strains with anionic capsules; however, LPS is the major contributor in non-encapsulated bacterial species.

Neutralization in part of the negative charges of LPS by metal cations helps to stabilize the membrane by decreasing the strong electrostatic repulsion between the highly negatively charged LPS molecules. Ca^{2+} and Mg^{2+} ions are primarily essential for the existence of the membrane. The phosphoryl groups on the LPS as well as the carboxyl group on one of the KDO units were shown to be involved in binding Ca^{2+} and Mg^{2+}. This was confirmed by metal binding studies conducted with heptoseless mutants of *E. coli* by ^{13}C and ^{31}P nuclear magnetic resonance [42].

3. Chemistry and biophysics of membrane proteins

3.1. Porins

The outer membrane of Gram-negative bacteria is perforated by a variety of different hydrophilic channels, that are formed by proteins called porins. The bacterial cell can express up to 10^5 copies of each different channel [4,6]. While some are constitutively expressed, others are inducible under certain growth conditions. Porins from many Gram-negative bacteria have been isolated and characterized [5,6]. They fall into two functional classes, the general diffusion porins, which are chemically non-specific although they may be weakly ion selective, and the specific porins, which contain substrate specific, saturable binding sites [4,6; see also Chapter 19]. While varying substantially in sequence, their physical properties are highly conserved [43]. Their monomeric molecular weight usually varies between 28 000 and 48 000, and they form trimeric arrangements in vivo. Most bacterial porins characterized to date have an acidic pI and a high content of beta-sheet structure (for structural information see Chapter 15).

3.2. Lipoproteins

Two different types of lipoprotein have been found in the outer membranes of *E. coli*: the Braun lipoprotein [44] and the so-called peptidoglycan-associated lipoproteins (PAL)

[45; see also Chapter 14]. The Braun lipoprotein is a small 7.2 kDa protein, existing in high copy numbers (7×10^5 per cell) in the outer membrane of *E. coli*. One-third of this protein is covalently linked to the peptidoglycan, while the remaining two-thirds are non-covalently associated [44,46]. The covalent linkage to the peptidoglycan occurs between the ε-amino group of the C-terminal lysine (or arginine) of the lipoprotein and every tenth to fifteenth carboxy group of diaminopimelic acid [44] of the peptidoglycan. The sulfhydryl group of the N-terminal cysteine is substituted with a diglyceride, while the amino group is substituted with a fatty acid through an amide linkage [44]. The polypeptide chain of both bound and free lipoprotein seems to be largely organized in α-helices as shown by Braun et al. [47]. Crosslinking studies of a hybrid lipoprotein lacking the lipid moiety showed that it exists as trimers [48] which may reflect the aggregation stage of the free lipoprotein. No clear evidence exists indicating the exposure of the lipoprotein on the cell surface. The hydrophilic amino acid composition of the lipoprotein indicates that possibly only the lipid portion penetrates into the outer membrane. Mutants in the structural gene for the lipoprotein are quite viable [49] and show normal diffusion rates of small hydrophilic solutes [50]. However, the cell wall structure of these mutants appears to be unstable based on their increased production of outer membrane vesicles, leakage of periplasmic enzymes and increased sensitivity to EDTA [51] demonstrating a structural role for lipoprotein. One proposed model predicts the peptidoglycan bound lipoprotein as a periplasmic space keeper, linking the outer membrane and the peptidoglycan at a fixed distance of 4.8 nm [44].

A broadly analogous lipoprotein in *P. aeruginosa* is the highly abundant low molecular-weight 9 kDa lipoprotein OprI, which shows 23–30% alignment with the *E. coli* major lipoprotein sequence [52]. Some strains of *Pseudomonas* apparently contain both a covalently bound and a free form of lipoprotein [53] whereas OprI from *P. aeruginosa* PAO1 is entirely non-covalently peptidoglycan associated [54]. Lipoproteins analogous to OprI have been found in other *Pseudomonas* species [55].

Protein 21K from *E. coli* and OprL (21 kDa) from *P. aeruginosa* are also lipoproteins. However, they are larger than Braun lipoprotein and are exclusively non-covalently associated with the peptidoglycan. They thus belong to the class of the so-called peptidoglycan associated lipoproteins (PAL) [45,56]. While OprL is a major protein in *P. aeruginosa*, the 21K protein is of low abundance in the cell wall of *E. coli* [45]. Similar lipoproteins are found among many Gram-negative [45,57] bacteria.

3.3. Protein/peptidoglycan association

Outer membrane proteins can be associated with the peptidoglycan either covalently or non-covalently. Examples of covalent interactions include the well known Braun lipoprotein of *E. coli* [44] and the major outer membrane protein (MOMP) of *Legionella* [58]. Alternatively certain proteins including porins and OmpA-like proteins have strong non-covalent associations with the peptidoglycan. The operational definition of non-covalent association is usually resistance to SDS solubilization at low to moderate temperatures, whereas covalently associated proteins resist extraction by boiling in SDS. Clearly even non-covalently associated proteins demonstrate strong associations with the peptidoglycan, and the necessity for heating in SDS (often to 56°C or greater) to release such proteins indicates that localized denaturation of the part of the protein in contact with the peptidoglycan may be required to free them.

These strong associations are probably important. In plasmolyzed cells, the peptidoglycan is aligned along the bottom of the outer membrane rather than shrinking with the cytoplasmic membrane. This association is partially uncoupled during septum formation and cell division, but it nevertheless appears to be important during cell division. Thus, *lkyD* mutants of *Salmonella*, lacking covalently bound lipoprotein, show outer membrane blebbing, particularly at the position of the division septum [59]. Another role of such associations may be in cell shape and osmotic stability determination (see below).

Porin OmpF associated with the peptidoglycan has been visualized as being arranged in a mosaic crystalline (hexagonal) array [60]. As described elsewhere in this book, OmpF trimers contain a triplet of water-filled channels [61]. While visualization of the hexagonal array of OmpF porin requires selective solubilization techniques, the native surface of *B. pertussis* has been shown to be completely covered with a crystalline structure resulting from the 40 kDa porin [62].

3.4. Multifunctional, structural proteins

OmpA is one of the most abundant and most widely studied outer membrane proteins in *E. coli* and many functions have been attributed to this 35 000 molecular weight, heat-modifiable protein (for review, see [4]). In addition to its role as a phage and colicin receptor, OmpA functions in stabilizing mating aggregates with F+ donor cells [63] and in formation of a non-specific diffusion channel [64]. The heat- and 2-mercaptoethanol-modifiable 35 000 molecular weight porin OprF from *P. aeruginosa* shows high homology to OmpA. The variant N-terminal domains of both OmpA and OprF have both been proposed to cross the membrane eight times in antiparallel beta-sheets [65,66], while the highly homologous C-terminal domains have been proposed to be periplasmic for OmpA [40] and transmembrane for OprF [65]. The two largest gaps in sequence alignment in the C-terminal domain are in the regions encompassing the four cysteines of OprF and near the region of the two cysteines of OmpA [67] with these cysteines forming disulphide bonded 'cysteine loops'. An important function of OmpA is in stabilizing the outer membrane and the cell wall. This was evident from studies using a *lpp ompA* double mutant of *E. coli* lacking Braun lipoprotein and OmpA. Such cells grow in an almost spherical form instead of the normal rod form, require high concentrations of divalent cations for growth, and show frequent blebbing. These properties were ascribed to the observed defect whereby peptidoglycan was no longer connected with the outer membrane [68]. OprF deficient mutants of *P. aeruginosa* showed similar defects [49,68]. The elongated morphology of the above *E. coli lpp ompA* mutant could partially be reconstituted by the cloned *P. aeruginosa oprF* gene [67].

OmpA and OprF both have an Ala-Pro rich region at residues 176–187 and 163–174, respectively, that separates the N- and C-terminal domains and resembles the trypsin sensitive 'hinge' regions of the IgG light chain [69] . However, there is as yet no proof that this region functions as a hinge. The demonstration of immunological cross-reactivity of OmpA with many other Enterobacteriaceae [70,71] and with *Haemophilus influenzae* and *Aeromonas salmonicida* [35] indicates that OmpA has been strongly conserved through evolution. In addition to the above-mentioned protein, protein PIII (the serum blocking protein) of *Neisseria gonorrhoeae* has significant sequence homology and cross-reacts immunologically with both OmpA [72] and protein OprF from *P. aeruginosa* (W.A. Woodruff and R.E.W. Hancock, unpublished data). It seems likely that in Gram-negative bacteria, there is a family of OmpA-like proteins, all of which have receptor and/or porin functions but which in addition have a major function in outer membrane stabilization through interactions with peptidoglycan.

Chlamydia uses a special strategy in outer membrane structural organization and stabilization. At all stages of the developmental cycle, the bacterium is surrounded by a double membrane; however, no significant amounts of peptidoglycan have been observed in *Chlamydia* at any stage of this cycle [73]. It is therefore apparent that *Chlamydia* requires a substitute to fulfil the role of peptidoglycan. The outer membrane of *Chlamydia* consist of up to 60% of a MOMP of approximate molecular weight 40 000. It acts as a trypsin-sensitive adhesin during infection [74], but also shows a porin function with an estimated pore radius of 0.65–0.9 nm. MOMP, however, is 200 times less efficient in porin formation than *E. coli* porins, and it becomes active only when treated with reducing agents [75]. MOMP contains at least three cysteines which are linked by disulphide bonds to other MOMP molecules and to two other cysteine-rich outer membrane proteins (12 kDa and 60 kDa), to form large aggregates [76]. These bonds play an important role in maintaining the structural integrity of the outer membrane of the infectious elementary body (EB) and they seem to replace the function of the missing peptidoglycan. The outer membrane of *Chlamydia* forms a hexagonal mesh [77] with depressions that can be seen by electron microscopy. A structural model proposed a hexagonal arrangement of six dimers of the MOMP arranged around the central depression [78] and freeze fracture experiments indicated a transmembrane channel [79] which has been proposed to correspond to the porin function of MOMP. However, given the evolutionary conservation of porins, it seems possible that the hexagonal arrangements instead reflect a trimeric porin unit. During the extracellular, inert stage, chlamydial elementary bodies (EB) are comparable to spores since they are resistant to osmotic pressure and sonication. After the EB is phagocytosed, it becomes exposed to the intraphagosomal reducing conditions. The cells subsequently change into reticulate bodies (RBs) which do not synthesize and are thus deficient in the cysteine-rich 12 kDa and 60 kDa outer membrane proteins and in which the intermolecular disulfide bonds of the MOMP are reduced. Under these conditions, the outer membrane becomes structurally pleiomorphic, and MOMP would open its pores, allowing uptake of ATP and other required nutrients. Lacking crosslinking by disulfide bonds, the RB becomes osmotically fragile. However, this is not a disadvantage in the high osmolarity intracellular environment.

Another special case is provided by spirochetes which contain outer membranes, called outer sheaths, which contain the periplasmic flagella that runs up the longitudinal axis of the cell giving these spirochetes their classical corkscrew motion. In one case, *Spirochaeta aurantia*, this outer sheath has no easily recognizable equivalent of an LPS molecule [80]. Two spirochetes examined to date [80,81] contain, as their predominant outer sheath proteins, porins which have by far the largest channel diameters of any porins observed to date. This has led to the suggestion that spirochetes are filter feeders and that the large porins ensure a continuous flow of nutrients through the periplasm during movement. One of these spirochete porins, the 53 kDa protein of *Treponema denticola* [81] joins the MOMPs from *Legionella pneumophila* [82], the *Chlamydia,* to form a selected group of porins which have a dual adhesin/porin function.

3.5. Stability of outer membrane proteins

Outer membrane proteins demonstrate remarkable stability to proteases and to detergent treatment [4]. Protease resistance is probably an appropriate feature given the surface localization of these proteins and the existence of some Gram-negative bacteria in environments in which they are often exposed to proteolytic attack (e.g. during infections). In the newly published *E. coli* OmpF structure [61], this protease resistance has been ascribed to the tight packing of the surface loop regions that separate adjacent transmembrane β-sheets. Presumably evolutionary selection based on deletion or alteration of susceptible amino acids has decreed such arrangements. Indeed this, together with antigenic drift, may be one of the driving forces that has led to sequence shuffling over evolution, thus limiting our ability to align the sequences of porins from different bacteria (see Chapter 17).

Detergent stability, on the other hand, would appear to be a property related to the predominant β-barrel structure of porins [61,83]. Thus, the ability of SDS to be inserted into this structure even after heating must be quite limited. As a result, we have proteins with amazing detergent stability. For example, OprF of *P. aeruginosa* retains substantial β-structure even after boiling in SDS [65], whereas most porins can reconstitute channels in lipid membranes even after SDS treatment at room temperature or greater.

4. Consequences of these properties

4.1. Exclusion properties of the outer membrane

The concept of the outer membrane as a molecular sieve provides a descriptive overview of its exclusion properties [4,6,84]. In general, one can state that the 'holes' of the sieve (i.e. the channels of porin proteins) define the size exclusion limit for most hydrophilic molecules (and ions) by limiting the size of molecules that can pass through these channels and by restricting the rate of passage of molecules of sizes approaching the diameters of the porin channels (as described by the Renkin correction, see Chapter 27 and [6]). Another generalization that would follow from this concept of a molecular sieve would be that the fabric of the sieve, in this case comprising LPS and various proteins, would be nearly impermeable to various molecules. As discussed below, this is a gross oversimplification. Several classes of molecules can pass across the outer membrane without accessing the channels of porins. These include polycations varying in size from trisaccharides through 30 amino acid peptides to polycationic proteins, under some circumstances DNA, certain classes of zwitterionic or uncharged antibiotics, and specific proteins, including antibacterial bacteriocins and hydrophobic compounds in some bacterial species. This is not to say that such molecules pass rapidly across the outer membrane. For example, black lipid bilayer experiments have indicated that small ions like K^+ and Na^+ can pass through the OmpF porin channel under an applied voltage of 10 mV (i.e. much less than the existing Donnan potential across the outer membrane [85]), at a rate exceeding 10^5 ions per channel per second. Given 10^5 OmpF channels per *E. coli* cell, the flux of Na^+ or K^+ across the outer membrane can exceed 10^{10} ions per second per cell. However, a far slower rate of passage for, for example, an antibacterial compound would suffice to give rise to physiological effects such as cell death. A rate of passage of only one molecule per second could build up a periplasmic concentration of the antibiotic of 2.4 μM within one (40-min) generation time. Thus, one must be cautious when applying the terms 'exclusion' or 'impermeability' with regards to outer membranes. Although a wide variety of molecules can pass across the outer membrane at slow rates, we use these terms here in the physiological sense. As discussed repeatedly in earlier reviews [4,35,86,87], the outer membranes of many bacteria are considered to exclude most hydrophobic substances, including detergents and hydrophobic antibiotics, as well as proteinaceous enzymes, including nucleases, phosphatases, kinases, proteases, peptidases, etc. This concept of 'exclusion' reflects the probability that these substances are not taken up across the outer membrane at a rate sufficient to give rise to physiological effects on cells (i.e. solubilization, inhibition of function or modification of bacterial macromolecules). This resultant barrier function, involving semi-selective exclusion of potentially harmful environmental molecules, is one of the most critical roles of outer membranes in Gram-negative bacteria, and affords these bacteria a generalized advantage in surviving in many ecological niches which contain high concentrations of potentially lethal substances.

4.2. LPS/LPS interactions: antibiotic uptake and interaction pathways.

As described in Section 2.3, a variety of data indicate that adjacent LPS molecules interact with one another. This is due to the partial neutralization of negative charges by monovalent, but more importantly divalent, cations. Removal of the divalent metal cations by chelators like EDTA [13] results in increased outer membrane permeability [88], structural perturbations [18,35] and, at higher concentrations, extraction of LPS and/or LPS–protein complexes from the cell surface [13]. Similar effects may be observed with various polycations [89] including polymyxins, aminoglycosides, etc. (see below) which competitively displace rather than chelate divalent cations and due to their bulky nature cause similar disruptions. Utilization of a polycationic fluorescent probe, dansyl polymyxin [90] has indicated that parallel interactions occur between the probe and purified LPS or the probe and intact cells [91]. However, certain important concepts must be recognized when one considers the nature of the cell surface. First, bacterial cell surfaces are highly negatively charged [41]. Thus, neutralization of LPS charges by divalent cations must be incomplete. In addition, the general concept of negatively charged LPS molecules bridged by divalent cations is an oversimplification. Indeed, the surface of bacterial cells can be best described as a Guoy–Chapman–Stern [92] layer with high negative electrostatic potential and with divalent and monovalent cations diffusing quite rapidly across this surface. Thus, the interaction of a large polycation with such a surface will first involve a localized neutralization of such a surface layer, including charge displacement or localized exclusion of cations, followed by integration of the polycation into the outer surface of the outer membrane bilayer. Probe displacement experiments have indicated that such polycations have a very high affinity for LPS (e.g. around 0.3–3 μM for polymyxin B) [90,93] although the affinity tends to decrease with decreasing cationic nature [93]. Evaluation of binding kinetics using both the fluorescent probe dansyl polymyxin [90], for intact cells and purified LPS, or the spin label probe CAT_{12} (4-dodecyl-dimethylammonium-1-oxyl-2,2,6,6-tetramethylpiperidine bromide) for purified LPS [93] have indicated that such interaction is cooperative. Thus, the interaction of one molecule of polycation with the outer membrane promotes the interaction of subsequent molecules.

Such interactions have substantial physiological importance since they explain two key properties of the outer membranes of bacteria such as *E. coli* and *P. aeruginosa*, namely their ability to exclude or resist attack by hydrophobic molecules and the existence of a specific pathway of uptake termed self-promoted uptake. Thus, exclusion of hydrophobic molecules including antibiotics, bile salts and anionic or neutral detergents reflects the inability of these molecules to pass across the area of strong negative electrostatic potential at the surface of the outer membrane. Consistent with this, disruption of this surface potential by treatment with polycations, or removal of divalent cations with EDTA causing charge repulsion amongst adjacent LPS molecules, causes enhanced susceptibility to hydrophobic probes [88] and antibiotics [94]. Bacteria that do not have such a strong electrostatic potential would presumably be more susceptible to such agents. For example, deep rough mutants of *Salmonella* and *E. coli* demonstrate enhanced susceptibility to such agents because of decreased surface potential due either to the abnormal presence of phospholipids in the outer monolayer of the outer membrane (in the view of [19] but not [16]), or the reduction of negatively charged groups on the LPS molecules of such bacteria [95] or both. Similarly, we hypothesize that other bacteria such as *Haemophilus influenzae*, *Neisseria* sp. etc. which contain a unique LOS species, instead of conventional LPS, could demonstrate a reduced surface potential (perhaps due to a requirement to interact with the negatively charged surfaces of eukaryotic cells). This would then explain their known increased susceptibility to hydrophobic agents [87].

Self-promoted uptake has been described in detail previously [87] and is only described in overview here. Basically, it represents a system by which bactericidal polycations and organic monovalent and divalent cations can interact with LPS binding sites, and cause permeabilization of the outer membrane to promote uptake of the permeabilizing antibiotic. Attack of Gram-negative bacterial cells by such compounds represents a conserved evolutionary theme (for review, see [88,96]) utilized by antibiotics such as aminoglycosides and polymyxins from certain microorganisms, and by peptides from insects or animal semen or the intracellular contents of eukaryotic cells, including phagocyte 'defensins'. It is known that the interaction of such compounds with cell surface LPS molecules (see above) is followed by structural perturbations to outer membranes and their increased permeability to probes including the β-lactam nitrocefin, the peptidoglycan-degrading enzyme lysozyme and hydrophobic compounds including the fluorescent probe 1-*N*-phenyl naphthylamine [97] and antibiotics. The relevance of such interactions to actual killing of cells has been demonstrated using mutants with increased susceptibility or resistance to such agents due to outer membrane alterations which influence the interaction of these compounds with the cell surface.

4.3. LPS–protein interactions

An area about which far less is known is the association of LPS with outer membrane proteins. It is well known that outer membranes upon purification are often associated with molar or greater quantities of LPS, as demonstrated by co-purification [15,98,99] and, in *P. aeruginosa*, by crosslinking [100] and crossed immunoelectrophoresis [101] experiments. These associations probably involve both ionic and hydrophobic interactions since procedures that disrupt both interactions must usually be applied to obtain LPS-free outer membrane proteins [e.g. 61]. In one case, OmpF porin of *E. coli*, the influence of cations on intrinsic tryptophan fluorescence, was interpreted as evidence for the presence of a divalent cation binding site on this protein that could be involved in interaction with LPS [102].

The relevance of such interactions is currently somewhat obscure. Data with phages that utilize outer membrane protein receptors have demonstrated that the presence of a normal LPS seems important for interaction of the phages with their receptors [103]. Conversely, Parr et al. [99] isolated a monoclonal antibody specific for LPS which preferentially recognized LPS in complex with OmpF or OmpC porin. Thus, we may assume that LPS stabilizes, anchors and/or orients proteins at the surface of the outer membrane.

Recently, Young and Hancock [104] demonstrated that overproduction of an outer membrane protein OprH in *P. aeruginosa* led to an 8–16-fold enhancement in supersusceptibility to quinolones, including ciprofloxacin and nalidixic acid, chloramphenicol and trimethoprim, whereas susceptibility to β-lactams and rifampicin were unaffected. Since data were presented that OprH was not functioning as a porin, we are left with the conclusion that either enhanced uptake of specific antibiotics occurs via sites created by LPS–OprH interactions, or that OprH somehow neutralizes the surface electrostatic potential. This indicates the possibility that protein–LPS interactions can mediate in antibiotic permeation pathways.

4.4. Fluidity, energization and hydrophobic permeability

Outer membranes have been traditionally viewed as quite rigid membranes due to their frequent intimate association with the underlying peptidoglycan and the bulky nature of LPS. However, this would appear to be an oversimplification. Two types of data indicate the fluidity of outer membranes. First, measurement of phase transitions in *E. coli* has indicated that when cells are grown at 37°C, the outer membranes are fluid above a transition temperature of 25°C [105]. Second, spin-label experiments and fluorine nuclear magnetic resonance spectroscopy data on *E. coli* vesicles demonstrated that the diffusion rate of lipids in the *E. coli* cell envelope is in the order of 10^{-8} cm/s which indicates that a given lipid molecule could move from one end of a bacterium to the other in less than a second (for review, see [106]).

The fluidity of outer membranes permits the passage of hydrophobic molecules under appropriate circumstances. However, one as yet unexplained phenomenon has been observed using hydrophobic fluorescent probes. When the outer membranes of *E. coli* and *P. aeruginosa* are permeabilized, they take up such probes transiently and then secrete them [97,107]. Administration of an inhibitor of cell energization prevents secretion such that net uptake is observed. This implies that cells contain an energized secretion system for hydrophobic compounds, perhaps one analogous to the tetracycline and quinolone secretion systems [108].

5. Conclusions

Understanding how even a single macromolecule achieves its function is quite difficult, as discussed elsewhere in this book. However, understanding how a large number of molecules integrate to give rise to a variety of general properties is far more complex. Nevertheless, studies of outer membranes have progressed to a stage where we can start to make educated guesses about the relationships between outer membrane organization and outer membrane function. Much has been learnt but much remains to be learnt.

Acknowledgement

The authors own research on outer membranes has been generously funded by the Canadian Cystic Fibrosis Foundation, Canadian Bacterial Diseases Network, Medical Research Council of Canada. C.B. was the recipient of a fellowship from the Swiss National Science Foundation.

References

1. Leive, L. (1974) Ann. N.Y. Acad. Sci. 235, 109–129.
2. Hirota, Y., Suzuki, H., Nishimura, Y. and Yasuda, S. (1977) Proc. Natl. Acad. Sci. USA 74, 1417–1420.
3. Inouye, M., ed. (1986) in: Bacterial Outer Membranes As Model Systems, Wiley–Interscience, New York.
4. Nikaido, H. and Vaara, M. (1985) Microbiol. Rev. 49, 1–32.
5. Benz, R. (1988) Annu. Rev. Microbiol. 42, 359–393.
6. Hancock, R.E.W. (1986) in: M. Inouye (Ed.), Bacterial Outer Membranes as Model Systems, Wiley–Interscience, New York, pp. 187–225.
7. Rietschel, E.T., Mayer, H., Wollenweber, H.W., Zahringer, U., Luderitz, O., Westphal, O. and Brade, H. (1984) in: J.Y. Homma, S. Kanegasaki, O. Luderitz, T. Shiba and O. Westphal (Eds.), Bacterial Endotoxin: Chemical, Biological and Clinical Aspects, Verlag Chemie, Weinheim, pp. 11–22.
8. Wu, L., Tsai, C.-M. and Frasch, C.E. (1987) Anal. Biochem. 160, 281–289.
9. Avrameas, S. (1969) Immunochemistry 6, 43–52.
10. Funahara, Y. and Nikaido, H. (1980) J. Bacteriol. 141, 1463–1465.
11. Lam, J.S., Lam, M.Y.C., Macdonald, L.A. and Hancock, R.E.W. (1987) J. Bacteriol. 169, 3531–3538.
12. Morrison, D.C. (1985) in: L.J. Berry (Ed.), Handbook of Endotoxin, Vol. 3, Elsevier, Amsterdam, pp. 25–55.
13. Rottem, S. and Leive, L. (1977) J. Biol. Chem. 252, 2077–2081.
14. Coughlin, R.T., Peterson, A.A., Huag, A., Pownall, H.J. and McGroarty, E.J. (1985) Biochim. Biophys. Acta 821, 404–412.
15. Van Alphen, L., Verkleij, A., Leurissen-Bijvelt, J. and Lugtenberg, B. (1978) J. Bacteriol. 134, 1089–1098.
16. Gmeiner, J. and Schlecht, S. (1980) Arch. Microbiol. 127, 81–86.

17. Schweizer, M., Hindennach, I., Garten, W. and Henning, U. (1978) Eur. J. Biochem. 82, 211–217.
18. Verkleij, A., Van Alphen, L., Bijvelt, J. and Lugtenberg, B. (1977) Biochim. Biophys. Acta 466, 269–282.
19. Smit, J., Kamio, Y. and Nikaido, H. (1975) J. Bacteriol. 124, 942–958.
20. Jann, K. and Jann, B. (1977) in: I.W. Sutherland (Ed.), Surface Carbohydrates of the Prokaryotic Cell, Academic Press, New York, pp. 247–287.
21. Wilkinson, S.G. (1988) in: C. Ratledge and S.G. Wilkinson (Eds.), Microbial Lipids, Vol. 1, Academic Press, San Diego, CA, pp. 299–553.
22. Rivera, M., Bryan, L.E., Hancock, R.E.W. and McGroarty, E.J. (1988) J. Bacteriol. 170, 512–521.
23. Le Dur, A., Caroff, M., Chaby, R. and Szabo, L. (1978) Eur. J. Biochem. 84, 579–589.
24. Whitfield, C., Perry, M.B., MacLean, L.L. and Yu, S.H. (1992) J. Bacteriol. 174, 4913–4919.
25. Palva, E.T. and Makela, H.P. (1980) Eur. J. Biochem. 107, 137–143.
26. Goldman, R.C. and Leive, L. (1980) Eur. J. Biochem. 107, 145–153.
27. Kenne, L. and Lindberg, B. (1983) in: G.O. Aspinall (Ed.), The Polysaccharides, Vol. 2, Academic Press, New York, pp. 287–363.
28. Prehm, P., Jann, B. and Jann, K. (1976) Eur. J. Biochem. 67, 53–56.
29. Kocharova, N.A., Knirel, Y.A., Kochetkov, N.K. and Stanislavsky, E.S. (1988) Bioorg. Khim. 14, 701–703.
30. Wilkinson, S.G. (1983) Rev. Infect. Dis. 5, S941–S949.
31. Parker, C.T., Kloser, A.W., Schnaitman, C.A., Stein, M.A., Gottesman, S. and Gibson, B.W. (1992) J. Bacteriol. 174, 2525–2538.
32. Radziejewska-Lebrecht, J. and Mayer, H. (1989) Eur. J. Biochem. 183, 573–581.
33. Mayer, H. and Weckesser, J. (1984) in: E.T. Rietschel (Ed.), Handbook of Endotoxin, Vol. 1, Elsevier, Amsterdam, pp. 221–247.
34. Raetz, C.R.H. (1978) Microbiol. Rev. 42, 614–659.
35. Lugtenberg, B. and Van Alphen, L. (1983) Biochim. Biophys. Acta 737, 51–115.
36. Melchoir, D.L. (1982) Current Top. Membr. Transport 17, 263–316.
37. Coughlin, R.T., Huag, A. and McGroarty, E.J. (1983) Biochim. Biophys. Acta. 729, 161–166.
38. Emmerling, G., Henning, U. and Gulik-Krzywicki, T. (1977) Eur. J. Biochem. 78, 503–509.
39. Van Alphen, L., Verkleij, A., Burnell, E. and Lugtenberg, B. (1980) Biochim. Biophys. Acta 597, 502–517.
40. Muhlradt, P.F. and Golecki, J.R. (1975) Eur. J. Biochem. 51, 343–352.
41. Bayer, M.E. and Sloyer, J.L. (1990) J. Gen. Microbiol. 136, 867–874.
42. Strain, S.M., Fesik, S.W. and Armitage, I.M. (1983) J. Biol. Chem. 258, 13466–13477.
43. Siehnel, R.J., Martin, N.L. and Hancock, R.E.W. (1990) Mol. Microbiol. 4, 831–838.
44. Braun, V. (1975) Biochim. Biophys. Acta 415, 335–377.
45. Mizuno, T. (1979) Biochemistry 86, 991–1000.
46. Inouye, M., Show, J. and Shen, C. (1972) J. Biol. Chem. 247, 8154–8159.
47. Braun, V., Rotering, H., Ohms, J.-P. and Hagenmeier, H. (1976) Eur. J. Biochem. 70, 601–610.
48. Yu, F., Yamada, H., Daishima and K. and Mizushima, S. (1984) FEBS Lett. 173, 264–268.
49. Gotoh, N., Wakebe, H., Yoshihara E., Nakae, T. and Nishino, T. (1989) J. Bacteriol. 171, 893–990.
50. Nikaido, H., Bavoli, P. and Hirota, Y. 1(977) J. Bacteriol. 132, 1045–1047.
51. Suzuki, H., Nishimura, Y., Yasuda, S., Nishimura, M., Yamada, M. and Hirota, Y. (1978) Mol. Gen. Genet. 167, 1–9.
52. Cornelius, P., Bouia, A., Belarbi, A., Guyonvarch, B., Kammerer, V., Hannaert, V. and Huber, J.C. (1989) Mol. Microbiol. 3, 421–428.
53. Mizuno, T. and Kageyama, M. (1979) J. Biochem. 85, 115–122.
54. Hancock, R.E.W., Irvin, R.T., Costerton, J.W. and Carey, A.M. (1981) J. Bacteriol. 145, 628–631.
55. Nakajima, K., Muroga, K. and Hancock, R.E.W. (1983) Int. J. Syst. Bacteriol. 33, 1–8.
56. Mizuno, T. (1981.) J. Biochem. 89, 1039–1049.
57. Mutharia, L.M. and R.E.W. Hancock. (1985) Int. J. Syst. Bacteriol. 35, 530–532.
58. Butler, C.A. and Hoffman, P.S. (1990) J. Bacteriol. 172, 2401–2407.
59. Fung, J., MacΛlister, T.J. and Rothfield, L.I. (1978) J. Bacteriol. 171, 983–990.
60. Rosenbusch, J.P. (1974) J. Biol. Chem. 249, 8019–8029.

278

61. Cowan, S.W., Schirmer, T., Rummel, G., Steiert, M., Ghosh, R., Pauptit, R.A., Jansonius, J.N. and Rosenbusch, J.P. (1992) Nature 358, 727–733.
62. Kessel, M., Brennan, M.J., Trus, B.L., Bisher, M.E. and Steven, A.C. (1988) J. Mol. Biol. 5, 275–278.
63. Achtmann, M., Schwochow, S., Helmuth, R., Morelli, G. and Manning, P.A. (1978) Mol. Gen. Genet. 164, 171–183.
64. Sugawara, E. and Nikaido, H. (1992) J. Biol. Chem. 267, 2507–2511.
65. Siehnel, R.J., Martin, N.L. and Hancock, R.E.W. (1990) in: S. Silver, A.M. Chakrabarty, B. Iglewski and S. Kaplan (Eds.), *Pseudomonas*: Biotransformations, Pathogenesis and Evolving Biotechnology, American Society for Microbiology, Washington, DC, pp. 328–342.
66. Morona, R., Klose, M. and Henning, U. (1984) J. Bacteriol. 159, 570–578.
67. Woodruff, W.A. and Hancock, R.E.W. (1989) J. Bacteriol. 171, 3304–3309.
68. Sonntag, I., Schwarz, H., Hirota, Y. and Henning, U. (1978) J. Bacteriol. 136, 280–285.
69. Poljak, R.J., Amzel, L.M., Avey, H.P., Chen, B.L., Phizackerley, R.P. and Saul, F. (1974) Proc. Natl. Acad. Sci. USA. 77, 3440–3444.
70. Beher, M.G. and Schnaitman, C.A. (1980) J. Bacteriol. 119, 906–913.
71. Overbeeke, N. and Lugtenberg, B. (1980) J. Gen. Microbiol. 212, 373–380.
72. Gotschlich, E.C., Seiff, M. and Blake, M.S. (1987) J. Exp. Med. 165, 471–482.
73. Barbour, A.G., Amano, K.-I., Hackstadt, T., Perry, L. and Caldwell, H.D. (1982) J. Bacteriol. 151, 420–428.
74. Su, H., Zhang, Y.-X., Barrera, O., Watkins, N.G. and Caldwell, H.D. (1988) Infect. Immun. 56, 2094–2100.
75. Bavoli, P., Ohlin, A. and Schachter, J. (1984) Infect. Immun. 44, 479–485.
76. Newhall, W.J. (1987) Infect. Immun. 55, 162–168.
77. Matsumoto, A. (1972) Med. J. 8, 149–157.
78. Chang, J.-J., Leonard, K., Arad, T., Pitt, T., Zhang, Y.-X. and Zhang, L.-H. (1982) J. Mol. Biol. 161, 579–590.
79. Louis, C., Nicolas, G., Eb, F., Fefebvre, J.-F. and Orfila, J. (1980) J. Bacteriol. 141, 868–875.
80. Kropinski, A.M., Parr, T.R., Angus, B., Hancock, R.E.W., Ghiorse, W.C. and Greenberg, E.P. (1987) J. Bacteriol. 169, 172–179.
81. Egli, C., Leung, W.K., Müller, K.-H., Hancock, R.E.W., McBride, B.C. (1993) Infect. Immun. 61, 1694–1699.
82. Payne, N.R. and Horwitz, M.A. (1987) J. Exp. Med. 166, 1377–1387.
83. Weiss, M.S., Abele, U., Weckesser, J., Welte, W., Schiltz, E., Schulz, G.E. (1991) Science 254, 1627–1630.
85. Stock, J.B., Rauch, B., Roseman, S. (1977) J. Biol. Chem. 252, 7850–7861.
86. Nikaido, H. (1976) Biochim. Biophys. Acta 433, 118–132.
87. Hancock, R.E.W. and Bell, A. (1988) Eur. J. Clin. Microbiol. Infect. Dis. 7, 713–720.
88. Hancock, R.E.W. and Wong, P.G.W. (1984) Antimicrob. Agents Chemother. 26, 48–52.
89. Gilleland, H.E. and Murray, R.G.E. (1976) J. Bacteriol. 125, 267–281.
89. Hancock, R.E.W. (1984) Annu. Rev. Microbiol. 38, 237–264.
90. Moore, R.A., Bates, N.C. and Hancock, R.E.W. (1986) Antimicrob. Agents Chemother. 29, 496–500.
91. Sawyer, J.G., Martin, N.C. and Hancock, R.E.W. (1988) Infect. Immun. 56, 693–698.
92. Chung, L., Kaloyanides, G., McDamel, R., McLaughlin, A. and McLaughlin, S. (1985) Biochemistry 24, 442–452.
93. Peterson, A.A., Hancock, R.E.W. and McGroarty, E.J. (1985) J. Bacteriol. 164, 1256–1261.
94. Vaara, M. and Vaara, T. (1983) Antimicrob. Agents Chemother. 24, 107–113.
95. Cox, A.D. and Wilkinson, S.G. (1991) Mol. Microbiol. 5, 641–646.
96. Hancock, R.E.W. (1991) ASM News 57, 175–182.
97. Loh, B.L., Grant, G. and Hancock, R.E.W. (1984) Antimicrob. Agents. Chemother. 26, 546–557.
98. Strittmatter, W. and Galanos, C. (1987) Microb. Pathogenesis 2, 29–36.
99. Parr, T.R., Poole, K., Crockford, G.W.K. and Hancock, R.E.W. (1986) J. Bacteriol. 165, 523–526.
100. Angus, B.L. and Hancock, R.E.W. (1983) J. Bacteriol. 155, 1042–1051.
101. Lam, J.S., Mutharia, L.M., Hancock, R.E.W., Høiby, N., Lam, K., Baek, L. and Costerton, J.W. (1983) Infect. Immun. 42, 88–98.

102. Hancock, R.E.W., Farmer, S., Li, Z. and Poole, K. (1991) Antimicrob. Agents Chemother. 35, 1309–1314.
103. Datta, D.B., Arden, B. and Henning, U. (1977) J. Bacteriol. 131, 821–829.
104. Young, M.L. and Hancock, R.E.W. (1992) Antimicrob. Agents Chemother. 36, 2566–2568.
105. Nakayama, H., Mitsui, T., Nishihara, M. and Kito, M. (1980) Biochim. Biophys. Acta 601, 1–10.
106. Cronan, J.E. (1979) in: M. Inouye (Ed.), Bacterial Outer Membranes. Biogenesis and Functions, Wiley–Interscience, New York, pp. 35–65.
107. Helgerson, S.L. and Cramer, W.A. (1977) Biochemistry 16, 4109–4117.
108. Cohen, S.P., Hooper, D.C., Wolfson, J.C., Sonza, K.S., McMurry, L.M. and Levy, S.B. (1988) Antimicrob. Agents Chemother. 32, 1187–1191.

J.-M Ghuysen and R. Hakenbeck (Eds.), *Bacterial Cell Wall*

CHAPTER 13

Biosynthesis and assembly of lipopolysaccharide

PETER REEVES

Microbiology Department, Bldg G08, University of Sydney, Australia NSW 2006

1. Introduction

Lipopolysaccharide (LPS) is a key component of the OM (outer membrane) which characterizes the Bacteria (previously known as Eubacteria and excluding the Archaebacteria) although lost in the Gram-positive line which appears to be a derivative of the basic Gram-negative form. LPS has thus been an integral component of the world's most numerous group of organisms for perhaps 3 billion years. It is a remarkable molecule, the inner Lipid A component replacing phospholipid as the major amphipathic molecule in the outer leaflet of the OM and the O polysaccharide component being amongst the most polymorphic of structures known. LPS is used as a receptor by bacteriophage; it also activates complement and some forms are potent toxins, leading to LPS often being called endotoxin [1,2].

LPS has been difficult to study and although much is known, it is possible that any structure given here may need at least minor revision. The general nature and role of LPS is discussed in Chapter 12, which should be used in conjunction with this chapter in which we look in detail at its structure, synthesis, genetics and evolution.

The subject has been reviewed many times with several books on the topic. Readers are referred in particular to a recent book [1] and reviews [3–5] and to some excellent earlier reviews [6–11].

2. Overview of structure, biosynthesis and genetics

2.1. Nomenclature for Escherichia, Shigella and Salmonella

We will have cause to compare structures and biosynthesis in different species and for a proper appreciation, it is essential that a natural classification be used. Unfortunately, in two important groups, the currently used classification is very unnatural but although known for years, there is considerable resistance to change. Sequence variation in *Salmonella* is only that expected for a species, although about 1000 serovars have been given full species status; we use here the single name *Salmonella enterica* [12] and use the old species names as serovar (sv) names. For similar reasons, the several species placed in the genus *Shigella* (*S. dysenteriae*, *S. sonnei*, *S. flexneri* and *S. boydii*) all properly belong in the species *E. coli* and are referred to as shigella, dysenteriae or sonnei etc. strains of *E. coli*.

We use *E. coli* strain K-12 and *S. enterica* strain LT2 as examples and refer to them simply as K-12 and LT2.

2.2. Overview of structure

LPS comprises several distinct regions: Lipid A, core (divided into inner and outer core) and O polysaccharide. LPS is extremely variable with the extent of variation increasing from Lipid A to O polysaccharide. The overall structure of Lipid A and core of LT2 is shown in Fig. 1. Lipid A was first observed as a fatty component after acid hydrolysis which cleaves the KDO–GlcN bond. The structure of Lipid A as extracted from LT2 (Fig. 2) is now fully established [3,13], after a long history [13]. The structure has been confirmed by in vitro synthesis [14,15] and the same structure was obtained for K-12. But note that the structure is that of Lipid A after cleavage from core by acid hydrolysis and

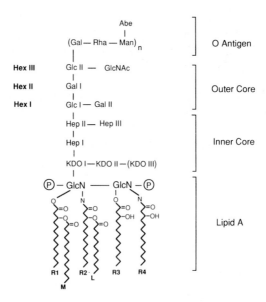

Fig. 1. LPS of *S. enterica* LT2 (sv Typhimurium) showing the regions of LPS. GlcN, Glucosamine: KDO, 3-deoxy-D-*manno*-octulosonic acid: Hep; L-*glycero*-D-*manno*-heptose: Glc; D-Glucose: Gal; D-Galactose: Rha; L-Rhamnose: Man; D-Mannose: Abe; Abequose: R1,2,3,4; R-3-hydroxymyristic acid: M; myristic acid: L; lauric acid. Note that KDOIII and the acyloxyacyl substitution on R1 are not stoichiometric and that non-stoichiometric PPEA substitution on HepI and KDO, and 4-amino-4-deoxy-L-arabinose phosphate on 4' are not shown. R595 (sv Minnesota) can have an acyloxyacyl substitution on R1. See refs. 13 and 16 for details of Lipid A, and the text for core and O polysaccharide.

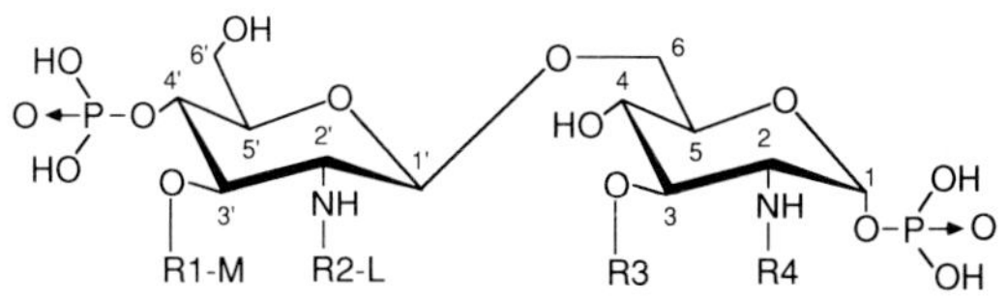

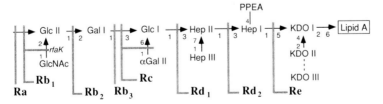

Fig. 2. Hepta-acyl Lipid A of *S. enterica* LT2 (sv Typhimurium) and R595 (sv Minnesota) and *E. coli* K-12. GlcNAc residues and acyl linkages R1, 2, 3, 4, M and L as Fig. 1.

there may be other substituents which were lost in preparation. The attachment of phosphoethanolamine and 4-amino-4-deoxy-L-arabinose in particular is thought to require further investigation [16].

The structures of the cores of K-12 and LT2 have been 'known' for many years but have undergone recent revision and it seems certain that further minor refinements will be made. Both comprise a KDO_2–Hep_3 inner core and an outer core of three hexoses in the main chain with a Gal substituent on the first and a GlcNAc or Hep substituent on the third. The nomenclature for each evolved separately and at times we use a common 'HexI, HexII, HexIII' nomenclature as proposed by Holst and Brade [17] and shown in Fig. 1. The structures have been reviewed recently [4,17] and new findings are that the GlcNAc in K-12 is attached to inner core heptose and not to HexIII as in LT2 and that there is a heptose attached to the HO-6 of terminal HexIII in K-12 in place of the GlcNAc in LT2 [17]. The GlcNAc on HexIII of K-12 previously observed by Jansson [18] was not seen by Holst and Brade [17] and its status is not clear as there is evidence for it presence on some molecules (see Section 5.3.2).

K-12 Core Structure

Fig. 3. Core structures of *E. coli* K-12 and *S. enterica* LT2. Overall structure are similar in both and all glycosidic linkages are α. The genes are essentially the same for both (see Section 3.1.7) but shown only for K-12. Mutants blocking formation of the bonds shown give LPS with truncated cores named Ra to Re as indicated for LT2. Similar mutants in K-12 have not been as extensively studied.

There is very little variation reported in Lipid A (see Section 3.2.1). The core is often divided into inner and outer cores as the outer core is more variable than the inner core, with five forms known for *E. coli* and two for *S. enterica* (see Section 3.2.2).

The O polysaccharide, a repeat unit polysaccharide [19] with units known as O units, is one of the most variable of cell constituents. This variation has been well documented but that reported so far is surely but a fraction of the total range; no attempt is made here

A

$$
\begin{array}{ccc}
\textbf{\textit{b}} & \textbf{\textit{a}} & \textbf{\textit{d}} \\
\boxed{\text{OAc}} \rightarrow \boxed{\text{DDH}} & & \boxed{\text{Glc}} \\
\alpha \big|^1_3 & & \downarrow \\
\rightarrow [\,\text{Man} - \boxed{c} \rightarrow \text{Rha} \rightarrow \text{Gal}\,] \; \boxed{\xrightarrow{f}} \\
& & \uparrow \\
& & \boxed{\text{OAc}} \\
& & \textbf{\textit{e}}
\end{array}
$$

B

$$
\begin{array}{l}
\overset{\text{Abe}}{\underset{}{\alpha\big|\,1,3}} \qquad \overset{\text{GlcOAc}}{\underset{}{\alpha\big|\,1,3}} \\
\xrightarrow[4]{}[\text{Rha}\xrightarrow[1,2]{\alpha}\text{Man}\xrightarrow[1,2]{\alpha}\text{Man}\xrightarrow[1,3]{\alpha}\text{Gal}]\xrightarrow[1]{\alpha} \qquad\qquad \textit{S. e.}\ \text{C2}
\end{array}
$$

$$
\xrightarrow[3]{}[\text{Man}\xrightarrow[1,2]{\alpha}\text{Man}\xrightarrow[1,2]{\alpha}\text{Man}]\xrightarrow[1]{\beta} \qquad\qquad \textit{E. c.}\ \text{O8}
$$

$$
\xrightarrow[3]{}[\text{Man}\xrightarrow[1,3]{\alpha}\text{Man}\xrightarrow[1,2]{\alpha}\text{Man}\xrightarrow[1,2]{\alpha}\text{Man}\xrightarrow[1,2]{\alpha}\text{Man}]\xrightarrow[1]{\alpha} \qquad\qquad \textit{E. c.}\ \text{O9}
$$

$$
\xrightarrow[3]{}[\text{Rha}\xrightarrow[1,3]{\alpha}\text{Rha}\xrightarrow[1,2]{\alpha}\text{Gal}\xrightarrow[1,3]{\alpha}\text{GlcNac}]\xrightarrow[1]{\alpha} \qquad \begin{array}{l}\textit{E. c.}\ \text{(shigella)}\\ \text{dysenteriae I}\end{array}
$$

$$
\xrightarrow[4]{}[\text{AltNAcUA}\xrightarrow[1,3]{\alpha}\text{NH}_2\text{-FucNAc}]\xrightarrow[1]{\alpha} \qquad \begin{array}{l}\textit{E. c.}\ \text{(shigella)}\\ \text{sonnei}\end{array}
$$

$$
\begin{array}{l}
\overset{\text{Col}}{\underset{}{\alpha\big|\,1,3}} \\
\underset{4}{}[\text{Glc}\xrightarrow[1,4]{\alpha}\text{Gal}\xrightarrow[1,3]{\alpha}\text{GlcNAc}]\xrightarrow[1]{\beta} \qquad\qquad \textit{E. c.}\ \text{O111}\\
\overset{}{\underset{}{\alpha\big|\,1,6}} \\
\text{Col}
\end{array}
$$

$$
\xrightarrow[3]{}[\text{Gal}\xrightarrow[1,3]{\beta}\text{Gal}]\xrightarrow[1]{\alpha} \qquad\qquad \textit{Kl.}\ \text{2a}
$$

$$
\underset{3}{}[\text{GlcNAc}\xrightarrow[1,5]{\beta}\text{Gal}]\xrightarrow[1]{\beta} \qquad\qquad \textit{Kl.}\ \text{2c}
$$

$$
\underset{3}{}[\text{D-Rha}\xrightarrow[1,3]{\alpha}\text{D-Rha}\xrightarrow[1,2]{\alpha}\text{D-Rha}]\xrightarrow[1]{\alpha} \qquad\qquad \textit{Ps.}\ \text{(A band)}
$$

$$
\begin{array}{l}
\overset{\text{OAc}}{\underset{}{\big\downarrow}} \\
\xrightarrow[4]{}[\text{Fuc}\xrightarrow[1,3]{\alpha}\text{Glc}\xrightarrow[1,3]{\beta}\text{Fuc}]\xrightarrow[1]{\beta} \qquad\qquad \begin{array}{l}\textit{E. c.}\\ \text{colanic acid}\end{array}\\
\qquad\qquad\qquad\qquad \beta\big|\,1,4 \\
\text{Gal}\xrightarrow[1,4]{\beta}\text{Glc}\xrightarrow[1,3]{\beta}\text{Gal}
\end{array}
$$

at a comprehensive listing but a representative sample is shown in Fig. 4. Many genetic and biosynthesis studies have used K-12 or LT2 and these are used as examples. Unfortunately, all surviving strains of K-12 lack O polysaccharide but Schnaitman's group have used a dysenteriae 1 *rfb* (+*rfp*) clone in K-12 to allow study of complete LPS in this well documented strain [20,21].

2.3. Overview of biosynthesis and genetics

LPS is in general synthesized as two separate components, Lipid A/core and O polysaccharide which are then ligated to give the complete molecule (Fig. 5), which itself may then be further modified.

LPS genetics was developed in *S. enterica* strain LT2 and the LT2 terminology has been adapted for other strains and species. Mutants which make incomplete LPS are known as rough mutants because colonies on agar plates are flatter, spread further and lack the normal 'smooth' appearance of wild type colonies. The mutations mapped at three loci first named rough A, rough B and rough C, but now *rfa*, *rfb* and *rfc*, which include genes for core synthesis, O unit synthesis and O polysaccharide polymerization, respectively. We can now see anomalies in this nomenclature; mutants affecting Lipid A or KDO were not detected in the early studies, probably because they were lethal, and those found later were named after the pathway affected and not as *rfa* mutants although one (*kdtA*, see below) maps at the *rfa* locus. Secondly, the existence of a separate *rfc* locus is now seen to be atypical as in most cases the O polysaccharide polymerase gene is in the *rfb* cluster, although still called *rfc*.

Synthesis of Lipid A is now best understood for K-12 but given the conserved nature of Lipid A, it is probable that it follows a similar course in many other species. The first

Fig. 4. (A) O units of groups *S. enterica* groups A, B, D1,2, E1–4. Residues and linkages shaded are variable and/or not present in all groups. a: DDH (dideoxyhexose) being paratose in group A, abequose in group B, tyvelose in groups D1,2 and absent in groups E1-4. b: O acetyl present in some group B strains (encoded by chromosomal gene *oafA*) where it confers epitope O5. c: linkage $\alpha 1$–4 in A, B, D1 and $\beta 1$–4 in E1–4, D2. d: In groups A, B, D $\alpha 1$–4 Glc is encoded by *oafR* and *oafE* genes on the chromosome and confers epitope 12_2; some strains carry phage P22 (or similar phage) with genes for a $\alpha 1$–6 Glc conferring epitope 1 (epitopes 1 and 12_2 both subject to form variation; see Section 4.3.3); see below for Glc in groups E1–4 and ref. 6 for review. e: group E1 see below. f: linkage between O units $\alpha 1$–2 in most strains of groups A, B, D1,2, $\alpha 1$–6 in groups E1,4 and $\beta 1$–6 in E2,3: phage P27 converts the linkage in groups B or D to $\alpha 1$–6 and confers epitope 27; there are naturally occurring strains of groups B and D2 with phage encoded epitope 27 [64]. Groups E1–4 all have the same *rfb* region [127] but may have the O antigen modified by presence of lysogenic phage; group E1 with epitopes 3 and 10 has the basic 3-sugar backbone and an O acetyl on Gal; phage ε_{15} carries genes to repress O acetyl expression and for a new O polysaccharide polymerase which gives a $\beta 1$–6 in place of $\alpha 1$–6 linkage thereby replacing epitope 10 with epitope 15; phage ε_{34}, which uses epitope 15 as receptor, leads to presence of $\alpha 1$–4 Glc on Gal and epitope 34; group E4 resembles E1 with an $\alpha 1$–6 Glc on Gal, but the genetic basis has not been reported. (B) O units of selected forms. Note that in many cases the structures are based on chemical studies only and the biological unit synthesized can be any linear permutation of the unit shown. *S.e*, *S. enterica*; *E.c*, *E. coli*; *Kl*, *Klebsiella*; *Ps*, *Pseudomonas aeruginosa*. Other symbols are O polysaccharide names except for colanic acid which is a capsule. *f*, furanose form (all others pyranose); AltNAcUA, 2-deoxy-2-acetamidoaltrose; NH2-FucNAc, 2-acetamido-4-amino-2,4,6-trideoxy-D-galactose; Col, colitose; Fuc, L-fucose; D-Rha, D-rhamnose; others as in Fig. 1. References given in reviews [7,8] except for *Klebsiella* [94] and *Pseudomonas* [130].

286

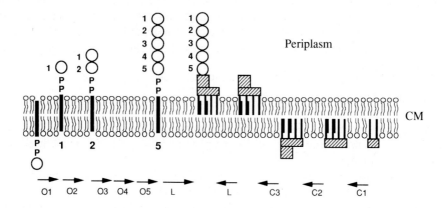

Fig. 5. Outline of LPS assembly on the cytoplasmic membrane. Circles represent O units assembled on UndPP. Both are numbered (to indicate order in which they enter the assembly process). Steps O1–O5, O polysaccharide assembly; C1–C3, Lipid A core assembly; L, ligation of O polysaccharide and Lipid A core. Step O1, flipping of UndPP-O; steps O2 etc addition of single O units to the chain at reducing end by transfer of growing chain to new O unit on UndPP carrier. C1 represents condensation of 2 substituted GlcNAc molecules to give lipid IVA; C2, the several steps in synthesis of core and addition of acyloxyacyl residues; and C3, flipping of Lipid A/core.

specific step is acylation of UDP-GlcNAc and proceeds through the steps described below, overlapping the sequential addition of core sugars, all as pyranose structures. From the first committed reaction to completion of Lipid A/core, the intermediates are anchored in the cytoplasmic membrane with the hydrophilic component presumably facing the cytoplasm where the substrates are.

O polysaccharide synthesis proceeds independently with each O unit assembled on the C_{55} isoprenoid carrier lipid, Undecaprenol (Und) phosphate (see below for details and the exception), the intermediates again embedded in the cytoplasmic membrane with the hydrophilic O unit on the cytoplasmic face of the membrane.

It is believed that Lipid A/core and O unit are separately translocated (flipped) to the outer (periplasmic) face of the CM (cytoplasmic membrane), where the O unit is polymerized to O polysaccharide before transfer to Lipid A/core by an enzyme called ligase. Nothing is known of the further steps in export of LPS to the outer face of the OM.

3. Lipid A/core

3.1. Lipid A/core of K-12 and LT2

3.1.1. Lipid A and core are synthesized as one unit
The distinction between Lipid A and core was made because the KDO-Lipid A linkage is acid labile [3] and acid hydrolysis gave water insoluble Lipid A and water soluble polysaccharide, both amenable to study, whereas complete LPS is insoluble in any convenient solvent. However, biosynthesis of Lipid A/core involves a continuous series of reactions

which we treat as one. The biosynthesis has been reviewed recently [16] and earlier reviews (see above) are also useful.

3.1.2. Synthesis of Lipid A/core precursors

L-*Glycero*-D-*manno*-heptose (referred to as heptose) and KDO are not generally present in non-LPS structures and synthesis of ADP-Hep and CMP-KDO is conveniently treated with LPS Synthesis. UDP-GlcNAc, UDP-Glc and UDP-Gal are present for other purposes and their biosynthetic pathways are encoded by housekeeping genes.

Heptose is added from its ADP derivative, thought to be synthesized in five steps from sedoheptulose-7-phosphate of the pentose phosphate pathway (for review, see [3]). In support of this proposal is the existence of the first enzyme of the proposed pathway, phosphoheptose isomerase [22] and the observation that a *tkt* (transketolase) mutant, blocked in sedoheptulose synthesis, produces heptose deficient LPS [23].

The last step, epimerization, is blocked in *rfaD* mutants known in both K-12 and LT2 and thought in K-12 to incorporate (at reduced level) D-*glycero*-D-*manno*-heptose into LPS in place of the L form [24]. A functional *rfaD* gene is required for viability at elevated temperatures [25] suggesting that this substitution is deleterious. The *rfaD* gene has been cloned and sequenced [25,26].

Genes for the other steps have not yet been defined, but *rfaE* (LT2 76 min) is presumably one of them as mutants produce Re LPS [6] (see Fig. 3). Extracts of *rfaE* (and rfaD) mutants can add heptose to KDO_2-IV_A in vitro if ADP-heptose is present showing that they are involved in synthesis of ADP-heptose [27]. Although the proposed heptose pathway seems reasonable, much of the supporting data needs to be confirmed; only two genes have been found although five are expected, and only *rfaD* has been characterized at all. The other genes are thought to map away from *rfa*.

KDO is transferred from CMP-KDO, synthesized from arabinose-5-P and PEP via a three-step pathway which is well characterized (for review, see [3]) with the genes (*kdsA* and *kdsB*) known for the first and last steps in both LT2 and K-12.

3.1.3. Synthesis of Lipid A/inner core in K-12

Synthesis of Lipid A (Fig. 6) has been reviewed recently [3,16]. The first committed step in assembly of Lipid A/core is transfer of *R*-3-hydroxymyristic acid to UDP-GlcNAc by UDP-*N*-acetylglucosamine transferase (Fig. 6). The product is membrane associated as are all subsequent intermediates. A ts mutant for this step enabled the gene, *lpxA*, to be identified [28].

The second step, deacetylation, is carried out by the product of *envA* [29], a gene known previously for an effect on septum separation during division [30]. The third step, N-acylation with *R*-3-hydroxymyristic acid, is carried out by the product of *lpxD* [31], separated from *lpxA* by one gene. *lpxD* was first identified (under the name *firA*) as a gene affecting rifampicin resistance [32] but more recently has been thought to have a general effect on OM function [33]. The acyl transferases involved in steps 1 and 3 are specific for *R*-3-hydroxymyristoyl-ACP [34].

The product, UDP-2,3-diacyl-GlcN, may then lose UMP to give 2,3-diacyl GlcN-1-P also known as lipid X, and one molecule each of these diacyl compounds are then condensed, with loss of UDP, to give a tetra-acyl disaccharide with the basic structure of

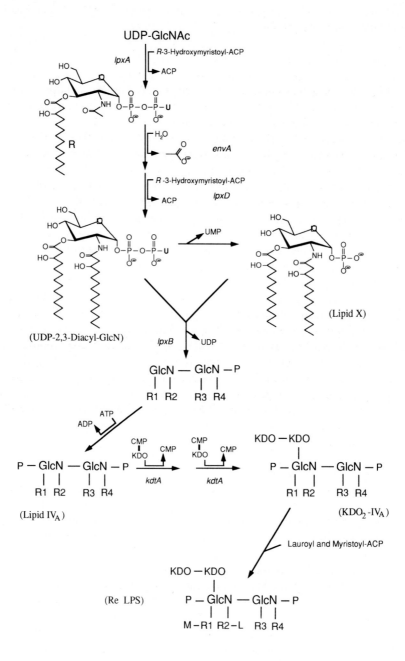

Fig. 6. Biosynthetic pathway of *E. coli* Re LPS showing genes involved. Symbols as Fig. 1. Modified after Raetz [3].

Lipid A. There are said to be two enzymes which release the UMP but little is known of them [3]. Lipid A disaccharide synthase, encoded by *lpxB*, carries out the condensation step and has been purified and characterized [35].

The 4' position is then phosphorylated by a kinase [36] (gene not known) to give lipid IV_A to which KDO is added from CMP-KDO by KDO transferase, a bifunctional enzyme which adds two KDO residues to give KDO_2-IV_A; the gene, *kdtA*, maps at one end of the *rfa* cluster. A third KDO residue is thought to be present although not stoichiometrically. A homologous *Chlamydia trachomatis* enzyme (GseA) is trifunctional adding three KDO residues and will incorporate three residues in the K-12 core or add an additional residue to a 2 KDO K-12 substrate in vitro. The presence of the third residue in a *gsaA+* K-12 strain prevents addition of heptose and the timing, distribution among LPS forms (see below) and genetics of KDOIII addition are not known [4].

KDO_2-IV_A is a substrate in vitro for addition of two further fatty acids as acyloxyacyl moieties or addition of two heptose residues, and the order in which the additions are made in vivo is not known. Mutants deficient in CMP-KDO synthesis accumulate lipid IV_A, showing that addition of KDO must precede addition of the two further fatty acids and that the enzymes possess KDO recognition domains. This is confirmed by the observation that lipid IV_A is not a substrate for in vitro fatty acid addition [37]. A lauroyl moiety is transferred to the hydroxyl of the 2N substituted *R*-3-hydroxymyristoyl residue, followed by transfer of a myristoyl moiety to the hydroxy group on the O3 substituted *R*-3 hydroxymyristoyl residue, both from ACP derivatives [37]. The lauroyl residue must be added before the myristoyl moiety can be added but these modifications of the lipophilic part of the molecule appear to be independent of addition of heptose and presumably subsequent sugar residues. There are no mutants defective in addition of the acyloxyacyl residues; in vitro studies showed only that addition of the heptose residues did not require their presence.

Null mutants in *lpxA*, *lpxD* or *kdtA* are probably lethal as only ts mutants have been isolated, and the same applies to *kdsA* and *kdsB*. These mutations define within limits the minimum structure required for cell viability. Unfortunately, because KdtA catalyses addition of two KDO residues, and this is followed by addition of two fatty acids by unidentified enzymes, it is not possible to precisely define the minimum structure which must include at least one KDO or fatty acid residue but possibly both KDO and both fatty acid residues for cell survival.

Mutants blocked in subsequent steps are viable and elucidation of LPS core structure was facilitated by a set of mutants of LT2 with specific blocks (see Fig. 3). In more recent studies, many of the equivalent K-12 genes have been identified [38].

HepI and HepII are added from ADP-Hep by heptosyltransferase I, encoded by *rfaC,* which has been analysed using an in vitro assay [36] and heptosyltransferase II encoded by *rfaF*.

3.1.4. K-12 and LT2 outer core (main chain)

This topic is discussed in detail by Schnaitman [4]. The first outer core sugar GlcI (HexI), is transferred from UDP-Glc by glucosyl transferase, RfaG, first defined in LT2 by Rd1 mutants and in *E. coli* by complementation of Rd1(LT2) mutants by clones of *rfa*(K-12)

[39] (Fig. 3). The gene has been cloned from LT2 [40] and cloned and sequenced from K-12 [41]. Synthesis of UDP-Glc requires *galU* and mutants also have the Rd phenotype.

From this point on, LT2 and K-12 structures and pathways differ. Addition of HexII(GalI) in LT2 requires the *rfaI* gene which has been cloned [40]. The K-12 gene maps at the corresponding position and is also known as *rfaI*, but the product adds a glucose (GlcII). It has also been cloned and sequenced [42] and will complement an LT2 mutant, probably adding a glucose residue.

Addition of HexIII (GlcII in LT2, GlcIII in K-12) requires gene *rfaJ* in LT2 and a similar gene is found in the corresponding place in K-12. However, the expected enzyme activity was not demonstrated in K-12 [40], although the K-12 gene can complement a *rfaJ* mutant of LT2 at low efficiency [42] but at high efficiency if *rfaI*(K-12) is also present suggesting specificity for addition to Glucose as HexII. Unlike mutants blocked at earlier stages, an *rfaJ* mutant produces multiple forms of LPS core [42] suggesting that once the terminal glucose is added, alternative non-stoichiometric modifications may occur.

3.1.5. K-12 and LT2 side chain residues

Additional residues are added to the main chain as synthesis progresses, and while there are a considerable amount of data, there is still much to learn about the reactions and genes involved.

A side chain Gal (GalI in K-12, GalII in LT2) is present on GlcI of both structures. Most *rfaB* (and *galE*) mutants have only GlcI with Rc LPS, suggesting that the side chain Gal must be present before the main chain is extended beyond GlcI [43]. However, one *rfaB* mutant had many LPS molecules extended beyond GlcI, some with O polysaccharide, although there was little or no GalII present [43]. Pradel et al. [42] suggest that RfaB protein in addition to its enzyme activity, engages in protein–protein interaction to allow further chain expansion and that transfer of GalII is not required for this additional activity; the Wollin mutant is then a missense mutant which does not transfer but does allow further chain extension. The raising of this possibility emphasizes how little we know about the organization of LPS synthesis. Kadam et al. [40] cloned the *rfaB* gene of LT2 and Pradel et al. [42] cloned and sequenced the K-12 gene.

rfaK and *rfaL* are present in both K-12 and LT2 but each has barely detectable amino acid level similarity although nearly identical predicted structures in the two strains. *rfaK* mutants of LT2 have long been known to lack the GlcNAc residue which is required by phage FO as part of its receptor. FO does not attack K-12 and now that it is known that K-12 has a heptose in place of the GlcNAc of LT2, the reason becomes clear [20]. Clones carrying *rfaKL*(LT2) confer FO sensitivity on a K-12 *rfaK* mutant [20] showing that RfaK has different functions in the two species, RfaK(K-12) probably adding GlcNAc to inner core heptose. The new structure also explains why K-12 is naturally resistant to LT2 phage Br2, which requires as receptor a core with at least GlcI present, as the GlcNAc residue on one of the K-12 heptose residues is in the Br2 receptor region of LPS. Klena et al. [20] find that an *rfaK* mutant of K-12 is sensitive to phage Br2 and that an *rfaK* (K-12) clone makes an LT2 strain resistant. These data strongly support the suggestion that RfaK(K-12) transfers a GlcNAc residue to Hep.

RfaL seems also to be involved in *rfaK* activity [20]. A plasmid carrying *rfaK* (K-12) and *rfaL* (K-12) partially complements a *rfaK* LT2 strain. However, if the plasmid is mutant in *rfaL* the *rfaK*(K-12) gene alone cannot complement, showing that RfaK(K-12) cannot work with RfaL(LT2) but only with RfaL(K-12). This defines a new and unexpected function for RfaL which is considered to be the ligase (see below). A reciprocal test 'showed' that RfaK and RfaL of LT2 also worked better together than if mutants were complemented by K-12 genes.

The rhamnose in the K-12 core reported many years ago has recently been located on KDOII [17]. The rhamnose biosynthetic pathway of K-12 is present only in its *rfb* gene cluster and a mutation (*rfbC*) in this pathway in several K-12 strains correlates with variation in banding pattern of LPS on SDS PAGE; *rfbC* mutants have three clear bands (B1, B2 and B3 in increasing mobility). The bands probably differ by one sugar for each interval and several LPS forms may have a given mobility. *rfbC*+ strains have a broad major band (B2) and a less abundant band (B3) [20,44] with B2 thought to include complete core and B1 to have GlcNAc added in place of O polysaccharide (see Section 5.2.3).

A K-12 *rfaK* mutant in a *rfbC* (no TDP-Rha) background lacks B1 and B2. *rfaK* is required for O polysaccharide addition (see Section 5.2.3) so B1 is not expected and B3 presumably lacks inner core GlcNAc.

Genes for other side chain moieties are less easy to identify. The genes available are *rfaP*, *rfaS*, *rfaQ*, *rfaY* and *rfaZ*. The pyrophosphorylethanolamine residue (PPEA) on HepI has been well characterized in *S. enterica* sv Minnesota [45]. Mutants at *rfaP* have been described for sv Minnesota and LT2 [46,47] and shown to have an Rc phenotype, being truncated at GlcI, and also lacking the phosphatidyl ethanolamine on HepI, the HepIII on HepII and the side chain GalII. RfaP may be the kinase which adds a phosphate to HepI to which phosphoethanolamine is then added to give pyrophosphoethanolamine [47] (but see [4]) and it appears that in the absence of this side group, in general the side chain HepIII and GalI residues are not added and the chain terminates at GlcI, although a few molecules are extended and also carry O polysaccharide. *rfaP*(K-12) complements LT2 mutants suggesting that they have the same structure and this was confirmed by structural studies [47]

rfaQ, *rfaS* and *rfaZ* mutations have relatively little effect on the three band pattern of an *rfbC* mutant [20] suggesting that several substituents are not added in the absence of rhamnose.

3.1.6. Lipo-oligosaccharide (LOS) of K-12

Klena et al. [20] used K-12 with a clone of the *rfb* gene cluster of (shigella) dysenteriae type 1 *E. coli*, which produced complete LPS (see Section 4.1.5) and observed that mutation in *rfaS* or *rfaZ* increased the mobility of the unsubstituted core band, but had no effect on the mobility of complete LPS. This showed that contrary to the usual assumption, unsubstituted Lipid A/core material differed from that which was part of complete LPS, as changes in the structure of the core component of complete LPS would affect the mobility of each band which would be detected at least for LPS with only few O units. *rfaS* and *rfaZ* must then affect an alternative pathway which leads to material destined for direct translocation to OM without addition of O polysaccharide. They named this material lipo-oligosaccharide (LOS) after the material found in *Neisseria* etc. which comprises core only without O polysaccharide (see Section 3.3).

Synthesis of LOS in K-12 requires *rfaS* and *rfaZ* but not *rfaL* [20,44]. In the presence of the dysenteriae 1 *rfb* gene cluster (which complements *rfbC* so TDP-Rha present), *rfaS* and *rfaZ*, there is an O polysaccharide ladder and LOS at B2. *rfaQ*, *rfaS*, and *rfaZ* mutants in this background affect the core bands but not O polysaccharide containing bands. *rfaQ* mutants have faster running LOS and in TDP-Rha minus strains it is B3 which is affected. Although there is as yet no direct evidence, a possibility [20,44] is that RfaQ adds HepIII. This addition is not stoichiometric [41] and also requires the presence of RfaP. *rfaZ* and *rfaS* mutants have even faster running LPS which perhaps lacks an additional sugar [20], possibly rhamnose [44]. *rfaZ* seems to best fit the requirements for the Rha transferase gene as it only affects band mobility in the presence of TDP-Rha [20] although the Rha transferase gene may be *rfaS* [20].

The provisional picture then is that Rha and HepIII are only present on LOS (B2) and in a normal K-12 strain the LPS core destined to receive O polysaccharide (also B2) is in part substituted with GlcNAc (B1) added in a *rfaL* dependent step. In a *rfbC* strain the LOS precursor (lacks Rha) runs as B3.

3.1.7. Maps of Lipid A and core genes

The *rfa* cluster (79 min in LT2, 81 min in K-12) shown in Fig. 7 includes the genes for all transferases for assembly of core (if one includes the adjacent *kdtA* gene as part of the cluster), together with the gene for the last step of ADP-heptose synthesis and *rfaL*, the ligase gene. Recent work mostly from the Sanderson and Schnaitman labs has given us complete *rfa* maps of both strains, complete sequence for K-12 and a substantial amount for LT2. The two maps were aligned by Schnaitman [38] and are extremely similar. There are three operons present as shown by sequence analysis [38] and insertional analysis [48], which also showed that insertions were generally polar indicating no strong internal promoters. One operon comprises *kdtA* and an ORF named *kdtB* [48] although apparently not required for KDO transferase (Anderson and Raetz, personal communication), the second contains the heptose transferases and the gene for the last step in heptose synthesis

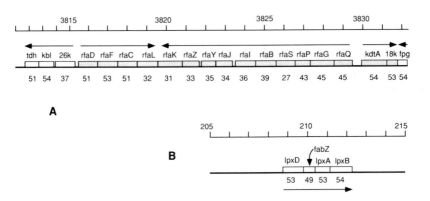

Fig. 7. Genetic maps of (A) *rfa* and (B) *lpx* regions of *E. coli* K-12. Core related genes of *rfa* region shown shaded. Scale above is the EcoliMap6 Kilobase scale (Rudd, personal communication) based on the Kohara scale [131] as [132,133]. G + C percentage of each gene shown below.

plus *rfaL* which acts very late in LPS synthesis and the third the genes for outer core synthesis.

The whole of the region has been sequenced in K-12 and much of it in LT2: The level of amino acid identity is 70% or more over most of the region but barely detectable for *rfaL* and *rfaK* which are adjacent but in separate operons [4,49]. The *rfaS* gene of K-12 is replaced in LT2 by what seems to be the fragmentary remains of an *rfaS* gene with very low sequence similarity [50]; it is non-functional, LT2 not producing the LOS form determined by the presence of *rfaS* in K-12. The only gene not discussed above is 'gene 15' identified by two *lacZ* fusions in K-12 [48] (see below). It has also been suggested that the gene upstream of *rfaD* may be a second Hep kinase [4] although *kbl* and *tdh* in the same operon have no obvious connection with LPS.

rfaH, a positive control gene acting through antitermination affects the large central outer core operon [51,52]. The promoters for the two divergent operons have been identified [53] and as expected for a positive control system, have no similarity to the consensus −35 region. Roncero and Casadaban [48] using translational fusions found a small 'gene' in this intergenic region which may be involved in regulation. There is as yet no convincing model for regulation of the *rfa* operons but possibilities are canvassed by Schnaitman [4].

Three of the genes for Lipid A synthesis (*lpxA*; *lpxD* and *lpxB* for steps 1, 3 and 5) map at 4 min (208–212 kb Kohara) with *fabZ* and several genes involved in macromolecular synthesis. *fabZ* (previously known as *orf17*) encodes a β-hydroxylacyl-ACP dehydratase [31] which uses the same substrate as *lpxA* but leads to unsaturated fatty acid synthesis. The other known gene, *envA* maps at 2 min (105 min kb Kohara).

The KDO pathway genes map separately with *kdsA* at 27 min in K-12 and 39 min in LT2; and *kdsB* at 85 min in K-12 and 16 min in LT2; the third gene is not yet known.

The gene for the last step in heptose synthesis, *rfaD*, maps within the *rfa* cluster, adjacent to *rfaC* and *rfaF* for the transferases for the two backbone heptose residues. *rfaE* maps separately at 76 min (LT2) and the other predicted genes are not known.

3.2. Variation in Lipid A and core

3.2.1. Variation in Lipid A

Lipid A is the most conserved part of LPS with most of the variation in the chain length of the GlcN substituted fatty acids and the number and chain length of acyloxyacyl substituted fatty acids. Only for *S. enterica* strains LT2 and R595(sv Minnesota), *E. coli* K-12 and *Rhodobacter sphaeroides* is the structure 'fully' known [13] (see caveat in Section 2.2). However, enough is known of many other structures to indicate the extent of variation over the broad reach of Gram-negative bacteria. For a detailed survey and analysis and primary references, readers are referred to Rietschel et al. [54] and Takayama [13].

A $\beta(1–6)$ linked disaccharide D-hexosamine is almost universal. It may be GlcN-GlcN disaccharide as in *S. enterica* and *E. coli*, or GlcN3N-GlcN or GlcN3N-GlcN3N. Man, GalA, Ara and AraN have also been reported. Where known, the sugars are in pyranose form and $\beta(1–6)$ linked. In a few cases, Lipid A contains a GlcN3N monosaccharide backbone.

There is much more variation in the fatty acids. In general C2, C3, C'2 and C'3 are all substituted, with amido linkages when appropriate, generally with hydroxy fatty acids or rarely 3-keto fatty acids [13]. These fatty acids are added before the two diacylated sugars are condensed and the 2,2' substituents are expected to be the same as the 3,3' substituents as is the case where data are available; for several species, it has been shown that the acyltransferases have appropriate fatty acid specificity [34].

The secondary acyloxyacyl substitutions are variable in number within and between strains and species, ranging from 1 to 3; it is interesting that in each well documented structure there is variation within a preparation in the number of acyloxyacyl substitutions with the full range of variation at this level occurring within *S. enterica* strain R595 which can have substitutions on R2, R1 and R2 or R1, R2 and R4 (palmitate). In most species, if there is only one acyloxyacyl substitution, it is usually on R2, with the second being on R1 or R4 with R1, R2 and R4 substituted if there are three substitutions. R3 is substituted only rarely.

Overall fatty acid composition alone is known for many more species and those with 10, 12 or 16 carbons are commonly reported, with both 3-hydroxy and normal fatty acids usually present, and 2-hydroxy or 3-keto-fatty acids occasionally present [13]. 3-Hydroxy fatty acids usually have 10, 12 or 14 carbons and the normal fatty acids are usually, but not always, saturated with 12–18 carbons. In both classes, a mixture of chain lengths is usual.

Particularly interesting is the presence of 27-hydroxy-octacosanoic acid in several bacteria from the α-2 subgroup of Proteobacteria [55]. This 28 carbon fatty acid is twice the normal length and presumably is exposed on the other (periplasmic) face of the OM.

3.2.2. Variation in core

Variation in core is much more substantial than in Lipid A, with outer core being more variable than inner core. Unfortunately, the structures of even the best known are still undergoing revision and further adjustments can be expected. That for *S. enterica* given in Fig. 3 is best documented and perhaps correct apart from non-stoichiometric substitution by phosphate and 2-aminoethyl phosphate. Most *S. enterica* strains are thought to have the same structure as LT2 but some lack the GlcNAc residue [56].

Five core types have been described for *E. coli* (Fig. 8) including some shigella strains which have the R1 form [17]. The R1 *rfa* regions from a classical *E. coli* strain and a sonnei strain have been cloned [57] and used in Southern hybridization to show that seven strains (including sonnei, flexneri and boydii) with R1 cores had general similarity although some variation in restriction sites. A weaker signal was picked up with K-12 suggesting only limited homology with *rfa* regions of strains with a different core structure. It will be very interesting to see more detailed comparison of the *rfa* regions of strains with R1 and K-12 cores. Core variation in *E. coli* has also been studied using monoclonal antibodies (MAb). MAb specific for different epitopes on the R1 core could pick out other R1 strains and development of this technique may enable more extensive analysis of core types than has been possible using only chemical studies.

Structures have been proposed for many other species but most are incomplete and should be treated as ranging from 'speculative' to 'may need minor revision'. The difficulties in determining structure include the lability of KDO, which is widely present; the

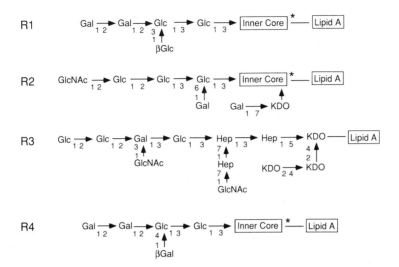

Fig. 8. Known LPS core structures of *E. coli* (other than K-12 shown in Figs. 1 and 3). All linkages α except where indicated. *, the Hep KDO region not fully determined (see [17] for references and details).

presence of phosphate substitutions, often non-stoichiometric, which interfere with analysis by GC; and the presence of multiple forms due to partial substitution of some residues. Holst and Brade [17] give an excellent review of the known structures and here only an overview is attempted.

The division into an inner core of KDO and heptose is close to universal, with one KDO (octulosonic acid in *Acinetobacter* [58]) and one or two Hep residues in the main chain being common.

The outer core generally contains only hexoses in the main chain, with hexoses, sometimes heptose, and often phosphate or 2-aminoethyl phosphate as side substitutions. The hexoses can include Glc, Gal, GalNAc, GlcNAc, Glc3NAc, GlcA, GlcN, Man, FucNAc-Me and GalNA, but Glc and Gal are particularly widespread with GlcNAc and GalNAc perhaps next [17]. However, it should be noted that this sample is biased towards organisms of medical interest. Many of the structures known are of R mutants and, in the absence of O polysaccharide, substitutions on the non-reducing terminal sugar of the main chain are not distinguished from extension of the main chain itself. The length of the main chain core can thus be exaggerated but three or less in the main chain is common, with up to about five reported. Linkages are generally α but β linkages are reported. Good compilations of structures can be seen in [17].

3.3. LPS without O polysaccharides

As we saw above, many bacteria easily lose the ability to make complete LPS by mutation. These 'rough' mutants frequently displace the wild type 'smooth' form during culture and laboratory strains may be known only in their secondary rough form, with *E. coli* K-12 being a well known example. In some genera, such as *Neisseria*, *Haemophilus*,

Chlamydia and *Bordetella*, it appears that the natural form lacks the repeat unit altogether and these have been called lipo-oligosaccharides (LOS) by some, although others [59] consider it premature to introduce a new terminology. *Neisseria* is perhaps the best studied example and several forms are known generally with quite complex outer core structures [17].

4. O unit structure and synthesis

4.1. The O unit

The O unit is by far the most variable part of LPS and a selection of forms is shown in Fig. 4. There can be more than 100 forms for one species and the total range must run into thousands with only a fraction characterized. Details of many structures are given by Kenne et al. [60] and Jann and Jann [8]. Biosynthesis has been studied in only a few cases. Characteristically, all genes for enzymes specific to O unit synthesis and also, except in our type case *S. enterica* group B and close relatives, the *rfc* gene for O polysaccharide polymerase are grouped in the *rfb* gene cluster, which in *E. coli*, *S. enterica* and related species maps close to *his* (histidine synthesis) and *gnd* (6-phosphogluconate dehydrogenase). There are several cases where there is interaction with the synthesis of other polysaccharides, with common steps sometimes but not always being duplicated. For example, the Enterobacterial common antigen (ECA) pathway uses the products of *rfbA* and *rfbB* in group B *S. enterica* but has its own genes for these steps in other strains which do not have these genes in their *rfb* cluster, whereas the mannose pathway genes are present in both the *rfb* and *cps* (colanic acid) capsule clusters [61]. We look below at some well studied O units of *S. enterica* and then a few selected examples from other species to illustrate specific aspects of structure or variation.

4.1.1. Salmonella enterica group B
The O unit of the group B *S. enterica* O polysaccharide has four sugars in its basic structure (Fig. 1). The biosynthetic pathway and genes are outlined in Fig. 9.

UDP-Gal is synthesized by housekeeping enzymes and 11 enzymes are required to synthesize the other three nucleotide sugars, with a further four transferases to synthesize the O unit on UndPP (Fig. 9). The GDP-mannose pathway serves to illustrate which genes are to be expected in an *rfb* cluster. *S. enterica* utilizes mannose by the pathway shown in Fig. 10. The *rfb* gene cluster (Fig. 11) includes the 15 genes shown in Fig. 9.

The mannose phosphate isomerase used in mannose catabolism is also used for mannose synthesis in the absence of exogenous mannose (Fig. 10) and only the genes for synthesis of GDP-mannose from mannose-6-P are specific to GDP-mannose synthesis and present in the *rfb* cluster.

The genes for each biosynthetic pathway are clustered (Fig. 11). Three of the transferases expected have been identified and the fourth tentatively [62]. Galactose transferase, which transfers galactose phosphate from UDP-Gal to UndP, is a membrane protein with several potential transmembrane segments seen in its sequence. The other three, which add sequentially rhamnose, mannose and abequose from nucleotide diphospho

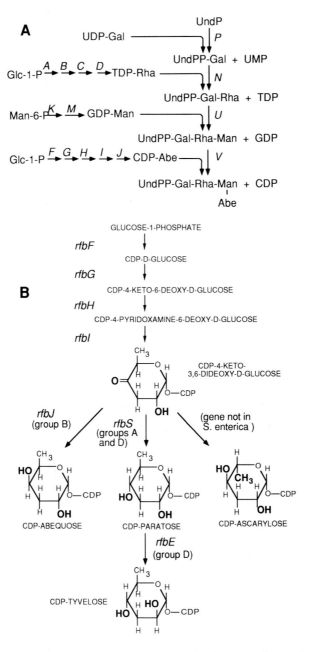

Fig. 9. Biosynthesis of O antigen of selected *S. enterica* groups. (A) Group B , *rfb* genes for biosynthetic pathways and O unit assembly shown in italics. Symbols as in Fig. 1 (see part B for details of CDP-Abe and Fig. 10
for details of GDP-Man pathways). (B) Biosynthetic pathway of dideoxyhexoses derived from CDP-Glc with
S. enterica groups and genes indicated.

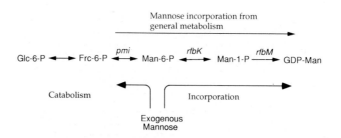

Fig. 10. Mannose metabolism in *E. coli* and *S. enterica* showing synthesis of GDP-man from either general metabolic pathways via fructose 6-P or from exogenous mannose. Note the central role of mannose-6-P and the *pmi* (mannose phosphate isomerase) gene.

sugars, are also found in the membrane fraction but lack potential transmembrane segments and are presumably peripherally associated.

In addition to genes allocated to predicted pathway enzymes and transferases, there is always one gene, named *rfbX*, which appears to encode an integral membrane protein. Recent reviews [4,63] should be consulted for details of *S. enterica rfb* regions.

4.1.2. O polysaccharides of other strains and species

The O polysaccharide is extremely variable with about 170 forms described for *E. coli* (including shigella strains) and about 60 for *S. enterica* [64]. This variation is in both linkages and sugars present as illustrated below (Section 4.1.3) for selected *S. enterica* forms. Similar levels of variation appear to be present in other species such as *Pseudomonas aeruginosa*, whereas other species are much less variable, with for example *Yersinia pseudotuberculosis* having only 12 forms, all but one including a dideoxyhexose [7,65,66]. Only representative cases can be discussed here.

4.1.3. Some S. enterica groups with related O units

Groups A, B, D, C2 and E1 have related O polysaccharides; the *rfb* gene clusters have all been sequenced [67–72] and here also there is considerable similarity as shown in Fig. 11. All the pathway genes have been identified (one only tentatively) and all transferase genes [62] other than the abequose transferases of groups B and C2 which by elimination are likely to be *orf14.1* of group B and *orf13.9* of group C2.

4.1.4. Other S. enterica

Only one other *S. enterica* O polysaccharide has been studied in detail. The group C1 O unit consists of one GlcNAc and four mannose residues in the main chain and a side chain glucose. It is linked by name with group C2, because they have antigen O6 in common. This is due to a common mannosyl-mannose linkage but otherwise the two have little in common in chemical structure or in organization of their *rfb* gene cluster. The groups studied represent only a fraction of the structural variation observed and presumably of the genetic variation present.

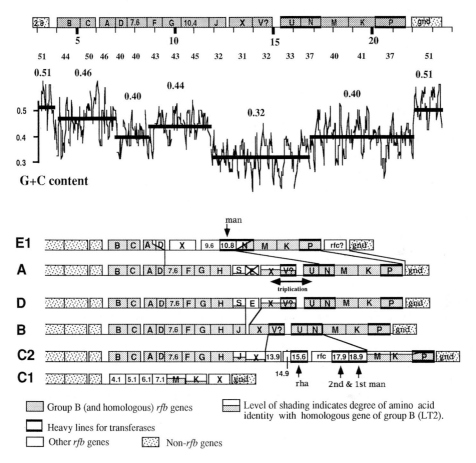

Fig. 11. Top: *Rfb* region of *S. enterica* strain LT2 (Gp B). *rfbB,C,A,D* rhamnose pathway genes; *orf7.6*, *rfbF,G,H,J* abequose pathway genes; *rfbM,K* mannose pathway genes; *rfbP,N,U,V* ; gal, rha, man and abe transferase genes, respectively (shown with heavy lines) which transfer sugar to oligosaccharide from nucleotide sugar: abequose transferase (*rfbV*) is allocated by elimination pending assay. The G + C content of each gene is shown below the bar, and the graph gives the content of 100 bp spans. Genes with letters (for *rfbA* etc) have in general been identified by subcloning and assay of gene products (for review, see [63]). Other genes are named by position in kb in the sequence; *orf7.6* is provisionally allocated to the unidentified gene of the dideoxyhexose pathway (*rfbI*) by correlation of presence of ORF and pathway product. Bottom: *rfb* region from six groups of *S. enterica*: The level of similarity is indicated by the level of shading, which reflects the level of amino acid identity to the homologous gene of group B. All groups have one gene (*rfbX*) with protein product predicted to have 12 transmembrane segments; function unknown but could be antigen export. Transferases of groups E1 and C2 indicated where not same as group B; Abe transferase of group C2 probably *orf13.9*. Group A is as group D but has a mutation in *rfbE* and one segment triplicated. Lines between groups connect breaks in homology (for review, see [63]).

4.1.5. E. coli

The structures of several *E. coli* O polysaccharides have been determined (see Fig. 4 for examples) and in 11 cases the *rfb* genes have been cloned (O1, O2, O4, O7, O9, O101, O111, flexneri Y2a, flexneri 6, dysenteriae type 1 and sonnei) [21,57,73–82]. Each was cloned into K-12 and some were then transferred to other *E. coli* strains or other species. In all but one case, a single clone gave full function confirming the conclusion drawn from recombination data with *S. enterica* that *rfb* genes are clustered. In three cases (O2, O7 and O111), the gene cluster was shown to map near *gnd*. In sonnei, the *rfb* cluster was on a plasmid and in dysenteriae type 1, one gene was on a plasmid but the others in a typical *rfb* cluster near *his* [83]. There is some disagreement on the *rfb* genes of O111 as the first publication [84] suggested that all or perhaps part of the *rfb* cluster was on a plasmid but a later study showed that a single clone of chromosomal DNA which mapped near *gnd* would give full function. The status of the earlier observation is not clear and may indicate variation in this serogroup. In six cases (O7, O101, O111, flexneri Y2a, flexneri 6 and sonnei), the *rfb* cluster was shown to function in an *rfb* delete strain and thus carried all genes specific to O unit synthesis.

In several cases, the cloned *rfb* DNA was used as a probe to test for homology with other strains [75,76,79,81,82,85]. Unfortunately, in general not much is known of the genes encoded by the DNA used as probe but the probes often hybridized with DNA from only some other strains, often known to have similar O polysaccharides. In total, these studies indicate that there is great diversity in the genes of the *rfb* cluster of *E. coli*.

Only for (shigella) dysenteriae type 1 has the entire *rfb* region been sequenced [21]. It contains eight genes: four rhamnose pathway and two rhamnose transferase genes, an *rfbX* gene and an *rfc* gene. The O unit structure is thought to be Rha-Rha-Gal-GlcNAc; *rfp*, a plasmid gene encoding a Gal transferase [86] is required for O polysaccharide synthesis and the *rfe* gene is also required, presumably for transfer of GlcNAc-P to UndP to initiate O unit synthesis (see Section 5.4). The *rfb* region of sonnei is present on a plasmid [87] and has also been cloned and sequenced [21].

4.1.6. Yersinia

Yersinia pseudotuberculosis is of particular interest as all but one serotype has a dideoxy-hexose in its O polysaccharide [7,65,66]. The O polysaccharides are otherwise quite diverse in their structure and the reasons for the predominance of dideoxyhexoses is not clear. The *rfb* cluster of serotype IIA has been cloned and the dideoxyhexose pathway region sequenced [88,89] and shown to have the same organization as in *S. enterica*. Fragments representing individual genes of the pathway were used as probes: all serotypes with abequose, paratose, tyvelose or ascarylose hybridized with the first four genes of the pathway, showing that the gene clusters were homologous. The *rfbJ* gene probed only abequose producing strains as expected as the paratose and ascarylose equivalents are expected to differ considerably. The group IIA strain sequenced had a defective copy of a gene very similar to *rfbE* of *S. enterica* group D, but with frame shift changes which would make it inactive; this was unexpected as *rfbE* converts CDP-paratose to CDP-tyvelose and has no potential role in an abequose producing strain. This '*rfbE*' gene probed all abequose and tyvelose, but not paratose producing strains. The inference is that formation of paratose is an original trait and not due to mutation of the *rfbE* gene of a tyvelose producing form as in *S. enterica* [63,68]. A further implication is that the abequose producing forms are derived from tyvelose producing forms by substitution of their *rfbS* gene with a *rfbJ* gene, the remnant *rfbE* gene giving away the origin; however, further study of several forms will be required to substantiate this.

4.1.7. Vibrio cholerae

Cholera is caused by *Vibrio cholerae* of 01 serotype, of which sub-serotypes Ogawa and Inaba have distinctive antigens plus a common specificity and a third, Hikojima, is a combination of the two. The only difference of significance in the *rfb* regions of Ogawa and Inaba is a single base change which leads to a stop codon in the Inaba strain in what is the *rfbT* gene of the Ogawa strain (no relation to *rfbT* of *S. enterica*) [90]. The *rfbT* gene of an Ogawa strain could convert an Inaba strain to Ogawa. These observations explained long standing observations on seroconversion. Ogawa strains convert easily to Inaba, presumably by loss of *rfbT* function and controversy over the ability of Ogawa strains to convert to Inaba is now seen to depend on the ability of the particular mutation in *rfbT* to revert.

4.2. Bacteria with two forms of O unit

We looked above at variation in the structure of LPS components but some strains have further complexity. There are for example several examples of multiple LPS forms.

4.2.1. Pseudomonas

Most strains of *Pseudomonas aeruginosa* have two distinct forms of LPS: one form called A band with a tri-D-rhamnose repeat unit present in most strains and B band which contains a typical serotype specific O antigen [91,92].

4.2.2. Klebsiella

Some *Klebsiella pneumoniae* strains also have two distinct forms of LPS. LPS of O1 and O2 strains was found to have two distinct D-galactan polymers [93,94]. Galactan I (subserogroup 2a, Fig. 4) was present in both strains with a wide range of polymer lengths and expression was linked to *his* in crosses with *E. coli*. Galactan II was only present as long chains and different forms, 2b and 2c (Fig. 4) were present in the two strains; the genes do not appear to be linked to *his*. Other *K. pneumoniae* O antigens also appear to be galactans but serological studies suggest that they do not carry the same Galactan I as the O1 and O2 strains.

4.2.3. E. coli capsules

Two types of capsule are found in *E. coli*. The type 1 capsule can be in part ligated to Lipid A/core [95] and since it is a repeat unit capsule, it is not easily distinguished from a second form of O unit. It may be significant that the *cps* (type I capsule) and *rfb* loci are very close, and the difference between the *E. coli* situation and that described above for *K. pneumoniae* and *P. aeruginosa* may only be in the amount and extent of ligation to Lipid A/ core of the second structure.

4.3. Modification of O polysaccharide

The O polysaccharide is sometimes modified after completion and in some cases at least after polymerization. Such modifications usually involve addition of glucose or O-acetyl groups or variation in linkage between O units.

4.3.1. Glucosylation
Some examples of glucosylation are shown in Fig. 4 and there are many other such cases. The glucosylation is often determined by genes on a phage or, if chromosomal, separate from the *rfb* genes. Groups B and D *S. enterica* can be glucosylated at C4 of Gal, determined by a chromosomal locus known as *oafR*, to add epitope 12_2 or at C6, determined by genes on phage P22, to add epitope 1, group E2 O polysaccharide can be glucosylated by genes on phage e34 to give epitope 34. The reactions for epitopes 12_2 [96–99] and 34 [96,97] have been studied in vitro; in both cases, the glucose is added from UndP-Glc, itself derived from UDP-Glc and in the case of epitope 34, the gene cluster includes genes for transfer of Glc to UndP and subsequent transfer to O polysaccharide. Interaction between genes for epitopes 1 and 12_2 lead to similar conclusions (see [6]), although only one gene was found in a *E. coli* flexneri X [100].

Glucosylation at C4 has been shown to affect the mobility only of LPS molecules with more than six O units, suggesting that glucose addition, known to occur before transfer of O polysaccharide to Lipid A/core [101], occurs only after the chain reaches this length. In support of non-glucosylation at C4 of shorter chain length molecules is the observation that the single O unit in LPS of an *rfc* mutant was not glucosylated [101] and there is indirect evidence that the glucose of *S. enterica* epitope 34 is added after polymerization [11].

If O units are flipped to the CM periplasmic face before polymerization, then Glc addition must also occur in the periplasm. The use of UndP-Glc as donor is presumably to allow flipping of the glucose end of the molecule to the periplasmic face of the CM. As yet no genes for this step have been reported but mutants may well not have been distinguished from blocks in transfer to O unit.

4.3.2. O acetylation
O acetylation is also observed in group B *S. enterica*, in this case on the Abe residue, the gene, *oafA*, being on the chromosome near *rfb*. Other examples include phage determined acetylation of rhamnose in group B [102] and flexneri strains of *E. coli*. In the latter case, sequencing shows that a single gene is involved [103]. As for glucosylation discussed above, this acetylation only affects mobility of longer chain molecules, but in this case the effect is attributed to the greater number of acetyl groups expected on the longer molecules being needed to have a detectable effect on mobility. In both cases, Western blot confirmation is needed of the presence or absence in shorter chains of the epitope involved.

The acetyl group of *S. enterica* epitope 10 is derived from acetyl-CoA with no intermediate detected [11]. This modification must then be carried out before the O unit is flipped.

Several examples of glucosylation and acetylation of shigella strains are discussed in a recent review [5]; flexneri strains are particularly interesting as 10 serovars differ in O acetyl and Glc substitutions on a basic 'Y' form [104]. Many may be due to lysogenic phage.

4.3.3. Form variation

Many of the modifications of O polysaccharide are subject to form variation in which expression is regulated to be either on or off for several generations; the mechanism is in general not understood for O polysaccharides but similar form variation of flagellar expression is well documented. Examples include the *S. enterica* chromosomal *oafR* gene and the gene(s) on phage 27 (for review, see [6]).

4.3.4. Linkage between O units

In several cases, a phage encoded function gives a new O polymerase activity which may replace or coexist with the original. Thus, *S. enterica* group E1 strains with α1–6 linkage are converted to E2 strains with β1–6 linkage, and phage P27 gives group B strains an α1–6 linkage in addition to the α1–2 (for review, see [6]).

5. Polymerization and assembly of components and export

5.1. Assembly of complete LPS and translocation to OM

We have looked at the synthesis of Lipid A/core and O unit and must now look at putting the pieces together. Most of the recent work on this topic comes from the Osborne group using LT2, who propose that polymerization of O unit to O polysaccharide and ligation both occur on the periplasmic face of the inner membrane, after flipping of UndPP-O unit and Lipid A/core across the CM.

Munford and Osborne [105] used ferritin-labelled antibody to locate O polysaccharide, using a *galE* mutant in which it was only made when galactose was provided in the medium. They found it on the periplasmic face of the CM and outer face of the OM when galactose was present but if galactose was withdrawn, then after 5 min the O polysaccharide level on the CM was greatly reduced and essentially absent by 10 min. It was concluded that O polysaccharide has a periplasmic stage during export although this interpretation has been questioned [3]. These data do not tell us if the LPS components are separately flipped to the periplasmic face of the CM or if complete LPS is flipped after ligation but there is indirect evidence for flipping of both UndPP-O unit and Lipid A/core, and thus for location on the periplasmic face of the subsequent steps of polymerization and ligation.

The evidence for O unit being flipped is the finding that the first step in O unit synthesis, transfer of Gal-P from UDP-Gal to UndP, is dependent on proton motive force in vivo but has no energy requirement in vitro [106]. It was proposed that the energy requiring step was most probably flipping of spent UndP(P) from the periplasmic to the cytoplasmic face. Similarly, whereas ligation in vitro has no energy requirement, proton motive force was required for ligation of preformed O polysaccharide to newly synthesized Lipid A/core [107]. This situation is achieved experimentally by blocking Lipid A/core synthesis in a *kdsA*(ts) strain by growth at 43°C, then adding DNP during shift down to 30°C. In the control without DNP, presynthesized O polysaccharide was ligated to core, but the reaction was blocked by DNP. DNP had no effect on the reaction in vitro. The observation is again taken to show the need for an energy dependent translocation, in this case of Lipid A/core from exposure on the cytoplasmic to the periplasmic face of the CM.

There is no simple alternative explanation for these energy requirements in vivo (but not in vitro) and the experiments lend considerable indirect support for the proposed pathway (shown in Fig. 5), but it awaits confirmation by studies of the enzymes and their reactions. Both *rfc* and *rfaL* have been cloned and sequenced, which will facilitate such studies.

There are no data on the transfer of O unit from exposure on the inner to the outer face of the CM. All *rfb* clusters studied thus far have a gene (*rfbX*) encoding a protein with about 12 transmembrane segments which may be involved. However, an *rfbX* mutant [21] had low amounts of LPS with a single O unit and none of the larger chain length, suggesting a block in polymerization and partial block in either flipping or ligation. *rfbT* mutants, once thought to be blocked in ligation, have been shown to map to *rfbP* and to accumulate what is probably a single O unit on UndPP [108], suggesting a block in flipping, although the location of the accumulated material has not been demonstrated. RfbP is thus bifunctional. The *rfe* gene has considerable similarity to the first half of *rfbP* and also adds a sugar (GlcNAc) plus phosphate to UndP in ECA subunit synthesis and in *E. coli* dysenteriae 1 O unit synthesis. *rfe* is followed by a gene, o349 [109], with similarity to the second half of *rfbP*. These two parts of *rfbP* may carry out the two described functions and correspond to separate genes in the *rfe* locus. We can expect considerable advance in our understanding of *rfbX* and '*rfbT*' in the next few years.

As noted above, there is also really nothing known of the translocation of LPS from the periplasmic face of the CM to the outer face of the OM.

5.2. Polymerization and its control

Polymerization itself seems to require the single enzyme, O antigen polymerase (Rfc). *rfc* mutants have LPS with only a single O unit, in addition to unsubstituted Lipid A/core and were first isolated in LT2, where they mapped well away from the *rfa* or *rfb* gene clusters. *rfc* mutants have since been found for *S. enterica* group C2 [71] and *E. coli* dysenteriae 1 [21] and, in both cases, they map to the *rfb* gene cluster. One would expect the O polysaccharide polymerases to be specific as the linkages formed vary widely. In the three cases in which the *rfc* gene has been identified, there is no K-12 gene able to complement and it seems reasonable to generalize and assume that K-12 is not able to provide a polymerase function for introduced *rfb* regions, and as there are now several cases of cloned *rfb* regions giving rise to long chain O polysaccharide in K-12 it seems that the *rfc* gene generally maps within *rfb*.

The polymerization reaction involves addition of a new O unit to a growing chain on undecaprenol to give elongation from the reducing end, probably by repeated transfer of the growing chain to new O units with change of UndPP molecule at each step [110]. The evidence for this mode of extension is good as labelled galactose, initially on the reducing end of the developing chain, is quickly chased into the body of the growing chain [111]. This model resembles the extension of a polypeptide chain or a fatty acid [110] and

Bastin et al. [112] have carried the analogy further by suggesting that Rfc has two sites, named R and D, which bind respectively the (receiving) UndPP carrying the new O unit and the UndPP carrying the O polymer which will be donated, with UndPP plus extended chain moving from the R to D site each cycle (Fig. 12).

O polysaccharide chain length is clustered around a 'preferred' modal length, which varies from one form to another. In some cases, there are secondary clusters generally at multiples of the major cluster length. This modal chain length distribution, which requires that the kinetics of either or both of ligation or polymerization vary in a chain length dependent manner [113], is dependent on a single gene called *cld* (or *rol*), [112,114] which has been sequenced from LT2 and two *E. coli* strains. The gene product has predicted transmembrane segments at each end and is presumably an integral membrane protein.

In the absence of the chain length determinant (CLD), at least for *E. coli* 0111, the distribution resembles that expected if the kinetics of polymerization and ligation are both independent of chain length with the probability (p) of an O chain being ligated to core

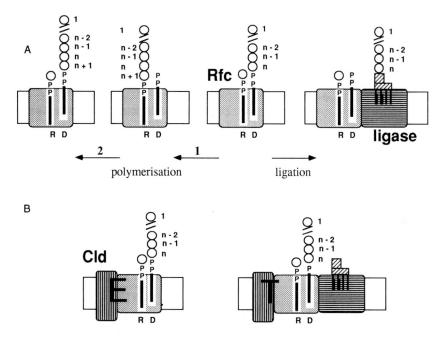

Fig. 12. Model for O antigen polymerase and CLD action. (A) Polymerase molecules are shown with two postulated sites for UndPP. Polymerization itself involves transfer of an O antigen chain of n O units (O_n) from UndPP-O_n to UndPP-O at the R site. It is proposed that the UndPP-O_{n+1} formed does not dissociate but moves to the D site as shown (step 2), to replace the spent UPP. After a new molecule of UndPP-O binds to the R site, the polymerization step can be repeated and long O antigen chains are built up on the polymerase. The O_n chain can also transfer or be committed to Lipid A core as shown (rightward arrow) but it is not known if there is any association between ligase and polymerase or if the UPP-O_n is released before interacting with ligase. (B) CLD is postulated to interact with polymerase to form a complex, and depending on the state of CLD, the polymerase preferentially catalyzes repeated polymerization cycles (state 'E') or preferentially commits the O antigen chain to ligation (state 'T'). The two states are shown diagrammatically below. From Bastin et al. [112].

being 0.065 and probability (q) of it undergoing further elongation being 0.935, although there was an excess of chains of lengths 1, 2 and 3.

The role of CLD then is to modify this intrinsic distribution to give the wild type distribution. The data fits a model in which p is reduced to less than 0.015 by chain length of 3 but between lengths 7 and 14 rises to 0.4 and remains there. With 40% of chains ligating each cycle, effectively all chains are transferred by length 20. In the model put forward (Fig. 12), CLD acts indirectly by contact with Rfc which retains contact with the growing chain in the ribosome and fatty acid synthetase manner. It was proposed that in the absence of CLD, Rfc operates with constant probabilities of transfer to ligase or further extension, but that in association with CLD, it changes first to reduced probability and then to increased probability of transfer, with modal chain length determined by the length reached during the period of reduced probability of transfer to ligase.

5.3. Ligation

5.3.1. Genetics

Until very recently, the ligation step seemed to be well understood at the genetic level although not well characterized at the biochemical level. Two genes, *rfaL* and *rfbT*, were thought to be involved, with mutation in either leading to accumulation of UndP associated O polysaccharide. Both observations now need reconsideration. In K-12, *rfaL* mutation can affect core in the absence of O polysaccharide [20]. At first sight this would imply that RfaL carries out a step in core synthesis, the block in ligation being secondary. However, an alternative explanation is that RfaL interacts with RfaK, the GlcNAc transferase. RfaL and RfaK are the two proteins which differ substantially between K-12 and LT2, and complementation studies suggest that there is an interaction between the two proteins; addition of O polysaccharide to core, the interaction being more effective if both proteins are from the same species [20].

In the case of '*rfbT*', now seen as a function of *rfbP* (see above), recent data suggest that mutants accumulate undecaprenol with a single O unit [108]; using the logic applied above, a possible explanation is that they are blocked in flipping UndPP-O unit and hence unable to carry out polymerization or ligation steps on the periplasmic face of the CM. Clearly more work is needed to determine the nature of the ligation step.

5.3.2. Specificity

O polysaccharides will often attach to core of unrelated species; clones of *rfb* clusters from several *E. coli* and *S. enterica* strains, *Y. pseudotuberculosis* and *Vibrio cholerae* have been expressed in K-12 and shown by SDS-PAGE to have a typical ladder pattern indicating attachment of O polysaccharide to K-12 Lipid A/core. However, the nature of the core does sometimes affect the ability of an O polysaccharide to be attached. In K-12 and some *S. enterica* strains, dysenteriae type 1 *rfb/rfp* genes were expressed and the O antigen attached to Lipid A/core. However, there was no attachment in sv Typhi unless the *rfa* genes of K-12 were added, but attachment in sv Dublin appeared normal. This requirement for the *rfa* region of K-12 can be met by clones of *rfaK*, suggesting that the *rfaK* genes of svs Dublin and Typhi are different (Brahmbhatt, personal communication): this reveals a new level of complexity in *S. enterica* as the two serovars have the same O antigen and no difference was expected in their *rfa* regions. A similar effect was observed with attachment of *V. cholerae* O antigen to core of sv Typhimurium but to sv Typhi only if the *rfa* region of K-12 were present [115]. Viret et al. [57] showed that the sonnei O polysaccharide expressed well in K-12 but that in *S. enterica* sv. Typhi or *V. cholerae* expression was poor with the O polysaccharide not properly anchored in the OM unless an *E. coli rfa* region was also present (both available R1 *rfa* forms were effective). Although we know very little of ligase specificity, it seems clear that, although there is some inter-strain variation, it can be quite wide at least for the O polysaccharide component. It remains to be seen if *E. coli* O polysaccharides are generally able to attach to any of the *E. coli* core forms, but as yet all seem to attach well to the K-12 core. The requirement for two genes for ligation (*rfaL* and *rfbT*) was thought to relate to the need for ligase to identify both core and O polysaccharide; however, more recent findings (see above) show that more study is needed to define even the components of ligase.

Mutants blocked in specific steps in the synthesis of dysenteriae 1 O unit make core with what appears to be partial core structures added [21]. The top band in normal K-12 LPS is only present in an *rfe+* strain suggesting that it contains GlcNAc transferred from UndPP-GlcNAc, presumably by ligase to HexIII; this would be the GlcNAc seen on HexIII in chemical studies. In appropriate strains, bands which could correspond to Gal-GlcNAc or Rha-Gal-GlcNAc on Lipid A/core were seen. No chemistry has been done as yet but it is difficult to escape the conclusion that partial core structures are added. The efficiency of transfer is less than for complete or polymerized O unit and intermediates which can be extended are not transferred suggesting that ligase has a strong preference for complete O unit but in its absence will transfer the intermediate which accumulates. The same must apply to flipping the intermediate across the CM and indeed this may be the major discriminatory step. This may well be a general phenomenon not observed before in part because resolution of core bands has only recently been easy and in part because mutants in *rfb* are often unstable and not easily studied.

5.4. Some O polysaccharides may be made without O unit intermediate

Polymerization of presynthesized O units as described above provides an elegant mechanism for synthesis of repeat unit polysaccharides. Surprisingly it appears that there may be an alternative solution to the problem. *E. coli* O8 and O9 O polysaccharides have a typical repeat unit structure (Fig. 4), consisting of three or five mannose residues, respectively, the repeat units being identified by the pattern of linkages. However, the complete polysaccharide is generally thought to be synthesized by a single chain mechanism in which each mannose is added to the non-reducing end of the growing chain. At the reducing end is a sugar, first reported to be Glc, through which the mannan is attached to UndPP (for review and references, see [8]).

Synthesis is *rfe* dependent and as *rfe* encodes a GlcNAc-1-P transferase [116] which makes UndPP-GlcNAc, the presence of Glc at the reducing end is now surprising. Recently, Rick (personal communication) has shown that O8 is made in mutants which cannot make UDPGlc but can make UDPGlcNAc. There was considerable evidence for Glc at the reducing end, but it now seems that the reducing end sugar is GlcNAc, although further study may show the situation to be more complex.

There is no direct evidence for growth at the non-reducing end but the failure to detect an oligosaccharide intermediate [117], the presence of Glc(NAc) on the reducing end and relative resistance to bacitracin have all been used to support the proposed model [8].

O9 LPS has a typical ladder on SDS-PAGE, which under the proposed model can only be explained if each mannose in the repeat unit has a specific transferase with one limiting the rate of extension such that most molecules end at the same residue; however, as we shall see, even that explanation is not available for a mutant form.

The O9 *rfb* cluster has been cloned and deletion of one part leads to synthesis of LPS with α(1–2) links only and a typical ladder pattern on SDS-PAGE with spacing appropriate to a repeat unit of three [77]; in this case, once synthesis has proceeded beyond enzyme 'sight' of the UndPP-Glc carrier, addition of another residue cannot change the appearance of the substrate as all linkages are the same, leaving no simple explanation for the ladder pattern. Perhaps the transferases form a complex [8] which once started always completes a round of transfers and the incomplete complex of the mutant adds three residues only per round.

An alternative explanation for the data which does not seem to have been considered is that synthesis is not continuous but follows the model of other O polysaccharides, except that the mannan only is transferred by the O9 polymerase by transglycosylation, leaving UndPP-Glc for recycling without further processing; this would account for the relative resistance observed to bacitracin. The failure to detect the oligosaccharide intermediate could be due to the kinetics of its synthesis and polymerization. This model would require that the O9 polymerase transglycosylation reaction occur in vitro without added energy source to account for the experimental data and that the UndPP-Glc acceptor be flipped back to the cytoplasmic face from periplasmic face of the CM for reuse. These are major caveats but the advantage of this model is that it allows all subsequent stages to be the same as for other lipopolysaccharides; under the current continuous synthesis model, we have to assume that the complete O polysaccharide is flipped across the CM if subsequent steps are as described for other LPSs. Clearly, further study of O8 and O9 synthesis is required.

Synthesis of several other O polysaccharides is *rfe* dependent, but in at least one case (see Section 4.1.5), the first sugar of the O unit is GlcNAc and Rfe simply adds this first residue of each O unit to UndPP. The other classic *rfe* dependent O polysaccharide, that of *S. enterica* group C1, also has GlcNAc in its O unit, so there is no need to postulate a single chain mechanism other than for O8 and O9.

6. Organization and biological function of LPS

6.1. Role of LPS

LPS is an integral part of the Gram-negative cell and its role in organization and interaction with other parts of the OM is discussed in Chapter 12. Rough mutants of pathogens, lacking O antigen and perhaps part of the core, are generally non-pathogenic and often serum sensitive. Clearly, the outer layers provide protection to the cell and the length of the O polysaccharide seems to be important in this regard [118], the suggestion being that the long chain sterically hinders access of complement to the OM (for discussion, see [85]). However, some pathogenic species lack polymerized O polysaccharide (see Section 3.3) and it cannot be necessary in all circumstances. The role of LPS in non-pathogens has not been explored in the same detail, but organisms such as *P. aeruginosa* have an extensive range of O unit structures and this must play an important part in adaptation for the polymorphism to have been maintained.

6.2. Tertiary structure of LPS

We have considered LPS structure largely in two dimensions as little information exists on the three-dimensional structure or its significance. There have been studies of phase transition for outer membranes (see Chapter 12) but we have only limited information on the arrangement of LPS in the membrane. In vitro studies (reviewed in 119] indicate that the amount of core present, presence or absence of O polysaccharide, number and nature of fatty acids present and nature of cation and pH are all important in determining phase transition temperature and by inference LPS molecular interactions. There have also been X-ray diffraction and ESR studies of extracted LPS of various types (for review, see [119]) but these tend to form non-lamellar structures which are not yet very helpful for determining the structure in the living membrane and the models shown in Fig. 13 are based on energy minimization.

A detailed model, based on energy minimization and well supported by NMR studies, has been proposed for the tertiary structure of O polysaccharide of groups A, B and D_1 of *S. enterica* [120,121] (Fig. 13). There is a quite rigid helical structure with ~3 repeat units per turn. The greatest flexibility is at the Rha-Gal linkage so the structural unit is Gal-Man(-DDH)-Rha rather than the Man(-DDH)-Rha-Gal unit which is made and polymerized. The 6-deoxy group of abequose or tyvelose (groups B and D1) is involved in an extensive hydrophobic surface which extends across the α face of Man and the 6-deoxy group of L-Rha. The group A (with DDH paratose) structure is similar, but less stable and substitution with DDH ascarylose changes the preferred conformer. The α-D-glucopyranosyl residue which may be linked to O-4 of Gal (Fig. 4) fits between the DDH residues of adjacent units with little overall change in gross geometry. However, the 1–6 linked Glc in phage P22 lysogenic strains and the change in the main chain linkage in P27 infected strains have a greater effect. The lesser stability of the group A structure is interesting as group A strains are much less common than group B or D1 strains and all seem to be recently derived by mutation from group D strains. This and the absence of an ascarylose form of this group of related O polysaccharides suggests that ability to form an appropriate stable tertiary structure may be important for proper function and if so this will limit the possibilities for O unit composition.

6.3. Biological effects

LPS has complex effects on animals including pyrogenicity, induction of endotoxic shock, activation of macrophages and other cells, and complement activation. These properties reside largely in Lipid A and require the presence of the β1–6 linked D-GlcN backbone, phosphorylation at positions 1 and 4', acylation and 3-acyloxyacylation for full

310

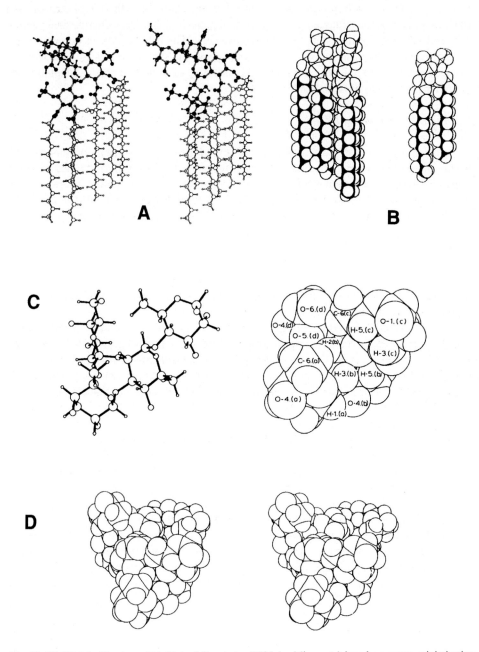

Fig. 13. (A) Model of hepta-acyl Lipid A of *S. enterica* R595 (sv Minnesota) based on energy minimization with views at 90° to each other [134]). (B) Space filling model of Re LPS of *E. coli* K-12 based on energy minimization with views at 90° to each other [134]. (C) Ball and stick and space filling models of an O unit of a group B *S. enterica* strain viewed from the reducing end of a Rha residue [120]: a, Abe; b, Man; c, Rha; d, Gal. (D) Stereo projection of three O units viewed as above [120].

effect. The details have been reviewed recently [122] as has the enhancing effect of core regions on these biological effects [123].

7. Evolution

The variation in O polysaccharides poses questions at several levels. All genetic variation in a species tends to be lost by random genetic drift. Polymorphisms are, however, widespread and due either to a balance between generation of variation by mutation and loss by genetic drift, or to some form of balanced selection. The variation in O polysaccharide core discussed above is clearly not in the former class and hence must be due to some form of balanced selection, and we can ask what that is.

At least in the Enterobacteriaceae, which are relatively well documented, related species may have quite different sets of O polysaccharide forms, and we can ask where these forms come from.

7.1. Maintenance of LPS variation

The enormous extent of variation in O polysaccharides is not easily accounted for by previously described forms of balanced selection and it is suggested [124] that it and other polymorphisms in bacteria are maintained by niche adaptation of clones together with niche-specific selection. If we consider E. coli and S. enterica, it is immediately apparent that specialized clones exist, examples being sv Typhi of S. enterica or the ETEC, EIEC etc. strains of E. coli. In the proposed model: (1) each clone is adapted to a particular niche, comprising host, mode of pathogenesis etc.; (2) natural selection operates to maintain adaptation to that niche; (3) the level of gene flow between clones of a species is low enough such that influx of non-adaptive alleles from other clones does not overwhelm the niche adaptation. In effect the level of genetic exchange between clones is low enough to enable each to maintain separate adaptations yet high enough for evolution and maintenance of species wide adaptations. The model has been tested against estimates of the parameters involved and seems quite capable of explaining the maintenance of O polysaccharide polymorphism if different O polysaccharides are adaptive in different niches [124]. A preferred allele can be maintained in a given clone if it has a selective advantage of 10^{-3} or 10^{-4}, that is if the preferred allele has a 1:0.999 advantage in contributing to the next generation. It seems highly plausible that variations in O polysaccharide could have selective advantage at this level, difficult though it would be to detect it directly.

7.2. Origin of the variation

We have referred above to the pattern of variation in the related S. enterica groups A, B, C2, D1 and E. The conclusions to be drawn have been discussed recently [63] and are summarized here. The five forms have rfb clusters in which genes common to two or more are always in the same order: rhamnose pathway, dideoxyhexose pathway, rfbX, group specific transferases, mannose pathway and rfbP, the common Gal transferase. Presumably the clusters had a common ancestor with this organization. The G + C content

312

is below the 0.50–0.53 characteristic of *S. enterica* coding sequence and furthermore is variable within each *rfb* cluster. We conclude from the variation that the cluster was assembled from several different sources with G + C contents in the range 0.32–0.45.

In general, genes which differentiate groups, being the group-specific transferases, *rfbX* and some of the dideoxyhexose pathway genes, are in a central region. Genes outside of this central region are usually identical in sequence if common to two groups with exceptions generally adjacent to the central group-specific region. The dideoxyhexose pathway is interesting in this regard as the genes common to groups D, C2 and B are adjacent to the common rhamnose pathway genes while the genes determining which dideoxyhexose is made are at the downstream end of the dideoxyhexose set in the group-specific region. It seems that during evolution, genes have been added or lost from the central region to give the range now seen. While this must have gone on over a very long period of time to account for the differences, there has been remarkably little divergence in the common sequences. We attribute this conservation of sequence to genetic exchange between forms such that random genetic drift occurs across the species.

Of the 170 forms of O polysaccharide reported for *E. coli* and the 60 forms reported for *S. enterica,* only two are shared by these two closely related species, and in the one such case which has been studied (*E. coli* O111 and *S. enterica* sv Adelaide), the two *rfb* regions are not sufficiently similar to hybridize on Southern blots [79]. This suggests that there is sufficient turnover of *rfb* regions for essentially all to have been replaced since the two species diverged.

We can now turn our attention to specific genes. *rfbJ* and *rfbS* are in the corresponding place in groups B and D, respectively; the products carry out a similar reaction on 4-keto-6-deoxy-CDP-glucose, reducing the 4-keto group to the galactose or glucose form, respectively (Fig. 9). We suggest that the two genes evolved in situ from a common ancestor. The difference is enormous indicating a very long period since divergence, but we have at present no indication as to which differences relate to the change of function and which are neutral.

The exceptional genes which are homologous but have diverged in sequence enable us to put some limits on time since divergence. For example, in the case of group C2 and group B, etc., we assume that at some time after evolution of functionally distinct central regions, the *rfbK* and *rfbM* genes of the two forms accumulated by chance sufficient difference to interfere with recombination, thereby reducing genetic exchange. The two forms of *rfb* cluster are maintained by niche adaptation, and reduction of recombination between the *rfbK* and *rfbM* genes of the two forms causes each to undergo genetic drift independently and the level of difference rises; if it reaches a level where recombination is effectively ruled out, then the two will continue to diverge, hitch-hiking with the central functionally important regions. We have no reason to believe that the differences between *rfbK* and *rfbM* of the two groups are adaptive and consider them in general to be neutral so differences should accumulate at the same rate as for neutral differences between species (accumulation of neutral base substitutions is determined by mutation rate and not affected by population size (see [125]). The level of difference observed is equivalent to that between genes of *Serratia marcescens* and *E. coli* which are thought to have diverged about 200 million years ago [71]. It seems that the *rfb* clusters of these related O polysaccharides diverged long before *E. coli* and *S. enterica* diverged an estimated 140 million years ago [126]. The difference between *rfbJ* of groups B and C2 is even greater and if we take this as our indicator, we look to *B. subtilis* and *E. coli* as having equivalent levels of difference and a divergence period of 1.5 billion years [71].

Groups A, B, C2, D and E1 are clearly related and, based on probing by rhamnose and mannose pathway genes, there are perhaps another nine groups in this family [127]. The group C1 *rfb* cluster is the only one outside of this family to have been studied in detail, and while there is no resemblance in detail, it too appears to be a mosaic of genes which arose outside of *S. enterica*, with G + C content ranging from 0.29 to 0.61. The same applies to the *rfb* cluster of *E. coli* dysenteriae I with G + C content ranging from 0.32 to 0.43 [21].

We can learn more of the evolution of *rfb* clusters from the CDP-dideoxyhexose pathway genes of a *Y. pseudotuberculosis* strain which has the same organization as in *S. enterica* and similar variation in G + C content [89], although the rest of the *rfb* cluster must be different as there is considerable variation in the O units. Similarly, the rhamnose and mannose pathways have the same organization wherever they occur, in both cases including occurrences in non-*rfb* clusters. The simplest explanation for these observations is that each pathway evolved once only and that there has been extensive gene capture and assembly of gene clusters from components from various sources.

It seems that we can generalize and say that evolution of O polysaccharide diversity has involved repeated gene transfer events, and one can perhaps look on the portfolio of O polysaccharides carried by a species as subject to continued turnover, such that related species like *E. coli* and *S. enterica* (nonetheless thought to have diverged 140 million years ago [128] have almost no overlap in O polysaccharide forms. These two species have two forms in common and we know that the *E. coli* gene cluster will barely probe the *S. enterica* genes [79] suggesting independent acquisition rather than retention of a component of the ancestral polymorphism. Many bacterial species comprise a series of clones which may be adapted to specific niches [124]. We can speculate that O polysaccharide specificity is part of that adaptation and as circumstances change, so the suitability of particular O polysaccharides changes, providing selective pressure for lateral transfer of new forms. We can only suggest a specific origin in one case; the *E. coli* (shigella) sonnei *rfb* genes have been cloned and will hybridize to DNA from a *Plesiomonas shigelloides* strain which produces an identical O polysaccharide [57]. The sonnei clone is particularly interesting as shigella clones are adapted to *H. sapiens*, a species which has only recently lived in densities suitable for its mode of pathogenesis [129]. Sonnei has its *rfb* genes on a plasmid which may well be an indication of relatively recent transfer and perhaps in sonnei, we see a *rfb* region soon after transfer maintained as part of its adaptation to a new niche.

We can extend the analysis to the *rfa* region as there are several forms in *E. coli* which are sufficiently different for us to conclude that they constitute a balanced polymorphism. It is of interest then that the *rfa* region is also of atypical G + C content. A G + C content scan of the *rfa* region is quite informative as the flanking genes and also the *kdtA* and heptose related *rfa* genes which are at the two ends of the *rfa* locus are of about 0.53 G + C content, whereas the rest of the *rfa* region stands out with G + C content ranging from 31 to 45 with *rfaS* being particularly low (Fig. 8). It appears that the genes for synthesis of the Lipid A and inner core are part of the common enterobacterial genome but

314

that the genes for the K-12 outer core were imported and incorporated later. The only other *rfa* region for which we have sequence information is *S. enterica*. The K-12 and LT2 *rfa* regions are very similar in organization [38] and in general have about 30% non-identity at amino acid level but differ substantially in *rfaK* and *rfaL* and *rfaS* [49]. RfaK and RfaL have only about 13% identity and only part of the *rfaS* gene is present in LT2 with about 25% identity. The two clusters appear to have diverged long before *E. coli* and *S. enterica* and must have been independently acquired or perhaps existed in the common ancestor as a polymorphism.

Acknowledgements

I thank those who contributed papers and other information prior to publication. Work carried out in my laboratory was supported by the Australian Research Council. I am indebted to those who have worked in the lab in various capacities and in particular Brian Neal who made most of the illustrations and Gordon Stevenson who helped with bibliographic work. Himanshu Brahmbhatt provided helpful criticism of the manuscript.

References

1. Morrison, D.C. and Ryan, J.L. (Eds.) (1992) Bacterial Endotoxic Lipopolysaccharides, Vol. 1, CRC Press, Boca Raton, FL.
2. Brade, H., Brade, L. and Ruetschek, E.T. (1988) Zbl. Bakteriol. Mikrobiol. Hyg. [A] 268, 151–179.
3. Raetz, C.R.H. (1990) Annu. Rev. Biochem. 59, 129–170.
4. Schnaitman, C.A. and Klena, J.D. (1993) Microbiol. Rev. 57, 655–682.
5. Brahmbhatt, H.N., Lindberg, A.A. and Timmis, K.N. (1992) Curr. Top. Microbiol. Immunol. 180, 45–64.
6. Mäkelä, P.H. and Stocker, B.A.D. (1984) in: E.T. Rietschel (Eds.), Handbook of Endotoxin, Vol. I, Elsevier, Amsterdam, pp. 59–137.
7. Kenne, L. and Lindberg, B. (1983) in: G.O. Aspinall (Eds.), The Polysaccharides, Vol. 2, Harcourt Brace Jovanovich, New York, pp. 287–363.
8. Jann, K. and Jann, B. (1984) in: E.T. Rietschel (Eds.), Elsevier, Amsterdam, pp. 138–186.
9. Rick, P.D. (1987) in: F.C. Neidhardt (Eds.), *Escherichia coli* and *Salmonella typhimurium* Cellular and Molecular Biology, Vol. 1, American Society for Microbiology, Washington, DC, pp. 648–662.
10. Osborn, M.J. (1969) Annu. Rev. Biochem. 38, 501–538.
11. Robbins, P.W. and Wright, A. (1971) in: G. Weinbaum, S. Kadis and S. Ajl (Eds.), Microbial Toxins, Vol. 4, Academic Press, New York, pp. 351–368.
12. Le Minor, L. and Popoff, M.Y. (1987) Int. J. Syst. Bacteriol. 37, 465–468.
13. Takayama, K. and Qureshi, N. (1992) in: D.C. Morrison and J.L. Ryan (Eds.), Bacterial Endotoxic Lipopolysaccharides, Vol. 1, CRC Press, Boca Raton, FL, pp. 44–65.
14. Imoto, M., Yoshimura, H., Sakaguchi, N., Kusumoto, S. and Shibe, T. (1985) Tetrahedron Lett. 26, 1545.
15. Kusimoto, S. (1992) in: D.C. Morrison and J.L. Ryan (Eds.), Bacterial Endotoxic Lipopolysaccharides, Vol. 1, CRC Press, Boca Raton, FL, pp. 81–105.
16. Raetz, C.R.H. (1992) in: D.C. Morrison and J.L. Ryan (Eds.), Bacterial Endotoxic Lipopolysaccharides, Vol. 1, CRC Press, Boca Raton, FL, pp. 67–80.
17. Holst, O. and Brade, H. (1992) in: D.C. Morrison and J.L. Ryan (Eds.), Bacterial Endotoxic Lipopolysaccharides, Vol. 1, CRC Press, Boca Raton, FL, pp. 135–170.

18. Jansson, P.E., Lindberg, A.A., Lindberg, B. and Wollin, R. (1981) Eur. J. Biochem. 115, 571–577.
19. Aspinall, G.O. (1983) in: G.O. Aspinall (Eds.), The Polysaccharides, Vol. 2, Academic Press, New York, pp. 1–9.
20. Klena, J.D., Asford II, R.S. and Schnaitman, C.A. (1992) J. Bacteriol. 174, 7297–7307.
21. Klena, J.D. and Schnaitman, C.A. (1993) Mol. Microbiol. 9, 393–402.
22. Eidels, L. and Osborne, M.J. (1974) Microbiol. Rev. 249, 5642–5648.
23. Eidels, L. and Osborn, M.J. (1971) Proc. Natl. Acad. Sci. USA 68, 1673–1677.
24. Coleman, W.G. and Leive, L. (1979) J. Bacteriol. 139, 899–910.
25. Raina, S. (1991) Nucleic Acids Res. 19, 3811–3819.
26. Pegues, J.C., Chen, L., Gordon, A.W., Ding, L. and W. G. Coleman, J. (1990) J. Bacteriol. 172, 4652–4660.
27. Sirisena, D.M., Brozek, K.A., MacLachlan, P.R., Sanderson, K.E. and Raetz, C.R.H. (1992) J. Biol. Chem. 267, 18874–18884.
28. Coleman, J. and Raetz, C.R.H. (1988) J. Bacteriol. 170, 1268–1274.
29. Young, K., Silver, L.I., Bramhill, D., Caceres, C.A., Stachula, S.A., Shelly, S.E., Raetz, C.R.H. and Anderson, M.S. (1993) in Abstracts of American Society for Biochemistry and Molecular Biology (ASBMB), San Diego, CA.
30. Normark, S., Boman, H.G. and Matsson, E. (1969) J. Bacteriol. 97, 1334–1342.
31. Mohan, S., Shelly, S.E., Raetz, C.R.H. and Anderson, M.S. (1993) in: Abstracts of American Society for Biochemistry and Molecular Biology (ASBMB), San Diego, CA.
32. Lathe, R., Lecocq, J.-P. and Resibois, A. (1977) Mol. Gen. Genet. 154, 43–51.
33. Vuorio, R. and Vaara, M. (1992) J. Bacteriol. 174, 7090–7097.
34. Williamson, J.M., Anderson, M.S. and Raetz, C.R.H. (1991) J. Bacteriol. 173, 3591–3596.
35. Radika, K. and Raetz, C.R.H. (1988) J. Biol. Chem. 29, 14859–14867.
36. Ray, B.L. and Raetz, C.R.H. (1987) J. Biol. Chem. 262, 1122–1128.
37. Brozek, K.A. and Raetz, C.R.H. (1990) J. Biol. Chem. 265, 15410–15417.
38. Schnaitman, C.A., Parker, C.T., Klena, J.D., Pradel, E.L., Pearson, N.B., Sanderson, K.E. and MacLachlan, P.R. (1991) J. Bacteriol. 173, 7410–7411.
39. Austin, E.A., Graves, J.F., Hite, L.A., Parker, C.T. and Schnaitman, C.A. (1990) J. Bacteriol. 172, 5312–5325.
40. Kadam, S.K., Rehemtulla, A. and Sanderson, K.E. (1985) J. Bacteriol. 161, 277–284.
41. Parker, C.T., Pradel, E. and Schnaitman, C.A. (1992) J. Bacteriol. 174, 930–934.
42. Pradel, E., Parker, C.T. and Schnaitman, C.A. (1992) J. Bacteriol. 174, 4736–4745.
43. Wollin, R., Creeger, E.S., Rothfield, L.I., Stocker, B.A.D. and Lindberg, A.A. (1983) J. Biol. Chem. 258, 3769–3774.
44. Klena, J.D. and Schnaitman, C.A. (1993) J. Bacteriol., submitted.
45. Lehmann, V., Luderitz, O. and Westphal, O. (1971) Eur. J. Biochem. 21, 339–347.
46. Helander, I.M., Vaara, M., Sukupolvi, S., Rhen, M., Saarela, S., Zahringer, U. and Mäkelä, P.H. (1989) Eur. J. Biochem. 185, 541–546.
47. Parker, C.T., Kloser, A.W., Schnaitman, C.A., Stein, M.A., Gottesman, S. and Gibson, B.W. (1992) J. Bacteriol. 174, 2525–2538.
48. Roncero, C. and Casadaban, M.J. (1992) J. Bacteriol. 174, 3250–3260.
49. Klena, J.D., Pradel, E. and Schnaitman, C.A. (1992) J. Bacteriol. 174, 4746–4752.
50. Klena, J.D., Pradel, E. and Schnaitman, C.A. (1992) J. Bacteriol. J. Bacteriol. 175, 1524–1527.
51. Farewell, A., Brazas, R., Davie, E., Mason, J. and Rothfield, L.I. (1991) J. Bacteriol. 173, 5188–5193.
52. Pradel, E. (1991) J. Bacteriol. 173, 6428–6431.
53. Clementz, T. (1992) J. Bacteriol. 174, 7750–7756.
54. Rietschel, E.T., Brade, L., Lindner, B. and Zähringer, U. (1992) in: D.C. Morrison and J.L. Ryan (Eds.), Bacterial Endotoxic Lipopolysaccharides, Vol. 1, CRC Press, Boca Raton, FL, pp. 3–41.
55. Bhat, U.R., Carlson, R.W., Busch, M. and Mayer, H. (1991) Int. J. Syst. Bacteriol. 41, 213–217.
56. Tsang, R.S.W., Schlecht, S., Aleksic, S., Chan, K.H. and Chau, P.Y. (1991) Res. Microbiol. 142, 521–533.
57. Viret, J.-F., Cryz Jr., S.J., Lang, A.B. and Favre, D. (1993) Mol. Microbiol. 7, 239–252.

316

58. Kawahara, K., Brade, H., Rietschel, E.T. and Zahringer, U. (1987) Eur. J. Biochem. 163, 489.
59. Hitchcock, P.J., Leive, L., Mäkelä, P.H., Rietschel, E.T., Strittmatter, W. and Morrison, D.C. (1986) J. Bacteriol. 166, 699–705.
60. Kenne, L., Lindberg, B., Soderholm, E., Bundle, D.R. and Griffith, D.W. (1983) Carbohydr. Res. 111, 289–296.
61. Stevenson, G., Lee, S.J., Romana, L.K. and Reeves, P.R. (1991) Mol. Gen. Genet 227, 173–180.
62. Liu, D., Haase, A.M., Lindqvist, L., Lindberg, A.A. and Reeves, P.R. (1993) J. Gen. Microbiol. 175, 3408–3413.
63. Reeves, P.R. (1993) Trends Genet. 9, 17–22.
64. Ewing, W.H. (1986) Edwards and Ewing's Identification of the Enterobacteriaceae, 4th edition, Elsevier, Amsterdam.
65. Komandrova, N.A., Gorshkova, R.P. and Ovodov, Y.S. (1986) Khimija Prirodnykh Soedinenij 5, 532–547.
66. Ovodov, Y.S., Gorshkova, R.P., Tomshich, S.V., Komandrova, N.A., Zubkov, V.A., Kalmykova, E.N. and Isakov, V.V. (1992) J. Carbohydr. Chem. 11, 21–35.
67. Verma, N.K., Quigley, N.B. and Reeves, P.R. (1988) J. Bacteriol. 170, 103–107.
68. Verma, V. and Reeves, P.R. (1989) J. Bacteriol. 171, 5694–5701.
69. Liu, D., Verma, N.K., Romana, L.K. and Reeves, P.R. (1991) J. Bacteriol. 173, 4814–4819.
70. Brown, P.K., Romana, L.K. and Reeves, P.R. (1991) Mol. Microbiol. 5, 1873–1881.
71. Brown, P.K., Romana, L.K. and Reeves, P.R. (1992) Mol. Microbiol. 6, 1385–1394.
72. Wang, L., Romana, L.K. and Reeves, P.R. (1992) Genetics 130, 429–443.
73. Ding, M.J., Svanborg, C., Haraguchi, G.E., Hull, R.A. and Hull, S.I. (1991) Microb. Pathogenesis 11, 379–385.
74. Neal, B.L., Tsiolis, G.C., Heuzenroeder, M.W., Manning, P.A. and Reeves, P.R. (1991) FEMS Microbiol. Lett. 82, 345–351.
75. Haraguchi, G.E., Hull, R.A., Krallmann-Wenzel, U. and Hull, S.I. (1989) Microb. Pathogenesis 6, 123–132.
76. Valvano, M.A. and Crosa, J.H. (1989) Infect. Immun. 57, 937–943.
77. Kido, N., Ohta, M., Iida, K.-I., Hasegawa, T., Ito, H., Arakawa, Y., Komatsu, T. and Kato, N. (1989) J. Bacteriol. 171, 3629–3633.
78. Heuzenroeder, M.W., Beger, D.W., Thomas, C.J. and Manning, P.A. (1989) Mol. Microbiol. 3, 295–302.
79. Bastin, D.A., Romana, L.K. and Reeves, P.R. (1991) Mol. Microbiol. 5, 2223–2231.
80. Bastin, D.A. and Reeves, P.R. (1993) unpublished.
81. Macpherson, D.F., Morona, R., Beger, D.W., Cheah, K. and Manning, P.A. (1991) Mol. Microbiol. 5, 1491–1499.
82. Cheah, K.-Y., Beger, D.W. and Manning, P.A. (1991) FEMS Microbiol. Lett. 83, 213–218.
83. Hale, T.L., Guerry, P., Seid, J., R. C., Kapfer, C., Wingfield, M.E., Reaves, C.B., Baron, L.S. and Formal, S.B. (1984) Infect. Immun. 46, 470–475.
84. Riley, L.W., Junio, L.N., Libaek, L.B. and Schoolnik, G.K. (1987) Infect. Immun. 55, 2052–2056.
85. Valvano, M.A. (1992) Can. J. Microbiol. 38, 711–719.
86. Sturm, S., Jann, B., Jann, K., Fortnagel, P. and Timmis, K.N. (1986) Microb. Pathogenesis 1, 299–306.
87. Sturm, S., Jann, B., Jann, K., Fortnagel, P. and Timmis, K.N. (1986) Microb. Pathogenesis 1, 307–324.
88. Kessler, A., Brown, P.K., Romana, L.K. and Reeves, P.R. (1991) J. Gen. Microbiol. 137, 2689–2695.
89. Kessler, A., Haase, A. and Reeves, P.R. (1993) J. Bacteriol. 175, 1412–1422.
90. Stroeher, U.H., Karageorgos, L.E., Morona, R. and Manning, P.A. (1992) Proc. Natl. Acad. Sci. USA 89, 2566–2570.
91. Lightfoot, J. (1991) J. Bacteriol. 173, 5624–5630.
92. Arsenault, T.L., Hughes, D.W., Maclean, D.B., Szarek, W.A., Kropinski, A.M.B. and Lam, J.S. (1991) Can. J. Chem. 69, 1273–1280.
93. Whitfield, C., Richards, J.C., Perry, M.B., Clarke, B.R. and Maclean, L.L. (1991) J. Bacteriol. 173, 1420–1431.
94. Whitfield, C., Perry, M.B., MacLean, L.L. and Yu, S.H. (1992) J. Bacteriol. 174, 4913–4919.
95. Jann, K., Dengler, T. and Jann, B. (1992) Zbl. Bakteriol. 276, 196–204.

96. Wright, A. (1971) J. Bacteriol. 105, 927–936.
97. Wright, A. and Barzilai, N. (1971) J. Bacteriol. 105, 937–939.
98. Nikaido, H., Nikaido, K. and Nakae, T. (1971) J. Biol. Chem. 246, 3902–3911.
99. Nikaido, K. and Nikaido, H. (1971) J. Biol. Chem. 246, 3912–3919.
100. Verma, N.K., Verma, D.J., Huan, P.T. and Lindberg, A.A. (1993) Gene 129, 99–101.
101. Takeshita, M. and Makela, P.H. (1971) J. Biol. Chem. 246, 3920–3927.
102. Wollin, R., Stocker, B.A.D. and Lindberg, A.A. (1987) J. Bacteriol. 169, 1003–1009.
103. Clark, C.A., Beltrame, J. and Manning, P.A. (1991) Gene 107, 43–52.
104. Carlin, N.A. and Lindberg, A.A. (1986) Infect. Immun. 53, 103–109.
105. Munford, C.A. and Osborn, M.J. (1983) Proc. Natl. Acad. Sci. USA 80, 1159–1163.
106. Marino, P.A., McGrath, B.C. and Osborn, M.J. (1991) J. Bacteriol. 173, 3128–3133.
107. McGrath, B.C. and Osborn, M.J. (1991) J. Bacteriol. 173, 3134–3137.
108. Wang, L. and Reeves, P.R. (1993) J. Bacteriol., submitted.
109. Daniels, D.L., Plunkett III, G., Burland, V. and Blattner, F.R. (1992) Science 257, 771–778.
110. Robbins, P.W., Bray, D., Dankert, A.W. and Wright, A. (1967) Science 158, 1536–1542.
111. Bray, D. and Robbins, P.W. (1967) Biochem. Biophys. Res. Commun. 28, 334–339.
112. Bastin, D.A., Brown, P.K., Haase, A., Stevenson, G. and Reeves, P.R. (1993) Mol. Microbiol. 7, 725–734.
113. Goldman, R.C. and Hunt, F. (1990) J. Bacteriol. 172, 5352–5359.
114. Batchelor, R.A., Haraguchi, G.E., Hull, R.A. and Hull, S.I. (1991) J. Bacteriol. 173, 5699–5704.
115. Tacket, C.O., Forrest, B., Morona, R., Attridge, S.R., Labrooy, J., Tall, B.D., Reymann, M., Rowley, D. and Levine, M.M. (1990) Infect. Immun. 58, 1620–1627.
116. Meier-Dieter, U., Barr, K., Starman, R., Hatch, L. and Rick, P.D. (1992) J. Biol. Chem. 267, 746–753.
117. Kopmann, H.J. and Jann, K. (1975) Eur. J. Biochem. 60, 587–601.
118. Grossman, N., Schmetz, M.A., Foulds, J., Klima, E.N., Jiminez, V., Leive, L.L. and Joiner, K.A. (1987) J. Bacteriol. 169, 856–863.
119. Seydel, U. and Brandenberg, K. (1992) in: D.C. Morrison (Eds.), Bacterial Endotoxic Lipopolysaccharides, Vol. 1, CRC Press, Boca Raton, FL, pp. 225–250.
120. Bock, K., Meldal, M., Bundle, D.R., Iversen, T., Pinto, M., Garegg, P.J., Norberg, T., Lindberg, A.A. and Svensen, S.B. (1984) Carbohydr. Res. 130, 23–34.
121. Bock, K., Meldal, M., Bundle, D.R., Iversen, T., Pinto, M., Gareg, P.J., Kvanstrom, I., Norberg, T., Lindberg, A.A. and Svensen, S.B. (1984) Carbohydr. Res. 130, 35–53.
122. Takada, H. and Kotani, S. (1992) in: D.C. Morrison and J.L. Ryan (Eds.), Bacterial Endotoxic Lipopolysaccharides, Vol. 1, CRC Press, Boca Raton, FL, pp. 107–134.
123. Cavaillon, J.-M. and Haeffner-Cavaillon, N. (1992) in: D.C. Morrison and J.L. Ryan (Eds.), Bacterial Endotoxic Lipopolysaccharides, Vol. 1, CRC Press, Boca Raton, FL, pp. 205–224.
124. Reeves, P.R. (1992) FEMS Microbiol. Lett. 100, 509–516.
125. Maynard Smith, J. (1989) Evolutionary Genetics, Oxford University Press, Oxford.
126. Ochman, H. and Wilson, A.C. (1987) in: F.C. Neidhardt (Eds.), Escherichia coli and Salmonella typhimurium Cellular and Molecular Biology, Vol. II, American Society for Microbiology, Washington, DC, pp. 1649–1654.
127. Xiang, S.-H., Haase, A.M. and Reeves, P.R. (1993) J. Bacteriol. 175, 4877.
128. Ochman, H. and Wilson, A.C. (1987) J. Mol. Evol. 26, 74–86.
129. Fenner, F. (1971) Med. J. Aust. 1, 1043–1102.
130. Yokota, S., Kaya, S., Sawada, S., Kawamura, t., Araki, Y. and Ito, E. (1987) Eur. J. Biochem. 167, 203–209.
131. Kohara, Y., Akiyama, K. and Isono, K. (1987) Cell 50, 495–508.
132. Rudd, K.E., Miller, W., Werner, C., Ostell, J., Tolstoshev, C. and Satterfield, S.G. (1991) Nucleic Acids Res. 19, 637–647.
133. Rudd, K.E. (1992) in: J. Miller (Eds.), A Short Course in Bacterial Genetics: A Laboratory Manual and Handbook for Escherichia coli and Related Bacteria, Cold Spring Harbor Laboratory Press, Cold Spring Harbor, NY, pp. 2.3–2.43.
134. Naumann, D., Schultz, C., Sabisch, A., Kastowsky, M. and Labischinski, H. (1989) J. Mol. Struct. 214, 213.

J.-M Ghuysen and R. Hakenbeck (Eds.), *Bacterial Cell Wall*

319

CHAPTER 14

Lipoproteins, structure, function, biosynthesis and model for protein export

V. BRAUN[1] and H.C. WU[2]

[1]*Mikrobiologie II, Universität Tübingen, Auf der Morgenstelle 28, D-72076 Tübingen, Germany and* [2]*Department of Microbiology, Uniformed Services University of the Health Sciences, 4301 Jones Bridge Road, Bethesda, MD 20814-4799, USA*

1. Introduction

Murein lipoprotein of *Escherichia coli* was, apart from the basic myelin protein, the first integral membrane protein for which the primary structure was determined [1]. It was the first protein in which a covalently bound lipid was demonstrated by determination of the structure of the lipid and its mode of attachment to the protein [2]. Both the structure of the lipid and the way it was bound to the protein turned out to be unique but later studies revealed many proteins, up to now more than 130 more or less well characterized lipoproteins, with the same lipid structure as that in the murein lipoprotein among Gram-negative and Gram-positive bacteria, and only in bacteria. Later, proteins carrying covalently bound lipids were also found in eukaryotic cells [3], but both the lipids and their attachment to the proteins differ considerably from the murein lipoprotein.

Murein was originally isolated from *Escherichia coli* B by a multistep procedure including the treatment with phenol which had been adapted from the isolation procedure of DNA and lipopolysaccharide. Electron microscopy of the murein revealed a thin layer which had the size and the shape of the *E. coli* cells, and was therefore designated the rigid layer. After treatment with lysozyme, the rigid layer was disintegrated, and globular material remained on the supporting colloid film. Treatment of the rigid layer with pepsin prior to incubation with lysozyme resulted in a thinner layer, and subsequent treatment with lysozyme left no visible structures on the film [4]. Apparently, protein material remained tightly associated with the murein. A much simpler procedure involving the solubilization of EDTA treated cell envelopes in boiling 4% sodium dodecylsulfate resulted in the isolation of a rigid layer of a similar morphology. The latter procedure applied to logarithmically growing cells yielded a single protein species covalently bound to the murein, the murein lipoprotein [5], also called Braun's lipoprotein, whereas stationary phase cells yielded a murein containing a mixture of poorly defined proteins at the murein, among them the lipoprotein.

Analytical work on the nature of the murein lipoprotein started in a fashion similar to finding a needle in a haystack. Incubation of *E. coli* cell envelopes with various proteases

320

resulted in a decrease in the optical density; the maximal decrease in OD was achieved with trypsin, the protease with the narrowest specificity. An exquisitely trypsin-sensitive protein, or even only a lysine or arginine bond, in the cell envelope was looked for which played an important role for the integrity of the cell envelope. It turned out that after trypsin treatment, no protein was left at the murein, and subsequent studies revealed that the bond between lipoprotein and murein was cleaved by trypsin [5].

2. Structure and subcellular location of murein lipoprotein (Braun's lipoprotein)

The mature murein lipoprotein consists of 58 amino acid residues in which no glycine, histidine, proline, phenylalanine and tryptophan is present [1]. The sequence is highly repetitive and suggests an evolution of the structural gene by a duplication of a 15-residue sequence followed by four successive duplications of the C-terminal 7 residues of the original 15 residues. The N-terminal and C-terminal sequences of the murein lipoprotein flank the repetitive motif, and they serve as the lipid (N-terminal) and murein (C-terminal) attachment sites (Fig. 1). The N-terminal cysteine is modified at the amino group mainly by a palmitate residue and at the SH group by a diacylglycerol residue with a fatty acyl composition very similar to that of phospholipids [2]. The structure of the lipid was confirmed by total chemical synthesis. The synthetic lipid S-[2,3-bis-(palmitoleyloxy)propyl]-N-palmitoleylcysteine can assume two diastereometric configurations (R,S and R,R) at the C-2 position of the propyl moiety of which the R (=L) configuration is the most likely one in the natural product [6]. Attachment to the murein occurs via the ε-amino group of the C-terminal lysine residue to the carboxyl group of the optical L-center of diaminopimelate where it replaces D-alanine. [7]. About every 10th to 12th murein subunit is substituted with lipoprotein [1].

The amino acid residues of murein lipoprotein are arranged such that all charged and hydrophilic side chains are located at one side, and all hydrophobic side chains at the opposing side of an α-helix [1]. Indeed, an α-helical content of over 80% was determined by circular dichroism measurements of lipoprotein samples isolated by using boiling so-

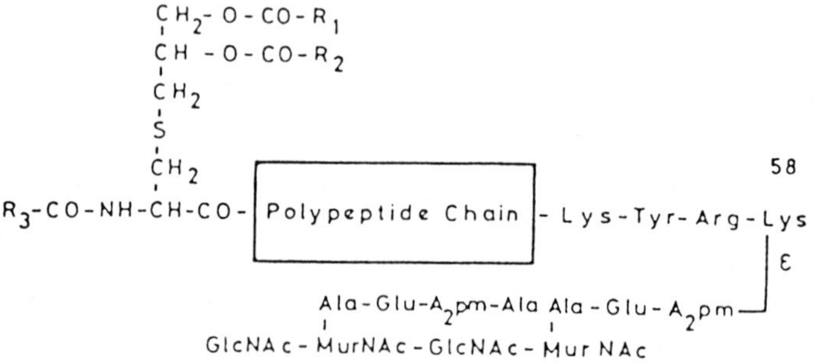

Fig. 1. Lipid and murein attachment sites at the lipoprotein (bound form). R_1, R_2, R_3 indicate fatty acyl chains at the amino-terminal end of the mature protein. The lipoprotein is bound via the epsilon amino group of the C-terminal lysine residue to the carboxyl group of the optical L-center of diaminopimelate (A_2pm) of murein (peptidoglycan). Two murein subunits are drawn to indicate replacement of D-alanine by lipoprotein.

dium dodecyl sulfate [8,9] as well as by employing a much milder procedure (3% octyl-polydisperse-oligoethylene at 37°C) [10]. A double or triple-helix-stranded coiled coil has been proposed as a possible arrangement of the lipoprotein polypeptide chain.

The lipid moiety of the murein lipoprotein causes aggregation of isolated lipoprotein. Replacement of the signal peptide of lipoprotein precursor by the signal peptide of the OmpF protein, and of the cysteine residue by Ala-Glu at the N-terminus of OmpF resulted in a nonacylated protein which, after cross-linking with dimethyl superimidate, was found in the periplasmic space as a trimer [11]. This result indicates the ability of the lipoprotein polypeptide chain to form trimers, as has been found for the OmpF and OmpC porins, and it shows that the polypeptide portion of lipoprotein is not integrated into the outer membrane. Since the acylated lipoprotein was exclusively found in the outer membrane fraction [12], it is attached to the outer membrane presumably via the insertion of the three acyl chains into the inner phospholipid leaflet of the outer membrane.

Twice the amount of the murein-bound form of the lipoprotein exists in the free form not covalently bound to murein [13]. However, it is found to be non-covalently associated with the murein [14] and presumably assumes the same location in the envelope as the bound form. Indeed, murein-bound form was chemically cross-linked to free form lipoprotein. It is very likely that like the bound form, the free form interacts with the outer membrane via the N-terminal lipid and with the murein non-covalently via the protein C-terminus. In fact, removal of the C-terminal lysine residue abolished the interaction of the free form with the murein [14].

Upon release of lipoprotein from the murein layer by cleaving the glycan strands with hen egg white lysozyme, two or three murein subunits remain bound to a single lipoprotein molecule. This finding suggests that lipoprotein is not concentrated at certain sites but is randomly distributed over the murein sacculus [1]. Immunogold-labeling of lipoprotein synthesized under *lac* control on isolated murein, and in cells in which lipoprotein became accessible to anti-lipoprotein antibodies by treatment with Tris-EDTA, revealed a homogeneous distribution of the label over the entire cell surfaces [15].

3. Structure and occurrence of lipoproteins

3.1. Lipoproteins of the murein lipoprotein type

Those lipoproteins which contain an N-terminal cysteine residue substituted with a glyceride residue through a thioether linkage will be defined as lipoproteins of the murein lipoprotein type. They are by far the most frequently occurring lipoproteins in bacteria (Table I). However, apart from the murein lipoprotein, only a few lipoproteins have been studied by chemical (Edman degradation, determination of glycerylcysteine, fatty acid analysis) and physical methods (electron spin and FAB mass spectroscopy) to identify the nature of the lipid substitution. Such a study for example revealed that the cytochrome subunit of the photosynthetic reaction center from *Rhodopseudomonas viridis* carries at the SH group of the N-terminal cysteine residue a glyceride residue containing 18:OH and 18:1 fatty acids but the amino group of the cysteine residue was found to be free [16]. Interestingly, high-resolution X-ray analysis failed to detect the two acyl groups due to

TABLE I

Lipoproteins in bacteria

Species	Lipoprotein	Ref.
Gram-negative bacteria		
Escherichia coli	Lpp (*lpp* = *mlpA*) bound form	5,104
	Free form	13
	Antigen 47	105
	RlpA (36K-lipoprotein)	106
	RlpB (19K-lipoprotein)	106
	PAL (= ExcC; peptidoglycan associated lipoprotein)	23,107
	NlpA (lipoprotein 28)	66,108
	NlpB	109
	F-TraT (surface exclusion)	110,111
	R100-TraT	112
	ColE1 lysis protein (CelA)	113,114
	ColE2 lysis protein (CelB)	115
	ColE3 lysis protein (CelC)	116
	ColA lysis protein (Cal)	117
	Cloacin lysis protein	118
	ColE8 lysis protein	119
	ColE9a lysis protein	120
	ColE9b lysis protein	120
	ColN lysis protein	121
	PBP3 (*ftsI*; penicillin) binding protein	122,123
	OsmB	124
	ColV2-K94 Iss (serum-resistance protein)	125
	lambda Bor (serum-resistance protein)	126
	EnvC	127
Shigella dysenteriae	Lpp	128
Shigella flexneri	MxiJ (Secretion of Ipa invasins)	43
	MxiM	43
Citrobacter freundii	Lpp	128
Salmonella typhimurium	Lpp	129
	TraT	130
Klebsiella pneumoniae	PulA (Pullulanase)	39,131
	PulS (protein secretion)	41
Klebsiella aerogenes	Lpp	128
	PulA	132
Enterobacter aerogenes	Lpp	128
Erwinia amylovora	Lpp	133

TABLE I (*continued*)

Species	Lipoprotein	Ref.
Serratia marcescens	Lpp	134
Morganella morganii	Lpp	135
Edwardsiella tarda	Lpp	128
Proteus mirabilis	Lpp	136
	PAL	22
Yersinia enterocolitica		137
	PCP$_{Ye}$	138
	YscJ (secretion factor)	44
Vibrio cholerae	TagA	125
	AcfD (colonization factor)	125
	TcpC (production of fimbriae)	139
Vibrio harveyi	Chitobiase	140
Haemophilus influenzae	OMP P6 (outer membrane protein P6)	141
	51K haemin-binding lipoprotein	142
	Protein D (IgD-binding lipoprotein)	143
	HbpA (haem binding protein)	144
	PAL (peptidoglycan-associated lipoprotein)	145
	PCP	145
Haemophilus somnus	LppA	146
Haemophilus ducreyi	HbpA	144
Pseudomonas aeruginosa	Protein I (OprI)	19,21
	Protein H	19
Pseudomonas solanacearum	Egl (endoglucanase)	147
Rhodopseudomonas viridis	Cytochrome C subunit	16
Rhodopseudomonas sphaeroides	Lpp	148
Xanthomonas campestris	XpsD (protein secretion)	42
Brucella abortus *Brucella melitensis* *Brucella ovis*	Lpp	149
Neisseria gonorrhoeae	Laz (azurin-like protein)	150
	PanI	151
Neisseria gonorrhoeae *Neisseria meningitidis*	Lip (H.8 antigen)	152,153

TABLE I (*continued*)

Species	Lipoprotein	Ref.
Treponema pallidum	Dpp (47K-protein) (detergent phase protein)	154,155
	35,5 K-protein	156
	TmpC (immunogenic membrane protein-C)	157
	TmpA	158,159
	TpD (34K-membrane immunogen	160
	15 K-protein (tpp15) (major membrane immunogen)	161
Treponema phagedenis	TmpA	158
Borrelia burgdorferi	OspA	162
	OspB	162
	OspD	163
	PCP-extracted lipoprotein	164
Serpulina hyodysenteriae	16K-Antigen	165
Francisella tularensis	17K protein	166
Gram-positive bacteria		
Bacillus licheniformis	Penicillinase (*blaP*)	35
Bacillus cereus	Penicillinase	167
	β-Lactamase III (*blaZ*)	168,169
Alkalophilic *Bacillus* sp. strain 170	LIPEN (penicillinase)	37
Bacillus subtilis	Slp (PAL-related Lpp)	170
	PrsA (protein secretion)	45
	OppA	33
	FhuD	34
	DciAE	171
Staphylococcus aureus	Penicillinase	38
Streptococcus pneumoniae	MalX	32
	AmiA	32
Streptococcus gordonii	SarA	48
Deinococcus radiodurans	HPI	172
Micrococcus luteus	38 K Protein	173
Lactococcus lactis	PrtM (maturing protein)	46
Mollicutes		
Mycoplasma hyorhinis	p37-Protein	32
	VlpA (antigenic variation)	27
	VlpB	27
	VlpC	27
	p120, p70, p23 (complement lysis)	174

TABLE I (*continued*)

Species	Lipoprotein	Ref.
Mycoplasma hyopneumoniae	4 lipoproteins	174
Mycoplasma arginini	About 20 lipoproteins	174
Mycoplasma capricolum	About 30 lipoproteins	175
Spiroplasma melliferum	Spiralin	176
Spiroplasma citri		
Acholeplasma laidlawii	D12, T_2, T_{4a}	177
	About 30 lipoproteins	178
Ureaplasma urealyticum	Antigen serotype 8	179
Rickettsia rickettsii	17K-Antigen	180
Bacteriophages		
T5	Llp	181

disorder in the crystals. In most cases, evidence for a lipoprotein of the murein lipoprotein type came from the examination of the amino acid sequences deduced from the nucleotide sequences which showed a cysteine residue as part of a lipoprotein consensus processing site (see Section 5), demonstration of an inhibition of the conversion of a larger precursor form to a smaller mature form by globomycin, which specifically inhibits signal peptide cleavage by signal peptidase II, incorporation of radioactively labeled palmitate and [2-^{3}H]glycerol into the protein in question, and in a few cases identification of glycerylcysteine. Sequence information exists for 76 lipoproteins containing signal peptidase processing sites of the murein lipoprotein type.

Enterobacteriaceae contain a lipoprotein of the same size and lipid structure as the *E. coli* murein lipoprotein [17]. Their sequence homology agrees with their known taxonomic distance. Among the enterobacterial lipoproteins sequenced, that of *Proteus mirabilis* is most distantly related to the *E. coli* murein lipoprotein. A covalent linkage to the murein and an amide as well as two ester-linked fatty acids were shown by chemical analysis [18]. The murein lipoprotein of *Pseudomonas aeruginosa* also occurs in bound and free forms [19]. Its amino acid sequence differs substantially from the murein lipoprotein of the Enterobacteriaceae (Fig. 2). Most notable is the replacement of the otherwise strictly conserved tyrosine residue by an alanine residue near the C-terminus [20,21].

A second lipoprotein which has been characterized by biochemical techniques was PAL of *P. mirabilis* [22]. This protein is associated but not chemically bound to murein, contains the same lipid structure as the murein lipoprotein and is associated with the outer membrane. The same protein occurs in *E. coli* [22] and consists of 173 amino acid residues including a signal sequence of 21 N-terminal residues [23]. PAL shows homologies to the murein lipoprotein, especially at the processing site. Additional lipoproteins have been found in *E. coli* and other Enterobacteriaceae which are listed in Table I. There are

326

```
1  M KATKLVLG A VI L GSTLLA GCSS NA K .....IDQLSSDVQTLN A KV D QLSNDVNAMRSDV Q A A KDD A AR AN Q R LDNMATKY.R K

2  M NRTKLVLG A VI L GSTLLA GCSS NA K .....IDQLSTDVQTLN A KV D QLSNDVTAIRSDV Q A A KDD A AR AN Q R LDNQAHSY.R K

3  M NRTKLVLG A VI L GSH.SA GCSS NA K .....IDQLSSDVQTLN A KV D QLSNDVNAMRSDV Q A A KDD A AR AN Q R LDNQAHAY.K K

4  M GRSKIVLG A VV L ASALLA GCSS NA K .....FDQLDNDVKTLN A KV D QLSNDVNAIRADV Q Q A KDE A AR AN Q R LDNQVRSY.K K

5  M N.NVLKFS A LA L AAVLAT GCSS HS K ETEARLTATEDAAARAQ A RA D EAYRKADEALGAA Q K A QQT A DE AN E R ALRMLEKASR K
```

Fig. 2. Comparison of the amino acid sequence of the lipoproteins of *Escherichia coli* (1), *Erwinia amylovora* (2), *Serratia marcescens* (3), *Morganella morganii* (4) and *Pseudomonas aeruginosa* (5). The residues in bold face are common to all five proteins.

certainly many more lipoproteins than are known today, and the greatest abundance of known lipoproteins in *E. coli* simply reflects our much more advanced state of knowledge on this organism compared to other bacterial species. For example, recent interests in the surface proteins of *Haemophilus influenzae* and *Treponema pallidum* as antigens and immunogens greatly increased the number of lipoproteins found in these organisms (Table I). Furthermore, two-dimensional gel electrophoresis of proteins labeled with radioactive palmitate frequently resolved more than 10 lipoproteins in a single species of which only a few were characterized further.

Many Gram-positive organisms also contain lipoproteins (Table I). Some are discussed in Section 4.

3.2. Other lipoproteins

A major outer membrane protein of *Rhodopseudomonas sphaeroides* contains a fatty acid residue in amide linkage at the N-terminal alanine [24]. The *E. coli* haemolysin is fatty acylated in an unknown manner [25]. Among 200 membrane proteins of *Acholeplasma laidlawii* resolved by two-dimensional electrophoresis, 23 were found to be covalently modified with acyl chains. The extent of modification and the composition of the fatty acyl chains somewhat depended on the exogenous and endogenous lipid supply but differed from that of the lipid fraction. In addition to the murein lipoprotein type, the fatty acylated proteins in *A. laidlawii* also contain one or two fatty acyl residues attached to the polypeptides by *O*-ester bonds [26]. Similar observations have been made with lipoproteins of other Mollicutes, for example *Mycoplasma hyorhinis* (see [27] and Table I).

4. Functions of lipoproteins

The lipid portion of the lipoproteins contributes mainly to the structural properties of the lipoproteins and may determine the membrane location of a protein. The protein portion can assume any of many functions so far identified (Table I). In a few cases, discussed at the end of this section, the lipid moiety plays a decisive role in the activity of a lipoprotein.

The murein lipoprotein is anchored to the outer membrane by the three fatty acyl chains [11]. A mutant of *E. coli* lacking the murein lipoprotein lysed when treated with

EDTA, and the lysis was prevented in the presence of 0.5 M sucrose [28]. Less severe reduction in Mg^{2+} resulted in the formation of outer membrane blebs at the cell surface. The mutants also leaked a considerable fraction of their periplasmic proteins, and were impaired in the invagination of the outer membrane during the formation of the nascent septum [29]. A similar phenotype was displayed in a mutant which contained the free form but contained a very reduced amount of the bound form of murein lipoprotein [30]. A mutant of *E. coli* lacking murein lipoprotein and the OmpA protein formed spherical cells and required increased concentrations of Mg^{2+} and Ca^{2+} for growth which only partially prevented cell lysis [31]. Both OmpA and murein lipoprotein can be chemically cross-linked to each other suggesting their close association in the cell envelope. These observations support the original proposal [1,5] that the murein lipoprotein serves an important function in maintaining the outer membrane structure.

An interesting function of the lipid moiety of lipoproteins has been found recently. Gram-negative bacteria contain nutrient transport systems with periplasmic binding proteins as essential components. Since Gram-positive bacteria are devoid of a periplasmic space they anchor the binding protein through the fatty acids of the N-terminal glyceride-cysteine to the outer surface of the cytoplasmic membrane [32–34].

Secreted proteins may undergo a cell-bound form prior to their release into the culture medium. Intermediary membrane fixation of the secreted protein may serve to concentrate their activities at the cell surface before they are diluted into the surrounding medium. The first examples were penicillinases of various species of *Bacillus* [35–37] and *Staphylococcus aureus* [38]. They are synthesized as pre-proforms which are first modified with lipid and the signal sequence is then cleaved during export through the cytoplasmic membrane; subsequently the lipid anchor of lipopenicillinase is removed by proteolysis, for example after residues 8 and 16 of processed form (proform) of the *B. licheniformis* penicillinase.

Pullulanase is one of about 20 lipoproteins identified in *Klebsiella pneumoniae* (now *K. oxytoca*). It contains a lipid of the murein lipoprotein type which binds the enzyme to the cell surface during logarithmic growth. The lipid moiety in pullulanase is not cleaved upon release of the enzyme into the culture medium at the end of exponential growth [39,40]. Thus, the lipid-modified enzyme forms large aggregates presumably because the fatty acyl chains form micelles. Another strain, *K. oxytoca* K21, secretes a pullulanase during logarithmic growth which is free of fatty acyl chains [40]. Therefore, neither secretion nor enzyme activity is related to lipid modification.

Another lipoprotein, PulS, is involved in pullulanase secretion. It was localized in the outer membrane and may be part of a secretion apparatus that translocates pullulanase through the outer membrane [41]. A protein, XpsD, required for the secretion of extracellular enzymes across the outer membrane of *Xanthomonas campestris* also seems to be a lipoprotein because it contains the typical processing site for signal peptidase II [42]. MxiJ is a lipoprotein which takes part in the secretion of *Shigella flexneri* Ipa invasins [43] and which exhibits homologies to YscJ, a lipoprotein of *Yersinia enterocolitica*, involved in the secretion of Yop proteins [44].

Two other homologous lipoproteins, PrsA of *Bacillus subtilis* [45] and PrtM of *Lactococcus lactis* [46], are required for the formation of active secreted α-amylase and PrtP proteinase, respectively, but their modes of action remain to be elucidated.

All small plasmids encoding colicins contain three genes which encode the colicin proper, the immunity protein and the lysis or colicin release protein. Lysis proteins are small lipoproteins which show a high sequence homology. The lysis proteins facilitate release of the colicins from the producing cells by pseudolysis. Acylation and processing is required for the lysis proteins to be active in colicin release. Directed mutagenesis resulted in unacylated forms which were inactive. Activation of phospholipase A in the outer membrane has been implicated in the colicin release process but the entire process seems to require more than phospholipase activity. For example replacement of the signal peptide of the cloacin DF13 lysis protein by the signal peptide of the murein lipoprotein yielded a hybrid protein which was inactive in cloacin release but provoked cell lysis and lethality [47].

Another case where the function of a protein is related to the presence of the lipid may be the hydrophobicity of the cell surface of oral streptococci conferred by lipoproteins (13 were identified by 2-D electrophoresis). These streptococci adhere to saliva-coated surfaces and co-aggregate with other bacteria [48].

Murein lipoprotein is a potent B-cell mitogen, and the stimulating activity resides in the lipopeptide moiety [49]. Based on this observation many lipopeptide derivatives were synthesized and their immunogenic activity was evaluated. These studies demonstrated that the lipopetide confers enhanced immunogenicity and antigenicity on proteins with low immunogenic activity [50, and refs. therein]. Lipopeptide mediates attachment to the macrophage cell membrane followed by internalization into the cytoplasm, and activates macrophages to secrete cytokines. Recently, virus-specific cytotoxic T-lymphocytes were primed with synthetic lipopeptide vaccine [50]. An octapeptide epitope of influenza virus nucleoprotein covalently linked to tripalmitoyl-S-glycerylcysteinyl-Ser-Ser induced high affinity influenza virus-specific cytotoxic T-lymphocytes in vivo as does infectious virus. Synthetic lipopeptides of the murein lipoprotein type have the potential to enhance humoral and cell-mediated immunity, and therefore may be useful for the development of new vaccines. This notion is supported by the finding that mice immunized with OspA lipoprotein were protected against challenge with infectious *Borrelia burgdorferi*, whereas lipid-free OspA did not induce a protective response and no serum immunoglobulin [51].

5. Biosynthesis of bacterial lipoproteins

The abundance of murein lipoprotein in *E. coli* and many unusual features of the biosynthesis of free form lipoprotein have made this protein one of the most extensively studied membrane proteins in *E. coli*. The unique features of the biosynthesis of murein lipoprotein include the abundance and stability of the mRNA for lipoprotein and the extensive modification of the translation product required to form the mature lipoprotein. The recognition that this protein is synthesized as a precursor form, the prolipoprotein, with N-terminal signal sequence led to the use of lipoprotein as a model protein in the study of protein export in *E. coli*. The exhaustive use of site-directed mutagenesis to alter the signal peptide of murein lipoprotein also represents the best studied structure/function relationship of a signal sequence in exported proteins.

329

Studies on the biosynthesis of murein lipoprotein began with the search for the do-
nor(s) of the glyceride moiety covalently attached to the N-terminal cysteine. The first
hint for the fatty acyl donors came from the fatty acid composition of murein lipoprotein
which was similar to that of bulk phospholipids of *E. coli* including the presence of cyclopropane fatty acids [52]. Inasmuch as cyclopropane fatty acids are synthesized only on
phospholipid-linked unsaturated fatty acids [53], this observation suggested the possibility that the fatty acids in lipoproteins were transferred from glycerophosphatides. Pulse-
chase experiments carried out to ascertain the nature of the lipid precursors for murein
lipoprotein indicated the existence of a large metabolic pool of the donor of the diglyceride moiety in murein lipoprotein [54,55]. Further in vivo labeling experiments suggested that the glyceryl moiety in murein lipoprotein was derived from the non-acylated
glycerol in phosphatidylglycerol (PG) [54,56], while both the O- and N-acyl moieties in
murein lipoprotein were derived from fatty acids in any of the major glycerophosphatides,
phosphatidylethanolamine (PE), phosphatidylglycerol (PG) or cardiolipin (CL) [57,58].

A major advance in our understanding of the biosynthesis of murein lipoprotein came
from the identification of its precursor form, the prolipoprotein, which contained a 20
amino acid signal sequence at its N-terminus [59]. This discovery was soon followed by
the isolation of an *E. coli lpp* (*mlpA*) mutant which produced an unmodified prolipoprotein due to a single amino acid substitution of Asp for Gly in the signal sequence of prolipoprotein [60] and by the discovery of globomycin, an antibiotic [61]. The use of globomycin in the study of lipoprotein biosynthesis in *E. coli* led to two important findings:
(i) it inhibited the maturation of lipoprotein at the step of the processing of lipoprotein
precursor, resulting in the accumulation of lipid-modified prolipoprotein in globomycin-
treated cells [62]; and (ii) in addition to lipid-modified precursor form of murein lipoprotein, the globomycin-treated cells accumulated the lipid-modified precursor forms of a
number of new lipoproteins of varying sizes localized in both the inner and outer membranes of *E. coli* [63]. These two observations had profound implications. It strongly suggested that lipid modification of prolipoprotein preceded, and might indeed be a prerequisite for, the processing of lipoprotein precursor to form mature protein, a postulate
consistent with the formation of an unmodified prolipoprotein in *lpp*D14 mutant of *E. coli*
[59]. Furthermore, accumulation of a number of lipid-modified prolipoproteins in globomycin-treated cells suggested the existence of a common biosynthetic pathway in *E.
coli* for structurally distinct but similarly lipid-modified proteins, the most abundant of
which is murein lipoprotein.

While the existence of the homologs of *E. coli* murein lipoprotein in other Gram-negative bacteria was known, the prevalence of lipid-modified proteins in other bacteria was
not suspected until the advent of recombinant DNA technology which greatly facilitated
the cloning and sequence determinations of genes of known or unknown functions. The
discovery of membrane-bound penicillinase in *Bacillus licheniformis* as a lipid-modified
protein like murein lipoprotein led to the identification of a consensus sequence (also
called lipobox) at the C-terminal region of the signal sequences of lipoprotein precursors
[35]. The list of lipid-modified proteins has been increasing steadily in recent years, and
an analysis of the −5 to +5 amino acid residues in 75 lipid-modified precursor proteins revealed a highly conserved -Leu-Ala/Ser-Gly/Ala-Cys- sequence at −3 to +1 positions
(Table II). The evidence for these proteins being lipid-modified was based on metabolic

TABLE II

Consensus sequence for prolipoprotein modification/processing enzymes[a]

Frequencies of amino acid residues at position

−5	−4	−3	−2	−1	+1	+2	+3	+4	+5
Leu(0.34)	Leu(0.34)	Leu(0.81)	Ala(0.36)	Gly(0.55)	Cys(1.00)	Ser(0.31)	Ser(0.28)	Asn(0.32)	Ser(0.23)
Ile(0.11)	Ala(0.18)	Ile(0.05)	Ser(0.30)	Ala(0.38)		Gly(0.19)	Ala(0.16)	Ser(0.17)	Asn(0.11)
Val(0.11)	Val(0.14)	Val(0.04)	Ile(0.12)	Ser(0.07)		Gln(0.13)	Asn(0.15)	Lys(0.08)	Gly(0.10)
Ala(0.11)	Thr(0.09)	Phe(0.03)	Thr(0.08)			Ala(0.11)	Gln(0.13)	Gly(0.07)	Ala(0.10)
Gly(0.09)	Ile(0.07)	Ala(0.03)	Val(0.07)			Asn(0.07)	Gly(0.07)	Met(0.07)	Thr(0.08)
Met(0.08)	Met(0.07)	Ser(0.03)	Gly(0.03)			Asp(0.05)	Val(0.04)	Glu(0.05)	Asp(0.07)
Phe(0.07)	Phe(0.05)	Thr(0.01)	Cys(0.01)			Lys(0.05)	Glu(0.04)	Asp(0.05)	Tyr(0.07)
Thr(0.04)	Gly(0.03)		Gln(0.01)			Thr(0.04)	Leu(0.03)	His(0.04)	Gln(0.05)
Ser(0.04)	Ser(0.03)		Met(0.01)			Leu(0.01)	Phe(0.03)	Thr(0.04)	Lys(0.05)
Asn(0.01)	His(0.01)					Val(0.01)	His(0.03)	Leu(0.03)	Leu(0.04)
						Phe(0.01)	Ile(0.01)	Gln(0.03)	Glu(0.04)
						Glu(0.01)	Lys(0.01)	Pro(0.03)	Pro(0.03)
							Asp(0.01)	Arg(0.01)	Val(0.01)
							Trp(0.01)	Trp(0.01)	Arg(0.01)
									His(0.01)

[a]The frequencies of amino acid residues at positions −5 to −5 in 75 lipoprotein precursors was calculated. No special consideration was given to the relative abundance of each amino acid in an averaged amino acid composition of proteins in a particular bacterial species. Nor was adjustment made for the number of homologous proteins used in the data bank; for example, a number of murein lipoproteins or colicin release proteins were treated as distinct proteins. The results are therefore somewhat biased in favor of more frequent representations of a particular lipoprotein or its homolog in the data bank.

labeling of cloned gene products with [³H]palmitate or [2-³H]glycerol and/or accumulation of the precursor form in globomycin-treated cells [64]. The high success rate of predicting the lipoprotein nature of new gene products based on deduced amino acids of cloned genes illustrates the usefulness of this approach in identifying new lipoproteins in bacteria.

Inspection of the amino acid sequences defining the modification and processing site of lipid-modified precursor proteins led to the following conclusions. The Cys residue at +1 contains the sulfhydryl group to which the diglyceride moiety is attached, and therefore by definition is invariant. Leu is favored at the −3 position in 81% of lipoprotein precursors; even Ile and Val are found at as low frequencies as those of Phe, Ala, Ser and Thr, each about 1–3%. Ala and Ser together account for 66% of −2 residues, with Gly at 3% only. In contrast, Gly and Ala account for 93% of all lipoprotein signal sequences at the −1 position with Ser at 7%. The first modification enzyme in this pathway seems to strongly prefer Leu at −3, Ala or Ser with a CH_3 or $OHCH_2$ side chain at −2, and Gly and Ala lacking OH substitution in the side chain at the −1 position. Hydrophobic amino acids are favored at −5 and −4, and no charged amino acids (with possible exception of His at −4 in one case) are found at −5 to +1.

The distribution of amino acids at +2 to +5 positions is more random. There appears to be an enrichment of amino acids (Asn, Gly and Ser) favoring the formation of β-turn. This is consistent with the requirement of a β-turn secondary structure in the major outer membrane prolipoprotein for the modification reactions [65]. Another consideration of the amino acid residues at the +2 to +5 region of the precursor proteins is the topogenic information at the N-terminus of the mature lipoproteins in the determination of the subcellular localization of the lipoproteins. Ser and Asp at +2 positions appear to determine the localization of the lipoproteins to the outer and inner membranes, respectively [66]. A relative abundance of outer membrane lipoproteins over inner membrane lipoproteins in the data used for the statistical analysis would contribute to a bias of Ser at +2 position instead of Asp. Inasmuch as the proposed biosynthetic pathway for bacterial lipoproteins stipulates that the modification of prolipoproteins with glycerol is the first reaction in this pathway, the consensus sequence of Leu-Ala/Ser-Gly/Ala-Cys defines the sequence recognized by the putative prolipoprotein glyceryl transferase. An assessment of the structural requirements for the succeeding enzymes in the pathway (O-acyl transferase or prolipoprotein signal peptidase) must be made with independent assays of the respective enzymes in vivo or in vitro.

A second approach to defining the structure and function relationship between the substrate (unmodified prolipoprotein) and the modification enzymes is the use of site-specific mutagenesis to vary the sequence of the prolipoprotein surrounding the modification and cleavage site [67]. Deletion or substitution of Gly_{20} with Val or Leu affected the modification of murein prolipoprotein, while substitution of Gly_{20} with Thr affected both the modification of prolipoprotein and the processing of lipid-modified prolipoprotein. Two other requirements for an efficient modification reactions were revealed by the studies of *lpp* mutants: the importance of a β-turn secondary structure immediately following the cleavage site (+2 to +5) (15), and the absence of a charged amino acid at the −7 position. Thus, *lpp* mutants with $Ala_{20}Ile_{23}Ile_{24}$ or $Ala_{20}Ile_{23}Lys_{24}$ substitution in prolipoprotein were defective in lipid modification, and these mutant prolipoproteins were predicted to have a reduced β-turn conformation in the junction of the signal sequence and the mature lipoprotein [15]. Substitutions of Gly_{14} with Asp, Glu, Arg and Lys affected to varying extents the modification of prolipoprotein by glyceryl transferase [59,68].

6. Biochemistry of prolipoprotein modification and processing

The biosynthetic pathway of lipoprotein was postulated based on in vivo and in vitro studies (Fig. 3) [69]. In vitro studies were initiated with the preparation of radiochemically pure unmodified prolipoprotein as the starting material which was converted by a solubilized cell envelope fraction to form mature lipoprotein. Subsequently, direct assays of individual reactions were established for some but not all of the multiple enzymes involved in this pathway.

The processing of lipid-modified prolipoprotein by a lipoprotein specific signal peptidase (also called signal peptidase II) is the best understood step in the overall pathway of lipoprotein maturation. The genes for this enzyme (lsp) have been cloned and sequenced from four different bacterial species [70–73]. The deduced amino acid sequences of these four SPase IIs exhibit a similar hydropathy profile suggestive of a similar membrane topology of this cytoplasmic membrane enzyme, with four transmembrane domains connected by two periplasmic and two cytoplasmic loops, and the positively charged N- and C-terminal portions of this enzyme remain in the cytoplasm, consistent with the inside-positive rule of integral membrane proteins [74]. Support for this computer-generated model was obtained from studies using SPase II-PhoA and SPase II-LacZ fusion proteins in E. coli [75]. Comparison of the primary structures of this enzyme from four distinct bacteria has revealed three highly conserved domains in polypeptides of similar sizes but low overall sequence homology [73]. The current thinking is that the basic features of the common topology of this enzyme in the membrane would allow the limited highly conserved regions at the periplasmic/cytoplasmic membrane interface to participate in the unique catalysis of proteolytic processing of lipid-modified prolipoproteins by SPase II.

The SPase II from E. coli has been purified [76]. It does not require phospholipids or metal ions for its activity, and is not inhibited by many of the common protease inhibitors [77]. Globomycin inhibits this enzyme non-competitively with a K_i of 36 nM, and the K_m for lipid-modified murein prolipoprotein is about $6\,\mu M$ [28]. While the E. coli enzyme is inhibited by sulfhydryl reagents and requires DTT for its stability [77], neither the enzyme from Pseudomonas fluorescens nor that from Staphylococcus aureus contains cysteine [72,73].

The processing of lipid-modified prolipoprotein by SPase II results in the formation of apolipoprotein lacking the amide-linked fatty acid, predominantly palmitate, in murein lipoprotein. This is followed by a transacylation reaction catalyzed by the apolipoprotein N-acyl transferase. An in vitro assay of this enzyme was developed based on the differential thermostability of SPase II and N-acyl transferase [79]. Radiochemically pure apolipoprotein was prepared by immunoprecipitation after incubation of membranes isolated from globomycin-treated cells at 80°C. Further incubation of apolipoprotein with a detergent solubilized cell envelope preparation resulted in the conversion of apolipoprotein to

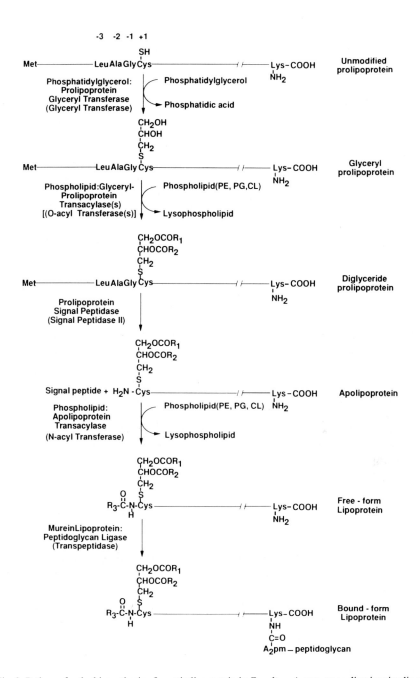

Fig. 3. Pathway for the biosynthesis of murein lipoprotein in *E. coli*. *m*-A$_2$pm, *meso*-diaminopimelic acid.

mature lipoprotein [80]. This enzyme has been located to the cytoplasmic membrane of *E. coli*, and requires neutral detergent for its activity. Using delipidated membranes as the enzyme source, it was shown that any of the major glycerophosphatides (PE, PG or CL) was active as the fatty acyl donor for this reaction [80]. This is consistent with the in vivo studies using fusion of [³H]palmitate-labeled phospholipid vesicles with intact cells as well as those using an *E. coli* mutant containing negligible amount of PE due to a null mutation in the gene encoding phosphatidylserine synthetase (*pss::kan*) [81]. It would appear that this enzyme catalyzes a transacylation reaction with a broad substrate specificity of phospholipids as the fatty acyl donor.

The first committed step in the lipoprotein biosynthetic pathway remains the least characterized reaction. This enzyme was first studied using an in vitro system which converts the unmodified prolipoprotein through successive modification and processing reactions to form mature lipoprotein based on SDS-PAGE analysis of the reaction products [69]. Recently, another in vitro system analogous to a prolipoprotein translocation assay, was developed using unmodified prolipoprotein synthesized in a cell-free transcription/translation coupled system [64,82]. Incubation of the nascent [³⁵S]methionine-labeled prolipoprotein with inverted membrane vesicles from *E. coli*, *B. subtilis*, and *Streptomyces* resulted in the conversion of unmodified prolipoprotein to fully mature lipoprotein [64,82; Hayashi and Wu, unpublished data]. The reaction products in fact contained various intermediates in the maturation pathway, including unmodified prolipoprotein, glyceride-modified prolipoprotein, apolipoprotein, and the mature lipoprotein, based on Tricine-SDS-PAGE analysis [64,83]. Using globomycin or vesicles prepared from an *E. coli* mutant defective in SPase II activity, the modification of prolipoprotein to form lipid-modified prolipoprotein has been separated from the subsequent processing and N-acylation reactions. However, separation of glyceryl prolipoprotein from unmodified prolipoprotein has yet to be achieved by any electrophoretic system, and current studies are limited to a coupled assay of the first two reactions, glyceryl transfer and O-acyl transfer. Further progress would require the development of simple and direct assays for each of these two enzyme activities.

7. Genetics of lipoprotein biosynthesis

The identification of the structural gene for SPase II (*lsp*) was relatively easy and straightforward due to the availability of a selective agent (globomycin) and a conditionally lethal mutant defective in the activity of this enzyme [84,85]. This gene was mapped to 0.5 min of the *E. coli* chromosome, and was part of a mixed-functions operon x-*ileS*-*lsp*-*orf149*-*orf316* [86,87]. The same genomic organization of *lsp* and its flanking genes was found in *Enterobacter aerogenes* and *Pseudomonas fluorescens* but not in *S. aureus* [71–73]. The physiological significance of the co-transcription of the *lsp* gene with *ileS* and other genes of unknown functions in Gram-negative bacteria remains unknown.

Little is known of the catalytic mechanism of this unique enzyme which requires lipid-modified prolipoprotein as the substrate. We have recently determined the amino acid substitution (Asp_{23} to Gly_{23}) in the *lsp* mutant mentioned above [88; Dai and Wu, unpublished data]. The mutated amino acid (Asp) is totally conserved in all four SPase IIs sequenced so far, and is located at the N-terminal region of the first periplasmic loop of this enzyme. The C-terminal cytoplasmic loops of SPase IIs contain multiple Lys residues which may serve as anchors for this enzyme to achieve its unique membrane topology in the cytoplasmic membrane. The importance of the C-terminal tail of this enzyme is supported by the observation that mutations affecting this region significantly reduced the activity of this enzyme [70].

Until recently, repeated attempts to isolate *E. coli* mutants defective in lipoprotein biosynthesis have been unsuccessful, either by a selection scheme based on isolation of suppressor mutations of signal sequence mutations or by brute-force screening of random mutant collections. However, two new mutants of *Salmonella typhimurium* have been isolated and partially characterized. One mutant was found to accumulate unmodified prolipoprotein at the non-permissive temperature, and in vitro assay confirmed that this *ts* mutation affected the combined activity of prolipoprotein glyceryl transferase and O-acyl transferases. Mapping, cloning and DNA sequence determinations have located the mutation to the 61 min of the *E. coli/S. typhimurium* map, immediately adjacent and 5' to the *thyA* gene [89]. The possibility that this *ts* mutation in *S. typhimurium* is allelic to the *umpA* gene in *E. coli* is confirmed by complementation analysis and by biochemical studies of three independent *umpA* alleles [90]. Accumulation of unmodified prolipoprotein was observed in *umpA* mutants under the non-permissive temperature. Furthermore, mutant cells transformed with the wild-type allele from *E. coli* or *S. typhimurium* restored the combined activity of prolipoprotein glyceryl and O-acyl transferase in vitro, and membrane vesicles from wild-type *E. coli* cells harboring the cloned *umpA*$^+$ gene exhibited higher activity. These results strongly suggest that the *umpA* gene corresponds to the structural gene for prolipoprotein glyceryl transferase. The second mutant was found to be defective in apolipoprotein N-acylation both in vivo and in vitro at the non-permissive temperature. Mapping and complementation test with cloned genes have led to the conclusion that this *ts* mutation in *S. typhimurium* is allelic to *cutE* of *E. coli*, located at 15 min of the *E. coli/S. typhimurium* map. This mutant was both temperature-sensitive and copper-sensitive (at the permissive temperature) for growth, a phenotype shared by the *cutE* mutant of *E. coli* [91,92]. The relationship of copper sensitivity and apolipoprotein N-acylation remains to be ascertained. It is worth noting that in each of these two mutants recently identified, the mutations affecting lipoprotein maturation are conditionally lethal. The biochemical basis for the temperature-sensitive phenotypes in both mutants remains to be elucidated.

The isolation of null mutations in structural genes encoding biosynthetic enzymes for phospholipids in *E. coli* has also contributed to a better understanding of lipoprotein maturation in bacteria. As mentioned in a previous section, a null mutation in the gene encoding phosphatidylserine synthetase, a key enzyme for the biosynthesis of major phospholipid PE, does not affect the maturation of lipoprotein either in vivo or in vitro [81]. This observation indicates that PE is not essential for the biosynthesis of lipoprotein either as a substrate (acyl donor) or for the activity of the membrane enzymes involved in this pathway. On the other hand, a null mutation in the structural gene encoding phosphatidylglycerol phosphate synthetase, a key enzyme in the biosynthesis of acidic phospholipids PG or CL, is not compatible with the maturation of murein lipoprotein [93–95]. The null mutation in *pgsA* is lethal in *E. coli* strains synthesizing normal lipoprotein, but is not lethal in *E. coli* mutants lacking murein lipoprotein or synthesizing non-modifiable mutant prolipoprotein due to mutations in the *lpp* gene. This observation is consistent with the proposed biosynthetic pathway of lipoprotein which requires PG as the glyceryl donor for lipoprotein modification. Inasmuch as murein lipoprotein represents the most abundant membrane lipoprotein in *E. coli*, mutations affecting lipid modification of murein prolipoprotein have a sparing effect on the limited supply of acidic phospholipids in *pgsA* mutants of *E. coli*. These observations suggest that *E. coli* cells require a minute amount of acidic phospholipids either for DNA replication, protein export or other essential cellular processes, and this requirement will be kept at a minimum in the absence of lipid modification of the major outer membrane (murein) lipoprotein.

8. Export of bacterial lipoproteins

A common feature of most exported proteins in bacteria is the presence of a transient signal sequence at the N-termini of their precursor forms. A canonical signal sequence contains three distinct domains, the N-terminal positively charged and hydrophilic domain (n), the hydrophobic segment devoid of charged amino acid residues (h), and the C-terminal region defining the cleavage site (c) [96]. The signal sequences of lipoprotein precursors have similar substructures as those in non-lipoprotein precursors, and the distinction between these two major groups of exported proteins lies in the c domain, with -Ala-X-Ala- defining the cleavage site recognized by signal peptidase I and -Leu-Ala/Ser-Gly/Ala-Cys- defining the modification site by prolipoprotein glyceryl transferase. The common structural features of the remaining part of the signal sequences suggest the possibility that these two groups of precursor proteins share a common export machinery prior to their divergence in the post-translocation modification/processing reactions. In vivo studies using conditionally lethal mutants of *E. coli* defective in protein export in general indicate that the export of murein lipoprotein requires functional SecA, SecE, SecY, SecD and SecF proteins [97,98]. Since each of these mutants was found to accumulate unmodified prolipoprotein at the non-permissive temperature, the translocation of unmodified prolipoprotein from the cytoplasm to the inner membrane catalyzed by the Sec proteins takes place prior to the encounter of the precursor proteins with the lipoprotein-specific modification and processing enzymes. These results indicate a common export machinery for both lipoprotein and non-lipoprotein precursors, and the divergence of the pathway is a late event in the export process. In fact, mutant forms of lipoprotein precursors may be exported to the outer membrane without modification or processing [59], or processed anomalously at alternate processing sites by SPase I [99–101], suggesting that export and translocation to the outer membrane does not require processing, and the common export pathway is not compartmentalized and is accessible to both SPase I and prolipoprotein modification and processing enzymes. The export of murein lipoprotein does not require functional SecB protein. On the other hand, other lipoprotein precursors may require SecB protein for their export [102].

9. Covalent assembly of murein lipoprotein to the peptidoglycan

Despite the discovery of covalent attachment of lipoprotein to murein in 1969, little progress has been made regarding the biochemistry and genetics of this process. The lack of significant progress is partly due to a lack of selectable phenotype of mutants lacking murein-bound lipoprotein and absence of a simple and sensitive assay for the putative lipoprotein peptidoglycan ligase.

Using site-directed mutagenesis to generate mutations in the *lpp* gene of *E. coli*, recent studies have demonstrated the broad substrate specificity of this enzyme [68,103a,b]. Neither lipid modification nor processing of prolipoprotein is essential for the attachment of lipoprotein molecules to the peptidoglycan [68]. Nor does an internal deletion of 21 amino acid residues in lipoprotein affect this reaction [94]. It is not surprising that the internal sequence is not essential for the formation of the murein-bound lipoprotein. Analysis of the amino acid sequences of prolipoproteins from *E. coli* and four other Enterobacteriaceae species including *Proteus mirabilis, Morganella morganii, Erwinia amylovora* and *Serratia marcescens* reveals a low degree of sequence homology (32%) in the internal sequence (Ser31-Asp61) as compared to highly conserved N- and C-termini (around 70%) [94,103]. Two kinds of mutations in the N-terminal region of prolipoprotein affected the formation of murein-bound lipoprotein: those resulting in the replacement of Gly_{14} with charged amino acid residues at the 14th position of prolipoprotein signal sequence, and those reducing the β-turn secondary structures at the junction of the signal sequence and the mature portion of lipoprotein [68]. On the other hand, certain structural features at the C-terminal region of lipoprotein are required for this reaction [103b]. Mutations resulting in an alteration of the predicted secondary structure from a random coil to a β-turn at the C-terminal region of the lipoprotein reduced the formation of murein-bound lipoprotein. In addition, a net charge of +2 at the C-terminus of the lipoprotein and the C-terminal sequence of -Tyr-Arg(or Lys)-Lys- are important for the recognition of the lipoprotein substrate by the putative lipoprotein peptidoglycan ligase [103a,b].

References

1. Braun, V. and Bosch, V. (1972) Proc. Natl. Acad. Sci. USA 69, 970–974.
2. Braun, V. and Hantke, K. (1974) Annu. Rev. Biochem. 43, 89–121.
3. Olson, E.N. (1988) Prog. Lipid Res. 27, 177–197.
4. Weidel, W. and Pelzer, H. (1964) Adv. Enzymol. 26, 193–232.
5. Braun, V. and Rehn, K. (1969) Eur. J. Biochem. 10, 426–438.
6. Wiesmüller, K.-H., Bessler, W. and Jung, G. (1983) Hoppe-Seyler's Z. Physiol. Chem. 364, 593–606.
7. Braun, V. and Bosch, V. (1972) Eur. J. Biochem. 28, 51–69.
8. Braun, V., Rotering, H., Ohms, J.-P. and Hagenmaier, H. (1976) Eur. J. Biochem. 70, 601–610.
9. Lee, N., Cheng, E. and Inouye, M. (1977) Biochim. Biophys. Acta 465, 650–656.
10. Manstein, D.J., Berriman, J., Leonard, K. and Rosenbusch, J.P. (1986) J. Mol. Biol. 189, 701–707.
11. Choi, D.-S., Yamada, H., Mizuno, T. and Mizushima, S. (1986) J. Biol. Chem. 261, 8953–8957.
12. Bosch, V. and Braun, V. (1973) FEBS Lett. 34, 307–310.
13. Inouye, M., Shaw, J. and Shen, C. (1972) J. Biol. Chem. 247, 8154–8159.
14. Choi, D.-S., Yamada, H., Mizuno, T. and Mizushima, S. (1987) J. Biochem. 102, 975–983.

338

15. Hiemstra, H., Nanninga, N., Woldringh, C.L., Inouye, M. and Witholt, B. (1987) J. Bacteriol. 169, 5434–5444.
16. Weyer, K.A., Schäfer, W., Lottspeich, F. and Michel, H. (1987) Biochemistry 26, 2909–2914.
17. Ching, G. and Inouye, M. (1985) J. Mol. Biol. 185, 501–507.
18. Gmeiner, J., Kroll, H.-P. and Martin, H.H. (1978) Eur. J. Biochem. 83, 227–233.
19. Mizuno, T. and Kageyama, M. (1979) J. Biochem. 85, 115–122.
20. Cornelis, P., Bouia, A., Belarbi, A., Guyonvarch, A., Kammerer, B., Hannaert, V. and Hubert, J.C. (1989) Mol. Microbiol. 3, 421–428.
21. Duchene, M., Barron, C. Schweizer, A. von Specht, B.-U. and Domedy, H. (1989) J. Bacteriol. 171, 4130–4137.
22. Mizuno, T. (1981) J. Biochem. 89, 1039–1049; 1051–1058; 1059–1066.
23. Chen, R. and Henning, U. (1987) Eur. J. Biochem. 163, 73–77.
24. Deal, C.D. and Kaplan, S. (1983) J. Biol. Chem. 258, 6530–6536.
25. Issartel, J.-P., Koronakis, V. and Hughes, C. (1991) Nature 351, 759–761.
26. Nyström, S., Wallbrandt, P. and Wieslander, A. (1992) Eur. J. Biochem. 204, 231–240.
27. Rosengarten, R. and Wise, K.S. (1991) J. Bacteriol. 173, 4782–4793.
28. Suzuki, H., Nishimura, Y., Yasuda, S., Nishimura, A., Yamada, M. and Hirota, Y. (1978) Mol. Gen. Genet. 167, 1–9.
29. Yem, D.W. and Wu, H.C. (1978) J. Bacteriol. 133, 1419–1426.
30. Fung, J., MacAlister, T.J. and Rothfield, L.I. (1978) J. Bacteriol. 133, 1467–1471.
31. Sonntag, I., Schwarz, H., Hirota, Y. and Henning, U. (1978) J. Bacteriol. 136, 280–285.
32. Gilson, E., Alloing, G., Schmidt, T., Claverys, J.-P, Dudler, R. and Hofnung, M. (1988) EMBO J. 7, 3971–3974.
33. Perego, M., Higgins, C.F., Pearce, S.R., Gallagher, M.P. and Hoch, J.A. (1991) Mol. Microbiol. 5, 173–185.
34. Schneider, R. and Hantke, K. (1993) Mol. Microbiol. 8, 111–121.
35. Lai, J.-S., Sarvas, M., Brammar, W.J., Neugebauer, K. and Wu, H.C. (1981) Proc. Natl. Acad. Sci. USA 78, 3506–3510.
36. Nielsen, J.B.K. and Lampen, J.O. (1982) J. Biol. Chem. 257, 4490–4495.
37. Kato, C., Nakano, Y. and Horikoshi, K. (1989) Arch. Microbiol. 151, 91–94.
38. McLaughlin, J., Murray, C.L. and Rabinowitz, J. (1981) J. Biol. Chem. 256, 11283–11291.
39. Pugsley, A.P., Chapon, C. and Schwartz, M. (1986) J. Bacteriol. 166, 1083–1088.
40. Kornacker, M.G., Boyd, A., Pugsley, A.P. and Plastow, G.S. (1989) Mol. Microbiol. 3, 497–503.
41. D'Enfert, C. and Pugsley, A.P. (1989) J. Bacteriol. 171, 3673–3679.
42. Hu, N.-T., Hung, M.-N., Chiou, S.-J., Tang, F., Chiang, D.-C., Huang, H.-Y. and Wu, C.-Y. (1992) J. Bacteriol. 174, 2679–2687.
43. Allaoui, A., Sansonetti, P.J. and Parsot, C. (1992) J. Bacteriol. 174, 7661–7669.
44. Michiels, T., Vanooteghem, J.-C., Lambert DE Rouvroit, C., China, B., Gustin, A., Boudry, P. and Cornelis, G.R. (1991) J. Bacteriol. 173, 4994–5009.
45. Kontinen, V.P., Saris, P. and Sarvas, M. (1991) Mol. Microbiol. 5, 1273–1283.
46. Haandrikman, A.J., Kok, J. and Venema, G. (1991) J. Bacteriol. 173, 4517–4525.
47. Cavard, D. and Oudega, B. (1992) in: R. James, C. Lazdunski and F. Pattus (Eds.), Bacteriocins, Microcins and Lantibiotics, NATO ASI Series H: Cell Biology, Vol. 65, Springer-Verlag, Berlin, pp. 297–305.
48. Jenkinson, H.F. (1992) Infect. Immun. 60, 1225–1228.
49. Melchers, F., Braun, V. and Galanos, C. (1975) J. Exp. Med. 142, 473–482.
50. Deres, K., Schild, H., Wiesmüller, K.-H., Jung, G. and Rammensee, H.-G. (1989) Nature 342, 561–564.
51. Erdile, L.F., Brandt, M.-A., Warakomski, D.J., Westrack, G.J., Sadziene, A., Barbour, A.G. and Mays, J.P. (1993) Infect. Immun. 61, 81–90.
52. Hantke, K. and Braun, V. (1973) Eur. J. Biochem. 34, 284–296.
53. Thomas, P.J. and Law, J.H. (1966) J. Biol. Chem. 241, 5013–5018.
54. Chattopadhyay, P.K. and Wu, H.C. (1977) Proc. Natl. Acad. Sci. USA 74, 5318–5322.
55. Schulman, H. and Kennedy, E.P. (1977) J. Biol. Chem. 252, 4250–4255.
56. Chattopadhyay, P.K., Lai, J-S. and Wu, H.C. (1979) J. Bacteriol. 137, 309–312.

57. Lai, J-S., Philbrick, W.M. and Wu, H.C. (1980) J. Biol. Chem. 255, 5384–5387.
58. Lai, J-S. and Wu, H.C. (1980) J. Bacteriol. 144, 451–453.
59. Inouye, S., Wang, S., Sekizawa, J., Halegoua, S. and Inouye, M. (1977) Proc. Natl. Acad. Sci. USA 74, 1004–1008.
60. Lin, J.J.C., Kanazawa, H., Ozols, H. and Wu, H.C. (1978) Proc. Natl. Acad. Sci. USA 75, 4891–4895.
61. Inukai, M., Takeuchi, M., Shimizu, K. and Arai, M. (1978) J. Antibiot. (Tokyo) 31, 1203–1205.
62. Hussain, M., Ichihara, S. and Mizushima, S. (1980) J. Biol. Chem. 255, 3707–3712.
63. Ichihara, S., Hussain, M. and Mizushima, S. (1981). J. Biol. Chem. 256, 3126–3129.
64. Hayashi, S. and Wu, H.C. (1992) in: N.M. Hooper and A.J. Turner (Eds.), Lipid Modification of Proteins: A Practical Approach, Oxford University Press, Oxford, UK, pp. 261–285.
65. Inouye, S., Duffaud, G. and Inouye, M. (1986) J. Biol. Chem. 261, 10970–10975.
66. Yamaguchi, K., Yu, F. and Inouye, M. (1988) Cell 53, 423–432.
67. Gennity, J., Goldstein, J. and Inouye, M. (1990). J. Bioenerg. Biomembr. 22, 233–269.
68. Zhang, W-Y., Inouye, M. and Wu, H.C. (1992) J. Biol. Chem. 267, 19631–19635.
69. Tokunaga, M., Tokunaga, H. and Wu, H.C. (1982) Proc. Natl. Acad. Sci. USA 79, 2255–2259.
70. Innis, M.A., Tokunaga, M., Williams, M.E., Loranger, J.M., Chang, S.Y., Chang, S. and Wu, H.C. (1984) Proc. Natl. Acad. Sci. USA 81, 3708–3712.
71. Isaki, L., Kawakami, M., Beers, R., Hom, R. and Wu, H.C. (1990) J. Bacteriol. 172, 469–472.
72. Isaki, L., Beers, R. and Wu, H.C. (1990) J. Bacteriol. 172, 6512–6517.
73. Zhao, X-J. and Wu, H.C.(1992) FEBS Lett. 299, 80–84.
74. von Heijne, G. (1986) EMBO J. 5, 3021–3027.
75. Munoa, F.J., Miller, K.W., Beers, R., Graham, M. and Wu, H.C. (1991) J. Biol. Chem. 266, 17667–17672.
76. Dev, I.K. and Ray, P.H. (1984) J. Biol. Chem. 259, 11114–11120.
77. Dev, K.I. and Ray, P.H. (1990) J. Bioenerg. Biomembr. 22, 271–290.
78. Dev, I.K., Harvey, R.J. and Ray, P.H. (1985) J. Biol. Chem. 260, 5891–5894.
79. Hussain, M., Ichihara, S. and Mizushima, S. (1982) J. Biol. Chem. 257, 5177–5182.
80. Gupta, S.D. and Wu, H.C. (1991) FEMS Microbiol. Lett. 78, 37–41.
81. Gupta, S.D., Dowhan, W. and Wu, H.C. (1991) J. Biol. Chem. 266, 9983–9985.
82. Tian, G., Wu, H.C., Ray, P.H. and Tai, P.C. (1989) J. Bacteriol., 171, 1987–1997.
83. Schägger, H. and von Jagow, G. (1987) Anal. Biochem. 166, 368–379.
84. Tokunaga, M., Loranger, J.M. and Wu, H.C. (1983) J. Biol. Chem. 258, 12102–12105.
85. Yamagata, H., Daishima, K. and Mizushima, S. (1983) FEBS Lett. 158, 301–304.
86. Regue, M., Remenick, J., Tokunaga, M., Mackie, G.A. and Wu, H.C. (1984) J. Bacteriol. 158, 632–635.
87. Miller, K.W., Bovier, J., Stragier, P. and Wu, H.C. (1987) J. Biol. Chem. 262, 7391–7397.
88. Yamagata, H., Ippolito, C., Inukai, M. and Inouye, M. (1982) J. Bacteriol. 152, 1163–1168.
89. Gan, K., Gupta, S.D., Sankaran, K., Schmid, M.B. and Wu, H.C. (1993) J. Biol. Chem. 268, 16544–16550
90. Williams, M.G., Fortson, M., Dykstra, C.C., Jensen, P. and Kushner, S.R. (1989) J. Bacteriol. 171, 565–568.
91. Gupta, S.D., Gan, K., Schmid, M.B. and Wu, H.C. (1993) J. Biol. Chem. 268, 16551–16556.
92. Rogers, S.D., Bhave, M.R., Mercer, J.F.B., Camakaris, J. and Lee, B.T.O. (1991) J. Bacteriol. 173, 6742–6748.
93. Asai, Y., Katayosa, Y., Hikita, C., Onta, A. and Shibuya, I. (1989) J. Bacteriol. 171, 6867–6869.
94. Zhang, W-Y., Dai, R-M. and Wu, H.C. (1992) FEBS Lett. 311, 311–314.
95. Heacock, P.N. and Dowhan, W. (1989) J. Biol. Chem. 264, 114972–14977.
96. von Heijne, G. (1985) J. Mol. Biol. 184, 99–105.
97. Watanabe, T., Hayashi, S. and Wu, H. C. (1988) J. Bacteriol., 170, 4001–4007.
98. Sugai, M., and Wu, H.C. (1992) J. Bacteriol. 174, 2511–2516.
99. Hayashi, S., Chang, S.Y., Chang, S. and Wu, H.C. (1984) J. Biol. Chem. 259, 10448–10454.
100. Ghrayeb, J., Lunn, C.A., Inouye, S. and Inouye, M. (1985) J. Biol. Chem. 260, 10961–10965.
101. Hayashi, S., Chang, S.Y., Chang, S. and Wu, H.C. (1986) J. Bacteriol. 165, 678–681.
102. Pugsley, A.P., Kornacker, M.G. and Poquet, I. (1991) Mol. Microbiol. 5, 343–352.

340

103. (a) Yu, F. (1987) in: M. Inouye (Ed.), Bacterial Outer Membranes as Model Systems, Wiley, New York, pp. 419–432. (b) Zhang, W-Y. and Wu, H.C. (1992) J. Biol. Chem. 267, 19560–19564.
104. Braun, V. (1975) Biochim. Biophys. Acta 415, 335–377.
105. Doherty, H., Yamada, H. Caffrey, P. and Owen, P. (1986) J. Bacteriol. 166, 1072–1082.
106. Takase, I., Ishino, F., Wachi, M., Kamata, H., Doi, M., Asoh, S., Matsuzawa, H. Ohta, T. and Matsuhashi, M. (1987) J. Bacteriol. 169, 5692–5699.
107. Lazzaroni, J.-C. and Portalier, R. (1992) Mol. Microbiol. 6, 735–742.
108. Yu, F., Inouye, S. and Inouye, M. (1986) J. Biol. Chem. 261, 2284–2288.
109. Bouvier, J., Pugsley, A.P. and Stragier, P. (1991) J. Bacteriol. 173, 5523–5531.
110. Perumal, N.B. and Minkley, Jr., E.G. (1984) J. Biol. Chem. 259, 5357–5360.
111. Harrison, J.L., Taylor, I.M., Platt, K. and O'Connor, C.D. (1992) Mol. Microbiol. 6, 2825–2832.
112. Ogata, R.T., Winters, C. and Levine, R.P. (1982) J. Bacteriol. 151, 819–827.
113. Waleh, N.S. and Johnson, P.H. (1985) Proc. Natl. Acad. Sci. USA 82, 8389–8393.
114. Oka, A., Nomura, N., Morita, M., Sugiaski, H., Sugimoto, K. and Takanami, M. (1979) Mol. Gen. Genet. 172, 151–159.
115. Cole, S.T., Saint-Joanis, B. and Pugsley, A.P. (1985) Mol. Gen. Genet. 198, 465–472.
116. Watson, R.J., Lau, P.C.K., Vernet, T. and Visentin, L.P. (1984) Gene 29, 175–184.
117. Cavard, D., Lloubes, R., Morlon, J., Chartier, M. and Lazdunski, C. (1985) Mol. Gen. Genet. 199, 95–100.
118. Hakkaart, M.J.J., Veltkamp, E. and Nijkamp, H.J.J. (1981) Mol. Gen. Genet. 183, 318–325.
119. Uchimura, T. and Lau, P.C.K. (1987) Mol. Gen. Genet. 209, 489–493.
120. James, R., Jarvis, M. and Barker, D.F. (1987) J. Gen. Microbiol. 133, 1553–1562.
121. Pugsley, A.P. (1988) Mol. Gen. Genet. 211, 335–341.
122. Hayashi, S., Hara, H., Suzuki, H. and Hirota, Y. (1988) J. Bacteriol. 170, 5392–5395.
123. Nakamura, M., Maruyama, I.N., Soma, M., Kato, J., Suzuki, H. and Hirota, Y. (1983) Mol. Gen. Genet. 191, 1–9.
124. Jung, J.U., Gutierrez, C. and Villarejo, M.R. (1989) J. Bacteriol. 171, 511–520.
125. Parsot, C., Taxman, E. and Mekalanos, J.J. (1991) Proc. Natl. Acad. Sci. USA 88, 1641–1645.
126. Barondess, J.J. and Beckwith, J. (1990) Nature 346, 871–874.
127. Klein, J.R., Henrich, B. and Plapp, R. (1991) Mol. Gen. Genet. 230, 230–240.
128. Nakamura, K., Pirtle, R.M. and Inouye, M. (1979) J. Bacteriol. 137, 595–604.
129. Lin, J. J.-C. and Wu, H.C.P. (1976) J. Bacteriol. 125, 892–904.
130. Rhen, M., O'Connor, C.D. and Sukupolvi, S. (1988) FEMS Microbiol. Lett. 52, 145–153.
131. Chapon, C. and Raibaud O. (1985) J. Bacteriol. 164, 639–645.
132. Katsuragi, N., Takizawa, N. and Murooka, Y. (1987) J. Bacteriol. 169, 2301–2306.
133. Yamagata, H., Nakamura, K. and Inouye, M. (1981) J. Biol. Chem. 256, 2194–2198.
134. Nakamura, K. and Inouye, M. (1980) Proc. Natl. Acad. Sci. USA 77, 1369–1373.
135. Huang, Y.-X., Ching, G. and Inouye, M. (1983) J. Biol. Chem. 258, 8139–8145.
136. Katz, E., Loring, D., Inouye, S. and Inouye, M. (1978) J. Bacteriol. 134, 674–676.
137. China, B., Michiels, T. and Cornelis, G.R. (1990) Mol. Microbiol. 4, 1585–1593.
138. Bäumler, A.J. and Hantke, K. (1992) J. Bacteriol. 174, 1029–1035.
139. Kaufman, M.R., Seyer, J.M. and Taylor, R.K. (1991) Genes Dev. 5, 1834–1846.
140. Jannatipour, M., Soto-Gil, R.W. and Zyskind, J.W. (1987) J. Bacteriol. 169, 3785–3791.
141. Weinberg, G.A., Towler, D.A. and Munson, JR., R.S. (1988) J. Bacteriol. 170, 4161–4164.
142. Hanson, M.S. and Hansen, E.J. (1991) Mol. Microbiol. 5, 267–278.
143. Janson, H., Heden, L.-O. and Forsgren, A. (1992) Infect. Immun. 60, 1336–1342.
144. Hanson, M.S., Slaughter, C. and Hansen, E.J. (1992) Infect. Immun. 60, 2257–2266.
145. Deich, R.A., Metcalf, B.J., Finn, C.W., Farley, J.E. and Green, B.A. (1988) J. Bacteriol. 170, 489–498.
146. Theisen, M., Rioux, C.R. and Potter, A.A. (1992) Infect. Immun. 60, 826–831.
147. Huang, J. and Schell, M.A. (1992) J. Bacteriol. 174, 1314–1323.
148. Baumgardner, D., Deal, C. and Kaplan, S. (1980) J. Bacteriol. 143, 265–273.
149. Gomez-Miguel, M.J., Moriyon, I. and Lopez, J. (1987) Infect. Immun. 55, 258–262.
150. Gotschlich, E.C. and Seiff, M.E. (1987) FEMS Microbiol. Lett. 43, 253–255.
151. Hoehn, G.T. and Clark, V.L. (1992) Infect. Immun. 60, 4704–4708.

152. Woods, J.P., Spinola, S.M., Strobel, S.M. and Cannon, J.G. (1989) Mol. Microbiol. 3, 43–48.
153. Baehr, W., Gotschlich, E.C. and Hitchcock, P.J. (1989) Mol. Microbiol. 3, 49–55.
154. Chamberlain, N.R., Brandt, M.E., Erwin, A.L., Radolf, J.D. and Norgard, M.V. (1989) Infect. Immun. 57, 2872–2877.
155. Weigel, L.M., Brandt, M.E. and Norgard, M.V. (1992) Infect. Immun. 60, 1568–1576.
156. Hubbard, C.L., Gherardini, F.C., Bassford, JR., P.J. and Stamm, L.V. (1991) Infect. Immun. 59, 1521–1528.
157. Schouls, L.M., Van Der Heide H.G.J. and Embden, J.D. (1991) Infect. Immun. 59, 3536–3546.
158. Yelton, D.B., Limberger, R.J., Curci, K., Malinosky-Rummell, F., Slivienski, L., Schouls, L.M., van Embden, J.D. and Charon, N.W. (1991) Infect. Immun. 59, 3685–3693.
159. Hansen, E.B., Pedersen, P.E., Schouls, L.M., Severin, E. and van Embden, J.D.A. (1985) J. Bacteriol. 162, 1227–1237.
160. Swancutt, M.A., Radolf, J.D. and Norgard, M.V. (1990) Infect. Immun. 58, 384-392.
161. Purcell, B.K., Swancutt, M.A. and Radolf, J.D. (1990) Mol. Microbiol. 4, 1371–1379.
162. Bergstroem, S., Bundoc, V. and Barbour, A.G. (1989) Mol. Microbiol. 3, 479–486.
163. Norris, S.J., Carter, C.J., Howell, J.K. and Barbour, A.G. (1992) Infect. Immun. 60, 4662–4672.
164. Katona, L.I., Beck, G. and Habicht, G.S. (1992) Infect. Immun. 60, 4995–5003.
165. Thomas, W., Sellwood, R. and Lysons, R.J. (1992) Infect. Immun. 60, 3111–3116.
166. Sjöstedt, A., Sandström, G., Tärnvik, A. and Jaurin, B.-A. (1990) J. Immunol. 145, 311–317.
167. Nielsen, J.B.K. and Lampen, J.O. (1982) J. Bacteriol. 152, 315–322.
168. Nielsen, J.B.K. and Lampen, J.O. (1983) Biochemistry 22, 4652–4656.
169. Hussain, M., Pastor, F.I.J. and Lampen, J.O. (1987) J. Bacteriol. 169, 579–586.
170. Hemilä H. (1991) FEMS Microbiol. Lett. 82, 37–42.
171. Mathiopoulos, C., Mueller, J.P., Slack, F.J., Murphy, C.G., Patankar, S., Bukusoglu, G. and Sonenshein, A.L. (1991) Mol. Microbiol. 5, 1903–1913.
172. Peters, J., Peters, M., Lottspeich, F., Schäffer, W. and Baumeister, W. (1987) J. Bacteriol. 169, 5216–5223.
173. Welby, M., De Bony, J. and Tocanne, J.-F. (1988) Biochim. Biophys. Acta 943, 190–198.
174. Bricker, T.M., Boyer, M.J., Keith, J., Watson-McKown, R. and Wise, K.S. (1988) Infect. Immun. 56, 295–301.
175. Dahl, C.E., Dahl, J.S. and Bloch, K. (1983) J. Biol. Chem. 258, 11814–11818.
176. Wroblewski, H., Nyström, S., Blanchard, A. and Wieslander, A. (1989) J. Bacteriol. 171, 5039–5047.
177. Nyström, S., Johansson, K.-E. and Wieslander, A. (1986) Eur. J. Biochem. 156, 85–94.
178. Dahl, C.E., Sacktor, N.C. and Dahl, J.S. (1985) J. Bacteriol. 162, 445–447.
179. Thirkell, D., Myles, A.D. and Russel W.C. (1991) Infect. Immun. 59, 781–784.
180. Anderson, B.E., Baumstark, B.R. and Bellini, W.J. (1988) J. Bacteriol. 170, 4493–4500.
181. Heller, K. (1992) personal communication.

J.-M Ghuysen and R. Hakenbeck (Eds.), *Bacterial Cell Wall*

CHAPTER 15

Structure-function relationships in porins as derived from a 1.8 Å resolution crystal structure

GEORG E. SCHULZ

Institut für Organische Chemie und Biochemie der Albert-Ludwigs-Universität, Albertstrasse 21, D-79104 Freiburg im Breisgau, Germany

1. General function and atomic structure

Gram-negative bacteria have a protective outer membrane containing channels that are permeable for nutrients. These channels are formed by porins which are usually homotrimeric proteins with subunit sizes ranging from 30 to 50 kDa and solute exclusion limits around 600 Da [1,2]. The channels are well permeable for polar solutes, but exclude non-polar molecules of comparable sizes. Porins have been subdivided into specific and non-specific porins. For a particular solute, specific porins show a comparatively large diffusion rate at low and saturation effects at high concentrations. In contrast, non-specific porins act like inert holes; their diffusion rates are always proportional to the solute concentration.

Electron microscopy studies resulted in a rather general description of the porin architecture [3,4]. The first porin crystals suitable for an X-ray analysis were obtained from *Escherichia coli* [5]. The first atomic structure was elucidated for the major porin of *Rhodobacter capsulatus* [6–10]. An analysis of other crystals using molecular replacement with the *R. capsulatus* porin structure showed that the 16-stranded β-barrel fold is present in quite a number of porins [11]. Now, three more porin structures are known in atomic detail [12,13]; they confirm the general picture derived from the first one.

The major porin of *R. capsulatus* yielded particularly good crystals after careful protein preparation [6]. As shown by amino acid sequence analysis [7], one subunit consists of 301 amino acids. The crystal structure has been solved at 1.8 Å resolution and refined to a crystallographic R-factor of 18.6% using a 97% complete data set [10]. The model of one subunit contains all polypeptide atoms, 274 water molecules, 3 calcium ions, 3 detergent molecules (octyltetraoxyethylene, C_8E_4) and 1 bound ligand. Since this ligand could not be identified, it was modeled as C_8E_4.

A porin subunit consists of a very large 16-stranded antiparallel β-barrel and 3 short α-helices. All β-strands are connected to their next neighbors. The loops at the bottom end of the barrel are short, this end is smooth. In contrast, the top end of the barrel is rough containing much longer loops. The longest loop has 43 residues and runs into the interior of the barrel where it constricts the pore to a small eyelet [10].

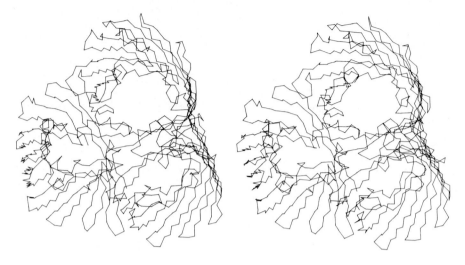

Fig. 1. Stereo view of the C$_\alpha$-backbone of trimeric porin taken from ref. 8. The view is from the external medium.

The C$_\alpha$ backbone chain of the trimeric porin (Fig. 1) shows that the β-barrel height in the trimer center is lower than the barrel height facing the membrane. This center forms a three-pronged star composed of all three subunits resembling the hub and spokes of a wheel. The mobilities of the atoms in this central star are the lowest of the whole molecule as determined by the crystallographic temperature factors [10]. The interior of this star is non-polar; 18 phenylalanines (6 from each subunit) interdigitate tightly, while the surface is polar. The star contains all six chain termini paired in salt bridges. Trimeric porin can thus be described as a rigid central core constructed like a water-soluble protein that is surrounded by three β-barrel walls fencing off the membrane.

2. Position in the membrane

2.1. Molecular orientation

The shape of the channel is illustrated by a cut through the center of the pore (Fig. 2). The aggregation to trimers connects the rear β-barrel walls at the height levels indicated by dashed lines. The central part is low, giving rise to a common channel formed by all three subunits in the upper half. The three pore eyelets limiting the diffusion are located between the barrel wall **1** close to the molecular threefold-axis and the 43-residue loop **3** inside the barrel.

The longitudinal position of porin in the membrane is clearly defined by the non-polar ring with a height of 24 Å that surrounds the trimer and fits the non-polar moiety of the membrane. The rough upper end of the β-barrel (Fig. 1) contains the larger loops with numerous charged side chains, whereas the smooth lower barrel end has only small loops with few polar residues. As shown by binding studies with antibodies and phages, the

345

large polar loops face the external medium [14]. Accordingly, the external medium is at the top of Figs. 1–5.

This orientation agrees well with the outer surface of the trimer (Fig. 3). While the non-polar surface moiety faces the non-polar interior of the lipid bilayer 5, the upper polar part with its numerous ionogenic side chains is most likely connected to the lipopolysaccharides (LPS) forming the external layer 6 of the bacterial outer membrane. Presumably, the numerous Asp and Glu side chains are glued by divalent cations (like calcium and magnesium) to the carboxylate groups of the LPS cores. This network integrates porin efficiently in the tough bacterial protection layer of crosslinked LPS molecules, avoiding a fragile protein–membrane interface.

Furthermore, Fig. 3 shows two girdles of aromatic residues along the upper and lower border lines between polar and non-polar residues. The upper girdle contains mostly tyrosines pointing with their hydroxyl groups to the upper polar moiety of the membrane, while the lower girdle has phenylalanines pointing to the non-polar membrane interior and also tyrosines pointing to the polar part of the periplasmic membrane layer. These patterns are obviously significant. They were also observed in the three recently elucidated porin structures [12,13].

In addition, there are four rare type-II' β-turns at the lower smooth end of the barrel. These four turns point with their peptide amides towards the membrane where they can form hydrogen bonds to the phosphodiesters of the periplasmic layer of the bacterial outer membrane.

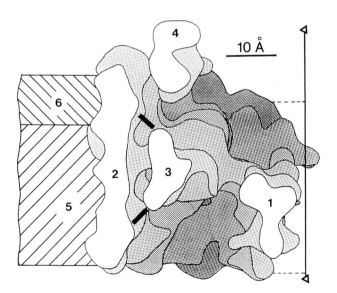

Fig. 2. Shape of a porin subunit represented by a sliced 6 Å resolution density map calculated from the high resolution model [9]. The density level is at 1σ, the distance from the viewer and is indicated by shading. The sectional areas 1, 2, 3, 4, 5 and 6 belong to the β-barrel wall at the trimer interface, to the β-barrel wall facing the membrane, to the pore size-defining the 43-residue loop inside the β-barrel, to a globular density facing the external medium, to the non-polar moiety of the membrane, and to the LPS core, respectively. As indicated by bars, there is no free space between sectional areas 2 and 3.

346

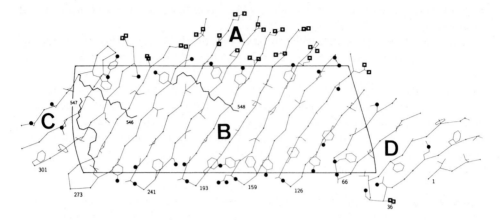

Fig. 3. Projection of the outer surface of the β-barrel onto a cylinder as taken from [10]. All loops at the perip-lasmic side are indicated. Surface areas A, B, C and D contain the ionogenic side chains connecting to the LPS core, the non-polar side chains facing the non-polar interior of the membrane, the left-hand and the right-hand parts of the interface, respectively. Ionogenic atoms are emphasized by squares and polar atoms by circles. The aromatic girdles are obvious.

2.2. Molecular packing in the crystal

The packing scheme (Fig. 4) in the exceptionally well ordered crystals of the *R. capsula-tus* porin [10] resembles the crystal packing of the photoreaction center [15] as the polypeptides form only polar contacts. Apart from the subunit interface, there exists merely one contact type, i.e. between head and tail, giving rise to a trimer arrangement as

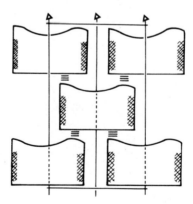

Fig. 4. Crystal packing arrangement of trimeric porin molecules, the molecular threefold axes ($\triangle$) are crystallographic. The space group $R3$ requires only one contact type for building up a three-dimensional network. This contact ($\equiv$) is head to tail and polar. There is no lateral contact between the non-polar surfaces (cross-hatched), which are presumably covered by detergent.

in a cubic closest packing. This contact contains five hydrogen bonds and one salt bridge and is obviously strong. The closest lateral distance between trimers is 15 Å, and it is between non-polar surfaces. Any conceivable contact through detergent molecules should therefore be very weak and should provide only a minor contribution to the crystal packing energy.

The molecular packing in the crystal corresponds to a stack of lipid bilayers containing porins, suggesting that such bilayers can also be formed in the crystal. This hypothesis can be tested by measuring the electric conductivity of the crystal along and perpendicular to the threefold rotation axis. Such measurements can also be done after changing the detergent in the crystal (possibly to lipids) or after crystallizing with other detergents, which would yield information about the detergent(lipid)–protein contact.

2.3. Suggested function of the aromatic girdles

Since the Tyr-Phe pattern (Fig. 3) has been observed in all the four known porin structures [10,12,13], it most likely reflects a function. These bulky aromatic side chains may prevent conformational damage of the protein on mechanical movements in the membrane as indicated in Fig. 5. Any transversal wave in the membrane or any knocking at a porin trimer immersed in the membrane causes rocking movements that would expose non-polar protein surface to a polar membrane layer and polar protein surface to the non-polar membrane interior. Both contact types give rise to a large surface tension which is likely to scramble the polypeptide conformation. Since the aromatic side chains rotate around their C_α–C_β-bonds much faster than the trimer can rock, they are able to shield the respective surfaces against the wrong counterpart (Fig. 5). The strong association between the upper polar protein surface and the LPS core units explains the absence of Phe from the upper aromatic girdle, as the crosslinked LPS core units prevent the non-polar membrane interior from reaching the upper polar protein surface, anyway.

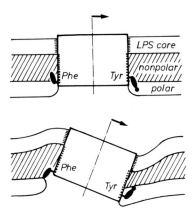

Fig. 5. Sketch describing the suggested shielding role of phenylalanines and tyrosines during relative movements of protein and membrane.

348

2.4. A conceivable folding pathway

The chain fold of the homotrimer with the central rigid star, the prongs of which are connected by the three β-barrel walls (see p. 344), suggests a folding pathway. Initially, the three-pronged star folds like a water soluble protein. Since this star contains all chain termini, the remaining chain parts form three large loops of about 200 amino acid residues each suspended between the three prongs. On contact with the non-polar membrane interior, the three 200-residue loops arrange themselves to the observed most simple β-barrel topology with all strands antiparallel and connected to their next neighbors. This folding process is straightforward as it requires no chain crossing or wrapping. On folding, the non-polar side chains of the β-barrel residues orient themselves to the outside and face the non-polar moiety of the membrane. Subsequently, the large 43-residue loop is inserted into the barrel. This loop supports the barrel at its center against the membrane pressure, and it defines the pore size.

3. Diffusion properties

3.1. The structure of the pore eyelet

The eyelet of the channel is almost exclusively lined by ionogenic groups that segregate into positively and negatively charged rims (Fig. 6). The positive rim is at the eyelet side closer to the trimer center and consists of half a dozen Arg, Lys and His side chains. The

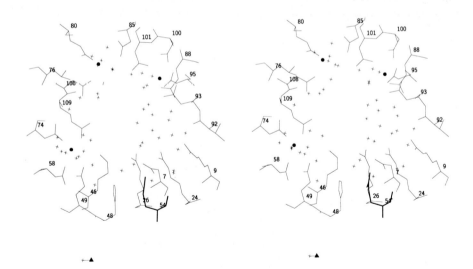

Fig. 6. Stereo view of a pore eyelet as taken from [10]. The pore eyelet is defined by negatively charged side chains at its upper rim and positively charged ones at the lower rim. There exists a strong transversal electric field between these charges as indicated by the rigidity of the neighboring arginines at the lower rim (see text). Water molecules (×) and bound Ca^{2+} (●) are given. The molecular threefold axis is indicated (▲).

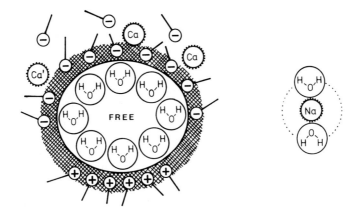

Fig. 7. Sketch of the pore eyelet indicating a ring of bound water molecules that are oriented in the transversal electric field. The rim of negatively charged side chains is strengthened by two strongly (Ca) and one weakly (Ca') bound Ca^{2+}. The depth of the pore eyelet is about 6 Å. For a hydrated ion, always half the hydration shell has an energetically unfavorable orientation in the electric field.

negative rim is further at the circumference of the trimer; it contains about a dozen Asp and Glu residues mostly located in the 43-residue loop that is inserted into the β-barrel. These abundant negative charges are partially compensated by two bound Ca^{2+} which tighten the rim structure appreciably. As a consequence, the removal of these Ca^{2+} should change permeabilities. Actually, it had been observed that the analyzed porin permits the diffusion of ATP only after the bacterium has been treated with EDTA [16].

3.2. The transversal field as an electric separator

The positive and negative rims juxtaposed across the eyelet cause a transversal electric field. The field strength is best estimated from two arginines participating in the positive rim. These arginines are at van der Waals distance in good electron density demonstrating that they are rigidly positioned in spite of their repelling positive charges (the crystals are at pH 7.2, the pK of Arg is 12.4). Such an arrangement is only possible if the charges are fixed by a strong electric field.

An X-ray structure analysis at 1.8 Å resolution allows the assignment of reasonably well fixed water molecules [10]. The eyelet contains quite a number of them (Fig. 6). They orient their dipoles in the electric field (Fig. 7). In the center of the eyelet there are mobile water molecules confined to a rather small cross section of about 4 Å by 5 Å. They form a hole through which a molecule with the cross section of an alkyl chain can permeate.

The diffusion of a small polar solute is illustrated in Fig. 8A. The solute is oriented by the transversal electric field formed by the charged rims of the eyelet and will remain oriented over the whole diffusion distance. The solute is oriented in the pore eyelet like a substrate on an enzyme. This reduces the entropy barrier of the diffusion process in the same way as this barrier is reduced for a chemical reaction on an enzyme surface. Without tumbling, the diffusion of polar solutes is appreciably accelerated.

Although the eyelet cross section that is free of fixed water molecules suffices for the penetration of alkyl chains, these still cannot permeate because they are blocked by the strong transversal field. Field and eyelet behave like a charged capacitor that stores an electric energy proportional to the high dielectric constant of 80 of oriented water molecules (Fig. 8B) multiplied by the volume. Any incoming molecule with a lower dielectric constant (the respective value for an alkyl chain is about 2), i.e. with a smaller dipole than water, causes a decrease in the capacitor energy and therefore generates a force expelling the intruder. Consequently, the transverse field acts as an electric separator, facilitating the permeation of polar solutes while blocking off non-polar ones.

A special situation seems to arise for ions. When diffusing through the pore, half their hydration shell enters in an energetically unfavorable orientation. In the *R. capsulatus* porin, there exists a possible separate pathway for positive ions at one corner of the eyelet. This pathway is outside the electric field and completely lined by carbonyl and carboxyl oxygens. It may explain the observed though low cation selectivity of this porin. Since this channel is blocked by a weakly bound Ca^{2+}, its importance can be tested by measuring cation currents as a function of Ca^{2+} concentration.

3.3. Structural changes caused by a voltage across the membrane

It has been observed that some porins incorporated in black lipid films reduce their electric conductivity on application of a voltage of around 100 mV across this film [2], i.e. there occurs voltage closure. Such behavior can be understood when considering the shape of the pore (Fig. 2). The membrane channel has a rather large cross section over its whole length except for the short distance of about 6 Å at the diffusion-defining eyelet at the center. This implies that any voltage across the membrane (say 100 mV) lies essentially across the eyelet with its small cross section (certainly more than 90 mV). Since the energy gained by moving a single charge (e.g. the protonated ε-amino group of a lysine)

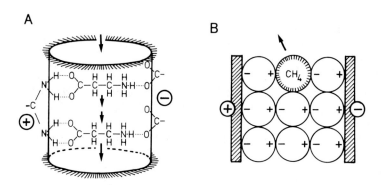

Fig. 8. Diffusion through the pore eyelet. (A) Sketch of a permeating zwitterion β-alanine which remains oriented in the field. Its diffusion is facilitated by entropy reduction. (B) Sketch of the pore eyelet as a capacitor, explaining the energetic exclusion of non-polar solutes from the diffusion through the pore.

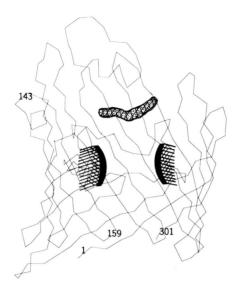

Fig. 9. Sketch of C_α backbone of one subunit with the pore eyelet and the observed solute binding site within the pore close to the eyelet. Note, that the solute binds at the external side of the eyelet.

over 100 mV is 10 kJ/mol and thus equivalent to the energy of a hydrogen bond, the strong longitudinal electric field along the eyelet can tear off any charged group that can be moved over the 6 Å length of the eyelet (which is possible for lysine). The applied voltage thus disrupts the eyelet structure diminishing the permeability.

3.4. Specificity of porins

Porins have been subdivided into two classes, specific and non-specific ones. The structurally known *R. capsulatus* porin has been previously classified as non-specific. In the crystal structure, however, this porin is ligated by a small molecule (Fig. 9). Moreover, efficient binding of tetrapyrrols to this porin has been reported [17]. In the crystal, the bound ligand cannot be identified. The binding site is formed between the 43-residue loop inserted into the barrel (where it forms the eyelet) and a little domain protruding to the external medium. Facing the external medium, this binding site can pick up solutes at very low concentrations. After being bound close to the eyelet (Fig. 9), the solute may subsequently dissociate and diffuse through the pore at a considerable rate. Binding and permeation should follow Michaelis–Menten kinetics and should show saturation effects as observed within the class of specific porins. Accordingly, the *R. capsulatus* porin previously assigned to the class of non-specific porins is actually both specific and non-specific, depending on the solute. Generalizing this observation, it is suggested that many porins belong to both classes, but that the specific solutes have been detected only in few cases.

352

Acknowledgement

I thank A. Kreusch, E. Schiltz, J. Weckesser, M.S. Weiss and W. Welte for their essential contributions to the structure analysis of the R. capsulatus porin.

References

1. Benz, R. and Bauer, K. (1988) Eur. J. Biochem. 176, 1–19.
2. Jap, B.K. and Walian, P.J. (1990) Q. Rev. Biophys. 23, 367–403.
3. Engel, A., Massalski, A., Schindler, H., Dorset, D.L. and Rosenbusch, J.P. (1985) Nature 317, 643–645.
4. Jap, B.K., Walian, P.J. and Gehring, K. (1991) Nature 350, 167–170.
5. Garavito, R.M. and Rosenbusch, J.P. (1980) J. Cell. Biol. 86, 327–329.
6. Kreusch, A., Weiss, M.S., Welte, W., Weckesser, J. and Schulz, G.E. (1991) J. Mol. Biol. 217, 9–10.
7. Schiltz, E., Kreusch, A., Nestel, U. and Schulz, G.E. (1991) Eur. J. Biochem. 199, 587–594.
8. Weiss, M.S., Kreusch, A., Schiltz, E., Nestel, U., Welte, W., Weckesser, J. and Schulz, G.E. (1991) FEBS Lett. 280, 379–382.
9. Weiss, M.S., Abele, U., Weckesser, J., Welte, W., Schiltz, E. and Schulz, G.E. (1991) Science 254, 1627–1630.
10. Weiss, M.S. and Schulz, G.E. (1992) J. Mol. Biol. 227, 493–509.
11. Pauptit, R.A., Schirmer, T., Jansonius, J.N., Rosenbusch, J.P., Parker, M.W., Tucker, A.C., Tsernoglou, D., Weiss, M.S. and Schulz, G.E. (1991) J. Struct. Biol. 107, 136–145.
12. Cowan, S.W., Schirmer, T., Rummel, G., Steiert, M., Ghosh, R., Pauptit, R.A., Jansonius, J.N. and Rosenbusch, J.P. (1992) Nature (London) 358, 727–733.
13. Kreusch, A., Neubüser, A., Weckesser, J. and Schulz, G.E., Protein Sci., in press.
14. Tommassen, J. (1988) in: J.A.F. Op den Kamp (Ed.), Membrane Biogenesis, Vol. H16, NATO ASI Series, Springer-Verlag, Berlin, pp. 351–373.
15. Deisenhofer, J. and Michel, H. (1989) Science 245, 1463–1473.
16. Carmeli, C. and Lifshitz, Y. (1989) J. Bacteriol. 171, 6521–6525.
17. Bollivar, D.W. and Bauer, C.E. (1992) Biochem. J. 282, 471–476.

J.-M Ghuysen and R. Hakenbeck (Eds.), *Bacterial Cell Wall*

CHAPTER 16

Structures of non-specific diffusion pores from *Escherichia coli*

SANDRA W. COWAN and TILMAN SCHIRMER

Department of Structural Biology, Biocentre, University of Basel, Klingelbergstrasse 70,
CH-4056 Basel, Switzerland

1. Introduction

The *Escherichia coli* porins have been studied extensively over the past 20 years and the results form a large part of what we know about porins in general, as can be seen in Chapters 12, 15, 17, 19 and 27) and in recent reviews [1–3]. In this section, the structures of two general diffusion pores from *E. coli* outer membranes are discussed in relation to their functional properties.

Matrix porin (OmpF) (encoded by the *ompF* gene) and phosphoporin (PhoE) (encoded by the *phoE* gene) allow the passive diffusion of hydrophilic solutes of up to a mass of ~600 Da across the outer membrane of *E. coli*. They are functional as trimers and in this state are extremely stable, being resistant to proteases, detergents and high concentrations of chaotropic agents [4]. This has allowed the collection of vast amounts of experimental data on their structure and function. Despite the high sequence homology between these two proteins (63% identity), they have distinct channel properties. OmpF is weakly cation selective whereas PhoE has a preference for the transport of anions [5] and has even been proposed to have an anion binding site [6]. The conductance of the OmpF pore [7] is larger than that of PhoE [8] (0.8 nS compared to 0.6 nS in 1 M salt) indicating that it probably has a larger pore size [5]. These differing properties are essential for the survival of cells in changing environments. Under normal laboratory conditions, OmpF is the most abundant porin in the outer membrane of *E. coli*. However, under conditions of phosphate starvation, PhoE is de-repressed and allows the cells to take up negatively charged nutrients such as phosphate more efficiently. When porins are reconstituted into artificial bilayers, voltage-gating can be observed [9]. The applied potential above which the OmpF pores close is ~90 mV [7] and for PhoE ~100 mV [8]. The physiological significance and the mechanism of pore closing are not yet clear.

The observation that OmpF porin can form hexagonal arrays within the outer membrane [4] immediately raised hopes that it could be crystallized and the structure determined by X-ray crystallography. In fact, three-dimensional crystals were obtained in 1980 [10], but the structure was determined only recently [11] with a new crystal form [12]. In this trigonal crystal form, there are hydrophobic contacts between the molecules and the threefold axis of the trimer coincides with a crystallographic threefold axis. The structure is now refined with data to a resolution of 2.4 Å. Virtually isomorphous crystals of PhoE porin have been obtained [13] and the structure of this protein has been determined to 3.0 Å resolution [11].

2. The atomic structures of OmpF and PhoE

2.1. Barrel topology

The structures of the OmpF and PhoE porins consist of a 16-stranded anti-parallel β-barrel enclosing the transmembrane pore (Fig. 1). This topology was first observed in the structure of a porin from *Rhodobacter capsulatus* [14] (see Chapter 15), and seems to be a common motif for general diffusion pores and more specific pores such as maltoporin [15], in outer membranes. In the *E. coli* porins, the amino and carboxy termini are connected by a salt bridge within one of the β-strands, making the structure pseudo-cyclic. Eight short β-hairpin turns define the smooth end of the barrel while the rough end is made up by long loops that are exposed at the cell surface. Six of these pack together and partially cover the entrance to the barrel. Another loop, which contains a short piece of α-helix, folds inside the barrel, constricting the width of the channel. These loop structures cause the channel to lie off-centre and at an angle of about 16° to the barrel axis. The remaining loop is involved in subunit interactions.

Both porins have very similar structures (279 $C\alpha$ atoms superimpose with a root mean square deviation of 0.74 Å), except in one short turn and some of the loops. All the insertions and deletions between the two proteins occur in the latter regions and only cause local differences in folding. This similarity is not surprising in light of their high sequence identity.

2.2. The trimer: structure and stability

The trimer is stabilized by both hydrophobic and hydrophilic interactions between the subunits (Fig. 2). The hydrophobic contacts are made by residues from the barrel walls. Along the trimer axis, large hydrophobic residues pack together to fill up the space completely leaving no room for water. Away from the trimer axis, the β-sheets of the barrel walls pack in an orthogonal manner with highly complementary surfaces. It is interesting to note that a substantial hydrophobic core is formed upon aggregation, but it remains to be seen whether there is a contribution to the stability of membrane proteins analogous to the hydrophobic effect for soluble proteins. Clearly, the complementarity of the surfaces of porin subunits (which can be partially illustrated by the interlocking of aromatic residues from both walls), is important for favouring subunit aggregation over unspecific protein–lipid interactions (a preference shown by model calculations even for single hydrophobic helices in a membrane [16]). For this effect, the contribution of hydrophilic interactions must also be relevant. These primarily involve a loop which reaches into the pore of a neighboring monomer where it makes extensive hydrogen bonds and a few salt bridges. This loop fills a gap in the wall of the adjacent subunit left by the loop which

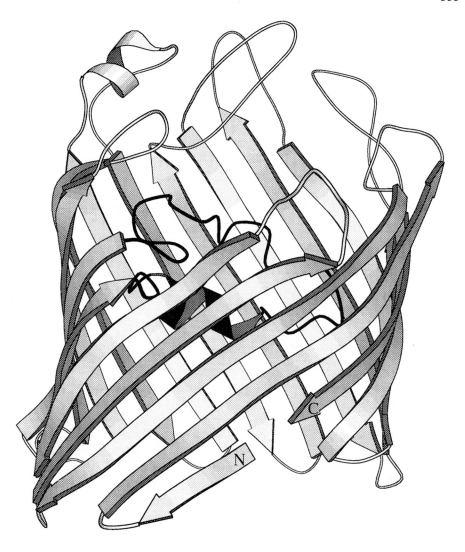

Fig. 1. Ribbon plot [34] of the OmpF monomer. The 16 β-strands are represented by arrows and the 2 α-helices by coils. The N- and C-termini are marked and the internal loop is shaded black for clarity. The periplasmic space is below the molecule in this view. (Adapted from [2]).

folds inside the barrel. For porin, the extensive, complementary contact areas and the combination of polar and apolar interactions between the subunits make it understandable that the trimers cannot be dissociated under non-denaturing conditions [4].

2.3. Structure in a lipid environment

The barrel is elliptical and has a cross section of 27 × 38 Å² defined by the main chain atoms. These dimensions agree very well with the results of investigations of two-dimen-

356

sional arrays of PhoE and OmpF using electron microscopy (EM) [17,18]. In fact, a projection of the models in the orientation shown in Fig. 2, superimposes very well on the EM projection maps, suggesting that the shape of the barrel in the three-dimensional crystals (which contain detergent) is the same as it is in the more natural lipid environment. Comparisons of three-dimensional EM reconstructions of negatively stained samples of OmpF at low resolution with images of the X-ray structure at a comparable resolution, also reveal no significant differences [19]. A more detailed comparison of the X-ray structure of PhoE with a three-dimensional map at 3.4 Å resolution [20] obtained from EM data of PhoE reconstituted with phospholipids [21] is underway (T. Earnest and B. Jap, personal communication). Initial results indicate that there are no gross structural deviations in the two environments. The availability of the OmpF and PhoE atomic structures allows the results from X-ray and EM crystallography of membrane proteins to be compared at high resolution for the first time and should provide information about the validity of the methods as well as their complementarity. This type of comparison will be even

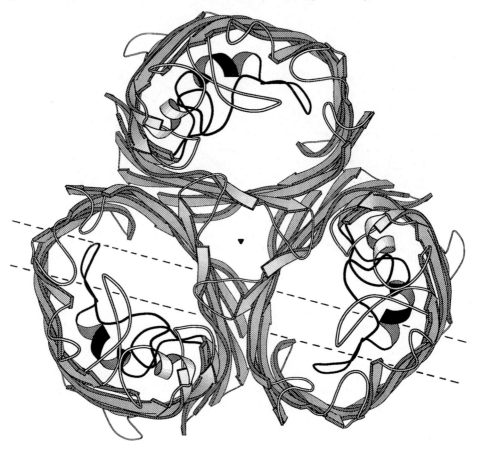

Fig. 2. Ribbon plot [34] of the OmpF trimer viewed down the threefold axis (indicated by a triangle). The internal loops are shaded black for clarity. The straight lines show the section viewed in Fig. 3. (Adapted from [2]).

more relevant when two-dimensional crystals of porin complexed with lipopolysaccharide (LPS) are analyzed. It has been shown that the presence of LPS affects the properties of porin and thus may be important for its function [7].

2.4. Orientation of porin in the membrane

A hydrophobic band around the trimer indicates the region of the protein that is in contact with the hydrophobic core of the membrane. Within its natural environment porin will, for symmetry reasons, have its threefold axis oriented normal to the membrane plane. The band is about 25 Å wide and consists mainly of small aliphatic residues with a marked distribution of aromatic residues on the edges. Many of the latter are tyrosine residues which have their hydroxyl groups directed into the hydrophilic regions on either side of the band. These aromatic residues probably serve to anchor the protein in the membrane, the flat aromatic surface being ideal for packing with fatty acyl chains (similar to the packing of arginine or lysine on aromatic rings often found in soluble proteins), thereby preventing leakage of the membrane which would be fatal for the cells. In tyrosine and tryptophan, as well as the combined properties of hydrophobicity and the ability to form hydrogen bonds, their polarizability may be favourable between two zones of vastly different dielectric constant. The importance of aromatic residues at this boundary is emphasized by a similar, although less pronounced, distribution of tryptophan residues in the bacterial photosynthetic reaction centre [22].

Mutations in PhoE porin which interfere with phage recognition and the binding of monoclonal antibodies directed against the cell surface [23,24] are found to occur in loops four, five, seven and eight. This indicates that the rough side of porin is exposed at the cell surface. This orientation is further confirmed by the result of a study with a β-lactamase-PhoE hybrid gene product, which showed that the amino terminus of porin lies on the periplasmic side of the membrane [25]. The solvent-exposed zones of the protein on either side of the hydrophobic band are quite dissimilar in size (Fig. 3). The large hydrophilic portion (~20 Å wide) of the protein on the extracellular side interacts (directly or indirectly via divalent cations) with the negatively charged head groups of LPS. Residues contributing to the hydrophilic surface on the periplasmic side of the membrane (~8 Å wide) come exclusively from the short β-hairpin turns.

3. The transmembrane pore

3.1. Pore architecture

The lining of the aqueous pore is similar to the external surface of soluble proteins, and has many well-ordered water molecules. Unlike porin from *R. capsulatus* [26], no calcium ions are incorporated in the structure (see Chapter 15). The shape of the pore varies on traversing the membrane and it can be divided up into three zones (Fig. 3). The pore entrance facing the extracellular surface (the mouth) is narrowed by the long loops at the rough end of the barrel to a diameter of between 11 and 19 Å. The solvent-accessible cross section decreases to 7×11 Å^2 where the internal loop and side chains from the

extracellular space

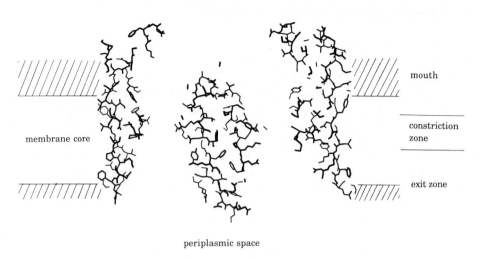

mouth

membrane core

constriction zone

exit zone

periplasmic space

Fig. 3. An 8 Å thick slice through the OmpF trimer (orientation indicated in Fig. 2). The relative location of the membrane is indicated schematically and the designation of the three different regions in the pores is shown on the right.

barrel wall near the threefold axis constrict the size of the pore at about half-way through the membrane (the constriction zone). The cross section of the channel increases abruptly (to 15×22 Å^2) just after the constriction zone since the pore size in this region (the exit zone) is simply delimited by the barrel walls. The shape of the pores is quite different to that found for the *R. capsulatus* porin where the entrance is much larger [27]. The major structural reason for this is that near the molecular threefold axis, the three pores in *R. capsulatus* porin run separately over a distance of only 20 Å, while in the *E. coli* porins, the pores are separated over a distance of 30 Å which spans the entire passage through the core of the membrane. Also, the loop structure at the entrance is more extensive in the slightly larger *E. coli* porins.

3.2. Selectivity filter

OmpF and PhoE filter solutes according to size and exhibit some charge selectivity [5]. Intuitively, one would expect this to occur at the narrowest part of the channel. The constriction zone (Fig. 4) is made up on one side by Asp$^{113/106}$ (sequential OmpF/PhoE residue numbering), Glu$^{117/110}$, and main chain carbonyl groups from the internal loop. In contrast, the side chains on the other side consist of basic residues (Lys$^{16/16}$, Arg$^{42/37}$, Arg$^{82/75}$ and Arg$^{132/126}$) resulting in a pronounced charge segregation across the pore. The physiological role of this polarity is unknown, but it seems to be important for the structure and/or function of porins since it is also observed in the pore of the *R. capsulatus* porin [27]. In addition, the residues mentioned above are strictly conserved between eight different porins from enteric bacteria [28]. The role of these residues as well as the inter-

A

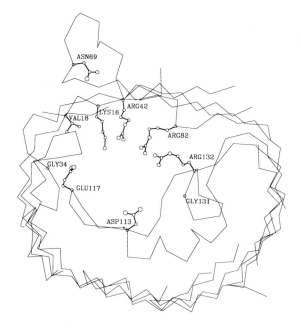

B

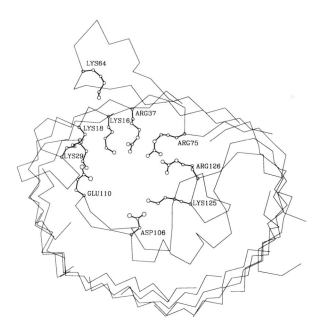

Fig. 4. Details of the constriction zones of (A) OmpF and (B) PhoE as seen from outside the cell. Residues mentioned in the text are represented by ball-and-stick models. The six loops at the entrance of the pores have not been included in the Cα trace for clarity. The C, N and O atoms are indicated by spheres of size increasing in this order.

nal loop in size exclusion has been demonstrated by the application of selection pressure; growing maltoporin-deficient cells on maltodextrin as the sole carbon source [29,30] should yield porin mutants with a larger pore size since most maltodextrins are too large to pass through the native OmpF and osmoporin (OmpC) (encoded by the *ompC* gene) pores. Indeed, OmpF mutants have been obtained [29] which have deletions of up to 15 residues in the internal loop or single site mutations replacing Asp^{113}, Arg^{42}, Arg^{82} or Arg^{132} by smaller uncharged residues. The same selection applied to OmpC resulted in mutations involving structurally equivalent residues [31]. That these changes involved residues solely from the constriction zone, implies that this region is critical for the determination of the size of solutes able to pass through the pore.

Since OmpF and PhoE have different charge selectivities, a comparison of their structures should reveal the features responsible. Hancock et al. [32] have shown that lysine residues are important for the anion selectivity of PhoE. Lysine residues found inside the pore of PhoE and not in OmpF include Lys^{18} (Val^{18} in OmpF) in the exit zone, Lys^{29} (Gly^{34}) in the mouth on loop one, Lys^{64} (Asn^{69}) in the mouth on loop two, and Lys^{125} (Gly^{131}) which protrudes into the constriction zone (Fig. 4). These residues have been mutated to glutamic acids [6] and exhibit effects on the anion selectivity of PhoE which are negligible for Lys^{18} and Lys^{29} and minor for Lys^{64}. The mutation of the constriction zone residue Lys^{125} to Glu, however, had a drastic effect. This PhoE mutant exhibits an inverted selectivity with a preference for cations similar to that of OmpF. It has been suggested that this residue is part of a 'hot spot' for phosphate binding, since PhoE pores are blocked by polyphosphates in the presence of magnesium [6]. These results indicate that residues in the constriction zone contribute to the selectivity of the pore concerned as well as the determination of solute exclusion size.

4. Concluding remarks

The 16-stranded antiparallel β-barrel which is observed in the structures of OmpF and PhoE porin as well as in that of *R. capsulatus*, seems to be a common feature of porins in the outer membrane of Gram-negative bacteria. The barrel provides the framework for an aqueous pore whose filtering properties are largely determined by the structure at the narrowest part of the channel, referred to as the constriction zone. The variation in channel dimensions allows efficient screening with rapid solute transfer. A preliminary screening for size and charge is provided by the mouth of the pore. The more stringent constriction zone is only 9 Å long and provides minimal (if any) interference before solutes pass into the large exit zone, which is like being released into bulk solvent. Of course, waste products must be able to leave the cell in the reverse direction and it is interesting to observe that the structures of the two different *E. coli* general diffusion pores on the periplasmic side are very similar both according to conformation and charge. The accumulation of solutes in the funnel on the periplasmic side will thus lead to a concentration-dependent and probably non-selective clearance of solutes through porin.

The crystal structures show only one conformation of the OmpF and PhoE pores. The dimensions at the narrowest part of the channels indicate that the pores are observed in an 'open' state. Applied voltage [9], osmotic pressure [7], and the lipid [7] and pH [33] environment of the protein affect the state, open or closed, of the channels. From the structures alone, it is not possible to determine which parts of the protein are involved in gating. This is an obvious subject for further site-directed mutagenesis experiments.

The sudden explosion of porin structure determinations over the past 2 years, allows several common features of their structures to be determined. In this way, rules can be developed which may help in the structure prediction of other hydrophilic outer membrane proteins. In the reverse sense, we can now re-evaluate the results of structure predictions based on theoretical methods and the immense amount of experimental data on the topology of OmpF and PhoE, and determine the relative merits of the methods used (see Chapter 17).

References

1. Jap, B.K. and Walian, P.J. (1990) Q. Rev. Biophys. 23, 367–403.
2. Cowan, S.W. (1993) Curr. Opin. Struct. Biol. 3, 501–507.
3. Nikaido, H. and Saier, Jr., H. (1992) Science 258, 936–942.
4. Rosenbusch, J.P. (1974) J. Biol. Chem. 249, 8019–8029.
5. Benz, R., Schmid, A. and Hancock, R.E.W. (1985) J. Bacteriol. 162, 722–727.
6. Bauer, K., Struyvé, M., Bosch, D., Benz, R. and Tommassen, J. (1989) J. Biol. Chem. 264, 16393–16398.
7. Buehler, L.K., Kusumoto, S., Zhang, H. and Rosenbusch, J.P. (1991) J. Biol. Chem. 266, 24446–24450.
8. Dargent, B., Hofmann, W., Pattus, F. and Rosenbusch, J.P. (1986) EMBO J. 5, 773–778.
9. Schindler, H. and Rosenbusch, J.P. (1978) Proc. Natl. Acad. Sci. USA 77, 1283–1285.
10. Garavito, R.M. and Rosenbusch, J.P. (1980) J. Cell. Biol. 86, 327–329.
11. Cowan, S.W., Schirmer, T., Rummel, G., Steiert, M., Ghosh, R., Pauptit, R.A., Jansonius, J.N. and Rosenbusch, J.P. (1992) Nature 358, 727–733.
12. Pauptit, R.A., Zhang, H., Rummel, G., Schirmer, T., Jansonius, J.N. and Rosenbusch, J.P. (1991) J. Mol. Biol. 218, 505–507.
13. Steiert, M., Ghosh, R., Schirmer, T. and Rosenbusch, J.P. (1992) Experientia 48, A22.
14. Weiss, M.S., Wacker, T., Weckesser, J., Welte, W. and Schulz, G.E. (1990) FEBS Lett. 267, 268–272.
15. Pauptit, R.A., Schirmer, T., Jansonius, J.N., Rosenbusch, J.P., Parker, M.W., Tucker, A.D., Tsernoglou, D., Weiss, M.W. and Schulz, G.E. (1991) J. Struct. Biol. 107, 136–145.
16. Wang, J. and Pullman, A. (1991) Biochim. Biophys. Acta 1070, 493–496.
17. Jap, B.K., Downing, K.H. and Walian, P.J. (1990) J. Struct. Biol. 103, 57–63.
18. Sass, H.J., Büldt, G., Beckmann, E., Zemlin, F., van Heel, M., Zeitler, E., Rosenbusch, J.P., Dorset, D.L. and Massalski, A. (1989) J. Mol. Biol. 209, 171–175.
19. Engel, A., Hoenger, A., Hefti, A., Henn, C., Ford, R.C., Kistler, J. and Zulauf, M. (1992) J. Struct. Biol. 109, 219–234.
20. Jap, B.K., Earnest, T.N., Walian, P. and Gehring, K. (1992) Program and abstracts of the American Crystallographic Association 1992 Annual meeting. A07, 29.
21. Jap, B.K., Walian, P.J. and Gehring, K. (1991) Nature 350, 167–170.
22. Schiffer, M., Chang, C.-H. and Stevens, F.J. (1992) Protein Eng. 5, 213–214.
23. Korteland, J., Overbeeke, N., de Graaff, P., Overduin, P. and Lugtenberg, B. (1985) Eur. J. Biochem. 152, 691–697.
24. van der Ley, P., Struyvé, M. and Tommassen, J. (1986) J. Biol. Chem. 261, 12222–12225.
25. Tommassen, J. and Lugtenberg, B. (1984) J. Bacteriol. 157, 327–329.
26. Weiss, M.S., Abele, U., Weckesser, J., Welte, W., Schiltz, E. and Schulz, G.E. (1991) Science 254, 1627–1630.
27. Weiss, M.S., Kreusch, A., Schiltz, E., Nestel, U., Welte, W., Weckesser, J. and Schulz, G.E. (1991) FEBS Lett. 280, 379–382.

28. Jeanteur, D., Lakey, J.H. and Pattus, F. (1991) Mol. Microbiol. 5, 2153–2164.
29. Benson, S.A., Occi, J.L.L. and Sampson, B.A. (1988) J. Mol. Biol. 203, 961–970.
30. Misra, R. and Benson, S.A. (1988) J. Bacteriol. 170, 528–533.
31. Misra, R. and Benson, S.A. (1988) J. Bacteriol. 170, 3611–3617.
32. Hancock, R.E.W., Schmidt, A., Bauer, K. and Benz, R. (1986) Biochim. Biophys. Acta 860, 263–267.
33. Todt, J.C., Rocque, W.J. and McGroarty, E. (1992) Biochemistry 31, 10471–10478.
34. Kraulis, P.J. (1991) J. Appl. Crystallogr. 24, 946–950.

J.-M Ghuysen and R. Hakenbeck (Eds.), *Bacterial Cell Wall*

CHAPTER 17

The porin superfamily: diversity and common features

DENIS JEANTEUR, JEREMY H. LAKEY and FRANC PATTUS

*European Molecular Biology Laboratory, Postfach 10.2209,
D-69012 Heidelberg, Germany*

1. Introduction

The name porin was first used by Taiji Nakae in 1976 [1] to describe a class of oligomeric proteins from the outer membrane of Gram-negative bacteria which 'form aqueous pores and penetrate through the thickness of the membrane in reconstituted vesicles as well as in native outer membrane, thus conferring hydrophilic permeability to that membrane'. Since that time, considerable progress has been made concerning the function, topology and, more recently, the atomic structures of these proteins. As a result, it is now clear that there exist two distinct classes of porins. The non-specific porins show limited selectivity between different ions and provide a general hydrophilic diffusion pathway with a size exclusion limit of 600–1400 Da depending on the porin. The term specific porins has been applied to the class of outer-membrane pore-forming proteins which display low selectivity for small molecules but quite selective facilitation of the diffusion of specific larger molecules such as maltose or nucleosides (LamB and Tsx porins) [2]. Both classes probably belong to the same structural class of β-barrel fold [3,4]. On the other hand, the TonB dependent receptors which are involved in the specific uptake of vitamin B12 or siderophores do not belong to the porin class. Although it was shown recently that they could be transformed into non-selective porins by the deletion of extracellular loops [5], only one specific class of compounds is able to cross the membrane through each of these receptors. The family of OmpA proteins were thought not to form pores in the outer membrane of Gram-negative bacteria. However, evidence that they may do so has recently been published [6]. In this chapter, we mainly focus on the so-called non-specific porin superfamily and, whenever necessary, discuss briefly the other outer-membrane protein families.

1.1. An unusual class of membrane proteins

According to their physicochemical properties and their localization in a lipid bilayer, the porins are unambiguously integral membrane proteins. However, when their sequences were determined by protein or DNA sequencing [7,8], it became apparent that they were a very peculiar class of membrane proteins. Their sequences are like soluble proteins, being

364

rather hydrophilic and devoid of long hydrophobic segments. Moreover, RAMAN, IR and CD spectroscopy revealed that they belong to the β-sheet class of protein structure in contrast to the then current dogma that membrane proteins could only traverse the bilayer as α-helices [8,9].

At present three porin structures are known, one at high resolution (1.8 Å, *Rhodobacter capsulatus* [10,11] and two at moderate resolution (OmpF and PhoE porins from *E. coli* at 2.4 and 3.0 Å resolution, respectively [12]) More detailed accounts of these results are presented in Chapters 15 and 16.

In this chapter, we restrict ourselves to the description of the salient features of these structures which are relevant to the prediction methods and conservation of important residues for pore function. For a more detailed analysis of porin channel function, the reader is referred to Chapters 19 and 27.

The three published crystal structures of porins reveal trimers of identical subunits. Each monomer consists of a 16-stranded anti-parallel β-barrel containing a pore. The barrel strands are connected by very short loops on the periplasmic face of the porin whereas the loops on the outside of the bacterium are long and of variable length. These long loops are often the sites for phage and colicin binding and even in intact cells can provide accessible epitopes for antibodies. The most striking and unexpected feature of all three structures is a long loop linking strands 5 and 6 which folds back into the pore to form (together with part of the inner barrel wall) the narrowest section of the channel.

Three other features of the known structures are worth mentioning in the current context:

- The regions of the hydrophobic protein–lipid interface of the β-barrel which are close to the lipid polar headgroups contain a ring of aromatic residues, consisting mainly of tyrosines and phenylalanines.
- At the level of the LPS headgroup on the outer face of the membrane, there is a large excess of negatively charged residues.
- The constriction zone of the channel has a very unusual distribution of charged groups, with carbonyl groups on one side and positively charged groups (mainly arginines) on the opposite side forming an electrostatic field parallel to the membrane plane.

1.2. Why do we need structure prediction?

Prior to the publication of these structures but with some low resolution information gained from electron crystallography [13–15], predictions of the porin polypeptide chain membrane topology were made initially in an attempt to correlate biochemical, immunological and mutational data with the porin functions of pore and receptor (for phage and colicins). This was done mainly on *E. coli* porins (see below). The structure of *R. capsulatus* porin, which was solved first [11], was a milestone in our understanding of porin function. However, in the absence of functional data on this particular porin and the lack of significant sequence homology with the well studied *E. coli* porins, prediction methods were still necessary to correlate the numerous *E. coli* functional data to the new β-barrel structure.

In the last few years, porin genes from many pathogenic Gram-negative bacteria have been cloned and sequenced. These porins are the focus of many studies because of their potential use as vaccines, for bacterial typing and their role in antibiotic resistance. Since their three-dimensional structures are unknown, prediction of their exposed loops is employed to define likely therapeutic antibody targets. Prediction methods are also useful for the study of other outer membrane protein families such as the TonB receptor family, the structures of which are unknown. It is hoped that we can improve these methods by analyzing their successes and failures with the known porin structures and thus aid further work with other families.

In the first part of this review, we present and analyze the different available prediction methods and update the alignment of the porin superfamily [16]. Use of the method on the OmpA family is also presented. In the second part, the conserved features of the alignment in relation to pore function are discussed. In the last part, we deal briefly with the diversity of the family regarding its evolution.

2. Sequence alignment and structure prediction

Many models, with a varying number and length of transmembrane segments, have been proposed which predict porin topology [17–28]. Although there are several methods which identify transmembrane segments, it is difficult to accurately predict the topology of an isolated single sequence. For well characterized porins such as those from *E. coli*, the prediction is made easier by our knowledge of the location of epitopes or specific residues involved in function. Even in such cases the predictions did not always fit with the structures that have appeared since. Thus, the problem was clear; how could we predict the topology of other less well characterized porins and at the same time be more confident about the result?

The simultaneous analysis of a set of porin sequences from distant species, instead of only one sequence, greatly improved the quality of the prediction [16]. This confirmed the results of Niermann and Kirschner [29] who showed with the well known 'TIM barrel' enzymes that prediction accuracy does indeed improve with increased sequence variability. Now that the structure of some porins is solved at high resolution [11,12], it is possible to analyze the prediction tools used in transmembrane topology determination.

Here, we show some of the sequence analysis tools that can be used for transmembrane topology prediction and show how the imperfections of these tools used alone can be compensated for by aligning the sequences from families of distantly related proteins. The current analysis, on 30 non-specific porin sequences from 5 distant families, allowed us to align the sequences and predict a transmembrane topology for each sequence with a higher confidence than was possible with one sequence alone.

2.1. Prediction tools for transmembrane segments

2.1.1. Turn prediction
Since the core of a membrane protein consists of alternating transmembrane elements, the first approach to determine transmembrane topology is to identify turns [28]. According

to Paul and Rosenbusch [28], amino acids can be separated into three groups: turn promoters (N, D, E, G, P, S), turn blockers (A, Q, I, L, M, F, W, Y) and the rest of the residues. Turns are then identified as a segment of three or more residues containing at least
one turn promoter and no turn-blockers. Turns are mainly predicted in loops by this
analysis but not exclusively. Short loops are difficult to localize and some loops are predicted to occur in known transmembrane regions. Similar results can be obtained with the
turn prediction criteria according to Chou and Fasman [30]. However, in spite of their
difficulties, these tools are important to confirm loops predicted by other methods. The
newly published porin structures have shown that periplasmic loops are very short, involving only a few residues [11]. As a consequence, the turns must be accurately defined
and it is possible that prediction methods of these specific turns based on a frequency
matrix of residue occurrence within these turns in *E. coli* and *Rhodobacter* porins may be
a good way of predicting the topology of other porins.

2.1.2. Hydrophobicity

Membrane spanning segments of non-pore-forming proteins can be derived by the analysis of hydrophobicity because they are involved only in hydrophobic interaction with
lipids or proteins [31]. Assuming that a transmembrane segment is made of $2n + 1$ amino
acids, then the mean hydrophobicity, $\overline{h}$ is defined as

$$\overline{h}_k(2n+1) = \frac{1}{2n+1} \sum_{j=k-n}^{k+n} h_j$$

where k is the number of the residue at the center of the window, h_j is the hydrophobicity
of residue j taken from a hydrophobicity scale. In this study, we used the PRIFT table
[32].

This method has been widely used for membrane proteins (for a review, see [30,33])
but is less efficient for pore-forming proteins. Since porin sequences are characterized by
the strong hydrophilicity of their sequence, this method is therefore useless for those
proteins.

2.1.3. Amphipathicity analysis

Transmembrane segments of pore-forming proteins are often not predicted with the hydrophobicity analysis alone because some pieces of secondary structure in pore-forming
proteins are amphipathic. They have one hydrophilic face towards the lumen of the pore
and one hydrophobic face towards either lipid or protein. As a consequence, the prediction of transmembrane segments of pore-forming proteins should be complemented by the
amphipathicity analysis.

The amphipathicity can be derived by calculation of the hydrophobic moment. The
vector hydrophobic moment $\langle u \rangle$ was defined by Eisenberg et al. [34,35] to quantify the
amphiphilicity of an α-helix but can also be used to quantify amphipathicity of other secondary structures. We have used a slight modification of Eisenberg's definition of the
hydrophobic moment as introduced by Finer-Moore and Stroud [36]:

$$\langle \mu_k \rangle(\nu, 2n+1) = \frac{1}{2n+1} \left| \sum_{j=k-n}^{k+n} \left[(h_j - \overline{h}_k) \exp(2\pi j \nu i) \right] \right|$$

$$= \frac{1}{2n+1} \left\{ \left[\sum_{j=k-n}^{k+n} (h_j - \overline{h}_k) \sin(2\pi j \nu) \right]^2 + \left[\sum_{j=k-n}^{k+n} (h_j - \overline{h}_k) \cos(2\pi j \nu) \right]^2 \right\}^{1/2}$$

where h_j and $2n+1$ are defined as above, $\overline{h}_k$ is the average hydrophobicity of the $2n+1$ residues from $k-n$ to $k+n$, and ν is the frequency of the secondary structure ($\nu = 1/3.6$ for α-helices and $1/2$ for a β-strand, and $1/2.25$ for a twisted β-strand). The subtraction of $\overline{h}_k$ from each h_j is to annul $\langle \mu \rangle$ for $\nu = 0$ (no secondary structure) instead of $\langle \mu \rangle = \overline{h}$ with the Eisenberg definition.

This prediction method has been used for topology determination of porins [9,21] but it can sometimes be confusing because it can predict more strands than actually exist. In OmpF, for instance, 18 strands are predicted using this kind of analysis [9].

2.1.4. Membrane criteria

These two tools (hydrophobicity and amphipathicity) can be combined in one. It has been shown by Eisenberg that α-helices could be classified with the aid of the hydrophobic moment plot, $\langle \mu \rangle$ versus $\overline{h}$ [35]. Segments of membrane proteins fall in different regions of the hydrophobic moment plot depending whether they are monomeric membrane anchors, 'surface seeking' or belonging to α-helical bundles. Those two criteria could be linearly combined and we found that the simplified value $\langle \mu \rangle + \overline{h}$ accurately determined the transmembrane segments [37]. By this analysis, a transmembrane segment would have $\langle \mu \rangle + \overline{h}$ above a certain threshold value which depends on the hydrophobicity table used. For the PRIFT table, we currently use a threshold of 1.5 to 2.

2.1.5. Alignments

Alignment of closely related sequences of porins like those from E. coli is quite easy with the usual alignment tools such as those available in the GCG package [38]. But for more distantly related sequences, those tools are not very accurate, mainly because those methods tend to introduce gaps which incur the same penalty all along the sequence. For membrane proteins, it is clear that loop regions are much more variable and in those regions even very long gaps may be easily introduced without problem. Conversely, insertion of gaps in the transmembrane regions should be more heavily penalized.

Multiple alignment and prediction of topology are complementary tasks. On the one hand, multiple alignment gives better accuracy to a prediction because, if the sequences are properly aligned, predictions of topological elements will be reinforced by being predicted in all aligned sequences. On the other hand, prediction of topology will help to align sequences because predicted topological elements can be lined up together. In addition, multiple sequence alignments highlight important conserved features of the sequences. As a result, structural information or biochemical information on one family of porins can be related to more distant porin families.

Sequence alignments were used for topology prediction as follows. Having s sequences aligned, we defined a new mean hydrophobicity $\overline{H}$ which is the mean hydrophobicity over a window of $2n + 1$ residues calculated from the average hydrophobicity of the s residues at each position:

$$\overline{H}_k(2n+1) = \frac{1}{(2n+1)s} \sum_{j=k-n}^{k+n} \left(\sum_{i=1}^{s} h_{j,i} \right)$$

where $h_{j,i}$ is the hydrophobicity of residue number j in the sequence number i whilst s is the total number of sequences. k and $2n + 1$ are as defined above.

We then determined two new hydrophobic moments $\langle M \rangle$ and $\langle \overline{\mu} \rangle$. $\langle M \rangle$ is the hydrophobic moment calculated from the average hydrophobicity of the s residues at each position:

$$\langle M_k \rangle(v, 2n+1, s) = \frac{1}{2n+1} \left| \sum_{j=k-n}^{k+n} \left\{ \left[\left(\sum_{i=1}^{s} \frac{h_{j,i}}{s} \right) - \overline{H}_k \right] \exp(2\pi j v i) \right\} \right|$$

whilst $\langle \overline{\mu} \rangle$ is the mean of the hydrophobic moment $\langle \mu \rangle$ calculated for each of the s sequences

$$\overline{\langle \mu_k \rangle}(v, 2n+1, s) = \frac{1}{2n+1} \left| \sum_{i=1}^{s} \left[\sum_{j=k-n}^{k+n} \left[\left(\frac{h_{j,i}}{s} - \overline{H}_k \right) \exp(2\pi j v i) \right] \right] \right|$$

The 'hydrophobic moment of the mean' $\langle M \rangle$ is not equal to the 'mean of the hydrophobic moment' $\langle \overline{\mu} \rangle$. $\langle M \rangle$ is more sensitive to the quality of the alignment because it is sensitive to the correct alignment of the repetitive features of the secondary structure. It is smaller than $\langle \overline{\mu} \rangle$ when the $\langle \mu \rangle$ of each sequence are misaligned. It is equal to $\langle \overline{\mu} \rangle$ when they are correctly aligned. The more sequences that are aligned and the more distant they are, the finer is the resolution when using $\langle M \rangle$ as the criterion, because in the variable regions its value is considerably reduced.

Identically, a membrane criteria was defined for the alignment parameters:

$\langle M \rangle + \overline{H} >$ threshold value

$\langle \overline{\mu} \rangle + \overline{H} >$ threshold value

Here, the threshold value is easily calculated by randomizing the order of the residues in each sequence and then inserting gaps at the same position as in the original sequences. The maximum value we obtained was 1.5.

Using the plot of $\langle M \rangle + \overline{H}$ and $\langle \overline{\mu} \rangle + \overline{H}$, it is possible to evaluate the quality of an alignment and, if the alignment is correct, it should indicate the transmembrane segments.

Finally the accuracy of the alignment could be checked by the conservation of certain residues. The existence of the ring of aromatic residues [11,12] at the water–lipid interface means that aromatic residues should concentrate at the end of predicted strands. Another characteristic of the structure is the charged residues in the constriction or eyelet of the pore [11,12]. Those important residues should be more or less conserved (Table I).

2.2. Results

In a previous paper, we aligned 14 porin sequences coming from two distant families and we predicted their topology [16]. We have now refined this alignment using the tools described above for the analysis of 30 sequences of non-specific porins which are now divided into five distantly related families. These families form the porin superfamily. We also have used this method for the topology determination of the OmpA and OprF group and for specific porins such as maltoporins and Tsx (see below).

The following criteria were used:

(1)   a β-strand is defined as:
- high value of $\langle M \rangle + \overline{H}$ (with $\nu = 1/2$ or $1/2.5$)
- no gaps
- no turn prediction
- sequence conservation

(2)   a loop is defined by:
- sequence variability
- turn promoters
- presence of gaps
- low value of $\langle M \rangle + \overline{H}$

TABLE I

Conservation of the eyelet residues[a]

OmpF	Neisseria	Haemophilus	Rhodobacter	Bordetella/Comamonas
K16	+	++	+[b]	I/I
R42	++[c]	++	+[b]	++
R82	++	++	E[b]	++
R132	+[c]	T	E	++
E62	++	++	K[b]	++
D113	+	+	+[b]	+
E117	+	L	+[b]	F/G

[a]The important eyelet residues are taken from [12] for OmpF and are fully conserved in the enteric porin family. The corresponding residues were taken from the alignment and are displayed as follows: ++, highly conserved (i.e. R = K and D = E); +, a conserved residue is within two spaces (this allows for the repetition of the β-strand). When no conservation is evident, the aligned residue is shown by its one letter symbol.
[b]Forms eyelet of Rhodobacter.
[c]Except NMPORA1.

2.2.1. The porin superfamily

Our last prediction [16] was found to be in good agreement with the structures of *E. coli* OmpF and PhoE that have since been solved [11,12].

Although it is very distant from other porin families, we have included the porin from *R. capsulatus* in our alignment because its structure has been solved and because the hydrophobicity and hydrophobic moment plot of this sequence alone is very clear.

Using this method we predict 16 strands unambiguously (Fig. 1) and we predict which residues are facing the membrane and which residues are facing the inside of the pore. Interestingly, the principal features, like aromatic ring or eyelet charged residues are quite conserved (Table I).

The length of our predicted β-strands are sometimes shorter that the one found by X-ray crystallography. That may be because bigger strands tend to have more hydrophilic residues at the exterior ends of the strand which we cannot predict with our method. To make the alignment clearer, we did not try to align the poorly conserved loop regions. They could clearly be better aligned but the required gaps would reduce the overall quality of the alignment. Similarly the 'PEFXG' motif we found previously [16] was not aligned (see below).

2.2.2. The OmpA and the LamB families

2.2.2.1. OmpA family.
The major outer membrane protein OmpA has been shown to function as a receptor for bacteriophage, to be required for the action of colicins K and L and to be necessary for efficient conjugation in recipient cells [39] (see also Chapter 12). This 325 residue polypeptide is thought to be a two domain protein with an N-terminal membrane domain carrying out all the above functions and a C-terminal periplasmic domain linked by a proline rich hinge region which is cleavable by pronase on isolated cell envelopes. A large number of OmpA sequences are known as are several mutations which affect phage and colicin binding.

A related family of outer membrane proteins is the OprF family from *Pseudomonas* [40,41]. Despite some debate, there is some good evidence these proteins form pores in the outer membrane. Like OmpA, OprF is responsible for the maintenance of the integrity of the cell envelope. Moreover, the two families can be easily aligned in their C-terminal half domain (see Fig. 2) and identical hydrophobicity and prediction patterns are obtained with the overall sequence. This similarity and the recent evidence that OmpA also forms channels in vesicles [6], and in planar bilayers [42] implies that the three-dimensional

---

Fig. 1. Alignment of porin sequences and predicted membrane-spanning strands. The $\langle M \rangle + \overline{H}$ was plotted using the set of 30 porin sequences shown on the phylogenic tree (Fig. 3). $\langle M \rangle$ was calculated with $\nu = 1/2$ or $\nu = 1/2.5$ in order to take into account untwisted and twisted β-strands. Each column represents the value of $\langle M \rangle + \overline{H}$ calculated with a window of 9 centered at the current position. The ten sequences aligned here are representative of the porin families. Three of them have their structures solved and we have boxed the 'true' strands revealed by X-ray crystallography. In order to show the alternation of hydrophobic residues, we have shaded the hydrophobic residues that are predicted to face the hydrophobic core of the membrane. Hydrophobic residues are lightly shaded, presenting also some polar properties (Y, H, P). Aromatic residues are shown in bold. Charged residues involved in the eyelet (but not involved in Ca^{2+} contact) are outlined. Conserved residues are printed slightly larger than less conserved. Under the alignment lines and numbers represents the predicted strands.

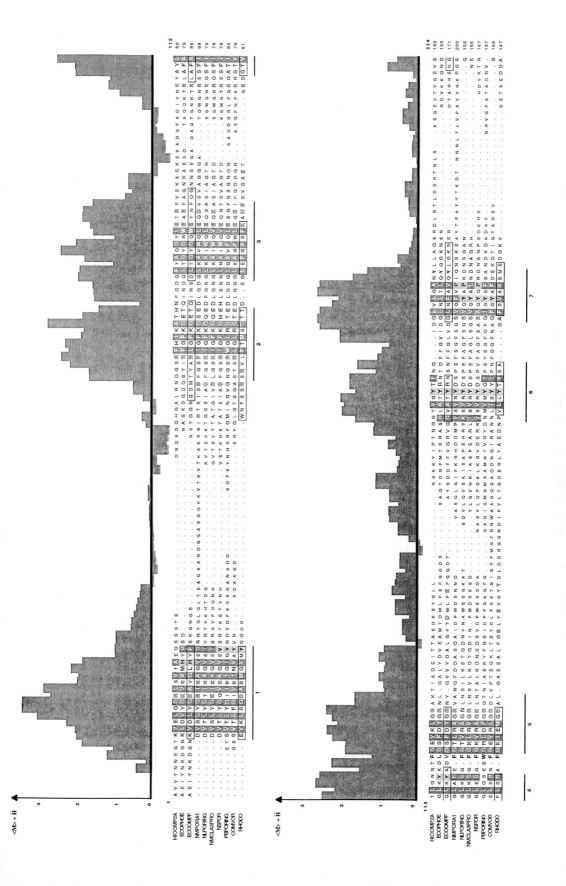

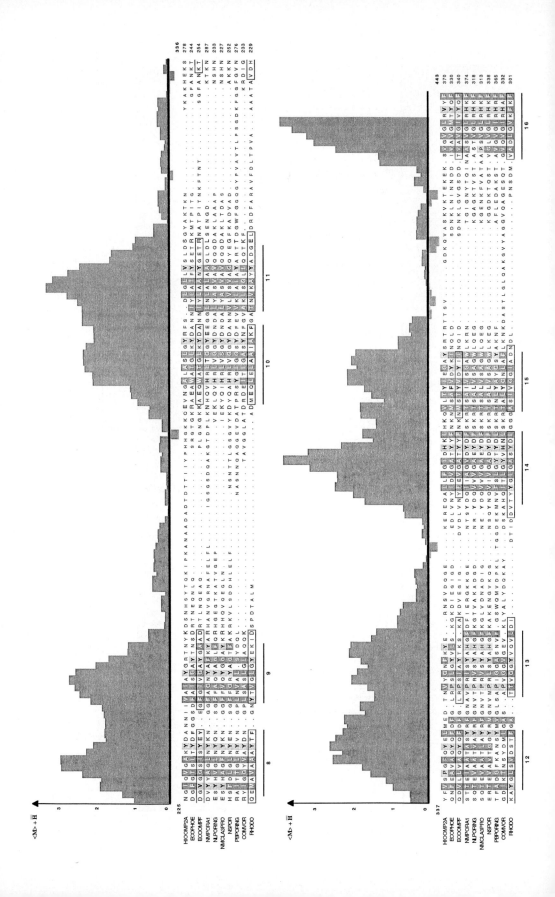

structure of these two classes of proteins is similar. However, in contradiction to current thinking on OmpA structure, Finnen et al. [43] have put forward evidence that the C-terminal domain of OprF protein is also transmembrane.

Vogel and Jähnig [9], have predicted eight transmembrane strands for the N-terminal domain of OmpA but did not comment on the possibilities of predicting other strands in the C-terminal domain (see [9, Fig 5]). Using our membrane criteria, we predict the same last seven transmembrane strands but not their strand 1 and strongly predict one strand just following the proline rich (OmpA) or cysteine rich region (OprF) and one other at the extreme C-terminal end of the polypeptide. In-between amphipathic α-helices are predicted instead of β-strands. Independent of which prediction is used, the biochemical data contains a contradiction. A mutation affecting phage binding has been located on the loop between strands 7 and 8 [44] of OmpA. This is followed by a strongly predicted single transmembrane strand which in turn indicates that the following loop is periplasmic. However, this region contains a disulfide bridge located on the surface of the OprF protein by epitope mapping [43]. One explanation could be that the disruption of a periplasmic disulphide bridge affects the conformation of an epitope located at the surface of the cell.

One striking sequence feature of the OmpA-OprF family may suggest that the transmembrane β-barrel terminates at the end of the N-terminal domain. This domain ends with a well conserved motif having a C-terminal phenylalanine and is linked to the hinge region by a glycine. This motif shows strong homology, and can be easily aligned with the 16th and last strand of the porin superfamily. It was shown by Klose et al. [45] that this small section is of crucial importance for the assembly of OmpA into the outer membrane. What is more, the C-terminal phenylalanine of PhoE has also been shown to be crucial in this respect [46].

Clear cut conclusions about the membrane fold of these proteins cannot therefore yet be established and this example illustrates the limits of confidence of such prediction methods.

2.2.2.2. LamB family. LamB porin (often called maltoporin) is a trimeric outer-membrane protein which form pores involved in the specific diffusion of maltose and maltodextrins across the outer membrane (see [27] and refs. therein). Despite the lack of any significant homology with the non-specific porins, it was shown from low resolution X-ray data that maltoporin adopts the same 16 stranded β-barrel fold as the *E. coli* and *Rhodobacter* porins [4].

Based on numerous biological and topological data, together with structure prediction, a 16 stranded model has been put forward and improved over the years [19,25,27]. More recently, the sequence of a sucrose-specific porin was determined which shows clear but low homology [47]. Using our prediction method on the four known highly homologous LamB sequences, 16 transmembrane strands are also predicted which nevertheless differ slightly from the present model of Charbit and Hoffnung [19]. Moreover, a common prediction could be obtained between the maltose and sucrose families but only at the N-terminal and C-terminal third of the sequences. More work remains to be done, perhaps with additional sequences, to obtain a consensus model for the two families. Again this is an example which underlines the limits of this prediction method.

```
                                                                                                                              100
OMPA  SHIDY   APKDNTWYTGAKLGWSQYHDTGF.....IDN..NGPTHENQLGAGAFGGYQVNPYVGFEMGYDWL..GRMPYKGSVENGAYKAQGVQLTAKLGYPITDDL    91
OMPA  ECOLI   APKDNTWYTGAKLGWSQYHDTGF.....INN..NGPTHENQLGAGAFGGYQVNPYVGFEMGYDWL..GRMPYKGSVENGAYKAQGVQLTAKLGYPITDDL    91
OMPA  SALTY   APKDNTWYAGAKLGWSQYHDTGF.....IHN..DGPTHENQLGAGAFGGYQVNPYVGFEMGYDWL..GRMPYKGSVENGAYKAQGVQLTAKLGYPITDDL    91
OMPA  ENTAE   APKDNTWYAGGKLGWSQFHDTGWYNSNLNN..NGPTHESQLGAGAFGGYQVNPYLGFELGYDWL..GRMPYKGVKVNGAFSSAQVQLTAKLGYPITDDL    95
OMPA  SERMA   APKDNTWYTGAKLGWSQYHDTGFYGNGYQNGIGNGPTHKDQLGAGAFLGYQANOYLGFEMGYDWL..GRMPYKGSVNGAFKAQGVQLAAKLSYPIADDL    98
PORF  PSEAE   ..QGQNSVEIEAFGKRYFTDSVRNMK.NADLY.ETGNKKVHGADTALDALYHFG.TPG.                                           81
PORF  PSESY   ..QGQGAVEIEGFAKKEMYDSARDFKNNGNLF.DDGKNIKGADTALDALYHFN.NPG.                                           79
ADH   PSEUFL  ..QGQGAVEGELFYKKQYNDSVKHIEDGFNP.GARIGYFLTDDLSLNLSYDKTNHTRSN.DGTGSQKIGGDTSLTAQYHFG.QAGV               82

                                                                                                                              200
OMPA  SHIDY   DVYTRLGGMVWRADTKAHNVTGESEKNHDTGVSPVFAGGVEWAITPE.....IATRLEYQWTNNIGDA..HTIGTRPDNGLLSLGVsYRFGQGE.AAPVVA   180
OMPA  ECOLI   DIYTRLGGMVWRADTKS.NVYGA..KNHDTGVSPVFAGGVEYAITPE.....IATRLEYQWTNNIGDA..HTIGTRPDNGMLSLGVsYRFGQGE.AAPVVA   185
OMPA  SALTY   DFYTRLGGMVWRADTKS.NVPGG.STKDHDTGVSPVFAGGIEYAITPE.....IATRLEYQWTNNIGDA..NTIGTRPDNGLLSLGVsYRFGQGE.AAPVVA   184
OMPA  ENTAE   DIYTRLGGMVWRADS.SNSIAGDNHDTGVSPVFAGGVEWAMTRD.....IATRLEYQWVNNIGDA..GTVGVRPDNGMLSVGVsYRFGQGE.EDNAPVVA   184
OMPA  SERMA   DIYTRLGGMVWRADSKANYGRTGQRLSDHDTGFVSNIGAAVGVEYALTKN..WATRLDYQGFVSNIGDA..GTVGARPDNTMLSLGVsYRFGQODDVVAP..A   191
PORF  PSEAE   VGLRPYVSAGLAHONITNINSNSDSO.GROOMT.MANIGAGLKYYFTENF.FAKASLDGQGYGLEKRDNGHO.                              164
PORF  PSESY   DMLRPYVSAGFSDDSIG.QNGRN.GRNGST.FANIGGGPKLYF.DNF.YARAGVEAQYNIDQGD.                                     160
ADH   PSEUFL  DSLRPYVEGGFGHQSRGNVKADGHSGRDQST.LAIAGAGAVYKYYFINNV.YARAGVEADYALDNGKW.                                 164

                                                                                                                              400
OMPA  SHIDY   PAPAP..........................APEVQTKHFTLKSDVLFNFNKATLKPEGQAALDQLYSQLSNLDPKDGSVVLGYTDRIGSD              243
OMPA  ECOLI   PAPAP..........................APEVQTKHFTLKSDVLFNFNKATLKPEGQAALDQLYSQLSNLDPKDGSVVLGYTDRIGSD              248
OMPA  SALTY   PAPAP..........................APEVTKTHFTLKSDVLFNFNKATLKPEGQQALDQLYTQLSNMDPKDGSVVLGYTDRIGSE              247
OMPA  ENTAE   PAPAP..........................APVVETKRFTLKSDVLFNFNKSTLKAEGQQALDQLYTQLSSMDPKDGSVVLGYTDAVGSD              247
OMPA  SERMA   AAPAP..EPVADVCSDSDNDGVCDNVDKCPDTPANVTVDANGCPAVAEVVRVQLDVKFDFDKSKVKENSYADIKNLADFMKQY.PSTSTTVEGHTDSVGTD   254
PORF  PSEAE   AAPAP..VAEVCSDSDNDGVCDNVDKCPDTPANVTVDADGCPAVAEVVRVELDVKFDFDKSVVKPNSYGDVKNLADFMAQY.QDRRVEIEVKGIKDVVTQPQA   259
PORF  PSESY   ..........TPAPAPEPTPEPEA..PVAQVVRVELDVKFDFDKSVVKPNSYGDVKNLADFMAQY.QDRRVEIEVKGIKDVVTQPQA                  253
ADH   PSEUFL  ..........................PATNVEVAGHTDSIGPD                                                          236

                                                                                                                              
OMPA  SHIDY   AYNQGLSERRAQSVVD.YLISKGIPADKISARGMGESNPVTGNTCDNVKQRA.ALIDCLAPDRRVEIEVKGIKDVVTQPQA
OMPA  ECOLI   AYNQGLSERRAQSVVD.YLISKGIPADKISARGMGESNPVTGNTCDNVKQRA.ALIDCLAPDRRVEIEVKGIKDVVTQPQA
OMPA  SALTY   QYNQKLSEQRAQSVVD.YLISKGIPSDKISARGMGESNPVTGNTCDNVKPRA.ALIDCLAPDRRVEIEVKGVKDVVTQPQA
OMPA  ENTAE   QYNQKLSEKRAQSVVD.YLVAKGIPANKISARGMGEADAVTGNTCGYKSGRATKAQIVCLAPDRRVEIEVKGIKDVVTQPQG
OMPA  SERMA   AYNQKLSERRANAVRDVLVNEYGVEGGRVNAVGYGESRPVADNATA.EGRAINRRVEAEVEAQAK.
PORF  PSEAE   AYNQKLSERRANAVRDVLVNQYGVGASRVNSVGYGESRPVADNATE.AGRAVNRRVEASVEAQAQ.
PORF  PSESY   AYNQKLSQRRADRVKQVLVKD.GVAPSRITAVGYGESRPVADNATE.
ADH   PSEUFL
```

3. Common functional features of the porin superfamily

3.1. The internal loop

A common characteristic of OmpF, PhoE, the porin from *R. capsulatus*, and perhaps of all true porins, is the presence of a peptide loop (loop 3) between β-strands 5 and 6 which enters into and constricts the channel. This creates the eyelet region which is largely responsible for the channel selectivity. The PEFXG motif has been shown to form the turn at the end of this loop in OmpF and PhoE and its strong conservation in *E. coli* suggests that the conformation of this turn is important. In *E. coli*, the individual residues involved in this eyelet are highly conserved and mutations allow maltodextrin permeation [24,48], destabilize the trimer, increase single channel conductance and increase voltage sensitivity [49]. However, Benson and co-workers [24,48] showed that deletions in this region could produce OmpF and OmpC proteins which not only were active but also allowed the permeation of maltodextrins. It follows from this that the loop, although important, is not essential for the β-barrel structure. Their conservation across the alignment presented in this paper is displayed in Table I. The major problem in examining the conservation of the residues of the loop is whether we allow the PEFXG motif to be aligned in *E. coli* and *Neisseria*. If we accept it as a valid motif then we are left with approximately 30 residues of unaligned sequence of *Neisseria* before the motif and 30 of *E coli* after. Within these 30 residues we find two predicted (and real; [12]) transmembrane β-strands in *E coli* which have no correlation in *Neisseria*. When examining the 30 extra residues of *Neisseria* we can predict two strands there as well. This result means that if the PEFXG motif is homologous in the two proteins, the internal loop must originate from strands 7 and 8 in *Neisseria* rather than from strands 5 and 6 which is the case in *E. coli* and *Rhodobacter*. With the enlarged sequence family presented here, we believe that alignment of the PEFXG motif in just two families is not a sufficient argument to introduce such large gaps and in so doing suggest that the internal loop of *Neisseria* is not loop 3. This makes the PEFXG motif in *Neisseria* a periplasmic loop between strands 6 and 7 as shown here and in the model of [22]. This model also predicts a 26 amino-acid long loop 3 which is not an exposed epitope, a situation that can be explained if it is an internal loop. In *Rhodobacter*, the sequence at the tip of a somewhat shorter loop is VGYTDL. In *Haemophilus*, the PEFXG motif is not present but the alignment before this loop is good. Srikumar et al. [26] have shown that an antibody to loop three of *H. influenzae* only binds to isolated protein and postulate that this epitope is part of the internal loop. Since the

Fig. 2. Alignment of OmpA family sequences and predicted membrane-spanning strands. There are two subfamilies, the OmpA family and the OprF family. Conserved residues throughout the entire set of sequences are shaded. Under the alignment lines represent putative transmembrane strands. The first seven strands are those predicted by Vogel and Jähnig [9]. The first strand (broken line) was not predicted by our algorithm. Positions of mutations affecting phage sensitivity are indicated by white letters on a dark background [44]. The long gap preceding the proline rich motif in the middle of the sequences separate the N-terminal, presumably transmembrane domain, from the C-terminal, presumably cytoplasmic domain. The residues from the 7th strand, presented in a larger and bolder font, form a pattern homologous to the well conserved motif of the 16th and last strand of the porin superfamily (see Fig. 1).

loop should determine the channel size, it is interesting to note that the conductance of *Haemophilus* (0.4 nS per monomer) [50] is lower than Omp F (0.7 nS) [51] or *Neisseria gonorrhoeae* (1.3–2 nS) [52,53] but the solute size exclusion limit is greater [50].

Predicted loops 3 and 4 of Class 2 meningoccocal porin become accessible to antibody binding only in outer membrane vesicles prepared by sarcosyl extraction. The large conductance of *Neisseria* porins may be explained by a small internal loop but their sensitivity to voltage invites comparison with *E. coli* where the internal loop is clearly involved in gating [49]. For these functional reasons, the accurate prediction of the internal loop is an important next step after assigning the main β-strands.

3.2. Antigenic determinants and vaccines

It has been assumed that the binding of an antibody to a membrane protein indicates the surface exposure of the relevant antigenic epitope. For outer membrane proteins of bacteria, there is the added possibility of phage binding sites to aid in topology studies. The knowledge of the PhoE and OmpF structures allows us to examine the quality of the previously gathered epitope mapping studies. Van der Ley et al. [17] identified three separate single residue antigenic mutants whilst it was shown that R158 was crucial for phage binding to PhoE [54,55]. These methods were successful because all these residues are exposed in the PhoE structure. Insertions or deletions were assumed to be non-disruptive only if they occurred in surface loops and this is generally true. However, the entire first β-strand, which is involved in monomer/monomer contacts as well as the channel wall, can be removed and still trimers are found in the outer membrane. Half of β-strand 5 can also be removed without affecting assembly [55,56].

In the case of pathogenic species, the analysis of surface epitopes also has direct relevance to vaccine production. The epitopes of the *Neisseria meningitidis* Class 1, 2 and 3 proteins and the *N. gonorrhoeae* A and B porins have been mapped [22,57,58]. All of these are predicted to be external loops by our method. Meningococcal class 1 porin has exposed antigens on loops 1, 4 and 5 and class 2 on loops 1 and 5. In the gonococci, only loop 5 for PIB and loop 6 for PIA were clearly surface exposed. No data are available for the class three topology but this meningococcal protein is closely related to PIA of *N. gonorrhoeae* [59,60]. The longest predicted loops were the ones which determine serotype specificity; loops 1 and 4 in class 1, loop 1 and 5 in class 2, loop 5 in PIB and loop 6 in PIA where loop 5 is 15 amino-acid residues shorter than in PIB. Many of these antibodies are bactericidal.

Haemophilus influenzae, which despite its name is a major cause of bacterial meningitis in children under 18 months, is clearly another target for vaccine development. Four regions provoked an antibody response and two of these, corresponding to loops 4 and 8 are exposed in whole cells [61]. Another antibody to loop 4 only binds to a peptide segment which includes part of predicted transmembrane segment 7. Since this epitope is not exposed in whole cells the accuracy of the loop prediction seems to hold.

Similarly loop 1 (residues 28–55) is not exposed in whole cells. Srikumar et al. [26,62] identified antibodies raised against *H. influenzae* porins by the use of overlapping synthetic hexapeptides, which provide a linear if not conformational epitope. They also identify loops 4 and 8 as being the most important surface exposed regions. They have also published a model topology using hydrophobic moment calculations and turn predictions. The results are largely similar to that above but the interested reader is encouraged to examine the original reference. Such exposed regions show the greatest variation between different serotypes with loop 4 being most variable in *H. influenzae* [63].

4. Evolution of the superfamily

In general, the phylogeny (Fig. 3) indicated by the porins agrees with that obtained by 16S RNA analysis [64]. Some measure of the evolution of porins can be seen in the sub-types of pathogenic species where the antigenic determinants in loop regions show greater variation than the highly conserved β-strands. Pressure of selection by antibodies and

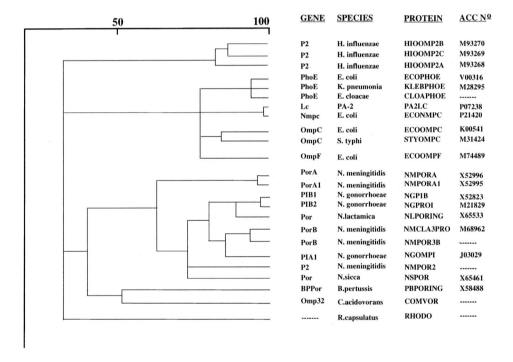

GENE	SPECIES	PROTEIN	ACC Nº
P2	H. influenzae	HIOOMP2B	M93270
P2	H. influenzae	HIOOMP2C	M93269
P2	H. influenzae	HIOOMP2A	M93268
PhoE	E. coli	ECOPHOE	V00316
PhoE	K. pneumonia	KLEBPHOE	M28295
PhoE	E. cloacae	CLOAPHOE	-------
Lc	PA-2	PA2LC	P07238
Nmpc	E. coli	ECONMPC	P21420
OmpC	E. coli	ECOOMPC	K00541
OmpC	S. typhi	STYOMPC	M31424
OmpF	E. coli	ECOOMPF	M74489
PorA	N. meningitidis	NMPORA	X52996
PorA1	N. meningitidis	NMPORA1	X52995
PIB1	N. gonorrhoeae	NGP1B	X52823
PIB2	N. gonorrhoeae	NGPROI	M21829
Por	N.lactamica	NLPORING	X65533
PorB	N. meningitidis	NMCLA3PRO	M68962
PorB	N. meningitidis	NMPOR3B	-------
PIA1	N. gonorrhoeae	NGOMPI	J03029
P2	N. meningitidis	NMPOR2	-------
Por	N.sicca	NSPOR	X65461
BPPor	B.pertussis	PBPORING	X58488
Omp32	C.acidovorans	COMVOR	-------
-------	R.capsulatus	RHODO	-------

Fig. 3. Possible phylogenetic tree derived from similarity of porin sequences. The aligned sequences were ana-lysed by the UWGCG DISTANCES programme which examines the pairwise similarity between different amino acid sequences [38]. The resulting matrix provides the similarity between all the components of the alignment. Each branching point shows the level of maximum similarity between the groups that it connects. The protein name chosen is the one most used in the literature describing the gene product. The species is that from which the gene was originally isolated. Where the accession number is provided, the sequence and gene name corresponds to that entry. Accession numbers beginning with the letter P are from the Swissprot database, others are from the Genbank/EMBL database. No accession number indicates that the entry was typed in by hand from the original reference. The baseline at the left of the diagram corresponds to the mean distance be-tween the randomized aligned sequences obtained from the programme SHUFFLE [38].

phage may play a role in forcing certain regions of porins to evolve at a higher rate than others. Another explanation is that external loops simply have more freedom to change without altering porin function.

In *H. influenzae*, the variations in porins are limited but apparently clonal with variations in porins being associated with defined serotypes and correlating well with differences in restriction fragment length polymorphisms of several genetic loci [63]. On the other hand, the more extensive family of *Neisseria* porins shows variations that do not correlate with species differences. At the most significant level, it can be seen that the meningococcal class 2 and 3 porins are more closely related to gonococcal porins than the class 1 meningococcal porin. At a more refined level, Feavers et al. [65] have shown that nucleotide differences of class 1 meningococcal porins were not consistent with the nucleotide sequence divergence of the whole chromosome, indicating that horizontal movement of genetic material is occurring. The gene for this porin also seems to be mosaic as smaller segments have clearly been swapped between different strains. Evolution of this type cannot only hamper attempts at vaccine production but may also provide a clue for more long-term porin evolution. In the alignment there is a region of homology between the *E. coli* strand 10 and *Haemophilus* strand 8 (GXKYDANN) which is shifted by the equivalent of a pair of β-strands and it may have occurred that such structural units can be transferred between mosaic genes to yield the variation we see now. It may provide the opportunity for evolution from the presumably eight stranded OmpA through the 16 stranded porins to the many stranded but possibly porin like selective transporters such as FepA [5]. Mosaicity of this type has been suggested elsewhere [18,66].The negligible sequence homology between the porins presented here and the structurally similar maltoporin [3,4] suggests that either the trimeric β-barrel has such advantages that it has arisen separately in bacterial evolution or that there was a common ancestor to all these porins. The sequence of the porin from *Thermotoga maritima* [67], which is considered ancestral to eubacterial species, may one day help in this respect.

References

1. Nakae, T. (1976) Biochem. Biophys. Res. Commun. 71, 877–884.
2. Nikaido, H. (1992) Mol. Microbiol. 6, 435–442.
3. Lepault, J., Dargent, B., Tichelaar, W., Rosenbusch, J.P., Leonard, K. and Pattus, F. (1988) EMBO J. 7, 261–268.
4. Pauptit, R.A., Schirmer, T., Jansonius, J., Rosenbusch, J.P., Parker, M.W., Tucker, A.D., Tsernoglou, D., Weiss, M.S. and Schulz, G.E. (1991) J. Struct. Biol. 107, 136–145.
5. Rutz, J.R., Lui, J., Goranson, J., Armstrong, S.K., McIntosh, M.A., Feix, J.B. and Klebba, P.E. (1992) Science 258, 471–475.
6. Sugawara, E. and Nikaido, H. (1992) J. Biol. Chem. 267, 2507–2511.
7. Mizuno, T., Chou, M.-Y. and Inouye, M. (1983) J. Biol. Chem. 258, 6932–6940.
8. Chen, R., Kramer, C., Schmidmayr, W., Chen-Schmeisser, U. and Henning, U. (1980) Biochem. J. 203, 33–43.
9. Vogel, H. and Jähnig, F. (1986) J. Mol. Biol. 190, 191–199.
10. Weiss, M.S., Kreusch, A., Schiltz, E., Nestel, U., Welte, W., Weckesser, J. and Schulz, G.E. (1991) FEBS Lett., 280, 379–382.
11. Weiss, M.S. and Shultz, G.E. (1992) J. Mol. Biol. 227, 493–509.

12. Cowan, S.W., Schirmer, T., Rummel, G., Steiert, M., Ghosh, R., Paupit, R.A., Jansonius, J.N. and Rosenbush, J.P. (1992) Nature 358, 727–733.
13. Jap, B.K., Walian, P.J. and Gehring, K. (1991) Nature 350, 167–170.
14. Dorset, D.L., Engle, A., Häner, M., Massalski, A. and Rosenbusch, J.P. (1983) J. Mol. Biol. 165, 701–710.
15. Engel, A., Massalski, A., Schindler, H., Dorset, D.L. and Rosenbusch, J.P. (1985) Nature 317, 643–645.
16. Jeanteur, D., Lakey, J.H. and Pattus, F. (1991) Mol. Microbiol. 5, 2153–2164.
17. van der Ley, P., Struyvé, M. and Tommassen, J. (1986) J. Biol. Chem. 261, 1222–1225.
18. Gerbl-Rieger, S., Engelhardt, H., Peters, J., Kehl, M., Lottspeich, F. and Baumeister, W. (1992) J. Struct. Biol. 108, 14–24.
19. Charbit, A., Ronco, J., Michel, V., Wert, C. and Hofnung, M. (1991) J. Bacteriol. 173, 262–275.
20. Francoz, E., Molla, A., Dassa, E., Saurin, W. and Hofnung, M. (1990) Res. Microbiol. 141, 1039–1059.
21. Tommassen, J. (1988) in: J.A.F. Op den Kamp (Ed.), Membrane Biogenesis, 16, Springer-Verlag, Berlin, pp. 352–364.
22. van der Ley, P., Heckels, J.E., Virji, M., Hoogerhout, P. and Poolman, J. (1991) Infect. Immun. 59, 2963–2971.
23. Klebba, P.E., Benson, S.A., Bala, S., Abdullah, T., Reid, J., Singh, S.P. and Nikaido, H. (1990) J. Biol. Chem. 265, 6800–6810.
24. Misra, R. and Benson, S.A. (1988) J. Bacteriol. 170, 3611–3617.
25. Francis, G., Brennan, L., Stretton, S. and Ferenci, T. (1991) Mol. Microbiol. 5, 2293–2301.
26. Srikumar, R., Dahan, D., Gras, M.F., Ratcliffe, M.J.H., van Alphen, L. and Coulton, J.W. (1992) J. Bacteriol. 174, 4007–4016.
27. Werts, C., Charbit, A., Bachellier, S. and Hofnung, M. (1992) Mol. Gen. Genet. 233, 372–378.
28. Paul, C. and Rosenbusch, J.P. (1985) EMBO J. 4, 1597–1985.
29. Niermann, T. and Kirschner, K. (1990) Protein Eng. 4, 137–147.
30. Fasman, G.D. and Gilbert, W.A. (1990) Trends Biochem. Sci. 15, 89–92.
31. Kyte, J. and Doolittle, R.F. (1982) J. Mol. Biol. 157, 105–132.
32. Cornette, J.L., Cease, K.B., Margalit, H., Spouge, J.L., Berzofsky, J.A. and DeLisi, C. (1987) J. Mol. Biol. 195, 659–685.
33. Jähnig, F. (1990) Trends Biochem. Sci. 15, 93–95.
34. Eisenberg, D., Weiss, R.M., Terwilliger, T.C. and Wilcox, W. (1982) Faraday Symp. Chem. Soc. 17, 109–120.
35. Eisenberg, D., Weiss, R.M. and Terwilliger, T.C. (1982) Nature 299, 371–374.
36. Finer-Moore, J. and Stroud, R.M. (1984) Proc. Natl. Acad. Sci. USA 81, 155–159.
37. Pattus, F., Heitz, F., Martinez, C., Provencher, S.W. and Lazdunski, C. (1985) Eur. J. Biochem. 152, 681–689.
38. Devereux, J., Haeberli, P. and Smithies, O. (1984) Nucleic Acids Res. 12, 387–395.
39. Morona, R., Tommassen, J. and Henning, U. (1985) Eur. J. Biochem. 150, 161–169.
40. Duchêne, M., Schweizer A., Lottspeich F., Krauss G., Marget, M., von Specht B.U. and Domdey, H. (1988) J. Bacteriol. 170, 155–162.
41. De Mot, R., Proost, P., Van Damme, J. and Vanderleyden, J. (1992) Mol. Gen. Genet. 231, 489–493.
42. Saint, N. (1992) Ph.D. Thesis, Pierre et Marie Curie University, Paris VI.
43. Finnen, R.L., Martin, N.L., Siehnel, R.J., Woodruff, W.A., Rosok, M. and Hancock, R.E.W. (1992) J. Bacteriol. 174, 4977–4985.
44. Morona, R., Klose, M. and Henning, U. (1984) J. Bacteriol. 159, 570–578.
45. Klose, M., MacIntyre, S., Schwarz, H. and Henning, U. (1988) J. Biol. Chem. 263, 13297–13302.
46. Struyvé, M., Moons, M. and Tommassen, J. (1991) J. Mol. Biol. 218, 141–148.
47. Schmid, K., Ebner, R., Jahreis, K., Lengeler, J.W. and Titgemeyer, F. (1991) Mol Microbiol. 5, 941–950.
48. Benson, S.A., Occi, J.L.L. and Sampson, B.A. (1988) J. Mol. Biol. 203, 961–970.
49. Lakey, J.H., Lea, E.J.A. and Pattus, F. (1991) FEBS Lett. 278, 31–34.
50. Vachon, V., Laprade, R. and Coulton, J. (1986) Biochim. Biophys. Acta 861, 74–82.
51. Lakey, J.H. and Pattus, F. (1989) Eur. J. Biochem. 186, 303–308.

380

52. Young, J.D., Blake, M., Mauro, A. and Cohn, Z.A. (1983) Proc. Natl. Acad. Sci. USA 80, 3831–3835.
53. Mauro, A., Blake, M. and Labarca, P. (1988) Proc. Natl. Acad. Sci. USA 85, 1071–1075.
54. Korteland, J., Overbeeke, N., de Graaff, P., Overduin, P. and Lugtenberg, B. (1985) Eur. J. Biochem. 152, 691–697.
55. Agterberg, M., Adriaanse, H., Van Bruggen, A., Karperien, M. and Tommassen, J. (1990) Gene 88, 37–45.
56. Agterberg, M., Adriaanse, H., Tijhaar, E., Resink, A. and Tommassen, J. (1989) Eur. J. Biochem. 185, 365–370.
57. McGuinness, B., Barlow, A.K., Clarke, I.N., Farley, J.E., Anilionis, A., Poolman, J.T. and Heckels, J.E. (1990) J. Exp. Med. 171, 1871–1882.
58. Butt, N.J., Lambden, P.R. and Heckels, J.E. (1990) Nucleic Acids Res. 18, 4258.
59. Butcher, S., Sarvas, M. and Runeberg, N.K. (1991) Gene 105, 125–128.
60. Wolff, K. and Stern, A. (1991) FEMS Microbiol. Lett. 67, 179–185.
61. Martin, D., Munson, R., Grass, S., Chong, P., Hamel, J., Zobrist, G., Klein, M. and Brodeur, B.R. (1991) Infect. Immun. 59, 1457–1464.
62. Srikumar, R., Chin, A.C., Vachon, V., Richardson, C.D., Ratcliffe, M.J.H., Saarinen, L., Käyhty, H., Mäkelä, P.H. and Coulton, J.W. (1992) Mol. Microbiol. 6, 665–676.
63. Munson, R., Bailey, C. and Grass, S. (1989) Mol. Microbiol. 3, 1797–1803.
64. Woese, C.R. (1987) Microbiol. Rev. 51, 221–271.
65. Feavers, I.M., Heath, A.B., Bygraves, J.A. and Maiden, M.C.J. (1992) Mol. Microbiol. 6, 489–495.
66. Gerbl-Rieger, S., Peters, J., Kellermann, J., Lottspeich, F. and Baumeister, W. (1991) J. Bacteriol. 173, 2196–2205.
67. Rachel, R., Engel, A.M., Huber, R., Setter, K.O. and Baumeister, W. (1990) FEBS Lett. 262, 64–68.

J.-M Ghuysen and R. Hakenbeck (Eds.), *Bacterial Cell Wall*
© 1994 Elsevier Science B.V. All rights reserved

Outer membrane proteins of *Escherichia coli*: mechanism of sorting and regulation of synthesis

ULF HENNING and RALF KOEBNIK

Max-Planck-Institut für Biologie, Corrensstrasse 38, D-72076 Tübingen, Germany

1. Introduction

For synthesis and localization of outer membrane (om) proteins three processes of basic interest are operating, using mechanisms which are still poorly understood: translocation across the plasma membrane, homing (sorting to om), and cross-regulation of their synthesis. Here we address the latter two mechanisms. We exclude the lipoproteins (see Chapter 14) and the TonB-dependent om receptors [1] for which the sorting mechanism has not been investigated. Also, we did not attempt to review the relevant literature exhaustively; we generally restrict ourselves to those cases which have been studied most thoroughly: the porins OmpC [2] and OmpF [3], the phosphoporin PhoE [4], the maltodextrin porin LamB [5] and the OmpA protein [6] (the references concern the primary structures of proteins). All are synthesized as precursors with a signal sequence and use the Sec system for translocation (for reviews, see [7,8]). A monograph has been published on protein targeting, which includes eukaryotic systems and covers the literature up to 1988 [9], another review discusses sorting of *E. coli* om proteins and covers the literature essentially also up to 1988 [10]. Figure 1 illustrates the topics in this chapter.

2. Sorting

2.1. Membrane incorporation of OmpA, OmpF and PhoE in vitro

A considerable body of experimental evidence suggests that the OmpA protein crosses the om from residues 1 to about 170 eight times with antiparallel β-strands [11–15] forming a typical amphiphilic β-barrel [16]; the C-terminal moiety is thought to be periplasmic. This type of membrane topology apparently had to be invented because anchoring in the om via typical lipophilic anchor sequences [17] is not possible; such a sequence, introduced into OmpA, caused the protein to become stuck in the plasma membrane [18]. The protein probably exists in the om as a monomer; oligomers have so far not been identified.

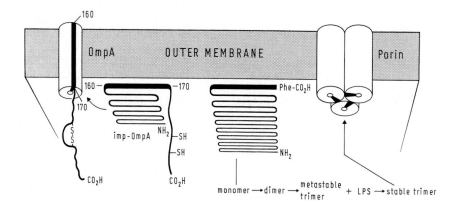

Fig. 1. Proteins of the om. The imp-OmpA and the porin monomer are thought to be attached to the inner side of the om. The prospective last trans-membrane strands of the imp-OmpA (residues 160–170) and of a porin monomer are indicated as heavy lines and as bars in the final products. The C-terminal phenylalanine of the porins is located at positions 346, 340, 330 and 421 of the OmpC, OmpF, PhoE and LamB proteins, respectively. The small circles at the center of OmpA and of a porin symbolize the channels. For all other details, see the following sections. The schemes are meant only to aid the reader in understanding amino acid positions, etc. mentioned in the text, they do not imply mechanisms, e.g. none of the processes may proceed within the periplasm.

It was also demonstrated that the polypeptide can act as a non-specific diffusion channel [19,19a]; there was no indication that OmpA had to oligomerize to exhibit this property. It had been shown earlier that denatured (SDS) OmpA refolded to its native conformation when lipopolysaccharide (LPS) was offered but not in the presence of synthetic dimyristoylphosphatidylcholine (DMPC) or total *E. coli* phospholipid [20]. It has recently been found that such denatured OmpA could very well assemble into DMPC vesicles provided octylglucoside was added first to the protein. The protein was already refolded in these SDS-octylglucoside micelles and could be transferred into DMPC vesicles [21]. In fact, the protein could be assembled directly into such vesicles under conditions which may be highly relevant to what is happening in vivo [22]. The vesicles had to be very small (diameter 30–50 nm), i.e. highly curved, and the lipid had to be in the fluid state. When the denatured (urea) OmpA was added to these vesicles, it refolded and inserted (yield 40–50% of added OmpA) in an oriented manner with the C-terminal moiety outside the vesicles. This was also achieved with a fragment of OmpA lacking most of its periplasmic part.

When the experiment was performed below the lipid phase transition temperature, the protein also folded and adsorbed to the vesicles but was not incorporated; raising the temperature also led in this case to incorporation. This adsorbed polypeptide, as judged by spectroscopic methods, possessed the same content of β-structure as the protein incorporated into the vesicles (or present in the om) but lacked two characteristic properties which it gained upon assembly in the vesicles: the so-called heat-modifiability and resistance to proteolytic degradation. (OmpA in the om or in these vesicles requires boiling in SDS to become completely unfolded; solubilization in the detergent at 50°C creates a species which migrates electrophoretically according to a molecular mass of 30–32 kDa. The boiled protein exhibits such a mass of 35–36 kDa [23 and refs. therein]. Trypsin acting on cell envelopes degrades the periplasmic part of the protein while the membrane moiety remains protected [6].) It is quite possible that this adsorbed, non-incorporated polypeptide corresponds to an assembly intermediate found in vivo (imp-OmpA, see Section 2.2.2.2.). When larger DMPC vesicles were used as a target for denatured OmpA, it was unable to assemble in their bilayer; instead, the protein aggregated and in this form was incompetent to insert into small vesicles. Assembly was achieved when these vesicles were prepared in the presence of octylglucoside.

These results show that a lipid bilayer is not required for OmpA to assume its native conformation and that it will spontaneously refold and insert into a membrane provided the latter possesses some defects (suboptimal packing of lipid in small vesicles or the presence of octylglucoside in large vesicles). As a note of caution regarding the relevance of these data to the in vivo process, it should be pointed out that it is quite unlikely that the protein will ever be in a random coil state in vivo. Nevertheless, it is clear that it is an intrinsic property of the polypeptide to be able to proceed from such a state into a lipid bilayer and to assume its native conformation.

Similar studies have been performed with the porin OmpF. The structures of several porins have been solved by X-ray crystallography [24–28], including those of OmpF and the closely related [2] phosphoporin PhoE [28]. OmpF is present in the om as a very stable trimer [29,30]. Each subunit consists of a 16-stranded antiparallel β-barrel. Trimeric OmpF has been isolated in biologically active state [31], i.e. able to form channels in planar lipid bilayers [32]. Such trimers were completely denatured by heating in guanidinium hydrochloride, the chaotropic agent was dialyzed away, the precipitated protein dissolved in SDS (OmpF remaining monomeric) and added to mixed micelles consisting of lecithin dissolved in octyl-pentaoxylethylene [33]. OmpF renatured with good yields (40–80%) and yielded trimers indistinguishable from the original trimers, including their ability to form channels. Thus, it is again clear that all information required for correct folding and trimerization is present within the OmpF protein and that for these processes, pre-existing membranes are not required. Yet, expression of this information needs a helper, e.g. in the form of detergent micelles.

Studies which use newly synthesized proteins are closer to the in vivo situation. It had been observed earlier that OmpF, when produced in spheroplasts, was lost in soluble form into the medium [34]. The system was later exploited [35] to show that what was lost was the mature protein as a monomer (or a very labile oligomer). This in itself is interesting because it demonstrates that spheroplasting (lysozyme-EDTA) causes the loss of something required for trimerization and membrane assembly of the porin. The monomer could be efficiently (70–80%) induced to trimerize and to become incorporated into cell envelope preparations provided Triton X-100 was present at low concentrations.

Trimerization was also achieved, in the absence of envelopes, by LPS or a mixture of Triton X-100 and lithium dodecylsulfate (LDS), this mixture probably representing Triton micelles carrying LDS. There was specificity regarding the structure of LPS; mutant forms missing most or all residues of the core oligosaccharide were ineffective [36]. The authors then used an OmpF synthesized in an in vitro transcription/translation system [37]. Probably the most telling results were obtained with a protein expressed from a gene lacking a segment coding for 16 out of 22 residues of the signal sequence, OmpF'. Using monoclonal antibodies directed against a surface epitope of the mature trimer or against the denatured monomer, it was shown that OmpF', although migrating electrophoretically as a monomer, reacted well with the former antibody but not so with the latter.

In contrast, the spheroplast generated protein was recognized by the latter and not efficiently by the former. OmpF' was able to trimerize, but much less efficiently (15–30%) than the spheroplast protein when tested under comparable conditions (except that Triton X-100 was without effect on OmpF'). When the in vitro protein synthesis was stopped before cell envelopes were added as a target for trimerization and membrane assembly, the amount of trimers formed decreased further. Clearly, OmpF' and the spheroplast OmpF differed in conformation, the former being less competent for trimerization and membrane assembly and losing still more such competence upon prolonged exposure to an aqueous environment.

The basis of the differences between OmpF' and spheroplast OmpF could reside in the fact that the latter was derived from a complete pre-OmpF. For a given precursor, processing can occur post- and/or co-translationally [38,39] and even when occurring co-translationally, processing is a late event during polypeptide chain elongation [38]. Translocation can apparently also occur co-translationally [40] and processing is no prerequisite for translocation [41,42]. Thus, most of the mature part of a precursor being exported may already be located at the periplasmic side of the membrane when the signal peptide is removed. Leader peptides modulate folding pathways of precursors [43–45]. Hence, a difference in conformation between the monomeric spheroplast OmpF and OmpF' could easily be due to 'imprinting' of pre-OmpF by the signal sequence.

Consistent with such a possibility is the behavior of hybrids between the precursor of the periplasmic phosphate binding protein PhoS and OmpF [46]. A fraction of them was localized to the om and exhibited OmpF functions but much of them remained in the monomeric state. It has also been shown that pre-OmpA possesses secondary and tertiary structure [47]. Further evidence suggesting that the conformation of pre-OmpA affects that of the processed form is given below (Section 2.2.2.2.). It has been shown that the TEM β-lactamase is translocated post-translationally [48]. Even in such cases, at least part of the conformation of a cytosolic precursor may be maintained during translocation. A polypeptide need not be in a linear form for successfully traversing the plasma membrane. OmpA possesses, within its periplasmic moiety, two cysteine residues which are 11 residues apart [6]. Intramolecular crosslinking these two did not inhibit import of the molecule into everted plasma membrane vesicles [49].

Similar experiments have been performed with the phosphoporin PhoE synthesized in vitro [50,51]. Using a gene missing an area encoding all of the hydrophobic core of the signal peptide, a PhoE was expressed with properties different from OmpF' or spheroplast OmpF. This monomer exhibited a heat-modifiability resembling that of OmpA and was recognized by monoclonal antibodies specific for cell surface-exposed conformational epitopes, i.e. this monomer (or very labile oligomer) was in a folded state. Trimerization was induced by the addition of cell envelope fractions, the most efficient being that of the om, but not by LPS alone. Trimerization was considerably less efficient (see Discussion in 37]) than reported for both OmpF species and the trimers were not assembled in the added membranes; they remained sensitive to proteinase K. It has been suggested [37] that this low efficiency of trimerization was due to the fact that a more or less oligomeri-zation-incompetent folded monomer was allowed to form by inhibiting protein synthesis before starting the trimerization reactions; from the in vitro transcription/translation sys-tem using OmpF', it appears that a large fraction of the trimers arose from nascent polypeptide chains. It should be added that with both the PhoE and OmpF' systems, dim-ers were also identified for OmpF that are known to be intermediates of porin assembly in vivo [52].

2.2. Sorting in vivo

2.2.1. Precursors
It has been assumed that signal peptides do not possess information for targeting [53–56], e.g. when such a peptide plus 158 residues of the periplasmic β-lactamase were fused to the complete PhoE protein, the porin ended up in the om [53]. This conclusion may need modification. A multivariate data analysis revealed that signal peptides from proteins with different final locations have clearly discernible different physico-chemical properties [57]. (It should be mentioned that the predictive power of this analysis is limited as the signal peptide of OmpA would localize the protein to the periplasm.) As suggested in the previous section (and see below), a signal peptide may influence the conformation of the processed protein and thus influence the efficiency of assembly in the om; such an effect may remain undetected unless the kinetics of assembly are followed carefully. An exam-ple for such a possibility is provided by the interaction of pre-OmpA with the cytosolic chaperone SecB. In cells with a *secB* null mutation [58], a decrease in the rate of process-ing could only be observed with pulse-chase experiments. No effect was visible when cells were analyzed at steady state growth levels, i.e. precursor accumulation was not de-tectable (unpublished data).

2.2.2. On the way to the outer membrane

2.2.2.1. The route. The most confusing situation concerns the pathway of proteins from the plasma membrane to the om. The morphological existence of adhesion sites between the two membranes [59; see also Chapter 21] and thus their involvement in protein export to the om has been questioned [60] (see rebuttal in [61]) and arguments have been put forward favoring a periplasmic route [10,37]; however, there are no hard facts proving any of these or other possibilities. There is conflicting evidence regarding the location of sites of protein newly incorporated into the om. For a porin of *Salmonella typhimurium*, it has been shown that it appears at the cell's surface in the form of randomly distributed, discrete patches [62]. The maltoporin LamB of *E. coli*, on the other hand, was thought to be integrated into the om predominantly in the vicinity of the forming septum [63]. If so, another type of adhesion site could play a role, the periseptal annuli which are involved in cell division [64,65]. Yet, it is obviously unlikely that for these two organisms and pro-teins, different mechanisms exist for the intercalation of the polypeptides into the om. There are also data concerning a property of sites of translocation in the plasma mem-brane which are not easily reconcilable. Using an in vitro system measuring import of pre-OmpA into membrane vesicles, it was found that vesicles of very low density were the most active [66]. Working with intact cells following translocation of the periplasmic maltose binding protein, it was demonstrated that nascent molecules of this protein were associated with a membrane fraction which had a density close to that of the om [67]. In summary, it is not yet known what is going on.

2.2.2.2. Folding and membrane incorporation. It is very well documented that folding intermediates of proteins exist after they are processed and have crossed the plasma membrane. It certainly appears that it is this folding which prepares proteins for entering the om or to become water soluble in the periplasmic space. As one example for periplasmic proteins, alkaline phosphatase may serve; it possesses two disulfide bonds [68] being formed only in the periplasm [69]. Using a null mutant of the recently discovered periplasmic disulfide oxidoreductase [70,71], two intermediates were identified [71]. One was the reduced protein and the other the oxidized form, both of which, in contrast to the enzyme in its final conformation, were trypsin-sensitive. However, sorting of proteins to the periplasm may involve more than simply acquiring water solubility; it could be that the translocation pathway across the plasma membrane is already participating in sorting of these polypeptides. Deletion mutants of β-lactamase [48] and maltose-binding protein [72–74] have been described which were translocated and processed but remained associated with the membrane. Such mutants had no effect on the level of proteins in the om. A distinct step of release from the periplasmic side of the plasma membrane has also been demonstrated for the wild type β-lactamase [75]. In addition, when the β-lactamase gene, placed under the control of the *lac* regulatory elements, was induced with isopropylthiogalactoside, enormous quantities of the enzyme were produced and its precursor accumulated but no block of export of OmpA, C or F was observed (G. Ried and U. Henning, unpublished). Thus, is seems that the translocation routes for periplasmic and om polypeptides diverge at some point.

Regions have been inserted into the genes of several om proteins, encoding foreign sequences at the turns of the β-barrels [13,14,76–81; 120, Section 3] and a remarkable tolerance of these polypeptides to such alterations was found. For example, a stretch of 54 non-LamB residues placed into a cell surface-exposed area of LamB had not much of an effect [79]. Insertion of 21 residue segments simultaneously into each of two such sites of OmpA did not prevent proper membrane incorporation [14] and similar multiple insertions did not inhibit the assembly process of PhoE [77]. With OmpA, such insertions have recently also been introduced into the three proposed turns at the periplasmic side of the om. In each case stretches of ca. 12 residues were perfectly tolerated, including the rather exotic sequence Arg-Arg-Arg-Val-Asp-Ala-Ser-Thr-Arg-Arg-Arg (G. Ried and U. Henning, unpublished). There are limitations [77]; for example, an accumulation of turn blocking residues [82,83] (Ile-Ala-Ile-Phe) at one 'inside' turn of OmpA conferred toxicity to the synthesis of the protein. In general, it is probably safe to conclude that all such turns do not possess topogenic information and that this information, the driving force for correct folding, is solely contained within the prospective transmembrane β-strands. In the following, we first discuss the least complicated case of OmpA (most likely no oligomerization involved) and then those of the trimeric porins LamB, OmpF and PhoE.

A folding intermediate, imp-OmpA (**im**mature **p**rocessed), has been identified which had passed the plasma membrane, was not yet assembled in the om and lacked heat-

modifiability [84]. Pulse-chase experiments followed by sucrose density gradient centrifugation located the protein, which was chased into the om, at the plasma membrane;
when overproduced, it was found to be associated with the om. As mentioned above
(Section 2.1), the conformation of this intermediate may well be identical with the OmpA
attached to small phospholipid vesicles at low temperature. Quite interestingly, a mutant
OmpA signal peptide was observed to restore a defect caused by a mutation in the mature
part of the protein [85]. The latter, an amino acid substitution, led to synthesis of an
OmpA that lacked heat-modifiability. Replacement of the signal sequence with another,
possessing an altered charge at its N-terminus (from +2 to −1), 'repaired' the mature
polypeptide, all of which was now heat-modifiable and thus apparently normally assembled in the om. The mutant signal peptide decreased the processing rate about fivefold
without affecting the total amount of OmpA produced. It is conceivable that, due to this
slowed rate, the conformation of the precursor differed from that of the mutant carrying
the alteration in the mature part only. If so, the result could be interpreted, as suggested
above, as an influence of the conformation of the precursor on that of the processed form.

How does the imp-OmpA find and enter the om? Ten overlapping deletions in *ompA*
have been constructed which miss segments of the area encoding the membrane moiety of
the protein [23,86]. The products were found in the periplasm in those five cases where
the region between residues 154 and 180 (the last trans-membrane β-strand consists of
residues ~160–170) was missing and the truncated proteins were seen associated with, but
not assembled in, the om in all other cases where this region was present. Unfortunately,
the nature of this association remains unknown and it is thus not clear whether it represents the normal sorting mechanism. If it does, there is no sequence specificity for the last
β-strand to function in assembly; it only needs to be amphiphilic or hydrophobic [15]. It
can also apparently be substituted for somehow at another site of OmpA. A remarkable
experiment was based on the facts that (i) a hybrid consisting of the signal peptide plus
the first nine N-terminal residues of the Braun lipoprotein fused to the mature periplasmic
β-lactamase directed the enzyme to the om [87] and (ii) a C-terminally truncated OmpA
ending at residue 159 (located at the last turn at the cell surface) was periplasmic (see
above). A tripartite gene was constructed encoding the pre-lipoprotein fragment mentioned fused to an OmpA fragment comprising residues 46–159 followed by the mature β-
lactamase [88]. A large fraction of the enzyme was located at the cells' surface. In this
protein, the OmpA part cannot form a β-barrel since it misses the first two and the last β-
strands. None of the partially deleted OmpAs mentioned above was assembled in the om,
including a protein missing residues 4–45; in fact, expression of this gene was very toxic.
It remains enigmatic how the OmpA part of the tripartite polypeptide could be arranged in
the om.

OmpA interacts specifically with LPS and it had been suggested earlier that it plays a
role in the cellular sorting of the protein [20,84,89]. This has become very unlikely since
de novo synthesis of LPS is not required for correct routing of OmpA [90] and since LPS
essentially occurs only in the outer leaflet of the om (for review, see [91]). Hence, the
protein apparently associates with LPS after it enters the om. How does it enter? The very
structure of the membrane part of the protein renders it very unlikely that the insertion
would proceed sequentially, i.e. one or two β-strands after another; the membrane
environment is not suitable for only parts of the β-barrel. (This may be the reason for the

high toxicity of the deletion mentioned above; the protein may initiate the intercalation process and then ruin the om by forcing regions into it which are not compatible with a lipid bilayer.)

In line with the results of the in vitro experiments, we suggest that the polypeptide, possessing a defined conformation, associates with the om followed by intercalation in toto. The role of the last β-strand, if any, may be to initiate this association (see Section 4). Not knowing anything about the route from the plasma membrane to the om, however, meaningful speculations regarding the way the om is found and offered are not possible. It may be added that the two cysteine residues within the periplasmic part of OmpA [6] form an intramolecular disulfide bond [92]. This bond is formed at a slower rate in the periplasmic oxidoreductase null mutant and it could be shown that this decreased rate is accompanied by a slower rate of folding of OmpA [71], perhaps only of the periplasmic moiety, but it could be that the efficiency of incorporation of the polypeptide into the om is also affected. If so, a periplasmic helper would exist and, as suggested before [22; 120, Section 3], other proteins of the chaperone type may be present in this compartment aiding om proteins in general.

A fairly consistent picture has evolved regarding the events following processing of the precursors of the trimeric porins. A set of overlapping deletions within the part of *phoE* which codes for the mature protein have been analyzed [41,93]. With a few notable exceptions, the corresponding proteins were found in the periplasm, not associated with the om. The authors concluded that sorting is probably determined by the overall conformation of the protein and not by relatively short stretches of amino acid residues. However, the lack of up to 14 N-terminal residues did not interfere with normal membrane incorporation [41], an understandable result in view of the now known 3D structure of the phosphoporin [28]. It was noted that at the C-terminus of 28 different om proteins, including a number of polypeptides of non-*E. coli*, Gram-negative organisms, a phenylalanine and, in two such proteins, a tryptophan is present [94]. In PhoE and OmpF, this position is located at the end of the last membrane spanning β-strand. Replacement of this residue in PhoE by other amino acid residues or a stop codon more or less drastically interfered with assembly in the om and led to an accumulation of monomers [94]. A phenylalanine is also present at the corresponding position of OmpA, but an aromatic residue at this location or nearby is not required for successful incorporation into the om [15]; OmpA does not trimerize and the primary effect in the *phoE* mutants could be a defective trimerization reaction. It has recently been stated, however, that in the in vitro system, trimerization was unaffected while membrane insertion was totally blocked [95]. Thus, the phenylalanine may be part of a sorting signal similar to that possibly existing in OmpA.

Several studies suggest that trimerization precedes membrane incorporation ('precedes' does not exclude the possibility that trimerization and assembly proceed concomitantly). It had been shown earlier that the subunits of OmpC and OmpF do not mix once present in the om as trimers [96]. Yet, the formation, in vivo, of heterotrimers between OmpC/F, OmpC/PhoE and OmpF/PhoE was readily detected; the amount of each heterotrimer corresponded to the level of expression of monomers [97]. Evidence for the existence of heterotrimers has also been presented for LamB in cells carrying a wild type and a mutant *lamB* gene [98]. Random mixing of monomers most likely then occurs after translocation across the plasma membrane.

In contrast to OmpA, the trimeric porins require LPS for success. The fatty acid analogue cerulenin blocks fatty acid and LPS syntheses [99,100] and was shown to cause a decrease in the cellular concentration of OmpC, OmpF and LamB but not of OmpA [101,102]. A more detailed study of this effect on OmpF revealed that the drug led to an accumulation of the monomer and, certainly indirectly, an inhibition of OmpF synthesis; this monomer remained unable to trimerize when the inhibitor was removed [103]. Similarly, in mutants producing a defective LPS (shortened core oligosaccharide), a drastic reduction in OmpC and OmpF concentrations was observed, caused by inhibition of their synthesis; no such effect on OmpA was observed [90] (the conclusion in this communication that the inhibition acted at the level of translation may be erroneous as detailed in Section 3). Most likely the effect of cerulenin and the LPS mutants have the same basis; the porins appear to require de novo LPS synthesis for efficient trimerization and continued synthesis. In good agreement with this proposal are results obtained with monoclonal antibodies directed against cell surface-exposed epitopes of OmpF [104]. The appearance of these epitopes during assembly of the processed protein was followed and four conformers were identified: a folded monomer, a metastable trimer (already dissociating at 50°C, instead of the >70°C required to monomerize wild type), a stable trimer and the final native trimer in the om. In contrast to the latter two, the appearance of the first two conformers was not inhibited in the presence of cerulenin; the monomer, metastable and stable trimers were found in the periplasmic fraction and the investigators suggested that trimerization and subsequent association with LPS occurred in this compartment. This fraction was defined as that being soluble after treatment of cells with sucrose, EDTA and lysozyme; all of the proteins may have been liberated from a peripheral association with the om by removal of divalent cations. Regardless of the exact location, all these data concerning OmpF, OmpC and probably also LamB again suggest that trimerization and association with LPS occur prior to membrane incorporation.

At variance with this conclusion was the interpretation of results obtained with a very interesting *tsf* (temperature sensitive folding) mutant of LamB [105]. A processed monomer and a metastable trimer have been described as folding intermediates of LamB [106]. The mutant, possessing an additional tyrosine in the mature LamB monomer [107], when grown at the non-permissive temperature, accumulated the monomer which was rapidly degraded unless cells missing the periplasmic protease DegP [108] were used. The effect was reversible; lowering the temperature allowed much of the monomer to be converted to stable trimers, hence, this monomer represented a true folding intermediate. The monomer was located (sucrose density gradient sedimentation and metrizamide density gradient floatation) at the om. The authors concluded that the monomer was behaving as an authentic om protein and that trimerization proceeded within this membrane. It could be, however, that the monomer was not incorporated into but only peripherally associated with the om (comparable to the partially deleted OmpA proteins discussed above; asking for accessibility of the mutant monomer to proteases may provide an answer) and that in the case of LamB, trimerization also precedes the final assembly process. If this is correct, membrane assembly of all trimeric porins may start with monomers already attached to the om. Association with newly synthesized LPS and trimerization should then occur at this site with trimerization presumably triggering membrane incorporation. Such a possibility would also be consistent with the fact that synthesis and transfer of LPS (mechanism of transfer also unknown) from the plasma membrane to the om [109,110] are not coupled with protein synthesis [111], i.e. the two types of molecules need not travel together.

3. Cross-regulation of synthesis of outer membrane proteins

The synthesis of these proteins is usually subject to multiple regulatory mechanisms, often one more individual and others involving groups of otherwise unrelated proteins. For example, synthesis of LamB, which is a member of the products of the maltose regulon, is under the control of the positive regulator for this regulon, MalT. The regulon, and thus the *lamB* gene, is also subject to catabolite repression, using the cyclic AMP-cyclic AMP-binding protein system (for review, see [112]). Finally, expression of the porin gene is subject to what we call cross-regulation here; *lamB* expression is coupled to that of other om proteins while the other members of the maltose system are not affected by this mechanism.

It has long been known that cells can sense the occupancy of the om by protein. For example, merodiploidy for the *ompA* gene did not cause a gene dosage effect [113] and induction of certain polypeptides of the om did not alter the concentration of total protein per unit surface [114,115]. This effect has recently been studied in some detail. Over-expression of the porin OmpC led to a rapid and almost complete block of synthesis of OmpA and LamB, but had no effect on the production of the periplasmic maltose-binding protein MalE [116]. As determined by pulse-labeling with [³H]uridine of cells over-producing OmpC, the concentration of *ompA* mRNA was not sufficiently reduced to explain the cessation of synthesis of this protein. In addition, expression of a *lamB-lacZ* fusion gene was not affected by overproduction of OmpC (in this protein, the first two residues of the LamB signal sequence were fused to most of the β-galactosidase). Essentially the same effects were observed in a strain carrying a mutant *ompC* gene coding for a protein missing residues 300 and 301 of the 346 residue porin [117]. The mutation caused a decrease in synthesis of the om proteins OmpA, C, F and Lc (the latter being a porin encoded by a lambdoid phage [118]), but apparently not of several unidentified minor proteins of this membrane. Expression of *ompF-lacZ* or *ompC-lacZ* operon fusions was not affected, and it was concluded that in this case and that of overproduction of OmpC, the mechanism of inhibition acted at the level of translation. For the mutant protein to be effective, it had to possess an intact signal peptide, destroying the continuity of its hydrophobic core (leading to cytosolic accumulation of the precursor) abolished the inhibitory action [117]. Most remarkably, however, a target protein did not need to be channelled into the export pathway to be subject to inhibition of its synthesis. Overproduction of OmpC inhibited production of LamB precursors with defective signal sequences not allowing them to be exported [119].

A very similar situation has been found for the OmpA protein. Overproduction of the polypeptide inhibited synthesis of OmpC and OmpF and its absence led to an increased concentration of OmpC and F (and vice versa). Furthermore, in 17 mutant proteins (insertions of 4 to 16 amino acids between residues 45 and 46), the expression (not over-expression) of two of them was found to inhibit synthesis of OmpC, OmpF and wild type

OmpA; there was no effect on the production of the periplasmic β-lactamase [120]. A correlation between sizes or sequences of the inserts and presence or absence of the inhibitory effect was not discernible, indicating that they acted indirectly by altering the conformations of the mutant OmpAs. The introduction of an internal deletion into one such effector protein abolished the inhibitory action. This deletion removed residues 154–227 and rendered the protein unable to assemble in the om; it remained periplasmic. A functional signal peptide was required for a mutant protein to be effective. Pulse-labeling with [³H]uridine showed that synthesis of *ompC* mRNA was not impaired when one of the effector proteins was produced which inhibited synthesis of OmpA, C, and F. It was concluded that also in these cases the feedback mechanism operated at the level of translation. (In none the cases described did precursors of the target proteins accumulate, i.e. a block of the export pathway was not the reason for their decreased concentrations.)

All these results allow a reasonable speculation to be made regarding the events involved in this regulatory mechanism [116,117,120]. The effector polypeptide (wild type or mutant) has to enter the export pathway and will bind to a component, presumably a protein of unknown location (plasma membrane, periplasm, om). As long as this component is bound to the effector, a signal is transmitted to the cytosol causing inhibition of synthesis of om proteins. An effective mutant protein would release this factor at slower rates than a wild type polypeptide; the internal deletion of OmpA abolishing the effector action may then, by virtue of an altered conformation, bind the factor only poorly.

However, there are conflicting results concerning the mechanism of inhibition [121]. Expression of a gene for an om protein from *Erwinia cloacae* in *E. coli* caused a reduction of OmpC and F concentrations. The effect of overproduction of this protein on the synthesis of OmpF was studied and it was found that it was inhibited and that *ompF* mRNA was not detectable when determined at steady state level. The authors also tested the influence of overproduction of the *E. cloacae* protein on the expression of *ompF*- and *ompC-lacZ* operon fusions. The activity of the β-galactosidase was reduced, but not as much as one might expect from the virtual absence of the *ompF* mRNA; this activity was reduced 4.4-fold or 2.5-fold when stemming from the *ompF-lacZ* or the *ompC-lacZ* fusion, respectively. The authors concluded that the inhibition occurred at the level of transcription. It is rather unlikely that the feedback mechanism in the case of the *E. cloacae* protein is different from that observed with OmpA or OmpC as effectors and we have reinvestigated the situation (M.-J. Lu and U.H., unpublished). It appears that all three groups have at least partially misinterpreted their results. We have now found that overproduction of OmpC causes an almost immediate decrease (about sixfold) in the stability of *ompA* mRNA; hence, the feedback control appears to influence mRNA decay rates. This will lead to lowered concentrations of target mRNA at steady state level and could easily have remained undetected in pulse label experiments; the response of protein or operon fusions would depend on whether the site(s) susceptible to mRNA degradation is(are) present in these fusions. These fusions used by the investigators cited were all different, and for most of them, the exact location of the fusion site has not been determined. The regulatory mechanism operates only for a subset of om proteins; many of those (unidentified) present at low concentrations were not affected. Hence, there should be a property common to the relevant mRNAs allowing them to be recognized as belonging to this subset.

It may well be that the same mechanism acts for periplasmic proteins [122,123]. A C-terminally truncated periplasmic phosphodiesterase was able to cross the plasma membrane but failed to be released into the periplasm. Synthesis of this protein inhibited the production of other (but not all) periplasmic polypeptides, including MalE, but not of several of the om, including LamB. On the basis of the effect of the incomplete phosphodiesterase on the expression of *malE-lacZ* operon or protein fusions, harboring only the part of *malE* encoding the signal peptide (no reduction of β-galactosidase activity), it was concluded that transcription and early translation were not affected. Yet, the control may not be translational (see above). The specific effect on the synthesis of a subclass of periplasmic proteins suggests, of course, that the inhibitory mechanism is the same as in case of the om proteins. If so, an epistasis experiment would suggest that the primary target of an effector protein is associated with the plasma membrane. The inhibition caused by the truncated esterase was dominant over the accumulation of pre-MalE in cells carrying a temperature sensitive *secA* mutation and grown at the non-permissive temperature [123]. (SecA acts at the inner side of the plasma membrane [124].)

4. Conclusions

The main obstacle for understanding the sorting mechanism is our lack of knowledge regarding the route from the plasma membrane to the om. Irrespective of this pathway, however, the presence of a receptor at the om, specific for its proteins, appears unlikely; it may create the chicken-and-egg problem. Should the route be periplasmic, a method of avoiding the plasma membrane has to be invoked (charge differences between the two membranes?) and the existence of a periplasmic chaperone(s), prohibiting folding to assembly-incompetence is likely. Comparable suggestions are not possible should adhesion sites be involved because their nature is unknown (membrane fusion has never been observed [61]).

None of the results discussed creates severe problems with the following suggested sequence of events. The imp-OmpA or a porin monomer associates with the om. It remains unknown whether a sorting signal is involved. (We feel that it does exist because membrane insertion occurs in one orientation and this probably requires an oriented association, e.g. with the last β-strand of OmpA facing the om. If so, such a signal may actually not serve sorting but a certain type of association.) The association may be accompanied by a conformational change, reminiscent of the membrane trigger hypothesis [125], causing non-sequential, spontaneous intercalation of the former into the membrane. The porin monomer trimerizes, picks up de novo synthesized LPS and assembles in the om as complete trimer. It should be stressed that this is speculative, at present it cannot be excluded that the monomer does enter the membrane. The LPS pick-up points to specialized sites of the om; it has been shown that newly synthesized LPS appears at the cell surface over sites of adherence of the two membranes [126], be this morphology per se an artefact of preparation or not. The existence of such sites is also suggested by the requirements for membrane insertion in vitro, e.g. highly curved vesicles for OmpA (Section 2.1). Can conditional, non-LPS mutants be isolated which are defective in sorting of more than one om protein?

The cross-regulation (Section 3) so far poses only questions. Which components are involved in transmitting which signal to the cytosol, how are mRNAs of a subset each of om and periplasmic proteins distinguished, is the feedback mechanism for both types of proteins basically the same? Answers could come from mutants (presumably conditional) with a defective feedback control.

Acknowledgement

We are grateful to Fritz Jähnig, Hiroshi Nikaido and Thomas Surrey for constructive criticism of this article.

References

1. Braun, V. and Hantke, K. (1991) in: G. Winkelmann (Ed.), CRC Handbook of Microbial Iron Chelates, Boca Raton, FL, pp. 107–138.
2. Mizuno, T., Chou, M.-Y. and Inouye, M. (1983) J. Biol. Chem. 258, 6932–6940.
3. Inokuchi, K., Mutoh, N., Matsuyama, S. and Mizushima, S. (1982) Nucleic Acids Res. 10, 6957–6968.
4. Overbeeke, N., Bergmans, H., van Mansfeld, F. and Lugtenberg, B. (1983) J. Mol. Biol. 163, 513–532.
5. Clément, J.M. and Hofnung, M. (1981) Cell 27, 507–514.
6. Chen, R., Schmidmayr, W., Krämer, C., Chen-Schmeisser, U. and Henning, U. (1980) Proc. Natl. Acad. Sci. USA 77, 4592–4596.
7. Randall, L.L., Hardy, S.J. and Thom, J.R. (1987) Annu. Rev. Microbiol. 41, 507–541.
8. Schatz, P. and Beckwith, J. (1990) Annu. Rev. Genet. 24, 215–248.
9. Pugsley, A. (1989) Protein Targeting, Academic Press, New York.
10. Nikaido, H. and Reid, J. (1990) Experientia 46, 174–180.
11. Morona, R., Klose, M. and Henning, U. (1984) J. Bacteriol. 159, 570–578.
12. Morona, R., Krämer, C. and Henning, U. (1985) J. Bacteriol. 164, 539–543.
13. Freudl, R., MacIntyre, S., Degen, M. and Henning, U. (1986) J. Mol. Biol. 188, 491–494.
14. Freudl, R. (1989) Gene 82, 229–236.
15. Klose, M., Jähnig, F., Hindennach, I. and Henning, U. (1989) J. Biol. Chem. 264, 21843–21847.
16. Vogel, H. and Jähnig, F. (1986) J. Mol. Biol. 190, 191–199.
17. von Heijne, G. (1981) Eur. J. Biochem. 120, 275–278.
18. MacIntyre, S., Freudl, R., Eschbach, M.-L. and Henning, U. (1988) J. Biol. Chem. 263, 19053–19059.
19. Sugawara, E. and Nikaido, H. (1992) J. Biol. Chem. 267, 2507–2511.
19a. Saint, N., De, E., Julien, S., Orange, N. and Molle, G. (1993) Biochim. Biophys. Acta 1145, 119–123.
20. Schweizer, M., Hindennach, I., Garten, W. and Henning, U. (1978) Eur. J. Biochem. 82, 211–217.
21. Dornmair, K., Kiefer, H. and Jähnig, F. (1990) J. Biol. Chem. 265, 18907–18911.
22. Surrey, T. and Jähnig, F. (1992) Proc. Natl. Acad. Sci. USA 89, 7457–7461.
23. Klose, M., Schwarz, H., MacIntyre, S., Freudl, R., Eschbach, M.-L. and Henning, U. (1988) J. Biol. Chem. 263, 13291–13296.
24. Weiss, M.S., Wacker, T., Weckesser, J., Welte, W. and Schulz, G.E. (1990) FEBS Lett. 267, 268–272.
25. Weiss, M.S., Kreusch, A., Schilz, E., Nestel, U., Welte, W., Weckesser, J. and Schulz, G.E. (1991) FEBS Lett. 280, 379–382.
26. Pauptit, R.A., Schirmer, T., Jansonius, J.N., Rosenbusch, J.P., Parker, M.W., Tucker, A.D., Tsernoglu, D., Weiss, M.S. and Schulz, G.E. (1991) J. Struct. Biol. 107, 136–145.
27. Weiss, M.S. and Schulz, G.E. (1992) J. Mol. Biol. 227, 493–509.
28. Cowan, S.W., Schirmer, T., Rummel, G., Steiert, M., Ghosh, R., Pauptit, R.A., Jansonius, J.N. and Rosenbusch, J.P. (1992) Nature (London) 358, 727–733.
29. Rosenbusch, J.P. (1974) J. Biol. Chem. 249, 8019–8029.

30. Nakae, T., Ishii, J. and Tokunaga, M. (1979) J. Biol. Chem. 254, 1457–1461.
31. Garavito, R.M. and Rosenbusch, J.P. (1986) Methods Enzymol. 125, 309–328.
32. Schindler, M. and Rosenbusch, J.P. (1984) FEBS Lett. 173, 85–89.
33. Eisele, J.-L. and Rosenbusch, J.P. (1990) J. Biol. Chem. 265, 10217–10220.
34. Metcalfe, M. and Holland, I.B. (1980) FEMS Microbiol. Lett. 7, 111–114.
35. Sen, K. and Nikaido, H. (1990) Proc. Natl. Acad. Sci. USA 87, 743–747.
36. Sen, K. and Nikaido, H. (1991) J. Bacteriol. 173, 926–928.
37. Sen, K. and Nikaido, H. (1991) J. Biol. Chem. 266, 11295–11300.
38. Josefsson, L.-G. and Randall, L.L. (1981) Cell 25, 151–157.
39. Freudl, R., MacIntyre, S., Degen, M. and Henning, U. (1988) J. Biol. Chem. 263, 344–349.
40. Smith, W.P. (1980) J. Bacteriol. 141, 1142–1147.
41. Bosch, D., Voorhut, W. and Tommassen, J. (1988) J. Biol. Chem. 263, 9952–9957.
42. Koshland, D., Sauer, R.T. and Botstein, D. (1982) Cell 30, 903–914.
43. Park, S., Liu, G., Topping, T.B., Cover, W.H. and Randall, L.L. (1988) Science 239, 1033–1035.
44. Liu, G., Topping, T.B. and Randall, L.L. (1989) Proc. Natl. Acad. Sci. USA 86, 9213–9217.
45. Laminet, A.A. and Plückthun, A. (1989) EMBO J. 8, 1469–1477.
46. Bolla, J.M., Bernadac, A., Lazdunski, C. and Pagès, J.-M. (1990) Biochimie 72, 385–395.
47. Lecker, S., Driessen, A.J.M. and Wickner, W. (1990) EMBO J. 9, 2309–2314.
48. Koshland, D. and Botstein, D. (1982) Cell 30, 893–902.
49. Tani, K. and Mizushima, S. (1991) FEBS Lett. 285, 127–131.
50. de Cock, H., Hekstra, D. and Tommassen, J. (1990) Biochimie 72, 177–182.
51. de Cock, H., Hendriks, R., de Vrije, T. and Tommassen, J. (1990) J. Biol. Chem. 265, 4646–4651.
52. Reid, J., Fung, H., Gehring, K., Klebba, P.E. and Nikaido, H. (1988) J. Biol. Chem. 263, 7753–7759.
53. Tommassen, J., van Tol, H. and Lugtenberg, B. (1983) EMBO J. 2, 1275–1279.
54. Hoffman, C.S. and Wright, A. (1985) Proc. Natl. Acad. Sci. USA 82, 5107–5111.
55. Takahara, M., Hibler, D.W., Barr, P.J., Gerlt, J.A. and Inouye, M. (1985) J. Biol. Chem. 260, 2670–2674.
56. Jackson, H.E., Pratt, J.E., Stocker, N.G. and Holland, I.B. (1985) EMBO J. 4, 2377–2383.
57. Sjöström, M., Wold, S., Wieslander, A. and Rilfors, L. (1987) EMBO J. 6, 823–831.
58. Kumamoto, C.A. and Beckwith, J. (1985) J. Bacteriol. 163, 267–274.
59. Bayer, M.H., Costello, G.P. and Bayer, M.E. (1982) J. Bacteriol. 149, 758–767.
60. Kellenberger, E. (1990) Mol. Microbiol. 4, 697–705.
61. Bayer, M.E. (1991) J. Struct. Biol. 107, 268–280.
62. Smit, J. and Nikaido, H. (1978) J. Bacteriol. 135, 687–702.
63. Ryter, A., Shuman, H. and Schwartz, M. (1975) J. Bacteriol. 122, 295–301.
64. Cook, W.R., MacAlister, T.J. and Rothfield, L.I. (1986) J. Bacteriol. 168, 1430–1438.
65. Cook, W.R., Kepes, F., Joseleau-Petit, D., MacAlister, T.J. and Rothfield, L.I. (1987) Proc. Natl. Acad. Sci. USA 84, 7144–7148.
66. Chen, L., Rhoades, D. and Tai, P.C. (1985) J. Bacteriol. 161, 973–980.
67. Thom, J.R. and Randall, L.L. (1988) J. Bacteriol. 170, 5654–5661.
68. Kim, E.E. and Wyckoff, H.W. (1991) J. Mol. Biol. 218, 449–464.
69. Derman, A.I. and Beckwith, J. (1991) J. Bacteriol. 173, 7719–7722.
70. Kamitami, S., Akiyama, Y. and Ito, K. (1991) EMBO J. 11, 57–62.
71. Bardwell, J.C.A., McGovern, K. and Beckwith, J. (1991) Cell 67, 581–589.
72. Ito, K. and Beckwith, J.R. (1981) Cell 25, 143–150.
73. Duplay, P., Smelcman, S., Bedouelle, H. and Hofnung, M. (1987) J. Mol. Biol. 194, 663–673.
74. Duplay, P. and Hofnung, M. (1988) J. Bacteriol. 170, 4445–4450.
75. Minski, A., Summers, R.G. and Knowles, J.R. (1986) Proc. Natl. Acad. Sci. USA 83, 4180–4184.
76. Agterberg, M., Adriaanse, H. and Tommassen, J. (1987) Gene 59, 145–150.
77. Agterberg, M., Adriaanse, H., van Bruggen, A., Karperien, M. and Tommassen, J. (1990) Gene 88, 37–45.
78. Charbit, A., Boulain, J.C., Ryter, A. and Hofnung, M. (1986) EMBO J. 5, 3029–3037.
79. Charbit, A., Molla, A., Saurin, W. and Hofnung, M. (1988) Gene 70, 181–189.
80. Charbit, A., Ronco, J., Michel, V., Werts, C. and Hofnung, M. (1991) J. Bacteriol. 173, 262–275.

395

81. Koebnik, R. and Braun, V. (1993) J. Bacteriol. 175, 826–839.
82. Paul, C. and Rosenbusch, J.P. (1985) EMBO J. 4, 1593–1597.
83. Chou, P.Y. and Fasman, G.D. (1978) Annu. Rev. Biochem. 47, 251–276.
84. Freudl, R., Schwarz, H., Stierhof, Y.-D., Gamon, K., Hindennach, I. and Henning, U. (1986) J. Biol. Chem. 261, 11355–11361.
85. Tanji, Y., Gennity, J., Pollit, S. and Inouye, M. (1991) J. Bacteriol. 173, 1997–2005.
86. Freudl, R., Schwarz, H., Klose, M., Rao Movva, N. and Henning, U. (1985) EMBO J. 4, 3593–3598.
87. Ghrayeb, J. and Inouye, M. (1984) J. Biol. Chem. 259, 463–467.
88. Francisco, J.A., Earhart, C.F. and Georgiu, G. (1992) Proc. Natl. Acad. Sci. USA 89, 2713–2717.
89. DiRienzo, J.M. and Inouye, M. (1979) Cell 17, 155–161.
90. Ried, G., Hindennach, I. and Henning, U. (1990) J. Bacteriol. 172, 6048–6053.
91. Nikaido, H. and Vaara, M. (1987) in: F.C. Neidhardt (Ed.), Escherichia coli and Salmonella typhimurium. Cellular and Molecular Biology, Vol. 1, American Society for Microbiology, Washington, DC, pp. 7–22.
92. Tani, K., Shiozuka, K., Tokuda, H. and Mizushima, S. (1990) J. Biol. Chem. 265, 17341–17347.
93. Bosch, D., Leunissen, J., Verbakel, J., de Jong, M., van Erp, H. and Tommassen, J. (1986) J. Mol. Biol. 189, 449–455.
94. Struyvé, M., Moons, M. and Tommassen, J. (1991) J. Mol. Biol. 218, 141–148.
95. Tommassen, J., Struyvé, M. and de Cock, H. (1992) Antonie van Leeuwenhoek J. Microbiol. 61, 81–85.
96. Ichihara, S. and Mizushima, S. (1979) Eur. J. Biochem. 100, 321–328.
97. Gehring, K.B. and Nikaido, H. (1989) J. Biol. Chem. 264, 2810–2815.
98. Marchal, C. and Hofnung, M. (1983) EMBO J. 2, 81–86.
99. Goldberg, I., Walker, J.R. and Bloch, K. (1973) Antimicrob. Agents Chemother. 3, 549–554.
100. Walenga, R.W. and Osborn, M.J. (1980) J. Biol. Chem. 258, 4257–4263.
101. Pagès, C., Lazdunski, C. and Lazdunski, A. (1982) Eur. J. Biochem. 122, 381–386.
102. Bocquet-Pagès, C., Lazdunski, C. and Lazdunski, A. (1981) Eur. J. Biochem. 118, 105–111.
103. Bolla, J.M., Lazdunski, C. and Pagès, J.-M. (1988) EMBO J. 7, 3595–3599.
104. Fourel, D., Mizushima, S. and Pagès, J.-M. (1992) Eur. J. Biochem. 206, 109–114.
105. Misra, R., Peterson, A., Ferenci, T. and Silhavy, T.J. (1991) J. Biol. Chem. 266, 13592–13597.
106. Vos-Scheperkeuter, G.H. and Witholt, B. (1984) J. Mol. Biol. 175, 511–528.
107. Heine, H., Francis, G., Lee, K. and Ferenci, T. (1988) J. Bacteriol. 170, 1730–1738.
108. Strauch, K. and Beckwith, J. (1988) Proc. Natl. Acad. Sci. USA 85, 1576–1580.
109. Osborn, M.J. (1984) The Harvey Lectures 78, 87–103.
110. MacGrath, B.C. and Osborn, M.J. (1991) J. Bacteriol. 173, 3134–3137.
111. Rothfield, L. and Pearlman-Kothencz, M. (1969) J. Mol. Biol. 44, 477–492.
112. Schwartz, M. (1987) in: F.C. Neidhardt (Ed.), Escherichia coli and Salmonella typhimurium. Cellular and Molecular Biology, Vol. 2, American Society for Microbiology, Washington, DC, pp. 1482–1502.
113. Datta, D.B., Krämer, C. and Henning, U. (1976) J. Bacteriol. 128, 834–841.
114. Boyd, A. and Holland, I.B. (1979) Cell 18, 287–296.
115. Diedrich, D.L. and Fralick, J.A. (1982) J. Bacteriol. 149, 156–160.
116. Click, F.M., McDonald, G.A. and Schnaitman, C.A. (1988) J. Bacteriol. 170, 2005–2011.
117. Catron, K.M. and Schnaitman, C.A. (1987) J. Bacteriol. 169, 4327–4334.
118. Blasband, A.J., Marcotte, W.R. and Schnaitman, C.A. (1985) J. Biol. Chem. 261, 12723–12732.
119. Click, E.M. and Schnaitman, C.A. (1989) J. Bacteriol. 171, 616–619.
120. Ried, G., MacIntyre, S., Mutschler, B. and Henning, U. (1990) J. Mol. Biol. 216, 39–47.
121. Stoorvogel, J., van Bussel, M.J.A.W.M. and van de Klundert, J.A.M. (1991) J. Bacteriol. 173, 161–167.
122. Hengge, R. and Boos, W. (1985) J. Bacteriol. 162, 972–978.
123. Hengge-Aronis, R. and Boos, W. (1986) J. Bacteriol. 167, 462–466.
124. Cunningham, K., Lill, R., Crooke, E., Rice, M., Moore, K., Wickner, W. and Oliver, D. (1989) EMBO J. 8, 955–959.
125. Wickner, W. (1980) Science 210, 861–868.
126. Mühlradt, P.F., Menzel, J., Golecki, J.R. and Speth, V. (1973) Eur. J. Biochem. 35, 471–481.

J.-M Ghuysen and R. Hakenbeck (Eds.), *Bacterial Cell Wall*
© 1994 Elsevier Science B.V. All rights reserved

CHAPTER 19

Uptake of solutes through bacterial outer membranes

ROLAND BENZ

*Lehrstuhl für Biotechnologie, Biozentrum der Universität Würzburg, Am Hubland,
D-97074 Würzburg, Germany*

1. Introduction

The cell envelope of Gram-negative bacteria consists of three different layers, the outer membrane, the murein and the inner membrane [1] (see also Chapter 1). The inner membrane represents a real diffusion barrier and contains the respiration chain, a large number of transport systems and the machinery for protein export [2]. Murein is a large heteropolymer that confers the rigidity of the cell envelope, the shape of the cell and protects it from osmotic lysis (see Chapter 2). The outer membrane of Gram-negative bacteria plays an important role in the physiology of these organisms. All nutrients or antibiotics either hydrophilic or hydrophobic have to cross this permeability barrier on the way into the periplasmic space and the cell, which means that it has special sieving properties. The active components of molecular sieving of the outer membrane are due to the presence of a few major proteins called 'porins' [3].

Most porins form transmembrane channels that sort solutes mainly according to their molecular masses and have little solute specificity. These porins represent general diffusion channels. Other porins contain binding sites for certain classes of solutes. These specific porins are very often induced under special growth conditions together with an inner membrane uptake system and a periplasmic binding protein. The primary structure of many porins has been resolved from the basis of their amino acid or their cDNA sequence. The sequences are not particularly hydrophobic, which means that the arrangement of the proteins in secondary and tertiary structure is responsible for their function as membrane channels. The porins are organized as trimers of three identical subunits of which each contains one channel [4,5]. The 3-D-structures of porin from *Rhodobacter capsulatus* [4] and of OmpF and PhoE from *Escherichia coli* [5] have been resolved from X-ray crystallography (see also Chapters 15 and 16). These porins and probably many if not all others are organized as hollow cylinders formed by 16 antiparallel β-barrels.

The function of the porins can be studied in intact cells and in vitro systems such as the liposome swelling assay [6] and the lipid bilayer technique [7,8]. Reconstitution experiments are necessary to re-establish the pore function and to show the integrity of the pore-forming complex after the isolation process. Furthermore, it is possible to perform a detailed study of the properties of the channels in the in vitro systems, which is very often not possible in in vivo systems. Reconstituted systems contain only a few components, and allow good control of the conditions. On the other hand, the main disadvantages of these systems are that artefacts are possible and that components are missing which may be necessary for the full activity of the porin pores.

The different reconstitution methods have their advantages and disadvantages. The lipid bilayer technique allows the resolution of molecular events, that is the study of single channels and the control of the integrity of the porin channels [7,8]. However, the estimation of the channel size from the single-channel conductance data is probably not very precise although not as bad as has been suggested recently [9]. The liposome swelling assay gives a better measure for the channel size and allows the measurement of solute flux through porin channels [6]. The basic disadvantage of this method is that it is not possible to control the integrity of the porin channels. The lipid bilayer technique allows also the evaluation of the stability constants for substrate binding to specific porins on the basis of titration experiments of lipid bilayer membranes with different solutes [10–12]. The binding site within specific porins represents a considerable advantage for the diffusion of defined classes of solutes through the outer membrane particularly at small solute concentrations. The binding proteins within the periplasmic space represent a sink for the solutes and allow vectorial transport and maximum presentation of them to the cytoplasmic membrane transport systems.

2. Structure and composition of the bacterial outer membranes

The murein of Gram-negative bacteria consists of a network of amino sugars and amino acids. The amino sugars (N-acetylglucosaminyl-N-actylmuramyl dimers) form long linear strands that are covalently linked between two muramyl residues by short tetrapeptides (see Chapters 2 and 3). Some of the components of the outer membrane are covalently bound to the peptidoglycan. Others appear to interact with the murein through ionic bridges in such a way that a tight network between murein and outer membrane is produced that is able to protect the cell from osmotic lysis [13]. This also means that bacterial outer membranes have a structure that is completely different to that of the inner membrane and most biological membranes.

2.1. Lipids and lipopolysaccharides

The outer membrane of Gram-negative bacteria is an asymmetric bilayer. The inner leaflet contains lipids that cover about 50% of the surface. The rest is occupied by proteins. The lipid composition is very similar to that of the inner membrane. The major lipid is the zwitterionic phosphatidylethanolamine [14]. In enteric bacteria, the outer monolayer is exclusively composed of lipopolysaccharides (LPS). Non-enteric bacteria or rough mutants of enteric bacteria may also contain phospholipids on the external surface [15].

LPS molecules are amphiphilic and exhibit structural similarities to lipids (see Chapter 13). Common to all LPS is the hydrophobic lipid A moiety (endotoxin). Its basic structure is a D-glucosaminyl-β-D-glucosamine backbone to which between five and seven saturated side chains are linked through ester and amide bonds. The hydrophilic polysaccharide core (O-antigen) consists of a variable number of identical saccharide subunits each composed of between 3 and 6 sugars (for review, see [16]). LPS molecules are associated with one another and proteins by ion bridges with divalent cations and form a rigid two-dimensional structure. In enteric bacteria, the lipopolysaccharides form a strong barrier to prevent the diffusion of hydrophobic molecules through the outer membrane [13].

2.2. Proteins and porins

The outer membrane of enteric and probably also of other Gram-negative bacteria contains approximately 50% protein (see Chapter 12). A large quantity of them belongs to a small number of 'major' proteins, which also include one or several porins with molecular mass between 30 and 50 kDa [17,18]. This means that the protein composition of the outer membrane of Gram-negative bacteria is relatively simple and SDS-PAGE of outer membrane proteins shows only a limited number of bands [19,20]. The most abundant protein in enteric bacteria is the murein-lipoprotein (see Chapter 14) with a molecular mass of 7.2 kDa and 7×10^5 copies per cell [21]. One-third of the lipoprotein molecules is covalently linked to the murein. Another major protein with a high copy number (10^5 per cell) in the outer membrane is OmpA with a molecular mass of 35 kDa (325 amino acids) [22]. Part of the OmpA molecule is localized in the periplasmic space (approximately 150 amino acids). The rest spans the outer membrane eight times [23]. Recently, it has been proposed that OmpA functions as an outer membrane channel similarly as the porins [24,25]. However, the data suggest that OmpA forms a channel with the same cross-section as OmpF [24] but much smaller activity (only 1% as compared with OmpF). This is difficult to understand since the diameter of the channel should only be 1.2 nm ($4.7 \times 8/\pi$ [4]) minus the length of the peptide side chains, which means that it is in reality probably only 0.6–0.7 nm wide. Similar considerations apply to the lipid bilayer data [25]. In this case, the conductance is larger than expected according to the dimensions of the channel. This means that the contribution of OmpA to the outer membrane permeability is questionable.

Besides the major proteins, the outer membrane also contains a large number of 'minor' proteins. The 'minor' proteins also have important roles in cell metabolism, for example in the export of toxins and colicins out of the cell [26,27]. TolC in *Escherichia coli*, recently identified as a porin, is responsible for this function. Other minor proteins such as FhuA or BtuB are involved in the uptake of iron [28] or vitamins [29], respectively. In the case of special starvation or growth conditions, some of these proteins such as PhoE [30,31] or LamB [32] of *Escherichia coli* become 'major' bands on SDS-PAGE because they are induced.

The outer membrane of any Gram-negative bacteria contains at least one porin in high copy number. Certain bacteria may contain several different constitutive porins. Others may be induced under special growth conditions. Porins from a large number of Gram-negative bacteria have been investigated (see [33] for a list of physical properties and channel characteristics of porins). The expression of porins in *E. coli* has been studied in detail. This organism is able to express a large number of porins in the outer membrane. Two general diffusion porins, OmpF and OmpC [34], are present in the outer membrane of *E. coli* K12. Their expression is influenced by the osmolarity of the culture medium and the nature of the carbon source. OmpF is preferentially expressed if the cells grow in media of low osmolarity or high cAMP levels [35,36]. OmpC is induced in media of high osmolarity. Change in the growth conditions may result in the formation of chimeric trimers, namely in trimers formed by two OmpC and one OmpF and vice versa [37]. The genes involved in the regulation of the osmolarity-sensitive expression of both porins are ompR (coding for a cytoplasmic 'regulator') and envZ (coding for a 'sensor' and located in the cell envelope) [38]. Another general diffusion porin, PhoE, is induced under conditions of phosphate limitations together with a periplasmic binding protein, an alkaline phosphatase and an inner membrane uptake system [39]. Several other porins may be expressed if the 'normal' porins are deleted or after phage attack [40].

Many porins of Gram-negative bacteria have similar properties as the general diffusion channels of E. coli. This means that most porins form wide water-filled channels in the outer membrane which sort solutes of different structures mostly according to their molecular weight. In fact, for these general diffusion porins, only minor interaction exists between the solutes and the pore interior. In many cases, porins appear to be closely associated with the peptidoglycan layer [3]. This offers an elegant method for their isolation and purification [41].

Besides the general diffusion pores, the outer membrane may also contain porins that are specific for one class of solutes. Prominent examples for these specific porins are LamB of E. coli [32,42] and Salmonella typhimurium [43] and OprP of Pseudomonas aeruginosa [44]

3. Investigation of the porin function in vitro

3.1. Study of porin function in vivo

The permeation of solutes through the outer membrane can be studied by the following several different methods. (i) the uptake of radioactively labelled substrates into the cell [32,45,46]; (ii) the use of the β-lactamase activity localized in the periplasmic space of certain Gram-negative bacteria [47,48]; (iii) cell growth [49]; (iv) susceptibility of bacteria against antibiotics [50]. From all these methods, the β-lactamase activity, that is the hydrolysis of β-lactam antibiotics in the periplasmic space, yields the most precise information on porin function. It has to be noted, however, that the absolute rates of hydrolysis in different strains of the same organisms are not meaningful because of different expressions of the porins. Similarly, the permeability properties of the outer membranes of different organisms cannot be compared using the same method. It is only possible to make a comparison on the basis of the uptake ratios of different β-lactam antibiotics (for instance between neutral, positively and negatively charged antibiotics [51], or between hydrophilic and hydrophobic antibiotics in a given system [52]. This means that only the selectivity of a given porin and selectivity changes after its side-directed mutagenesis can be evaluated from these experiments [51,52]. All the methods mentioned above have a considerable drawback since they give only precise information on the outer membrane permeability if the flux through this membrane is rate-limiting and not the β-lactamase activity or the inner membrane transport system [53]. Furthermore, it is essential that the solute flux occurs only through one porin. Substitute methods for the study of porin function are provided by in vitro techniques.

3.2. Isolation of bacterial porins

The porins of most Gram-negative bacteria are tightly associated with the peptidoglycan layer [3,41]. This allows the rapid isolation of the porins through the preparation of the murein-protein complex [41]. The cells are harvested in the late logarithmic phase. They are washed and resuspended in a small volume before they are passed several times through a French pressure cell. The pellet of a subsequent centrifugation step contains the cell envelope. Most components of the cell envelope are soluble in detergents. The insoluble material is composed of murein and a few proteins either covalently bound to or associated with the peptidoglycan. The porins can be released by standard methods either by digestion of the murein or by the salt extraction method [3,41,54]. After the release, pure porin may be obtained by column chromatography using gel filtration or affinity chromatography.

If the porin is not peptidoglycan-associated or if the channel-forming activity is destroyed by the SDS-treatment, the outer membrane has to be isolated first. The procedure follows basically the method first proposed by Miura and Mizushima [55]. The cells are disrupted and the envelope fraction is layered on top of a two-step sucrose gradient (70% and 54% [12]). After centrifugation in an ultracentrifuge for about 12 h at 80 000 $\times$ g, the outer membrane is in the pellet because of its higher density compared with the inner membrane.

3.3. Reconstitution methods for porins

Several different methods have successfully been used to reconstitute porin pores into model systems. Here, only two methods are described in detail, the liposome swelling assay and the lipid bilayer technique. These two methods have been used in the majority of reconstitution experiments. For the other methods, the interested reader is referred to recent reviews [18,33,56]. The liposome swelling method was introduced in the study of the permeability properties of porin channel by Nikaido and co-workers [6,52,57,58] following a method established earlier by Bangham et al. [59]. Porin-containing multilamellar liposomes are formed from lipid and protein in a buffer containing a large molecular mass solute such as dextran or stachyose that is not able to penetrate the porin channels and is entrapped inside the liposomes. The liposomes are added under rapid mixing to an isotonic solution of a test solute. If this solute can penetrate the pores, the total concentration of solutes inside the liposomes increases because stachyose or dextran is retained. The liposomes are ideal osmometers, which means that the influx of the test solute occurs together with water. The liposome swelling, that is the influx of the test solute, can be detected by a measurement of the optical density, which is measured about 10 s after the mixing [6]. The initial swelling rate is a measure of the penetration rate of the solute through the porin channels. It is possible to compare the penetration rates of different solutes by using the same liposome preparation and to calculate the effective diameter of the porin channel from the initial swelling rates according to the theory of Renkin [60]. It has to be noted that the time between the mixing of the liposomes with the test solutes and the measurement of the optical density has a certain influence on the channel size (Zembke and Benz, unpublished results). Larger channel diameters, as observed by Nikaido and Rosenberg [6,52], are obtained when this time is below 10 s.

Two slightly different methods have been used for the reconstitution of porins into lipid bilayer membranes. The first uses solvent-containing membranes formed according to the Mueller–Rudin method [7], while the other uses virtually solvent-free membranes [8]. The simplest method consists of the addition of detergent-solubilized porin, in very small concentrations (1–100 ng/ml), to the aqueous phase bathing a painted (i.e. solvent-containing) lipid bilayer membrane. Subsequently, the membrane conductance (the membrane current per unit voltage) increases in step-like fashion [7].

Figure 1 shows a typical reconstitution experiment of this type. LamB of *E. coli* K12 was added in final concentration of about 10 ng/ml to the aqueous phase bathing a lipid bilayer membrane from diphytanoyl glycerophosphocholine/n-decane. The addition of the porin results in a stepwise increase of the membrane conductance. Two types of steps are observed. Almost all steps had a single-channel conductance of 150 pS. Only one step in Fig. 1 has an amplitude of about 1.5 nS, which means that it is about ten times larger than the other steps. This could mean that a large LamB aggregate is inserted into the membrane. It is more likely, however, that the large step represents the insertion of an OmpF channel into the membrane. This means that porin preparations may contain contaminants, which cannot be detected in SDS-PAGE, but can be observed in lipid bilayer experiments. On the other hand, it is possible to sort different porin channels using this method. It has to be noted that a similar sorting of channels is not possible by the liposome swelling assay. Using the same approach, multichannel experiments (experiments with a large number of channels) can be performed. A maximum of between 10^6 and 10^8 pores per cm^2 can be incorporated into lipid bilayer membranes. This means that the reconstitution of porin pores is not a rare event.

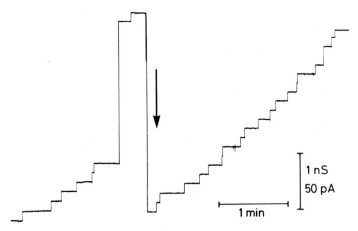

Fig. 1. Single channel record of a membrane from diphytanoyl glycerophosphocholine/n-decane in the presence of 5 ng/ml LamB of *E. coli* and 1 M KCl. A voltage of 50 mV was applied through calomel electrodes with salt bridges; $T = 25°C$. The arrow indicates the shift of the base line. Reproduced from [10] with permission.

The experiments with solvent-free membranes were performed in the following way [8]. Vesicles reconstituted from lipid and protein were spread on the surface of the aqueous phase on both sides of thin Teflon foil, which has a small circular hole (less than 100 μm diameter) above the initial water levels. The surfaces of both aqueous compartments were covered by lipid multilayers containing protein. The water levels on both sides of the membrane are now raised and a folded lipid bilayer membrane is formed across the small hole. So far it is not clear how the porin pores are incorporated into the lipid bilayers since no channels could be detected after the formation of the membranes. The pores were always activated as multiples of the single conductance unit at large voltages, presumably because of the fusion of porin-containing vesicles with the membrane [8]. When the porins were added to the aqueous phase after formation of the membranes from pure lipids [61], the pores were also incorporated into the membranes as single conductive units.

As far as single-channel conductance and channel selectivity are concerned, the results obtained with both types of lipid bilayer membranes are identical. Substantial differences were only observed for the voltage-dependence of the porin channels. Whereas porin channels incorporated into solvent-containing membrane show only a minor voltage-dependence [7,10], a strong voltage-dependence has been observed for channels reconstituted into solvent-free bilayers at voltages above 100 mV [8,61]. The reason for this discrepancy between the results obtained from both methods is discussed in some detail in the next section.

4. Properties of the general diffusion porins

The general diffusion channels of the outer membrane of Gram-negative bacteria sort hydrophilic solutes mainly according to their molecular mass. Molecules with larger molecular mass than a well-defined exclusion limit are excluded from the pores and cannot enter the periplasmic space, whereas smaller solutes seem to permeate freely. The exclusion limit of porins from *E. coli* [62], *S. typhimurium* [3] and *P. aeruginosa* [63] have been measured using radioactive labelled solutes. The exclusion limits are 600, 800 and about 5000 Da, respectively. The diameters of other general diffusion pores have been derived from the liposome swelling assay. These diameters range between 1 and 1.5 nm [6,52,64–66]. The majority of porin studies have been performed with the lipid bilayer technique (see [33,67] for lists). Wide water-filled channels have also been found in these experiments [33]. This means that a large number of ions are permeable through general diffusion porins without any detectable interaction with the channel interior. Table I shows the single channel conductances, G, measured with *Rhodobacter capsulatus* porin. The single channel conductance is a linear function of the bulk aqueous conductivity, σ. This means that despite a considerable variation in G for different salts and concentrations, the ratio G/σ varied only slightly. This is also demonstrated in Fig. 2, in which the average single channel conductances, G, of OmpF from *E. coli* [68], of the porin from *Rhodobacter capsulatus* [69] and of the porin of the anaerobic Gram-negative *Pelobacter venetianus* [70] are given as functions of the bulk aqueous conductivity σ. The difference

TABLE I

Average single-channel conductance, G, of *Rhodobacter capsulatus* porin in different salt solutions

Salt	c (M)	G (nS)	G/σ (nm)
KCl	0.003	0.012	0.29
	0.01	0.038	0.29
	0.03	0.12	0.29
	0.1	0.35	0.29
	0.3	1.1	0.29
	1.0	3.3	0.30
	3.0	9.5	0.35
RbCl	1.0	3.4	0.30
LiCl	1.0	1.4	0.20
K-acetate	1.0	2.3	0.33
Tris–HCl	0.5	0.52	0.17

The membranes were formed from diphytanoyl glycerophosphocholine/n-decane. $V_m = 20$ mV; $T = 25°C$. Taken from [69].

between OmpF and the *R. capsulatus* porin probably reflects the larger diameter of the latter channel [4,5].

The single channel conductance data of Table I suggest, in principle, that the *R. capsulatus* porin is selective for cations. Nevertheless, anions must have also a certain permeability through the porin channel since potassium chloride and potassium acetate do not have the same single channel conductance, which would be expected for a selective channel. This result suggests that the anions influence the movement of the cations through the porin (i.e. anions are also present inside the channel), otherwise it is difficult to understand why the single channel conductance in KCl is higher than in potassium acetate (see also below). The influence of the anions on the cation mobility is easy to understand on the basis of the 3D structure of the porin [4], since the channel interior contains both negatively and positively charged amino acids. The presence of only one type of charged group inside a channel results in charge effects, which means that the single channel conductance is not a linear function of the bulk aqueous ion concentration. Instead, a dependence on the square root of the concentration has been observed for the porins of *Pelobacter venetianus* ([70]; also included in Fig. 2) and *Acidovorax delafildii* [71]. After correction for the charge effects, a linear conductance– concentration relationship may be obtained [71].

Most porins of enteric bacteria are cation selective, probably because bile acids and other anionic tensides are present in the intestine. The ionic selectivity of porins can be measured in vivo by using the β-lactamase activity and positively or negatively charged cephalosporins [72,73]. From the two in vitro methods, the lipid bilayer technique offers considerable advantage over the liposome swelling assay, since Donnan potentials may influence the movement of charged solute into the liposome. In fact, most selectivity measurements in vitro have been performed with the latter method and only a few with the liposome swelling assay.

The lipid bilayer assay allows the evaluation of the ionic selectivity by measuring the membrane potential under zero-current conditions. From the measured V_m and the con-

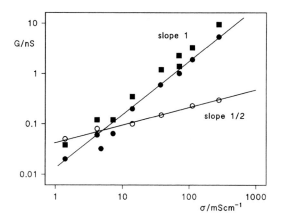

Fig. 2. Average single channel conductance, *G*, of OmpF of *E. coli* [68] (full circles), of porin from *R. capsulatus* [69] (full squares) and of porin from *Pelobacter venetianus* [70] (open circles) given as a function of the specific conductance of the aqueous salt solutions. The membranes were formed from diphytanoyl glycerophosphocholine/n-decane. The lines have a slope of 1 or 1/2.

centration gradient c''/c' across the membrane, the ratio P_c/P_a of the permeabilities (P_c for cations and P_a for anions) can be calculated using the Goldman–Hodgkin–Katz equation [68]. Figure 3 shows the results of selectivity measurements with different porins from *E. coli* reconstituted into lecithin membranes [74]. As mentioned above, most porins are cation-selective. In these cases, a positive potential is obtained at the more dilute side of

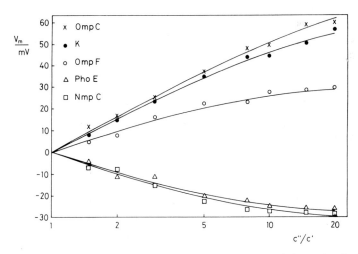

Fig. 3. Zero-current membrane potentials V_m of membranes from diphytanoyl glycerophosphocholine/n-decane. in the presence of five different porins of *E. coli* as a function of the KCl-gradient c''/c' across the membranes. The potential corresponds to the more dilute side of the membrane. The lines were drawn according to the Goldman–Hodgkin–Katz equation [68] with the following values for P_{cation}/P_{anion}: 26 (OmpC), 12 (K), 3.6 (OmpF), 0.3 (PhoE) and 0.26 (NmpC). Reproduced from [74] with permission.

TABLE II

Permeability ratios P_{cation}/P_{anion} of different porins from *E. coli*, *R. capsulatus* and *P. aeruginosa* for three different salts

Species	Porin	P_{cation}/P_{anion}		
		LiCl	KCl	KCH$_3$COO
E. coli	OmpF	1.5	3.6	8.2
	OmpC	15	26	54
	PhoE	0.11	0.30	0.65
	K	7.6	15	34
	NmpC	0.055	0.27	0.53
R. capsulatus		2.2	6.1	17
P. aeruginosa	OprP	≪0.01	≪0.01	≪0.01

The permeability ratios were calculated from experiments similar to those of Fig. 3 by using the Goldman–Hodgkin–Katz equation [68]. The data were taken from refs. [74] (*E. coli*), [69] (*R. capsulatus*) and [84] (*P. aeruginosa*).

the membrane. For two porins (PhoE and NmpC), the more dilute side is negative. Table II shows the permeability ratios P_a/P_c for different porins of *E. coli* [74] and of porin from *R. capsulatus* [69]. OmpF and OmpC are present in the strain K12. Their expression is regulated by the osmolarity of the growth media [34] and other conditions [35]. PhoE is induced in *E. coli* under the conditions of phosphate starvation [30,39]. NmpC appears in revertants of porin-deficient mutants [40] and porin K was found in *E. coli* strains that form capsules [75].

The results of Table II demonstrate that the selectivity of the general diffusion porins is not an absolute one, i.e. these channels are not permeable for one class of ions alone. This means that the mobility of the counterions also plays an important role in channel selectivity. This result is consistent with the results obtained from X-ray crystallography of OmpF and PhoE of *E. coli* [5] and of porin from *R. capsulatus* [4] (see also Chapters 15 and 16). In all these cases, negatively and positively charged amino acids are present inside the channels. The positively charged groups are preferentially located towards the side of the channels, at which the monomers are connected within the trimer [4,5]. The overall selectivity is caused by an excess of one class of charges, either positive or negative as chemical modifications of porins clearly indicate [76,77].

4.1. The eyelet controls the properties of general diffusion pores

The architecture of the general diffusion channels of different Gram-negative organisms, such as *E. coli* [5], *R. capsulatus* [4], *Haemophilus influenza* [78] and others are probably very similar (see also Chapter 12). The channel monomer is formed by 16 antiparallel transmembrane β-strands forming a hollow cylinder with a length of approximately 4 nm and a diameter of approximately 24 nm ($= 16 \times 4.7$ nm/π [5]). From the diameter of the cylinder formed by the polypeptide chain, approximately 0.4–0.6 nm has to be subtracted for the polypeptide side chains that limit the passage of the hydrophilic solutes. The porin channels are further influenced by the third external loop (the eyelet) between β-strands 6 and 7 (from the N-terminal end) which is folded back inside the channel and decreases its diameter to approximately 0.8×1 nm [4,5]. This eyelet is also very important for the channel selectivity. It has been demonstrated by side-specific mutagenesis of PhoE that Lys 125 (localized within this third external loop) plays a substantial role in the preference for anionic solutes and the binding of phosphate and other anions [51]. The insertion of a nine amino acid long stretch into the third loop of PhoE (the eyelet) results in a dramatic decrease of the cross-section of the channel and the single channel conductance decreases by more than a factor of ten [50].

Some porins form voltage-dependent channels when they are reconstituted into lipid bilayer membranes [8,79]. This voltage dependence probably represents a reconstitution artefact since porin channels in intact cells are not affected by Donnan potentials, which represents the only possibility of a potential across the outer membrane in vivo [80]. The eyelet is probably responsible for the voltage dependence since the structure of the porin channel itself should be very stable as it is stabilized by many hydrogen bonds [4,5] and considerable energy would be required to break them down. The voltage dependence is much higher in virtual solvent-free membranes than it is in solvent-containing membranes (painted bilayers). The reason for this is still questionable. It may be caused by the structural difference of solvent-free and solvent-containing bilayers. The latter have a thickness of the hydrocarbon core of approximately 4 nm [81] which is very close to that of the outer membrane of Gram-negative bacteria and the length of the outer rim of porin trimers [4,5]. Solvent-free membranes are much thinner and have a thickness of 2.5–2.8 nm [82]. This also means that the porin is not completely buried in the membrane which in turn may influence the stability of the location of the third loop within the channel and the channel becomes voltage-dependent.

5. Properties of the specific porins

5.1. Mechanism of solute transport through specific porins

For the transport of substrates through specific porins, it is assumed that it can be explained by a simple two-barrier one site-model [83–85]. This model assumes a binding site in the centre of the channel. The rate constant k_1 describes the jump of the sugars from the aqueous phase (concentration c) across the barrier to the central binding site, whereas the inverse movement is described by the rate constant k_{-1}. The stability constant of the binding between a substrate molecule and the binding site is $K = k_1/k_{-1}$. Furthermore, it is assumed that only one substrate can bind to the binding site at a given time and that no substrate or other molecule [84,85] can pass the channel when the binding site is occupied. This means that a substrate can enter the channel only when the binding site is free. The probability, p, that the binding site is occupied (identical concentrations on both sides) and does not conduct ions is given by

$$p = Kc/(1 + Kc) \tag{1}$$

and that it is free and the channel conducts ions is given by

$$1 - p = 1/(1 + Kc) \tag{2}$$

G_{max} is the conductance of a large number of specific porins (number n; single channel conductance G) reconstituted into a membrane in the absence of the substrate:

$$G_{max} = nG \tag{3}$$

The conductance, $G(c)$, of the n channels in the presence of a substrate (concentration c on both sides of the membrane) decreases since the substrate molecules bind with the stability constant, K, to the binding-site. $G(c)$ is given by the probability that the binding site is free:

$$G(c) = G_{max}/(1 + Kc) \tag{4}$$

or by the following equation:

$$(G_{max} - G(c))/G_{max} = Kc/(Kc + 1) \tag{5}$$

which means that Lineweaver–Burke plots of the titration experiments could be used for the evaluation of the substrate binding.

The concentration gradient $c'' - c'$ across the membrane results in a net flux through the channel under stationary conditions. The net flux of substrate molecules, Φ, through such a one-site two-barrier channel is given by the net movement of molecules across one barrier:

$$\Phi = k_1 c''/(1 + K') - k_{-1} K'/(1 + K') \tag{6}$$

K' is given by

$$K' = K(c' + c'')/2 \tag{7}$$

In eqn. (6), the rate constants k_1 and k_{-1} are multiplied by the probabilities that the binding site is free $(1 - p = 1/(1 + K'))$ or occupied $(p = K'/(1 + K'))$, respectively. The reason for this is that the binding site can only be occupied by a substrate molecule from the aqueous phase when it is free and a substrate can only leave the site when it is occupied. Equation (7) has the following form in the case $c'' = c, c' = 0$:

$$\Phi = k_1 c/(2 + Kc) \tag{8}$$

The latter form may be used for the discussion of the concentration dependence of the substrate flux through a specific porin. Using instead the following expression (with P permeability):

$$\Phi = Pc \tag{9}$$

the permeability of a one-site two barrier channel is given by

$$P = k_1/(2 + Kc) \tag{10}$$

which means that the maximum permeability, P_{max}, is obtained at very small substrate concentration:

$$P_{max} = k_1/2 \tag{11}$$

Similar to enzyme reactions, the maximum turnover number, T, of the channel (i.e. the maximum flux at very high substrate concentration on one side of the membrane) is given by

$$T = \Phi_{max} = k_{-1} \qquad (12)$$

5.2. Sugar transport through LamB of Escherichia coli

The LamB-proteins of *E. coli* [32] and *S. typhimurium* [43] are inducible specific porins, which are encoded in the maltose-inducible mal-regulon of Enterobacteriaceae. This regulon consists of two regions, malA and malB [86–88]. Both regions contain a number of genes coding for proteins involved in the uptake of maltose and maltodextrins into the cell and their degradation. The properties of LamB of *E. coli* and *S. typhimurium* have been studied in detail in vivo [32,43,46] and in vitro [10,57,61,85] and the sequences of both proteins are known [89,90]. The LamB channel is different from the general diffusion porins and does not show any immunologic cross reaction with them [91]. The results of liposome swelling experiments in the presence of maltoporin showed that the rate of penetration of maltose was much larger than that of sucrose or lactose despite similar molecular weight [57]. This facilitated diffusion has been explained by the assumption of a specific binding site for maltose and maltooligosaccharides inside the channel [92].

Lipid bilayer experiments are not very well suited for permeability measurements with uncharged solutes. Experiments with LamB are no exception. However, lipid bilayer experiments are able to give insight into the binding of maltose and maltooligosaccharides to this channel since the channel conductance is influenced in the presence of sugars. LamB has a single-channel conductance of 155 pS in 1 M KCl. The addition of maltotriose reduces the single-channel conductance, and large concentrations completely block the pore [10]. In fact, it is possible to titrate the LamB conductance with maltose, maltooligosaccharides and other sugars and to calculate the binding constant of these substances to the channel interior as described in the previous section [85]. Figure 4 shows an experiment of this type. LamB was added to a black lipid bilayer membrane, and the increase in the conductance was measured as a function of time. Thirty minutes after the addition, the membrane conductance was almost stationary. Different concentrations of maltotriose were added to the aqueous phase with stirring (arrows). The membrane conductance decreased as a function of the maltotriose concentration (Fig. 4).

Using the formalism described above, a binding constant of about 2×10^3 l/mol for maltotriose to the pore interior can be calculated from the conductance decrease [10,85]. This value and other stability constants (see Table III) for the binding of different sugars show excellent agreement with the binding constant of maltotriose to whole cells [42]. These results indicate that it is possible to study the specificity of channels for neutral solutes in lipid bilayer membranes. Table III shows the stability constant for the binding of a large variety of sugars to the binding site inside the LamB channel. The stability constant usually increased with the number of residues in a sugar chain. For example, in the series glucose, maltose to maltopentaose, the binding constant increased about 1800-fold, whereas there was no further increase between 5 and 7 glucose residues in the maltooligosaccharides. This may be explained by assuming that the binding site has approximately

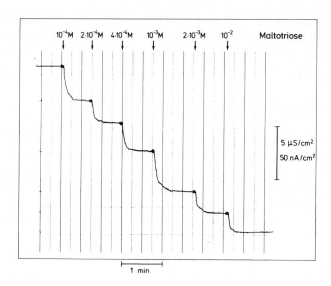

Fig. 4. Titration of LamB-induced membrane conductance with maltotriose. The membrane was formed from diphytanoyl glycerophosphocholine/n-decane. The aqueous phase contained 50 ng/ml LamB, 1 M KCl and maltotriose at the concentrations shown at the top of the figure; $V_m = 10$ mV. Reproduced from [10] with permission.

the length of five glucose units, i.e. it is about 2.5 nm long. All disaccharides such as maltose, lactose, sucrose and others had stability constants between 18 and 250 l/mol for the binding to LamB. This means that sucrose and maltose had approximately the same affinity to LamB. The relative rates of permeation of these sugars as derived in vitro [57] and in vivo [93], differed substantially. This can be explained by the kinetics of the sugar transport and means that widely separated k_1 and k_{-1} may have the same ratio K (i.e. the

TABLE III

Stability constants, K, for the binding of different sugars to the LamB channel

Sugar	K (l/mol)	K_S (mmol/l)
D-Glucose	9.5	110
Maltose	100	10
Maltotriose	2500	0.40
Maltotetraose	10000	0.10
Maltopentaose	17000	0.059
Maltohexaose	15000	0.067
Maltoheptaose	15000	0.067
Trehalose	46	22
Lactose	18	56
Sucrose	67	5
Gentibiose	259	4

K was calculated from titration experiments similar to that shown in Fig. 4. K_S is the half saturation constant. Taken from [85].

same stability constant K; see below). Further studies on mutant LamB [46,94–97] and the investigation of the primary sequence of the mutants [95,97] and the evaluation of sugar flux and binding [94,96] may give further insight into the three-dimensional structure of the binding site and may help to complete the model of LamB [98].

5.3. Analysis of sugar-induced current noise of LamB channels

The results of the liposome swelling assay, together with the binding data, allow only a qualitative description of the kinetics of sugar movement through the LamB channel [85,99]. This means that the rate constants k_1 and k_{-1} can only be given with respect to the movement of maltose [85]. Furthermore, the rate constants cannot be derived from the analysis of the sugar-induced block of the LamB channels in single-channel experiments because of the pure time resolution of this method. Therefore, an alternative method was used, the analysis of the sugar-induced current noise of the LamB channel [100,101]. The experiments were performed in the following way: black lipid bilayer membranes were formed and LamB was added to both sides of the membrane. Instead of 30 min (as in the case of the titration experiments), a wait of about 60 min was necessary to avoid any further conductance increase during the experiments since a change in the membrane conductance had a large influence on the noise measurements. Subsequently, the power density spectrum of the current noise was measured at a voltage of 25 mV (see Fig. 5, trace 1). Then maltopentaose was added at a concentration of 0.037 mM and the power density spectrum was measured again and the background noise subtracted (trace 2). At another concentration of maltopentaose (c = 0.077 mM), the power density spectrum (minus the background) corresponded to trace 3. The power density spectra of traces 2 and 3 are of the Lorentz type expected for a random switch with unequal on and off probabilities [99]. Both could be fitted to the equation

$$S(f) = S_0/(1 + (f/f_c)^a) \tag{13}$$

where $S(f)$ is the power density of the current given as a function of the frequency (in A^2

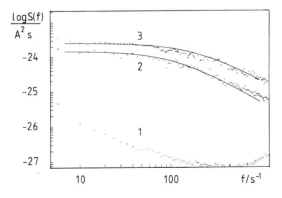

Fig. 5. Power density spectra of sugar-induced current noise of 380 LamB channels. Trace 1 shows the control (1 M KCl). Trace 2: the aqueous phase contained 0.039 mM maltopentaose (f_c = 106 1/s) and trace 3, 0.077 mM maltopentaose (f_c = 177 1/s); T = 25°C; V_m = 25 mV. The data were fitted to eqn. (13).

$\times$ s), S_0 is the plateau value at very small frequencies f (in 1/s) and f_c is the corner frequency. The exponent a was approximately 1.6 in the experiments given in Fig. 5. The corner frequency obviously increased with increasing sugar concentration. According to a previously published formalism [102], $2\pi f_c$ was set to the rate, $1/\tau$, of the current-modulating process, that is the binding of the sugar to the binding site inside the channel. The corner frequencies of the experiments shown in Fig. 5 and of other maltopentaose concentrations (not shown) could be reasonably well fitted to the equation

$$2\pi f_c = 1/\tau = k_1 c + k_{-1} \qquad (14)$$

The values for both rate constants derived from the experimental results of Fig. 5 and similar experiments with other maltopentaose concentrations were $k_1 = 8.1 \times 10^6$ l/(mol s) and $k_{-1} = 310$ s^{-1}. The stability constant was $K = 2.6 \times 10^4$ M^{-1}, which agrees reasonably well with the results of the titration experiments (see Table III). Similar experiments were performed with a variety of sugars but not with maltose and glucose since the corner frequency, f_c, in these cases was outside the time resolution of the experimental setup. The rate constants for the on- (k_1) and off-processes (k_{-1}) for the binding of these sugars are given in Table IV [100]. The data for glucose and maltose are calculated on the basis of the relative rates of permeation of glucose and maltose taken from Luckey and Nikaido [57] with respect to other maltooligosaccharides by using eqn. (8) and the binding constants of Table III. The results suggest that the rate constants for the on-process is more or less the same for the maltooligosaccharides. Substantial differences were observed for the movement of the sugars out of the channel, probably because the interaction between binding site and sugar increases with the number of glucose residues.

One interesting result of the noise analysis is that the transport kinetics of maltose and sucrose can be compared. The data of Table IV clearly indicate that sucrose is an extremely bad substrate for the LamB channel since the on- and the off-rates for its transport are considerably smaller than those for maltose. No wonder that enteric bacteria may contain a plasmid that codes for proteins for an uptake and degradation system for sucrose [103–105]. This plasmid (pUR400) also contains the gene scrY that codes for an outer

TABLE IV

Rate constants k_1 (on) and k_{-1} (off) of the transport of different sugars through LamB of E. coli

Sugar	K (l/mol)	k_1 (l/mol)	k_{-1} (s^{-1})
G_1	9.5	1×10^6	1.1×10^5
G_2	100	8×10^5	8000
G_3	8000	1×10^7	1300
G_4	12000	8×10^6	670
G_5	13000	7×10^6	530
G_6	20000	5×10^6	240
G_7	30000	4×10^6	120
Sucrose	80	4×10^3	50

The indices refer to the number of glucose subunits within the sugars. The data for G_3 to G_7 and for sucrose were taken from the analysis of the current noise [100]. The rate constants for glucose (G_1) and maltose (G_2) were calculated using eqn. (8), the binding constants and the relative permeation rates of the sugars were taken from Luckey and Nikaido [57].

membrane protein which has a similar function as LamB [106,107]. ScrY exhibits a high homology to LamB but has an N-terminal extension of approximately 70 amino acids [106,107]. The results of in vitro experiments with ScrY have demonstrated that it is a specific porin and contains a binding site for sugars [106,108]. Again the binding constants are very similar for sucrose and maltose [108] which probably means that, in the case of ScrY, the kinetics of maltose transport could be slower than that of sucrose, i.e. the opposite situation as in LamB.

5.4. Tsx of Escherichia coli

The *tsx* gene of *E. coli* encodes a minor outer membrane protein (Tsx) that is the receptor for colicin K and bacteriophage T6 [109]. Synthesis of Tsx is co-regulated with the systems for nucleoside uptake and metabolism, suggesting that it plays an important role in the permeation of nucleosides across the outer membrane [28,110,111]. Indeed, *E. coli* strains lacking Tsx are impaired in the uptake of all nucleosides, with the exception of cytidine and deoxycytidine [28,110]. This Tsx-mediated permeation of nucleosides across the outer membrane is most clearly detected when the exogenously provided nucleosides are present at low substrate concentration, i.e. when the outer membrane is the limiting factor for the overall transport process [112]. At high substrate concentrations, the Tsx protein becomes dispensable, and the nucleosides diffuse across the outer membrane primarily through OmpF. Investigation of the Tsx-mediated translocation of nucleosides across the outer membrane has revealed a number of remarkable properties of Tsx. Interestingly, the Tsx-channel apparently discriminates between the closely structurally related pyrimidine nucleosides, cytidine and thymidine [28,110]. Furthermore, the uptake of deoxynucleosides is more strongly dependent on Tsx than that of the corresponding nucleosides [28,110,112]. On the other hand, Tsx plays no role in the uptake of the free bases or in the permeation of nucleoside monophosphates in vivo [110,113].

Tsx forms ion-permeable channels upon reconstitution in lipid bilayer membranes. These channels have an extremely small single channel conductance of about 10 pS [12,114]. This is more than 100 times less than the conductance of OmpF (1500 pS) and more than 10 times less than that of LamB (155 pS) under otherwise identical conditions. Similar to the experiments with LamB and sugars, it is possible to block the flux of ions through Tsx with nucleosides. This means that it is possible to measure the binding of different nucleosides, deoxynucleosides and free bases to the binding site by using titration experiments with lipid bilayer membranes. Table V shows the results of these titration measurements. From the different pyrimidines, thymine had the largest stability constant for binding to the Tsx-channel, whereas the half saturation constant for the binding of cytidine could not be given and was possibly much larger than 20 mM. A comparison of the structures of the different pyrimidines shows that thymine contains an additional methyl-group in the 5-position compared to uracil. This group obviously results in a larger stability constant for the binding. Cytosine has an amino group in the 4-position instead of the carbonyl group compared to uracil. This results in a considerable decrease in the binding affinity of this compound, probably because of the missing hydrogen bound between an OH-group of the protein and the carbonyl or because of the increased hydrophilic nature of the whole molecule.

414

TABLE V

Stability constants, K, for the binding of nucleobases, nucleosides, and deoxynucleosides to the Tsx-channel (half saturation constant $K_S = 1/K$)

Compound	K (l/mol)	K_S (mmol/l)
Adenine	500	2.0
Adenosine	2000	0.50
Deoxyadenosine	7100	0.14
Guanine	nd	nd
Guanosine	1000	1.0
Deoxyguanosine	3100	0.32
Thymine	170	5.8
Thymidine	5000	0.20
5'-Deoxythymidine	20000	0.050
Cytosine	nd	nd
Cytidine	46	22
Deoxycytidine	100	10
Uracil	50	20
Uridine	1900	0.54
Deoxyuridine	19000	0.053

The data were taken from [114]. K was calculated from titration experiments similar to that shown for LamB (Fig. 4).

Similar considerations also apply to a comparison of the binding affinities of adenine and guanine. However, the relationship between binding affinity and the structure of the purines and the pyrimidines cannot be understood from of the data of Table V alone. The binding between Tsx and purines and pyrimidines is stabilized if a ribofuranose is bound to the 9-position in the case of the purines or to the 1-position in the case of the pyrimidines. This increase in the stability constant could in principle be caused by the formation of additional hydrogen bounds. However, the removal of the hydroxyl-group in the 2'-position (deoxynucleosides) resulted in an even larger stability constant, which makes it rather unlikely that hydrogen bonds are the reason for the larger stability constants. It seems, moreover, that the binding is also dependent on some kind of hydrophobic interaction between the binding site and the different molecules. This hypothesis may be supported by the rather small single-channel conductance (10 pS) of the Tsx-channel in 1 M KCl [12,114].

5.5. Function of TolC of Escherichia coli

Besides the major outer membrane proteins, the cell wall of Gram-negative bacteria also contains a number of minor proteins that are also important for the physiology of the cells. One of them, TolC, is an outer membrane protein with a molecular mass of 52 kDa which is believed to be involved in outer membrane protein regulation and in protein uptake (for a recent review, see [115]). Mutations in the tolC locus greatly reduce the expression of OmpF and NmpC by reduction of the corresponding subscripts [116]. This effect seems to be regulated through an increased expression of the *mic*F gene in the absence of TolC [117]. The effect of TolC on outer membrane protein expression has probably nothing to do with the function of the mature protein itself since outer membrane proteins influence the expression of one another. On the other hand, it has been demonstrated recently that TolC is essential for secretion of hemolysin (HlyA) (see also Chapter 20) and of colicin V out of *E. coli* cells [26,27].

Interestingly, TolC also forms oligomers in the outer membrane [118,119]. Reconstitution experiments with TolC by using the lipid bilayer technique show that TolC is also a pore-forming component (see Fig. 6). Only the oligomeric but not the monomeric form of the protein is able to increase the specific conductance of artificial lipid membranes [119]. The channels formed by TolC have a small single-channel conductance of 80 pS in 1 M KCl, similar to LamB but much smaller than the general diffusion pores OmpF and OmpC. Titration of TolC-induced membrane conductance with peptides suggests that the ion flux through TolC is blocked partially or completely in the presence of small polypeptides [119]. This result indicated binding of the peptides to some sort of binding site inside the TolC channel. The stability constant for the binding of H-Gly-Gly-Leu-OH to the binding site as derived from titration experiments is 20 l/mol [119] which means that the half saturation constant is 50 mM. The half saturation constant for the binding of different peptides is dependent on their amino acid composition and on their length [119].

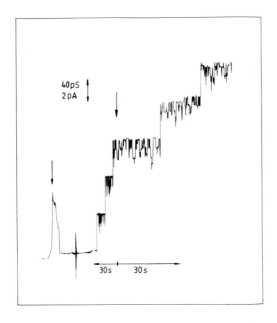

Fig. 6. Single-channel recording of a diphytanoyl glycerophosphocholine/n-decane membrane in the presence of 10 ng/ml TolC of *E. coli*. The aqueous phase contained 1 M KCl. The applied membrane potential was 20 mV; $T = 20°C$. The arrows indicate the blackening of the membrane and the change in the chart speed.

These results suggest that TolC is an outer membrane channel selective for peptides. Small peptides may penetrate the outer membrane of E. coli through the general diffusion channels. However, starting with a certain molecular mass of the peptides, the permeation through OmpF and the other porins may be rather slow or impossible. This means that the outer membrane should contain a pathway for peptides and this pathway could be TolC, i.e. it is a porin specific for peptides, similar to LamB or Tsx for sugars and nucleosides, respectively. This is consistent with the observation that TolC is arranged as an oligomer, probably as a trimer, in the outer membrane and that the monomer is inactive in the lipid bilayer assay.

TolC is involved in permeation of colicin E1 through the outer membrane [115]. Furthermore, TolC is essential for the secretion of HlyA and colicin V out of the cell [26,27]. Deletion of tolC inhibits export of both proteins and leads to an accumulation in the cells. The proteins necessary for the export of Bordetella pertussis cyclolysin (CyaE, see ref. [120]) and of Erwinia chrysanthemi protease (PrtF; see [27,121,122]) out of the cells exhibit some sequence homologies with TolC [123] which suggests that these proteins have a similar function and act probably also as outer membrane channels for proteins.

5.6. OprP of Pseudomonas aeruginosa

The outer membrane of P. aeruginosa has very special sieving properties, which make this organism quite resistant to most antibiotics [124]. It contains a variety of different porins, which are mostly specific porins and are involved in the uptake of different classes of substrates. Protein D1 is involved in sugar uptake and is induced in the presence of glucose [58]. Protein D2 plays probably an important role in the permeation of amino acids and peptides since it has been demonstrated that it contains a binding site for basic amino acids and peptides [125] . The general diffusion pore of the P. aeruginosa outer membrane is probably OprF, which has a large exclusion limit but only a low channel-forming activity in vivo and in vitro [63,124,126,127]. It has to be noted that the role of OprF as an outer membrane porin is still a matter of debate since its primary sequence exhibits some homology with the primary structure of OmpA of E. coli and some investigations question its role as a porin.

OprP (protein P) with a molecular mass of 48kDa [128] is another outer membrane porin of P. aeruginosa. This porin is induced when the organisms grow in media with low phosphate concentrations together with an alkaline phosphatase, a periplasmic binding protein and the inner membrane transport system [44,128]. In this respect, it shares some similarities with PhoE of E. coli. However, in contrast to PhoE [74], OprP and the closely related OprO (which is probably a channel for polyphosphate [129,130]) form highly anion-selective channels in lipid bilayer membranes which are at least 100 times more permeable for chloride than for potassium [44,77,84,129]. The OprP pore has the largest single channel conductance in 0.1 M chloride solution [84]. A variety of other anions also permeate through the channel, whereas in the presence of salts with large organic anions, such as HEPES (N-2-hydroxyethylipiperazine-N'-2-ethanesulfonic acid), no conductance fluctuations could be observed [84]. Interestingly, the single channel conductance was extremely small in 100 mM phosphate (pH 6) although OprP should be a channel for

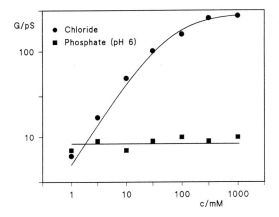

Fig. 7. Single-channel conductance of the OprP channel of *P. aeruginosa* as a function of the KCl and the phosphate concentration in the aqueous phase (pH 6). The membranes were formed from diphytanoyl glycerophosphocholine/n-decane; $T = 20°C$; $V_m = 50$ mV. The solid lines were drawn with eqn. (15) and the corresponding parameters given in Table VI. The data were taken from [131].

phosphate. Furthermore, no concentration dependence could be observed in monobasic phosphate down to 1 mM as Fig. 7 clearly indicates.

The conductance versus concentration curve for chloride given in Fig. 7 can be explained by the one-site two-barrier model [84] using the same formalism as above (i.e. assuming a single file channel and that the stability constant of the binding between an ion and the binding site is $K = k_1/k_{-1}$). Under these assumptions, the dependence of the single-channel conductance, $G_0(c)$, on the aqueous anion concentration c is given by [84]

$$G_0(c) = G_{0,max}Kc/(1 + Kc) \qquad (15)$$

where $G_{0,max}$ is the maximum single channel conductance at very high ion concentration:

$$G_{0,max} = e^2k_{-1}/(2kT) \qquad (16)$$

e (= 1.6×10^{-19} A s) is the elementary charge, k (= 1.38×10^{-23} J/K) is the Boltzmann constant and T is the absolute temperature. Table VI shows the maximum single channel conductances, $G_{0,max}$, of OprP for a variety of different anions [84,131].

The binding constants for phosphate could not be derived from the concentration dependence of the single channel conductance because of the strong saturation. It is possible, however, to study the influence of phosphate on the conductance of the OprP channel by measuring single-channel conductances in aqueous solutions containing 0.1 M KCl and increasing concentrations of phosphate [131]. The addition of phosphate to the KCl solutions has a similar effect on the conductance as described above for sugars and nucleosides although phosphate contributes somewhat to the single-channel conductance. This also means that, in these cases, the binding constants could be evaluated using the one-site two-barrier model. Table VI shows the results for the binding constants and for the on- (k_1) and the off-rate constants (k_{-1}) derived either from the concentration dependence of the single-channel data or from titration experiments [84,131]. The on-rate k_1

TABLE VI

Parameters of OprP-mediated anion transport

Anion	K (l/mol)	$G_{0,max}$ (pS)	k_1 (10^7 l/(mol s))	k_{-1} (10^7 s^{-1})
Fluoride	3.5	515	60	17
Chloride	20	280	180	9.0
Bromide	4.7	265	40	8.5
Iodide	1.3	110	4.6	3.5
Phosphate (pH 6)	11000	9	5700	0.44
Sulfate (pH 6)	500	18	310	0.62
Citrate (pH 6)	400	21	260	0.64

The stability constants, K, for the binding of different anions to OprP and the maximum single channel conductance, $G_{0,max}$, were taken from [84] and [131] by fitting the concentration dependence of the single channel conductances with eqn. (15) or from a fit of the phosphate-induced block of chloride conductance data, respectively. k_1 and k_{-1} were calculated from K and $G_{0.max}$ using eqn. (16) and $K = k_1/k_{-1}$.

(which is equal to twice the maximum permeability) has a maximum for monobasic phosphate followed by chloride, whereas the inverse reaction rate k_{-1} (i.e. the maximum turnover number of ions through a one-site two-barrier channel) was largest for fluoride. This means that at very small anion concentrations, the permeability of the channel is highest for phosphate, whereas it conducts F$^-$ or Cl$^-$ best at large anion concentrations. This is again consistent with the in vivo-situation of phosphate-starvation at which the channel should have a maximum permeability for phosphate.

6. Role of periplasmic binding proteins in solute uptake

The periplasmic space between inner and outer membranes represents an additional cellular compartment and occupies between 5% and 20% of the total cell volume according to different estimations [18,132]. It plays an important role in the physiology of Gram-negative bacteria because binding proteins for classes of solutes such as sugars, amino acids and phosphate and the β-lactamase activity are located there [2,47,133–135]. Furthermore, it contains enzymes some of which function as processing enzymes and convert non-transportable metabolites to transport substrates [2,136]. The periplasmic space is almost iso-osmotic with the cytoplasm (around 300 mosM). This means that the osmotic pressure, normally around 3.5 bar and at maximum about 7–8 bar [1], is maintained across the outer membrane and not across the inner membrane. The periplasmic space is strongly anionic compared to the external medium, mainly caused by the anionic membrane-derived oligosaccharides (MDO) present in the periplasmic space to maintain part of the osmolarity. MDO also contribute to the Donnan potential across the outer membrane, which can be as large as 100 mV (inside negative) in media of low ionic strength [80,132]. Recently, it has been suggested that MDO may be also involved in the porin channel gating [137].

The periplasmic space has a gel-like structure [138] which means that the proteins localized there have only limited mobility. It represents a sink for solutes because of the

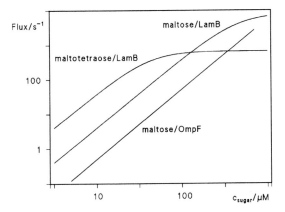

Fig. 8. Flux of maltose and maltotetraose through one single LamB channel as a function of the corresponding sugar concentration on one side of the channel. The concentration on the other side was set to zero. The flux was calculated using eqn. (8) and the rate constants given in Table IV. The flux of maltose through OmpF was calculated on the basis on its dimensions [5] and using the Renkin correction factor [60].

presence of the binding proteins. A solute molecule bound inside one of these specific porins discussed here has two possibilities. It can move to the external surface or to the periplasmic space. The presence of the binding proteins in a high concentration (up to 1 mM) inside the periplasmic space [2,138] reduces the probability that the solute molecules move back to the outer membrane channels and may leave the cell. This means that the high affinity binding proteins reduce the concentration of unbound solutes inside the periplasmic space and lead to a vectorial transport from the outer membrane channels to the cytoplasmic membrane transport systems.

7. Conclusions

The outer membrane of Gram-negative bacteria contains different pathways for the permeation of solutes. The permeation of vitamins and siderophores occurs through TonB-dependent highly specified channels that are probably gated [139]. One always open pathway is provided by the general diffusion pores which form more or less structured holes with defined exclusion limits and little solute specificity. The flux of sugars (and other substrates) through a general diffusion pore is linearly dependent on the concentration. Another class of permeability pathways is given by the specific porins. These channels are highly structured and contain binding sites for one class of specific solutes. The presence of a binding site leads to the saturation of solute flux above the half saturation constant $1/K$. However, the presence of a binding site offers a considerable advantage for the flux of solutes across the outer membrane at small external solute concentrations, the in vivo situation, at which the general diffusion pores are rate-limiting [53].

Figure 8 shows the maximum flux of maltose and maltotetraose across LamB calculated on the basis of eqn. (8) under the assumption that the concentration of the sugars on one side (the periplasmic side) is zero. The data were calculated using the rate constants given in Table IV. The flux of maltose through a general diffusion pore (dotted line) was estimated according to the dimensions of OmpF [5]. It was corrected by using the Renkin correction factor [60] for the hit of the maltose to the rim of the OmpF channel, which leads to the reflection of the sugar. Specific porins have their maximum permeability in the linear range of Fig. 8 and it is proportional to the rate constant for the movement from the external solution to the binding site inside the channels (i.e. by $k_1/2$). The saturation of the flux at very high substrate concentrations is given by the rate constant k_{-1}, which is the maximum turnover number of ions through a one-site two-barrier channel. The comparison of the different fluxes of Fig. 8 again demonstrates the advantage of a binding site for the maximum scavenging of substrates. Furthermore, it demonstrates that the flux through a general diffusion pore can exceed that through a specific porin at very high substrate concentration.

Acknowledgements

I would like to thank Winfried Boos, Erhard Bremer, Christine Egli, Tom Ferenci, Robert E.W. Hancock, Taiji Nakae, Jan Tommassen, Greetje H. Vos-Scheperkeuter and Jürgen Weckesser for the fruitful and excellent collaboration during many years and my collaborators Katharina Bauer, Elke Maier, Stefan Nekolla, Angela Schmid and Katrin Schülein for their excellent work with outer membrane porins on which this review is largely based. My own research was funded by the Deutsche Forschungsgemeinschaft (SFB 176) and the Fonds der Chemischen Industrie.

References

1. Beveridge, T.J. (1981) Int. Rev. Cytol. 72, 229–317.
2. Ros, B.P. (1978) Bacterial Transport, Marcel Dekker, New York.
3. Nakae, T. (1976) J. Biol. Chem. 251, 2176–2178.
4. Weiss, M.S., Abele, U., Weckesser, J. Welte, W., Schiltz, E. and Schultz, G.E. (1991) Science 254, 1627–1630.
5. Cowan, S.W., Schirmer, T., Rummel, G., Steiert, M., Gosh, R., Pauptit, R.A., Jansonius, J.N. and Rosenbusch, J.P. (1992) Nature 356, 727–733.
6. Nikaido, H. and Rosenberg, E.Y. (1981) J. Gen. Physiol. 77, 121–135.
7. Benz, R., Janko, K., Boos, W. and Läuger, P. (1978) Biochim. Biophys. Acta 511, 305–319.
8. Schindler, H. and Rosenbusch, J.P. (1978) Proc. Natl. Acad. Sci. USA 75, 3751–3755.
9. Nikaido, H. (1992) Mol. Microbiol. 6, 435–442.
10. Benz, R., Schmid, A., Nakae, T. and Vos-Scheperkeuter, G.H. (1986) J. Bacteriol. 165, 978–986.
11. Dargent, B., Rosenbusch, J. and Pattus, F. (1987) FEBS Lett. 220, 136–142.
12. Maier, C., Bremer, E., Schmid, A. and Benz, R. (1987) J. Biol. Chem. 263, 2493–2499.
13. Lugtenberg, B. and van Alphen, L. (1983) Biochim. Biophys. Acta 737, 51–115.
14. Cronan, J.E. and Vagolos, P.R. (1972) Biochim Biophys. Acta 265, 25–60.
15. Resch, C.M. and Gibson, J. (1983) J. Bacteriol. 155, 345–350.
16. Galanos, C., Lüderitz, O., Rietschel, E.T. and Westphal, O. (1977) Int. Rev. Biochem.14, 239–335.
17. Benz, R. (1985) CRC Crit. Rev. Biochem. 19, 145–190.
18. Nikaido, H. and Vaara, M. (1985) Microbiol. Rev. 49, 1–32.
19. Henning, U. and Haller, I. (1975) FEBS Lett. 55, 161–164.

20. Uemura, J. and Mizushima, S. (1975) Biochim. Biophys. Acta 413, 163–176.
21. Braun, V. (1975) Biochim. Biophys. Acta 415, 335–377.
22. Chen, R., Schmidmayr, W., Krämer, C., Chen-Schmeisser, U. and Henning, U. (1980) Proc. Natl. Acad. Sci. USA 77, 4592–4596.
23. Vogel, H. and Jähnig, F. (1986) J. Mol. Biol. 190, 191–199.
24. Sugawara, E. and Nikaido, H. (1992) J. Biol. Chem. 267, 2507–2511.
25. Saint, N., De, E., Julien, S., Orange, N. and Molle, G. (1993) Biochim. Biophys. Acta 1145, 119–123.
26. Wandersman, C. and Delepelaire P. (1990) Proc. Natl. Acad. Sci. USA 87 4776–4780.
27. Fath, M.J., Skvirsky R.C. and Kolter, R. (1991) J. Bacteriol. 172 7549–7556.
28. Hantke, K. (1976) FEBS Lett. 70, 109–112.
29. Reynolds, P.R., Mottur, G.P. and Bradbeer C. (1980) J. Biol. Chem. 255, 4313–4319.
30. Overbeeke, N. and Lugtenberg, B. (1980) FEBS Lett. 112, 229–232.
31. Tommassen, J., de Geus, P., Lugtenberg, B., Hackett, J. and Reeves, P. (1982) J. Mol. Biol. 157, 265–274.
32. Szmelcman, S. and Hofnung, M. (1975) J. Bacteriol. 124, 112–118.
33. Benz, R. and Bauer, K. (1988) Eur. J. Biochem 176, 1–19.
34. Van Alphen, W. and Lugtenberg, B. (1977) J. Bacteriol 131, 623–630.
35. Kawaji, H., Mizumo, T. and Mizushima, S. (1979) J. Bacteriol. 140, 843–847.
36. Schnaitman, C.A. (1974) J. Bacteriol. 118, 454–464.
37. Sen, K. and Nikaido, H. (1991) J. Biol. Chem. 266, 11295–11300.
38. Mizuno, T., Wurtzel, E.T. and Inouye, M. (1982) J. Biol. Chem. 257, 13692–13698.
39. Tommassen, J. and Lugtenberg, B. (1981) J. Bacteriol. 143, 151–157.
40. Blasband, A.J., Marcotte Jr., W.R. and Schnaitman, C.A. (1986) J. Biol. Chem. 261, 12723–12732.
41. Nikaido, H. (1983) Methods Enzymol. 97, 85–100.
42. Ferenci, T., Schwentorat, M., Ullrich, S. and Vilmart, J. (1980) J. Bacteriol. 142, 521–526.
43. Palva, E.T. (1978) J. Bacteriol. 136,286–294.
44. Hancock, R.E.W., Poole, K. and Benz, R. (1982) J. Bacteriol. 150, 730–738.
45. Sonntag, I., Schwarz, H., Hirota, Y. and Henning, U. (1978) J. Bacteriol. 136, 280–285.
46. Heine, H.-G., Kyngdon, J. and Ferenci, T. (1987) Gene 53, 287–292.
47. Zimmermann, W. and Rosselet, A. (1977) Antimicrob. Agents Chemother. 12, 368–372.
48. Sawai, T., Matsuba, K. and Yamagishi, S. (1977) J. Antibiot. 90, 1134–1136.
49. Misra, R. and Benson, S. (1988) J. Bacteriol. 170, 528–533.
50. Struyvé, M., Visser, J., Adriaanse, H., Benz, R. and Tommassen, J. (1992) Mol. Microbiol. 7, 131–140.
51. Bauer, K., Struyvé, M., Bosch, D., Benz, R. and Tommassen, J. (1989) J. Biol. Chem. 264, 16393–16398.
52. Nikaido, H. and Rosenberg, E.Y. (1983) J. Bacteriol. 153, 241–252.
53. West, I.C. and Page, M.G.P. (1984) J. Theor. Biol. 110, 11–19.
54. Nakamura, K. and Mizushima, S. (1976) J. Biochem. 80, 1411–1422.
55. Miura, T and Mizushima, S. (1968) Biochim. Biophys. Acta 150, 156–164.
56. Hancock, R.E.W. (1987) J. Bacteriol. 169, 929–933.
57. Luckey, M. and Nikaido, H. (1980) Proc. Natl. Acad. Sci. USA 77, 165–171.
58. Trias, J., Rosenberg, E.Y. and Nikaido, H. (1988) Biochim. Biophys. Acta 938, 493–496.
59. Bangham, A.D., De Gier, J. and Greville G.D. (1967) Chem. Phys. Lipids 1, 225–246.
60. Renkin, E.M. (1954) J. Gen. Physiol. 38, 225–253.
61. Dargent, B., Hofmann, W., Pattus, F. and Rosenbusch, J.P. (1986) EMBO J. 5, 773–778.
62. Nakae, T. (1976) Biochem. Biophys. Res. Commun. 71, 877–884.
63. Hancock, R.E.W., Decad, G.M. and Nikaido, H. (1979) Biochim. Biophys. Acta 554, 323–331.
64. Flamman, H.T. and Weckesser, J. (1984) J. Bacteriol. 159, 410–412.
65. Weckesser, J., Zalman, L.S. and Nikaido, H. (1984) J. Bacteriol. 159, 199–205.
66. Vachon, V., Lyew, D.I. and Coulton, J.W. (1985) J. Bacteriol. 162, 918–924.
67. Hancock, R.E.W. (1986) in: M. Inouye (Ed.), Bacterial Outer Membranes as Model Systems, Wiley, New York, pp. 187–225.
68. Benz, R., Janko, K. and Läuger, P. (1979) Biochim. Biophys. Acta 551, 238–247.
69. Benz, R., Woitzik, D., Flammann, H.T. and Weckesser, J. Arch. Microbiol. 148, 226–320.

422

70. Schmid, A., Benz, R. and Schink, B. (1991) J. Bacteriol. 173, 4909–4913.
71. Brunen, M., Engelhardt, H., Schmid, A. and Benz, R. (1991) J. Bacteriol. 173, 4182–4187.
72. Nikaido, H., Rosenberg, E.Y. and Foulds, J. (1983) J. Bacteriol. 153, 232–240.
73. Van der Ley, P., Burm, P., Agterberg, M., Van Meersbergen, J. and Tommassen, J. (1987) Mol. Gen. Genet. 209, 585–591.
74. Benz, R., Schmid, A. and Hancock, R.E.W. (1985) J. Bacteriol. 162, 722–727.
75. Whitfield, C., Hancock, R.E.W. and Costerton, J. W. (1983) J. Bacteriol. 156, 873–879.
76. Darveau, R.P., Hancock, R.E.W. and Benz, R. (1984) Biochim. Biophys. Acta 74, 69–74.
77. Hancock, R.E.W., Schmid, A., Bauer, K. and Benz, R. (1986) Biochim. Biophys. Acta 860, 263–267.
78. Srikumar, R., Dahan, D., Gras, M.F., Ratcliffe, M.J.H., van Alphen, L. and Coulton, J.W. (1992) J. Bacteriol. 174, 4007–4016.
79. Lakey, J.H. (1987) FEBS Lett. 211, 1–4.
80. Sen, K., Hellman, J. and Nikaido, H. (1988) J. Biol. Chem. 263, 1182–1187.
81. Benz, R. and Janko, K. (1976) Biochim. Biophys. Acta 455, 712–738.
82. Benz, R., Fröhlich, O., Läuger, P. and Montal, M. (1975) Biochim. Biophys. Acta 394, 323–334.
83. Läuger, P. (1973) Biochim. Biophys. Acta 300, 423–441.
84. Benz, R. and Hancock, R.E.W. (1987) J. Gen. Physiol. 89, 275–295.
85. Benz, R., Schmid, A. and Vos-Scheperkeuter, G.H. (1988) J. Membr. Biol. 100, 21–29.
86. Debarbouille, M. and Schwartz, M. (1979) J. Mol. Biol. 132, 521–534.
87. Raibaud, O. and Schwartz, M. (1980) J. Bacteriol. 143, 761–771.
88. Raibaud, O., Roa, M., Braun-Breton, C. and Schwartz, M. (1980) Mol. Gen. Genet. 147, 241–248.
89. Clement, J.M. and Hofnung, M. (1981) Cell 27, 507–514.
90. Francoz, E., Molla, A., Saurin, W. and Hofnung, M. (1990) Res. Microbiol. 141, 1039–1059.
91. Bloch, M.-A. and Desaymard, C. (1985) J. Bacteriol. 163, 106–110.
92. Luckey, M. and Nikaido, H. (1980) Biochem. Biophys. Res. Commun. 93, 166–171.
93. Schwartz, M. (1983) Methods Enzymol. 97, 100–112.
94. Nakae, T., Ishii, J. and Ferenci, T. (1986) J. Biol. Chem. 261, 622–626.
95. Ferenci, T. and Lee, K.-S. (1989) J. Bacteriol. 171, 855–861.
96. Benz, R., Francis, G., Nakae, T. and Ferenci, T. (1992) Biochim. Biophys. Acta 1104, 299–307.
97. Charbit, A., Gehring, K., Nikaido, H., Ferenci, T. and Hofnung, M. (1988) J. Mol. Biol. 201, 487–493.
98. Ferenci, T., Saurin, W. and Hofnung, M. (1988) J. Mol. Biol. 201, 493–496.
99. Benz, R. (1988) Annu. Rev. Microbiol. 42, 359–393.
100. Nekolla, S. (1989) Diplom Thesis Universität Würzburg
101. Benz, R. and Nekolla, S. (1990) in: E. Quagliariello, S. Papa, F. Palmieri and C. Saccone (Eds.), Structure, Function and Biogenesis. Developments in Biochemistry, Vol. 28, Elsevier, Amsterdam, pp. 307–310.
102. Verween, A.A. and DeFelice, L.J. (1974) Prog. Biophys. Mol. Biol. 28, 189–265.
103. Lengeler, J.W., Mayer, R.J. and Schmid, K. (1982) J. Bacteriol. 151, 468–471.
104. Schmid, K., Ebner, R., Altenbuchner, J., Schmitt, R. and Lengeler, J.W. (1988) Mol. Microbiol. 2, 1–8.
105. Hardesty, C., Colon, G., Ferran, C. and DiRienzo, J.M. (1987) Plasmid 18, 142–155.
106. Hardesty, C., Ferran, C. and DiRienzo, J.M. (1991) J. Bacteriol. 173, 449–456.
107. Schmid, K., Ebner, R., Jahreis, K., Lengeler, J.W. and Titgemeier F. (1991) Mol. Microbiol. 5, 941–950.
108. Schülein, K. Schmid, K. and Benz, R. (1991) Mol. Microbiol. 5, 2233–2241.
109. Manning, P.A., Pugsley, A.P. and Reeves, P. (1977) J. Mol. Biol. 116, 285–300.
110. Krieger-Brauer, H.J. and Braun, V. (1980) Arch. Microbiol. 124, 33–242.
111. Bremer, E., Gerlach, P. and Middendorf, A. (1988) J. Bacteriol. 170, 108–116.
112. Munch-Petersen, A. and Mygind, B. (1983) in: A. Munch-Petersen (Ed.), Metabolism of Nucleotides, Nucleosides and Nucleobases in Microorganisms, Academic Press, London, pp. 259–305.
113. van Alphen, W. , van Selm, N. and Lugtenberg, B. (1978) Mol. Gen. Genet. 159, 75–83.
114. Benz, R., Schmid, A., Maier, C. and Bremer, E. (1989) Eur. J. Biochem. 699–705.
115. Webster, R.E. (1991) Mol. Microbiol. 5, 1005–1011.
116. Morona, R. and Reeves, P. (1982) J. Bacteriol. 150, 1016–1023.
117. Misra, R. and Reeves, P. (1987) J. Bacteriol. 169, 4722–4730.

118. Morona, R., Manning, P.A. and Reeves, P. (1983) J. Bacteriol. 153, 693–699.
119. Benz, R., Maier, E. and Gentschev, I. (1993) Zlb. Bakteriol. 278, 187–196.
120. Glaser, P., Sakamoto, H., Bellalou, J., Ullmann, A. and Danchin, A. (1988) EMBO J. 7, 3997–4004.
121. Delepelaire, P. and Wandersman, C. (1991) Mol. Microbiol. 5, 2427–2434.
122. Guzzo, J, Duong, F., Wandersman, C., Murgler, M. and Lazdunski, A. (1991) Mol. Microbiol. 5, 447–453.
123. Niki, H., Imamura, R., Ogura, T. and Hiraga, S. (1990) Nucleic Acids Res. 18, 5547.
124. Hancock, R.E.W., Siehnel, R. and Martin, N. (1990) Mol. Microbiol. 4, 1069–1075.
125. Trias, J. and Nikaido, H. (1990) J. Biol. Chem. 265, 15680–15684.
126. Yoshimura, F., Zalman, L.S. and Nikaido, H. (1983) J. Biol. Chem. 258, 2308–2314.
127. Benz, R. and Hancock, R.E.W. (1981) Biochim. Biophys. Acta 646, 298–308.
128. Siehnel, R.J., Martin, N.L. and Hancock, R.E.W. (1990) Mol. Microbiol. 4, 831–838.
129. Hancock, R.E.W., Egli, C., Benz, R. and Siehnel, (1992) J. Bacteriol. 174, 471–476.
130. Siehnel, R.J., Egli, C., Wong, G. and Hancock, R.E.W. (1992) Mol. Microbiol. 6, 2319–2326.
131. Benz, R. Egli, C. and Hancock, R.E.W. (1993) Biochim. Biophys. Acta 1149, 224–230.
132. Stock, J.B., Rauch, B. and Roseman, S. (1977) J. Biol. Chem. 252, 7850–7861.
133. Szmelcman, S., Schwartz, M. Silhavy, T.J. Boos, W. (1976) Eur. J. Biochem. 65, 13–19.
134. Poole, K. and Hancock, R.E.W. (1984) Eur. J. Biochem. 144, 607–612.
135. Duplay, P., Bedouelle, H., Fowler, A., Zabin, I., Saurin, W. and Hofnung, M. (1984) J. Biol. Chem. 259, 10606–10613.
136. Freundlieb, S. and Boos, W. (1986) J. Biol. Chem. 261, 2946–2953.
137. Delcour, A.H., Adler, J., Kung, C. and Martinac, B. (1992) FEBS Lett. 304, 216–220.
138. Brass, J.M. (1986) Curr. Top. Microbiol. Immunol. 129, 1–92.
139. Rutz, J.M., Liu, J., Lyons, J.A., Goranson, J., Armstrong, S.K., McIntosh, M.A., Feix, J.B. and Klebba, P.E. (1992) Science 258, 471–475.

J.-M Ghuysen and R. Hakenbeck (Eds.), *Bacterial Cell Wall*
© 1994 Elsevier Science B.V. All rights reserved

Secretion of hemolysin and other proteins out of the Gram-negative bacterial cell

VASSILIS KORONAKIS and COLIN HUGHES

Cambridge University Department of Pathology, Tennis Court Road, Cambridge CB2 1QP, UK

1. Introduction

Like all other cells, bacteria need to move newly synthesized proteins to specific cellular locations. The minority of proteins which are destined to traverse the energized hydrophobic cytoplasmic membrane (CM) of Gram-positive and Gram-negative bacteria are handled by the conventional export machinery, which recognizes an N-terminal signal peptide in the precursor substrate protein and comprises several secretion (Sec) proteins located in the cytoplasm and the membrane. Some proteins of the Gram-negative bacterial cell are directed further than the periplasm, reaching the exterior surface of the outer membrane (OM) and the extracellular medium. Such complete secretion is especially well characterized among pathogenic bacteria, in which cell surface structures or diffusible proteins determine the specific interactions with eukaryotic cells that underlie the ability to colonize, invade and survive in mammalian or plant hosts. This chapter sets out the distinct and common features of these secretion mechanisms.

1.1. Export of proteins across the cytoplasmic membrane

The Sec proteins and their interactions have been identified by extensive biochemical and genetic analyses (reviewed in [1–3]) and these now allow mechanistic studies of the CM protein translocation process, which is indicated schematically in Fig. 1. In bacteria, protein export can be entirely post-translational and in order to maintain a partially unfolded, secretion-competent form, the newly synthesized proteins first need to bind to cytoplasmic chaperones. The chaperone SecB maintains the loosely folded state of cytosolic proteins destined for translocation across the membrane [4], although some protein substrates may be handled by alternative chaperones such as trigger factor, GroEL or Ffh, which has homology with a component of the eukaryotic signal recognition particle [5]. Recognition of the substrate (preprotein)–SecB complex by the membrane complex is determined primarily by SecA, the peripheral subunit of the membrane translocase, which has affinities for SecB and both the signal and mature domains of the preprotein substrate [6]. In addition to SecA, the preprotein translocase comprises integral membrane components of acidic phospholipid and the proteins SecE/Y/(band 1), indeed crosslinking studies suggest

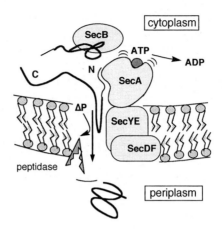

Fig. 1. Components of the conventional signal peptide-dependent export of proteins across the cytoplasmic membrane.

that SecA and SecY together form the proteinaceous pathway through which substrate proteins pass during membrane transit [7]. The later stages in translocation and release of the substrate protein involve the membrane proteins SecD [8] and possibly SecF. Substrate proteins secreted by this conventional pathway reach the periplasm as mature proteins following removal of the N-terminal signal peptide by the signal (leader) peptidase. Extensive in vivo and in vitro studies of the energy required during this process have shown that both the proton motive force, ΔP, and ATP are needed (reviewed in [9]). The early stage up until the precursor interacts with the SecA and SecE/Y components of the membrane translocase, does not seem to require energy, but initial limited translocation of an N-terminal loop and the dissociation from SecA consume energy from the binding and the hydrolysis of ATP by SecA [10]. The subsequent translocation of the precursor protein through the membrane is driven by the total proton motive force or electrochemical potential, ΔP, acting either (or both) directly on the translocation substrate and/or indirectly by affecting the membrane phospholipid and protein components of the secretion apparatus [9–11]. Recycling of SecA and re-association with distal sequences of the substrate protein still located in the cytoplasm allow the translocation cycle to continue until complete, the vectoral movement itself also generating energy to drive further unfolding [12].

1.2. Secretion beyond the periplasm and out of the Gram-negative cell

Extracellular proteins secreted by Gram-negative bacteria are listed in Table I. In most cases, protein translocation to the outside of the cell comprises two apparently distinct stages, using as substrates for OM translocation the periplasmic intermediates, which are either proven or assumed to be generated by the conventional SecA-based export across the cytoplasmic membrane, i.e. they are signal peptide-dependent. This type of secretion is exemplified by the processes which direct secretion of immunoglobulin A (IgA) protease, pullulanase (PulA), cholera toxin, aerolysin and the *Serratia/Proteus* hemolysin

(ShlA/HpmA), and also the assembly of cell surface structures such as the fimbrial extensions bearing specific mammalian cell adhesins such as the pyelonephritis-associated (Pap) or P-type pili. Nevertheless, not all examples of secretion out of the cell require conventional cytoplasmic membrane transfer, and in particular the *E. coli* hemolysin protein and a large number of other toxins and proteases are translocated across the outer membrane in a process that is signal peptide-independent. The interest in these secretion processes, both signal peptide-dependent and -independent, has to a large extent been prompted originally by their central role in bacterial pathogenicity and in many cases by the hope that the secretion mechanisms might be exploited for the presentation of foreign antigens by live attenuated vaccine strains. But their study is now of fundamental importance in understanding the principles of membrane targetting and translocation.

2. Signal peptide-dependent secretion out of the cell

The release of certain bacteriocins (e.g. cloacin and related colicins) into the medium by *Klebsiella* and *E. coli* is associated with 'permeabilization', i.e. loss of integrity of the outer, and possibly also inner, membranes [13]. The bacteriocin substrate protein itself

TABLE I

Extracellular proteins of Gram-negative bacteria

Extracellular protein	Bacterium	Surface/ medium	Signal peptide	Helper proteins
IgA protease	*Neisseria*	m	+	−
Serine protease	*Serratia*	m	+	−
Endoglucanase	*Pseudomonas*	m	+	−
ST$_A$ enterotoxin	*E. coli*	m	+	−?
Cloacin, colicin	*Klebsiella, E. coli*	m	(+)	+?
ShlA/HpmA hemolysin	*Proteus, Serratia*	m?	+	+
Pullulanase	*Klebsiella*	m	+	+
'Exoenzymes'	*Pseudomonas, Erwinia* *Xanthomonas*	m	+	+
Aerolysin	*Aeromonas*	m	+	+
CTX toxin	*Vibrio*	m	+	+
Type IV pilin	*Pseudomonas, Neisseria*	S	+	+
P-type pilin	*E. coli*	S	+	+
Type I pilin	*E. coli*	S	+	+
OM proteins (Yops)	*Yersinia*	S	−	+?
Flagellin	*E. coli, Caulobacter*	S	−	+
Invasion proteins	*Shigella, Salmonella*	S	?	+
HlyA hemolysin, Leukotoxin	*E. coli, Proteus, Morganella* *Actinobacillus, Pasteurella*	m	−	+
CyaA cyclolysin	*Bordetella pertussis*	m	−	+
Proteases	*Erwinia, Serratia, Proteus*	m	−	+
NodO (nodulation)	*Rhizobium*	m	−	+
ColV colicin	*E. coli*	m	−	+

428

does not appear to have a signal peptide, but the process is nevertheless signal peptide-dependent as such a peptide is cleaved from the ancillary 'lysis' or 'release' protein, which is located in the cell envelope. There is no evidence of stable periplasmic intermediates in this process and it is suggested that secretion occurs in some way directly from the cytoplasm to the OM, involving during release the activation of the OM phospholipase PldA. Having mentioned this apparently disruptive method of reaching the outside of the cell, we focus in this section on examples of signal peptide-dependent protein secretion in which the OM remains intact and the cell maintains its integrity. These 'two-stage' mechanisms are outlined schematically in Fig. 2. The periplasmic intermediates released after removal of the N-terminal signal peptide can be translocated across the outer membrane either with or without the assistance of multiple helper proteins. Furthermore, one can usefully draw a distinction between the secretion of proteins to the cell-free medium and the related processes by which subunit components are assembled to form cell-surface structures such as fimbrial adhesins. Molecular genetic studies are now revealing important common features between them, most obviously by defining closely related export helper proteins.

2.1. Without helper proteins: the IgA protease

In perhaps the most straightforward and well-characterized case of extracellular secretion, the IgA-cleaving protease of mucosal pathogens [14], in particular *Neisseria gonorrhoeae*, is translocated across the OM by a 45-kDa C-terminal helper domain (the β domain) within the 170-kDa *iga* gene product itself, apparently without any extra helper proteins [15]. The IgA protease precursor gains access to the periplasm via the conven-

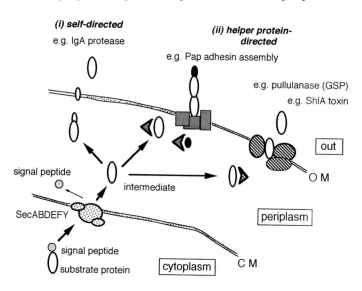

Fig. 2. Secretion out of the cell by mechanisms that are dependent upon conventional export across the cytoplasmic membrane. Processes are divided into those requiring helper proteins for outer membrane secretion, and those in which substrate proteins are apparently able to direct their own secretion.

tional export process, during which its N-terminal signal peptide is removed. Immediately, without forming a stable periplasmic pool, the C-terminal β domain inserts into the OM and forms an export 'pore' via which the bulk of the protease is translocated. Following this looping through, the helper domain is removed by autocleavage by the IgA proteolytic activity, at proline-rich sequences which, with elegant simplicity, match those recognized and cleaved in the mammalian host IgA target molecule. The mature enzyme of ca. 120 kDa is thus released into the external medium. This 'self-governed' OM translocation process appears to be used for the export of a similar IgA-specific enzyme by *Haemophilus influenza* [16] and a broader spectrum serine protease by *Serratia marcescens* [17]. By studying the ability of a fused IgA protease C-terminus to export heterologous peptides to the surface of the OM, the functional core of the β export domain has been identified as ca. 30 kDa, with a primary structure conserved among the secreted IgA proteases, and the data indicate that the conformation of the passenger protein is a critical factor in the export mechanism [18,19].

At first glance, the secretion of the small heat-stable enterotoxin (STA) into the medium shares some features with the above. Release of this polypeptide is also dependent upon a cleaved signal peptide, which directs the 53 residue proST to the periplasm and also maintains the substrate in a stable conformation [20] and OM translocation appears to be directed by the pro domain of the periplasmic protoxin. The proST translocated to the medium is autoproteolytically matured to remove the pro domain and yield the heat and protease-resistant 19-residue toxin. However, experimental deletion of the pro domain of the gene product does not actually prevent secretion of the artificially truncated toxin, indeed it accelerates secretion and shortens the life of the periplasmic intermediate. Moreover, the pro domain can itself be secreted from the periplasm. Such early data have suggested tentatively that small peptides can gain exit across the OM by more than one mechanism, and it may be that a 'background' process recognizes peptides on the basis of their size. Although no helper proteins have been identified in ST secretion, it cannot be excluded that the mechanism involves some helper function.

2.2. With a single helper protein: the ShlA/HpmA toxins

A further example of signal peptide-dependent secretion out of the cell, again in pathogenic bacteria, is the export of the pore-forming toxin ShlA by *Serratia marcescens* (and its homologue HpmA by *Proteus mirabilis*) [21]. In this case, a membrane-associated periplasmic intermediate of the ca. 165 kDa ShlA is believed to be secreted by a single cosynthesized integral outer membrane protein ShlB [22]. The ShlB protein is, like ShlA, exported by the signal peptide-dependent inner membrane pathway, and the expression of recombinant *shlA* and *shlB* genes is sufficient for export of the hemolysin from *E. coli*. Remarkably, however, ShlB is apparently also necessary for the post-translational modification of the *shlA* gene product such that it becomes hemolytic [23]. This unusual activation apparently takes place in the periplasm as loss of the ShlB helper protein leads to the periplasmic accumulation of inactive ShlA product [22], but it is still not clear whether the two functions of activation and secretion are mechanistically linked. The information for complete secretion is located at the N-terminus of the mature substrate ShlA, and close similarity between this region of the ShlA/HpmA sequence and that of the N-terminus of fimbrial hemagglutinin of *Bordetella pertussis* [24] suggests that this adhesin protein may be secreted in a similar way.

2.3. With multiple helper proteins

2.3.1. The general secretion pathway: enzymes, toxins and type IV pili

A distinct example of export across the OM that now appears to be extremely widespread was first described for the secretion of pullulanase, the polysaccharide-degrading lipoprotein enzyme, by *Klebsiella oxytoca* [25]. In this process, cytoplasmic membrane-associated periplasmic intermediates are generated by the Sec pathway and a cleaved N-terminal signal, but in contrast to the above examples, the translocation across the OM and release of this N-terminally acylated mature protein into the medium require the assistance of at least 12 helper proteins, for the most part encoded as part of the maltose regulon [26–28]. The fatty acylation is not a requirement for secretion but it has been suggested that the modification may enhance the efficiency of translocation [29].

The Pul proteins are sufficient to direct extracellular secretion of pullulanase by *E. coli* and these are proposed following initial investigations to be located in several subcellular compartments, primarily in the inner membrane (PulC, F, G, H, J, K, L, M, N, O) cytoplasm (PulE, B), or outer membrane/periplasm (PulD, S). The requirement for the cytoplasmic protein PulE in the secretion to the extracellular medium has suggested that the first stage of the SecA-dependent translocation across the CM might not be entirely separate from the second stage of OM translocation. The putative ATP-binding protein PulE might for example act as a specific cytoplasmic chaperone, or 'activate' other Pul secretion proteins by phosphorylation [30]. Nevertheless, this possibility of 'overlapping' stages may not be supported entirely by subsequent experiments in which stable periplasmic intermediates have been demonstrated [29].

Homologues of many of the *pul* genes in other Gram-negative bacteria have been shown to be required for secretion of various proteins to the extracellular medium, in particular the cellulases and pectinases of the plant pathogens *Erwinia chrysanthemi* [31] and *Xanthomonas campestris* [32], and exotoxin, elastase and other proteins of the human pathogen *Pseudomonas aeruginosa* [33]. Mutation of the secretion genes, termed *out* or *xcp*, result in a pullulanase-like accumulation of periplasmic or membrane-located intermediates in these organisms. The *pul*, *xcp* and *out* genetic determinants are closely conserved in structure and organization [34–36], although the secretion systems seem to have substrate-specific components. The emerging commonality of this pathway suggests that it represents in effect the second half of the conventional (general) secretion pathway from Gram-negative bacteria with the exception of *E. coli* and its closest relatives. The intensive study of *E. coli* has thus in one sense appeared to delay the concept of a complete general secretion pathway (GSP).

The GSP concept has been subsequently extended to other less similar secretion processes and this has provided a substantial step forward in understanding the underlying mechanism. The GSP secretion genes have been shown to share substantial similarity with Pil proteins required for the biogenesis of type IV cell surface fimbrial (pili) structures which determine bacterial colonization of mammalian cells by *Pseudomonas aeruginosa* (and also *Neisseria gonorrhoeae*) [37]. In particular, PulE(XcpR) and PulF (XcpS) are related to the pilus assembly proteins PilB and PilC, while XcpA and PulO resemble closely the PilD prepilin endopeptidase [38], suggesting a requirement for processing and assembly of GSP components. Indeed, components of the *Pseudomonas* secretion apparatus are processed by type IV prepilin peptidase [39] and the PulO protein is able to process correctly the type IV prepilin protein [40]. Processing sites have been identified within the 'pilin-like' XcpTUVW (PulGHIJ) proteins and the absence of processing in *xcpA* mutants causes accumulation of the secretion components themselves in the inner membrane [41]. The close similarity of the protein translocation processes to a mechanism for pilus assembly has suggested that the GSP could in every case involve a pilus-like structure as a means of generating inner-outer membrane connections [41].

The export of aerolysin by *Aeromonas hydrophila* and other species involves a stable periplasmic intermediate generated by the signal peptide-dependent conventional system, but in this case conversion of proaerolysin to the mature active pore-forming toxin requires C-terminal proteolytic processing on the cell-surface [42,43]. This previously anonymous secretion event has now also been brought under the umbrella of the pullulanase-type GSP, as the mutant gene, *exeE*, that causes defective aerolysin secretion and outer membrane assembly has been shown to be homologous to *pulE* and to be located in a *pulC-O*-like operon [44]. It has been reported that the second stage of transfer to the extracellular medium requires ΔP [45], a surprising feature which perhaps may be re-evaluated, particularly as PulE is the putative ATP-binding cytoplasmic protein that has been tentatively proposed as a transducer of ATP-derived energy to other GSP secretion components. It is possible that disruption of ΔP by uncouplers would deplete ATP and thus compromise the source of ExeE-derived energy used in the late stage of secretion to the extracellular medium. The ExeE(PulE) protein has gained further relatives in the ComG and VirB proteins that are required, respectively, for DNA uptake and DNA transfer in *Bacillus subtilis* [46] and *Agrobacterium tumefaciens,* the latter having been shown to bind and hydrolyze ATP [47]. A further relative is the PilT 'twitching' gene, which is involved in bacterial movement across solid surfaces [48].

It is inevitable that other signal peptide-dependent examples of extracellular secretion will be added to the GSP family. Periplasmic intermediates are for example substrates in the secretion of *Vibrio cholera* enterotoxin (CTX) to the cell-free medium, in a process in which the mature oligomeric toxin structure appears also to be assembled in the periplasm [49]. The folding and assembly of the CTX (or *E. coli* homologue ETX) is maintained by a periplasmic disulphide isomerase homologous to DsbA [50], although it is not known whether this has consequences for oligomerization or secretion or both events. The OM targetting component of the toxin resides in the B subunits but as yet unidentified OM or periplasmic helper proteins are believed to be required for assembly and also secretion. This view is supported primarily by the periplasmic location of CTX when synthesized in *E. coli* [51], whereas the normally periplasmic CTX-like heat-labile enterotoxin (ETX) of *E. coli* is exported out into the medium when made in *Vibrio cholera* [52].

2.3.2. Assembly of P-type fimbrial adhesins

A different type of signal peptide-dependent process is utilized in the assembly of most of the well-characterized fimbrial adhesins on the surface of the bacterial cell. The best studied system is the export of the Pap (or P-type) protein filaments needed for colonization of the kidney via binding to Gal-Gal moieties on the host epithelial cells [53], and genetic comparisons confirm that this serves as a model for the assembly of other fimbrial adhesins. The translocation of pilin subunits (PapAFX) including the tip adhesin (54) to the OM surface occurs via assembly of mature periplasmic subunits that are maintained in an appropriate conformation so that they can be presented effectively to a specific OM assembly complex (including PapC, the 'translocation platform') by a specific periplasmic chaperone, or 'usher' (PapD), one of a family of periplasmic chaperones that share conserved immunoglobulin-like structural features [55]. The second stage of the process is thus governed by protein–protein recognition involving specific helper proteins in the periplasm and the OM. If the critical role of the chaperone protein family is limited to pilin assembly, this might reflect a specific requirement in assembly processes for several substrates to 'queue' orderly as stable periplasmic intermediates rather than interacting rapidly with the membrane [53].

3. Signal peptide-independent secretion out of the cell

Several proteins which are important to bacterial pathogenicity during infection of animals and plants are secreted out of the cell by a process which is signal peptide-independent, and this appears not to require a periplasmic translocation intermediate. The primary example of this type of mechanism is the secretion by *E. coli* of the 110-kDa hemolysin, a toxin that forms pores in eukaryotic membranes and disrupts host cell functions [56–58].

3.1. Secretion of hemolysin

Secretion of hemolysin transfers the 110-kDa hydrophilic protein toxin (HlyA) across both cytoplasmic and outer membranes without employing an N-terminal targetting signal or the cellular SecA-dependent machinery. The process is directed by the *hlyB* and *hlyD* genes [59,60] which are contiguous and co-expressed with the *hlyC* and *hlyA* genes that are required for the synthesis of protoxin and the acyl carrier protein-dependent fatty acylation that matures it to cytolytically active toxin [56,57,61] (Fig. 3). The specific secretion proteins, HlyB and HlyD, are 80 kDa and 54 kDa, respectively, and are present at relatively low levels in cell membrane fractions, primarily due to down regulation of secretion gene expression within the *hlyCABD* operon. This is achieved by uncoupling operon transcription of the synthesis and secretion genes via transcript termination at a *rho*-independent terminator within the operon [62,63], an effect probably accentuated by differential stability of the truncated and elongated transcripts. Expression is dependent on structural elements, including the *hlyR* locus, which are positioned upstream of the *hly* operon promoter sequences and thought to act as binding sites for regulatory proteins involved in both *hly* transcription activation and antitermination [64,65]. A recently identified activator of *hlyCABD* gene expression is the 18-kDa product of the *rfaH* (*sfrB*) gene which positively regulates transcript initiation and possibly antitermination in the operons encoding synthesis of sex pilus and lipopolysaccharide of *E. coli* and *Salmonella* [66]. The suggestion that the *hly* operon is part of a coordinately activated regulon encoding virulence and fertility is supported by evidence of shared 5' sequences in the RfaH-acti-

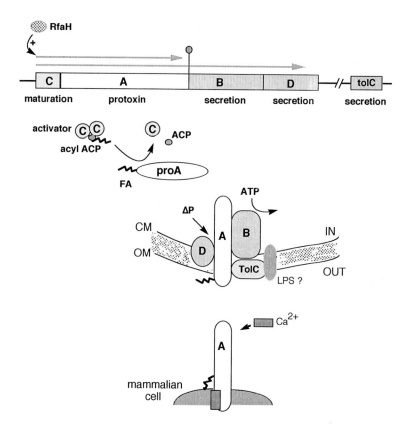

Fig. 3. Hemolysin synthesis, maturation and secretion. Expression of the *hlyCABD* operon is governed by *cis*- and *trans*-acting components including those upstream of the *hly* promoter region and the activator protein RfaH. It generates two *hly* transcripts, truncated (Tt) and elongated (Te), as a result of transcription termination and antitermination at the *hlyA*-*hlyB* intergenic space. Gene *hlyA* encodes the 1024 residue inactive proHlyA which is activated to the toxic HlyA by acyl carrier protein (ACP)-dependent HlyC-directed fatty acylation. Mature HlyA is secreted across both membranes in a process dependent upon the inner membrane proteins HlyB, HlyD and also the minor outer membrane component TolC. The toxin binds Ca^{2+} and changes conformation; its initial interaction with the mammalian cell membrane is suggested to be mediated by the fatty acid (FA), and may be followed by a specific interaction with a putative receptor. The toxin then forms cation selective pores in the mammalian cell.

vated operons and also in others such as those for capsule synthesis (Bailey and Hughes, unpublished data). The discovery of a role in *hlyCABD* expression for the LPS (*rfa*) operon transcriptional activator RfaH is perhaps not so surprising as LPS has now been suggested to substantially influence both the secretion [67] and toxic activity [68] of the toxin.

3.1.1. A model for export of toxins, proteases and nodulation proteins

The unconventional signal peptide-independent process by which HlyA protein is secreted is a model for the secretion of many important proteins. These include other toxins, such as the hemolysins and leukotoxins of *Proteus, Morganella, Actinobacillus* and *Pasteurella* [69–71], and the adenylate cyclase-hemolysin bifunctional protein (cyclo-lysin) of *Bordetella pertussis* [72], certain proteases of *Erwinia, Pseudomonas* and *Serratia* [73–75] and a bacteriocin, colicin V [76], by different Gram-negative bacterial pathogens of man, animals and plants, and also nodulation factors [77] by bacterial plant symbionts (Table II; reviewed in [57,78–80]). The substrate proteins that are exported range in size from 11 kDa (ColV), through 48–55 kDa (the proteases) up to 102–177 kDa (cytolytic toxins); in the case of the toxins they comprise a structurally related family, primarily defined by a common Ca^{2+}-binding repeat domain [80,81]. Table II also lists other Gram-negative HlyB-homologues (sometimes termed ABC exporters, see later) which are associated with the export of non-protein substrates [78] such as polysaccha-rides, antibiotics or substrates which have yet to be defined, and also examples of related exporters in Gram-positive bacteria and in medically important eukaryotic translocation processes such as those which determine multiple drug resistance in human tumour cells and in parasites.

TABLE II

HlyB-type exporter proteins

Protein	Species	Substrate
(i) Gram-negative bacteria		
Hly/LktB	*E. coli, Morganella, Proteus, Pasteurella, Actinobacillus*	HlyA hemolysin/leukotoxin
CyaB	*Bordetella pertussis*	CyaA cyclolysin
PrtD	*Erwinia chrysanthemi Serratia marcescens*	Prt B,C proteases, metalloprotease
	Pseudomonas aeruginosa	APR alkaline protease
?	*Rhizobium leguminosum*	NodO nodulation protein
CvaB	*E. coli*	ColV bacteriocin
McbF	*E. coli*	MccB17 peptide antibiotic
BexA	*Haemophilus influenza*	Capsular polysaccharide
KpsT	*E. coli*	Capsular polysaccharide
NdvA	*Rhizobium meliloti*	Glucan polysaccharide
ChvA	*Agrobacterium tumefaciens*	Glucan polysaccharide
HelA	*Rhodobacter capsulatus*	Putative heme
(ii) Other organisms		
CylB	*Enterococcus faecalis* (Gram-+)	Cytolysin
SpaB	*Bacillus subtilis* (Gram-+)	Peptide antibiotic
MsrA	*Staphylococci* (Gram-+)	Macrolide antibiotics
ComA	*Streptococcus pneumoniae* (Gram-+)	Putative competence factor
STE6	Yeast	Pheromone
WB	*Drosophila*	Pteridine
PFMDR	*Plasmodium*	Chemotherapeutic drugs
MDR	Human	Anti-tumour drugs
CFTR	Human	Ions
HAM, Mtp	Human	Peptides

HlyB-directed secretion apparently requires only two specific accessory proteins. All the well-characterized examples of HlyB-dependent secretion in Gram-negative species also require a close relative of the co-synthesized HlyD, although this protein is apparently not obligatory for the CylB-dependent export of a cytolysin across the single Gram-positive membrane [82], nor for transporters which export smaller non-proteinaceous substrates across either one or two membranes. Export of hemolysin and other proteins out of the Gram-negative cell also requires TolC, an outer membrane porin encoded by an 'unlinked' chromosomal gene [83,84]. The direct involvement of TolC is also implied by the identification of *tolC* homologues as additional secretion genes, *prtF* and *cyaE*, respectively, in the protease and cyclolysin operons of *Erwinia* and *Bordetella* [83,85] and it may be that the apparent requirement for LPS in secretion [67] reflects a role in the correct insertion of outer membrane proteins such as TolC, as may also be the case with OmpC, OmpF and flagellin [86]. As HlyB homologues can evidently export proteins and other substrates across the cytoplasmic membrane in a signal peptide-independent manner, without the assistance of HlyD and TolC, it is likely that the HlyD and TolC proteins specifically influence the later stages of secretion. The secretion process is defined primarily by the function of HlyB, and it is this protein which we now look at in some detail.

3.1.2. A central role for HlyB: a superfamily transporter ATPase

HlyB is a 707 residue cytoplasmic membrane protein [87,60] which belongs to the ATP-Binding Cassette (ABC) superfamily of prokaryotic and eukaryotic transporters. HlyB is believed to be central to energy generation, substrate recognition and the structure of the bacterial membrane translocator, and it combines both cytoplasmic (ABC) and membrane domains fused in a single polypeptide, as shown in Fig. 4 [78,88]. It therefore resembles closely in structure and exporter function the eukaryotic transporters, most particularly the multidrug resistance P-glycoproteins (Mdr, pfMdr) which pump chemotherapeutic drugs out of human tumour cells and parasites, but also others such as the cystic fibrosis trans-membrane conductance regulator and the peptide transporter encoded by the major histocompatibility locus, as indicated in Fig. 5 [88,89]. The fused structure of the exporters contrasts that of the bacterial importers in which autonomous ABC components interact with distinct hydrophobic integral membrane components to import small molecules such as amino acids, ions, peptides, vitamins and sugars bound by specific periplasmic binding

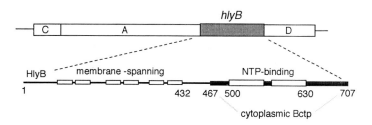

Fig. 4. The 707 residue HlyB secretion protein indicating the N-terminal membrane-associated region, and the cytoplasmic C-terminal peptide sequence (Bctp) encompassing the sequences needed for nucleotide binding and hydrolysis.

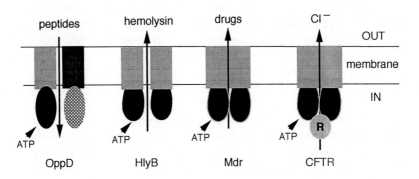

Fig. 5. ATP-binding transporter homologues of the HlyB secretion protein. OppD is an example of one of the many components of bacterial importer complexes, while the mammalian exporters Mdr and CFTR closely resemble HlyB in structure and function, combining cytoplasmic and integral membrane domains. The HlyB 'translocase' is assumed to form a dimeric structure directly analogous to the mammalian exporters by combining two copies of the HlyB protein.

proteins [88]. These importers are characterized by their requirement for specific periplasmic binding proteins which provide the initial site for substrate recognition, a function not needed by the exporters.

Hydrophobicity analysis of the HlyB amino acid sequence suggests that the 500 N-terminal amino acids determine interaction with the bacterial membrane(s) while the C-terminal 200 residues form a hydrophilic domain located in the cytoplasm [71,90,91]. The hydrophobic stretches are predicted to span the bacterial CM which, in the absence of structural data, has been tested by generating fusion proteins between HlyB and the reporter protein β-lactamase. The findings from an analysis of a limited number of random HlyB-β-lactamase protein fusions suggested eight membrane-spanning segments connected by substantial periplasmic and cytoplasmic loops [90]. A similar analysis [91], in this case of a similarly small number of random HlyB-LacZ and HlyB-PhoA hybrids, also suggested eight transmembrane segments, although not identical to those of the initial study. The topological predictions derived from the fusion data do not correspond unequivocally to the hydrophobicity predictions. The use of antisera raised against HlyB homologue PrtD, the exporter of protease, has shown that the ABC exporters are cytoplasmic membrane proteins [85], and protease accessibility studies support the view that at least one large domain of the polypeptide is exposed to the cytoplasm or periplasm, and suggest that the most likely arrangement is six transmembrane sequences (Fig. 4). This proposal would seem to agree with genetic and immunological studies of the better characterized OppB/C components of the enterobacterial oligopeptide importer, which predicted six membrane-spanning sequences [92]. A more confident appraisal of the HlyB/PrtD exporter topology awaits site-directed creation of reporter fusions and the use of monoclonal antibodies directed in particular against the sequences which are suggested to be looping out or buried in the membrane. In view of the requirements of most ABC transporters for two transmembrane domains and two ATP-binding domains, it is tempting to speculate the HlyB exporter is a homodimer comprised of two HlyB monomers.

Within the cytoplasmic domain of HlyB, closest identity to the ABC superfamily spans a region of approximately 200 amino acids, the 'ATP-binding cassette' [88], containing the diagnostic features of (i) a glycine rich domain beginning at residue 502, the 'Walker A' motif, thought to be responsible for nucleotide binding, (ii) a similar short glycine rich repeat domain starting at residue 603, and (iii) an aspartate at residue 630, part of the 'Walker B' motif, which is thought to be essential for Mg^{2+} coordination [93,94]. As none of the cytoplasmic components of transport ATPases has been crystallized, only speculative structural models have been proposed based on tertiary structure predictions and comparison with the adenylate kinase 'skeleton' [88,93]. Nevertheless, neither HlyB-directed ATP binding nor hydrolysis early in secretion has yet been demonstrated, although amino acid substitutions in the HlyB ATP-binding cassette do debilitate secretion [71]. A number of ABC transporters have been shown to bind ATP analogues, and replacement of conserved residues within the putative nucleotide binding site (A motif) affect ATP binding in the bacterial histidine uptake protein HisP and the mammalian CFTR [95]. In bacterial periplasmic binding protein-dependent maltose import, ATP hydrolysis occurs concomitantly with in vivo sugar uptake [96] and ATP is required for uptake into reconstituted proteoliposomes [97,98]. ATP hydrolysis by importers in reconstituted liposomes appears to require the presence of substrate bound by the specific periplasmic binding protein, suggesting that substrate-induced conformational interaction between the integral membrane domain and cytoplasmic ABC domain influences hydrolysis of bound ATP [99].

It has not been possible until recently to purify a soluble ABC protein to a high concentration, either as an autonomous bacterial cytoplasmic/peripheral membrane importer component, or as the larger bacterial and eukaryotic ABC exporters in which isolation of purified soluble protein is made more unlikely due to the fusion of the nucleotide-binding domain to the integral membrane component. Separation and overexpression of the MalK-like C-terminal 240 amino acid cytoplasmic domain of the HlyB exporter generates only aggregated protein in inclusion bodies which must be solubilized with urea or detergent (V. Koronakis unpublished data), as was also encountered with the overexpressed MalK protein [100].

It seemed likely that direct recovery of soluble active HlyB might be achieved by separating the ABC domain from the integral membrane sequences and at the same time encouraging a dimeric tertiary structure, so we recently generated a fusion of the ABC domain of the HlyB exporter (Fig. 4) with the protein glutathione S-transferase (GST) which is extremely soluble and believed to be active as a dimer [101]. The fusion protein GST-Bctp retained wild-type transferase activity and demonstrated clear ATP-binding and a high level of substrate-independent ATPase activity comparable with maximum levels of ABC homologues assayed in vivo [102].

This suggests that the two protein domains were folded independently, and molecular weight chromatography supported the view that the bifunctional fusion protein was active as a dimer. In the intact HlyB transporter (Figs. 4 and 5), a dimeric structure is likely to be determined and/or stabilized by the N-terminal integral membrane domain, possibly upon interaction with the substrate. GST-Bctp bound specifically to ADP-agarose and was eluted only by ATP and ADP, affinity behaviour which was confirmed in both the full length HlyB and the unfused HlyB cytoplasmic domain synthesized in vitro. Mg^{2+}

enhanced GST-Bctp binding of ATP and ADP analogues, and the stoichiometry in both cases was close to equimolar. Binding of MgATP or MgADP induced substantial conformational change in the protein, as demonstrated by resultant protection from proteolysis. The ATPase activity of the GST-Bctp protein was Mg^{2+}-dependent with a V_{max} and K_m comparable to the activity of bacterial importers (MalK) and human Mdr proteins reconstituted into proteoliposomes, and over an order of magnitude higher than in vitro measurements of disaggregated MalK purified from inclusion bodies. The ABC components of both the importing and exporting members of this family of transport systems are located on the cytoplasmic side of the membrane.

As well as transmitting a conformational change directly to the associated transmembrane domains/components with the effect of opening or closing a pore structure, the ABC domains could also be involved in establishing one part of a 'ratchet'-mechanism [103] in which the energy of ATP hydrolysis is used to strip the translocating protein from bound chaperones on the cytoplasmic side of the membrane or for inducing a translocation competent folding state in the substrate protein. Our results are compatible with the view that both conformational change and enzymatic hydrolysis by HlyB could be involved in protein secretion. In the homologous mammalian exporters, it is not known how the NTP-binding domains interact or complement each other, and in the CFTR protein ATP hydrolysis by one of the domains is sufficient to open the channel [104].

3.1.3. Targetting of hemolysin: association with the HlyB translocase
Hemolysin is secreted as the mature acylated form of the *hlyA* gene product proHlyA, following the covalent attachment of a fatty acid moiety in a cytoplasmic maturation mechanism directed by the dimeric HlyC activator, a putative acyl transferase, and dependent upon the acyl carrier protein (Fig. 3) [56,61]. This specific and novel HlyC-directed fatty acylation is required to target the hemolysin toxin to mammalian cell membranes, prior to forming cation-selective pores and disrupting the host cell [58,105,106]. The HlyB-dependent secretion process readily accepts and translocates the non-acylated proform of hemolysin (i.e. toxin maturation is not a requirement for HlyBD-dependent secretion of HlyA). However, the presence of the fatty acid may modulate the kinetics of toxin association with the bacterial membrane, an interesting possibility considering the lipophilic nature of substrates exported by certain eukaryotic ABC transporters, i.e. lipophilic drugs by human P-glycoproteins, and the acylated peptide (farnesylated yeast a mating factor) by STE6 protein [107].

The initial targetting of the 1024 residue hydrophilic HlyA protein to the cytoplasmic membrane is directed by an unprocessed secretion signal located within its C-terminal 48 amino acids, and stepwise truncation of this sequence progressively debilitates secretion of the (pro)HlyA protein into the cell-free medium, as shown in Fig. 6 [108–110]. Despite a lack of primary sequence similarity between the C-terminal targetting signals of the toxins and proteases secreted by HlyB homologue-dependent pathways, they do appear to share higher-order structural similarities since the *E. coli* HlyB and HlyD proteins recognize and secrete the hemolysins and leukotoxins of *P. vulgaris*, *M. morganii*, *A. pleuropneumoniae*, *P. haemolytica*, and the bifunctional adenyl cyclase-hemolysin of *B. pertussis* (the so-called RTX (repeat toxins) family [57,79,80]), and also the proteases of *E. chrysanthemi* [111] and *P. aeruginosa*, the *E. coli* bacteriocin ColV and the *Rhizobium*

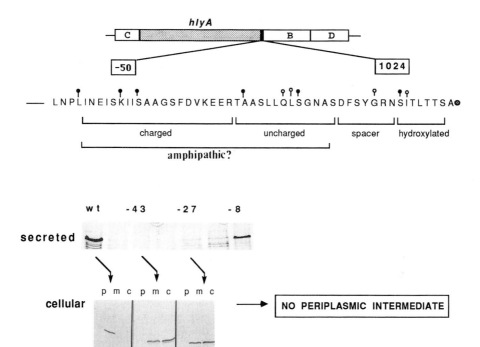

Fig. 6. The C-terminus of the *E. coli* 1024 residue HlyA protein showing putative features of the uncleaved targetting signal, derived from primary sequence analyses of the HlyB/D-secreted proteins and extensive site-directed mutagenesis (see text). C-terminal truncation of HlyA (shown as the black lollipops at –8, –20, –27, –40, –49 residues) leads to a dramatic loss in extracellular toxin, as demonstrated by the Coomassie-stained secreted protein in the SDS-polyacrylamide gel. Fractionation and immunoblotting of HlyA protein from cells containing the wild-type (wt) HlyA and the –43 and –27 truncated derivatives shows that while secretion-defective substrate protein can accumulate in the membrane (m) and cytoplasm (c), like the wild-type they do not accumulate in the periplasm (p).

NodO protein. Comparison of the C-terminal signal sequences reveals that they contain three regions [109]: charged, uncharged and hydroxylated, the last two separated by a spacer sequence (Fig. 6). Although typically at the C-terminus of secreted polypeptides, similar features have been identified in the N-terminal signal of the secreted ColV protein [112; P. Stanley personal communication].

Extensive mutagenesis of the C-terminus of *E. coli* HlyA supports the functional importance of each of these three regions, in particular the presence and distribution of acidic and basic residues in the charged region and the hydroxylated residues at the extreme C-terminus [109]. Results of subsequent mutagenesis analysis [113] do not appear to conflict with this view. The charged region is 20–25 amino acids long, including 6–8 acidic and basic residues, while the uncharged region is 7–13 amino acids long, rich in small, hydroxylated residues and usually has aspartic acid and a basic residue at either end. Plotted as an α-helix, the charged region of the C-terminal signal possesses an imperfect amphipathic character [108,109], which as the helix is stretched (as a 3_{10}-helix) continues into the uncharged region (Fig. 6). This suggests that the HlyA C-terminal secretion signal might adopt a largely helical conformation, which is also felt to be important to the function of N-terminal bacterial leader peptide sequences [114] and mitochondrial import presequences [115] in determining lipid association and influencing translocation in response to transmembrane potential. The hydroxylated cluster is located within the 11 C-terminal amino acids and generally includes a S/T-L/V-S/T tripeptide motif. Its conservation at the extreme C-terminus suggests a common role such as the initial interaction with the translocation machinery (specifically HlyB). Such an interaction could precede an induced conformational change, ensuring that the C-terminal 50 amino acids enter a closer association with both the CM and HlyB.

These views are incorporated in the three-step model outlined in Fig. 7, the sequential formation of (i) a membrane lipid-signal intermediate, (ii) an initial signal-translocator complex, and (iii) a complete HlyA-translocator complex [57,109]. In contrast to genes central to conventional protein secretion in bacteria and mitochondrial import [114,116], *hlyB* and *hlyD* can be deleted without causing pleiotropic effects on the cell, thus opening the way to experimental differentiation between binding of translocation substrates to the lipid bilayer and binding to the specific secretion machinery formed by HlyB and HlyD. The proposal that the HlyA signal interacts with first the lipid bilayer and subsequently the secretion protein(s), is supported by investigations of targetting to the plasma membrane with fusions of the HlyA C-terminal signal to non-translocatable globular proteins (V. Koronakis, unpublished results).

3.1.4. Translocation of hemolysin
Neither intact HlyA, HlyA C-terminal peptides, nor HlyA carrying defective secretion signals appear in the periplasm, as is shown in Fig. 6 [108,110,117[. This supports a model in which HlyA is exported directly into the medium without a periplasmic intermediate, analogous to the import of protein to the mitochondrial matrix, in which a junc-

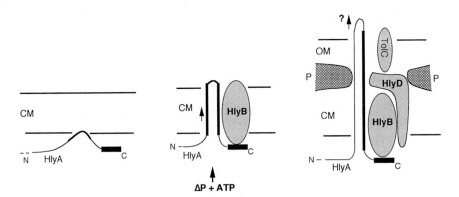

Fig. 7. Possible early and late steps in the HlyB-dependent translocation of HlyA across the CM and OM during export without a periplasmic (P) intermediate. The N-terminal and C-terminal ends of HlyA are labelled N and C, respectively. Filled box at the HlyA C-terminus indicates hydroxylated cluster.

tion between the two membranes during translocation has been identified by electron microscopy [118]. If the initial stage of HlyB-dependent secretion does involve interaction of substrate with the lipid bilayer to gain access to the secretion machinery, this provokes interesting analogies with the 'flippase' model proposed for the export of lipophilic drugs by the mammalian ABC protein, the multidrug resistance P-glycoprotein [119,120]. In this case, the substrate is predicted to interact directly with membrane lipids prior to interaction with a substrate binding site on the transporter protein. This model suggests that the transporter is not simply a hydrophilic hole which shields the substrate from the lipid bilayer [119]. This is especially appealing for the HlyB secretion system since such a pore with set dimensions is unlikely to be able to accommodate toxin/protease substrates up to 1000 amino acids and larger, in contrast to a structure able to open and close ('breathe') as sections of HlyA pass through.

By disrupting the proton motive force, or its components, during in vivo HlyB/D-dependent export of a C-terminal segment of HlyA, the secretion process has been divided into two energetically distinct stages [117]. The early stage of secretion requires the total protonmotive force (ΔP), while a late stage, characterized by strong dependence upon pH and temperature, does not. The early ΔP-requirement can be met by either of the ΔP components, the membrane potential ($\Delta \Phi$) or the pH gradient (ΔpH). This is also the case for conventional processing and transfer of β-lactamase across the E. coli CM by the Sec pathway [121], but not for mitochondrial import across two membranes, which requires the membrane potential specifically [122].

The ΔP requirement for the early stage of the HlyBD-directed secretion is suggested to reflect the needs of one or more steps at the CM, such as substrate binding, its subsequent release [117], or perhaps the cytosolic association of substrate with HlyB/(HlyD) protein which then inserts into the CM in a ΔP-dependent manner. The first possibility would be analogous to the requirement defined for conventional SecA-dependent bacterial secretion. In vitro and in vivo experiments have indicated that during conventional protein translocation across the CM, membrane association of the pre-protein only requires the energy of the SecA ATPase, while ΔP is required for the subsequent entrance to the periplasm once the translocation intermediate has been released from SecA [10]. Secretion of M13 procoat, which does not utilize the SecA/Y translocase, also requires ΔP but in this case apparently for substrate insertion into the membrane [123].

It seems likely that one or more of the conventional requirements for ΔP is shared by the early stage of hemolysin export, but in the latter case, subsequent entry into some form of energetically primed membrane fusion complex may replace conventional release of protein into the periplasm (Fig. 7). In this way, the OM barrier is not confronted directly, in contrast to the movement of periplasmic aerolysin which is suggested, surprisingly, also to require ΔP to cross this membrane [45]. In such a model, HlyA could bypass the periplasm and cross into the external medium following the formation of a complete proteinaceous bridge, and possibly by fusing the CM and OM (Fig. 7). Fractionation of cells has shown that, when HlyB is absent, HlyA is associated with the CM, whereas in the presence of HlyB and the N-terminal two-thirds of HlyD, HlyA is associated with both membranes with part of it exposed on the cell surface [124]. One might postulate that the truncated HlyD is sufficient to stabilize the structure crossing the two membranes, but cannot continue the translocation. On the other hand, cell fractionation of the

HlyB/D/TolC homologues PrtD/E/F have suggested that the three proteins are not in intimate contact in translocation complexes [85]. This would argue against the role of a stable 'pore' or stabilized adhesion zone and suggests a dynamic translocation machinery, perhaps involving only a transient contact between inner and outer membranes. This remains to be resolved.

A translocation intermediate in the ΔP-independent late stage of export has been shown to be inaccessible to trypsin in whole cells and spheroplasts [117], indicating that the substrate is still located at the inside of the cytoplasmic membrane, or possibly in a HlyB/HlyD-induced or stabilized membrane contact site. The requirement for ΔP is therefore not a result of translocation to the OM and beyond. While this late stage could be driven directly by HlyB, e.g. by ATP binding and/or hydrolysis, but it is also possible that this movement needs no additional energy. The persistence of late-stage secretion in the presence of high levels of ΔP-uncouplers [117] supports the latter possibility as does analogy with the bacterial signal peptide-dependent secretion process [9–11] and protein import into microsomes and mitochondria [125,126], all of which require ATP hydrolysis early in the process. Further parallel with the conventional SecA-dependent translocation [9,121] suggests that the early requirement for total protonmotive force (ΔP) in the HlyB-dependent mechanism probably reflects the needs of substrate association with the CM translocation machinery and transfer across the CM.

3.2. An alternative signal peptide-independent mechanism? Proteins involved in cell invasion and motility

The hemolysin secretion process is as yet the only clear example of complete secretion which is independent of the signal peptide-directed CM translocation, but recent work indicates that further signal-peptide independent processes may account for important features of flagella assembly and also the export of a wide range of surface proteins involved in bacterial invasion of mammalian host cells and possibly bacterial cell differentiation. There are for example no cleaved signal peptides on the secreted outer membrane proteins of *Yersinia enterolytica* and related species [127] (termed Yops, e.g. YopH, YopQ YopE etc.) which are essential for invasive virulence, e.g. cytotoxicity, platelet aggregation and antiphagocytic activity, although targetting information may be present at the N-terminus. The gene *yscC* of the *virC* locus (*yscA-yscM*), has been shown to be involved in Yop secretion and its product appears to share a relationship with the *Klebsiella* PulD protein needed for the pullulanase (GSP) secretion. The YscC protein has been shown to be closely related to the MxiD protein of *Shigella flexneri* which is required for the proper membrane localization of Ipa (invasion) proteins which are involved in many aspects of the invasive process, entry intro eukaryotic cells, contact-mediated hemolysis and lysis of the phagocytic vacuole [128–130]. A further homologue of the GSP proteins YscC/PulD/MxiD is HrpA (hypersensitive response and pathogenicity), a protein of the plant pathogen *Pseudomonas solanacearum* that is believed to be involved in eliciting host plant responses, presumably by means of bacterial surface or cell-free molecules [131]. That this signal peptide-independent process may share characteristics with the signal peptide-dependent GSP is intriguing in that it suggests that certain helper proteins may be involved in junction points connecting the different secretion pathways.

Homologies continue to be revealed throughout the invasion proteins of *Yersinia, Shigella* and *Salmonella* [128,130], and it seems that a novel and common secretion process may be crucial to the extracellular localization of virulence and motility related proteins in these bacteria. Central to the production of Yops by *Yersinia* is the *IcrD* gene product, the regulator of the 'low calcium response' during which surface proteins and virulence are induced by physiological controls, particularly calcium and temperature. LcrD is a ca. 70-kDa inner membrane protein predicted to have eight amino terminal transmembrane segments that anchor a large cytoplasmic carboxy-terminal domain to the inner membrane. Two close relatives are the invasion proteins VirH (MxiA) of *Shigella* and InvA of *Salmonella,* which are required for the presentation of analogous virulence-related proteins on the cell surface [132,133]. Other known members of this emerging family include the FlbF and FlhA proteins which are needed for the appearance of flagella on the surface of *Caulobacter* and *E. coli,* respectively [134,135]. A further member of this family, the SwaA protein of migratory *Proteus mirabilis,* has been found to be central to flagella production, invasion of eukaryotic cells and also the differentiation from vegetative to aseptate swarm filaments (Gygi, Bailey, Allison and Hughes, unpublished results). The secretion mechanism associated with this novel family of proteins apparently shares features with both HlyB-directed signal peptide-independent secretion and also the GSP. It will be of particular interest to discover what functional relationships underlie these secretion mechanisms.

Acknowledgements

We thank our colleagues in the laboratory, particularly Eva Koronakis and Peter Stanley, for their contributions to the work on hemolysin secretion, and also the MRC and Wellcome Trust for financial support.

References

1. Bieker, K.L., Philips, G.J. and Silhavy, T. (1990) J. Bioenerg. Biomembr. 22, 291–310.
2. Schatz, P.J. and Beckwith, J. (1990) Annu. Rev. Genet. 24, 215–248.
3. Wickner, W., Driessen, A.J.M. and Hartl, F. (1991) Annu. Rev. Biochem. 60, 101–124.
4. Kumamoto, C.A. (1989) Proc. Natl. Acad. Sci. USA 86, 5320–5324.
5. Rapoport, T.A. (1991) Nature 349, 107–108.
6. Hartl, F.U., Lecker, S., Schiebel, E., Hendrick, J.P. and Wickner, W. (1990) Cell 63, 269–279.
7. Joly, J.C. and Wickner, W. (1993) EMBO J. 12, 255–263.
8. Matsuyama, S., Fujita, Y. and Mizushima, S. (1993) EMBO J. 12, 265–270.
9. Geller, B.L. (1991) Mol. Microbiol. 5, 2093–2098.
10. Schiebel, E., Driessen, A.J., Hartl, F.-U. and Wickner, W. (1991) Cell 64, 927–939.
11. Driessen, A.J.M. (1992) EMBO J. 11, 847–853.
12. Arkowitz, R.A., Joly, J.C. and Wickner, W. (1993) EMBO J. 12, 243–253.
13. De Graaf, F.K. and Oudega, B. (1986) Curr. Top. Microbiol. Immunol. 125, 183–205.
14. Kornfield, S.J. and Plaut, A.G. (1981) Rev. Infect. Dis. 3, 521–534.
15. Pohlner, J., Halter, R., Bayreuther, K. and Meyer, T.F. (1987) Nature 325, 458–462.
16. Poulsen, K., Brandt, J., Hjorth, J.P., Thogersen, H.C. and Killian, M. (1989) Infect. Immun. 57, 3097–3105.

444

17. Miyazaki, H., Yanagida, N., Horinuchi, S. and Beppu, T. (1989)J. Bacteriol.171,6566-6572.
18. Klauser, T., Pohlner, J. and Meyer, T.F. (1990) EMBO J. 9, 1991–1999.
19. Klauser, T., Pohlner, J. and Meyer, T.F. (1992) EMBO J. 11, 2327–2335.
20. Yang, Y., Gao, Z., Guzman-Verduzco, L.-M., Tachias, K. and Kupersztoch, Y.M. (1992) Mol. Microbiol. 6, 3521–3529.
21. Uphoff, S.T. and Welch, R.A. (1990) J. Bacteriol. 172, 1206–1216.
22. Schiebel, E., Schwarz, H. and Braun, V. (1989) J. Biol. Chem. 264, 16311–16320.
23. Ondraczek, R., Hobbie, S. and Braun, V. (1992) J. Bacteriol. 174, 5086–5094.
24. Delisse-Gathoye, A.-M., Locht, C., Jacob, F., Raaschou-Nielsen, M., Heron, I., Ruelle, J.-L., De Wilde, M. and Cabezon, T. (1990) Infect. Immun. 58, 2895–2901.
25. Pugsley, A., d'Enfert, C., Reyss, I. and Kornacker, M.G. (1990) Annu. Rev. Genet. 24, 67–90.
26. Pugsley, A.P. and Kornacker, M.G. (1991) J. Biol. Chem. 266,13640–13645.
27. Pugsley, A.P., Kornacker, M.G. and Poquet, I. (1991) Mol. Microbiol. 5, 343–352.
28. Pugsley, A.P., Poquet, I. and Kornacker, M.G. (1991) Mol. Microbiol. 5, 865–873.
29. Poquet, I., Faucher, D. and Pugsley, A.P. (1993) EMBO J. 12, 271–278.
30. Possot, O., d'Enfert, C., Reyss, I. and Pugsley, A.P. (1992) Mol. Microbiol. 6, 95–105.
31. He, S., Lindeberg, M., Chatterjee, A.C. and Collmer, A. (1991) Proc. Natl. Acad. Sci. USA 88, 1079–1083.
32. Dums, F., Dow, J.M. and Daniels, M.J. (1991) Mol. Gen. Genet. 229, 357–364.
33. Bally, M., Ball, G., Badere, A. and Lazdunski, A. (1991) J. Bacteriol. 173, 479–486.
34. Condemine, G., Dorel, C., Hugovieux-Cotte-Pattat, N. and Robert-Baudouy, J. (1992) Mol. Microbiol. 6, 3199–3213.
35. Lindeberg, M. and Collmer, A. (1992) J. Bacteriol. 174, 7385–7397.
36. Filloux, A., Bally, M., Ball, G., Akrim, M., Tommassen, J. and Lazdunski, A. (1990) EMBO J. 9, 4323–4329.
37. Lory, S. (1992) J. Bacteriol. 174, 3423–3428.
38. Nunn, D. and Lory, S. (1991) Proc. Natl. Acad. Sci. USA 88, 3281–3285.
39. Nunn, D. and Lory, S. (1992) Proc. Natl. Acad. Sci. USA 89, 47–51.
40. Dupuy, B., Taha, M.-K., Possot, O., Marchal, C. and Pugsley, A.P. (1992) Mol. Microbiol. 6, 1887–1894.
41. Bally, M., Filloux, A., Akrim, M., Ball, G., Lazdunski, A. and Tommassen, J. (1992) Mol. Microbiol. 6, 1121–1132.
42. Howard, P.S. and Buckley, J.T. (1985) J. Bacteriol. 161, 1118–1124.
43. Howard, P.S., Garland, W.J., Green, M.J. and Buckley, J.T. (1987) J. Bacteriol. 169, 2869–2871.
44. Jiang, B. and Howard, S.P. (1992) Mol. Microbiol. 6, 1351–1361.
45. Wong, K.R. and Buckley, T.J. (1989) Science 246, 654–656.
46. Albano, M., Breitling, R. and Dubnau, D. (1989) J. Bacteriol. 171, 5386–5404.
47. Christie, P.J., Ward, J.E., Gordon, M.P. and Nester, E.W. (1989) Proc. Natl. Acad. Sci. USA 86, 9677–9681.
48. Whitchurch, C.B., Hobbs, M., Livingstone, S.P., Krishnapillai, V. and Mattick, J.S. (1991) Gene 101, 33–44
49. Hirst, T.R. and Holmgren, J. (1987) J. Bacteriol. 169, 1037–1045.
50. Yu, J., Webb, H. and Hirst, T.R. (199Z) Mol. Microbiol. 6, 1949–1958.
51. Pearson, G.D.N. and Mekalanos, J. (1982) Proc. Natl. Acad. Sci. USA 79, 2976–2980.
52. Neil, R.J., Ivins, B.E and Holmes, R.K. (1983) Science 221, 289–291.
53. Jones, C.H., Jacob-Dubuisson, F., Dodson, K., Kuehn, M., Slonim, L., Striker, R. and Hultgren, S.J. (1992) Infect. Immun. 60, 4445–4451.
54. Kuehn, M.J., Heuser, J., Normark, S. and Hultgren, S.J. (1992) Nature 356, 252–255.
55. Holmgren, A., Kuehn, M.J., Branden, C.-I. and Hultgren, S.J. (1992) EMBO J. 11, 1617–1622.
56. Issartel, J.-P., Koronakis, V. and Hughes, C. (1991) Nature 351,759–761.
57. Hughes, C., Stanley, P. and Koronakis, V. (1992) Bioessays 14, 519–526.
58. Bhakdi, S. and Martin, E. (1991) Infect. Immun. 59, 2955–2962.
59. Felmlee, T., Pellett, S., Lee, E.-Y. and Welch, R . (1985) J. Bacteriol. 163, 88–93.
60. Wagner, W., Vogel, M. and Goebel, W. (1983) J. Bacteriol. 154, 200–210.

61. Hardie, K.R., Issartel, J.-P., Koronakis, E., Hughes, C. and Koronakis, V. (1991) Mol. Microbiol. 5, 669–679.
62. Koronakis, V., Cross, M. and Hughes, C. (1989) Mol. Microbiol. 3, 1397–1404.
63. Koronakis, V., Cross, M. and Hughes, C. (1988) Nucleic Acids Res. 16, 4789–4799.
64. Vogel, M., Hess, J., Then, I., Juarez, A. and Goebel, W. (1988) Mol. Gen. Genet. 212, 76–84.
65. Cross, M.A., Koronakis, V., Stanley, P. and Hughes, C. (1990) J. Bacteriol. 172, 1217–1224.
66. Bailey, M.J.A., Koronakis, V., Schmoll, T. and Hughes, C. (1992) Mol. Microbiol. 6, 1003–1012.
67. Wandersman, C. and Letoffe, S. (1993) Mol. Microbiol. 7, 141–150.
68. Stanley, P., Diaz, P., Bailey, M.J.A., Juarez, A. and Hughes, C. (1993) Mol. Microbiol., in press.
69. Gygi, D., Nicolet, J., Frey, J., Cross, M., Koronakis, V. and Hughes, C. (1990) Mol. Microbiol. 4, 123–128.
70. Koronakis, V., Cross, M., Senior, B., Koronakis, E. and Hughes, C. (1987) J. Bacteriol. 169, 1509–1515.
71. Koronakis, V., Koronakis, E. and Hughes, C. (1988) Mol. Gen. Genet. 213, 551–555.
72. Glaser, P., Sakamoto, H., Bellalou, J,. Ullmann, A. and Danchin, A. (1988) EMBO J. 7, 3997–4004.
73. Letoffe, S., Delepelaire, P. and Wandersman, C. (1990) EMBO J. 9, 1375–1382.
74. Létoffé, S., Delepelaire, P. and Wandersman, C. (1991) J. Bacteriol. 173, 2160–2166.
75. Guzzo, J., Duong, F., Wandersman, C., Murgier, M. and Lazdunski, A. (1991) Mol. Microbiol. 5, 447–453.
76. Gilson, L., Mahanty, H.K. and Kolter, R. (1990) EMBO J. 9, 3875–3884.
77. Scheu, A.K., Economou, A., Hong, G.F., Ghelani, S., Johnston, A.W.B. and Downie, J.A. (1992) Mol. Microbiol. 6, 231–238.
78. Koronakis, V. and Hughes, C. (1993) Semin. Cell Biol. 4, 7–15.
79. Coote, J.G. (1992) FEMS Microbiol. Rev. 88, 137–162.
80. Welch, R.A. (1991) Mol. Microbiol. 5, 521–528.
81. Ludwig, A.T., Jarchau, T., Benz, R. and Goebel, W. (1988) Mol. Gen. Genet. 214, 553–561.
82. Gilmore, M.S., Segarra, R.A. and Booth, M.C. (1990) Infect. Immun .58, 3914–3923.
83. Wandersman ,C. and Delepelaire, P. (1990) Proc. Natl. Acad. Sci. USA 87, 4776–4780.
84. Benz, R., Maier, E. and Gentschev, I. (1993) in: Molecular Pathogenesis of Bacteria – Basic and Applied Aspects, Zbl. Bakteriol. Suppl., Gustav Fischer Verlag, Stuttgart, in press.
85. Delepelaire, P. and Wandersman, C. (1991) Mol. Microbiol. 5, 2427–2434.
86. Ried, G. ,Hindennach, I. and Henning, U. (1990) J. Bacteriol. 172, 6048–6053.
87. Felmlee, T., Pellett, S. and Welch, R.A. (1985) J. Bacteriol. 163, 94–105.
88. Hyde, S.C., Emsley, P., Hartshon, M.J., Mimmack, M., Pearce, S.R., Gallagher, M.P., Gill, D.R., Hubbard, R.E. and Higgins, C.F. (1990) Nature 346, 362–365.
89. Deverson, E.V., Gow, I.R., Coadwell, W.J., Monaco, J.J., Butcher, G.W. and Howard, J.C. (1990) Nature 348, 738–741.
90. Wang ,R., Seror, S.J., Blight, M., Pratt, J.M., Broome–Smith, J.K. and Holland, I.B. (1991) J. Mol. Biol. 217, 441-454.
91. Gentschev, I. and Goebel, W. (1992) Mol. Gen. Genet. 232, 40–48.
92. Pearce, S.R., Mimmack, M.L., Gallagher, M.P., Gileadi, U., Hyde, S.C. and Higgins, C.F. (1992) Mol. Microbiol. 6, 47–57.
93. Mimura, C.S., Holbrook, S.R. and Ames, G.F.-L.(1991) Proc. Natl. Acad. Sci. USA 88, 84–88.
94. Schlichting, I., Almo, S.C., Rapp, G., Wilson, K., Petratos, K., Lentfer, A., Wittinghofer, A., Kabsch, W., Pai, E.F., Petsko, G.A. and Goody, R.S. (1990) Nature 345, 309–315.
95. Shyamala, V., Baichwal, V., Beall, E. and Ames, G.F.-L. (1991) J. Biol. Chem. 266, 18714–18719.
96. Mimmack, M.L., Gallagher, M.P., Hyde, S.C., Pearce, S.R., Booth, I.R. and Higgins, C.F. (1989) Proc. Natl. Acad. Sci. USA 86, 8257–8261.
97. Davidson, A.L. and Nikaido, H .(1990) J. Biol. Chem. 265, 4254–4260.
98. Davidson, A.L., Shuman, H.A. and Nikaido, H. (1992) Proc. Natl. Acad. Sci. USA 89, 2360–2364.
99. Petronilli, V. and Ames, G.F.-L. (1991) J. Biol. Chem. 266, 16293–16296.
100. Walter, C., Hoener zu Bentrup, K. and Schneider, E. (1992) J. Biol. Chem. 267, 8863–8869.
101. Reinemer, P., Dirr, H.W., Ladenstein, R., Schaeffer, J., Gallay, O. and Huber, R. (1991) EMBO J. 10, 1997–2005.

102. Koronakis, V., Hughes, C. and Koronakis, E. (1993) Mol. Microbiol. 8, 1163–1175.
103. Simon, S., Peskin, C. and Oster, G.F. (1992) Proc. Natl. Acad. Sci. USA 89, 3770–3774.
104. Anderson, M.P., Berger, H.A., Rich, D.P., Gregory, R.J., Smith, A.E. and Welsh, M.J. (1991) Cell 67, 775–784.
105. Benz, R., Schmid, A., Wagner, W. and Goebel, W. (1989) Infect. Immun. 57, 887–895.
106. Menestrina, G., Mackman, N., Holland, I.B. and Bhakdi, S. (1987) Biochim. Biophys. Acta 905, 109–117.
107. Anderegg, R.J., Betz, R., Carr, S.A., Crabb, J.W. and Duntze, W. (1988) J. Biol. Chem. 263, 18236–18240.
108. Koronakis, V., Koronakis, E. and Hughes, C. (1989) EMBO J. 8, 595–605.
109. Stanley, P., Koronakis, V. and Hughes, C. (1991) Mol. Microbiol. 5, 2391–2403.
110. Mackman, N., Baker, K., Gray, L., Haig, R., Nicaud, J.-M., and Holland, I.B. (1987) EMBO J. 6, 2835–2841.
111. Delepelaire, P. and Wandersman, C. (1990) J. Biol. Chem. 265, 17118–17125.
112. Fath, M.J., Skvirsky, R.C. and Kolter, R. (1991) J. Bacteriol. 173, 7549–7556.
113. Kenny, B., Taylor, S. and Holland, I.B. (1992) Mol. Microbiol. 6, 1477–1489.
114. Schatz, P.J. and Beckwith, J. (1990) Annu. Rev. Genet. 24, 215–248.
115. Roise, D., Theiler, F., Horvath, S.J., Tomich, J.M., Richards, J.H., Allison, D.S. and Schatz, G. (1988) EMBO J. 7, 649–653.
116. Baker, K.P., Schaniel, A., Vestweber, D. and Schatz, G. (1990) Nature 348, 605–609.
117. Koronakis, V., Hughes, C. and Koronakis E. (1991) EMBO J. 10, 3263–3272.
118. Schleyer, M. and Neupert, W. (1985) Cell 43, 339–350.
119. Higgins, C.F. and Gottesman, M.M. (1992) Trends Biochem. Sci. 17, 18–21.
120. Devaux, P.F. (1991) Biochemistry 30, 1163–1173.
121. Bakker, E.P. and Randall, L.L. (1984) EMBO J. 3, 895–900.
122. Pfanner, N. and Neupert, W. (1985) EMBO J. 4, 2819–2825.
123. Date, T., Goodman, J.M. and Wickner, W. (1980) Proc. Natl. Acad. Sci. USA 77, 4669–4673.
124. Oropeza-Wekerle, R.L., Speth, W., Imhof, B., Gentshev, I. and Goebel, W. (1990) J. Bacteriol. 172, 3711–3717.
125. Chirico, W.J., Waters, M.G. and Blobel, G .(1988) Nature 332, 805–810.
126. Pfanner, N., Tropschug, M. and Neupert, W. (1987) Cell 49, 815–823.
127. Michiels, T., Vanoothegan, J.-C., DeRouvroit, C.L., China, B., Gustin, A., Boudry, P. and Cornelis, G.R. (1991) J. Bacteriol. 173, 4994–5009.
128. Allaoui, A., Sansonetti, P.J. and Parsot C. (1993) Mol. Microbiol. 7, 59–68.
129. Hakansson, S., Bergman, T., Vanooteghem, J.-C., Cornelis, G. and Wolf-Watz, H. (1993) Infect. Immun. 61, 71–80.
130. Allaoui, A., Sansonetti, P.J. and Parsot, C. (1992) J. Bacteriol. 174, 7661–7669.
131. Gough, C.L., Genin, S., Zischek, C. and Boucher, C.A. (1992) Mol. Plant-Microbe Interact. 5, 384–389.
132. Andrews, G.P. and Maurelli, A.T. (1992) Infect. Immun. 60, 3287–3295.
133. Galan, J.E., Ginocchio, C. and Costeas, P. (1992) J. Bacteriol. 174, 4338–4349.
134. Sanders, L.A., Way, S.V. and Mullin, D.A. (1992) J. Bacteriol. 174, 857–866.
135. Macnab, R.M. (1993) Annu. Rev. Genet,, in press.

J.-M Ghuysen and R. Hakenbeck (Eds.), *Bacterial Cell Wall*
© 1994 Elsevier Science B.V. All rights reserved

CHAPTER 21

Periplasm

MANFRED E. BAYER and MARGRET H. BAYER

Fox Chase Cancer Center, Institute for Cancer Research, 7701 Burholme Avenue, Philadelphia,
PA 19111, USA

1. Introduction: definition of the periplasm and its activities

This chapter addresses current views on the structural and functional organization of the periplasm. Selected topics are meant to serve as examples for models of pathways in the macromolecular transfer across envelope membranes and periplasm.

Periplasm is a concept pertaining to the envelope of Gram-negative bacteria, and comprises the molecules and ions that are localized within the space between the inner membrane (IM) and outer membrane (OM). The periplasmic space can be considered as a trans-shipment region carrying out the traffic between the interior and exterior of the cell. The periplasmic space is accessible in the intact bacterium from the cell's environment via three classes of transport pathways: (1) the non-specific protein assemblies, pores of the OM which allow passage of small, hydrophilic solutes; (2) via specific protein channels [1]; (3) whereas larger molecular weight substrates are bound with high affinity and specificity, and are actively transported across the OM [2,3]. From the cytoplasmic side, the periplasm is accessed with the help of a variety of transport mechanisms. Molecular transport across the IM from inside the cell may have different requirements for insertion of N- or C-termini of proteins such as leader peptidase [4] and may either terminate with parts of the exported molecule exposed at the outer surface of the IM, or the excreted molecules will remain in the periplasm. Furthermore, the molecule may be targeted for insertion into the OM or be destined for release from the cell surface [5]. To achieve this latter translocation, specific signal motifs within the polypeptide or on the folded mature molecule will be recognized by an export machinery [6]. An alternative pathway, namely a temporary or continuing contact of the newly synthesized molecule with a membranous environment can be provided by localized membrane connections between IM and OM, such as the membrane adhesion sites (Fig. 1) [7,8]. As discussed later, these sites appear to serve in both macromolecular import and export. The question of whether the adhesions can be described as contacts (Fig. 2(1)), fusions of IM and OM (Fig.2(2)) or pores (Fig. 2(3)) will have to be addressed by further research. Current evidence points towards the existence of contacts between IM and OM (Fig. 2(1)). Extracellular secretion of enzymes and toxins, without affecting the envelope membranes, appears to use translocation machineries of diverse pathways which are shared by a variety of Gram-negative

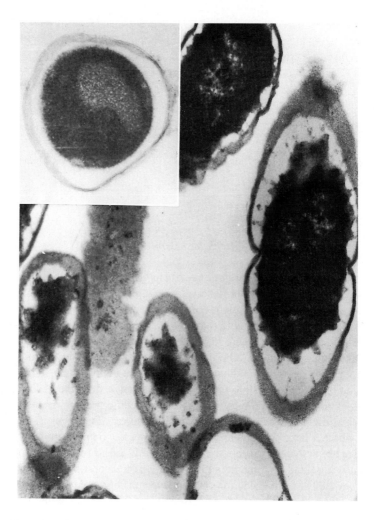

Fig. 1. Ultrathin section of plasmolized and conventionally processed *E. coli*. Numerous IM/OM adhesion sites are visible. Insert: Plasmolized, impact-frozen and cryo-substituted *E. coli*. Note the absence of adhesion sites.

bacteria. The possibility of a periplasmic transport via vesicle budding from the IM and fusion with the OM (Fig. 2(4)) has not been substantiated experimentally.

The periplasm contains a large number of proteins that have been classified in major groups [9]: (a) binding proteins for amino acids, carbohydrates and vitamins; (b) enzymes involved in scavenging and detoxifying activities; and (c) a group of other diverse proteins. The conventional mechanism of entry from the intracellular compartment into the periplasm occurs either post-translationally or co-translationally with the cleavage of the protein by signal peptidases. Other exported proteins are not cleaved and do not appear to use the periplasmic route [5,10]. Periplasmic binding proteins interact with incoming substrate [11] and facilitate its further translocation to the target at the IM.

Thus, the periplasm comprises not only the mixture of resident molecules, but it also houses those which are temporarily present during their import into the cell as well as during their export to the outer membrane, where they are assembled to form multimeric products such as fimbriae, protein pores and receptors for viruses, vitamins and colicins.

For the import of larger molecules (cobalamines, siderophores), pathways across the OM are available which catalyze active transport and release the incoming substrate from the OM into the periplasm [3]. Furthermore, substrates may be targeted to take the route via bridge-like structures, the adhesions between OM and IM, thereby circumventing the periplasm altogether. It has been well established that receptors for uptake of some nutrients are shared with the sites of interaction of colicins and bacteriophages. According to electron microscopic and kinetic studies of bacteriophage adsorption [8,12], virus particles appear to adsorb irreversibly to membrane adhesions and use these areas as DNA injection sites.

Two polymer species are permanent residents of the periplasm: (a) peptidoglycan (PG) and (b) membrane-derived oligosaccharides (MDO). Peptidoglycans form the shape-determining structure in the cell wall of a bacterium and include a group of closely related structures termed mureins, which comprise a large sacculus entirely made of sugar chains and peptide side chains. Although the chemical structure and biosynthesis of PG is well known, its physical arrangement in the periplasm is not unambiguously established.

The MDOs contain glucose as sole sugar substituted with phosphoglycerol, ethanolamine and succinyl ester groups. Their synthesis is controlled by the osmolarity of the environment of the cell, with low osmotic pressure of the growth medium enhancing their synthesis significantly [13]. MDOs exhibit a variable negative charge and have also been described as affecting the shutter functions of OM pores [14]. Another oligosaccharide group deserves to be mentioned here. The genus *Rhizobium* produces periplasmic cyclic

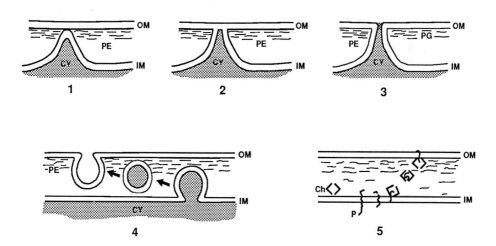

Fig. 2. Diagram showing modes of membrane contact, (1–3) periplasmic vesicle formation plus fusion (4) and diffusion (5) as mechanisms for translocation of proteins and polysaccharides. CY, ctyoplasm; PG, peptidoglycan; PE, periplasm (or space opened during plasmolysis); Ch, chaperone; IM, inner membrane; P, protein; OM, outer membrane.

glucans which are made of 11–13 glucose units per ring, with a phosphocholine association [15]. The function of these glucans as a periplasmic component has not been established.

2. Osmotic pressure; release of periplasmic proteins

If the periplasm is isotonic with the cytoplasm, the high concentration of cations in the periplasm relative to that in the environment will cause a hydrostatic pressure which is higher than that of the cell's environment. Thus, an outwards force is exerted against the shape-determining peptidoglycan sacculus. Without an intact peptidoglycan, the OM and IM would yield to the pressure, especially under shock conditions; large hernia-like extrusions of membranes will therefore develop [15], an effect also encountered when the peptidoglycan is weakened as a consequence of unbalanced autolytic activity or antibiotic action [16]. The osmotic pressure within a cell depends largely on the osmotic condition of the environment [17]. High osmolarity of the growth medium induces an increased uptake of ions such as potassium, glutamate and proline and a water loss from the cytoplasmic compartment. E. coli tolerates short periods of relatively high osmotic pressure. Brief exposure to 30% (w/v) sucrose in L-broth has no effect on the colony formation, whereas higher sucrose concentrations were found to reduce colony counts [18]. Sudden increase in external osmotic pressure causes plasmolysis, an outflow of water from the cytoplasm and the retraction of the IM from the OM, accompanied by an opening up and widening of the periplasmic space [7,8]. The bridge-like adhesion sites have been observed to rupture at high osmolarities, for example in 35% sucrose [18].

Sudden reduction of the external osmotic pressure (hypo-osmotic shock) of cells having previously adjusted to high osmolarities, for example, from 20% sucrose into water, may cause a large rate of cell death with 1% or fewer colony-forming survivors [15], whereas relatively light shocks allow most of the cells to survive. The sudden increase in pressure, exerted by the swelling cytoplasm on the peptidoglycan-encased periplasmic compartment, causes periplasmic components to be ejected through the OM into the medium. Aiding this release are pretreatments, such as a partial removal of lipopolysaccharide by EDTA and lysozyme from the OM [19]. A release of periplasmic proteins occurs also in envelope-defective mutants which lack proteins essential for the integrity of the OM [20] or in spheroplasts, as well as after chloroform treatment [11]. Periplasmic-binding proteins may undergo conformational (hinge-bending) changes after binding with the substrate. This has been shown in X-ray crystallographic studies of maltodextrin-binding protein complexes [21] and suggests an IM receptor differentiation between bound and ligand-free protein which would affect targeting to the IM and subsequent translocation into the cytoplasm. In addition to their function in transport, periplasmic-binding proteins play key roles in the chemotactic responses to the powerful attractants aspartate/serine; glucose/galactose; ribose/maltose. Secondary receptors represented by integral IM proteins bind specifically the complexes of protein and chemotactic molecules and link periplasmic signals to the flagellar motors [22]. Furthermore, they serve as detectors of other environmental conditions such as pH and temperature.

3. Molecular access to the periplasm; diffusion events

For a charged molecule to reach the periplasm from the environment of the cell, the first barriers are the strong negative charges of the molecules of the bacterial capsule and of the O-antigen and core region of the LPS. Added to the charge effects is the reduction in the diffusion rate of the entering molecule due to steric hindrance by these macromolecules. The second barrier consists of the variety of hydrophilic pores whose tertiary structure defines the pore openings [23] and provides channels open to molecules with molecular sizes not exceeding approximately 700 Da [24]. Pores can assume open and closed states. Depending on the area density of open pores, a molecule that is already positioned near the cell surface may collide with the cell many times before a successful entry into an open pore can be gained. Calculations showed (Litwin and Bayer, unpublished) that closing of 50% or 75% of the randomly distributed pores of a computer-simulated *E. coli* surface increases the time of random walk, which ends in successful entry of an open pore, by factors 10^2 to 10^3 times. Furthermore, the presence of exopolysaccharides (capsular and O-antigens) will impede the free movement of the walk along the cell surface and thus prolong the travel time significantly.

Once a molecule has gained entry into the periplasmic domain, it will encounter the structural elements of the murein sacculus. The peptidoglycan has been viewed as a sheet of murein [25], which is associated with the OM by two types of linkages: (1) via the covalently linked lipoproteins [26] that reach into the OM lipid domain and (2) via tight associations to the OM porins such as OmpF and OmpC [27]. The total number of linkages has been estimated to several hundred thousand contacts between the murein and the OM.

4. Periplasm and periplasmic space

With the aid of the electron microscope, the periplasmic space was observed early on in ultrathin sections of fixed, dehydrated and resin-embedded eubacteria [28], as well as in cryo-fixed, freeze-fractured enterobacteria [29,30] and was seen as a space between the boundaries of the OM and IM with a width of 7–15 nm. In other preparations, a separate thin layer of 2.5 nm thickness was visible stretching more or less parallel to a wavy OM [31]. Due to its sensitivity to lysozyme treatment [32], this structure was interpreted as representing the peptidoglycan sacculus. The peptidoglycan layer was often found to be adjacent to a layer of globular protein localized to the inner contour of the OM [28,32] (Fig. 3A). In most of these studies, the periplasmic space was found to be a more or less featureless domain of generally low contrast.

In more recent studies, two procedural systems were employed that were designed to minimize cell extraction and shrinkage, both of which had been shown to occur during fixation and cell dehydration [30,33]. One of these procedures employs aldehyde fixation in combination with a progressive lowering of the temperature during dehydration (PLT method) and subsequent low temperature embedding [34]. In the other group of procedures, the initial chemical fixation is circumvented by rapid freezing of the specimen and subsequent cryo-sectioning [35] or cryo-substitution followed by embedding in suitable resins [36,37]. Use of either of these methods revealed a different aspect of the periplas-

452

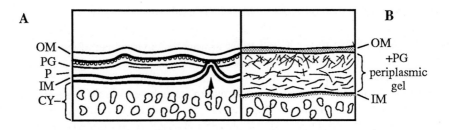

Fig. 3. Envelope structures after different preparation methods. (A) Cross-sectioned cell envelope after conventional preparation of slightly plasmolized cell. Arrow indicates adhesion site. (B) Cross-sectioned cell envelope after low temperature fixation and cryo-substitution. Periplasmic gel and peptidoglycan are shown in one compartment. A peptidoglycan layer is no longer identifiable. OM, outer membrane profile plus particulate layer; PG, (1–3 layered) peptidoglycan (its relation to adhesion sites is not established); IM, inner membrane profile; CY, cytoplasm plus ribosomes; P, periplasm spanning the area between PG layer and IM.

mic space. Instead of a low density domain of variable thickness between OM and IM, the area was now observed as being filled with an electron-scattering (contrast-generating) substance. These results were more pronounced when preparations were examined with a scanning transmission microscope operating in Z-contrast mode, in which the contrast is largely dependent on the atomic composition and concentration [38]. From these results, Hobot et al. [34] derived a model of the periplasm in which the peptidoglycan forms a diffuse gel (the 'periplasmic gel') with zones of varying degrees of hydration (Fig. 3B). A study of a variety of similarly prepared Gram-negative bacteria showed that the thickness of the periplasmic domain seems to depend on the bacterial species and ranges from approximately 10–25 nm [37,39]. Using a different method of rapid cryo-fixation, the width of the periplasm of *E. coli* B and *E. coli* K29 was measured to be 13.7 < 2.4 nm, but varied in some cells to reveal a space of more than 25 nm [18].

Based on these results and on assuming an average cell diameter of 1 μm (and a length of the bacterium between 2.2 and 3 μm), the volume of the periplasmic space can be estimated as being approximately 10% of the total cell volume; however, higher estimates were also reported [40]. The cell width has been determined after freeze-fracturing of unfixed *E. coli* B, grown in L-broth with a generation time of 23–25 min [30]. Other growth conditions will result in smaller cells and smaller envelope dimensions [41]. This result is in general agreement with measurements of a sucrose-permeable space which is correlated to the periplasmic volume and which was estimated to be 5% of the cell volume [24].

The importance of such spatial considerations becomes clear when the requirements are considered for the biochemical and physical behavior of periplasmic components such as electron carriers, enzyme complexes, binding proteins, and oligosaccharides. A group of Gram-negative bacteria are known to contain in their periplasm large quantities of redox proteins constituting up to 15% of the soluble bacterial protein; to accommodate these proteins, a periplasmic space of ~50 nm width had been suggested [42]. Although these latter biochemical estimates appear to be high, one might expect a variation among different cell strains. Recent electron microscopic results of cryo-prepared bacteria point towards an average periplasmic width of 10–25 nm [37]. Such measurements are averages only, since cells cryo-fixed with a different procedure show a variable, uneven width of the periplasm and also exhibit areas of contact between IM and OM [18]. The contribution of the peptidoglycan to the spatial arrangement between OM and IM has been addressed in a study using neutron scattering of isolated peptidoglycan sacculi. Labischinski et al. [43] measured the surface of the murein sacculus to be 75–80% single-layered and 20–25% triple-layered. The thickness of the single layered part was estimated to be 2.5 nm, that of the triple-layered part 7.5 nm. The results would be not incompatible with recent electron microscopic data: a 'somewhat fuzzy' region of 5–7 nm was measured in thin sections of isolated purified peptidoglycan [34], and Leduc et al. [44] found, after special staining, a sacculus thickness of 6.6 < 1.5 nm in ultrathin sections of *E. coli.*

The contribution of the 75–80% of single-layered peptidoglycan of 2.5 nm thickness to the entire periplasmic space is relatively small. Therefore, the contribution to the contrast in some of the electron micrographs, which seems to be 'filling' the space (see, e.g. [37]), would have to come only partially from the 'triple-layered' peptidoglycan, but largely from the periplasmic proteins which contribute to approximately 4% of the total protein content of *E. coli* [45]. Preliminary electron microscope data from our laboratory suggest that the membrane-derived oligosaccharides fail to provide a significant contrast enhancement in either PLT or cryo-fixed preparations.

5. Plasmolysis; osmotic condition of the periplasm

An osmotic pressure of 5–6 kg/cm^2 has been measured in *E. coli* [46]. This turgor forces the cytoplasmic contents plus the surrounding flexible plasma membrane (IM) into the shape provided by the peptidoglycan. Without the maintenance of a controlled synthesis of peptidoglycan during growth and extension of the cell envelope ([47], see also Chapter 7), the hydrostatic pressure from within the cell will drive the IM plus cytoplasm through any weakened area of the cell wall, and will cause eventually disruption of the membrane and cell death. Plasmolysis occurs when the osmotic pressure of the environment exceeds the pressure that maintains IM and OM in apposition. In the last century, plasmolysis was observed in plant cells [48] and in bacteria [49] and revealed plasmolysis bays as well as the separation of the envelope into plasma membrane and cell wall. To induce plasmolysis, *E. coli* is exposed to concentrated solutions of saccharides (sucrose, mannose) to which the IM is impermeable, or to NaCl concentrations of, for example, 0.25 M. In *E. coli* and *Salmonella anatum*, plasmolysis is dominated by two structural features: (1) the cell shrinks away from the OM and forms several inward cavities or bays, most often at one, sometimes at both cell poles; these bays can be readily observed in the light microscope; (2) at much of the cylindrical part of their envelope, plasmolyzed cells show a varying degree of separation of the IM from the OM. This can be seen in the electron microscope after conventional fixation (Fig. 1) or plunge-freezing and cryo-substitution (Figs. 4 and 5).

Plasmolysis is a relatively short-lived reversible condition. The cell achieves the reversal by an uptake of inorganic ions such as potassium, or by the uptake of proline, polyols and polyamines. Within limits, the uptake re-establishes a positive turgor pressure. A recovery is achieved in moderately strong osmolarities only (in 700–850 mOsm of sucrose in growth medium, for example), whereas higher concentrations cause loss of viability in *E. coli*, as well as structural disruption of the zones of membrane adhesions [18]. It has been shown that the effect of osmotic stress correlates with the water activity (an expression of the ratio of vapor pressure of the solution over that of water). For *E. coli* to grow, a limiting numerical value of 0.93 was found; extreme pressure represents a value of 0.75 found in halobacteria, and bacterial life does not seem to be possible at values below 0.66 [50].

Stock et al. [40] concluded that the osmotic pressures in the cytoplasm and periplasm are equal; for this to occur, molecules larger than those able to pass the pores of the OM will provide the osmotic conditions of the periplasm. A contributor to this function is the group of membrane-derived oligosaccharides [13] whose synthesis is regulated by the osmotic pressure of the growth media. At low osmotic strength, their synthesis was found to be 16 times as high as in 0.4 M Na concentrations. The contribution of the murein to periplasmic osmotic conditions at either of its proposed structural aspects, as highly hydrated gel or as more condensed, largely monolayered sheet, is not known.

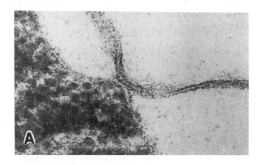

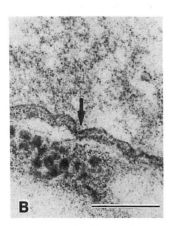
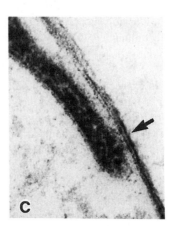

Fig. 4. Electron micrographs of plasmolized and subsequently cryo-fixed and cryo-substituted *E. coli* B. Crossing the periplasm are membrane adhesions differing in size. (A) Typical adhesion. Note the tight contact between inner and outer membrane (see Fig. 5b). (B) Small contact of outer and inner membrane (see Fig. 5a). (C) Membrane contact at rim of polar bay.

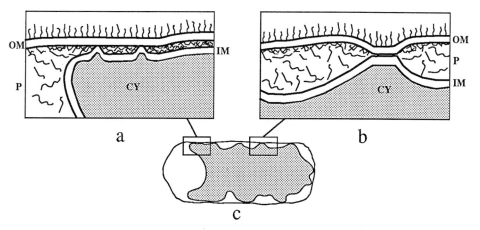

Fig. 5. Envelope membranes in plasmolized *E. coli* after cryo-fixation and cryo-substitution. (a) Represents membranes near polar bay; (b) adhesion site; (c) overview of cell. P, periplasmic space; OM, outer membrane; IM, inner membrane, CY, cytoplasm.

6. The physical state of the periplasm; viscosity

A molecule entering the periplasm is confronted with a large variety of molecular species. If one assumes that a viscous periplasmic gel exists in the periplasmic space, the molecular travel in the periplasm would be quite restricted, especially within the suggested denser parts of the peptidoglycan. The average dimension of openings in highly crosslinked murein has been estimated to be approximately 5×2 nm [51]. A varying degree of hydration of the murein and its partially triple-layered organization, as well as the number of periplasmic binding proteins and their large copy number (for example, *E. coli* may have 30 000 copies of maltose binding proteins per cell) will generate a relatively close packing with restricted freedom of molecular motion. Experiments in which labeled proteins were inserted in the periplasmic space seem to support this view. When the lateral diffusion rate of maltose binding protein was measured by photo bleaching recovery, a diffusion coefficient of 0.9×10^{-10} cm^2 s^{-1} was observed [51]. This value is 10^3 times lower than that for water, and about 10^2 times lower than that for a similar-sized protein in the cytoplasm. Thus, the lateral diffusion through the periplasm is extremely slow. This finding does to some degree cause difficulties in the interpretation of data showing relatively rapid responses of the cell, for example to adsorption of bacteriophages or to chemotactic signals. Thus, molecular transport will have to overcome the high viscosity of the periplasm. Several mechanisms to bridge the periplasmic space have been suggested, including those which would imply the structurally well-defined OM/IM contacts, the adhesion sites.

Photo bleaching recovery measures diffusion in the lateral direction, but fails to measure movement across the periplasm. This fact may still allow for the existence of periplasmic compartments in which the diffusion rate could be high, possibly approaching that of water. These compartments would have to be sealed off from the rest of the periplasm for example by tight-fitting periplasmic annuli [52]. However, this is unlikely to be the case since a lack of tightness of membrane sealing along the annuli has been reported [53]. A different mechanism of transport across the periplasm was suggested by Brass et al. [51], who envisaged a semi-ordered chain of binding proteins that would facilitate a transfer of molecules from one binding protein to the adjacent neighbor. Ultrastructural evidence for such aggregates is not available; rapid immobilization of molecules by cryo-microscopy might provide the answer, however, with due concern regarding the pitfalls in the interpretation of results of various rapid freezing procedures [18].

7. Contacts between outer and inner membrane

The electron microscope and, to some extent, also the light microscope, have provided solid evidence for the presence of connections between IM and OM in plasmolized Gram-negative bacteria. Several types of membrane associations can be considered, with their main feature being a close contact between outer and inner membrane (Fig. 2). Contacts between adjacent membranes have been amply observed in eukaryotic organelles and are best exemplified by mitochondrial membranes. Inner and outer membrane contacts were first described by Hackenbrock [54] in thin sections of liver mitochondria and by VanVenetie and Verkleij [55] in freeze-fractured mitochondrial preparations. The number of contacts was found to increase during oxidative phosphorylation, and an outer membrane pore which binds hexokinase (in brain and liver mitochondria) was localized to contact sites [56]; from lipid composition and dynamics of liver mitochondria, two populations of contact sites were described, constituting, respectively, the IM and OM portion of the contact zone [57]. An integral membrane protein of the pea chloroplast was identified as the receptor for protein import into chloroplasts and was localized to areas of IM/OM contact [58]. Although the envelope of Gram-negative bacteria such as *E. coli* B appears to have structural features which are in many aspects similar to those of mito-chondria, the major functional difference is that in the bacterial cell, all envelope compo-nents are synthesized within the cell, and may, after passage through the inner membrane, be incorporated in the outer membrane or released from the cell into the environment.

8. Structures crossing the periplasmic domain

8.1. Flagella

The structural design of a flagellum of *E. coli* and *Salmonella* has been studied exten-sively by both electron microscopy and by genetic procedures. The flagellum is anchored with its basal body in the cytoplasmic membrane. The rotational motor is localized at this site. The basal body shows two inner rings which are positioned in the periplasm and two outer rings in the OM and IM, respectively [59]. The flagellar motor system will react to periplasmic signals from energy-transducing and switching components which affect its rotational direction. Growth of the flagellum is proposed to occur by passing monomeric flagellin from the cytoplasm through the central hole of the filament to assemblage points at the distal tip [22]. During plasmolysis of flagellated *E. coli*, the outer and inner membranes are held together at the flagellar insertion site as seen in ultrathin sections [60]. In other bacteria such as spirochetes, the flagella are polar and remain localized largely in the periplasmic space.

8.2. F pili

E. coli HfrH forms approximately one F pilus per cell. Ultra-thin sections of plasmolyzed cells revealed the insertion site of the pilus at an inner/outer membrane adhesion [8]. In these preparations, the F pilus was made visible by adsorbing the pilus-specific RNA phages (either MS2 or R17) before induction of plasmolysis and fixation.

8.3. Membrane adhesion sites

Plasmolysis of Gram-negative *E. coli* reveals two distinctive types of structures that bridge the periplasm: (1) Discrete domains exist at which OM and IM remain attached to each other forming 100–200 connections [7]. At the site of adhesion, the inner membrane is closely associated with the inner contour of the outer membrane (Fig. 4). The size of the contact zone varied from approximately 20 nm to over 100 nm. The cross-sectional view of the extended portion of the IM at the smaller adhesion zones (before the IM makes contact with the OM) is more or less circular, with a central opening. Figure 1 shows that cells whose widened space between IM and OM is sectioned in a grazing plane, reveal more or less circular IM portions of adhesion sites. (2) Freshly plasmolized cells also show large vacuole-like invaginations of the inner membrane; these 'bays' are visible in the light microscope [16,49,49a,52,61] and are placed either at one or on both poles of the cell and along the side of the cell. The location of the 'bay' areas has been associated with current and future sites of cell division [62]. In this model, the adhesions along the bay are proposed to be continuous annuli, sealing off the contents of the bay from the rest of the periplasmic space. However, our data [18] and those of Anba et al. [53] suggest that the contact is not tight but rather incomplete along the adhesion areas surrounding a polar or a lateral bay. Therefore, it appears unlikely that the contents of a 'plasmolysis bay' are effectively separated from the rest of the periplasm.

8.4. Bacteriophage infection

When bacteriophages adsorb to either plasmolized or unplasmolized *E. coli*, they do so by adsorbing to areas that coincide with membrane adhesion sites [12]. Thus, the virus is seen to inject its DNA into a structure that might protect the nucleic acid from exposure to the periplasmic nucleases. The phages that adsorb preferentially to adhesion sites were *E. coli* phages T1, T2 to T7, PhiX174 and BF23, as well as the capsule-penetrating *E. coli* phages K26, K29 and K1 [12,63,64,65]; and LPS degrading *Salmonella* phages P22 [66], epsilon 15 and epsilon 34 [8].

The data suggest that the sites of membrane adhesions are also sites for entry of the DNA of very different classes of bacteriophages. In vitro experiments showed that a forced contact of OM vesicles with IM lipids can trigger a release of phage T4 DNA [67]. The notion that the adsorbing bacteriophages may generate adhesion zones in vivo is not supported by the finding that the number of adhesion sites in *E. coli* was not increased following adsorption of either phage T2, T5 or PhiX174. Furthermore, the capability of phage particles to 'find' adhesion sites is largely lost in super-infecting phages; this exclusion effect is already observed within 1–5 min after primary infection of the cell with homologous phage T2 [63]. More than 50% of super-infecting phage T2 are not positioned over adhesion sites and remain adsorbed with partially filled heads indicating a defective signal for DNA release; furthermore, DNA that has been released from most of the super-infecting phages is degraded in the periplasm. A significant difference in the number of adhesions before and after T2-super-infection was not observed.

9. Preservation of periplasmic structures

Membrane adhesions were visualized by electron microscopy of a variety of conventionally fixed plasmolized *E. coli* strains as well as in *Salmonella anatum* and *S. typhimurium* strains [8,65,66]. When a group of cryo-methods such as PLT or impact cryo-fixation, were employed [34,68,69], the membrane adhesions and even the large plasmolysis 'bays' were not observed (see also Fig. 1, insert). However, a subsequent comparative study of a variety of freezing methods revealed significant differences in the structural preservation of cryo-fixed cell envelopes, depending on the type of freezing process used [18]. While both the freezing by impact of the specimen on a cold metal surface as well as the PLT method failed to maintain plasmolysis bays and adhesion sites, other methods of rapid freezing, namely plunging the cells into a cryo-liquid, were found to preserve the plasmolysis bays and membrane adhesions. A combination of UV-flash photo-crosslinking [18] and plunge-freezing plus cryo-substitution increased the preservation of well-defined membrane adhesions, with their number approaching that of approximately 50 per cell. High resolution micrographs of these cryo-fixed sites revealed the distance at the contact between OM and IM to be less than 5 nm, leaving a narrow space stainable with uranium ions (Fig. 4A) or lanthanides [70]. Also, broader areas of membrane apposition were observed which were mainly at the borders of large bays. There the two membranes are separated by a gap of 10–50 nm; apparently at these areas the membranes are held together by small intermittent 'spotwelds', which also represent sites of membrane contact as close as those of the more obvious adhesion sites (Fig. 4B). Our recent data indicate that periseptal annuli represent a 'multiple spot-welded' zone of IM–OM contacts with many zones of non-adhering membrane portions between the contact areas [18]. These data suggest that the periplasmic space is interwoven with at least 50, and possibly several hundred, membrane bridges that form contacts of varying structural stability (Figs. 1 and 5).

10. Examples of export pathways: periplasm and membrane adhesions

The molecular architecture of the cell envelope requires that OM constituents and molecules destined for export will either have to be transported through the periplasmic space or via membrane adhesion sites or other, as yet undiscovered conduits.

10.1. Polysaccharides; rate of LPS conversion

Export of the oligo- and polysaccharides of the cell surface is a vectorial translocation process, in which multi-enzyme systems deliver the macro molecules from their place of synthesis, the inner membrane [71] to the exterior plane of the outer membrane. Localization of newly synthesized LPS after induction of LPS production in galactose-epimerase-less mutants had revealed patches of the exported antigen on the cell surface; the patches coincided with adhesion sites, whereas periplasmic label was not observed [72a]. We used another approach, namely a conversion of LPS, whose change in serotype was induced with bacteriophage epsilon 15. We observed 20–40 cell surface domains which exhibited new antigenic reactivity and localized to adhesion sites [72b]. The new LPS also represents the receptor for phage $\varepsilon 34$. When phage $\varepsilon 34$ adsorption was used to measure the presence of newly exported LPS, the phage conversion was observed to commence 70 s after the phage $\varepsilon 15$ had been added to the culture of *S. anatum*. The time needed for adsorption and the response of the cell membrane to attachment of phage $\varepsilon 15$ is 45–55 s under these experimental conditions [73]. During the following 15–25 s, DNA injection has to take place, and subsequently the enzymes necessary to synthesize the new O-antigen have to be synthesized and become functional. Thereafter, the converted LPS will have to be synthesized and transported to the OM. In summary, it takes possibly a few seconds to transcribe and translate proteins followed by protein folding [74a,b] of the group of (at least three) enzymes, plus a time span for synthesis and export of the new O-antigen. The timescale for each of these steps within the 10–25 s is difficult to establish. However, the alternative pathway for LPS to diffuse across a highly viscous periplasm might require considerably more time than that involving membrane adhesions. The electron microscope showed that not only immunolabel specific for the newly synthesized LPS is accumulating over adhesion sites, but also $\varepsilon 34$ virus particles were positioned over adhesion sites. Significant label of the periplasm was not observed.

A similar localization was found for membrane domains involved in the export of capsular antigens. In these studies, the periplasmic domain did not reveal capsular antigen either [65,75].

10.2. Peptidoglycan

In order to allow for a cell to grow, to extend its surface and to form a septum, the bag-shaped murein polymer has to be opened to provide acceptor sites for subsequent insertion of new building blocks [33,47] (see also Chapters 4, 7). A controlled action of enzymes such as the murein hydrolases is responsible for shaping the sacculus. Localization of the corresponding hydrolytic enzymes has been performed at a wide range of resolution by a variety of methods [76]. More recently, for example, immuno-electron microscopy was used to locate the prevailing autolysin, the soluble lytic transglycosylase; most of the enzyme was localized to the outer membrane domain and, in isolated murein sacculi, to the outer surface of the peptidoglycan [77]. This would support a model predicting a layered surface growth of the sacculus. Interestingly, although the enzyme is classified as periplasmic, it is strongly bound to the substrate and is not released by osmotic shock. Such a strong muramidase binding to murein has also been observed in *Bacillus subtilis*

[78] and in *Streptococcus faecium* [79]. The data support the notion that the periplasmic domain in *E. coli* contains significant amounts of immobile constituents.

10.3. Fimbrial proteins

A typical example for periplasmic transport are fimbrial proteins which assemble to the filamentous organelles responsible for bacterial attachment to the glycolipid and gly-coprotein receptors of eukaryotic cells. The export from the cytoplasmic compartment to the OM involves a proton-motive force and an export machinery consisting of products of the *sec* genes [9,80], signal peptidases, plus a translocation ATPase [81]. Once in the periplasm, a periplasmic chaperone, for example PapD in the Ppap system will protect the fimbrial protein (PapA) monomer [82,83] from enzyme attack. After diffusion through the periplasm and the peptidoglycan layer(s), the protein complex is inserted and polymerized in the OM, involving the function of protein PapC [84]. Other fimbrial systems such as the 987p system appear to work with analogous pathways [85].

10.4. F pilus

A set of proteins involved in F pilus function is found in IM pools and in the cell enve-lope [86]. The transfer region of the *E. coli* K12 plasmid F encompasses many genes re-quired for plasmid transfer. Among these, the F-pilus protein encoded by the *tra*W gene undergoes signal sequence processing [87a] and is rapidly assembled at the OM to form the pilus. According to Silverman [87b], F pili grow from their base at the IM. Together with other proteins, such as the recently discovered product of the *trb*I gene [88] which localizes to the IM, the assembly as well as the retraction of the F pilus seems to involve both the periplasm plus the inner membrane. Electron microscopic data indicate that a structure resembling a membrane adhesion is present at the F-pilus insertion of *E. coli* HfrH [60]. However, in sections of cryo-fixed clusters of mating cells and transfer-blocked *E. coli*, structures such as assembled pili were not found [89]. Reasons for this could be either the low contrast of the pilus or the retracted pilus has been fully depolym-erized when cell–cell contact is made in a mating cluster.

10.5. OM pores

A temporary periplasmic residence has been demonstrated for another group of proteins targeted to the OM, the transport proteins [1] that form the variety of OM pores. The pro-teins mediate passage of hydrophilic substrates of limited size, either as unspecific porins or as porins such as LamB, which provides facilitated diffusion of maltose [90]. In addi-tion, outer membrane proteins are proposed to form specific transport systems by com-plexing IM proteins such as TonB with the OM receptors to allow the uptake of vitamin B_{12}, B group colicins and iron chelators [91]. Studies of the topography of insertion sites of OM pore formers have shown that newly induced LamB protein was inserted within 40 s over the entire OM surface with a peak of insertion at the division site of the cell [92]. In most of these studies, the periplasmic location was established with outer membrane proteins generated by overproducing or by polypeptide fusion; however, one should be aware that in these experiments protein transfer is often incomplete and may result in periplasmic or cytoplasmic localization, whereas in the wild-type or in in vivo conditions the target is the OM. Smit and Nikaido [93] localized the insertion of porin to membrane adhesion sites; more recently, however, Sen and Nikaido [94] re-assigned the pathway for protein OmpF to the periplasmic route. The authors caution that having obtained the data from mutant spheroplasts in vitro, the necessary presence of detergents could have caused an in vitro environment mimicking that of an in vivo adhesion site. Fractionation studies of cell envelopes revealed fractions of intermediate density containing most of the newly made OM proteins [95]. The pitfalls of protein localization by fractionation of envelope membranes have been demonstrated [96], leading to the conclusion that in situ analysis such as immuno-electron microscopy of cells might be advantageous for studies of transport intermediates.

10.6. TonB

TonB, an IM protein, appears to extend across the periplasm to form a temporary connection to the OM. The protein forms a complex with proteins ExbB and ExbD [97,98]. It is an energy transducer which couples energy from the inner membrane to the active transport of siderophores and vitamin B12 across the outer membrane. The transport system consists of a high affinity OM receptor, periplasmic binding proteins and IM transport proteins. Studies of the core of TonB by NMR suggested that the molecule can stretch to 10 nm in length, potentially allowing the protein to span the periplasmic space [99]. Although the presence of TonB in adhesion sites has been questioned [91], in favor of such an association might be the finding that phage BF23, a phage that shares the vitamin B12 receptor, adsorbs preferentially to membrane adhesion sites of *E. coli* [60]. The model of Holroyd and Bradbeer [100] predicts that TonB interacts transiently with the OM receptor to release the ligand into the periplasm. Subsequently, TonB returns to its uncoupled state whereby a continued depolarization of the IM is prevented and the OM receptor is recycled to its functional conformation [91].

11. Extracellular secretion of proteins

E. coli and other bacteria secrete a number of cytotoxic proteins whose export circumvents the periplasm. One group contains proteins synthesized with a signal sequence, whose determinants for extracellular export appear to be part of the mature peptide chain. Examples are the multi-subunit toxins (cholera, pertussis) and pullulanase [6]. The other group of secreted proteins lacks typical N-terminal signal sequences and is classified as RTX family, whose members include *E. coli* hemolysin, colicin V and metallo-proteases. The secretion of *E. coli* hemolysin requires three to four additional proteins (HlyB, D, TolC and HlyC). HlyB is involved in the transport of a wide variety of unrelated proteins, including the provision of transmembrane activity across both, bacterial plasma membranes as well as eukaryotic membranes, and functioning in drug expulsion, pheromone transport and ion channel action. A model of Wandersman and Delepelaire

462

[101] envisages a selective pore complex of HlyB and D at the IM, contacting the OM protein TolC. Such a bridge would bypass the periplasm completely.

It has been discussed [5] whether the export machinery for these proteins might be located at the membrane adhesion sites, which may translocate the protein to the OM and also deliver energy to the secretion apparatus. Some of these extracellular excretion pathways use the periplasm, but the molecule might still be sorted at the adhesions for the final export process. At present, a structural correlation of adhesions to these classes of protein export has not been made. While short-lived proteinaceous bridge-like connections can be envisaged, some of the above-mentioned studies concerning the sites of adsorption of bacteriophages T5 or BF23 suggest an involvement of close IM–OM contacts in the functions of surface receptors and export sites. Obviously, these secretion processes add a highly dynamic aspect to the array of trans-periplasmic transport mechanisms.

Acknowledgements

This work was supported by grants from the National Science Foundation (MCB-9206674) and the National Institutes of Health (CA-06927, RR-05539) and by an appropriation from the Commonwealth of Pennsylvania. The authors thank R. Diehl and G. Zygmunt for expert editing and typing of the manuscript.

References

1. Nikaido, N. and Saier, M.H. (1992) Science 258, 936–942.
2. Kadner, R.J. (1990) Mol. Microbiol. 4, 2027–2033.
3. Braun, V., Günter, K. and Hantke, K. (1991) Biol. Metals 4, 14–22.
4. Lee, J.I., Kuhn, A. and Dalbey, R.E. (1992) J. Biol. Chem. 267, 938–943.
5. Lory, S. (1992) J. Bacteriol. 174, 3423–3428.
6. Pugsley, A.P., D'Enfert, C. and Kornacker, M.G. (1990) Annu. Rev. Genet. 24, 67–90.
7. Bayer, M.E. (1968a) J. Gen. Microbiol. 53, 395–404.
8. Bayer, M.E. (1981) Int. Rev. Cytol. (Suppl.) 12, 39–70.
9. Oliver, D. and Beckwith, J. (1981) Cell 25, 765–772.
10. Pugsley, A.P. (1992) Proc. Natl. Acad. Sci. USA 89, 12058–12062.
11. Ames, G.F.L. (1986) Annu. Rev. Biochem. 55, 397–405.
12. Bayer, M.E. (1968b) J. Virol. 2, 346–356.
13. Kennedy, E.P. (1982) Proc. Natl. Acad. Sci. USA 79, 1092–1095.
14. Delcour, A.H., Adler, J., Kung, C. and Martinac, B. (1992) FEBS Lett. 15, 167–169.
15. Bayer, M.E. (1967) J. Bacteriol. 93, 1104–1112.
16. Bayer, M.E. (1967) J. Gen. Microbiol. 46, 237–246.
17. Csonka, L.N. (1989) Microbiol. Rev. 53, 121–147; Csonka, L.N. (1981) Mol. Gen. Genet. 182, 82–86.
18. Bayer, M.E. (1991) J. Struct. Biol. 107, 268–280.
19. Neu, H.L. and Heppel, L.A. (1965) J. Biol. Chem. 240, 3685–3692.
20. Levengood-Freyermuth, S.K., Click, E.M., Webster, R.E. (1993) J. Bacteriol. 175, 222–228.
21. Sharff, A.J., Rodseth, L.E., Spurlino, J.C. and Guiocho, F.A. (1992) Biochemistry 31, 10657–10663.
22. Macnab, R.M. (1987) in: F. Neidhardt (Ed.), *Escherichia coli* and *Salmonella typhimurium*, American Society for Microbiology, Washington, DC, pp. 70–83.
23. Klebba, P.E., Benson, S.A., Bala, S., Abdullah, T., Reid, J., Singh, S.P. and Nikaido, H. (1990) J. Biol. Chem. 265, 6800–6810.

24. Nikaido, N. and Nakae, T. (1979) Adv. Microb. Physiol. 20, 163–250.
25. Weidel, W.H. and Pelzer, H. (1964) Adv. Enzymol. 26, 193–232.
26. Braun, V. (1975) Biochim. Biophys. Acta 415, 335–377.
27. Rosenbusch, J.P. (1974) J. Biol. Chem. 249, 8019–8029.
28. De Petris, S. (1967) J. Ultrastruct. Res. 19, 45–83.
29. Nanninga, N. (1970) J. Bacteriol. 101, 297–303.
30. Bayer, M.E. and Remsen, C.C. (1970) J. Bacteriol. 101, 304–313.
31. Beveridge, T.J. (1981) Int. Rev. Cytol. 72, 229–317.
32. Murray, R.G.E., Steed, P. and Elson, H.E. (1965) Can. J. Microbiol. 11, 547–560.
33. Nanninga, N. (1985) in: N. Nanninga (Ed.), Molecular Cytology of *Escherichia coli*, Academic Press, London.
34. Hobot, J.A., Carlemalm, E., Villiger, W. and Kellenberger, E. (1984) J. Bacteriol. 160, 143–152.
35. Dubochet, J.A., McDowall, W., Menge, B., Schmid, E.N. and Lickfeld, K.G. (1983) J. Bacteriol. 155, 381–390.
36. Graham, L.L. and Beveridge, T.J. (1990) J. Bacteriol. 172, 2150–2159.
37. Graham, L.L., Beveridge, T.J. and Nanninga, N. (1991) Trends Biochem. Sci. 16, 328–329.
38. Carlemalm, E. and Kellenberger, E. (1982) EMBO J. 1, 63–67.
39. Graham, L.L., Harris, R., Villiger, W. and Beveridge, T.J. (1991b) J. Bacteriol. 173, 1623–1633.
40. Stock, J.B. Rauch, B. and Roseman, S. (1977) J. Biol. Chem. 252, 7850–7861.
41. Woldringh, C.L. and Nanninga, N. (1985) in: N. Nanninga (Ed.), Molecular Cytology of *Escherichia coli*, Academic Press, New York, pp. 161–197.
42. Van Wielink, J.E. and Duine, J.A. (1990) Trends Biochem. Sci. 15, 136–137.
43. Labischinski, H. (1991) J. Bacteriol. 173, 751–756.
44. Leduc, M., Frehel, C., Siegel, E. and van Heijenoort, J. (1989) J. Gen. Microbiol. 135, 1243–1254.
45. Nossal, N.G. and Heppel, L.A. (1966) J. Biol. Chem. 241, 3055–3062.
46. Mitchell, P. and Moyle, J. (1956) Symp. Soc. Gen. Microbiol. 6, 150–180.
47. Ghuysen, J.M. and Shockman, G.D. (1973) in: L. Leive (Ed.), Bacterial Membranes and Walls, Marcel Dekker, New York, pp. 37–130.
48. Naegeli, C. (1855) in: C. Naegeli and C. Cramer (Ed..), Pflanzenphysiologische Untersuchungen Fr. Schulthess, Zürich.
49. Fischer, A. (1891) Ber. Kgl. Sächs. Ges. Wissensch. Math.-Phys. CL 43, 53–74.
49a. Mulder, E. and Woldringh, C.L. (1993) J. Bacteriol. 175, 2241–2247.
50. Kushner, D.J. (1978) in: D.J. Kushner (Ed.), Microbiol Life in Extreme Environments, Academic Press, London, pp. 317–368.
51. Brass, J.M., Higgins, C., Foley, M., Rugman, P.A., Birmingham, J. and Garland, P.B. (1986) J. Bacteriol. 165, 787–794.
52. MacAlister, T.J., MacDonald, B. and Rothfield, L.I. (1983) Proc. Natl. Acad. Sci. USA 80, 1372–1376.
53. Anba, J., Bernadac, A., Pages, J.M. and Lazdunski, C. (1984) Biol. Cell 50, 273–277.
54. Hackenbrock, C.R. (1968) Proc. Natl. Acad. Sci. USA 61, 598–605.
55. VanVenetie, R. and Verkleij, A.J. (1982) Biochim. Biophys. Acta 692, 379–405.
56. Kottge, M., Adams, V., Riesinger, I., Bremm, G., Bosch, W., Brdiczka, D., Sandri, G. and Panfili, E. (1988) Biochim. Biophys. Acta 395, 807–832.
57. Ardail, D., Privat, J.P., Egret-Charlier, M., Levrat, C., Lerme, F. and Louisot, P. (1990) J. Biol. Chem. 265, 18797–18802.
58. Pain, D., Kanwar, Y.S. and Blobel, G. (1988) Nature 331, 232–236.
59. DePamphilis, M.L. and Adler, J. (1971) J. Bacteriol. 105, 396–407.
60. Bayer, M.E. (1979) in: M. Inoye (Ed.), Bacterial Outer Membranes, Biosynthesis, Assembly, Function, Wiley, New York, pp. 167–202.
61. Olijhoek, A.J.M., VanEden, C.G., Trueba, F.J., Pas, E. and Nanninga, N. (1982) J. Bacteriol. 152, 479–484.
62. Cook, W.R., MacAlister, T.J. and Rothfield, L.I. (1986) J. Bacteriol. 168, 1430–1438.
63. Bayer, M.E. (1975) in: A. Tzagoloff (Ed.), Membrane Biogenesis, Plenum, New York, pp. 393–427.
64. Bayer, M.E., Thurow, H. and Bayer, M.H. (1979) Virology 94, 95–118.
65. Kröncke, K.-D. Golecki, J.R. and Jann, K. (1990) J. Bacteriol. 172, 3469–3472.

464

66. Crowlesmith, I. Schindler, M. and Osborn, M.J. (1978) J. Bacteriol. 135, 259–269.
67. Furukawa, H. and Mizushima, S. (1982) J. Bacteriol. 150, 916–924.
68. Armbruster, B.L., Carlemalm, E., Chiovetti, R., Garavito, R.M., Hobot, J.A., Kellenberger, E. and Villiger, W. (1982) J. Microsc. 126, 77–85.
69. Kellenberger, E. (1990) Mol. Microbiol. 4, 697–705.
70. Bayer, M.E. and Bayer, M.H. (1991) J. Bacteriol. 173, 141–149.
71. Osborn, M.J., Gander, J.E. and Parisi, E. (1972) J. Biol. Chem. 247, 3973–3986.
72. (a) Muhlradt, P., Menzel, J., Golecki, J. and Speth, V. (1973) Eur. J. Biochem. 35, 471–481. (b) Bayer, M.E. (1974) Ann. N.Y. Acad. Sci. 235, 6–28.
73. Bayer, M.E. and Bayer, M.H. (1981) Proc. Natl. Acad. Sci. USA 78, 5618–5622.
74. (a) King, J. (1989) Chem. Eng. News 10, 32–54. (b) Randall, L.L., Hardy, S.J. and Thom, J.R. (1987) Annu. Rev. Microbiol. 41, 507–541.
75. Bayer, M.E. and Thurow, H. (1977) J. Bacteriol. 130, 911–936.
76. Höltje, J.V. and Schwarz, U. (1985) in: N. Nanninga (Ed.), Molecular Cytology of *Escherichia coli*, Academic Press, London, pp. 77–119..
77. Walderich, B. and Höltje, J.-V. (1991) J. Bacteriol. 173, 5668–5676.
78. Hobot, J.A. and Rogers, H.J. (1991) J. Bacteriol. 173, 961–967.
79. Barrett, J.F. (1984) J. Biol. Chem. 259, 11818–11827.
80. Lecker, S., Lill, R., Ziegelhoffer, T., Georgopoulos, C., Bassford Jr., P.J., Kumamoto, C. and Wickner, W. (1989) EMBO J. 8, 2703–2709.
81. Eisenstein, B.I. (1987) in: F. Neidhardt et al. (Eds.), *Escherichia coli* and *Salmonella typhimurium*, Vol. 1, American Society for Microbiology, Washington, DC, pp. 84–89.
82. Lindberg, F., Lund, B., Johansson, L. and Normark, S. (1987) Nature 328, 84.
83. Kuehn, M.J., Normark, S. and Hultgren, S.J. (1991) Proc. Natl. Acad. Sci. USA 88, 10586–10590.
84. Norgren, M., Baga, M., Tennent, J.M. and Normark, S. (1987) Mol. Microbiol. 1, 169–178.
85. Schifferli, D.M., Beachey, E.H. and Taylor, R.K. (1991) J. Bacteriol. 173, 1230–1240.
86. Willetts, N. and Skurray, R. (1987) in: F. Neidhardt (Ed.), *Escherichia coli* and *Salmonella typhimurium*, American Society for Microbiology, Washington, DC, pp. 1110–1133.
87. (a) Ippen-Ihler, K., Moore, D., Laine, S., Johnson, D.A. and Willets, N.S. (1984) Plasmid 11, 116–129. (b) Silverman, P.M. (1987) in: M. Inouye (Ed.), Bacterial Outer Membranes as Model Systems, Wiley, New York, pp. 277–309.
88. Maneewannakul, S., Maneewannakul, K. and Ippen-Ihler, K. (1992) J. Bacteriol. 174, 5567–5574.
89. Dürrenberger, M.B., Villiger, W. and Bächi, T. (1991) J. Struct. Biol. 107, 146–156.
90. Nikaido, N. and Reid, J. (1990) Experientia 46, 174–180.
91. Postle, K. (1990) Mol. Microbiol. 4, 2019–2025.
92. Vos-Scheperkeuter, G.H., Pas, E., Brakenhoff, G.J., Nanninga, N. and Witholt, B. (1984) J. Bacteriol. 159, 440–447.
93. Smit, J. and Nikaido, H. (1978) J. Bacteriol. 124, 687–702.
94. Sen, K. and Nikaido, H. (1990) Proc. Natl. Acad. Sci. USA 87, 743–747.
95. deLeij, L., Klingma, J. and Withold, B. (1979) Biochim. Biophys. Acta 553, 224–234.
96. Tommassen, J. (1986) Microbiol. Pathogenesis 1, 225–228.
97. Braun, V. (1989) J. Bacteriol. 171, 6387–6390.
98. Kampfenkel, K. and Braun, V. (1992) J. Bacteriol. 174, 5485–5487.
99. Evans, J.S., Levine, B.A., Trayer, I.P., Dorman, C.G. and Higgins, C.F. (1986) FEBS Lett. 208, 211–216.
100. Holroyd, C. and Bradbeer, C. (1984) in: L. Leive and D. Schlesinger (Ed.), Microbiology – 1984, American Society for Microbiology, Washington, DC, pp. 21–23.
101. Wandersman, C. and Delepelaire, P. (1990) Proc. Natl. Acad. Sci. USA 87, 4776–4780.

J.-M Ghuysen and R. Hakenbeck (Eds.), *Bacterial Cell Wall*

465

CHAPTER 22

Transmembrane signal transducing proteins

MICHAEL G. SURETTE and JEFFRY B. STOCK

Department of Molecular Biology, Princeton University, Princeton, NJ 08544, USA

1. Introduction

The life of a bacterium is characterized by changing environmental conditions. To compete and survive, these cells have evolved signal transduction pathways that allow them to adapt to different surroundings. Responses range from changes in gene expression leading to altered metabolism and differentiation to behavioral responses such as chemotaxis that allow cells to move to more favorable environments. Adaptive responses are generally mediated by signal transduction pathways initiated by receptors within the cell envelope that communicate with signal transduction proteins in the cytoplasm [1]. Each receptor is generally composed of two functional domains: a sensory domain that interacts with stimuli within the periplasm and a signaling domain that modulates effector activities in the cell. This review focusses on the structure and function of the cell surface receptors.

2. Overview of membrane receptors

The sensory inputs that initiate signal transduction pathways in bacteria are diverse and often quite complex. Cells are sensitive to virtually every aspect of their environment including the concentrations of a large number of different small molecules such as peptides, amino acids, sugars, oxygen, and phosphate, as well as parameters such as osmotic pressure, pH, temperature, and light. Environmental changes are monitored by sensory receptors, and this information is used to regulate appropriate adaptive response mechanisms.

The best characterized sensory apparatus is the system of chemotaxis receptors in *Escherichia coli* that monitors levels of attractant and repellent chemicals [2–5]. *E. coli* sense small molecules that diffuse through outer membrane porins into the periplasm where they interact with chemoreceptor proteins. Attractants such as aspartate and serine, and repellents such as Co^{2+} and Ni^{2+} bind directly to the sensory domains of transmembrane receptor proteins. Chemo-attractants such as ribose, galactose, maltose, and oligopeptides first bind to stereospecific binding proteins in the periplasm, and the binding protein-attractant complexes then interact with the sensory domains of chemoreceptors in the inner membrane. These binding proteins are members of a large family of

466

soluble periplasmic components that function primarily to deliver solutes to transport proteins in the inner membrane [6]. Only a few are known to serve a dual role in chemosensing.

There are several other examples of sensory receptors that receive information by interacting with solute transport systems. Regulation of gene expression in response to phosphorus availability in *E. coli* is coupled to the phosphate specific transport system (PST) [7]. The periplasmic binding protein, PhoE, that delivers periplasmic phosphate to the membrane components of the PST system does not appear to be directly involved, however. Instead, an auxiliary cytoplasmic protein, PhoU, functions to transfer information concerning the activity of the PST system to a transmembrane receptor, PhoR.

Sensing of extracellular hexose phosphates in *E. coli* represents still another example of the close relationship between transporters and sensors. The receptor in this case, UhpB, does not have a periplasmic sensory domain. In its place is an extremely hydrophobic domain that crosses the membrane several times. Sensing of hexose phosphates requires an additional membrane component, UhpC, that is homologous to the hexose phosphate transporter, UhpT. It is thought that hexose phosphates bind to UhpC which in turn interacts with the integral membrane domain of UhpB to induce a response from the signalling domain of UhpB [8]. In this case, a transport protein has apparently evolved to become a protein with purely sensory function.

Another example of a family of proteins with clear homology to transport proteins are the receptors that mediate phototaxis responses in *Halobacterium halobium*. The photoreceptors, sensory rhodopsins, are integral membrane proteins with seven transmembrane helices that are homologous to the light driven ion pumps, bacteriorhodopsin and halorhodopsin [9]. Sensory rhodopsin is not a pump, however. Instead, it passes information to an associated transmembrane receptor that is homologous to the chemoreceptors that mediate chemotactic responses in *E. coli* [10].

In all of these cases, the sensory system involves a transmembrane protein that functions as a receptor-transducer to pass information into the cytoplasm. There are many other instances, however, where no receptor-transducer is involved. Instead, the transporter functions to directly pass information to cytoplasmic components. The best example of this phenomena is provided by the phosphotransferase system (PTS) that mediates the uptake and phosphorylation of sugars (for recent reviews, see [11–14]). PTS transport proteins regulate the activities of signal transduction proteins in the cytoplasm that control motility and gene expression in response to the rate of sugar uptake.

3. Chemotaxis receptors

The best characterized sensory receptors in bacteria are those that mediate chemotaxis responses. The sensitivities of these proteins are generally modulated by the methylation and demethylation of glutamate γ-carboxyl groups. This modification appears to be unique to this family of membrane proteins, and accordingly the receptors have been termed the methylated chemoreceptor proteins or MCPs.

3.1. MCP family of receptor/transducer proteins

The bacterial chemoreceptors comprise a large family of proteins. MCP or MCP-like proteins have been identified directly in a number of different bacterial species (*Bacillus subtilis, Spirochaeta auranta, Pseudomonas aeruginosa, P. putida, Caulobacter crescentus, Rhodospirillum rubrum, Agrobacterium tumefaciens, Halobacterium halobium* and *Myxococcus xanthus*) and indirectly by cross-reactivity to an antibody directed against an *E. coli* MCP in several others (see [15] and refs. therein). The best characterized MCPs are those from *E. coli* and *Salmonella typhimurium*. These have been studied extensively and the recent development of in vitro systems to analyze the biochemical events associated with signal transduction in chemotaxis [16–18] have made them attractive model systems for basic research on sensory receptor function.

The MCPs all have a similar structural organization [19]. The MCP that mediates chemotaxis to aspartate in *E. coli*, Tar, is by far the most studied and best understood (Fig. 1). Tar monomers are 60-kDa proteins with two sequences of hydrophobic amino acids that form membrane spanning α-helices, TM1 and TM2. A short N-terminal sequence of six residues within the cytoplasm precedes TM1; TM1 leads to the 158-residue periplasmic sensory domain at the outer surface of the cytoplasmic membrane. The second transmembrane helix, TM2, leads from the sensory domain to a cytoplasmic signaling domain at the inner surface of the membrane. The signaling domain interacts with at least four signal transduction proteins in the cytoplasm (Fig. 2): two, CheA and CheW, are directly involved in the chemotaxis signaling pathway, and two, CheR and CheB, are involved in feedback regulation of receptor activity. A ternary complex formed with the receptor, CheA and CheW acts as a kinase to phosphorylate the chemotaxis response regu-

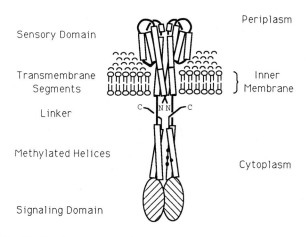

Fig. 1. Structural model of the Tar chemoreceptor. The receptor is a homodimer spanning the inner cytoplasmic membrane. The structure of the sensory domain is derived from the crystal structure of Milburn et al. [24]. The positioning of the transmembrane segments is based on predictions from the sensory domain crystal structure [24] and disulfide crosslinking studies [33,34]. The coiled-coil domain is predicted from amino acid sequence analysis [4,100]. The sites of methylation within the coiled-coil domain and the N- and C-termini of the subunits are indicated.

468

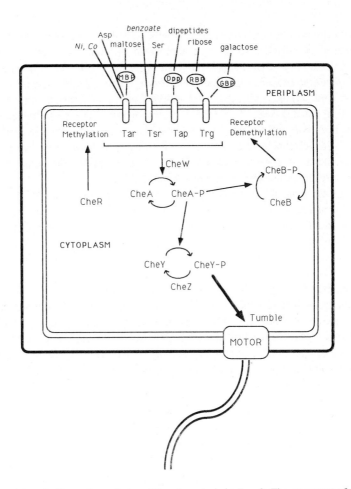

Fig. 2. The signal transduction pathway that mediates chemotaxis in *E. coli*. The movement of these bacteria alternates between smooth swimming and tumbling. Tumbles serve to randomize the direction of smooth swimming. Chemotaxis is achieved by suppressing the tumble frequency when moving towards attractants or away from repellents. The signaling pathway involves a family of membrane bound chemoreceptors that sense attractants and repellents and six interacting cytoplasmic components (the Che proteins) that transduce information from the receptors to the flagella motors. The receptors together with CheW regulate the activity of a histidine kinase, CheA. Activation of CheA by the receptors (with CheW) results in autophosphorylation of CheA on a histidine residue. CheA can then donate this phosphate to aspartyl carboxylates in two proteins CheB and CheY. Phospho–CheY binds to components at the flagellar motor to cause a tumble. The CheZ protein accelerates the rate of CheY–P dephosphorylation. CheB–P catalyzes the demethylation of methyl-glutamyl residues in the receptors. CheR is an *S*-adenosylmethionine-dependent methyltransferase that catalyzes the methyl esterification of specific receptor glutamyl residues. Receptor methylation is involved in the desensitization of the receptors to attractants and repellents. Periplasmic binding proteins that interact with the chemoreceptors are also illustrated. These are the maltose binding protein (MBP) which is a ligand for the Tar chemoreceptor, the dipeptide binding protein (Dpp) that interacts with the Tap receptor, and the ribose binding protein (RBP) and galactose binding protein (GBP) both of which interact with the Trg chemoreceptor.

lator, CheY [17,18,20]. The activity of this kinase is negatively regulated by the chemo-attractant aspartate, and positively regulated by methyl esterification of four specific glu-tamyl side chains within the cytoplasmic domain. CheR is an S-adenosylmethionine-de-pendent enzyme that catalyzes the methylation reaction [21], and CheB is an esterase that functions to remove methyl groups [22].

To begin to define the mechanism of Tar signaling it is important to understand the na-ture of the interactions between Tar monomers within the plane of the membrane. The wild-type Tar receptor lacks cysteine residues, and several investigators have used site-directed mutagenesis to add cysteine residues at defined positions and measure the for-mation of disulfide crosslinks between monomers. This 'site-directed crosslinking' ap-proach was used to demonstrate that in the absence of aspartate, Tar is predominantly a dimer with a slow rate of subunit exchange suggesting the existence of a small pool of the monomeric species. Aspartate appeared to stabilize the dimer, since it prevented the ex-change reaction [23]. Heterodimers between closely related MCP receptors were not formed. This high degree of specificity even applies to the very closely related Tar pro-teins from $E.\ coli$ and $S.\ typhimurium$ [23].

3.2. Sensory domain

The structure of the periplasmic domain of the *Salmonella* Tar protein has been solved by standard X-ray crystallographic methods [24]. Structures with and without aspartate bound were obtained using a disulfide crosslinked dimer of the periplasmic domain. The monomeric subunit is a four helix bundle with helices A and D predicted to extend through the membrane as TM1 and TM2, respectively. The ligand binding site is located at the interface of the two monomers with asymmetric contacts with each subunit. This is analogous to the ligand binding site of the human growth hormone receptor (hGHR) whose structure has also recently been determined [25]. hGHR is a class I receptor tyro-sine kinase and ligand binding promotes dimerization [26].

One very interesting feature of the chemoreceptors is the ability of a single receptor protein to integrate signals from several different stimulatory ligands. The best character-ized example of this is the Tar-mediated responses to aspartate and maltose in $E.\ coli$ [27]. Whereas aspartate binds directly to the sensory domain of Tar, maltose binds to a periplasmic maltose binding protein (MBP). The X-ray crystal structure of $E.\ coli$ MBP has been determined, and Stoddard and Koshland [28] have used a molecular docking algorithm to predict the site of binding of MBP in the Tar sensory domain. Their results, which are consistent with an extensive genetic analysis [29,30], indicate that the maltose-MBP complex binds at the interface of the two subunits at a site that is close to, but dis-tinct from, the site of aspartate binding.

Genetic studies of Trg, the MCP in $E.\ coli$ that interacts with binding protein homologs of MBP that bind ribose (RBP) and galactose (GBP) have led to the identification of point mutations in the Trg sensory domain that affect binding protein interactions [31]. These mutant Trg proteins fall into two general classes: those that are unable to respond to stimuli, and those that behave as though they are being continually stimulated even in the absence of stimulatory ligands. The mutations map to a region which corresponds to the part of Tar that forms the sensory domain dimer interface. Thus, genetically induced perturbations of the dimer interface can mimic the effects of stimulatory ligands to cause transmembrane signaling.

3.3. Membrane spanning domains

The transmembrane segments of the chemoreceptors are central to the process of signal transduction across the membrane and hence have been the focus of several studies. Oosawa and Simon [32] introduced two mutations in TM1 of the Tar receptor (A19K, and a seven amino acid deletion, $\Delta7$–13) that result in receptors that remain associated with the membrane and still bind aspartate, but are deficient in stimulus-response coupling. Neither mutant was modified by CheR or CheB. Several second site suppressors were found for the A19K mutant. Four of eighteen suppressors mapped to the second transmembrane segment and one to the first transmembrane segment. All but one of these second site suppressors results in a negatively charged residue that may form an ion pair with the positively charged lysine. This suggests a close association of the two transmembrane segments. Surprisingly, all 13 of the remaining mutations mapped to a sequence of 40 amino acids in the cytoplasmic domain that links TM2 to the first set of methylated glutamate residues.

Site directed cysteine crosslinking studies have been used to study interactions between transmembrane segments. Lynch and Koshland [33] introduced disulfides at specific positions and measured rates of disulfide bond formation. Pakula and Simon [34] developed a method to randomly insert cysteines in the two transmembrane segments, and then screen for pairs that produce disulfide bonds. The two approaches gave very similar results. Both groups concluded that the TM1s of each monomer are closely associated, whereas the TM2s are not in contact. They proposed that TM1 and TM2 form a distorted, relatively planar, four helix bundle in the receptor dimer.

3.4. Cytoplasmic domains

The cytoplasmic portion of the chemoreceptors has been divided into several distinct functional domains. Two regions that contain the sites of reversible methylation are predicted to be extended α-helices [4]. They are referred to as K1 and R1 in reference to tryptic fragments of the Tar receptor that include the sites of methylation [35,36]. Methylation plays a central role in adaptation and modulates the kinase activity of the ternary Tar-CheW-CheA complex [18,37]. The methylation of specific glutamate residues is catalyzed by the CheR methyltransferase using S-adenosylmethionine as a methyl donor [21]. Each receptor has 3–5 glutamates that can be methylated [38], some of which are genetically encoded as glutamines and subsequently deamidated by the CheB esterase/amidase to yield substrate glutamates for the methyl transferase [39]. CheB also catalyzes the demethylation reaction [22]. The CheB protein has a response regulator domain that is homologous to CheY [40,41]. When CheB is phosphorylated, the esterase/amidase activity is turned on [42]. Regulation of CheB by the MCP-linked CheA kinase provides a feedback mechanism to modulate kinase activity. A model has been presented where α-helices corresponding to the methylated K and R sequences form a coiled coil [4]. In the receptor dimer, the coiled coil regions from each monomer would coalesce to form a four helix bundle [43].

The region from the end of the second transmembrane segment to the start of K1 is called the linker region [44]. Some suppressors of a point mutation in TM1 clustered in this region and several mutations that were 'locked' in either of the two extreme signaling states are also found in this region. In addition, a conserved portion of the linker region was recently identified in an unrelated receptor [45]. These results suggest that conformations of this region can influence the signaling state of the receptor and this region may play an important role in signal transmission from the sensory to the signaling domain.

The amino acid sequence between the two methylated helices is the most highly conserved region among the chemoreceptor family. Most missense mutations that lock the receptors in a kinase on (repellent) or kinase off (attractant) state are located to this region [44], as are allele specific second site suppressors of CheW mutants [46]. These results are consistent with this being the site of interaction of the receptors with CheW and CheA. Overexpression of a gene that encodes only this region of the Tar receptor results in production of a stable peptide that can be purified from *E. coli* cells. Circular dichroism of this purified fragment indicates a significant α-helical content (Surette and Stock, unpublished data). The ability to express and purify this region of the protein as a stable peptide supports the notion that this is a discrete structural domain within the cytoplasm.

The extreme C-terminus of the MCPs is the least conserved region. Nevertheless, truncation of the C-terminus severely compromises signaling ability. Deletion of 35 amino acids [47] or 84 amino acids [48] from the C-terminus of Tar interferes with methylation. These truncated proteins bind aspartate and are capable of generating signals in response to attractant [47,48] and repellents [48], but are not able to adapt to stimuli.

Another region of the receptors that is found in the cytoplasmic domain is the N-terminal six amino acids that precede TM1. This sequence is non-contiguous with the rest of the cytoplasmic domain. This region may simply play a structural role necessary for the anchoring of TM1, however, the apparent importance of TM1 in the signaling process suggests that the cytoplasmic N-terminus may play a functional role as well. One recent study has targeted this region of the protein for investigation using site-directed crosslinking of introduced cysteine residues [49]. These experiments demonstrate that residue 4 from each Tar monomer (4 and 4') are in close proximity and that attractant binding stimulates the formation of a disulfide bond between cysteines at this location. Furthermore, the crosslinked receptor is a better substrate for methylation.

3.5. Methylation and adaptation

Chemotaxis in bacteria is dependent not only on the ability of the receptor to regulate a cascade of phosphorylation reactions in response to a stimulus but also on the capacity to adapt or desensitize to that stimulus. This process is achieved, at least in part, by modulating the level of methylation of the chemoreceptors [50–52].

Both the methylation and demethylation reaction are dependent on the signaling state of the receptor. An increase in attractant (or removal of repellent) leads to an increase in receptor methylation and addition of repellent or removal of attractant leads to a decrease in methylation [50,53–56]. The change in methylation level results from two processes.

First, attractant binding increases the accessibility of methylation sites to the methyltransferase [55,56]. This is presumably a direct consequence of the conformational changes induced by the stimulatory ligand. Second, the changes in kinase activity associated with stimulus binding cause changes in CheB phosphorylation [18,42,57]. Thus, the addition of attractant leads to an increase in receptor methylation because of a decrease in overall CheB activity due to a decrease in CheB phosphorylation.

The relationship between methylation and the binding of a stimulus suggests that the reciprocal relationship might also be true. That is, an increase in methylation might result in a decrease in affinity for ligand. Several studies have attempted to test this idea. It had been demonstrated earlier that a glutamine residue produces about the same effect on receptor conformation as a methyl glutamate at the same position [58]. Dutten and Koshland [59] found no significant effects of ligand binding with receptors that differed in the number of glutamines at positions of glutamate methylation. These measurements were carried out using membranes containing overexpressed receptor. The results imply that the binding of ligand and the level of methylation are independent. Different results were obtained when ligand binding was measured in membranes containing wild-type levels of Tar and Tsr receptors [60]. Receptors that were fully methylated or fully unmethylated were prepared from CheB and CheR strains, respectively. For the Tsr receptor, the dissociation constant was at least 100 times higher for the methylated receptor, whereas only a 10-fold difference was observed for the Tar receptor. The differences observed in these studies may be due to the experimental methods used to prepare receptors and measure ligand binding. In addition, the signaling state of the receptor is a ternary complex with CheA and CheW [17,18,20] and the methylation observed in the absence of these complexes may not reflect the adaptation process in vivo.

An increase in Tar receptor methylation was found to increase CheA kinase activity in a reconstituted system with Tar and CheW [18]. The increased activity was offset by aspartate binding. The effect of receptor methylation level on the ability of receptors to activate the CheA kinase was further studied by Borkovich et al. [37]. They found only a sevenfold difference in aspartate binding in unmethylated compared to fully methylated receptors. More importantly, they demonstrated that the level of methylation affected the ability of the receptor to activate and inhibit the kinase. Increased methylation correlated with increased kinase activity both in the presence and absence of aspartate. The fully methylated receptor was unable to inactivate the kinase even at high concentrations of aspartate.

3.6. Mechanism of transmembrane signaling

The simplest mechanism for transmembrane signaling is based on reversible oligomerization induced by ligand binding [4,19,26]. According to this idea, ligand-induced oligomerization of the extracellular domain causes oligomerization of the intracellular domain. For receptors that do not signal through this mechanism (i.e. monomeric or stable oligomeric receptors), several possibilities exist [61]. These may be divided into two categories, intramolecular and intermolecular, depending on whether the signaling event is within or between monomers. In the case of intramolecular models, conformational changes induced by stimuli are transmitted across the membrane within the monomer. For monomers with single transmembrane segments, one might imagine that this could be achieved by a pulling or pushing of the transmembrane segment perpendicular to the membrane. Receptors with more than one transmembrane segment, could use a similar piston-like action, or any rotational movement of the transmembrane segments relative to one another. For models involving intermolecular signaling, changes in the relative positioning of the subunits are seen to occur upon ligand binding. This can be accomplished by a number of motions including rotational or lateral movements of the subunits. Signaling by reversible oligomerization can be thought of as an extreme example of this type of mechanism. These mechanisms are not necessarily mutually exclusive and it is not unexpected that transmembrane signaling by some receptors will be the result of more than one type of conformational change.

Disulfide crosslinking studies suggest that the Tar receptor of *S. typhimurium* is a stable dimer and thus aspartate binding is not necessary for dimer formation [23]. This would appear to rule out oligomerization as a mechanism of transmembrane signaling. Aspartate binding to the Tar sensory domain causes a rotation of subunits with respect to one another [24]. This result is consistent with a mechanism of transmembrane signaling that involves ligand-induced changes in the relative positioning of subunits. The small changes observed in the sensory domain would be amplified by a rigid transmembrane segment. This has been described as a scissor-like motion with the center of rotation in the periplasmic domain [24].

This mechanism has been challenged by Milligan and Koshland [62] who generated disulfide crosslinked heterodimers of Tar that contained one subunit of full length Tar and a second subunit with TM2 and the cytoplasmic domain deleted. The periplasmic domain remained intact in these heterodimers and ligand binding was essentially unaffected. The full length subunit could be methylated by CheR, albeit at a relatively slow rate compared to the full length homodimer. When assayed in vitro, aspartate caused a small but significant increase in receptor methylation of the heterodimers implying intramolecular transmembrane signaling.

The observation that a stable C-terminal proteolytic fragment could be isolated during the purification of the Tar receptor [63,64] led to experiments to test the ability of this domain to function in vivo. Oosawa et al. [65] expressed C-terminal constructs of wild-type and two mutant Tar receptors and examined their effects. The wild-type C-terminal fragment was poorly modified by CheR and CheB and the mutation that produced attractant-like effects was not a substrate for these enzymes in vivo. The mutation that produced repellent-like effects was modified by both CheR and CheB. Expression of this mutant C-terminus in strains lacking all other chemoreceptors caused repellent-induced behaviors indicating that this construct retained some capacity to activate the kinase. These same proteins were subsequently purified and characterized by size exclusion chromatography [66]. All displayed molecular weights of ~100 000 (predicted monomer molecular weight of 31 000), however the attractant-like mutant eluted in two forms (~100 kDa and 225 kDa). These results indicate that significant conformational differences and oligomeric states may be exhibited by these mutants. This study has been extended by Long and Weis [67] to include wild-type C-terminus and nine mutants that caused either attractant or repellent phenotypes when present in the full length receptor. They combined gel permeation chromatography and light scattering to demonstrate that the C-terminal fragment has an extended non-spherical shape and the 110 kDa size predicted from the chromatography experiments corresponds to a monomer of 31 kDa. Furthermore, they demonstrated the ability of the fragments to oligomerize. This oligomerization is increased at higher protein concentration and lower pH. In addition, five of the six 'smooth swimming' mutants had a greater tendency to form oligomers. These results have been used to establish a model of transmembrane signaling that involves receptor oligomerization [67]. Disulfide crosslinking studies, however, suggest that the receptor exists as a stable dimer both in the presence and absence of ligand [23]. Monomer-dimer equilibria of isolated domains may not accurately reflect the state of the intact receptor where additional contacts through other domains and the constraints of two dimensional diffusion within the membrane may contribute significantly to dimerization.

The different studies outlined above suggest different mechanisms of transmembrane signaling. First, based on crystallographic studies of the apo- and liganded forms of the sensory domain, a mechanism of intersubunit signaling was presented [24]. Second, methylation studies on heterodimers of full length and truncated receptors were consistent with an intrasubunit mechanism of transmembrane signaling [62]. Third, a ligand induced oligomerization model has been revived with the results of the experiments of Long and Weis on the properties of cytoplasmic constructs of the Tar receptor [67].

4. Histidine kinase receptors

The largest family of signal transducing receptors in bacteria are the histidine kinase receptors. It has been predicted that there may be as many as 50 members of this family in E. coli [1,68]. These histidine kinases are part of the 'two component' family of signal transducing systems. The basic components of these systems are a histidine kinase and a response regulator. The histidine kinase autophosphorylates itself on a histidine residue using ATP as a phospho-donor. The phosphate is subsequently transferred to an aspartate residue on the response regulator. Figure 3 illustrates the basic features of two component systems. The reader is directed to several comprehensive reviews on this subject [1,5,68–70].

The response regulators are usually composed of a conserved domain that contains the site of phosphorylation and an associated effector domain. The 'activity' of the effector domain is regulated by phosphorylation of the response regulator domain. For example, CheB which catalyzes the removal of methyl groups from the chemoreceptors is regulated by phosphorylation of a response regulator domain [42]. This domain normally inhibits the activity of the methylesterase, probably by steric hindrance [40]. Proteolytic or genetic removal of the response regulator domain results in constitutive activation of the enzyme [40,42]. Many of the response regulator domains are coupled to a DNA-binding domain and phosphorylation alters the DNA binding activity of the protein leading to changes in transcription.

Cross-talk, phosphorylation of a response regulator by a kinase other than its usual partner, has been demonstrated in vitro [71,72] but these reactions are inefficient compared to the reactions between normal pairs of proteins and their significance in vivo remains obscure [73]. The specificity for the phospho-transfer from kinase to response

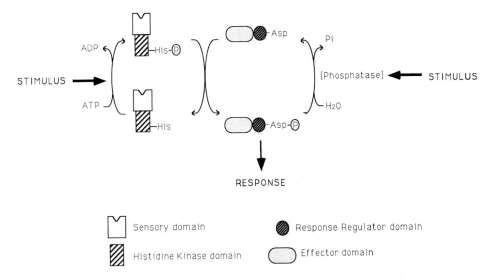

Fig. 3. The two component paradigm of bacterial signal transduction. The histidine kinase is composed of a kinase domain and a sensor domain. Activation of the kinase leads to autophosphorylation on a histidine residue and this phosphoryl group is subsequently transferred to an aspartate residue of a response regulator domain. The activity of the effector domain is regulated by phosphorylation of the response regulator domain. Stimuli can affect the autophosphorylation reaction of the histidine kinase or the dephosphorylation of the response regulator.

regulator and the phosphatase activity must arise from interactions at the interface between the two proteins. Recent experiments demonstrate that the transfer of phosphate from the phospho-histidine of the kinase to the aspartate residue of the response regulator may be catalyzed by the response regulator rather than the kinase. The response regulators can be phosphorylated by small molecule phospho-donors in vitro (phosphoramidate, acetyl phosphate and carbamoyl phosphate; but not nucleotide phosphates) [74]. This implies that the ability to carry out the phospho-transfer reaction is an intrinsic feature of the response regulator domain and not the histidine kinase.

Some of the small molecule phospho-donors are metabolic intermediates and at least one of these (acetyl phosphate) accumulates to high enough levels in vivo under certain growth conditions to contribute to the phosphorylated pool of some of the response regulators (McCleary and Stock, unpublished data). Not all response regulators can utilize these small molecule donors with equal efficiency indicating some specificity for these reactions [74]. The phosphorylation of response regulators by metabolic intermediates such as acetyl phosphate may represent an important mechanism to couple the metabolic state of the cell with signaling pathways that sense the external environment.

The histidine kinase typically has both kinase and phosphatase activities. The output is thus a balance of these two opposing reactions. The level of phosphorylated response regulator can be elevated by increasing kinase activity, decreasing phosphatase activity or both. Under conditions when phosphorylation of the response regulator may be independent of the kinase (e.g. high intracellular acetyl phosphate concentrations), the phosphatase activity of the histidine kinase domain can play an important role in regulating the level of phosphorylated response regulator.

4.1. Histidine kinase superfamily

The generic histidine kinase receptor has a periplasmic sensory domain and a cytoplasmic histidine kinase domain. The histidine kinase EnvZ is the best characterized example of this class of receptors. EnvZ regulates the phosphorylation levels of the transcriptional activator OmpR. The EnvZ/OmpR system regulates porin expression in response to changing medium osmolarity [75,76]. The topology of EnvZ is similar to the Tar chemoreceptor. It consists of periplasmic sensory domain flanked by two transmembrane segments. The sensory domain is connected to the cytoplasmic histidine kinase domain by a single transmembrane segment [77]. The nature of the environmental signal detected by the sensory domain of EnvZ remains to be established.

In addition to the receptors with architecture like EnvZ, several other variations exist in which additional domains are present (for comprehensive lists, see [68]). VirA of *Agrobacterium tumefaciens* [78], ArcB [79] and BarA [80] are examples of proteins similar to EnvZ but with a response regulator domain at the C-terminus. The function of the response regulator domain on these receptors remains to be determined but could be involved in regulation of the kinase domain. Phosphorylation of this potential regulatory domain by other kinases or small molecule phospho-donors may be a mechanism to integrate signals from other pathways. Other histidine kinase receptors lack a periplasmic sensory domain, but have extensive hydrophobic regions that may span the membrane several times (UhpB [8] , DivJ [81] ArcB [79]). These receptors may be regulated through interactions with other membrane proteins (e.g. UhpB/UhpC). KdpD appears to lack a periplasmic domain but is predicted to have two large cytoplasmic domains [82]. The KdpD receptor may represent another example of a mechanical sensor, in this case responding to turgor pressure at the inner membrane. Some of the different domain organizations of histidine kinase receptors are illustrated in Fig. 4.

4.2. Mechanism of signaling by histidine kinases receptors

EnvZ has been the most extensively studied of the members of this family of receptors. Recent experiments by Inouye and co-workers have shed some light on a potential mechanism of transmembrane signaling. The results reported by this group on the nature of the kinase autophosphorylation reaction has important implications for the nature of the signaling mechanism. First they demonstrated that a fusion of the N-terminal domain of the chemotaxis receptor Tar and the C-terminal kinase domain of EnvZ (called Taz1) was able to activate OmpC expression in response to aspartate [83]. Since this requires phosphorylated OmpR, the implication is that aspartate binding was causing activation of the histidine kinase domain or inhibition of the phosphatase activity. Phosphorylation of OmpR by Taz1 has been demonstrated in vitro [84]. This implies that the aspartate binding to Taz1 mimics the effect of increased osmolarity on EnvZ.

The second set of experiments utilized two defective kinase mutants. The first was mutated at the site of autophosphorylation (His243 to Val). The second mutant was deleted

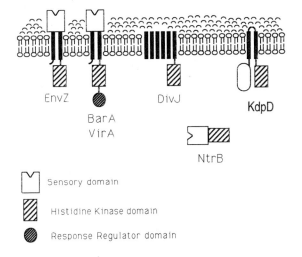

Fig. 4. Histidine kinase receptors. The EnvZ receptor has a histidine kinase domain connected to a periplasmic sensory domain by a single transmembrane segment [77]. The BarA [80] and VirA [78] have a topology similar to EnvZ but have an additional cytoplasmic domain that is homologous to the response regulator domain. DivJ [81] and UhpB [8] have an extensive hydrophobic region that potentially traverses the bilayer several times. KdpD is predicted to have two cytoplasmic domains [82]. NtrB [101] is an example of a histidine kinase receptor that is not membrane bound but is soluble within the cytoplasm.

at the C-terminus and no longer included the ATP binding site. Both these proteins were unable to autophosphorylate themselves, but when the two proteins were mixed together, phosphorylation of the second mutant was observed [84]. Similar results were observed when the same mutations were introduced into Taz1 [84]. These findings imply that the autophosphorylation reaction can occur in trans between subunits in a receptor dimer. This may be a general feature of the histidine kinases. This transphosphorylation of histidine kinases has recently been demonstrated for CheA [85] and NtrC (A. Ninfa, personal communication), two soluble cytoplasmic members of the histidine kinase family.

These results suggest that signal transduction by EnvZ may occur through dimerization. The experiments with Taz1 may lend support to this model. Recently, the resolution of the crystal structure of the ligand binding domain of Tar indicate that the ligand binds at the interface of the two domains [24]. In the Taz1 construct, this may result in dimerization although biophysical studies to establish this have not yet been reported. This is analogous to the class I receptor tyrosine kinases of eukaryotes [26]. Here, ligand binding induces dimerization and the juxtaposing of the intracellular kinase domains allows them to carry out transphosphorylation. The extrapolation of this simple model to EnvZ is complicated, however, by the observation that truncated EnvZ without its sensory domain is able to autophosphorylate [86–88] and the transphosphorylation reaction of Taz1 mutants occurs in the absence of aspartate [84]. These mutants are also defective in phosphatase activity. The level of OmpR-phosphate accumulating in the cell is a result of the balance of the kinase and phosphatase activities, and low levels of endogenous kinase activity would be accentuated in phosphatase mutants. The demonstration that deletion of much of the periplasmic domain results in kinase+/phosphatase− mutants, emphasizes the importance of regulation of the phosphatase activity [89]. Indeed, regulation of phosphatase activity may be responsible for changing levels of OmpR-phosphate. The regulation of porin expression by EnvZ has recently been further complicated by the identification of a second receptor capable of complementing EnvZ deletions [80]. This protein (BarA) is homologous to both EnvZ and OmpR, having both a histidine kinase and a response regulator domain similar. This protein complements the EnvZ deletion only when expressed from a high copy plasmid. The autophosphorylation of BarA has been demonstrated in vitro but no OmpR phosphorylation was detected under the in vitro conditions used.

5. Receptor-effector elements

Whereas chemoreceptors and histidine kinase receptors initiate complex signal transduction cascades within the cell, a more direct connection between sensing and response elements can occur at the membrane. The ToxR protein of *Vibrio cholerae* provides a good example of this principle (for a recent review, see [90]). ToxR is a transmembrane protein with an amino terminal cytoplasmic domain and a carboxy terminal domain in the periplasmic space [91]. These two domains are connected by a single transmembrane segment. The cytoplasmic region has a site-specific DNA binding domain [91]. The DNA binding domain is homologous to the conserved DNA binding domain found in several members of the two component family of response regulators. The periplasmic domain is thought to act as an environmental sensor monitoring changes in osmolarity, pH, temperature and the presence of certain amino acids. In this system, the receptor and effector components of the signaling pathway have been combined in a single protein.

The ToxR protein regulates transcription by binding a tandemly repeated element upstream of target genes [91]. The generation of fusions between this receptor and alkaline phosphatase that were still capable of activating transcription suggest that the mechanism of transmembrane signaling may be dimerization [91]. Since PhoA is active as a dimer, the PhoA-ToxR fusions that retain both PhoA and ToxR activity are likely to be dimers. Dimerization may thus be a prerequisite for DNA binding.

As with many of the signaling receptors, the ligand(s) for ToxR remain to be identified. Recent studies suggest that the sensing occurs indirectly through the ToxS protein, predicted by PhoA fusions to be a periplasmic protein anchored to the inner membrane by a single domain [92]. It is thus likely that ToxR-ToxS interactions occur in the periplasm (and possibly in the transmembrane regions). Replacing the periplasmic domain of ToxR with PhoA results in ToxS independent activation of ToxR dependent transcription [92]. Other fusions in which most of the periplasmic domain of ToxR are retained, protect ToxS from proteolytic degradation and ToxS decreases PhoA activity. These results suggest direct interaction of these two proteins and are consistent with a model in which ToxS binding to ToxR causes the dimerization of ToxR and consequent activation of DNA binding.

Only a few other bacterial signal transducing receptor have been characterized that are analogous to ToxR. One example appears to be CadC, a transcription regulator of genes induced under acidic media conditions. The gene encoding the CadC protein of *E. coli* has recently been characterized and sequence analysis predicts a protein of similar topology to ToxR with an amino terminal DNA binding domain homologous to ToxR, a single transmembrane sequence and a large C-terminal domain [93]. The C-terminus may be localized to the periplasm where it may act as a pH sensor. Biochemical analysis of CadC DNA binding and membrane topology have yet to be reported.

Two other membrane bound receptors have been described that may regulate transcription directly through a DNA binding domain in response to environmental signals. The NosR protein of *Pseudomonas stutzeri* is a positive transcriptional activator of the denitrification gene cluster (nos) [94]. The gene is predicted to encode an integral membrane protein of ~82 kDa with seven to nine transmembrane segments. The protein is not homologous to any known protein but is predicted to have a helix-turn-helix DNA binding motif consistent with its predicted role as a transcriptional regulator. In addition, there are two cysteine rich sequences in the C-terminal hydrophilic domain that are homologous to some bacterial ferrodoxins. This motif may be involved in redox sensing.

A membrane bound transcriptional activator (FecI) has been identified as a component of a system for the regulation of iron uptake in *E. coli* by extracellular Fe^{III}-dicitrate [95]. This system is proposed to consist of a periplasmic binding protein (FecR) and FecI. The 19-kDa FecI contains a helix-turn-helix DNA binding motif. Analysis of the primary sequence fails to predict any typical transmembrane segments, even though the protein is localized to the inner membrane fraction.

6. The sensory rhodopsins

Phototaxis by *Halobacterium halobium* is mediated by two retinal-based receptors that are homologous to the light activated ion pumps, bacteriorhodopsin and halorhodopsin (Fig. 5). These sensory rhodopsins mediate the phototaxis response by generating signals that regulate the direction of flagella rotation (for reviews, see [9]). Sensory rhodopsin I (SRI) mediates phototaxis towards green-orange light and away from blue light [96]. The second receptor (SRII) also mediates a repellent response to blue light [97]. The sensory rhodopsins do not cause changes in membrane potential and the signals generated are integrated with the signals from the chemoreceptors of this organism presumably through common cytoplasmic signaling components. Recently, a gene encoding an additional signaling component of the SRI dependent phototaxis system has been cloned [10]. The protein product (HtrI, halobacterial transducer I) of this gene is predicted to consists of two transmembrane segments with a small periplasmic domain. The cytoplasmic domain is very homologous to *E. coli* chemoreceptors such as Tar, but has an additional segment of about 200 amino acids between the second transmembrane segment and the C-terminal chemoreceptor domain. This later region is postulated to interact with SRI. Presumably, a similar protein exists for SRII. The transmembrane signaling system in phototaxis may be thought of as comprised of two membrane proteins: a sensory protein (sensory rhodospin) and a signaling protein (HtrI). The C-terminal domain of HtrI presumably regulates the activity of the soluble cytoplasmic components of the chemotaxis system. Methylation of

480

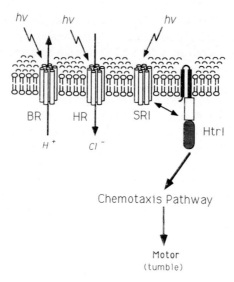

Fig. 5. Rhodopsins of *Halobacterium halobium*. The family of bacteriorhodopsins are all predicted to have seven transmembrane segments. Bacteriorhodopsin (BR) and halorhodopsin (HR) are light activated ion pumps, for H^+ and Cl^-, respectively. The closely related sensory rhodopsins SRI and SRII (not illustrated) do not act as ion pumps but are sensors for the phototactic response. This response for SRI requires a homolog of the MCP chemoreceptors called HtrI [10] which is thought to interact with the soluble components of the chemotactic signal transduction pathway.

this protein has been demonstrated in vivo suggesting an adaptation mechanism analogous to that of the MCP chemoreceptors of *E. coli*.

7. Conclusions

The past few years have witnessed a rapid expansion in our knowledge of bacterial signal transduction pathways. Even an organism as well characterized as *E. coli* continues to provide novel mechanisms. As new methodology removes the barriers to genetic analysis of other organisms, the diverse world of bacterial sensory receptors will continue to be revealed. We can expect to see more and more diversity, in addition to the continued emergence of a few common themes. We might also expect to see some bacterial homologs of important components of eukaryotic sensory transduction. A tyrosine kinase activity has been detected in *Pseudomonas solanacearum* [98] and a eukaryotic-like serine/threonine protein kinase involved in differentiation of *Myxococcus xanthus* has also been identified [99].

Acknowledgments

This work was supported in part by a grant from the National Institutes of Health (AI20980). M.G.S. was supported by a postdoctoral fellowship from the Medical Research Council of Canada.

References

1. Stock, J.B., Ninfa, A.J. and Stock, A.M. (1989) Microbiol. Rev. 53, 450–490.
2. Macnab, R.M. (1987) in: F.C. Neidhardt et al. (Eds.), *Escherichia coli* and *Salmonella typhimurium*: Cellular and Molecular Biology, American Society for Microbiology, Washington, DC, pp. 732–759.
3. Stewart, R.C. and Dahlquist, F.W. (1987) Chem. Rev. 87, 997–1025.
4. Stock, J.B., Lukat, G.S. and Stock, A.M. (1991) Annu. Rev. Biophys. Biophys. Chem. 20, 109–136.
5. Bourret, R.B., Borkovich, K.A. and Simon, M.I. (1991) Annu. Rev. Biochem. 60, 401–441.
6. Furlong, C.A. (1987) in: F.C. Neidhardt et al. (Eds.), *Escherichia coli* and *Salmonella typhimurium*: Cellular and Molecular Biology, Vol. 1, American Society for Microbiology, Washington, DC, pp. 768–796.
7. Torriani, A. (1990) BioEssays 12, 371–376.
8. Island, M.D., Wei, B.-Y. and Kadner, R.J. (1992) J. Bacteriol. 174, 2754–2762.
9. Spudich, J.L. and Bogomolni, R.A. (1988) Annu. Rev. Biophys. Biophys. Chem. 17, 193–215.
10. Yao, V.J. and Spudich, J.L. (1992) Proc. Natl. Acad. Sci. USA 89, 11915–11919.
11. Meadow, N.D., Fox, D.K. and Roseman, S. (1990) Annu. Rev. Biochem. 59, 497–542.
12. Saier, Jr., M.H. (1993) J. Cell. Biochem. 51, 62–68.
13. Titgemeyer, F. (1993) J. Cell. Biochem. 51, 69–74.
14. Chen, Y., Fairbrother, W.J. and Wright, P.E. (1993) J. Cell. Biochem. 51, 75–82.
15. Morgan, D.G., Baumgartner, J.W. and Hazelbauer, G.L. (1993) J. Bacteriol. 175, 133–140.
16. Borkovich, K.A., Kaplan, N., Hess, J.F. and Simon, M.I. (1989) Proc. Natl. Acad. Sci. USA 86, 1208–1212.
17. Borkovich, K.A. and Simon, M.I. (1990) Cell 63, 1339–1348.
18. Ninfa, E.G., Stock, A., Mowbray, S. and Stock, J. (1991) J. Biol. Chem. 266, 9764–9770.
19. Krikos, A., Mutoh, N., Boyd, A. and Simon, M.I. (1983) Cell 33, 615–522.
20. Gegner, J.A., Graham, D.R., Roth, A.F. and Dahlquist, F.W. (1992) Cell 70, 975–982.
21. Springer, W.R. and Koshland, Jr., D.E. (1977) Proc. Natl. Acad. Sci. USA 74, 533–537.
22. Stock, J.B. and Koshland, Jr., D.E. (1978) Proc. Natl. Acad. Sci. USA 75, 3659–3663.
23. Milligan, D.L. and Koshland, Jr., D.E. (1988) J. Biol. Chem. 263, 6268–6275.
24. Milburn, M.V., Prive, G.G., Milligan, D.L., Scott, W.G., Yeh, J., Jancarik, J., Koshland, D.E.J. and Kim, S.-H. (1991) Science 254, 1342–1347.
25. DeVos, A.M., Ultsch, M. and Kossiakoff, A.A. (1992) Science 255, 306–312.
26. Ullrich, A. and Schlessinger, J. (1990) Cell 61, 203–212.
27. Mowbray, S.L. and Koshland, Jr., D.E. (1987) Cell 50, 171–180.
28. Stoddard, B.L. and Koshland, Jr., D.E. (1992) Nature 358, 774–776.
29. Manson, M.D. and Kossman, M. (1986) J. Bacteriol. 165, 34–40.
30. Gardina, P., Conway, C., Kossman, M. and Manson, M. (1992) J. Bacteriol. 174, 1528–1536.
31. Yaghmai, R. and Hazelbauer, G.L. (1992) Proc. Natl. Acad. Sci. USA 89, 7890–7894.
32. Oosawa, K. and Simon, M. (1986) Proc. Natl. Acad. Sci. USA 83, 6930–6934.
33. Lynch, B.A. and Koshland, D.E.J. (1991) Proc. Natl. Acad. Sci. USA 88, 10402–10406.
34. Pakula, A.A. and Simon, M.I. (1992) Proc. Natl. Acad. Sci. USA 89, 4144–4148.
35. Kehry, M.R. and Dahlquist, F.W. (1982) J. Biol. Chem. 257, 10378–10386.
36. Terwilliger, T.C. and Koshland, Jr., D.E. (1984) J. Biol. Chem. 259, 7719–7725.
37. Borkovich, K.A., Alex, L.A. and Simon, M.I. (1992) Proc. Natl. Acad. Sci. USA 89, 6756–6760.
38. Clarke, S. (1985) Annu. Rev. Biochem. 54, 479–506.
39. Kehry, M.R., Bond, M.W., Hunkapiller, M.W. and Dahlquist, F.W. (1983) Proc. Natl. Acad. Sci. USA 80, 3599–3603.
40. Simms, S.A., Keane, M.G. and Stock, J. (1985) J. Biol. Chem. 260, 10161–10168.
41. Stock, A., Koshland, Jr., D.E. and Stock, J. (1985) Proc. Natl. Acad. Sci. USA 82, 7989–7993.
42. Lupas, A. and Stock, J. (1989) J. Biol. Chem. 264, 17337–17342.
43. Stock, J.B., Surette, M., McCleary, W.R. and Stock, A.M. (1992) J. Biol. Chem. 267, 19753–19756.
44. Ames, P. and J.S. Parkinson (1988) Cell 55, 817–826.
45. Collins, L.A., Egan, S.M. and Stewart, V. (1992) J. Bacteriol. 174, 3667–3675.

482

46. Liu, J. and Parkinson, J.S. (1991) J. Bacteriol. 173, 4941–4951.
47. Russo, A.F. and Koshland, D.E. (1983) Science 220, 1016–.
48. Krikos, A., Conley, M.P., Boyd, A., Berg, H.C. and Simon, M.I. (1985) Proc. Natl. Acad. Sci. USA 82, 1326–1330.
49. Stoddard, B.L., Bui, J.D. and Koshland, D.E. (1992) Biochemistry 31, 11978–11983.
50. Springer, M.S., Goy, M.F. and Adler, J. (1979) Nature 280, 279–284.
51. Stock, J. and Stock, A. (1987) Trends Biochem. Sci. 12, 371–375.
52. Stock, J.B. (1990) in: W.K. Paik and S. Kim (Eds.), Protein Methylation, CRC Press, Boca Raton, FL, pp. 275–284.
53. Goy, M.F., Springer, M.S. and Adler, J. (1977) Proc. Natl. Acad. Sci. USA 74, 4964–4968.
54. Borczuk, A., Staub, A. and Stock, J. (1986) Biochem. Biophys. Res. Commun. 141, 918–923.
55. Springer, M.S., Zanolari, B. and Pierzchala, P.A. (1982) J. Biol. Chem. 257, 6861–6866.
56. Stock, J.B. and Koshland, Jr., D.E. (1981) J. Biol. Chem. 256, 10826–10833.
57. Hess, J.F., Oosawa, K., Kaplan, N. and Simon, M.I. (1988) Cell 53, 79–87.
58. Stock, J., Kersulis, G. and Koshland, Jr., D.E. (1985) Cell 42, 683–690.
59. Dunten, P. and Koshland, Jr., D.E. (1991) J. Biol. Chem. 266, 1491–1496.
60. Yonekawa, H. and Hayashi, H. (1986) FEBS Lett. 198, 21–24.
61. Pakula, A. and Simon, M. (1992) Nature 355, 496–497.
62. Milligan, D.L. and Koshland, Jr., D.E. (1991) Science 254, 1651–1654.
63. Mowbray, S.L., Foster, D.L. and Koshland, Jr., D.E. (1985) J. Biol. Chem. 260, 11711–11718.
64. Foster, D.L., Mowbray, S.L., Jap, B.K. and Koshland, Jr., D.E. (1985) J. Biol. Chem. 260, 11706–11710.
65. Oosawa, K., Mutoh, N. and Simon, M.I. (1988) J. Bacteriol. 170, 2521–2526.
66. Kaplan, N. and Simon, M.I. (1988) J. Bacteriol. 170, 5134–5140.
67. Long, D.G. and Weis, R.M. (1992) Biochemistry 31, 9904–9911.
68. Parkinson, J.S. and Kofoid, E.C. (1992) Annu. Rev. Genet. 26, 71–112.
69. Stock, J.B., Stock, A.M. and Mottonen, J.M. (1990) Nature 344, 395–400.
70. Bourret, R.B., Hess, J.F., Borkovich, K.A., Pakula, A.A. and Simon, M.I. (1989) J. Biol. Chem. 264, 7085–7088.
71. Ninfa, A.J., Ninfa, E.G., Lupas, A.N., Stock, A., Magasanik, B. and Stock, J. (1988) Proc. Natl. Acad. Sci. USA 85, 5492–5496.
72. Igo, M.M., Ninfa, A.J., Stock, J.B. and Silhavy, T.J. (1989) Genes Dev. 3, 1725–1734.
73. Wanner, B.L. (1992) J. Bacteriol. 174, 2053–2058.
74. Lukat, G.S., McCleary, W.R., Stock, A.M. and Stock, J.B. (1992) Proc. Natl. Acad. Sci. USA 89, 718–722.
75. Csonka, L.N. (1989) Microbiol. Rev. 53, 121–147.
76. Igo, M.M., Slauch, J.M. and Silhavy, T.J. (1990) New Biol. 2, 5–9.
77. Forst, S., Comeau, D., Norioka, S. and Inouye, M. (1987) J. Biol. Chem. 262, 16433–16438.
78. Leroux, B., Yanofsky, M.F., Winans, S.C., Ward, J.E., Ziegler, S.F. and Nester, E.W. (1987) EMBO J. 6, 849–856.
79. Iuchi, S., Matsuda, Z., Fujiwara, T. and Lin, E.C.C. (1990) Mol. Microbiol. 4, 715–727.
80. Nagasawa, S., Tokishita, S., Aiba, H. and Mizuno, T. (1992) Mol. Microbiol. 6, 799–807.
81. Ohta, N., Lane, T., Ninfa, E., Sommer, J.M. and Newton, A. (1992) Proc. Natl. Acad. Sci. USA 89, 10297–10301.
82. Polarek, J.W., Williams, G. and Epstein, W. (1992) J. Bacteriol. 174, 2145–2151.
83. Utsumi, R., Brissette, R.E., Rampersaud, A., Forst, S.A., Oosawa, K. and Inouye, M. (1989) Science 245, 1246–1249.
84. Yang, Y. and Inouye, M. (1992) Proc. Natl. Acad. Sci. USA 88, 11057–11061.
85. Wolfe, A.J. and Stewart, R.C. (1993) Proc. Natl. Acad. Sci. USA 90, 1518–1522.
86. Aiba, H., Mizuno, T. and Mizushima, S. (1989) J. Biol. Chem. 264, 8563–8567.
87. Forst, S., Delgado, J. and Inouye, M. (1989) Proc. Natl. Acad. Sci. USA 86, 6052–6056.
88. Igo, M.M. and Silhavy, T.J. (1988) J. Bacteriol. 170, 5971–5973.
89. Tokishita, S.-I., Kojima, A., Aiba, H. and Mizuno, T. (1991) J. Biol. Chem. 266, 6780–6785.
90. DiRita, V.J. (1992) Mol. Mirobiol. 6, 451–458.

91. Miller, V.L., Taylor, R.K. and Mekalanos, J.J. (1987) Cell 48, 271–279.
92. DiRita, V.J. and Mekalanos, J.J. (1991) Cell 64,
93. Watson, N., Dunyak, D.S., Rosey, E.L., Slonczewski, J.L. and Olson, E.R. (1992) J. Bacteriol. 174, 530–540.
94. Cuypers, H., Viebroc-Sambale, A. and Zumft, W.G. (1992) J. Bacteriol. 174, 5332–5339.
95. van Hove, B., Staudenmaier, H. and Braun, V. (1990) J. Bacteriol. 172, 6749–6758.
96. Bogomolni, R.A. and Spudich, J.L. (1982) Proc. Natl. Acad. Sci. USA 79, 6250–6254.
97. Marwan, W. and Oesterhelt, D. (1987) J. Mol. Biol. 195, 333–342.
98. Atkinson, M., Allen, C. and Sequeira, L. (1992) J. Bacteriol. 174, 4356–4360.
99. Zhang, W., Munoz-Dorado, J., Inouye, M. and Inouye, S. (1992) J. Bacteriol. 174, 5450–5453.
100. Lupas, A., VanDyke, M. and Stock, J. (1991) Science 252, 1162–1164.
101. Keener, J. and Kustu, S. (1988) Proc. Natl. Acad. Sci. USA 85, 4976–4980.

J.-M Ghuysen and R. Hakenbeck (Eds.), *Bacterial Cell Wall*

485

CHAPTER 23

Mechanisms of chromosomal β-lactamase induction in Gram-negative bacteria

STAFFAN NORMARK[1], EVELINE BARTOWSKY[1], JAY ERICKSON[1], CHRISTINE JACOBS[4], FREDERIK LINDBERG[2], SUSANNE LINDQUIST[3], KATHLEEN WESTON-HAFER[1] and MIKAEL WIKSTRÖM[3]

Departments of [1]Molecular Microbiology and [2]Medicine, Washington University, Medical School, Box 8230, 660 S. Euclid Ave., St. Louis, MO 63110, USA, [3]Department of Microbiology, Umeå University, Umeå, Sweden and [4]Université de Liège, Centre d'Ingénierie des Protéines, Liège, Belgium

1. Introduction

Chromosomally-mediated β-lactamase production is a major determinant of β-lactam resistance seen among many pathogenic and opportunistic *Enterobacter* spp., *Citrobacter freundii*, *Serratia* spp., *Morganella* spp., indole-positive *Proteus* spp. and *Pseudomonas aeruginosa*. These inducible enzymes, present in most clinical isolates of these species, generally convey resistance to β-lactamase-labile agents such as benzylpenicillin, ampicillin, and the first-generation cephalosporins. The susceptibility of these organisms to a β-lactam antibiotic is influenced by both the magnitude of β-lactamase production (degree of antibiotic-stimulated induction or constitutive overproduction), the drug's ability to resist hydrolysis by the β-lactamase, the rate of penetration through the outer membrane and the binding affinity for the penicillin-binding proteins. Naturally occurring *Enterobacter cloacae* isolates which exhibit only basal, low-level β-lactamase production may be moderately susceptible to ampicillin whereas the more common β-lactamase-inducible isolates are ampicillin-resistant [1]. When constitutive or semi-constitutive β-lactamase hyperproduction is present, as with stably derepressed isolates, resistance is seen even to relatively β-lactamase-stable, weakly β-lactamase-inducing agents such as the extended-spectrum penicillins and third-generation cephalosporins.

Gram-negative bacilli producing inducible, chromosomally mediated β-lactamases are common pathogens in hospitalized patients. *Pseudomonas aeruginosa*, *Enterobacter* spp., *Citrobacter* spp. and *Serratia marcescens* accounted for 22% of all infectious isolates from participating hospitals during 1986–1989 in the National Nosocomial Infections Surveillance System [2]. These organisms were responsible for 33% and 11% of nosocomial pneumonias and bacteremias, respectively. If an estimated 3.6 million nosocomial infections occur in the United States annually [3], these species account for nearly 800 000 infections in patient-care facilities, each year.

Gram-negative isolates resistant to multiple, 'β-lactamase-stable' β-lactams have been reported to emerge during β-lactam therapy. A recent review of studies addressing this issue indicated that resistance emerged in 14–56% of patients with infections caused by these Gram-negative species and when present, resulted in relapse or treatment failure in 25–75% of cases [4]. Isolates acquiring resistance typically exhibited stably derepressed β-lactamase production [4–6].

Factors associated with this emergence of resistance include bone, soft tissue and lower respiratory infections (especially cystic fibrosis), neutropenia and third-generation cephalosporin therapy [4,7,8]. Attempts to limit the emergence of resistance using combination antibiotic therapy have been disappointing and the question of a need for more judicious third-generation cephalosporin usage has been raised [8]. Broadly, β-lactam-resistant Gram-negative bacilli typically are as virulent as the parent strains and can spread from patient to patient even in the absence of the selective pressure of β-lactam antibiotics [4]. It is not surprising that these organisms remain serious nosocomial pathogens.

In this chapter, we summarize our current understanding of the molecular mechanisms by which β-lactam antibiotics may induce expression of chromosomal β-lactamase in Gram-negative organisms as well as the genetic basis for the appearance of so-called stably derepressed mutants.

2. Components of the inducible β-lactamase regulatory system in Gram-negative organisms

2.1. The amp regulon

Most Gram-negative enterobacteria carry a chromosomal β-lactamase gene denoted *amp*C [9,10]. This gene is constitutively expressed at low levels in *E. coli*. The *amp*C expression in this species is governed by a relatively weak promoter and by transcriptional termination at an attenuator site immediately 5' of the *amp*C gene [11,12]. Mutants of *E. coli* overexpressing chromosomal β-lactamase are either due to up promoter mutations, mutations decreasing transcriptional attenuation, *amp*C gene amplification, or insertion elements such as IS2 providing a novel hybrid promoter [11,13–17]. Combinations of the above genetic mechanisms may provide an appreciable β-lactam resistance [18]. However, since the frequency by which each mutation occurs is low (10^{-7}–10^{-9}), β-lactam resistance is unlikely to emerge during ongoing therapy. In a number of Gram-negative species such as *Citrobacter freundii*, *Enterobacter cloacae*, *Pseudomonas aeruginosa*, *Serratia marcescens*, *Yersinia enterocolitica* and indole-positive *Proteus* spp., the chromosomal *amp*C gene is inducible by β-lactam antibiotics. Available information suggests that the induction mechanism is the same for each of these organisms.

The main genetic difference between species such as *E. coli* expressing β-lactamase constitutively and those that are inducible is the presence in the latter of a regulatory gene, *amp*R, that is located close to *amp*C but transcribed in the opposite orientation [19]. *E. coli* not only lacks *amp*R but also the site between *amp*R and *amp*C to which this regulatory protein binds. When the *ampR*, *ampC* genes from either *C. freundii* or *E. cloacae* are introduced on a plasmid into *E. coli*, the cloned *ampC* gene is inducible by β-lactams. A genetic inactivation of *ampR* abolished inducibility demonstrating that AmpR function was required for this process.

In both *C. freundii* and *E. cloacae*, β-lactam-resistant, so-called stably derepressed mutants arise at an appreciable frequency (10^{-6}–10^{-7}) that overproduce the AmpC β-lactamase in the absence of β-lactam inducer [20]. Similar β-lactam-resistant mutants were selected in *E. coli* K12 harboring the *C. freundii* *ampR* and *ampC* genes on a multicopy plasmid [21]. These *E. coli* mutants were either fully constitutive or semiconstitutive for β-lactamase expression, and all mapped to a single locus at 2.6 min on the *E. coli* chromosome denoted *ampD* encoding a 20.5 kDa cytosolic protein [22]. Likewise, transformation with plasmids carrying *E. coli* *ampD* could restore the wild-type phenotype of inducible β-lactamase production in all stably derepressed clinical isolates and spontaneous laboratory mutants of *E. cloacae* and *C. freundii* tested (F. Lindberg, unpublished data). This indicated that AmpD acts as a negative regulator of β-lactamase production and mutations in this gene can lead to clinical β-lactam resistance [21]. Nucleotide sequencing of *ampD* from *E. coli*, *C. freundii* and *E. cloacae* demonstrate that they encode proteins showing 75–89% identity to one another [23].

A specific signal transducer has been postulated to exist transmitting the β-lactam-induced signal across the cytoplasmic membrane. The *ampG* gene is currently the best candidate to encode for a signal transducing element since mutants in this gene are totally non-inducible [24,25]. A summary of the genes and gene-products in the *amp* regulon is given in Table I.

2.2. Involvement of penicillin-binding proteins and FtsZ in β-lactamase induction

It is very likely that other genes besides *ampD*, *ampG* and *ampR* are involved in *ampC* β-lactamase induction, but mutations in such genes may be lethal. A number of *E. coli* mutants in penicillin-binding protein genes have been tested for β-lactamase induction. β-Lactams acting as good inducers usually have high affinity for the low molecular weight PBPs. However, mutants inactivated for PBP4, PBP5 or PBP6 are still inducible and express low levels of β-lactamase in the absence of β-lactam inducer. Likewise, double and triple mutants in PBP4, PBP5 and PBP6 remain inducible. Single mutants affected in either PBP1A or PBP1B are not affected in β-lactamase expression and minicells prepared from a PBP3 *ts* mutant are still inducible when treated with β-lactams at 42°C (S. Lindquist, H. Martin and K. Weston-Hafer, unpublished data). However, a mutant in PBP2 has been reported to be non-inducible in the reconstituted *E. coli* system [26]. Also, conditional *ftsZ* mutants are non-inducible at the restrictive temperature [27]. There are as yet no data showing whether a loss of FtsZ or PBP2 activity affects AmpG function or if expression of these two former proteins are required to elicit a β-lactam-induced signal. The prevailing hypothesis is that β-lactams affect a more global regulatory network to elicit the inducible signal. Since FtsZ is required for bacterial cell division, the β-lactam-induced signal may in some way be linked to the septation process.

TABLE I

Genes and proteins involved in β-lactamase induction in Gram-negative bacteria

Gene	Function	Size (kDa)	Protein location	Gene location (in E. coli)	Species sequenced from	Ref.	Mutational phenotype when inactivated
ampC	Structural gene for Class C β-lactamase	39.7	Periplasm	94.3'	C. freundii E. cloacae E. coli Ps. aeruginosa Y. enterocolitica	55 56 11 42 57	* No β-lactamase produced
ampR	Transcriptional activator of ampC	32	Cytoplasm	deleted	C. freundii E. cloacae Y. enterocolitica Rhodops. capsulata	29 58 57 43	* Low expression of non-inducible β-lactamase
ampD	Negative regulator	20.5	Cytoplasm	2.6'	E. coli E. cloacae C. freundii	28 22 23 23	* AmpR dependent hyperproduction of β-lactamase, altered crosslinking pattern in murein, and increased release of labelled DAP from the murein
ampE	ATPase?	32.1	Inner membrane	2.6'	E. coli	22	* not required for β-lactamase induction
ampG	ligand transporter ?	55	Inner membrane	9.9'	E. coli E. cloacae (in progress)	25	* Low expression of non-inducible β-lactamase at the basal level

3. The ampD operon

The *amp*D gene is the first gene in an operon also containing *amp*E. AmpE encodes a 32-kDa transmembrane protein carrying a putative ATP-binding site [22]. AmpE was initially thought to be involved in β-lactamase regulation and hypothesized to act as a signal transducer able to bind β-lactam at the periplasmic side of the plasma membrane [28]. A clean chromosomal 'knockout' mutation has recently been generated in the *E. coli amp*E gene. This *amp*E mutant expressed the same basal level of cloned *C. freundii* AmpC β-lactamase as the wild-type and β-lactamase expression was inducible by β-lactams provided AmpR was also expressed (M. Wikström, unpublished data). Therefore AmpE has no detectable effect on β-lactamase expression and is not acting as a signal transducer. Since the stop codon of *amp*D overlaps the initiation codon of *amp*E in both *E. coli* and *C. freundii*, these two genes may be translationally coupled to ensure a stoichiometric relationship between the two proteins. Thus, AmpD and AmpE may well functionally interact without affecting β-lactamase expression.

Point mutations, deletions, as well as insertions have been isolated in *amp*D yielding a similar phenotype suggesting that a functional inactivation of the AmpD protein results in an elevated β-lactamase expression [22,23,28]. The AmpD protein therefore acts as a negative regulator for β-lactamase expression. An extract containing AmpD does not result in a gel mobility shift of a DNA fragment carrying the intercistronic region between *amp*R and *amp*C suggesting that AmpD does not function as a conventional repressor [22; C. Jacobs, unpublished data]. Since AmpD expression in the absence of AmpR has no effect on AmpC expression, AmpD might act by binding to AmpR. However, the gel shift caused by AmpR is not affected and the DNase I footprint generated by AmpR is not affected by the presence or absence of AmpD in the extract [29]. We are therefore left with the possibility that AmpD affects β-lactamase expression in an indirect manner that is discussed below.

An *amp*D mutant of *E. coli* in the absence of β-lactamase is slightly more sensitive to β-lactam antibiotics than the wild-type [22]. It is unlikely that this hypersensitivity is due to an increased permeability through the outer membrane but likely reflects some alterations in the response to β-lactams. AmpD mutants also exhibited a higher release of incorporated [^{3}H]diaminopimelic acid from the murein compared to wild-type cells. AmpD mutants grown in the presence of diaminopimelic acid also differ from the wild-type in their murein fragment pattern as deduced by reverse phase high performance liquid chromatography [30], reinforcing the notion that AmpD might affect murein metabolism.

4. The ampG operon

The *amp*G locus was identified by isolating a β-lactam-sensitive mutant of an *E. cloacae* strain derepressed for chromosomal β-lactamase [24]. An *E. cloacae* DNA fragment cloned from the parental strain restored the β-lactam resistance to the mutant. This DNA fragment was subsequently used to clone the corresponding *amp*G locus from *E. coli*. In a parallel study, a series of β-lactam- sensitive mutants were isolated from a β-lactam-resistant *E. coli amp*D mutant strain harboring the *C. freundii amp*C and *amp*R genes. Three of these mutants were expressing AmpC β-lactamase at low basal levels not affected by β-lactam inducer. The cloned *ampG* gene from either *E. coli* or *C. freundii* complemented the three *E. coli* mutations suggesting that they all carried lesions in the *ampG* locus [25]. This was subsequently shown by sequencing the mutations and comparing these sequences to the wild-type *E. coli ampG* gene. Each mutant carried a Gly→Asp amino acid replacement in AmpG at three different locations in the protein. The AmpG amino acid sequence deduced from the nucleotide sequence suggests that it is a transmembrane protein, 491 amino acids in size, carrying several hydrophobic putative transmembrane sequences. The *ampG* gene product is not essential for growth of *E. coli* since an *E. coli* mutant carrying a *kan*^{r} insert early in the *ampG* gene shows the same growth rate as the parent. Such a clean 'knock out mutation' will express cloned *C. freundii* AmpC β-lactamase at low levels in a non-inducible manner. The finding that several independently isolated mutants in different species affected the same gene strongly argues that, besides mutations in *ampR*, *ampG* mutations are unique in providing a non-inducible non-conditional phenotype. Mutations in *ampD* only affect β-lactamase expression in *ampG* wild-type strains, showing that the AmpG protein is required not only for β-lactam-mediated induction but also for the activation of β-lactamase expression resulting from a genetic inactivation of *ampD*.

The *ampG* gene of *E. coli* is located at 9.8 min on the chromosome and is the second gene in an operon also encoding a putative lipoprotein [25]. An insertional inactivation of this upstream gene abolishes β-lactamase induction. This effect is probably due to transcriptional polarity on *ampG* since inducibility is restored by only providing *ampG* transcribed from a vector promoter. The *ampG* operon is also located close to the morphogene *bolA*. Overproduction of the *bolA* gene product results in rounded cells and in an overexpression of PBP6 [31]. The *bolA* transcription is low during logarithmic growth. Transcriptional *lacZ* fusions in either *ampG* or in the preceding lipoprotein gene were used to monitor transcriptional activity in different *E. coli* backgrounds and during different growth conditions. The two genes in the *ampG* operon are transcribed at a higher rate in logarithmically grown cells as compared to stationary-phase cells. It has been shown that the neighboring *bolA* morphogene is under the control of the RpoS starvation-induced sigma factor [32]. One possibility is therefore that RpoS acts on *bolA* and that BolA acts as a transcriptional regulator for the *ampG* operon. However, transcription of the *ampG* operon was not significantly affected by either a mutation in *bolA* or *rpoS* [25]. Moreover, AmpC β-lactamase is still inducible in *rpoS* and *bolA* mutants. Thus, it seems unlikely that the *bolA* gene controls expression from the *ampG* operon. Furthermore, expression from the *ampG* operon does not require the RpoS sigma factor. Finally, transcriptional activity over the *ampG* operon as measured by transcriptional fusions was not affected in an *E. coli* strain deleted for the *ampDE* operon. Therefore AmpD is unlikely to affect *ampC* transcription by repressing expression of AmpG.

The 491 amino acid AmpG protein carries no sequence motifs typical of β-lactam-binding proteins [33] and therefore differs markedly from the membrane-bound BlaR1 signal transducer in *Bacillus licheniformis* [34]. In addition, AmpG carries no homology to histidine kinases known to act as transmembrane sensors and signal transducers in many so-called two-component regulatory systems [35]. AmpG with its many putative membrane spanning regions may instead function as a permease allowing transport across

TABLE II

Some members of the LysR family of transcriptional activators

Protein	Bacterial species	Size (aa)	Inducer	Gene(s) regulated	Target pathway	Ref.
AmpR	*Citrobacter freundii*	291	β-Lactam	*ampC*	β-Lactamase	29
	Enterobacter cloacae	291	β-Lactam	*ampC*	β-Lactamase	58
	Yersinia enterocolitica	294	β-Lactam	*ampC*	β-Lactamase	57
	Rhodopseudomonas capsulata	289	β-Lactam	ORF1	β-Lactamase	43
CatR	*Pseudomonas putida*	289	*cis-cis* muconate	*catBC*	Benzoate utilization	59
CysB	*Escherichia coli*	324	O-Acetyl-L-serine	*cys*	Cysteine biosynthesis	60
	Salmonella typhimurium	324	O-Acetyl-L-serine	*cys*	Cysteine biosynthesis	60
IlvY	*Escherichia coli*	297	Acetolactate (L-isoleucine)	*ilvC*	Isoleucine biosynthesis	61
LysR	*Escherichia coli*	311	DAP (diaminopimelate)	*lysA*	Lysine biosynthesis	62
MetR	*Escherichia coli*	317	Homocysteine	*metJ/H*	Methionine biosynthesis	63
	Salmonella typhimurium	276	Homocysteine	*metJ/H*	Methionine biosynthesis	64
MleR	*Lactococcus lactis*	291	L-Malate	*metJ/H*	Malolactic acid fermentation	65
MprR	*Streptomyces coelicolor* 'Müller'	316	?	*mprA*	Metalloprotease	66
NahR	*Pseudomonas putida*	374	Salicylate	*nah, sal*	Degradation of naphthalene	67
NodD	*Rhizobium leguminosarum*	308	Flavonoid	*nodABC*	Nodulation genes	68
	Rhizobium meliloti	310	Flavonoid	*nodABC*	Nodulation genes	69
OxyR	*Escherichia coli*	305	Oxidative stress	H_2O_2	Inducible genes	70
	Salmonella typhimurium	305	Oxidative stress	H_2O_2	Inducible genes	70
RbcR	*Chromatium vinosum*	302	?	*rbcAB*	Carbon fixation	38
TrpI	*Pseudomonas aeruginosa*	293	Indole glycerol phosphate	*trpBA*	Tryptophan biosynthesis	71

the cytoplasmic membrane of ligands interacting with an intracellular regulator such as AmpR.

The deduced *ampG* amino acid sequence shows a 38% identity to *orf* 3 in the *lic* 3 locus of *Haemophilus influenzae* [36]. The *lic* 3 locus is involved in variable lipopolysaccharide expression but no specific role in this process has yet been assigned to the *orf* 3 gene product.

5. Mechanisms of converting AmpR into a transcriptional activator for ampC

5.1. AmpR belongs to the LysR family of transcriptional activators

AmpR of enterobacteria with inducible β-lactamase belongs to a large family of transcriptional activator proteins referred to as the LysR family [37]. To date over 30 proteins have been found to belong to this family [38], some of which are summarized in Table II. These proteins show a significant homology to one another along the entire sequence, but are most similar in their amino terminal regions containing a helix-turn-helix motif. An AmpR^{S35F} mutant affected in the second helix of this motif is unable to bind to its target DNA and cannot affect *ampC* transcription [39]. Most of these regulatory proteins are activated by low molecular weight effector molecules (Table II) but can bind target DNA both in the absence and presence of activating ligand. In the case of CysB, both an inducer O-acetyl-L-serine and an anti-inducer thiosulphate can affect the ability of the transcriptional regulator, to activate transcription [40]. The only exception to this rule is OxyR, a regulator responding to oxidative stress [41]. Activation in this case seems to involve direct oxidation of the OxyR protein rather than ligand binding. None of the members of the LysR family has been shown to be covalently modified and they have no significant homology to response regulators in two-component regulatory systems. The significant homology between AmpR and TrpI (Table III) which is known to be activated by indolglycerol phosphate argues that β-lactam induction is caused by ligand binding to AmpR converting it into a transcriptional activator.

5.2. AmpR homologues and AmpR like proteins in other species

Pseudomonas aeruginosa is a Gram-negative non-fermenting organism which, like many enterobacterial species, expresses an inducible chromosomal β-lactamase. This β-lactamase shows significant homology to the AmpC enzymes of enterobacteria. The *ampC* gene of *P. aeruginosa*, PAO1 is preceded by an open reading frame encoding a protein belonging to the AmpR family (Fig. 1, Table III). Unfortunately, only part of this AmpR sequence has been determined [42].

Rhodopseudomonas capsulata, a photosynthetic Gram-negative bacterium expresses an inducible chromosomal class A β-lactamase. Interestingly, this β-lactamase gene is preceded by an open reading frame showing high homology to the *ampR* genes preceding class C (*ampC*) β-lactamase genes in enterobacteria [43] (Fig. 1, Table III).

Streptomyces cacaoi, a Gram-positive organism with an inducible β-lactamase carries three open reading frames upstream from the β-lactamase gene that are transcribed in the

TABLE III

Evolutionary distance of AmpR, AmpR-like proteins and TrpI[a]

	C.f	E.c	Y.e	R.c	P.a	S.c	TrpI
C.f	1.0000	0.8832	0.7423	0.5017	0.7407	0.1993	0.3746
E.c		1.0000	0.7560	0.5052	0.7111	0.2062	0.3883
Y.e			1.0000	0.4983	0.7630	0.2475	0.3652
R.c				1.0000	0.6815	0.2249	0.3495
P.a					1.0000	0.2148	0.5407[b]
S.c						1.0000	0.1945
TrpI							1.0000

Sequence comparisons of the six different AmpR proteins (Distances program of the GCG package).
[a]C.f, *Citrobacter freundii* [29]; E.c, *Enterobacter cloacae* [58]; Y.e, *Yersinia enterocolitica* [57]; P.a, *Pseudomonas aeruginosa* [42]; R.c, *Rhodopseudomonas capsulata* [43]; S.c, *Streptomyces cacaoi* [44]; TrpI, *P. aeruginosa* [71].
[b]Only 135 aa in the *P. aeruginosa* AmpR protein sequenced thus far. The value is higher since the comparison concerns the more conserved amino terminal parts.

opposite orientation. The first open reading frame encodes a protein possibly belonging to the LysR family [44]. The homology with the AmpR proteins is, however, quite low (Table III). Interestingly, the β-lactamase gene *bla*A is followed by an open reading frame for a protein containing all motifs typical of a β-lactam-binding protein [44]. The precise role of this protein in β-lactamase regulation is not known.

Taken together, β-lactamase induction in *C. freundii, E. cloacae, Y. enterocolitica, P. aeruginosa, Rps. capsulata* and *S. cacaoi* seems to require an AmpR or AmpR-like protein. Moreover, the regulator seems more conserved in these organisms than the β-lactamase it regulates. The helix-turn-helix motifs are particularly conserved among the AmpR homologues suggesting that they may recognize similar operator sequences (Fig. 1). Table III gives the evolutionary distances for the respective AmpR homologues. These distances seem to reflect the overall genetic distances between the different species suggesting that they all, with the probable exception of the AmpR-like protein from *S. cacaoi*, have evolved as gene products from one ancestral gene.

5.3. AmpR is required for β-lactam-mediated induction as well as for ampD mutant mediated β-lactamase hyperproduction in C. freundii and in the heterologous E. coli system

In *C. freundii*, the *amp*R gene has been inactivated by a *kan* insertion resulting in a totally non-inducible mutant expressing the same low levels of AmpC β-lactamase as the uninduced, wild-type parent (in the absence of β-lactam inducer) (F. Lindberg, unpublished data). Inducibility is restored by transforming only *amp*R on a multicopy plasmid into such a mutant. In the heterologous *E. coli* system, *amp*C induction is also dependant on *amp*R. In *E. coli*, β-lactamase expression is twice the basal level in the absence of AmpR suggesting that AmpR might act as a repressor in the absence of inducer. As in the reconstituted *E. coli* system, AmpR expression in *C. freundii* is also required for the high level

```
                    1                                                 50
AmpRcf              MTRSYIPLNSLRAFEAAARHLSFTRAAIELNVTHSAISQHVKSLEQQLNC
AmpRec              MTRSYLPLNSLRAFEAAARHLSFTHAAIELNVTHSAISQHVKTLEQHLNC
AmpRye              MVRSYIPLNSLRAFEAAARQLSFTKAAIELNVTHAAISQQVKALEQRLNC
AmpRrc              MDRPDLPLNALRVFEVMMRQGSFTKAAIELRVTQAAVSHQVARLEDLLGT
AmpRpa              MVRPHLPLNALAAFEASARHLSFTRAAIELCVTQAAVSHQVKSLEERLGV
AmpR Identity*      M-R---PLN-L--FE---R--SFT-AAIEL-VT--A-S--VK-LE--L--
LysR consensus#     M------LR-L--F-----------AA--L---QP--S-Q---LE--LG-
                                             [--helix-turn-helix--]

                    51                                               100
AmpRcf              QLFVRGSRGLMLTTEGESLLPVLNDSFDRMAGMLDRFATKQTQEKLKIGV
AmpRec              QLFVRVSRGLMLTTEGENLLPVLNDSFDRIAGMLDRFANHRAQEKLKIGV
AmpRye              RLFIRISRGLVLTTEGENLLPILNDSFDRIADTLDRFSTGIIREKVRVGV
AmpRrc              ALFLRTSQGLIPTDEGRLLFPVLEHGFDAMSRVLDRLGGRRDIEVLKVGV
AmpRpa              ALFKRLPRGLMLTHEGESLLPVLCDSFDRIAGLLERFEGGHYRDVLTVGA
AmpR Identity       -LF-R-SRGL-LT-EGE-LLPVL-DSFDR-A--LDRF------E-L-VGV
LysR consensus      -LF-R--R----T--G-----------------------------L-I--

                    101                                              150
AmpRcf              VGTFAIGCLFPLLSDFKRSYPHIDLHISTHNNRVDPAAEGLDYTIRYGGG
AmpRec              VGTFATGVLFSQLEDFRRGYPHIDLQLSTHNNRVDPAAEGLDYTIRYGGG
AmpRye              VGTFATGYLLSRLRDFQQHSPHVDILLSTHNNRVDVVAEGLDYAIRYGNG
AmpRrc              NTTFAMCWLMPRLEAFRQAHPQIDLRISTNNNRVEILREGLDMAIRFGTG
AmpRpa              VGTFTVGWLLPRLEDFQARHPFIDLRLSTHNNRVD...............
AmpR Identity       --TF----L---L--F----P--D---ST-NNRV----EGLD--IR-G-G
LysR consensus      --------LP----------P-----L--------L-----D--------

                    151                                              200
AmpRcf              AWHDTDAQYLCSALMSPLCSPTLASQIQTPADILKFPLLRSYRRDEWALW
AmpRec              AWHGTEAEFLCHAPLAPLCTPDIAASLHSPADILRFTLLRSYRRDEWTAW
AmpRye              ALAWHESHFMYAPPLAQLCAPSISKRFTPPTDLQRFMLLGSYRAMNWSAW
AmpRrc              GWTGHDAIPLAEAPMAPLCAPGLASRLLHPSDLGQVTLLRSYRSAEWPGW
AmpRpa              ..................................................
AmpR Identity       -----------------LC-P--------P-D-----LL-SYR---W--W
LysR consensus      -------------------------------L------------------

                    201                                              250
AmpRcf              MQAAGEAPPSPTHNVMVFDSSVTMLEAAQAGMGVAIAPVRMFTHLLSSER
AmpRec              MQAAGEHPPSPTHRVMVFDSSVTMLEAAQAGVGIAIAPVDMFTHLLASER
AmpRye              FAAAGGSVPSPSQQIMMFDSSVSMLEAAQAEIGIALAPPAMFMHLLRSER
AmpRrc              FEAAG..VPCPPVTGPVFDSSVALAELATSGAGVALLPISMFESYIAQGR
AmpRpa              ..................................................
AmpR Identity       --AAG---P-P------FDSSV---E-A----G-A--P--MF-------R
LysR consensus      ---------------------------V--G-G----P------------

                    251                                         295
AmpRcf              IVQPFLTQIDLGSYWITRLQSRPETPAMREFSRWLTGVLHK...
AmpRec              IVQPFATQIELGSYWLTRLQSRAETPAMREFSRWLVEKMKK...
AmpRye              IIQPFSTTVSLGGYWLTRLQSRTETPAMRDFALWLLSEMKSEGE
AmpRrc              LAQPFGVTVSVGRYYLAWPSDRPATSAMSTFSRWLTGQSAE...
AmpRpa              .............................................
AmpR Identity       --QPF------G-Y-------R--T-AM--F--WL----------
LysR consensus      ---------------------------------------------
```

semi-constitutive expression of β-lactamase seen in *amp*D mutants. A *C. freundii amp*D, *amp*R double mutant expresses only low basal levels of AmpC β-lactamase, whereas the *amp*D single mutant expresses a 100-fold higher level of β-lactamase in the absence of an inducer. This can be further increased by a factor of two after β-lactam induction. Also, β-lactamase hyperproduction can be restored to such *C. freundii* double mutants by introducing the *amp*R gene on a plasmid (C. Jacobs et al., unpublished data). The AmpR dependence of the *amp*D mutant phenotype suggests that AmpD affects AmpC expression via AmpR.

5.4. Can more than one signal activate AmpR?

Recently, an AmpR mutant was generated in vitro, AmpR^{G102D} (Fig. 2), that when introduced into an *amp*D$^-$, *amp*R$^-$ double mutant of *C. freundii* resulted in an only 9-fold elevation of basal β-lactamase expression as compared to 100-fold for AmpR^{wt} (C. Jacobs et al., unpublished data). Interestingly, *C. freundii amp*D$^-$ expressing this AmpR mutant protein was fully inducible by β-lactams. The AmpR^{G102D} mutant phenotype was not specific for any particular *amp*D allele. Even in an *amp*D null background, AmpR^{G102D} provided an inducible phenotype. If one positively acting ligand is accumulating in an AmpD$^-$ background and a separate but similar ligand accumulates after β-lactam treatment, then the AmpR^{G102D} mutant phenotype could be explained by a lower ability to interact with the ligand provided in the AmpD$^-$ mutant while retaining the ability to interact with the β-lactam-induced ligand.

5.5. AmpR is kept in a 'locked' or repressed conformation in wild-type cells in the absence of β-lactam inducer

The AmpR^{G102D} mutant provides a higher basal level of β-lactamase compared to wild-type AmpR in the presence of AmpD (9-fold in *C. freundii* and 13-fold in an *E. coli* background). Since this AmpR^{G102D}-mediated activation of the system is independent of the AmpG protein, we would have to suggest that AmpR^{G102D} in the absence of positively acting ligands is conformationally altered such that it can activate *amp*C transcription.

An AmpR^{G102E} mutant (Fig. 2) was obtained by selecting for high β-lactam resistance in an *amp*G::*kan*^{R} *E. coli* background [39]. The AmpR^{G102E} protein mediates a 30-fold increase in basal β-lactamase expression. Inactivation of *amp*D has no significant effect on the expression level in either *C. freundii* or *E. coli*. Activation does not require AmpG

Fig.1. Alignment of the AmpR proteins sequenced thus far from Gram-negative bacteria. Only 135 amino acids of the *P. aeruginosa* AmpR have been sequenced. * AmpR identity represents homology between all 5 protein sequences until aa^{135} and then all 4 proteins sequence until the end. # LysR consensus is from [38]. The helix-turn-helix motif which is postulated to be the region of the protein that makes contact with the DNA is indicated. AmpRcf, *C. freundii*, AmpRec, *E. cloacae*, AmpRye, *Y. enterocolitica*, AmpRrc, *Rps. capsulata*, AmpRps, *P. aeruginosa*.

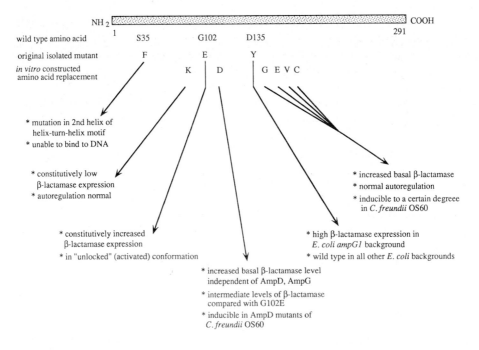

Fig. 2. Properties of *C. freundii* AmpR mutant proteins. Mutants of AmpR that were isolated by chemical mutagenesis [39] and in vitro mutagenesis [46] are shown.

and expression is not affected by β-lactams. Thus, a change from an aspartic acid to a glutamic acid residue at position 102 makes AmpR totally non-responsive to β-lactams. Our current hypothesis is that in wild-type cells in the absence of a β-lactam inducer, AmpR is kept in such a conformation that a domain involved in transcriptional activation is shielded or blocked. We interpret the AmpR[G102E] mutant as being 'unlocked' now exposing the region responsible for transcriptional activation. An analogous mutant phenotype in the related CysB protein has been interpreted in a similar manner [45]. Since AmpR[G102E] is not affected by β-lactam inducer or by *amp*D mutant-induced signalling, these signals might act to 'unlock' the AmpR[wt] protein by affecting a region in AmpR involving residue 102.

5.6. AmpR binds DNA and represses ampR transcription both in its repressed and activated state

DNase I footprinting with AmpR containing extracts shows that the protein protects a 38 bp region located immediately upstream of the *amp*C promoter [29]. The protected region covers the *amp*R promoter directing transcription in the opposite direction. Using an *amp*R::*lacZ* transcriptional fusion, it has been possible to show that AmpR in *trans* represses *amp*R transcription. The AmpR[G102D] and AmpR[G102E] mutants repressed *amp*R transcription to the same extent as the wild-type protein showing that these mutations at position 102 do not affect repression of the *amp*R promoter but specifically affect transcriptional activation of the *amp*C promoter. AmpR transcription is also unaffected by mutations in *amp*D, and *amp*G and does not respond to β-lactam inducers. Thus, AmpR must be able to bind to the *amp*R, *amp*C intercistronic region also in its repressed state.

5.7. AmpR^{G102K} has lost the ability to activate the ampC promoter while retaining its ability to repress ampR transcription

Since the glycine at amino acid position 102 in AmpR is seemingly located in a region of the protein affecting its activation state it was altered into a Lys residue by in vitro mutagenesis. The resulting AmpR^{G102K} mutant (Fig. 2) exhibited the same repression of *amp*R transcription showing that the mutated protein had not affected autoregulation. However, AmpR^{G102K} did not respond to β-lactam inducer and was not affected by an *amp*D mutant background either in *E. coli* or *C. freundii*. Interestingly, AmpR^{G102K} gives a faster migrating complex in gel mobility shifts with fragments carrying the entire *amp*R-*amp*C intercistronic region as compared to AmpR^{wt}, AmpR^{G102D} and AmpR^{G102E}, suggesting that its interaction with DNA was altered [46].

5.8. AmpR interaction with target DNA

The region protected from DNaseI cleavage is large enough (38 bp) to contain more than one operator binding site for AmpR. To test this hypothesis, fragments were synthesized carrying essentially each half of the 38 bp region (Fig. 3). A fragment carrying only the 5'

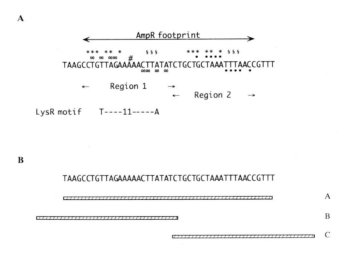

Fig. 3. The sequence for the 38 bp DNase I footprinted region of *C. freundii* (C.f) as well as 4 bp flanking on either side. The coding strand for *amp*C is shown. Regions 1 and 2 are indicated and, ∞ and • indicate palindromic sequences within each region. # is the axis of symmetry for the palindromic sequence in Region 1. * and § are repeated sequences from each half of the palindromic sequence in the two regions. Fragments A, B and C were used in gel shift mobility assays. A and B will bind AmpR whereas fragment C will not in gel retardation assays.

498

half (relative to the *ampC* promoter) could compete out AmpR binding to a fragment carrying the entire 38 bp region and was retarded by an AmpR-containing extract in a gel mobility shift. The 3' half of the AmpR-binding region on the other hand, did not compete out AmpR binding to a fragment carrying the entire binding region, and was not retarded by extracts containing AmpR[wt] [46]. Insertions of 2 bp, 5 bp or 10 bp into the center of the 38 bp region did not affect the ability of AmpR to bind, showing that it does not constitute a continuous binding region. Finally, two point mutations in the 5' region dramatically decreased AmpR binding whereas four mutations affecting the sequence of the 3' half of the 38 bp region did not. Therefore, the 38 bp region seems to contain one operator binding site within the 5' half of the 38 bp region (Region 1). It is not known how AmpR interacts with the 3' half of the 38 bp region (Region 2). One possibility is that AmpR binds to Region 1 via its helix-turn-helix motif and at this site interacts with Region 2 via some other part of AmpR. At present we cannot exclude the possibility that Region 2 contains a weak operator binding site that only can be saturated after full occupancy at the operator site in Region 1.

A comparison between the intercistronic regions in *C. freundii*, *E. cloacae* and *Y. enterocolitica* (Fig. 4) reveals two totally conserved stretches within the 38 bp region,

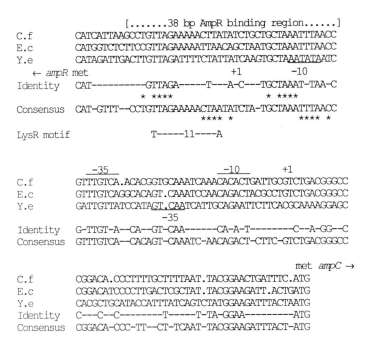

Fig. 4. Alignment of the intercistronic region between *ampR* and *ampC* from *C. freundii* (C.f), *E. cloacae* (E.c) and *Y. enterocolitica* (Y.e). Only the coding strand for *ampC* is indicated. The −10 and −35 as well as the transcriptional start site (+1) are indicated for both *ampR* and *ampC*. Identity represents total homology between the three sequences. Consensus is a 2 out of 3 match. The LysR binding motif is from [48]. * indicates palindromic sequences within the 38 bp AmpR binding region.

GTTAGA- and -TGCTAAAT-, separated by 15 nucleotides containing three conserved nucleotides. It has been shown that *C. freundii* AmpR can activate the *E. cloacae ampC* gene and vice versa suggesting that the respective AmpR protein can bind to the heterologous intercistronic region despite the sequence divergence of the central part [47]. Region 1 contains palindromic base pairs around an A/T rich core (Fig. 3). Moreover, it contains the T-N11-A motif found in the operator for most members of the LysR family [48]. Region 2 also contains palindromic base pairs but carries no LysR motif. Furthermore, Region 1 shows some sequence homology with Region 2 (Fig. 3).

5.9. Several in vitro generated mutations at position 135 in AmpR convert the regulator into a transcriptional activator for ampC

If an unlocking mechanism may expose an interactive surface on AmpR, one would expect several unrelated mutations in AmpR to open up this surface and convert AmpR into a transcriptional activator. In vitro mutagenesis at position 135 in AmpR has yielded five mutants (Fig. 2), four of which still bind DNA as evidenced by repression of *ampR* transcription. Interestingly, all four of these mutants produced greater expression of *ampC* than AmpR^{wt}. Thus, conservative as well as non-conservative replacements at either residue 102 or residue 135 may convert AmpR into a transcriptional activator for *ampC*. We propose that a similar unlocking of AmpR occurs in response to β-lactam inducers or in *ampD* mutant backgrounds. So far we have been unable to convert AmpR^{wt} to its 'unlocked' or activated state by addition of β-lactams and other putative ligands. However, if ligand binding is weak and the locked conformation energetically favourable, ligand loss during electrophoresis may result in AmpR converting to its 'locked' or repressed state.

5.10. AmpR binding to target DNA induces DNA bending

In contrast to many other transcriptional activators, members of the LysR family bind target DNA both in the absence and presence of activating ligand. One member of this family, OccR was recently shown to induce DNA bending [49]. Binding of octopine, the activating ligand, to OccR partially relaxed the bend of target DNA, which was associated with a reduced size of the DNaseI footprint. In contrast, the footprint for TrpI, was extended by adding the inducer indole glycerol phosphate [50]. In the latter study, it was not shown if TrpI in the presence of inducer affected DNA bending or if the activated form of TrpI could recognize a second operator binding sequence closer to the regulated promoter. It was recently shown that AmpR can induce DNA bending to target DNA [46]. The AmpR^{G102E} mutant mediating high constitutive *ampC* transcription induced DNA bending to the same extent as the wild-type protein. In contrast, AmpR^{G102K} mediating low constitutive *ampC* transcription caused no detectable bending of target DNA. AmpR^{G102K} as well as AmpR^{G102E} could bind to the operator sequence in Region 1. It may be that AmpR^{G102K} binds as a dimer to DNA whereas AmpR^{wt} and AmpR^{G102E} bind as a tetrameric complex. Such a difference could explain the faster migrating gel retardation complex with extracts containing AmpR^{G102K}. Perhaps DNA bending requires binding of an AmpR tetramer.

6. In search of the AmpR binding ligands

There are several arguments supporting an indirect role of the β-lactam inducer in AmpR activation. It has been demonstrated that β-lactamase expressed in the cytoplasm does not prevent *ampC* induction whereas the same enzyme directed towards the periplasm may block induction completely by efficiently hydrolyzing the β-lactam antibiotic [51]. Also, cytoplasmic β-lactamase cannot measurably provide any increased β-lactam resistance to the cell, suggesting that β-lactams cannot enter the bacterial cytoplasm [52]. The fact that AmpR can be activated in the absence of β-lactams in *ampD* mutants has also been taken as an evidence for an endogenous ligand activating AmpR.

AmpC induction can occur in non-growing but metabolically active *E. coli* minicells [53]. Moreover, such minicells retain inducibility despite pretreatment with high concentrations of cycloserine (F. Lindberg, unpublished data). Therefore, an active peptidoglycan biosynthesis is seemingly not a requirement for β-lactam induction. Since β-lactams can induce transcription from an *ampC::lacZ* transcriptional fusion in an *E. coli* background expressing only very low levels of chromosomal β-lactamase, it may be concluded that β-lactam hydrolysis is not required for induction. In fact, β-lactam hydrolyzed by β-lactamase is inactive as an inducer. The intact β-lactam ring is therefore a requirement for inducer activity.

Certain β-lactams such as imipenem and cefoxitin are excellent inducers whereas for example, cefazolin, moxalactam and aztreonam are not. These differences may depend either on different binding affinities to certain PBPs or to different abilities to enter the cytosol and activate AmpR. In general, the most potent inducers have a high affinity for the low molecular weight PBPs [54]. However, if inhibition of carboxypeptidase activity in the cell is generating an endogenous signal, then one would expect mutants in PBP4, PBP5 and PBP6 to behave as an *ampD* mutant, i.e. overproduce the β-lactamase in the absence of inducer. This is, however, not the case. It has not been possible to activate β-lactamase expression by any available PBP mutation. Thus, β-lactam induction does not appear to be associated with the inhibition of a particular PBP.

Gram-negative bacteria such as *Escherichia coli* normally recycle 40–50% of the peptidoglycan each generation [72,73]. It has been shown that the muramyl tripeptide, L-Ala-D-Glu-Dap is allowed to enter the cell via the oligopeptide transport system [74]. Mutants defective in the *opp* system are still inducible by β-lactams (unpublished). Recently, Park [73] presented evidence for a separate low affinity uptake system for muramyl peptides. It may be that AmpG constitutes that system. If so, the activating ligands for AmpR may be muramyl peptides derived from the peptidoglycan.

7. Model for chromosomal β-lactamase induction in enterobacteria

Based on the considerations given above, we propose the following model for β-lactamase induction (Fig. 5). The AmpR regulator expressed in wild-type cells is kept in a 'locked' conformation such that a region of the protein involved in dimer-dimer and possibly AmpR–RNA polymerase interaction is buried by a region of the protein encompassing amino acid residues 102 and 135. In the 'locked' conformation, AmpR binds the op-

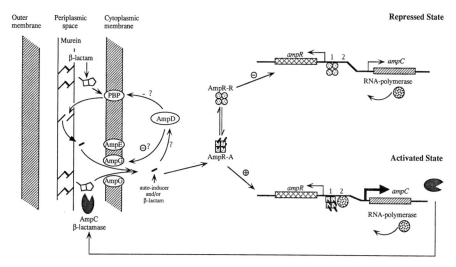

Fig. 5. Model for β-lactamase induction in Gram-negative bacteria. Explanation is presented in the text.

erator. This interaction induces a bend to target DNA preventing RNA polymerase from binding to the *amp*C promoter. This interaction also leads to a repression of *amp*R transcription. This 'locked' or repressed conformation of AmpR can be opened up by one or more auto-inducer; elicited by β-lactam action or constitutively present in AmpD-mutant cells. These ligands may be degradation products from the peptidoglycan and require the transmembrane AmpG protein to enter the cell and interact with AmpR. AmpG therefore most likely acts as a permease for AmpR-binding ligands. The 'unlocking' of AmpR is believed to expose an interactive surface allowing an altered interaction with target DNA such as a relaxation of the induced DNA bend allowing for a productive interaction with RNA-polymerase at the *amp*C promoter. Clearly, this model can only be proved or disproved by the identification of AmpR-binding ligands and in vitro demonstration that ligand binding causes a conformational change in AmpR allowing productive interaction of RNA polymerase with the *amp*C β-lactamase promoter.

References

1. Livermore, D.M. (1987) Eur. J. Clin. Microbiol. 6, 439–445.
2. Schaberg, D.R., Culver, D.H. and Gaynes, R.P. (1991) Am. J. Med. 91 (Suppl. 3B), 72S–75S.
3. Haley, R.W., Culver, D.H., White, J.W., Morgan, W. M. and Emori, T.G. (1985) Am. J. Epidemiol. 121, 159–167.
4. Sanders, W.E. and Sanders, C.C. (1988) Rev. Infect. Dis. 10, 830–838.
5. Dworzack, D.L., Pugsley, M.P., Sanders, C.C. and Horowitz, E.A. (1987) Eur. J. Clin. Microbiol. 6, 456–459.
6. Follath, F., Costa, E., Thommen, A., Frei, R., Burdeska, A. and Meyer, J. (1987) Eur. J. Microbiol. 6, 446–450.
7. Bryan, C.S., John, J.F., Pai, M.S. and Austin, T.L. (1985) Am. J. Dis. Child. 139, 1086–1089.
8. Chow, J.W., Fine, M.J., Shlaes, D.M., Quinn, J.P., Hooper, D.C., Johnson, M.P., Ramphal, R., Wagener, M., Miyashiro, D.K. and Yu, V.L. (1991) Ann. Intern. Med. 115, 585–651.

502

9. Bergström, S., Olsson, O. and Normark, S. (1982) J. Bacteriol. 150, 528–534.
10. Jaurin, B. and Grundström, T. (1981) Proc. Natl. Acad. Sci. USA 78, 4897–4901.
11. Jaurin, B., Grundström, T., Edlund, T. and Normark, S. (1981) Nature 290, 221–225.
12. Olsson, O., Bergström, S. and Normark, S. (1982) EMBO J. 11, 1411–1416.
13. Normark, S., Edlund, T., Grundström, T., Bergström, S. and Wolf-Watz, H. (1977) J. Bacteriol. 132, 912–922.
14. Edlund, T. and Normark, S. (1981) Nature 292, 269–271.
15. Jaurin, B., Grundström, T. and Normark, S. (1982) EMBO J. 7, 875–881.
16. Jaurin, B. and Normark, S. (1983) Cell 32, 809–816.
17. Normark, S., Bergström, S., Edlund, T., Grundström, T., Jaurin, B., Lindberg, F.P. and Olsson, O. (1983) Ann. Rev. Genet. 17, 499–525.
18. Olsson, O., Bergström, S., Lindberg, F.P. and Normark, S. (1983) Proc. Natl. Acad. Sci. USA 80, 7556–7560.
19. Lindberg, F, Westman, L. and Normark, S. (1985) Proc. Natl. Acad. Sci. USA 82, 4620–4624.
20. Normark, S., Lindquist, S. and Lindberg, F. (1986) J. Infect. Dis. 49, 38–45.
21. Lindberg, F., Lindquist, S. and Normark, S. (1986) J. Antimicrob. Chem. 18 (Suppl C), 43–50.
22. Lindquist, S., Galleni, M., Lindberg, F. and Normark, S. (1989) Mol. Microbiol. 3, 1091–1102.
23. Kopp, U., Wiedemann, B., Lindquist, S. and Normark, S. (1993) Antimicrob. Agents Chemother. 37, 224–228.
24. Korfmann, G. and Sanders, C.C. (1989) Antimicrob. Agents Chemother. 33, 1946–1951.
25. Lindquist, S., Weston-Hafer, K., Schmidt, H., Pul, C., Korfmann, G., Erickson, J., Sanders, C., Martin, H.H. and Normark, S. (1993) Mol. Microbiol. 9, 703–715.
26. Oliva, B., Bennett, P.M. and Chopra, I. (1989) Antimicrob. Agents Chemother. 33, 1116–1117.
27. Ottolenghi, A.C. and Ayala, J.A. (1991) Antimicrob. Agents Chemother. 35, 2359–2365.
28. Honoré, N., Nicholas, M.-H. and Cole, S.T. (1989) Mol. Microbiol. 3, 1121–1130.
29. Lindquist, S., Lindberg, F. and Normark, S. (1989) J. Bacteriol. 171, 3746–3753.
30. Tuomanen, E., Lindquist, S., Sande, S., Galleni, M., Light, K., Gage, D. and Normark, S. (1991) Science 251, 201–204.
31. Aldea, M. Garrido, T., Herández-Chico, C., Vicente, M. and Kushner, S.R. (1989) EMBO J. 8, 3923–3931.
32. Lange, R. and Hengge-Aronis, R. (1991) J. Bacteriol. 173, 4474–4481.
33. Joris, B., Ghuysen, J.-M., Dive, G., Renard, A., Dideberg, O., Charlier, P., Frére, J.-M., Kelly, J.A., Boyington, J.C., Moews, P.C. and Knox, J.R. (1988) Biochem. J. 250, 313–324.
34. Kobayashi, T., Zhu, Y.F., Nicholls, N.J. and Lampen, J.O. (1987) J. Bacteriol. 169, 3873–3878.
35. Ronson, C.W., Nixon, B.T. and Ausubel, F.M. (1987) Cell 49, 579–581.
36. Maskell, D.J., Szabo, M.J., Butler, P.D., Williams, A.E. and Moxon, E.R. (1991) Mol. Microbiol. 5, 1013–1022.
37. Henikoff, S., Haughn, G.W., Calvo, J.M. and Wallace, J.C. (1988) Proc. Natl. Acad. Sci. USA 85, 6602–6606.
38. Viale, A.M., Kobayashi, H., Akazawa, T. and Henikoff, S. (1991) J. Bacteriol. 173, 5224–5229.
39. Bartowsky, E. and Normark, S. (1991) Mol. Microbiol. 5, 1715–1725.
40. Hryniewicz, M.H. and Kredich, N.M. (1991) J. Bacteriol. 173, 5876–5886.
41. Storz, G., Tartaglis, L.A. and Ames, B.N. (1990) Science 248, 189–194.
42. Lodge, J.M., Minchin, S.D., Piddock, L.J.V. and Busby, S.J.W. (1990) Biochem. J. 272, 627–631.
43. Campbell, J.I.A., Scahill, S., Gibson, T. and Ambler, R.P. (1989) Biochem. J. 260, 803–812.
44. Urabe, H. and Ogawara, H. (1992) J. Bacteriol. 174, 2834–2842.
45. Kredich, N.M. (1992) Mol. Microbiol. 6, 2747–2753.
46. Bartowsky, E. and Normark, S. (1993) Mol. Microbiol. 10, in press.
47. Lindberg, F. and Normark, S. (1987) J. Bacteriol. 169, 758–763.
48. Goethals, K., van Montagu, M. and Holsters, M. (1992) Proc. Natl. Acad. Sci. USA 89, 1646–1650.
49. Wang, L., Helmann, J.D. and Winans, S.C. (1992) Cell 69, 659–667.
50. Chang, M. and Crawford, I.P. (1989) Nucleic Acids Res. 18, 979–988.
51. Everett, M.J., Chopra, I. and Bennett, P.M. (1990) Antimicrob. Agents Chemother. 34, 2429–2430.
52. Broome-Smith, J.K. and Spratt, B.G. (1986) Gene 49, 341–349.

53. Lindquist, S. (1992) Thesis, Umeå University, Umeå, Sweden.
54. Martin, H.H., Schmidt, B., Bräutigam, S., Noguchi, H. and Matsuhashi, M. (1988) in: P. Actor et al. (Eds.), Antibiotic Inhibition of Bacterial Cell Surface Assembly and Function, American Society for Microbiology, Washington, DC, pp. 494–501.
55. Lindberg, F. and Normark, S. (1986) Eur. J. Biochem. 156, 441–445.
56. Galleni, M., Lindberg, F., Normark, S., Cole, S., Honore, N., Joris, B. and Frere, J-M (1988) Biochem. J. 250, 753–760.
57. Seoane, A., Franca, M.V. and Lobo, J.M.G. (1992) Antimicrob. Agents Chemother. 36, 1049–1052.
58. Honoré, N., Nicholas, M.H. and Cole, S.T. (1986) EMBO J. 5, 3709–3714.
59. Rothmel, R.K., Aldrich, T.L., Houghton, J.E., Coco, W.M., Ornston, L.N. and Chakrarty, A.M. (1990) J. Bacteriol. 172, 922–931.
60. Ostrowski, J., Jagura-Burdzy, G. and Kredich, N.M. (1987) J. Biol. Chem. 262, 5999–6005.
61. Wek, R.C. and Hatfield, G.W. (1986) J. Biol. Chem. 261, 2441–2450.
62. Stragier, P. and Patte, J.-C. (1983) J. Mol. Biol. 168, 333–350.
63. Maxon, M.E., Wigboldus, J., Brot, N. and Weissbach, H. (1990) Proc. Natl. Acad. Sci. USA 87, 7076–7079.
64. Plamann, L.S. and Stauffer, G.V. (1987) J. Bacteriol. 169, 3932–3937.
65. Renault, P., Gaillardin, C. and Heslot, H. (1989) J. Bacteriol. 171, 3108–3114.
66. Dammann, T. and Wohleben, W. (1992) Mol. Microbiol. 6, 2267–2278.
67. Schell, M.A. and Sukordhaman, M. (1989) J. Bacteriol. 171, 1952–1959.
68. Rossen, L., Sherman, C.A., Johnston, A.W.B. and Downie, J.A. (1985) EMBO J. 4, 3369–3373.
69. Egelhoff, T.T., Fisher, R. T., Jacobs, R. T., Mulligan, J.T. and Long, S.R. (1985) DNA 4, 241–248.
70. Christman, M.F., Storz, G. and Ames, B.N. (1989) Proc. Natl. Acad. Sci. USA 86, 3484–3488.
71. Chang, M., Hadero, A. and Crawford, I.P. (1989) J. Bacteriol. 171, 172–183.
72. Goodell, E.W. (1985) J. Bacteriol. 163, 305–310.
73. Park, J.T. (1993) J. Bacteriol. 175, 7–11.
74. Goodell, E.W. and Higgins, C.F. (1987) J. Bacteriol. 169, 3861–3865.

J.-M Ghuysen and R. Hakenbeck (Eds.), *Bacterial Cell Wall*
© 1994 Elsevier Science B.V. All rights reserved

505

Induction of β-lactamase and low-affinity penicillin binding protein 2' synthesis in Gram-positive bacteria

BERNARD JORIS, KARIN HARDT and JEAN-MARIE GHUYSEN

*Centre d'Ingénierie des Protéines, Université de Liège, Institut de Chimie, B6,
B-4000 Sart Tilman (Liège 1), Belgium*

1. Introduction

Resistance to β-lactam antibiotics can be β-lactamase- and/or penicillin-binding protein (PBP)-mediated. PBP-mediated resistance in methicillin-resistant *Staphylococcus aureus* and penicillin-resistant *Enterococcus hirae* strains is considered to occur by acquisition of an additional PBP which has a low affinity for the drug and, apparently, can take over the functions required for wall peptidoglycan synthesis under conditions where the other PBPs are inactivated by a β-lactam antibiotic (see Chapter 25).

The aim of this chapter is to present a review on the molecular mechanisms by which a β-lactam compound can induce β-lactamase synthesis in *Bacillus licheniformis* and both β-lactamase and low-affinity PBP2' synthesis in *Staphylococcus aureus*. The mechanisms differ markedly from those responsible for chromosomal β-lactamase induction in Gram-negative bacteria (see Chapter 23). The mechanisms involve specialized repressors and sensory-transducers.

2. The regulons

The class A β-lactamases BlaP of *B. licheniformis* and BlaZ of *S. aureus* have high similarity in both their primary and three-dimensional structures (see Chapter 6). In *B. licheniformis*, the β-lactamase-encoding *bla*P is chromosomal [1]. In *S. aureus*, the β-lactamase-encoding *bla*Z is either chromosomal, in which case it is part of the transposon Tn552 [2,3], or plasmid-borne, in which case the left-hand half of the transposon sequence has been lost [4]. In both *B. licheniformis* and *S. aureus*, β-lactamase expression is under the control of two open reading frames: *bla*I encodes the repressor BlaI and *bla*R1 encodes the penicillin-sensory transducer BlaR [3–7]. In the absence of a β-lactam, β-lactamase synthesis is maintained at a low basal level by the repressor BlaI. Binding of a β-lactam (e.g. the inducer) to the sensory-transducer BlaR causes derepression and a high level of β-lactamase synthesis [8–12].

The low affinity PBP2' of S. aureus is a multi-domain protein that is anchored in the plasma membrane at the N-terminus of the polypeptide chain (see Chapter 6). The PBP2'-encoding *mec*A is found in the same chromosomal location in all the methicillin-resistant staphylococcal strains [13–16]. Regulation of PBP2' expression can be of two types. Transcription of *mec*A may be under the control of *mec*I and *mec*R whose encoded proteins, the repressor MecI and the penicillin-sensory transducer MecR, are homologous to the corresponding β-lactamase regulators BlaI and BlaR, respectively [15,17,18]. Alternatively, if *mec*I and *mec*R are altered or absent, transcription of *mec*A is dependent on the presence of a penicillinase plasmid and controlled by the plasmid-borne regulators BlaI and BlaR of β-lactamase production [19–21].

The three regulons are shown schematically in Fig. 1. They have common features. The genes encoding the regulators BlaI, BlaR, MecI and MecR are transcribed in the direction opposite to the genes encoding the proteins responsible for β-lactam resistance, BlaP, BlaZ and PBP2'. The pair *bla*I-*bla*R1 and the pair *mec*I-*mec*R are transcribed as polycistronic mRNAs and the expression of the regulators is autoregulated [3,10,18]. The three regulons have also their own features. In *B. licheniformis*, *bla*I is located upstream from *bla*R1 (the two ORFs are separated by a single base pair). In *S. aureus*, *bla*I and *mec*I are located downstream from and overlap *bla*R1 and *mec*R (11 and 1 base pairs, respectively). Repression of *bla*P in *B. licheniformis* is more strict than that of *bla*I-*bla*R1 because of the presence of two repressor binding sequences for *bla*P [5,9,11,12].

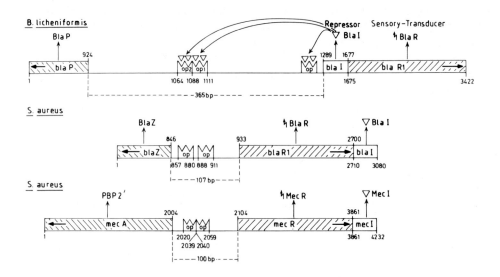

Fig. 1. Schematic representation of the loci involved in the regulation of the synthesis of the β-lactamase BlaP of *B. licheniformis*, the β-lactamase BlaZ of *S. aureus* and the penicillin-binding protein (PBP)2' of *S. aureus*. Op, operator. The arrows indicate the direction of the transcription. The scale of the intergenic regions is 10-fold that of the structural genes.

3. The operators-repressors

Regulation of β-lactamase and PBP2' synthesis is at the transcriptional level. Many bacterial repressors are dimeric molecules that recognize specific palindromic sequences of the DNA helix [22,23]. The significance of the palindromic nature of these sequences is that the two DNA binding sites are related by an approximate twofold symmetry axis and that each subunit of the dimeric repressor has a helix-turn-helix (HTH) motif whose second (recognition) helix interacts at the major groove of the DNA in its B-form. The two recognition helices of the dimeric repressor are separated by a distance corresponding to one turn of the B-DNA. When one recognition helix binds to the DNA major groove, the second recognition helix also binds to the major groove one helical turn along the DNA [24].

The nucleotide sequence alignments shown in Fig. 2 highlight the approximate twofold symmetry axis of the seven operators under comparison. The figure also identifies the nucleotides of the *B. licheniformis* operators that the repressor protects against hydroxyl radical attack and/or methylation [12]. Such a nucleotide distribution on the DNA strands is expected to occur if, indeed, each subunit of the dimeric repressor binds to half of the DNA binding site on the same face of the B-DNA helix. As derived from Fig. 2, the seven operators have a high degree of similarity. The sequences that overlap the symmetry axis have seven conserved base pairs for ten aligned base pairs.

The HTH motif of the bacterial transcription regulation proteins is found embedded in domains of remarkably varied structures [23]. The *Bacillus* and staphylococcal repressors are ~125 amino acid residues long. Based on amino acid alignments (not shown), the *S. aureus* BlaI and MecI repressors are 60% identical when compared to each other and 31–41% identical when compared to the *Bacillus* BlaI repressor. Each repressor possesses a conserved 20 amino acid polypeptide stretch (Fig. 3) which bears the signature of the HTH motif of the *Escherichia coli* lactose operon repressor (LacI) family [25,26] except that a glutamine or serine residue replaces the conserved alanine residue of helix 1. As also shown in the bottom part of Fig. 3, the distribution of the hydrophobic residues along the amino acid sequences is such that helix 2 of the HTH motif of the *B. licheniformis* and *S. aureus* repressors has the characteristic features of an amphipathic helix, as found in the repressors of the LacI family. However, putative domains involved in dimerization or cofactor binding are not detected.

4. The sensory-transducers

β-Lactam antibiotics do not penetrate through the bacterial plasma membrane [27] and, therefore, cannot interact directly with the repressor to switch on gene expression. Specific inducibility requires a transducer specialized in the transmission of a chemical signal produced by the presence of a β-lactam in the environment, to the interior of the cell.

Transducers contain an extracellular domain located on the outer face of the membrane which is responsible for signal reception, and a cytosolic domain which is responsible for the generation of an intracellular signal. Ligands either bind directly to the extra-

508

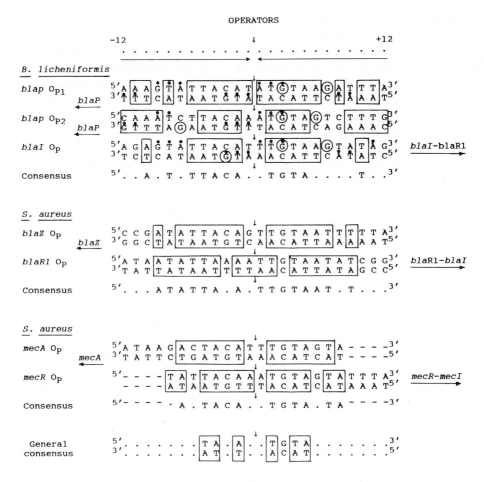

Fig. 2. Nucleotide sequences of the operator regions recognized by the BlaI and MecI repressors. Boxes, bases involved as palindromic sequences and symmetry dyad; filled circles, nucleotides whose sugar is protected by the repressor from hydroxyl radical attack; circled G, guanine residues protected by the repressor from methylation.

cellular domain of the transducer or, alternatively, they bind first to soluble binding proteins (located in the periplasm of Gram-negative bacteria) and the ligand; ligand-binding protein then binds to the extracellular domain of the transducer [28]. The transducer BlaR of *B. licheniformis* (Fig. 4) and the transducers BlaR and MecR of *S. aureus* have a conventional membrane topology, except that the penicillin (ligand)-binding protein, i.e. the sensor, is fused to the transducer by means of an additional transmembrane segment.

The membrane topology of the penicillin sensory-transducer BlaR (Fig. 4) involved in the specific inducibility of β-lactamase synthesis in *B. licheniformis* has been established by random gene fusion experiments [29] with a reporter β-lactamase gene (unpublished data). Among the five polypeptide stretches la, 1, 2, 3a and 3 that exhibit hydrophobic potential, stretches 1, 2 and 3 are transmembrane segments defining two intracellular domains A and C and two extracellular domains B and D (Fig. 4). The intracellular domain A and extracellular domain B are relatively short polypeptide stretches, 40–50 amino acid

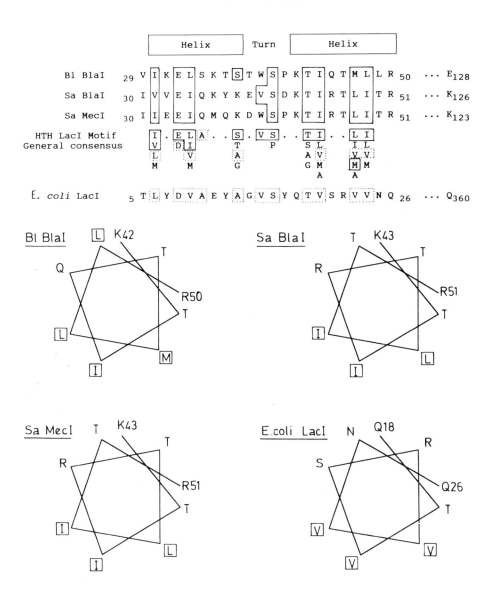

Fig. 3. Amino acid alignments of the putative helix-turn-helix (HTH) motifs of the repressors BlaI of *B. licheniformis* (B.l.) and BlaI and MecI of *S. aureus* (S.a.), and schematic representation of the amphipathic helix 2. The comparison is made by reference to the HTH motif consensus of the repressors of the LacI family. The hydrophobic residues of the amphipathic helix 2 are boxed.

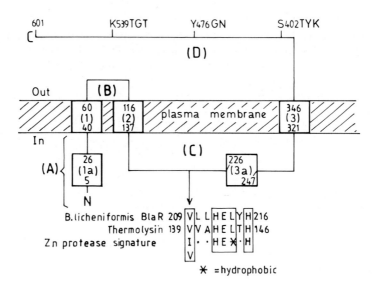

Fig. 4. Membrane topology of the sensory-transducer BlaR of *B. licheniformis*. Hydrophobic segments are numbered 1a, 1, 2, 3a, 3. Domains A, B, C, D are drawn to approximately the same scale. The Zn peptidase motif of domain C is indicated. The amino acid groupings S402TYK, Y476GN and K539TGT of domain D, characteristic of the penicilloyl serine transferases, are also indicated.

residues long. The intracellular domain C and extracellular domain D are much larger polypeptide stretches, approximately 180 and 250 amino acid residues long, respectively.

The extracellular domain D bears the motifs characteristic of the penicilloyl serine transferases [30] (Fig. 4). It has been overexpressed independently from the rest of the protein in the periplasm of *E. coli* [31]. Although it has similarity, in primary structure, with the β-lactamases of class D [32], the isolated domain D is a high-affinity penicillin-binding protein and the adduct formed by reaction with penicillin is not dissociated by hot SDS [31]. Based on these properties, it is proposed that the extracellular domain D is the penicillin sensor of BlaR; the transducer consists of three domains A, B, C and two transmembrane segments 1 and 2; and the transducer is fused to the sensor via the transmembrane segment 3.

The intracellular domains A and C of BlaR have no site of methylation/demethylation [31]. They also have no site that could be associated with that of a histidine kinase and to which generation of an intracellular signal via phosphorylation/dephosphorylation reactions might be attributed [33]. However, the intracellular domain C bears, in its amino acid sequence, the amino acid grouping V209$\cdot\cdot$HEL$\cdot$H216 which is the signature of a neutral zinc metallopeptidase [34,35] (Fig. 4). In thermolysin, the model of the zinc peptidases, H142 and H146 serve as zinc ligands and E143 acts as the required nucleophile [34]. The presence of a HE(X2)H motif is sufficient to identify a member of this family of peptidases [26]. Note also that domain C of BlaR has two glutamic residues at positions 245 and 266 (not shown). One of them might be equivalent to E166, the third zinc ligand of thermolysin.

The sensory transducers BlaR of *B. licheniformis*, BlaR of *S. aureus* and MecR of *S. aureus* have very similar molecular organization (Figs. 5 and 6). Based on the optimal amino acid alignments shown in Fig. 5, their hydrophobicity profiles are virtually super-imposable (Fig. 6, upper part). Moreover, the zinc-peptidase motif of domain C and the penicilloyl serine transferase motifs of domain D occur at equivalent places along the amino acid sequences (Figs. 5 and 6). This high similarity in structural design is con-served in spite of considerable variations in the extents of identity along the sequences (Fig. 6, lower part). The percentages of highly conserved amino acid residues (Fig. 5) in the transducers are 21 for the pair *B. subtilis* BlaR–*S. aureus* BlaR, 23 for the pair *B. subtilis* BlaR–*S. aureus* MecR and 30 for the pair *S. aureus* BlaR–*S. aureus* MecR. The corresponding values for the sensors are 33, 40 and 44, respectively.

Among the regions of the transducers that have significant identity scores (Fig. 6, lower part), one is the transmembrane segment 1 and the others are several peptide seg-ments of domain C, among which the zinc-peptidase motif-bearing peptide has the highest score. Domain A, domain B and transmembrane segments 2 and 3 lack similarity in their amino acid sequences. Hence, similarity between the transmembrane segment 2 and the second transmembrane segment of the Tsr and Tar transducers involved in chemotaxis in *E. coli* and *Salmonella typhimurium* [31] is restricted to *B. licheniformis* BlaR. The G124 →D mutation affecting transmembrane 2 causes loss of inducibility of β-lactamase syn-thesis in *B. licheniformis* [37].

Among the regions of the sensors that have high identity scores (Fig. 6, lower part), three possess the STYK, YGN and KTGT motifs of the penicilloyl serine transferases. The active-site serine of the tetrad STYK is at the amino end of a segment of high hydro-phobicity. In the penicilloyl serine transferases of known three-dimensional structure, this serine residue is at the amino end of an α-helix that is buried in the protein structure (see Chapter 6).

5. Unanswered questions and future prospects

The central question is: upon receipt of the message generated by penicillin binding, how does a presumed conformational change of the sensors lead to derepression of β-lac-tamase and PBP2' synthesis? At this time, a definite answer cannot be offered but a test-able hypothesis can be proposed which rests upon the observation that intracellular do-main C of the transducers bears the signature of a neutral metallopeptidase. According to this view, the 'peptidase' site of the transducers would be in an inactive form in its resting state, e.g. in the absence of inducer. Upon binding to the sensor, penicillin would act as a trigger setting the peptidase in motion. Peptidase activity might then lead to derepression either by direct proteolysis of the repressor or by proteolysis of a cytosolic compound with generation of an antirepressor.

Microbiologists also seek answers to questions regarding the mechanisms through which resistance via PBP2' synthesis is expressed homogeneously or heterogeneously in *S. aureus* strains (see Chapter 25); the role(s) played by *bla*R2 in the regulation of β-lactamase synthesis in *B. licheniformis* [38] and *S. aureus* [39]; and the marked differ-ences observed in the kinetics of inducibility of β-lactamase and PBP2' synthesis.

Transducer ⟶

```
                 1                            1a                                   1              60
Bl BlaR1   MSSSFFIPFL VSQILLSLFF SIIILIKKLL RTQITVGTHY YISVISLLAL IAPFIPFHFL  60
Sa BlaR1   .MAKLLIMSI VS...FCFIF LLLLFFRYIL KRYFNYMLNY KVWYLTLLAG LIPFIPI.KF  55
Sa MecR    VLSSFLMLSI IS...SLLTI CVIFLVRMLY IKYTQNIMSH KIWLLVLVST LIPLIPFYKI  57
Consensus  ----L----- -S-------- ---------- ---------- ------L--- --P-IP-J--

                 61                                          120
Bl BlaR1   KSHHFDWIL. ..NLGGAQSA LSQTHSTDKT TEAIGQHVNW VQDFSLSIEQ SSSKMIDSAF 117
Sa BlaR1   SLFKFNNV.. .NNQAPTVES KSHDLN.HNI NTT.....KP IQEFATDIHK FNWDSIDNIC 106
Sa MecR    SNFTFSKDMM NRNVSDTTSS VSHMLDGQQS SVT....... .KDLAINVNQ FETSNITYMI 109
Consensus  ----F----- --N------- -S-------- ---------- ---------- -----I----

                 121        2                                          180
Bl BlaR1   FAVWILGVAV MLLATLYSNL KIGKIKK... .....NLQIV NNKELLSLFH TCKEEIRFHQ 169
Sa BlaR1   TVIWIVLVII LSFKFLKALL YLKYLKKQSL ...YLNENEK NKIDTILFNH QYKKNI.... 159
Sa MecR    LLIWVFGSLL CLFYMIKAFR QIDVIKSSSL ESSYLNERLK VCQSKMQF.. .YKKHI.... 162
Consensus  ---W------ ---------L -----K---- -----N---- ---------- --K--I----

                 181                                         240
Bl BlaR1   KVILSRSPLI KSPITFGVIR PYIILPKD.I SMFSADEMKC VLLHELYHCK RKDMLINYFL 228
Sa BlaR1   ..VIRKAETI QSPITFWYGK YIILIPSSYF KSVIDKRLKY IILHEYAHAK NRDTLHLIIF 217
Sa MecR    ..TISYSSNI DNPMVFGLVK SQIVLPTVVV ETMNDKEIEY IILHELSHVK SHDLIFNQLY 220
Consensus  ---------I --P--F---- --I--P---- ---------- --LHE--H-K --D--N----

                 241  3a                                     300
Bl BlaR1   CLLKIVVWFN PLVWYLSKEA KTEMEISCDF AVLKTLDKKL HLKYGEVILK FTSIK.QRTS 287
Sa BlaR1   NIFSIIMSYN PLVHIVKRKI IHDNEVEADR FVLNNINKNE FKTYAESIMD .SVLNVPFFN 276
Sa MecR    VVFKMIFWFN PALYISKTMM DNDCEKVCDR NVLKILNRHE HIRYGESILK CSILKSQHIN 280
Consensus  ---------N P--------- ----E---D- -VL------- ---Y-E-I-- ----------

                 301                                    3     354
Bl BlaR1   SLLAASEFSS SYKHIKRRIV TVVNFQTASP LLKAKSALVF TLVLGAILAG TPSV  341
Sa BlaR1   KNILSHSFNG KKSLLKRRLI NIKE.ANLKK QSKLILIFIC IFTFLLMVIQ SQFL  329
Sa MecR    .NVAAQYLLG FNSNIKERVK YIALYDSMPK PNRNKRIVAY IVCSISLLIQ APLL  333
Consensus  ---------- -----K-R-- ---------- ---------- ---------- ----
```

355 Sensor ⟶

```
                                                                 414
Bl BlaR1   SILAMQKETR FLPGTNVEYE DYSTFFDKFS A.SGGFVLFN SNRKKYTIYN RKESTSRFAP 400
Sa BlaR1   MGQSIT.DYN YKKPLHNDYQ ILDK.SKIFG SNSGSFVMYS MKKDKYYIYN EKESRKRYSP 387
Sa MecR    SAHVQQ.D.. .KYETNVSYK KLNQLAPYFK GFDGSFVLYN EREQAYSIYN EPESKQRYSP 389
Consensus  ---------- --------Y- --------F- ---G-FV--- -----Y-IYN --ES--R--P

                 415                                    474
Bl BlaR1   ASTYKVFSAL LALESGIITK NDSHMTWDGT QYPYKEWNQD QDLFSAMSSS TTWYFQKLDR 460
Sa BlaR1   NSTYKIYLAM FGLDRHIIND ENSRMSWNHK HYPFDAWNKE QDLNTAMQNS VNWYFERISD 447
Sa MecR    NSTYKIYLAL MAFDQNLLSL NHTEQQWDKH QYPFKEWNQD QNLNSSMKYS VNWYYENLNK 449
Consensus  -STYK---A- ---------- ------W--- -YP---WN-- Q-L---M--S --WY------

                 475                                    534
Bl BlaR1   QIGEDHLRHY LKSIHYGNED FSVPADYWLD GSLQISPLEQ VNILKKFYDN EFDFKQSNIE 520
Sa BlaR1   QIPKNYTATQ LKQLNYGNKN LGSYKSYWME DSLKISNLEQ VIVFKNMMEQ NNHFSKKAKN 507
Sa MecR    HLRQDEVKSY LDLIEYGNEE ISGNENYWNE SSLKISAIEQ VNLLKNMKQH NMHFDNKAIE 509
Consensus  ---------- L----YGN-- ------YW-- -SL-IS--EQ V---K----- ---F------

                 535                                    594
Bl BlaR1   TVKDSIRLEE SNGRVLSGKT GTSVINGELH AGWFIGYVET ADNTFFFAVH IQGEKRAAGS 580
Sa BlaR1   QLSSSLLIKK NEKYELYGKT GTGIVNGKYN NGWFVGYVIT NHDKYYFATH L.SDGKPSGK 566
Sa MecR    KVENSMTLKQ KDTYKYVGKT GTGIVNHKEA NGWFVGYVET KDNTYYFATH LKGEDNANGE 569
Consensus  ----S----- ------GKT GT---N---- -GWF-GYV-T ------FA-H --------G-

                 595        615
Bl BlaR1   SAAEIALSIL DKKGIYPSVS R 601
Sa BlaR1   NAELISEKIL KEMGVLNGQ. . 585
Sa MecR    KAQQISERIL KEMELI.... . 585
Consensus  -A--I---IL ---------- -
```

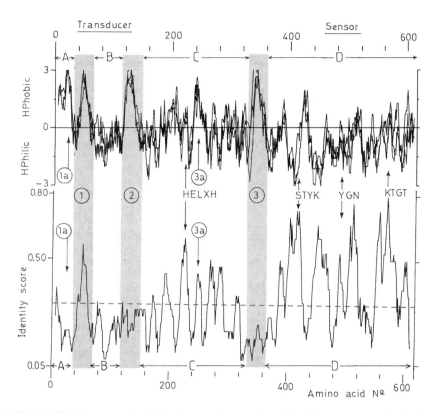

Fig. 6. Hydrophobicity (upper part) and identity score (lower part) profiles of the sensory-transducers BlaR of *B. licheniformis*, BlaR of *S. aureus* and MecR of *S. aureus*. The plots were obtained by using the GCG program (Pepplot and Plotsimilarity software) using a window of 10 amino acid residues [36]. The amino acid numbering is based on the standard scheme shown in Fig. 5. The domains A, B, C and D, the hydrophobic segments la, 1, 2, 3a and 3, and the zinc peptidase and penicilloyl-serine transferase motifs are indicated.

*bla*R2 is always chromosomal and unlinked to the *bla* regulon. Mutations in the locus *bla*R2 generate varying phenotypes. *S. aureus* magno-constitutive mutants produce large quantities of β-lactamase in the absence of inducer [39]. Other mutants are meso-consti-tutive; induction causes only a two-fold increased β-lactamase production [39]. The *B. licheniformis* Pen31 strain is micro-constitutive; it produces half the amount of enzyme that is produced by the uninduced wild-type strain and the synthesis is not inducible [40]. Whether the mutations responsible for these phenotypes reside in the same locus remains to be established. Moreover, the nature of the *bla*R2 gene product(s) is unknown.

Fig. 5. Optimal amino acid alignments of and standard numbering scheme for the sensory-transducers BlaR of *B. licheniformis*, BlaR of *S. aureus* and MecR of *S. aureus*. The hydrophobic segments la, 1, 2, 3a and 3 are boxed. The zinc peptidase and penicilloyl-serine transferase motifs are underlined. The standard numbering is shown above the aligned sequences.

Maximal synthesis of the *S. aureus* β-lactamase occurs within 15 min after induction and is maintained at this high level only in the continuous presence of the inducer [39]. Synthesis of the *Bacillus* β-lactamase is maximal 90 min after induction. The enzyme level remains several-fold higher than that of the uninduced strain for 2–3 h and continued presence of the inducer is not required. However, the amplitude of the response is a function of the initial inducer concentration [8,38]. Synthesis of PBP2' by a *S. aureus* strain bearing the *mec* cluster is a slow process. Full development of resistance on methicillin-containing agar plates is observed after 48 h [15]. In contrast, induction of PBP2' synthesis by a strain whose *mec*A is under the control of a penicillinase plasmid is as rapid as the induction of β-lactamase synthesis [15]. These different behaviours may be in some way related to the affinity of the β-lactam inducer for the sensor, the stability of the adduct formed, the affinity of the repressor for its operator and the organization of the regulons. The presence of two operators downstream from *bla*P (Fig. 1) may explain the relatively slow rate of induction of β-lactamase synthesis in *B. licheniformis*. During the first 30 min after induction, transcription from *bla*I promoter gives rise to polycistronic mRNAs. Subsequently, short transcripts encoding only the repressor accumulate and, after 1 h, they disappear progressively. The presence of inverted repeats in the putative Shine–Dalgarno sequence of *bla*R1 might explain the observed phenomenon [3,7].

Specific inducibility of β-lactamase synthesis among Gram-positive bacteria other than *B. licheniformis* and *S. aureus* may proceed through different mechanisms. *Streptomyces cacaoi* has two genes *bla*U (U for Umeå) and *bla*L (L for Liège) each encoding a distinct β-lactamase [41]. Transcription of *bla*L is under the control of at least two regulatory genes [42]. Inactivation of ORF1 or inactivation of ORF2 causes a 30–60-fold decreased basal level of β-lactamase synthesis and loss of inducibility. ORF1 encodes a Lys-R type DNA-binding protein which is related to the AmpR involved in the inducibility of β-lactamase synthesis in *E. coli* (see Chapter 23). ORF2 encodes a protein that has features of the penicilloyl serine transferases (but is not detected as a PBP in the plasma membrane) and also possesses a peptide sequence that is compatible with a phosphorylation site.

Finally, non-specific inducibility of β-lactamase synthesis by phosphate, molybdate, vanadate, tungstate and carbodiimides in *B. licheniformis* and by ferrous ions in *S. aureus* is well documented [38]. The underlying mechanism is unknown.

Acknowledgements

This work was supported in part by the Belgian programme on Inter-University Poles of Attraction initiated by the Belgian State, Prime Minister's Office, Science Policy Programming (PAI no. 19) and by the Fonds de la Recherche Scientifique Médicale (contract no. 3.4531.92). B.J. is Chercheur qualifié du Fonds National de la Recherche Scientifique (FNRS, Brussels).

References

1. Neugebauer, K., Sprengel, R. and Schaller, H. (1981) Nucleic Acids Res. 11, 2577–2588.
2. Rowland, S.J. and Dyke, K.G.H. (1989) EMBO J. 8, 2761–2773.

3. Rowland, S.J. and Dyke, K.G.H. (1990) Mol. Microbiol. 4, 961–975.
4. Wang, P.Z., Projan, S.J. and Novick, R.P. (1991) Nucleic Acids Res. 19, 4000.
5. Himeno, T., Imanaka, T. and Aiba, S. (1986) J. Bacteriol. 168, 1128–1132.
6. Imanaka, T., Himeno, T. and Aiba, S. (1987) J. Bacteriol. 169, 3867–3872.
7. Kobayashi, T., Zhu, Y.F., Nicholls, N. and Lampen, J.O. (1987) J. Bacteriol. 169, 3873–3878.
8. Salerno, A.J. and Lampen, J.O. (1986) J. Bacteriol. 166, 769–778.
9. Grossman, M.J. and Lampen, J.O. (1987) Nucleic Acids Res. 15, 6049–6062.
10. Salerno, A.J. and Lampen, J.O. (1988) FEBS Lett. 227, 61–65.
11. Wittman, V. and Wong, H.C. (1988) J. Bacteriol. 170, 3206–3212.
12. Wong, H.C., Lin, H.C., Chang, S. and Wittman, V. (1990) Genet. Biotechnol. Bacilli 3, 115–122.
13. Hackbarth, C.J. and Chambers, H.F. (1989) Antimicrob. Agents Chemother. 33, 995–999.
14. Patel, A.H., Foster, T.J. and Pattee, P.A. (1989) J. Gen. Microbiol. 135, 1799–1807.
15. Ryffel, C., Kayser, F.H. and Berger-Bächi, B. (1992) Antimicrob. Agents Chemother. 36, 25–31.
16. Tesch, W., Ryffel, C., Strassle, A., Kayser, F.H. and Berger-Bächi, B. (1990) Antimicrob. Agents Chemother. 34, 1703–1706.
17. Tesch, W., Strassle, A., Berger-Bächi, B., O'Hara, D., Reynolds, P.E. and Kayser, F.H. (1988) Antimicrob. Agents Chemother. 32, 1494–1499.
18. Hiramatsu, K., Asada, K., Suzuki, E., Okonogi, K. and Yokota, T. (1992) FEBS Lett. 298, 133–136.
19. Reynolds, P.E. and Brown, D.F.J. (1985) FEBS Lett. 192, 28–32.
20. Murakami, K., Nomura, K., Doi, M. and Yoshida, T. (1987) Antimicrob. Agents Chemother 31, 1307–1311.
21. Ubukata, K., Nonoguchi, R., Matsuhashi, M. and Konno, M. (1989) J. Bacteriol. 171, 2882–2885.
22. Harisson, S.C. and Aggarwal, A.K. (1990) Annu. Rev. Biochem. 59, 933–969.
23. Harisson, S.C. (1991) Nature 353, 715–719.
24. Brennan, R.G. (1991) Curr. Opinions Struct. Biol. 1, 80–88.
25. Vartak, N.B., Reizer, J., Gripp, J.T., Groisman, E.A., Wu, L.F., Tomich, J.M. and Saier, Jr, M.H. (1991) Res. Microbiol. 142, 951–963.
26. Bairoch, A. (1992) Dictionary of Protein Sites and Patterns (Prosite), release 9.2.
27. Nikaido, H. (1989) Antimicrob. Agents Chemother. 33, 1831–1836.
28. Furlong, C.E. (1987) in: F.C. Neidhardt, J.L. Ingraham, K.B. Low, B. Magasanik, M. Schaechter and H.E. Umbarger (Eds.), *Escherichia coli* and *Salmonella typhimurium.* Cellular and Molecular Biology, American Society for Microbiology, Washington, DC, pp. 768–796.
29. Broome-Smith, J.K. and Spratt, B.G. (1986) Gene 49, 341–349.
30. Joris, B., Ghuysen, J.-M., Dive, G., Renard, A., Dideberg, O., Charlier, P., Frère, J.M., Kelly, J., Boyington, J., Moews, P. and Knox, J.R. (1988) Biochem. J. 250, 313–324.
31. Joris, B., Ledent, P., Kobayashi, T., Lampen, J.O. and Ghuysen, J.-M. (1990) FEMS Microbiol. Lett. 70, 107–113.
32. Zhu, Y.F., Curran, I.H.A., Joris, B., Ghuysen, J.-M. and Lampen, J.O. (1990) J. Bacteriol. 172, 1137–1141.
33. Stock, J.B., Ninfa, A.J. and Stock, A.M. (1989) Microbiol. Rev. 53, 450–490.
34. Christianson, D.W. (1991) Adv. Protein Chem. 42, 281–329.
35. Vallee, B.L. and Auld, D.S. (1992) Faraday Discuss. 93, 47–65.
36. Devereux, J., Haeberli, P. and Smithies, O. (1984) Nucleic Acids Res. 12, 387–395.
37. Zhu, Y.F., Englebert, S., Joris, B., Ghuysen, J.-M., Kobayashi, T. and Lampen, J.O. (1992) J. Bacteriol. 174, 6171–6178.
38. Collins, J.F. (1979) in: J.M.T. Hamilton-Miller and J.T. Smith (Eds.), β-Lactamases, Academic Press, London, pp. 351–368.
39. Dyke, K.G.H. (1979) in: J.M.T. Hamilton-Miller and J.T. Smith (Eds.), β-Lactamases, Academic Press, London, pp. 291–310.
40. Dubnau, D.A. and Pollock, M.R. (1965) J. Gen. Microbiol. 41, 7–21.
41. Magdalena, J., Forsman, M., Lenzini, M.V., Brans, A. and Dusart, J. (1992) FEMS Microbiol. Lett. 99, 101–106.
42. Lenzini, V.M., Magdalena, J., Fraipont, C., Joris, B., Matagne, A. and Dusart, J. (1992) Mol. Gen. Genet. 235, 41–48.

J.-M Ghuysen and R. Hakenbeck (Eds.), *Bacterial Cell Wall*
517

Resistance to β-lactam antibiotics

BRIAN G. SPRATT

Microbial Genetics Group, School of Biological Sciences, University of Sussex, Falmer, Brighton BN1 9QG, UK

1. Introduction

β-Lactam antibiotics exert their lethal effects by inhibiting the final stages of peptidoglycan synthesis. Their enzymatic targets are the high molecular weight penicillin-binding proteins (high-M_r PBPs) which catalyze the final transpeptidation steps in peptidoglycan synthesis [1,2]. As discussed in Chapter 6, the high-M_r PBPs, together with the low-M_r PBPs, and the active-site serine classes of β-lactamases, are members of a superfamily of penicillin-interacting enzymes that are believed to have a very ancient common evolutionary origin. All three classes of enzymes interact with penicillin by a homologous mechanism via the formation of a covalent acyl-enzyme involving an active-site serine residue [2]. In the interaction of penicillin with PBPs, the acyl-enzyme is not hydrolyzed at a significant rate, and β-lactams act as irreversible inhibitors. In contrast, the acyl-enzyme formed between a β-lactamase and penicillin is a transient species on the pathway of hydrolysis, and β-lactamases thus catalyze the destruction of penicillin.

2. Susceptibility and resistance to β-lactam antibiotics

The high-M_r PBPs are minor components of bacterial cytoplasmic membranes [3]. They possess an amino-terminal hydrophobic sequence that acts as a non-cleaved, signal-like sequence, which acts to translocate the rest of the protein across the cytoplasmic membrane, and to anchor the protein in the membrane [1]. High-M_r PBPs are thus believed to extend from the outer surface of the cytoplasmic membrane towards their peptidoglycan substrate. Consequently, β-lactam antibiotics have essentially free access to their targets in Gram-positive species; there is little evidence that the cell walls of Gram-positive bacteria significantly hinder the access of β-lactams to the high-M_r PBPs and there are no convincing reports of increased resistance to β-lactams in Gram-positive species arising through alterations of cell wall structures. However, in Gram-negative bacteria, β-lactam antibiotics have to pass through the outer membrane to gain access to the PBPs. The outer membrane acts as a barrier to large β-lactams that are above the exclusion limit of the outer membrane porins and, although these antibiotics may have high affinities for PBPs, they have no activity against Gram-negative bacteria. β-Lactams that are below the exclusion limit of the porins permeate at rates that are dependent on both their size and overall charge [4] (see Chapter 27).

In the simplest case, the MIC of a β-lactam approximates to the concentration that is required to inactivate the high-M_r PBP that has the greatest affinity for the antibiotic. This situation probably applies to a few Gram-positive species, like *Streptococcus pneumoniae,* where there is unhindered access of the antibiotic to the PBPs, and no β-lactamase activity. In most cases, the MIC is considerably higher than the concentration required to inactivate the high-M_r PBPs as a result of the presence of β-lactamase activity and, in Gram-negative bacteria, the permeability barrier afforded by the outer membrane.

Recently, there has been considerable success in predicting the MICs of β-lactam antibiotics for Gram-negative bacteria [5,6]. These methods calculate the concentration of β-lactam outside of the bacteria that is required to achieve a concentration in the periplasm that inactivates the high-M_r PBP with the highest affinity for the antibiotic. The estimates must take into account the rate at which the β-lactam enters the periplasm, and the rate at which it is destroyed by β-lactamase. They depend on reasonably accurate estimates of the affinities of the antibiotic for the PBPs, its rate of permeation across the membrane, the number of molecules of β-lactamase in the periplasm, and the K_m and V_{max} for its hydrolysis by β-lactamase. These methods have in most cases been rather successful in predicting the MICs of β-lactam antibiotics, and have been very useful in focusing attention on the interplay between outer membrane permeability and β-lactamase hydrolysis in determining the MICs of laboratory mutants and clinical isolates that are resistant to β-lactam antibiotics.

Resistance to β-lactam antibiotics can emerge in three main ways [7,8]. By far the most widespread mechanism is the destruction of the antibiotic by a β-lactamase. Resistance can also occur by the development of high-M_r PBPs that have reduced affinity for β-lactam antibiotics or, in Gram-negative bacteria, by a reduction in the permeability of the outer membrane. Resistance can either be an intrinsic property of the species, or it can be acquired. Thus, all isolates of *Pseudomonas aeruginosa* are intrinsically-resistant to most of the older penicillins and cephalosporins, as a result of their poor penetration through the outer membrane, and their susceptibility to the chromosomally encoded β-lactamase. However, this species is intrinsically susceptible to imipenem, although individual isolates of the species can acquire resistance during therapy. Unfortunately, 'intrinsic resistance' has been use to describe any type of resistance that is not mediated by β-lactamases (e.g. PBP-mediated resistance). The term 'intrinsic resistance' is unsatisfactory in this context, and should not be used, as it implies that it is a natural feature of all members of the species, whereas PBP-mediated resistance is usually acquired. Consequently, the terms β-lactamase-mediated resistance and PBP-mediated resistance are used here. It should, however, be stressed that resistance is often multifactorial, involving an interplay between β-lactamase hydrolysis, altered permeability and altered PBPs.

3. PBP-mediated resistance

PBP-mediated resistance can be of two main types [7]. In methicillin-resistant *Staphylococcus aureus,* and some penicillin-resistant enterococci, an additional low affinity PBP is found that is absent in susceptible isolates (see below). In these cases, the new high-M_r PBP can apparently take over the function of the normal high-M_r PBPs when these are inactivated by a β-lactam antibiotic. In other cases, e.g. *Neisseria gonorrhoeae*, *N. meningitidis*, *S. pneumoniae*, viridans group streptococci, and *Haemophilus influenzae*, there have been reductions in the affinities of some of the normal PBPs for β-lactam antibiotics [7]. In those species that have been examined, the development of low affinity forms of normal PBPs has occurred, unexpectedly, by inter-species recombinational events that replace parts of the normal PBP gene with the corresponding parts from the homologous PBPs of closely related species [9]. It is probably not a coincidence that all of the species in which low affinity forms of normal PBPs have emerged are naturally transformable.

Bacteria possess multiple high-M_r PBPs that catalyze subtly different reactions in the biosynthesis of the peptidoglycan during the cell division cycle [1,2,10]. In most cases, the emergence of penicillin resistance will therefore require the stepwise development of low affinity forms of multiple PBPs. A simple example is provided by *N. gonorrhoeae* which has only two high-M_r PBPs [11]. The affinity of PBP2 for penicillin is about 10-fold higher than that of PBP1. The emergence of resistance first requires the development of a low affinity form of PBP2. However, even if PBP2 loses all affinity for penicillin, it will provide only a 10-fold increase in resistance to penicillin as killing will then occur by the inactivation of the unaltered PBP1. Higher levels of PBP-mediated resistance can only occur if isolates that produce low affinity forms of PBP2 also develop low affinity forms of PBP1 (in practice, reductions in outer membrane permeability also contribute to high level penicillin resistance in gonococci).

PBP-mediated resistance of this type is therefore often characterized by a slow increase over the years in the MIC of penicillin of some isolates of a species as, for example, has happened in *N. gonorrhoeae* and *S. pneumoniae*, and is occurring now in *N. meningitidis*. In the following sections, the molecular basis of PBP-mediated resistance in *Staph. aureus*, enterococci, pathogenic *Neisseria*, and *S. pneumoniae*, the species in which the phenomenon has been most thoroughly studied, is briefly described. PBP-mediated resistance in *H. influenzae* is not discussed as there has been no significant progress on the mechanism since the phenomenon was reviewed previously [7].

3.1. PBP-mediated resistance in Staphylococcus aureus and coagulase-negative Staphylococci

The emergence of penicillinase-producing isolates of *Staph. aureus* led to the development of the penicillinase-stable, methicillin. Isolates with high levels of resistance to methicillin appeared rapidly (MRSA) and have since caused considerable problems worldwide. MRSA strains vary in their expression of resistance. A few strains are homogeneously resistant to methicillin, such that most bacteria survive when plated on methicillin, whereas the majority of strains show heterogeneous expression of resistance where only a small proportion of the population survives exposure to the antibiotic [12].

Resistance in MRSA is due to the production of a novel low affinity PBP (PBP2'), which is not found in normal *Staph. aureus* isolates, and which can apparently take over the enzymatic functions of the normal high-M_r PBPs when these are inactivated by methicillin [13–15]. The PBP2' gene (*mec*A) has been found in all MRSA isolates and its inactivation by the insertion of *Tn*551 results in loss of methicillin resistance [16]. Furthermore, the introduction of a plasmid expressing PBP2' into a susceptible *Staph. aureus* strain renders it methicillin-resistant [17]. Although the role of PBP2' in resistance is beyond doubt, there is a poor correlation between the amount of PBP2' and the level of resistance, and there is evidence that additional genes (see below) are important for determining the level of methicillin resistance, and whether resistance is expressed homogeneously or heterogeneously. PBP2' is presumed to be a methicillin-resistant peptidoglycan transpeptidase but direct evidence that PBP2' functions as a transpeptidase is lacking and, perhaps surprisingly, there is no obvious effect on peptidoglycan structure in cells expressing PBP2' [18]. The *mec*A gene has been found in all MRSA although there are low level methicillin-resistant isolates that lack *mec*A where resistance appears to be due to the development of low affinity forms of the normal high-M_r PBPs [19,20].

In most heterogeneous MRSA strains, the expression of PBP2' is inducible by methicillin [21,22]. Regulation of the expression of PBP2' has been shown to act at the transcriptional level [23] and two different types of regulatory system have been established. In some strains, inducibility of PBP2' is independent of the presence of the penicillinase plasmid. These strains appear to have an intact *mec* region, consisting of *mec*A, the structural gene for PBP2', and another two open reading frames (*orf1* and *orf2*) that are transcribed in the opposite direction to *mec*A [24]. *Orf1* and *orf2* encode proteins that are clearly homologous to the regulators of penicillinase production in *Bacillus licheniformis* (BlaR1 and BlaI) and *Staph. aureus* [25,26]. BlaR1 is believed to be a transmembrane PBP that detects the presence of β-lactams in the environment and transmits a signal to the penicillinase repressor, BlaI, resulting in the induction of PBP2' [26,27].

In many MRSA strains, inducibility of PBP2' correlates with the presence of the penicillinase plasmid; loss of the plasmid results in constitutive expression of PBP2' and the appearance of homogeneous expression of methicillin resistance [21]. These strains have a truncated *mec* region lacking most the *blaR1* homologue and all of *blaI* [24]. Inducibility of expression of PBP2' occurs in strains that carry the penicillinase plasmid since the plasmid-encoded penicillinase repressor can recognize the regulatory sequences upstream of *mec*A [22].

The origin of the *mec* region is unknown. The region is not present in normal *Staph. aureus* isolates and has presumably been introduced from an unknown source by an illegitimate recombinational event [15]. Song et al. [15] proposed that a gene encoding a low affinity PBP became fused to the regulatory system of the *Staph. aureus* penicillinase plasmid. This is supported by the very similar genetic organization of the genes in the *mec* region, and those required for the expression of the *Staph. aureus* penicillinase. However, there is only a very low level of sequence similarity between the *blaI* and *blaR1* genes of the *Staph. aureus* penicillinase plasmid and their homologues in the *mec* region [24]. The recent fusion of a low affinity PBP gene to the regulatory genes on the widespread staphylococcal penicillinase plasmid is thus ruled out. It seems more likely that the whole *mec* region has been introduced from an unknown source.

The induction of PBP2' is slower, and the induced level of PBP2' is less, in MRSA strains containing the complete *mec* region compared to those in which *mec*A expression is under the control of the regulatory products of the penicillinase plasmid. It has been suggested that the complete *mec* region appeared first and that subsequently, under the selective pressures of methicillin usage, a deletion occurred that removed the *mec* regulatory genes, and put *mec*A under the control of the homologous genes of the penicillinase plasmid [24].

The *mec* genes are found in the same chromosomal location in all MRSA strains and it is likely that the introduction of *mec* occurred only once [28]. Methicillin-resistant isolates of coagulase-negative *Staphylococci*, including *Staph. epidermidis, Staph. haemolyticus* and *Staph. simulans,* also possess a *mec*A gene that is almost identical to that of *Staph. aureus* [29,30]. MRSA isolates show considerable diversity when analyzed by multilocus enzyme electrophoresis [31] and it appears that subsequent to its introduction, the *mec* region has been distributed horizontally, both within the *Staph. aureus* population, and also into the coagulase-negative *Staphylococci*.

Auxiliary genes *(fem)* that are essential for the full expression of methicillin resistance, and which alter peptidoglycan turnover and autolytic activity, have been identified using *Tn*551 mutagenesis [32–34]. The best characterized of these genes, *femA*, encodes a protein of 47 000 Da, which appears to be essential for the synthesis of pentaglycyl components of the peptidoglycan [35,36].

3.2. PBP-mediated resistance in enterococci

Enterococci are a major cause of serious nosocomial infections. In recent years, the emergence of high level resistance to aminoglycosides, vancomycin and β-lactam antibiotics has caused serious concerns about our ability to treat severe enterococcal infections. β-Lactamase-mediated resistance to penicillin appeared in *Enterococcus faecalis* in 1984 [37], and in *Enterococcus faecium* in 1992 [38], and is still relatively rare. However, even in the absence of β-lactamase activity, enterococci often have relatively low susceptibility to penicillin compared to most other Gram-positive pathogens, which has been correlated with the presence of a very low affinity PBP of about 71 000–77 000 Da [39,40]. The amount of this low affinity PBP (confusingly called either PBP5 or PBP3^r) has been shown to increase in laboratory mutants selected for increased resistance to penicillin [39,41]. The presence of a single low affinity PBP that can take over the functions of the other normal PBPs in the presence of penicillin is reminiscent of the situation in methicillin-resistant *Staph. aureus* strains. Recently, it has been shown that the low affinity PBP3^r and PBP5 are closely related in amino acid sequence (78% sequence identity), and that they are about 30% similar in sequence to PBP2' of *Staph. aureus* [41,42]. Interestingly, there is evidence that the gene encoding the low affinity PBP3^r of the clinical isolate S185 is on a large plasmid, linked to an erythromycin resistance gene, probably within a transposon (J. Coyette, personal communication). This idea is consistent with earlier work that showed the instability of penicillin resistance in enterococci [43]. Thus, the low affinity PBP(s) of enterococci, like PBP2' of *Staph. aureus*, may be an extra PBP that has recently been acquired from another species.

3.3. PBP-mediated resistance in Neisseria gonorrhoeae and Neisseria meningitidis

Penicillin resistance in *N. gonorrhoeae* is either due to the acquisition of the TEM-1 β-lactamase, or is due to a combination of the development of low affinity forms of both

PBP1 and PBP2, combined with a reduction in the permeability of the outer membrane [44–46].

Genetic analysis using transformation has shown that PBP-mediated penicillin resistance in gonococci is due to alterations of at least four genes [45]. Mutations in these genes individually provide only small increases in resistance to penicillin, but their combined effect is to increase the MIC for penicillin by a factor of about 1000. Two of the resistance genes (*pon* and *pen*A), encoding PBP1 and PBP2, respectively (the two high-M_r PBPs of this species), provide resistance only to β-lactam antibiotics [45], whereas the other two genes (*mtr* and *pen*B) affect permeability and give increased resistance to both β-lactams and unrelated antibiotics.

The *pen*B locus is probably identical to *por*, which encodes the major outer membrane protein I (PI) which is an anion-selective porin [47]. Gonococci express one of two major forms of the porin, PIA or PIB. Isolates expressing PIB have increased levels of resistance to penicillin and some other antibiotics compared to those expressing PIA. For unknown reasons, the difference in antibiotic resistance resulting from expression of the two forms of the PI porin is only observed in gonococci that possess *pen*A and *mtr* mutations. The *mtr* gene has been sequenced and shown to encode a protein of M_r 24 000 that has a helix-turn-helix DNA-binding motif and homology to several repressor proteins (W. Pan and B.G.S., unpublished results). Interestingly, *mtr* mutations result in increased levels of an outer membrane protein [45] suggesting, perhaps, that the *mtr* gene product controls the expression of this protein.

The development of low affinity forms of PBP2 was probably the first step in the emergence of resistance in gonococci and similar events have now occurred to produce low level penicillin-resistant isolates of *N. meningitidis*. The molecular basis of the development of low affinity forms of PBP2 has been studied in both gonococci and meningococci [48–50]. In both species, the PBP2 (*pen*A) genes from penicillin-susceptible isolates are very uniform, but those from resistant isolates have a mosaic structure, consisting of regions that are essentially identical to those in susceptible strains, and regions that are very different in sequence. This mosaic gene structure appears to have arisen by inter-species homologous recombinational events, presumably mediated by genetic transformation, that have replaced parts of the *pen*A gene of susceptible strains with the corresponding regions from the *pen*A genes of the closely related commensal species, *N. flavescens* or *N. cinerea* [48–50].

How do these inter-species recombinational events result in the production of forms of PBP2 with decreased affinity for penicillin? The answer is clearest for the recombinational events involving *N. flavescens*. Isolates of this species, including those obtained in the pre-antibiotic era, are considerably more resistant to penicillin than susceptible isolates of *N. gonorrhoeae* or *N. meningitidis*. The higher level of intrinsic resistance to penicillin of *N. flavescens* is due, at least in part, to the production of a low affinity PBP2. In these naturally transformable species, under the pressures applied by the use of penicillin, those rare inter-species recombinational events that replace the *pen*A genes of gonococci and meningococci (or the relevant parts of them) with the 'penicillin-resistant' *pen*A gene of *N. flavescens* have been selected [49]. Laboratory experiments can mimic the events that are proposed to have occurred in nature. Thus, chromosomal DNA from a pre-antibiotic era *N. flavescens* isolate can transform *N. meningitidis* to an increased level of penicillin resistance, and the resulting transformants have been shown to have replaced their *pen*A genes with that from the *N. flavescens* DNA donor (L.D. Bowler and B.G. Spratt, unpublished experiments).

Only three different classes of altered *pen*A genes have been found in penicillin-resistant gonococci [50]. The mosaic *pen*A genes of penicillin-resistant meningococci are much more diverse: 30 different mosaic *pen*A genes have been identified in the 78 penicillin-resistant meningococci that have been examined [51,52]. Some of these mosaic genes may have a common origin, with variation subsequently being introduced by a number of possible mechanisms, but others have clearly arisen by independent inter-species recombinational events. The *pen*A genes of penicillin-resistant gonococci and meningococci are distinct implying that meningococci have not obtained their altered *pen*A genes from penicillin-resistant gonococci [50].

3.4. PBP-mediated resistance in Streptococcus pneumoniae and viridans group Streptococci

β-Lactamase-producing isolates of pneumococci or viridans group streptococci have not so far been encountered. Isolates of these species have, however, developed resistance to β-lactam antibiotics. Penicillin-resistant pneumococci were first reported in the late 1960s and highly penicillin-resistant strains, and multiply-antibiotic-resistant strains, were reported from South Africa in the late 1970s [53]. Subsequently, isolates with intermediate level penicillin resistance (MICs of 0.1–1 μg/ml), or high level resistance (MICs > 1 μg/ml), have been isolated in most countries where adequate surveys have been carried out. In some regions, 30–65% of isolates from pneumococcal infections, or carriers, have intermediate or high level resistance to penicillin [53].

Resistance in pneumococci appears to be entirely due to the development of low affinity PBPs. In high level penicillin-resistant isolates, there have been reductions in the affinity of at least four of the five high-M_r PBPs [7,54]. The PBP genes of penicillin-resistant pneumococci are very different in sequence to those of truly penicillin-susceptible isolates [55,56]. As in the case of the PBP2 gene of penicillin-resistant meningococci and gonococci, the PBP1A [57], PBP2X [58] and PBP2B genes [56] of resistant pneumococci have a mosaic structure, consisting of regions that are similar to those in susceptible pneumococci, and regions that are as much as 23% diverged. Pneumococci, like meningococci and gonococci, are naturally transformable and the mosaic structure is believed to have arisen by the replacement of parts of the PBP genes with the corresponding regions from closely related streptococcal species that possess PBPs with lower affinity than those of their pneumococcal homologues [56].

Unlike the situation in *Neisseria*, it has been very difficult to identify unambiguously the origins of the diverged regions in the PBP genes of penicillin-resistant pneumococci, although the obvious sources are among the closely related viridans group streptococci. Penicillin resistance mediated by PBP changes has also occurred in viridans group streptococci and, at least in some cases, resistance appears to be due to the spread of altered PBP genes from penicillin-resistant pneumococci into viridans isolates [59]. The most likely scenario is that pneumococci have gained increased resistance to penicillin by recruiting parts of the PBP genes of those viridans species that by chance encode low affinity PBPs, and have subsequently donated the resulting mosaic PBP genes to other viridans species that have high affinity PBPs. Careful sequence analysis of the PBP genes of properly speciated viridans group streptococci, isolated in the pre-antibiotic era, will be required to establish critically the events that have occurred during the formation of mosaic PBP genes in pneumococci.

The high levels of resistance to penicillin in some regions has resulted in a switch from penicillins to third generation cephalosporins for the treatment of pneumococcal infections. However, even those penicillin-resistant pneumococci that were isolated before the introduction of third generation cephalosporins show considerable cross resistance to these compounds. Thus, the MIC of cefotaxime and ceftriaxone in penicillin-resistant pneumococci is typically about half that of benzylpenicillin, but the cephalosporins retain useful activity as a result of their improved tissue levels and pharmacokinetics. It might be expected that the relatively high MICs of third generation cephalosporins against penicillin-resistant pneumococci would result in the rapid emergence of strain with slight further increases in MIC which would lead to treatment failure. Such isolates have now been reported both in the United States and Spain [60,61]. Resistance to ceftriaxone and cefotaxime in a clinical isolate has been shown to be due only to alterations of PBP2X and PBP1A [61]. Presumably, the re-modelling of the active sites of PBPs that occurred to reduce greatly their affinities for penicillins has also resulted in decreased affinities for most other β-lactam structures. Unfortunately, the conversion of penicillin-resistant pneumococci that have cross resistance to cefotaxime and ceftriaxone, but which are still treatable with these agents, into isolates that have clinically significant resistance probably only requires slight further reductions in the affinities of PBP1A and PBP2X for these antibiotics.

3.5. Molecular basis of the re-modelling of high-M_r PBPs

The three-dimensional structures of a low-M_r PBP, and of Class A and Class C β-lactamases, have been determined [2] but there are no structures of high-M_r PBPs. The slight sequence similarities between all of these classes of active-site serine enzymes, which are largely confined to a small number of conserved motifs [1,2], and the overall similarity in the distribution of secondary structure elements between low-M_r PBPs and serine-β-lactamases, supports the idea that they form a homologous superfamily [2].

In almost all cases, the amino acid sequences of the low affinity PBPs of penicillin-resistant clinical isolates differ considerably from those in susceptible isolates and an analysis of the contribution of particular amino acid differences to the reduced affinity is difficult. The amino acid substitutions that occur in the PBPs of laboratory mutants selected for increased resistance to β-lactams have also been examined [62–64], but these substitutions are less interesting than those found in resistant clinical isolates as they often result in reduced growth rate, or increased thermosensitivity of the bacteria, in contrast to the robust growth of clinical isolates that have low affinity PBPs. Attempts to understand the molecular basis of the greatly reduced affinity for penicillin of the PBPs from penicillin-resistant isolates are also limited by the lack of a three-dimensional structure for high-M_r PBPs. Any understanding of the mechanism by which particular amino acid substitutions reduce affinity depends on the identification of the corresponding residue in a member of the superfamily whose three-dimensional structure is known. In many cases this cannot be achieved with the required accuracy as a result of ambiguities in the alignment of high-M_r PBPs with those enzymes whose structures have been determined.

The alterations in PBP2 of penicillin-resistant gonococci provide one situation where a single amino acid alteration has clearly been shown to contribute to a decrease in affinity for penicillin. An additional Asp-345A residue is found in PBP2 of all penicillin-resistant clinical isolates but is not found in PBP2 from penicillin-susceptible strains. Insertion of Asp-345A is known to be the main cause of the reduced affinity of PBP2 [65]. Asp-345A is located between the active-site serine residue (Ser-310) and the conserved Ser-X-Asn motif (residues 362–364). Alterations at the corresponding amino acid residue (Val-344) have been found in laboratory-generated mutant forms of PBP3 of *E. coli* that have decreased affinity for cephalexin [62].

The contribution of the amino acid differences between PBP2B of a penicillin-susceptible isolate (strain R6, MIC of 0.006 μg benzylpenicillin/ml) and a penicillin-resistant clinical isolate (strain 64147, MIC of 6 μg/ml) of *S. pneumoniae* has been carried out [66; C.G. Dowson and B.G. Spratt, unpublished data]. The penicillin-sensitive transpeptidase domain of the susceptible and resistant isolates differ at 17 residues but the substitutions that cause the difference in affinity could be localized to two positions. At one of these positions, there were seven contiguous amino acid alterations in the 'resistant' PBP2B. Replacement of this seven residue segment in stain R6 with the segment from the resistant strain 64147 resulted in a large decrease in the affinity of PBP2B and provided a 100-fold increase in penicillin resistance in an appropriate genetic background (a pneumococcal strain that contained a normal penicillin-susceptible form of PBP2B but penicillin-resistant forms of all of the other high-M_r PBPs).

This region in pneumococcal PBP2B (residues 425–431) is between the active-site serine residue (Ser-385) and the Ser-X-Asn motif (residues 442–444). There are thus three examples where alteration of the amino acid sequence in this region reduce affinity for β-lactam antibiotics. Unfortunately, an understanding in molecular terms of the reduction in affinity caused by alterations within this region must await the determination of the three-dimensional structure of the transpeptidase domain of a high-M_r PBP.

The second substitution in pneumococcal PBP2B that influences affinity for penicillin is located immediately carboxy-terminal to the Ser-X-Asn conserved motif, where there is a substitution of Thr-445 by Ala within the sequence Ser^{442}-Ser-Asn-Thr^{445}. This substitution has been found in PBP2B from all resistant pneumococci and its introduction into PBP2B of the susceptible strain R6 resulted in an approximately 100-fold decrease in affinity for benzylpenicillin. When this substitution was combined with the seven-residue substitution (see above), the effects of the two substitutions on affinity were additive, and the resulting low affinity PBP2B could provide a 200-fold increase in resistance to penicillin in the appropriate genetic background.

4. β-Lactamase-mediated resistance

β-Lactamases have been classified in several ways [67–70]. The simplest classification, based largely on amino acid sequence data, divides β-lactamases into four major groups.

The active site serine β-lactamases are classified into Classes A, C and D. The Class A enzymes are usually plasmid-encoded and include the TEM-1 enzyme, which is now found in about 50% of enterobacterial isolates, and the SHV-1 enzyme which is particularly common in *Klebsiella pneumoniae*. The β-lactamases that are found in most *Staph. aureus* isolates are Gram-positive examples of Class A enzymes. The Class C enzymes are typically chromosomally encoded and thus, unlike most Class A enzymes, are present in all members of the species. Typical Class C enzymes are the chromosomal β-lactamases of the enterobacteria and *P. aeruginosa*. The Class C enzymes are typically inducible by β-lactam antibiotics but in some species, notably *E. coli* and *K. pneumoniae*, the regulatory region has been deleted and β-lactamase is expressed at a very low constitutive level. The mechanistically distinct group of enzymes that require zinc for activity are placed in Class B, although this does not imply that all zinc-β-lactamases are homologous.

The existence of enzymes that inactivate β-lactam antibiotics has been recognized since the 1940s, and the relentless increase in the frequency of β-lactamase-mediated resistance has been discussed extensively elsewhere [71–73]. During the last decades, there has been an inexorable spread of β-lactamase genes into species that previously were not known to possess them. β-lactamases were detected for the first time in *N. gonorrhoeae* and *H. influenzae* in the 1970s [73], in *Enterococcus faecalis* [37] and *N. meningitidis* [73] in the 1980s and in *Ent. faecium* in the 1990s [38]. Molecular studies have provided convincing evidence for the origins of the β-lactamase genes in these species. Thus, the TEM-1 gene appears to have found its way from enterobacteria into *Haemophilus* and from there into *N. gonorrhoeae* [74], whereas the enterococcal gene has come from *Staph. aureus* [75]. β-Lactamases are now so widespread that there are very few examples of major bacterial pathogens in which β-lactamase-producing isolates are absent.

The increasing problems caused by β-lactamase-producing bacteria has been countered by the development of β-lactam antibiotics that are 'resistant' to enzymatic hydrolysis. β-Lactam antibiotics that combine potent broad-spectrum antibacterial activity with a high degree of stability to hydrolysis by most β-lactamases have been available since about 1980. These include the third generation cephalosporins (e.g. cefotaxime, ceftazidime, ceftriaxone), the 7-α-methoxy group (e.g. cefoxitin and moxalactam), the carbapenems (e.g. imipenem and meropenem), the fourth generation cephalosporins (e.g. cefepime and cefpirome), as well as the narrow-spectrum monobactams (e.g. aztreonam). As an alternative strategy, β-lactamase inhibitors (e.g. clavulanic acid, sulbactam and tazobactam) have been developed and used to protect β-lactamase-susceptible penicillins from hydrolysis by β-lactamases. Resistance to all of these 'β-lactamase-stable' β-lactam antibiotics, and to inhibitor/β-lactam combinations, has now emerged in the relatively short time since they were introduced.

Two major classes of resistance mechanisms have been described: those that extend the substrate specificity of the β-lactamase, and those that result from synthesis of greatly increased amounts of normal β-lactamases. In each of these resistance mechanisms, but particularly in the latter, reductions in permeability often contribute crucially to resistance. No attempt is made to survey the wide variety of β-lactamases that are now encountered; only the developments arising from the introduction of the 'β-lactamase-stable' β-lactam antibiotics and the inhibitors are described.

4.1. Extended spectrum β-lactamases

Transferable resistance to third generation cephalosporins was first reported from Germany and France in the early 1980s, less than 3 years after these compounds were introduced [76]. Resistance was initially found in *K. pneumoniae* isolates and was shown to be due to the emergence of variants of plasmid-encoded Class A β-lactamases that have increased rates of hydrolysis of third generation cephalosporins [76]. Resistance due to extended-spectrum β-lactamases is still mainly encountered in *K. pneumoniae,* and to a lesser extent in *E. coli,* but has also been found in several other enteric bacteria.

The nucleotide sequences of over 30 extended-spectrum β-lactamases have been reported [77,78]. Most of these enzymes are variants of the TEM-1, TEM-2 or SHV-1 β-lactamases, which are by far the commonest Class A β-lactamases among enterobacteria. The amino acid sequences of extended-spectrum β-lactamases differ from those of their normal parent β-lactamases at between one and four positions. In both TEM and SHV β-lactamases, the substitutions are found at only eight different residues (corresponding to Gln-37, Glu-102, Arg-162, Gln-203, Ala-235, Gly-236, Glu-237 and Thr-261 in the TEM-1 sequence). Differences at three of these positions (e.g. Gln-37, Gln-203 and Thr-261) are almost certainly due to polymorphisms that have no effect on substrate specificity, but those at the other five positions alter specificity [77–79]. The five causative substitutions are at residues that are within, or adjacent to, the conserved sequence motifs [2] that are known to be located within the active centre of TEM β-lactamase [80]. The recent progress in our understanding of the structure of the enzyme-substrate complex, and the catalytic mechanism of TEM β-lactamase [80], suggests that a convincing explanation for the altered specificity resulting from these amino acid substitutions will soon be forthcoming. Hopefully, this may lead to new ideas for the design of β-lactamase-stable structures that are less susceptible to the development of extended-spectrum variants.

Extended-spectrum β-lactamases have been found in many countries over the last few years. This is probably partly due to the ease with which they arise under the intense selective pressures of antibiotic usage. However, rapid local dissemination of the variant TEM and SHV β-lactamase genes within enteric bacteria probably occurs readily as they are usually carried, together with other resistance genes, on large conjugal plasmids [81]. Unlike most TEM and SHV β-lactamase genes, the variant genes appear almost invariably to be non-transposable [81]. Isolates expressing extended-spectrum TEM and SHV β-lactamases cause considerable therapeutic problems as they tend to be resistant to several other antibiotics, in addition to third generation cephalosporins and most other β-lactam antibiotics. However, they remain susceptible to carbapenems and α-methoxy-cephalosporins [78]. At least in France, over 10% of *K. pneumoniae* from hospitals carry extended-spectrum TEM or SHV β-lactamases, and such strains are now increasingly encountered outside the hospital environment.

Extended-spectrum β-lactamases have so far only been characterized from members of the Enterobacteriaceae. However, TEM-1 β-lactamase is common in isolates of both *N. gonorrhoeae* and *H. influenzae* and the increasing use of third generation cephalosporins to treat infections caused by these species is likely to result in the emergence of extended-spectrum enzymes. At present there appear to be no reports of extended-spectrum β-lactamases in these species, although a few β-lactamase-producing gonococcal isolates with

MICs for ceftriaxone of >0.5 μg/ml, that might be due to such enzymes, were detected in recent surveys from Thailand and The Philippines [82,83].

The emergence of extended-spectrum β-lactamases is perhaps not surprising given the ability of enzymes to alter their substrate specificity under strong selection [84,85]. What is perhaps surprising is that it has not happened more often. For example, the *Staph. aureus* Class A β-lactamase has apparently not evolved the ability to hydrolyze methicillin, although this enzyme is widespread in the species, and methicillin has been used extensively over many years for the treatment of staphylococcal infections.

4.2. Plasmid-mediated resistance to α-methoxy-cephalosporins, carbapenems and β-lactamase inhibitors

The extended-spectrum Class A enzymes derived from TEM and SHV β-lactamases do not provide resistance to 7-α-methoxy cephalosporins or carbapenems, and remain susceptible to inhibition by clavulanic acid [78]. The Class C chromosomal β-lactamases are more efficient at hydrolyzing the newer β-lactams than class A enzymes and their overexpression (see below) is known to provide resistance to 7-α-methoxy compounds, as well as to third generation cephalosporins. Furthermore, they are not inhibited by clavulanic acid or other β-lactamase inhibitors. A worrying development is the appearance of chromosomal Class C β-lactamase genes on plasmids. The best documented example is the plasmid-encoded MIR-1 enzyme, encountered in *K. pneumoniae*, which provides resistance to third generation cephalosporins and 7-α-methoxy compounds, but not to carbapenems. In this case, it appears that a chromosomal Class C β-lactamase gene, from an enterobacterial isolate that is about 90% related in nucleotide sequence to *Enterobacter cloacae*, has been translocated onto a plasmid and transferred into *K. pneumoniae* [86]. A similar transferable plasmid-encoded enzyme, apparently derived from a Class C β-lactamase, has recently been reported from an *E. coli* strain [87].

The carbapenems are among the most β-lactamase-stable of the newer β-lactam antibiotics and clinically significant resistance due to extended-spectrum Class A or Class C β-lactamases has not yet been reported. β-Lactamases that hydrolyze carbapenems efficiently have been reported, but so far all are zinc enzymes [88]. Most of these are chromosomal β-lactamases that were originally identified in species that were not major pathogens and which until relatively recently have not been considered to be of much significance. However, with the increasing use of imipenem, the carbapenem-hydrolyzing enzymes are becoming a cause for concern. In recent years, they have been reported from *Aeromonas* species, *Bacteroides fragilis*, *Serratia marcescens* and *Enterobacter cloacae* [88]. Some of these enzymes are potentially very dangerous as they provide resistance to essentially all β-lactams, and are not inhibited by any of the β-lactamase inhibitors, which only inactivate serine-β-lactamases.

The increasing use of imipenem and meropenem is also likely to apply strong selective pressures for the translocation of the zinc-β-lactamase genes onto plasmids and their spread into new species. A possible example of this phenomenon is the identification in *P. aeruginosa* of a plasmid-encoded zinc-β-lactamase that provides transferable resistance to carbapenems, third generation cephalosporins and 7-α-methoxy compounds [89].

Increased resistance to β-lactam/β-lactamase inhibitor combinations (e.g. amoxycillin plus clavulanic acid; augmentin) has also been reported and appears to involve increased expression of normal TEM-1 or SHV-1 β-lactamase [90]. Laboratory studies with both TEM and Ohio-1 (a member of the SHV family) Class A β-lactamases have shown that single point mutations can provide increased resistance to inhibition by clavulanic acid, tazobactam and sulbactam. In both cases, resistance was due to alterations of Met-69, immediately adjacent to the active-site Ser-70 [91,92]. The recent report of a derivative of TEM β-lactamase that is 100-fold less sensitive to inhibition by clavulanic acid suggests that similar mutations may have occurred in nature [93].

4.3. Resistance due to chromosomal β-lactamases and decreased permeability

The very low rates of hydrolysis of third generation cephalosporins compared to first and second generation cephalosporins, under standard assay conditions using high concentrations of substrate, led to their designation as β-lactamase-stable cephalosporins [94]. However, at the low substrate concentrations likely to occur in the periplasmic space of Gram-negative bacteria, the rates of hydrolysis of third generation cephalosporins by the chromosomal class C β-lactamases are appreciable, and in many cases are not so different from those of earlier cephalosporins [95,96]. This arises because although the V_{max} for hydrolysis of third generation cephalosporins is very low, they have a very high affinity for the Class C β-lactamases, such that V_{max}/K_m is significant. In contrast, the older cephalosporins have a higher V_{max} but also a higher K_m. Thus, at the low concentrations of β-lactams in the periplasm, the third generation cephalosporins are hydrolyzed at close to their V_{max}, whereas the earlier compounds will be hydrolyzed at a rate well below their V_{max} [95,96].

The improved efficacy of third generation cephalosporins against those enterobacterial species that contain inducible class C β-lactamases is now largely ascribed not to their β-lactamase stability but to the fact that they fail to induce the β-lactamase. It is not surprising, in hindsight, that the initial enthusiasm for third generation cephalosporins has been tempered by the frequent appearance during therapy of isolates of *Enterobacter cloacae*, *Citrobacter freundii*, *P. aeruginosa*, and other species that possess inducible class C β-lactamases, of resistant mutants that produce very high levels of the β-lactamase constitutively [94]. The appreciable rates of hydrolysis of third generation cephalosporins by massively overproduced Class C β-lactamases can provide resistance, particularly as many of these compounds are poor at permeating the outer membranes of these species [95].

The MICs of β-lactams that have extremely low rates of hydrolysis are only slightly increased in mutants that constitutively produce large amounts of Class C β-lactamase. In these cases, neither β-lactamase hyper-production, nor decreased permeability, are alone predicted to be sufficient to result in a substantial increase in MIC. According to the target access index of Nikaido and Normark [5], resistance to extremely poorly hydrolyzed β-lactams is predicted to emerge only if very high level β-lactamase production is combined with a large decrease in permeability (or with decreased affinities of the PBPs). Laboratory mutants and clinical isolates of *Enterobacter cloacae* that are resistant to the poorly hydrolyzed imipenem and meropenem support this view, as they produce high levels of β-lactamase and have reduced permeability [97,98].

Greatly reduced permeability of the outer membrane, in the absence of any hydrolysis by a β-lactamase, usually provides little increase in the MIC of β-lactam antibiotics [4]. However, as seen above, even a very low rate of hydrolysis, combined with greatly reduced permeability, provides substantially increased levels of resistance. β-lactams that are totally resistant to hydrolysis should therefore be largely insensitive to the combined effects of high level constitutive β-lactamase production and greatly reduced permeability. This situation is approximated by some of the methoxyimino-cephalosporins (cefepime, cefpirome and particularly cefaclidine), which are even more stable to hydrolysis by Class C β-lactamases than the carbapenems, mainly due to their much lower affinity for these enzymes. These 'fourth generation' cephalosporins therefore retain activity against strains of *Enterobacter cloacae* that produce high constitutive levels of β-lactamase and have decreased outer membrane permeability [99,100].

The emergence of resistance to third generation cephalosporins has focused attention on the desirable attributes of improved β-lactam antibiotics for Gram-negative infections. Firstly, they should be extremely resistant to hydrolysis by β-lactamases, by having very high K_m values, and very low V_{max} values. Secondly, they should have high rates of penetration through the outer membrane and, thirdly, they should have high affinities for the high-M_r PBPs. At present, some of the fourth generation cephalosporins approach these ideals, having significantly increased β-lactamase stability and improved penetration. These features ensure that fourth generation cephalosporins retain activity against hyper-β-lactamase-producing strains, as described above. However, the MICs of these compounds for normal *E. coli* or *Enterobacter cloacae* strains producing basal levels of β-lactamase are not very different from those of third generation cephalosporins. This may imply that the decreased affinity for β-lactamases has been gained at the cost of a decreased affinity for the high-M_r PBPs [100] although, at least in *E. coli,* this appears not to be the case [101]

Another interesting approach has been the development of catechol-containing β-lactam antibiotics that can permeate the Gram-negative outer membrane through the specific transport systems for the uptake of iron-siderophore complexes [102]. One potential advantage of this approach is that the ability to transport iron is crucial to the pathogenicity and survival of bacteria in vivo. Thus it is hoped that the development of resistance by mutational loss of iron transport systems will be less likely to occur as it would severely compromise bacterial survival in vivo.

5. Resistance to imipenem in Pseudomonas aeruginosa

P. aeruginosa is intrinsically resistant to many β-lactam antibiotics as a consequence of the low permeability of the outer membrane and the presence of an inducible class C β-lactamase. Imipenem is one of the few β-lactam antibiotics that has useful activity against this species as it is β-lactamase-stable and, unlike other β-lactams which permeate through the general porin pathway, can pass efficiently through the D2 protein which apparently forms a specific channel for the uptake of basic amino acids [103,104]. The use of imipenem for the treatment of *P. aeruginosa* infections has been limited by the frequent emergence of resistance during therapy [105]. The resistant mutants lack the D2 outer membrane protein and have higher MICs because the antibiotic can now only enter the periplasm using the much less effective general porin pathway [105,106]. These mutants lack cross resistance to other β-lactams (except closely related carbapenems) since these are unable to use the D2 basic amino acid transporter.

Although it has been suggested that reduced permeation of imipenem is the explanation of the increased resistance to imipenem, the major effect of the loss of the D2 protein may be to make the chromosomal β-lactamase (which hydrolyzes imipenem very poorly) much more effective at protecting the PBPs from acylation. Thus, the loss of the D2 protein in a β-lactamase-negative mutant of *P. aeruginosa* results in only a twofold increase in MIC, whereas loss of the protein in a normal β-lactamase-producing strain results in an eightfold increase [107].

References

1. Spratt, B.G. and Cromie, K.D. (1988) Rev. Infect. Dis. 10, 699–711.
2. Ghuysen, J.-M. (1991) Annu. Rev. Microbiol. 45, 37–67.
3. Spratt, B.G. (1977) Eur. J. Biochem. 72, 341–352.
4. Nikaido, H. (1989) Antimicrob. Agents Chemother. 33, 1831–1836.
5. Nikaido, H. and Normark, S. (1987) Mol. Microbiol. 1, 29–36.
6. Waley, S.G. (1987) Microbiol. Sci. 4, 143–146.
7. Spratt, B.G. (1989) in: L.E. Bryan (Ed.), Microbial Resistance to Drugs, Springer-Verlag, Berlin, pp. 77–100.
8. Sanders, C.C. and Sanders, W.E. (1992) Clin. Inf. Dis. 15, 824–839.
9. Spratt, B.G., Dowson, C.G., Zhang, Q.-Y., Bowler, L.D., Brannigan, J.A. and Hutchison, A. (1991) in: J. Campisi, D.D. Cunningham, M. Inouye and M. Riley (Eds.), Perspectives on Cellular Regulation: From Bacteria to Cancer, Wiley-Liss, New York, pp. 73–83.
10. Spratt, B.G. (1975) Proc. Natl. Acad. Sci. USA 72, 2999–3003.
11. Dougherty, T.J., Koller, A.E. and Tomasz, A. (1980) Antimicrob. Agents Chemother. 18, 730–737.
12. Chambers, H.F. (1988) Clin. Microbiol. Rev. 1, 173–186.
13. Brown, D.F.J. and Reynolds, P.E. (1980) FEBS Lett. 122, 275–278.
14. Hayes, M.V., Curtis, N.A.C., Wyke, A.W. and Ward, J.B. (1981) FEMS Microbiol Lett. 10, 119–122.
15. Song, M.D., Wachi, M., Doi, M., Ishino, F. and Matsuhashi, M. (1987) FEBS Lett. 221, 167–171.
16. Mathews, P. and Tomasz, A. (1990) Antimicrob. Agents Chemother. 34, 1777–1779.
17. Inglis, B., Mathews, P.R. and Stewart, P.R. (1988) J. Gen. Microbiol. 134, 1465–1469.
18. de Jonge, B.L.M., Chang, Y.-S., Gage, D. and Tomasz, A. (1992) J. Biol. Chem. 267, 11248–11254.
19. de Lencastre, H., Figueiredo, A.M., Urban, C., Rahal, J. and Tomasz, A. (1991) Antimicrob. Agents Chemother. 35, 632–639.
20. Tomasz, A., Drugeon, H.B., de Lencastre, H., Jabes, D., McDougal, L. and Bille, J. (1989) Antimicrob. Agents Chemother. 33, 1869–1874.
21. Ubukata, K., Yamashita, N. and Konno, M. (1985) Antimicrob. Agents Chemother. 27, 851–857.
22. Opal, S.M., Boyce, J.M., Medeiros, A.A., Mayer, K.H. and Lyhte, L.W. (1989) J. Antimicrob. Chemother. 23, 315–325.
23. Ryffel, C., Kayser, F.H. and Berger-Bachi, B. (1992) Antimicrob. Agents Chemother. 36, 25–31.
24. Hiramatsu, K., Asada, K., Suzuki, E., Okonogi, K. and Yokota, T. (1992) FEBS Lett. 298, 133–136.
25. Kobayashi, T., Zhu, .Y.F., Nicholls, N.J. and Lampen, J.O. (1987) J. Bacteriol. 169, 3873–3878.
26. Rowlands, S.J. and Dyke, K.G.H. (1990) Mol. Microbiol. 4, 961–975.
27. Zhu, Y.F., Curran, I.H.A., Joris, B., Ghuysen, J.-M. and Lampen, J.O. (1990) J. Bacteriol. 172, 1137–1141.
28. Pattee, P.A., Lee, H.C. and Bannantine, J.P. (1990) in: R.P. Novick (Ed.), Molecular Biology of the Staphlyococci, VCH, New York, pp. 41–58.

532

29. Ryffel, C., Tesch, W., Birch-Machin, I., Reynolds, P.E., Barberis-Maino, L., Kayser, F.H. and Berger-Bachi, B. (1990) Gene 94, 137–138.
30. Ubukata, K., Nonoguchi, R., Song, M.D., Matsuhashi, M. and Konno, M. (1990) Antimicrob. Agents Chemother. 34, 170–172.
31. Musser, J.M. and Kapur, V. (1992) J. Clin. Microbiol. 30, 2058–2063.
32. Berger-Bachi, B., Barberis-Maino, L., Strassle, A. and Kayser, F.H. (1989) Mol. Gen. Genet. 219, 263–269.
33. Tomasz, A. (1991) in: R.P. Novick (Ed.), Molecular Biology of the Staphylococci, VCH, New York, pp. 565–583.
34. Berger-Bachi, B., Strassle, A., Gustafson, J.E. and Kayser, F.H. (1992) Antimicrob. Agents Chemother. 36, 1367–1373.
35. Maidhof, H., Reinicke, B., Blumel, P., Berger-Bachi, B. and Labischinski, H. (1991) J. Bacteriol. 173, 3507–3513.
36. De Jonge, B.L.M., Chang, Y.-S., Gage, D. and Tomasz, A. (1992) J. Biol. Chem. 267, 11255–11259.
37. Murray, B.E. and Mederski-Samoraj, B. (1983) J. Clin. Invest. 72, 1168–1171.
38. Coudron, P.E., Markowitz, S.M. and Wong, E.S. (1992) Antimicrob. Agents Chemother. 36, 1125–1126.
39. Fontana, R., Cerini, R., Longoni, P., Grossato, A. and Canepari, P. (1983) J. Bacteriol. 155, 1343–1350.
40. Williamson, R., Le Bourguenec, C., Gutman, L. and Horaud, T. (1985) J. Gen. Microbiol. 131, 1933–1940.
41. Piras, G., El Kharroubi, A., Van Beeumen, J., Coeme, E., Coyette, J. and Ghuysen, J.-M. (1990) J. Bacteriol. 172, 6856–6862.
42. Piras, G., Raze, D., El Kharroubi, A., Hastir, D., Englebert, S., Coyette, J. and Ghuysen, J.-M. (1993) J. Bacteriol. 175, 2844–2852.
43. Eliopoulos, G.M., Wennersten, C. and Moellering, R.C. (1982) Antimicrob. Agents Chemother. 22, 295–301.
44. Cannon, J. and Sparling, P.F. (1984) Annu. Rev. Microbiol. 38, 111–133.
45. Faruki, H. and Sparling, P.F. (1986) Antimicrob. Agents Chemother. 30, 856–860.
46. Dougherty, T.J. (1986) Antimicrob. Agents Chemother. 30, 649–652.
47. Carbonetti, N., Simnad, V., Elkins, C. and Sparling, P.F. (1990) Mol. Microbiol. 4, 1009–1018.
48. Spratt, B.G. (1988) Nature 332, 173–176.
49. Spratt, B.G., Zhang, Q.-Y., Jones, D.M., Hutchison, A., Brannigan, J.A. and Dowson, C.G. (1989) Proc. Natl. Acad. Sci. USA. 86, 8988–8992.
50. Spratt, B.G., Bowler, L.D., Zhang, Q.-Y., Zhou, J. and Maynard Smith, J. (1992) J. Mol. Evol. 34, 115–125.
51. Zhang, Q.-Y., Jones, D.M., Saez Nieto, J.A., Perez Trallero, E. and Spratt, B.G. (1990) Antimicrob. Agents Chemother. 34, 1523–1528.
52. Campos, J., Fuste, M.C., Trujillo, G., Saez-Nieto, J., Vazquez, J., Loren, J.G., Vinas, M. and Spratt, B.G. (1992) J. Infect. Dis. 166, 173–177.
53. Appelbaum, P.C. (1992) Clin. Inf. Dis. 15, 77–83.
54. Hakenbeck, R., Briese, T., Chalkley, L., Ellerbrok, H., Kalliokoski, R., Latorre, C., Leinonen, M. and Martin, C. (1991) J. Infect. Dis. 164, 313–319.
55. Dowson, C.G., Hutchison, A. and Spratt, B.G. (1989) Mol. Microbiol. 3, 95–102.
56. Dowson, C.G., Hutchison, A., Brannigan, J.A., George, R.C., Hansman, D., Linares, J., Tomasz, A., Maynard Smith, J. and Spratt, B.G. (1989) Proc. Natl. Acad. Sci. USA 86, 8842–8846.
57. Martin, C., Sibold, C. and Hakenbeck, R. (1992) EMBO J. 11, 3831–3836.
58. Laible, G., Spratt, B.G. and Hakenbeck, R. (1991) Mol. Microbiol. 5, 1993–2002.
59. Dowson, C.G., Hutchison, A., Woodford, N., Johnson, A.P., George, R.C. and Spratt, B.G. (1990) Proc. Natl. Acad. Sci. USA 87, 5858–5862.
60. Bradley, J.S. and Connor, J.D. (1991) Pediatr. Infect. Dis. J. 10, 871–873.
61. Munoz, R., Dowson, C.G., Daniels, M., Coffey, T.J., Martin, C., Hakenbeck, R. and Spratt, B.G. (1992) Mol. Microbiol. 6, 2461–2465.
62. Hedge, P.J. and Spratt, B.G. (1985) Eur. J. Biochem. 151, 111–121.
63. Hedge, P.J. and Spratt, B.G. (1985) Nature 318, 478–480.

64. Laible, G. and Hakenbeck, R. (1991) J. Bacteriol. 173, 6986–6990.
65. Brannigan, J.A., Tirodimos, I.A., Zhang, Q.-Y., Dowson, C.G. and Spratt, B.G. (1990) Mol. Microbiol. 4, 913–919.
66. Maynard Smith, J., Dowson, C.G. and Spratt, B.G. (1991) Nature 349, 29–31.
67. Ambler, R.P. (1980) Philos. Trans. R. Soc. London Ser. B. 289, 321–331.
68. Bush, K. (1989) Antimicrob. Agents Chemother. 33, 259–263.
69. Bush, K. (1989) Antimicrob. Agents Chemother. 33, 264–270.
70. Bush, K. (1989) Antimicrob. Agents Chemother. 33, 271–276.
71. Medeiros, A.A. (1989) in: L.E. Bryan (Ed.), Microbial Resistance to Drugs, Springer-Verlag, Berlin, pp. 101–127.
72. Sanders, C.C. (1989) in: L.E. Bryan (Ed.), Microbial Resistance to Drugs, Springer-Verlag, Berlin, pp. 129–149.
73. Sanders, C.C. (1992) Clin. Infect. Rev. 14, 1089–1099.
74. Brunton, J., Meier, M., Erhman, N., Clare, D. and Almawy, R. (1986) J. Bacteriol. 168, 374–379.
75. Zscheck, K.K. and Murray, B.E. (1991) Antimicrob. Agents Chemother. 35, 1736–1740.
76. Kliebe, C., Nies, B.A., Meyer, J.F., Tolxdorff-Neutzling, R.M. and Wiedemann, B. (1985) Antimicrob. Agents Chemother. 28, 302–307.
77. Philippon, A., Labia, R. and Jacoby, G. (1989) Antimicrob. Agents Chemother. 33, 1131–1136.
78. Jacoby, G.A. and Medeiros, A.A. (1991) Antimicrob. Agents Chemother. 35, 1697–1704.
79. Collatz, E., Labia, R. and Gutmann, L. (1990) Mol. Microbiol. 4, 1615–1620.
80. Strynadka, N.C.J., Adachi, H., Jensen, S.E., Johns, K., Sielecki, A., Betzel, C., Sutoh, K. and James, M.N.G. (1992) Nature 359, 700–705.
81. Jacoby, G.A. and Sutton, L. (1991) Antimicrob. Agents Chemother. 35, 164–169.
82. Clendennen, T.E., Hames, C.S., Kees, E.S., Price, F.C., Rueppel, W.J., Andrada, A.B., Espinosa, G.E., Kabrerra, G. and Wignall, F.S. (1992) Antimicrob. Agents Chemother. 36, 277–282.
83. Clendennen, T.E., Eccheverria, P., Saengeur, S., Kees, E.S., Boslego, J.W. and Wignall, F.S. (1992) Antimicrob. Agents Chemother. 36, 1682–1687.
84. Clarke, P.H. (1978) in: L.N. Ornston and J.R. Sokatch (Eds.), The Bacteria, Vol. 6, Academic Press, New York, pp. 137–218.
85. Hall, A. and Knowles, J.R. (1976) Nature 264, 803–804.
86. Papanicolaou, G.A., Medeiros, A.A. and Jacoby, G.A. (1990) Antimicrob. Agents Chemother. 34, 2200–2209.
87. Woodford, N., Payne, D.J., Johnson, A.P., Weinbren, M.J., Perinpanayagan, R.M., George, R.C., Cookson, B.D. and Amyes S.G.B. (1990) Lancet 336, 253.
88. Livermore, D.M. (1992) J. Antimicrob. Chemother. 29, 609–616.
89. Watanabe, M., Iyobe, S., Inoue, M. and Mitsuhashi, S. (1991) Antimicrob. Agents Chemother. 35, 147–151.
90. Martinez, J.L., Cercenado, E., Rodriguez-Creixems, M., Vicente-Perez, M.F., Delgado-Iribarren, A. and Baquero, F. (1987) Lancet ii, 1473.
91. Oliphant, A.R. and Struhl, K. (1989) Proc. Natl. Acad. Sci. USA 86, 9094–9098
92. Bonomo, R.A., Currie-McCumber, C. and Shlaes, D.M. (1992) FEMS Microbiol. Lett. 92, 79–82.
93. Thomson, C.J. and Amyes, S.G.B. (1992) FEMS Microbiol. Lett. 91, 113–118.
94. Sanders, C.C. (1987) Annu. Rev. Microbiol. 41, 573–593.
95. Vu, M. and Nikaido, H. (1985) Antimicrob. Agents Chemother. 27, 393–398.
96. Livermore, D.M. (1985) J. Antimicrob. Chemother. 15, 511–514.
97. Lee, E.H., Nicolas, M.H., Kitzis, M.D., Pialoux, G., Collatz, E. and Gutmann, L. (1991) Antimicrob. Agents Chemother. 35, 1093–1098.
98. Raimondi, A., Traverso, A. and Nikaido, H. (1991) Antimicrob. Agents Chemother. 35, 1174–1180.
99. Nikaido, H., Liu, W. and Rosenberg, E.Y. (1990) Antimicrob. Agents Chemother. 34, 337–342.
100. Bellido, F., Perchere, J.-C. and Hancock, R.E.W. (1991) Antimicrob. Agents Chemother. 35, 73–78.
101. Pucci, M.J., Boicesowek, J., Kessler, R.E. and Dougherty, T.J. (1991) Antimicrob. Agents Chemother. 35, 2312–2317.
102. Curtis, N.A.C., Eisenstadt, R.L., East, S.J., Cornford, R.J., Walker, L.A. and White, A.J. (1988) Antimicrob. Agents Chemother. 32, 1879–1886.

534

103. Trias, J. and Nikaido, H. (1990) J. Biol. Chem. 265, 15680–15684.
104. Satake, S., Yoshihara, E. and Nakae, T. (1990) Antimicrob. Agents Chemother. 34, 685–690.
105. Quinn, J.P., Dudek, E.J., DiVincenzo, C.A., Lucks, D.A. and Lerner, S.A. (1986) J. Infect. Dis. 154, 289–294.
106. Yoneyama, H. and Nakae, T. (1991) FEBS Lett. 283, 177–179.
107. Livermore, D.M. (1992) Antimicrob. Agents Chemother. 36, 2046–2048.

J.-M Ghuysen and R. Hakenbeck (Eds.), *Bacterial Cell Wall*
© 1994 Elsevier Science B.V. All rights reserved

CHAPTER 26

Resistance to glycopeptide antibiotics

REGINE HAKENBECK

Max-Planck Institut für Molekulare Genetik, Ihnestrasse 73, D-14195 Berlin 33, Germany

1. Introduction

The history of glycopeptides represents a perfect example of the various circumstances that provoke medical and scientific interest in an antibiotic. Due to increased isolation of *Staphylococcus aureus* strains resistant against antibiotics such as erythromycin, tetracyclin and penicillin in the early 1950s, a large-scale screening program aimed at the identification of antibiotics with high anti-staphylococcal activity was initiated. As a result, the first glycopeptide-producing microorganism *Streptomyces orientalis* was found in a soil sample from the jungle in Borneo. The compound isolated from the fermentation broth (early lots of which were named 'Mississippi mud' due to its appearance, later it was named vancomycin) was shown to be highly active against Gram-positive bacterial species [1,2]. Laboratory-induced levels of staphylococcal resistance against vancomycin were almost negligible [3,4] and resistance remained of no clinical significance for a long time. Vancomycin was soon introduced clinically, but problems with toxicity and the introduction of β-lactamase-resistant penicillins in the early 1960s caused a decline in the use of vancomycin. The appearance of methicillin-resistant staphylococci and other, multiply resistant Gram-positive organisms revived interest in glycopeptide antibiotics especially since improved purification procedures largely reduced early toxicity problems of vancomycin [5,6]. Other glycopeptides were used as animal feed additives (see [7] and refs. therein), whereas ristocetin, isolated for the first time from *Nocardia lurida* [8] in the 1950s, causes aggregation of blood platelets [9,10] and could therefore not be used therapeutically. Recently, teicoplanin (produced by *Actinoplanes teichomyceticus* nov. sp. [11]) has been introduced for the treatment of serious infections in human due to resistant Gram-positive bacteria or in cases of β-lactam allergy. The emergence of resistance to glycopeptide antibiotics in staphylococci and enterococci confronts scientists with a bacterial defence mechanism which is unexpected and had been considered to be highly improbable. It results in a severe clinical problem and a demand for new antibacterial agents.

2. Structure of glycopeptides

The structures of vancomycin and teicoplanin (Fig. 1) reveal several features common to all members of the glycopeptide group [7,12]. They contain a central heptapeptide back-

536

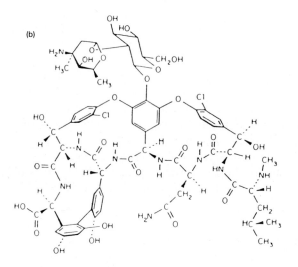

Fig. 1. Structures of (a) teicoplanin and (b) vancomycin.

bone with five highly conserved amino acids. The amino acids are linked via several phenolic acid residues, resulting in the formation of complex tetra- or tricyclic structures. The compounds have seven aromatic rings (except vancomycin which has five) carrying various substituents. The variable presence of chlorine groups poses interesting biosynthetic questions. Attached to the core structure, the aglycone, are a variable number of sometimes unusual sugars or amino sugars which are not essential for antibiotic activity. The source of these sugars is largely unknown. In the teicoplanin complex, an amino sugar carries the various fatty acids characteristic for each member, rendering the molecule more hydrophobic than vancomycin [13,14].

3. Mode of action

It is important to realize that the early biochemical studies on the mode of action of glycopeptides were carried out before the structure of these antibiotics was known. The elucidation of the spatial arrangement of glycopeptides then confirmed and illustrated the unique interaction between these compounds and their target, a specific bacterial cell wall component. Much of this work has been reviewed in several articles [7,12,15 and refs. therein], and the following summarizes the major aspects.

Glycopeptide antibiotics are too large (>1 kDa) to cross the outer membrane of Gram-negative organisms and are hence effective only against Gram-positive bacteria. Initial experiments suggesting that vancomycin interferes with cell wall synthesis were reported by Reynolds and Jordan who showed that vancomycin-treated bacteria accumulate UDP-N-acetylmuramyl-peptides [16,17]. Vancomycin and ristocetin form a 1:1 complex with UDP-MurNAc-pentapeptide [18], and bind not only to bacterial cells but also to cell walls [19–21]. The specificity of this binding was elucidated by Perkins and collaborators, including the important observations that ristocetin or vancomycin binding could be reversed by addition of peptides ending with acyl-D-Ala-D-Ala [22], and that prerequisites of this binding were the D-configuration of both alanine residues and a free terminal carboxyl group [23–26].

A key step in elucidating the interaction of glycopeptides with the target site was the determination of the three-dimensional structure of a vancomycin degradation product, CDP-1, and its complex with acetyl-D-Ala-D-Ala by X-ray analysis [27]. Further structural refinement and stereochemical details were obtained by applying mass spectrometry and ^{1}H NMR methods, and from analysis of nuclear Overhauser effects [7,12]. Figure 2 illustrates the hydrogen bond interactions between the aglycone of a glycopeptide and its target peptide, and a space-filling model of the complex derived from computer analysis is represented in Fig. 3. The acyl-D-Ala-D-Ala terminus fits tightly into a cleft of the antibiotic ristocetin or aridicin; vancomycin which is a more flexible molecule undergoes a large conformational change on binding to target peptides, and the bound form closely resembles the ristocetin/aridicin complex.

Ward [28] and Johnston and Neuhaus [29] have suggested that the main target peptide of glycopeptides is represented by the growing glycan chain rather than the lipid intermediate. In the presence of glycopeptides, transpeptidase and D,D-carboxypeptidase activities that depend on accessibility of C-terminal D-Ala-D-Ala residues will be prevented.

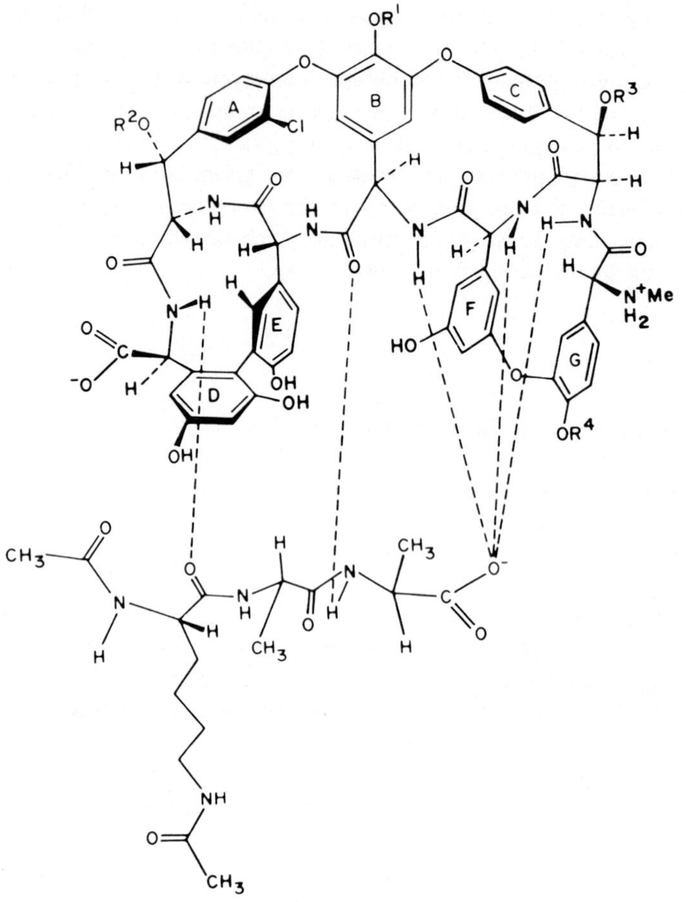

Fig. 2. Representation of an aridicin-like Ac$_2$-L-Lys-D-Ala-D-Ala complex showing interpeptide hydrogen bonds (from [12]).

This has been shown for the soluble D,D-carboxypeptidase of *Streptomyces albus* G [30] and for *Escherichia coli* enzymes [31,32]. It is also conceivable that a transglycosylase reaction may be inhibited by glycopeptides by steric hindrance although the disaccharide portion of the muropeptide is not masked in the glycopeptide complex [33]. This is in agreement with early in vitro studies where murein synthesis (due to polymerization of uncrosslinked murein) was blocked by vancomycin and ristocetin [34,35], and later experiments where the polymerization reaction was studied directly [36]. Interference with multiple critical steps in murein assembly may well explain why it was impossible to reach significant resistance levels in the laboratory. Resistance to glycopeptides due to alterations in the target site was therefore believed to be highly unlikely; it would require alterations in the specificity of a whole set of enzymes involved in biosynthesis of the murein layer.

4. Phenotypes of glycopeptide resistant bacteria

Several Gram-positive species are intrinsically constitutively resistant to glycopeptides. This includes high level resistant species like actinomycetes (the glycopeptide producers), and leuconostoc, pediococcus and lactobacillus species [37–42]. Low level intrinsically resistant strains are *Enterococcus gallinarum* and *Enterococcus casseliflavus* [44–45]. Although still uncommon, the clinical isolation of those strains is reported more frequently [46 and refs. therein]. Glycopeptide resistant strains of other Gram-positive species have rarely been observed [47 and refs. therein, 48], but they have not been investigated in further detail.

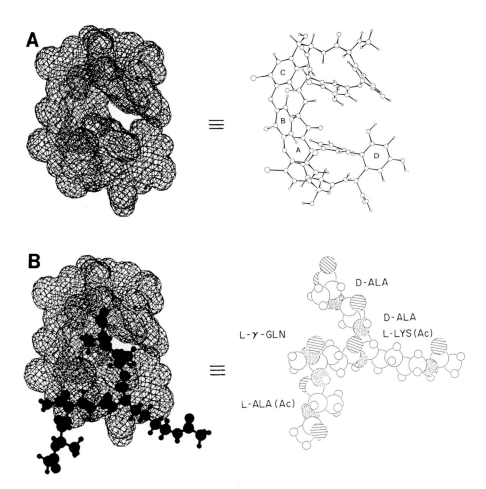

Fig. 3. Computer-generated model showing space-filling representation of (A) aridicin aglycone and (B) aridicin aglycone-diacetyl-L-Ala-(γ)-D-Gln-L-Lys(Ac)-D-Ala-D-Ala complex. Structures were obtained using distance geometry with subsequent energy minimization (from [12]).

Since the late 1980s, glycopeptide-resistant coagulase-negative staphylococci [49–51 and refs. therein] and enterococci [52 and refs. therein] have been isolated in clinical settings on several occasions. Resistance is considered to be an acquired property; the incidence of glycopeptide resistant strains is still relatively low. Recently, glycopeptide resistant *E. faecium* were also found in environmental samples such as waste water of sewage treatment plants, demonstrating the spread of the resistant determinants [53]. In none of these cases was inactivation of the antibiotic reported.

In enterococci, three classes of glycopeptide resistance VanA, VanB and VanC have been distinguished phenotypically on the basis of the resistance level, cross-resistance between vancomycin and teicoplanin, and whether resistance is inducible (Table I) [44,54]. Class C strains, represented by *E. gallinarum*, show low level intrinsic, constitutive resistance to vancomycin and not to teicoplanin [43–45]. *E. faecium* and *E. faecalis* strains with an acquired, inducible resistance are either of class A or class B. The high-level resistant class A phenotype includes resistance to teicoplanin; only this resistance has been demonstrated to be transferable; class B strains achieve only a moderate resistance level to vancomycin [55–61]. The incidence of class B strains may be underestimated because this phenotype is difficult to detect by disk diffusion assays [61].

5. Biochemical and genetic basis of glycopeptide resistance in enterococci

5.1. Class A vancomycin resistance

5.1.1. Transfer of vancomycin resistance determinants
The presence of plasmids transferring class A resistance has been documented in several *E. faecium* strains [55,56,62], whereas transfer was unrelated to plasmid DNA in other strains of *E. faecium* or *E. faecalis* [57,58,60]. Four of the plasmids were distinct in terms of size, restriction profiles, other resistance determinants and transfer properties, but apparently related at the DNA sequence level [55,56]. The host range differed considerably, one plasmid being readily transferred into a variety of Gram-positive species including

TABLE I

Glycopeptide resistance in enterococci

Relevant property	vanA	vanB	vanC
Resistance	Acquired	Acquired	Intrinsic
MIC (μg/ml)			
Vancomycin	≥64	16–32	8–16
Teicoplanin	≥16	0.5	0.5
Transferability	+	–	–
Inducibility (van/tei)	+/(+)	+/–	–/–
Resistance protein (kDa)	40	39.5	nd
Species	*E. faecium*	*E. faecium*	*E. gallinarum*
	E. faecalis	*E. faecalis*	

E. faecium, E. faecalis, Streptococcus pyogenes, Streptococcus lactis, and *Listeria monocytogenes.* Resistance was inducible in the transconjugants to varying degrees. Attempts to transfer this plasmid to *B. subtilis* or *S. aureus* failed. The plasmid pIP816 was not self-transferable but could be introduced into a *S. sanguis* strain via transformation. A DNA fragment from pIP816 carrying a resistance determinant was cloned into a Gram-positive/Gram-negative shuttle vector [63]. When introduced into *E. faecalis* and *B. thuringiensis* (but not in *E. coli* or *B. subtilis*), it also conferred inducible glycopeptide resistance.

Sequence homology between DNA of a large number of highly resistance *E. faecium* isolates and a pIP816 probe encoding a vancomycin resistance determinant strongly suggested that the resistance phenotype is due to dissemination of a gene specifically associated with resistant strains rather than of a bacterial clone or a single plasmid [54]. Typing of a number of strains including environmental samples confirmed the unrelatedness of highly resistant *E. faecium* [53]. Recently, sequence determination of the regions that flank the vancomycin resistance gene cluster (see below) of pIP816 revealed that the *van* gene cluster in pIP816 is carried by a 10.8 kb transposon designated Tn*1546* with similarity to transposons of the Tn*3* family. Evidence was obtained of elements related to Tn*1546* in other highly resistant enterococci, and further results indicated that transposition plays a role in dissemination of the vancomycin resistance gene cluster [64]. The origin of the *van* resistance transposon is still unknown. So far, the *van* genes have been detected only in enterococci, but transposition of the vancomycin transposon into a broad range plasmid may change the situation dramatically. Astoundingly, experiments have been carried out under laboratory conditions showing that indeed conjugative transfer of glycopeptide resistance from *E. faecalis* to *S. aureus* is possible [65].

5.1.2. The role of VanA and VanH

In several strains, subinhibitory concentrations of vancomycin induced the synthesis of a membrane-associated protein that appeared to be required for expression of vancomycin resistance [57,58,60]. The *vanA* gene encoding the inducible resistance protein was identified on the 34 kb plasmid pIP816 of *E. faecium* BM4147 using a two-stage cloning procedure based on the use of a Gram-positive/Gram-negative shuttle vector [66]. First, *Eco*RI restriction fragments of pIP816 were cloned and transformed into *E. coli,* and characterized by restriction profiles. Recombinant plasmids were then cloned into an *E. coli* strain containing a conjugative plasmid which could mobilize the shuttle vector derivatives. These constructs served as donors in mating experiments with different Gram-positive species which were tested for expression of glycopeptide resistance [63]. A 1.7-kb derivative encoded a 40-kDa protein in *E. coli* designated VanA. The deduced amino acid sequence of the *vanA* gene revealed a high degree of homology to D-Ala-D-Ala ligases of *E. coli* and *S. typhimurium,* and *vanA* could transcomplement an *E. coli* mutant with a thermosensitive D-Ala-D-Ala ligase activity [66]. The VanA protein purified from *E. coli* contained ligase activity of altered substrate specificity. It preferentially catalyzed the ester bond formation between D-Ala and other residues such as D-2-hydroxy acids [67].

A gene located upstream from *vanA* encodes a dehydrogenase VanH that has D-keto acid reductase activity, and can provide the D-2-hydroxyacid substrate of VanA [68,69].

VanH shows some sequence similarity with D-lactate dehydrogenases found in many lactobacilli and leuconostoc, i.e. bacteria with intrinsic vancomycin resistance. When D-lactate was added to the culture medium of a strain with an insertionally inactivated *vanH* gene, it restored the vancomycin resistance phenotype, suggesting that the depsipeptide D-alanyl-D-lactate is the in vivo product of VanA activity [70]. A novel murein precursor was found after induction of vancomycin resistance [71]. Analysis of the structure of murein precursors confirmed that D-lactate is the substrate of VanA in vivo [72–74].

5.1.3. The van gene cluster

In addition to *vanA* and *vanH*, three other genes contribute to vancomycin resistance. This five gene cluster appears to be sufficient for murein synthesis in the presence of vancomycin [75]. The distal part includes *vanA, vanH* and a third gene *vanX* of unknown function. These genes are regulated at the transcriptional level by a two-component regulatory system *vanS-vanR* which is located in the proximal part of the cluster. These two proteins are homologous to the two families of signal-transducer/response-regulator proteins (for review, see [76]) and are probably involved in transmitting a signal in response to the presence of vancomycin outside the membrane [75]. Insertional inactivation of the regulator component VanR suppressed vancomycin resistance completely [75,77].

A sixth gene *vanY* which is located downstream of the *vanX* gene was not required for vancomycin resistance in the genetic construct used, although it is inducible in response to the addition of glycopeptide antibiotics [78]. In addition to release of D-Ala, VanY also hydrolyzed muropeptide terminating with D-lactate, i.e. the precursor synthesized in vancomycin-induced cells [79]. It thus corresponds to the inducible D,D-carboxypeptidase described earlier in other strains [80,81]. No similarity to known D,D-carboxypeptidases was detectable at the sequence level [78], and the enzyme did not display transpeptidase or β-lactamase activity. Its physiological role remains to be clarified.

5.2. Class B and class C vancomycin resistance

VanB strains produce a 39.5 kDa membrane protein termed VanB [59] which is immunologically unrelated to the 39–40 kDa protein described in class A strains [46,82]. Internal fragments of the *vanB* genes of an *E. faecalis* and *E. faecium* strain, respectively, could be amplified using degenerate oligonucleotides that had been shown to prime amplification of the *vanA* gene and other D-Ala-D-Ala ligases [83]. Comparisons of the *E. faecalis* sequence confirmed that VanB was highly similar to VanA (77% identity) and to the other ligases tested (<40%); similarities of the N-terminal sequence of VanB to D-Ala-D-Ala ligases had been noted before [82]. *VanB* related sequences were detected by hybridization in a number of *E. faecium* and *E. faecalis* strains displaying the vanB phenotype but with different vancomycin resistance levels, indicating that a single class of resistance determinants is represented in class B strains [84]. Consistent with the proposed role of the VanB protein are reports on murein precursor analyses of a high and a low vancomycin-resistant strain that demonstrate the appearance of a new peak after vancomycin induction [71]. An inducible carboxypeptidase activity was also found in class B strains [81], but its role in the glycopeptide resistance mechanism is still unknown.

The *vanC* gene of *E. gallinarum* was cloned with the help of the same pair of oligonucleotides used for identification of the *vanB* gene fragment; it encodes a D-Ala-D-Ala ligase related protein [85]. The *vanC* probe was specific for *E. gallinarum*; no homology with DNA to *E. casseliflavus* strains was detected which express a glycopeptide resistance phenotype very similar to that of *E. gallinarum* [86].

6. Concluding remarks

In all three classes of glycopeptide resistance, a D-Ala-D-Ala ligase with modified substrate specificity appears to be an important component of the resistance mechanism. The relatedness of other genes involved in resistance is not known but may account for the different modes of inducibility. In all cases, resistance is most likely achieved by the synthesis of a novel murein subunit which will not complex with glycopeptides, allowing maturation of the murein network in the presence of those antibiotics. The impact of the change in murein composition on other enzymes acting on the murein network, especially penicillin-binding proteins (PBPs), is not known in the induced resistance phenotype, nor has the specificity of PBPs in intrinsically resistant species been analyzed. PBPs can interact with Ac_2-L-Lys-D-Ala-D-Lac and other esters and thioesters [87,88], and reaction constants of the PBPs are probably as diverse as with β-lactams. Different PBPs may be responsible for transpeptidation of muropentapetide and the D-lactate-containing homologue, respectively. Such a change in the role of individual PBPs due to glycopeptide treatment would also result in a shift in β-lactam sensitivity as has been proposed by Al-Obeid et al. [89] to account for the striking synergy between glycopeptides and certain β-lactam antibiotics of low- and high-level vancomycin resistant enterococci [44,90].

Acknowledgements

I thank P. Courvalin and L. Gutmann for providing manuscripts prior to publication, M. Achtmann for critical reading of the manuscript, and F. Parenti for contributing Fig. 1.

References

1. McCormick, M.H., Stark, W.M., Pittenger, G.E., Pittenger, R.C. and McGuire, J.M. (1956) Antibiot. Ann. 1955–1956, 606–611.
2. Griffith, R.S. (1981) Rev. Infect. Dis. 3, S200–S204.
3. Ziegler, D.W., Wolfe, R.N. and McGuire, J.M. (1956) Antibiot. Ann. 1955–1956, 612–618.
4. Griffith, R.S. and Peck, Jr., F.B. (1956) Antibiot. Ann. 1955–1956, 619–622.
5. Riley, Jr., H,D. (1970) Med. Clin. N. Am. 54, 1277–1289.
6. Farber, B.F. and Moellering, R.C. (1983) Antimicrob. Agents Chemother. 23, 138–141.
7. Barna, J.C.J. and Williams, D.H. (1984) Annu. Rev. Microbiol. 38, 339–357.
8. Grundy, W.E., Sinclair, A.C., Theriault, R.J., Goldstein, A.W., Rickher, C.J., Warren, Jr., H.B., Oliver, T.J. and Sylvester, J.C. (1957) Antibiot. Ann. 1956–1957, 687–692.
9. Coller, B.S. and Gralnick, H.R. (1977) J. Clin. Invest. 60, 301–312.
10. Howard, M.A. and Firkin, B.G. (1971) Thromb. Diath. Haemorrh. 26, 362–369.
11. Parenti, F., Beretta, G., Berti, M. and Arioli, V. (1978) J. Antibiot. 31, 276–281.
12. Jeffs, P.W. and Nisbet, L.J. (1988) in: P. Actor, L. Daneo-Moore, M. Higgins, M.R.J. Salton and G.D.

544

Shockman (Eds.), Antibiotic Inhibition of Bacterial Cell Surface Assembly and Function, American Society for Microbiology, Washington, DC, pp. 509–530.

13. Parenti, F. (1986) J. Hosp. Infect. 2 (Suppl. A), 79–83.
14. Somma, S., Gastaldo, L. and Corti, A. (1984) Antimicrob. Agents Chemother. 26, 917–923.
15. Perkins, H.R. (1987) in: D.J. Tipper (Ed.), Antibiotic Inhibitors of Bacterial Cell Wall Biosynthesis, Pergamon Press, New York, pp. 115–132.
16. Jordan, D.C. (1961) Biochem. Biophys. Res. Commun. 6, 167–170.
17. Reynolds, P.E. (1961) Biochim. Biophys. Acta 52, 403–405.
18. Chatterjee, A.N. and Perkins, H.R. (1966) Biochem. Biophys. Res. Commun. 24, 489–494.
19. Best, G.K. and Durham, N.N. (1965) Arch. Biochem. Biophys. 111, 685–691.
20. Sinha, R.K. and Neuhaus, F.C. (1968) J. Bacteriol. 96, 374–382.
21. Perkins, H.R. and Nieto, M. (1970) Biochem. J. 116, 83–92.
22. Nieto, M., Perkins, H.R. and Reynolds, P.E. (1972) Biochem. J. 126, 139–149.
23. Perkins, H.R. (1969) Biochem. J. 111, 195–205.
24. Nieto, M. and Perkins, H.R. (1971) Biochem. J. 123, 773–787.
25. Nieto, M. and Perkins, H.R. (1971) Biochem. J. 123, 789–803.
26. Nieto, M. and Perkins, H.R. (1971) Biochem. J. 124, 845–852.
27. Sheldrick, G.M., Jones, P.G., Kennard, O., Williams, D.H. and Smith, G.A. (1978) Nature 271, 223–225.
28. Ward, J.B. (1974) Biochem. J. 141, 227–241.
29. Johnston, L.S. and Neuhaus, F.C. (1975) Biochemistry 14, 2754–2760.
30. Leyh-Bouille, M., Ghuysen, J.-M., Nieto, M., Perkins, H.R., Schleifer, K.H. and Kandler, O. (1970) Biochemistry 9, 2971–2975.
31. Izaki, K. and Strominger, J.L. (1968) J. Biol. Chem. 243, 3193–3201.
32. Bogdanovsky, D., Bricas, E. and Dezelé, P. (1969) C.R. Acad. Sci., Series D 269, 390–393.
33. Reynolds, P.E. (1989) Eur. J. Clin. Microbiol. Infect. Dis. 8, 943–950.
34. Anderson, J.S., Matsuhashi, M., Haskin, M.S. and Strominger, J.L. (1965) Proc. Natl. Acad. Sci. USA 53, 881–889.
35. Anderson, J.S., Matsuhashi, M., Haskin, M.S. and Strominger, J.L. (1967) J. Biol. Chem. 242, 3180–3190.
36. van Heijenoort, Y., Derrien, M. and van Heijenoort, J. (1978) FEBS Lett. 89, 141–144.
37. Lerner, P.I. (1974) Antimicrob. Agents Chemother. 5, 302–309.
38. Bayer, A.S., Chow, A.W., Betts, D. and Guze, L.B. (1978) Am. J. Med. 64, 808–813.
39. Vescovo, M., Morelli, L. and Bottazzi, V. (1982) Appl. Environ. Microbiol. 43, 50–56.
40. Orberg, P.K. and Sandine, W.E. (1984) Appl. Environ. Microbiol. 48, 1129–1133.
41. Colman, G. and Efstratiou, A. (1987) J. Hosp. Infect. 10, 1–3.
42. Buu-Hoi, A., Branger, C. and Acar, J.F. (1985) Antimicrob. Agents Chemother. 28, 458–460.
43. Swenson, J.M., Hill, B.C. and Thornsberry, C. (1989) J. Clin. Microbiol. 27, 2140–2142.
44. Shlaes, D.M., Etter, L. and Gutmann, L. (1991) Antimicrob. Agents Chemother. 35, 776–779.
45. Vincent, S., Knight, R.G., Green, M., Sahm, D.F. and Shlaes, D.M. (1991) J. Clin. Microbiol. 29, 2335–2337.
46. Nicas, T.I., Cole, C.T., Preston, D.A., Schabel, A.A. and Nagarajan, R. (1989) Antimicrob. Agents Chemother. 33, 1477–1481.
47. Watanakunakorn, C. (1981) Rev. Infect. Dis. 3, S210–S215.
48. Shlaes, D.M., Marino, J. and Jacobs, M.R. (1984) Antimicrob. Agents Chemother. 25, 527–528.
49. Goldstein, F., Coutrot, A., Sieffer, A. and Acar, J.F. (1990) Antimicrob. Agents Chemother. 34, 899–900.
50. Moore, E.P. and Speller, D.C.E. (1988) J. Antimicrob. Chemother. 21, 417–424.
51. Schwalbe, R.S., Stapleton, J.T. and Gilligan, P.H. (1987) N. Eng. J. Med. 316, 927–931.
52. Leclercq, R., Dutka-Malen, S., Brisson-Noël, A., Molinas, C., Derlot, E., Arthur, M., Duval, J. and Courvalin, P. (1992) Clin. Infect. Dis. 15, 495–501.
53. Klare, I., Heier, H., Claus, H. and Witte, W. (1992) FEMS Microbiol. Lett. 106, 23–30.
54. Dutka-Malen, S., Leclercq, R., Coutant, V., Duval, J. and Courvalin, P. (1990) Antimicrob. Agents Chemother. 34, 1875–1879.

55. Leclercq, R., Derlot, E., Duval, J. and Courvalin, P. (1988) N. Engl. J. Med. 319, 157–161.

56. Leclercq, R., Derlot, E., Weber, M., Duval, J. and Courvalin, P. (1988) Antimicrob. Agents Chemother. 33, 10–15.

57. Nicas, T.I., Wu, C.Y.E., Hobbs, Jr., J.N., Preston, D.A. and Allen, N.E. (1989) Antimicrob. Agents Chemother. 33, 1121–1124.

58. Shlaes, D.M., Bouvet, A., Devine, C., Shlaes, J.H., Al-Obeid, S. and Williamson, R. (1989) Antimicrob. Agents Chemother. 33, 198–203.

59. Williamson, R., Al-Obeid, S., Shlaes, J.H., Goldstein, F.W. and Shlaes, D.M. (1989) J. Infect. Dis., 159, 1095–1104.

60. Shlaes, D.M., Al-Obeid, S., Shlaes, J.H., Biosivon, A. and Williamson, R. (1989) J. Antimicrob. Chemother. 23, 503–508.

61. Sahm, D.F., Kissinger, J., Gilmore, M.S., Murray, P.R., Mulder, R., Solliday, J. and Blarke, B. (1989) Antimicrob. Agents Chemother. 33, 1588–1591.

62. Handwerger, S., Pucci, M.J. and Kolokathis, A. (1990) Antimicrob. Agents Chemother. 34, 358–360.

63. Brisson-Noël, A., Dutka-Malen, S., Molinas, C., Leclercq, R. and Courvalin, P. (1990) Antimicrob. Agents Chemother. 34, 924–927.

64. Arthur, M., Molinas, C., Depardieu, F. and Courvalin, P. (1993) J. Bacteriol. 175, 117–127.

65. Noble, W.C., Virani, Z. and Cree, R.G.A. (1992) FEMS Microbiol. Lett. 93, 195–198.

66. Dutka-Malen, S., Molinas, C., Arthur, M. and Courvalin, P. (1990) Mol. Gen. Genet. 224, 364–372.

67. Bugg, T.D.H., Dutka-Malen, S., Arthur, M., Courvalin, P. and Walsh, C.T. (1991) Biochemistry 30, 2017–2021.

68. Arthur, M., Molinas, C., Dutka-Malen, S. and Courvalin, P. (1991) Gene 103, 133–134.

69. Bugg, T.D.H., Wright, G.D., Dutka-Malen, S., Arthur, M., Courvalin P. and Walsh, C.T. (1992) Biochemistry 30, 10408–10415.

70. Arthur, M., Molinas, C., Bugg, T.D.H., Wright, G.D., Walsh, C.T. and Courvalin, P. (1992) Antimicrob. Agents Chemother. 36, 867–869.

71. Billot-Klein, D., Gutmann, L., Collatz, E. and van Heijenoort, J. (1992) Antimicrob. Agents Chemother. 36, 1487–1490.

72. Handwerger, S., Pucci, M.J., Volk, K.J., Liu, J and Lee, M.S. (1992) J. Bacteriol. 174, 5982–5984.

73. Allen, N.E., Hobbs, Jr., J.N., Richardson, J.M. and Riggins, R.M. (1992) FEMS Miocrobiol. Lett. 98, 109–116.

74. Messer, J. and Reynolds, P.E. (1992) FEMS Microbiol. Lett. 94, 195–200.

75. Arthur, M., Molinas, C. and Courvalin, P. (1992) J. Bacteriol. 174, 2582–2591.

76. Surette, M.G. and Stock, J.B. (1994) in: J.-M. Ghuysen and R. Hakenbeck (Eds.), Bacterial Cell Wall, New Comprehensive Biochemistry, Vol. 27, Elsevier, Amsterdam, pp. 465–483.

77. Handwerger, S., Discotto, L., Thanassi, J. and Pucci, M.J. (1992) FEMS Microbiol. Lett. 92, 11–14.

78. Arthur, M., Molinas, C. and Courvalin, P. (1992) Gene 120, 111–114.

79. Wright, G.D., Molinas, C., Arthur, M., Courvalin, P. and Walsh, C.T. (1992) Antimicrob. Agents Chemother. 36, 1514–1518.

80. Al-Obeid, S., Collatz, E. and Gutmann, L. (1990) Antimicrob. Agents Chemother. 34, 252–256.

81. Gutmann, L., Billot-Klein, D., Al-Obeid, S., Klare, I., Francoual, S., Collatz, E. and van Heijenoort, J. (1992) Antimicrob. Agents Chemother. 36, 77–80.

82. Al-Obeid, S., Gutmann, L., Shlaes, D.M., Williamson, R. and Collatz, E. (1990) FEMS Microbiol. Lett. 70, 101–106.

83. Evers, S., Sahm, D.F. and Courvalin, P. (1993) Gene 124, 143–144.

84. Quintiliani, R., Evers, S. and Courvalin, P. (1993) J. Infect. Dis. 167, 1220–1223.

85. Dutka-Malen, S., Molinas, C., Arthur, M. and Courvalin, P. (1992) Gene 112, 53–58.

86. Leclercq, R., Dutka-Malen, S., Duval, J. and Courvalin, P. (1992) Antimicrob. Agents Chemother. 36, 2005–2008.

87. Rasmussen, J.R. and Strominger, J.L. (1978) Proc. Natl. Acad. Sci. USA 75, 84–88.

88. Adam, M., Camblon, C., Christiaens, L. and Frère, J.-M. (1990) Biochem. J. 270, 525–529.

89. Al-Obeid, S., Billot-Klein, D., van Heijenoort, J., Collatz, E. and Gutmann, L. (1992) FEMS Microbiol. Lett. 91, 79–84.

90. Leclercq, R., Bingen, E., Su, Q.H., Lambert-Zechovski, N., Courvalin, P. and Duval, J. (1991) Antimicrob. Agents Chemother. 35, 92–98.

J.-M Ghuysen and R. Hakenbeck (Eds.), *Bacterial Cell Wall*
© 1994 Elsevier Science B.V. All rights reserved

CHAPTER 27

Diffusion of inhibitors across the cell wall

HIROSHI NIKAIDO

Department of Molecular and Cell Biology, University of California, Berkeley, CA 94720, USA

1. Introduction

One major function of the bacterial cell envelope is to prevent, or at least slow down, the influx of deleterious compounds from the environment. The outer membrane (OM) of Gram-negative bacteria is very effective in this function. In addition, the cell wall of at least some Gram-positive species, such as mycobacteria, acts as an effective permeability barrier. This chapter summarizes the current knowledge in this area. No attempt has been made to cover the literature exhaustively.

2. Outer membrane of Gram-negative bacteria

2.1. Introduction

Gram-negative bacteria are covered by the OM, which is located outside the peptidoglycan layer. Its basic continuum is a lipid bilayer containing both lipopolysaccharides (LPS) and the more common glycerophospholipids, and most of the intrinsic OM proteins appear to function in providing selective permeability. Most detailed studies have so far been carried out with *Escherichia coli* and *Salmonella typhimurium*. However, *Pseudomonas aeruginosa* has also received much attention as a prototype of organisms that lack the classical high-permeability porins, and therefore show a high degree of intrinsic resistance to a large number of agents. Reviews exist on OM permeability in general [1] and in relation to antibiotic resistance [2,3].

2.2. Porin pathway

The 'classical' trimeric porins of *E. coli* and *Rhodobacter capsulatus* were discussed in Chapter 12. The most constricted portion of their channels has dimensions of about 7 Å × 10 Å. This allows the passage of many small antibiotics, as described below. There are reviews on the permeability properties of porin channels [4,5].

Enteric bacteria including *E. coli* produce at least two porins, OmpF and OmpC, in the usual media. Conditions prevailing in the body of host animals (high osmotic pressure, high temperature, lower pH) favor the synthesis of OmpC, which is a more desirable porin in such an environment because it produces a slightly smaller channel [1]. PhoE, with preference for anionic compounds, may be produced in some environments where the inorganic phosphate concentration is low. In reconstituted systems, the size of porin channels becomes smaller at acidic pH [4,6], but it has not yet been shown that this alteration also occurs in intact cells.

Most β-lactam compounds are small, and have an elongated shape. Thus, they are predicted to diffuse through the *E. coli* (and *S. typhimurium*) porin channels without much difficulty. That this does occur was shown using several approaches. (a) This diffusion process can be studied quantitatively by the method of Zimmermann and Rosselet [7], which analyzes the rate of hydrolysis of β-lactam compounds by intact cells of Gram-negative cells as a combination of the spontaneous diffusion process across the OM according to Fick's law, and the enzymatic hydrolysis in the periplasm according to Michaelis–Menten kinetics. By applying this method to porin-deficient strains of *S. typhimurium*, it was shown that more than 90% of the influx of cephaloridine occurs through the porin channels [8]. (b) By using liposome vesicles containing purified porins in the bilayer, most cephalosporins were shown to penetrate through the porin channel at substantial rates [9,10]. In fact, cephaloridine (M_r 415), the fastest penetrating cephalosporin so far tested, diffused about twice as rapidly through the *E. coli* OmpF channel than lactose (M_r 342), although these rates were about two orders of magnitude slower than the diffusion rate of a much smaller molecule, arabinose (M_r 150). (c) Porin-deficient mutants are significantly more resistant at least to some of the β-lactams [3,11–13]. Although there are claims that penicillins diffuse mainly through the lipid bilayer region, this is unlikely at least in the enteric bacteria of the wild type [2].

Quantitative determination of β-lactam diffusion rates in intact cells [14] and in reconstituted liposomes [10] showed that the rates are influenced by several parameters. (a) There is an inverse correlation between the lipophilicity of the monoanionic cephalosporin molecule and the permeation rate, and a tenfold increase in the octanol/water partition coefficient (of the protonated, uncharged forms) causes a four- to fivefold decrease in the penetration rate [10,14]. (b) With the OmpF and OmpC porins of *E. coli*, dipolar ionic cephalosporins diffuse much more rapidly than monoanionic cephalosporins, which in turn diffuse more rapidly than dianionic compounds [10,14]. This is consistent with the slight cation preference of these channels (Chapter 12). In contrast, the PhoE porin of *E. coli* prefers anionic cephalosporins [14]. Perhaps the anion preference of the gonococcal porin [15] contributes to the very high susceptibility of this organism to anionic benzylpenicillin. (c) The penetration of anionic compounds is impeded even more in intact cells, because of the presence of Donnan potential across the OM, with the interior negative [16]. (d) Finally, oxyiminocephalosporins, such as cefotaxime, ceftazidime or ceftriaxone, diffuse much more slowly than expected from their overall size, hydrophobicity and charge [11]. Possibly this is caused by the steric hindrance due to the oxyimino substituents, which protrude out of the average plane of the cephalosporin molecule.

Permeation rates of various β-lactam compounds through the *E. coli* OmpF porin span a wide range, the fastest penetrating compound (imipenem) diffusing at a rate at least forty times higher than some of the slower compounds [11]. Examination of these rates indicates that the permeability cannot be the only parameter that determines the efficacy of the drug. In fact, the influx of β-lactam molecules is counterbalanced by their hydrolysis by periplasmic β-lactamases, which are encoded by chromosomal genes of almost any Gram-negative bacterial species. In addition, many clinical isolates now contain R plasmids that code for additional β-lactamases. The interaction between permeability, enzymatic hydrolysis and the binding affinity to the target (penicillin-binding proteins) can be treated quantitatively, and the efficacy of any β-lactam can be predicted reasonably accurately for *E. coli* through this theoretical treatment [17].

Penetration mechanisms are probably similar in other organisms belonging to Enterobacteriaceae. However, the porins are very different in *Pseudomonas aeruginosa*. This organism, which shows an extremely low non-specific permeability in its OM (for review, see [18]), seems to lack entirely the 'classical' trimeric porins of the enteric bacteria [19]. (Although some OM proteins of this species have been reported to exist as trimers, the trimeric structure, if any, appears to be quite unstable, unlike the very stable, tight trimeric assembly found in classical porins.) The major non-specific porin of this organism is OprF [19,20], which is a homologue of OmpA protein of enteric bacteria [21]. Both OprF and OmpA exist as monomers, and produce pores that allow only very slow penetration of solutes [19,22]. The large pore size of OprF (about 2 nm in diameter) was surprising, but it is possible either that the channel offers much more resistance to solutes during their passage, or that only a fraction of the channels is open at any given time [18]. Although one laboratory has claimed that OprF is not a porin, and that other 'true' porins of this organism produce channels much narrower than the classical *E. coli* porins, these conclusions are based on incorrect interpretation of the data [19]. Cephalosporins that utilize the non-specific OprF porin show very slow penetration rates as expected, and this fact certainly explains the high general resistance of *P. aeruginosa* against β-lactams [23,24]. Most of the compounds with strong antipseudomonad activity are likely to utilize permeation pathways other than the non-specific porins (see below).

Much less direct evidence is available for the importance of porin pathways in the penetration of other classes of agents. This is because (a) direct assays of permeability of the Zimmermann–Rosselet type are difficult for most other agents, and (b) reconstitution assays utilizing liposomes are also impossible for lipophilic agents that can diffuse through the lipid bilayer of liposomes. Furthermore, mutational loss of porins does not always result in large increases in minimal inhibitory concentration (MIC), because the effect of penetration rates becomes prominent only when there is a degradation or modification of the drug after its penetration through the OM [2,3]. Nevertheless, it appears reasonably certain that the following classes of agents traverse the OM of enteric bacteria predominantly through porin channels.

(a) *Aminoglycosides*. Because of their extremely hydrophilic nature, it is difficult to imagine that they use anything other than porin channels, although liposome assays are not possible owing to the net positive charge of these molecules [2]. Some of the aminoglycosides are fairly large in size, but the positive charge should help in their influx. The mutational loss of porins does not usually increase the MIC of aminoglycosides, but it is possible that they penetrate rapidly enough through the residual permeation pathways (for example, remaining porin channels).

(b) *Chloramphenicol*. Its small size and absence of charge suggests an easy passage through the porin channels. Loss of porin increases the MIC of this agent [25], and indeed one class of the classical chloramphenicol-resistance mutations in *E. coli* K12, *cmlB*, was later identified as mutations in the porin structural gene *ompF*. In *Hemophilus influenzae*, a chlorampenicol-resistant clinical isolate was reported to lack a 40 000 Da OM protein [26] which presumably corresponds to the major non-specific porin of this organism [27]. A chloramphenicol-resistant mutant of *Pseudomonas cepacia* with decreased permeation rate for this drug has also been reported, but in this case, there was no obvious change in the OM protein pattern [28].

(c) *Tetracyclines*. They are fairly compact molecules with a number of hydrophilic groups on the surface. The MIC is increased slightly upon the mutational loss of porins [25]. However, diffusion through the bilayers may make a significant contribution with more lipophilic derivatives (see below).

(d) *Fluoroquinolones*. They are again compact molecules with some hydrophilic groups. The MIC is slightly increased in porin-deficient mutants [29].

2.3. Specific channels

Porin channels often prefer certain broad classes of solutes on the basis of their gross physicochemical properties. However, these channels do not show specificity in the traditional sense used in biochemistry. In contrast, there are porin-like channels with true specificity. In *E. coli*, the examples are LamB (phage lamda receptor) and Tsx (phage T6 receptor) proteins, which produce channels with internal binding sites that bind specifically oligosaccharides of maltose series [30–32] and nucleosides [33], respectively. Because of the presence of the specific binding sites, these channels accelerate the diffusion of specific ligands when they are present at low concentrations, but their diffusion becomes slowed down at high concentrations. Thus, they exhibit a saturation kinetics for diffusion [34], in contrast to the diffusion through the non-specific porin, where the rate shows a linear dependence on the concentration difference across the membrane.

The importance of the specific pathways in antibiotic diffusion was first shown with the protein D2 (OprD) channel in *P. aeruginosa*. Quinn and co-workers [35] found that imipenem-resistant mutants of *P. aeruginosa*, found in clinical material, were unusual among β-lactam-resistant mutants in that their susceptibility to other β-lactams was unaltered, and that they did not overproduce any β-lactamase. Further, they showed that the mutant was lacking a 45 000 Da OM protein, although they still contained normal amounts of the non-specific porin OprF [35]. It was then shown that the missing protein was protein D2 [36], and that the influx of imipenem through this channel occurred according to saturation kinetics, with an apparent K_m of about 0.1 mM [37]. Although the true nature of the physiological substrate for this channel is a matter for conjecture, the diffusion of imipenem is inhibited competitively by basic amino acids, and possibly this channel functions in the selective uptake of either basic amino acids or peptides containing these amino acids [38]. Imipenem and other carbapenems were also found to be less active in media containing high concentrations of basic amino acids [39].

When we consider that *P. aeruginosa* lacks entirely the efficient trimeric porins present in many other Gram-negative organisms, it is not surprising that this organism produces specific OM channels in order to take up the essential nutrients efficiently.

There are data suggesting that other β-lactams showing exceptional potency against *P. aeruginosa* might also be taken up by specific channels [40], but the identity and specificity of these channels are still not clear. In contrast, we have seen that most β-lactams utilize the non-specific porin channel for influx into enteric bacteria, and imipenem is not an exception [41].

2.4. TonB-dependent uptake pathways

For the uptake of compounds that exist in very low concentrations in the environment, for example vitamin B_{12} and iron-chelator complexes, specific channels are less than ideal because the dissociation constants of their binding sites are usually in the range of 0.1–1 mM, and because they can achieve only passive equilibration of solutes across the membrane. Thus, pathways based on a different mechanism are required. These pathways are composed of OM proteins and a cytoplasmic membrane protein, TonB [42,43]. The OM proteins bind the specific ligands at an affinity far higher than those encountered in the specific channels, and therefore are often called 'receptors'. Nevertheless, they appear to be constructed essentially as gated channels [44]. The important feature of these systems is that they catalyze an uphill accumulation of ligands in the periplasm [45,46], and the TonB protein is thought to effect this by physically interacting with OM channels and thereby somehow transmitting the energy stored in the cytoplasmic membrane to the outer membrane receptors.

Attempts have been made to utilize these pathways, especially the iron-chelator pathways, for improving the influx of agents into Gram-negative bacteria. Many of these chelators that are produced by microorganisms are called 'siderophores' and usually contain multiple catechol groups or hydroxamate groups for the chelation of Fe^{3+} at very high affinity [47]. One siderophore-antibiotic hybrid, CGP 4832, contains a morpholine substituent at the 3-position of rifamycin and an *N*-methylpiperidine substituent at the 25-position. This compound is at least 200 times more active than rifampicin against *E. coli* [48]. Analysis of resistant mutants indicated that the drug crossed the OM via the *FhuA* receptor, whose physiological substrate is ferrichrome [48], a hydroxamate-type siderophore produced by fungi. Interestingly, structural similarity between CGP 4832 and ferrichrome is not obvious, and this indicates that we know very little about how the siderophore receptors recognize their cognate ligands.

Many cephalosporin derivatives with catechol groups or its analogs [49–57], or with hydroxamate groups [58], have been synthesized. Most of these compounds contain monomeric catechol, in contrast to many of the natural siderophores (e.g. enterobactin) that contain multiple catechol groups arranged to face a central ferric ion. Many of these cephalosporins are very effective, sometimes showing MIC values as low as 1 ng/ml for enteric bacteria, consistent with their uphill accumulation in the periplasm. Some also show exceptional activity against *P. aeruginosa*. That these compounds are transported by siderophore receptor(s) is suggested by the observation that adding excess Fe^{3+} to the medium represses the production of these receptors, thereby making the drug less active [50–55]. Furthermore, resistant mutants selected in vitro by these compounds were often found to be defective in the TonB protein [50]. Use of defined mutants of *E. coli* showed that FepA (receptor for the endogenous tricatechol siderophore, enterobactin), FecA (receptor for ferric-citrate complex), FhuA (receptor for a hydroxamate siderophore, ferrichrome) and FhuE (receptor for another hydroxamate siderophore, rhodotorulic acid) were not involved in the influx of catechol-type cephalosporins, but defects in Fiu or Cir made the mutants more resistant especially with cephalosporins susceptible to enzymatic hydrolysis, and double defects in both genes produced very resistant strains [51,54]. With one catechol-cephalosporin, the resistance of the *cir fiu* double mutant was indeed shown to be due to a drastically decreased influx of the drug across OM [54]. There is some evidence suggesting that Cir is involved in the uptake of monomeric catechol compounds [59], and it has been proposed that catechol-cephalosporins are taken up more in an unliganded form through these channels [54]. The potential utilization of two alternative pathways gives a strong advantage to these compounds, because the antibiotic will not select for receptor mutants, and the only mutants selected for, *tonB*, are not a problem in a clinical setting, as they are avirulent as a result of their inability to take up iron in a severely iron-depleted environment in the tissues of host animal.

2.5. Diffusion through lipid bilayer

Lipid bilayers are usually quite permeable to uncharged, lipophilic molecules [60]. This is because the lateral mobility of the phospholipid molecules and hydrocarbon chains creates transient lacunae within the membrane, into which the solute molecules can partition. However, the bilayer in the OM is unusual in its construction, the outer leaflet being composed almost entirely of LPS, at least in enteric bacteria [1]. LPS is much larger than glycerophospholipids, containing 6–7 covalently linked fatty acids, and therefore interacts with its neighbors much more strongly [61]. The lateral interaction between LPS molecules is further strengthened by divalent cation bridges between the negatively charged groups of LPS, and perhaps by hydrogen bonds between carbohydrate chains. Moreover, the hydrocarbon interior of LPS leaflet is more rigid (nearly crystalline) [62,63], because all fatty acid residues in LPS are saturated. Thus, the partitioning of external solute molecules into the interior of LPS is expected to be more difficult than into the interior of glycerophospholipids, and experiments with several probes indeed showed that the partition coefficient into LPS was lower by a factor of about ten [64]. Experimental determination of diffusion rates of steroids, used as hydrophobic probes, across the OM of *E. coli, S. typhimurium*, and some other Gram-negative species showed that these molecules cross the OM bilayer more slowly, perhaps by a factor of 50–100, than through the typical glycerophospholipid bilayers [65]. This means that although OM slows down the diffusion of lipophilic molecules, it is not an absolute barrier. In contrast, it functions as a nearly perfect barrier for amphiphilic molecules, for example, nafcillin [66], because only an insignificant fraction of the drug exists in the protonated, uncharged form that is capable of partitioning into the lipid interior [65].

In fact, there is much evidence that OM allows the influx, at a significant rate, of hydrophobic agents. Thus, *E. coli* does produce intracellular enzymes for the metabolism of bile acids [67], a finding suggesting that small amounts of bile acids do penetrate into the cytoplasm all the time. As an another example, although *E. coli* is generally thought to be resistant to erythromycin, its MIC of 50–100 μg/ml suggests that significant penetration does occur. In fact, intact cells of *E. coli* containing erythromycin esterase presumably in the cytoplasm were shown to hydrolyze erythromycin in the medium rather rapidly [68], and from this we can estimate that a minimal permeability coefficient of 10^{-6} cm/s for erythromycin through the OM bilayer (this can be compared to the permeability coefficient of 10^{-5} cm/s for cephalothin through the porin channels of *E. coli* OM [9]).

A hallmark of compounds diffusing through the lipid interior of OM is that their influx is increased drastically in 'deep rough' mutants that synthesize extremely deficient forms of LPS [66], because the OM of these mutants contains some glycerophospholipid bilayer domains [1]. Many antiseptics and disinfectants, such as quaternary ammonium compounds, triphenylmethane dyes and butylparaben, which are all hydrophobic, belong to this class and inhibit the deep rough mutants much more effectively [69]. Chlorhexidine, presumably because of its positive charge, constitutes an exception and works equally well on *E. coli* and the Gram-positive *Staphylococcus aureus* [69].

The behavior of compounds with multiple protonation sites, such as tetracyclines and fluoroquinolones, is more complex [70]. When titration is carried out, for example by adding alkali to an acidic solution of a typical fluoroquinolone, say norfloxacin, much of the removal of the protons follows the pathway shown in the upper half of Fig. 1 (i.e. from A^0B^+ to A^-B^+ to A^-B^0). The macroscopic dissociation constants (K_1 and K_2) obtained in such an experiment are thus fairly close to the microscopic dissociation constants K_A and K_C of Fig. 1. However, the major species that diffuses through the bilayer is likely to be the uncharged species, A^0B^0. In order to predict the abundance of this species, one needs the microscopic dissociation constants K_B and K_D, which are usually far from K_1 and K_2. For norfloxacin and tetracycline, the misuse of macroscopic constants in this prediction led to the idea that the uncharged species correspond to only 0.6% and 0.0001% of the total drug population around pH 7, and to the frequently expressed notion that this species is too scarce to contribute much to diffusion across the membrane. However, use of correct microscopic constants [71,72] shows that a substantial fraction

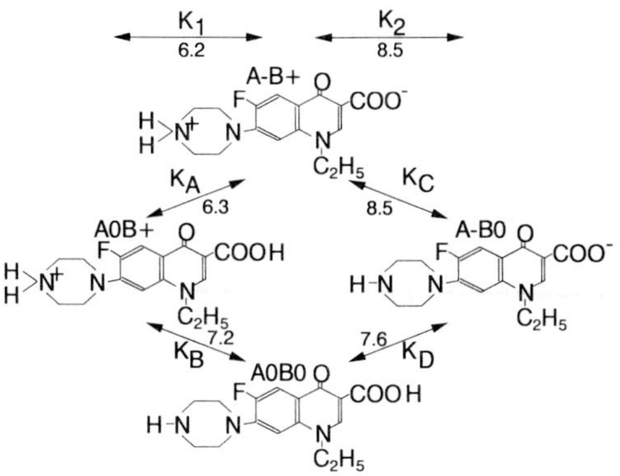

Fig. 1. Protonation behavior of norfloxacin. When titration is carried out in the usual manner, two macroscopic acid dissociation constants, K_1 and K_2, are observed. Their magnitudes are shown as pK_a values underneath the arrows. Norfloxacin, however, shows a complex protonation behavior as defined by the four microscopic acid dissociation constants, K_A to K_D, the magnitudes of which are again shown as pK_a values next to the arrows. Each species is denoted by the charge state of the acid group (A) and the base group (B), e.g. A^0B^+. The acid dissociation constants are from ref. 71.

554

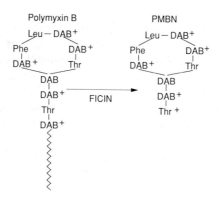

Fig. 2. Preparation of polymyxin nonapeptide (PMBN). DAB denotes diaminobutyric acid.

(10% and 9%, respectively) of the drugs is in the uncharged form. Thus, it is not surprising that diffusion through the OM bilayer seems to contribute significantly in the influx of more hydrophobic tetracyclines [73,74] and probably of some fluoroquinolones [70].

Because LPS contains a large number of anionic groups, the LPS leaflet of OM must be stabilized by divalent cation bridges. Competition with these divalent cations, or their removal, is an effective way of destabilizing the OM bilayer, thereby increasing the chances for solute partition and permeation. Polycations with strategically placed positive charges, such as polymyxin nonapeptide [75] (Fig. 2) dramatically permeabilizes the OM bilayer, making *E. coli* up to 100 times more susceptible to lipophilic and amphiphilic agents such as novobiocin, rifampin, vancomycin and fusidic acid. Significantly, molecules containing the same number of positive charge as polymyxin nonapeptide, but more flexible, such as pentalysine, are much less effective, because their modest permeabilizing action is totally abolished by such concentrations of monovalent and divalent cations as are present in body fluids [76]. Many cationic peptides and proteins of animal origin also increase the permeability of the OM bilayer, and the action of these agents has been reviewed [76].

Aminoglycosides contain multiple positive charges. This led to the concept that these drugs also destabilize the OM, thereby increasing their own penetration rate across the outer membrane ('self-promoted uptake'[77]). In fact, it was possible to show that *P. aeruginosa* OM became more permeable to nitrocefin (a fairly lipophilic cephalosporin) and to NPN (a very lipophilic fluorescent probe) upon treatment with gentamicin [78]. This model, however, has been criticized because unlike polymyxin nonapeptide, which remains very active in the presence of 100 mM Na^+ and 4 mM Mg^{2+}, the OM-permeabilizing action of aminoglycosides was totally abolished by just 1 mM Mg^{2+} [78]. From comparison with pentalysine, it appears that the aminoglycoside effect is also extremely sensitive to the concentrations of Na^+ [76]. These data suggest that the 'self-promoted uptake' is unlikely to play a major role in our body fluids. However, the efficacy of aminoglycosides in controlling *P. aeruginosa* infections of animals and humans does suggest some efficient mechanism of entry; possibly the self-promoted uptake pathway becomes resistant to the Na^+ and Mg^{2+} effect under certain local conditions prevailing at the site of infection.

3. Cell wall of Gram-positive bacteria

3.1. Introduction

Most Gram-positive cell walls consist of peptidoglycan, polysaccharides and similar polymers such as teichoic acids. That is, they are devoid of lipidic material that is the hallmark of the Gram-negative cell wall. The porosity of these 'typical' Gram-negative cell wall preparations does not appear to have been investigated with modern technology, but it is reasonable to assume that they are quite porous, as *Bacillus megaterium* cell wall for example allows a substantial penetration by 100 000-Da dextrans [79].

3.2. Mycobacterial cell wall

The organisms of the *Corynebacterium–Nocardia–Mycobacterium* group make a striking exception to the statement above, because their cell wall contains substantial amount of lipids of unusual structure. At least the mycobacterial cell wall has been suspected to function as an efficient permeability barrier, because of the generalized intrinsic resistance of this group of organisms towards a number of antimicrobial agents.

It was only in 1990, however, that the permeability of mycobacterial cell wall was determined in a quantitative manner [80]. Use of the Zimmermann–Rosselet method showed that the cell wall of *Mycobacterium chelonae* indeed had an exceptionally low permeability to cephalosporins. The permeation rate of cephaloridine, for example, was about 10 000-fold lower than through the OM of *E. coli*. It was more than ten times slower even when compared with the rate of diffusion through the *P. aeruginosa* OM.

The effect of solute parameters and temperature on this slow penetration process suggested that hydrophilic solutes traverse the mycobacterial cell wall through water-filled channels [80]. What appears to be the major pore-forming protein in the cell wall was recently identified by reconstitution methods [81]. This protein produces only a low level of permeability presumably because (a) the cell wall contains only a very small amount of protein, and (b) a unit amount of protein produces much lower permeability in comparison with the classical OmpF porin of *E. coli* [81].

The exceptionally low permeability of mycobacterial cell wall also suggests that the basic continuum of this structure must have an unusual structure. Thus, if the rest of the cell wall had the permeability typical of the glycerophospholipid bilayer, some of the monoanionic cephalosporins must have penetrated through this region of the cell wall at least 100 times faster than actually observed. Recent X-ray diffraction studies showed that the hydrocarbon chains in the cell wall of *M. chelonae* are arranged in a parallel, paracrystalline array, in a direction perpendicular to the plane of the cell wall [82]. This is not surprising because much of the lipids in mycobacterial cell wall are composed of mycolic acids, with their extremely long hydrocarbon chains (up to 50 carbons in one of the 'arms') with only one or two double bonds. Such lipids will produce an exceptionally tightly packed, rigid structure, that would be even less permeable than the LPS-containing leaflet of the Gram-negative OM. Although the covalent linkage of most of the mycolic acid residues to an underlying polysaccharide molecule was often thought to impair the tight organization of mycolic acid chains, the monosaccharide units in the cell wall arabinogalactan appears to be connected to each other in a manner that ensures the maximal flexibility of the chains, i.e. galactose exists as galactofuranose connected usually 1–6 [83], and this will certainly facilitate the tight packing of the mycolic acid residues. The cell wall was also shown to contain unexpectedly large amounts of lipids with shorter fatty acid residues, and these lipids may comprise the outer leaflet of the structure. Thus, the mycobacterial cell wall, at least in principle, appears to be constructed as an asymmetric lipid bilayer, as originally proposed by Minnikin [84]. It is interesting to see that two very distant groups of bacteria, Gram-negative bacteria and Gram-positive mycobacteria, arrived at similar solutions for the purpose of building effective permeability barriers, although the less impermeable part of the structure is located in the inner leaflet in the proposed model of the mycobacterial cell wall. Nevertheless, being a lipid bilayer, even mycobacterial cell wall cannot be totally impervious. Indeed, agents with increased lipophilicity appear to penetrate through the mycobacterial cell wall at a significant rate [85].

References

1. Nikaido, H. and Vaara, M. (1985) Microbiol. Rev. 49, 1–32.
2. Nikaido, H. (1989) in: L.E. Bryan (Ed.), Handbook of Experimental Pharmacology, Vol. 91, Springer-Verlag, Berlin, pp. 1–34.
3. Nikaido, H. (1989) Antimicrob. Agents Chemother. 33, 1831–1836.
4. Benz, R. (1988) Annu. Rev. Microbiol. 42, 359–393.
5. Nikaido, H. (1992) Mol. Microbiol. 6, 435–442.
6. Todt, J.C., Rocque, W.J. and McGroarty, E.J. (1992) Biochemistry 31, 10471–10478.
7. Zimmermann, W. and Rosselet, A. (1977) Antimicrob. Agents Chemother. 12, 368–372.
8. Nikaido, H., Song, S.A., Shaltiel, L. and Nurminen, M. (1977) Biochem. Biophys. Res. Commun. 76, 324–330.
9. Nikaido, H. and Rosenberg, E.Y. (1983) J. Bacteriol. 153, 241–252.
10. Yoshimura, F. and Nikaido, H. (1985) Antimicrob. Agents Chemother. 27, 84–92.
11. Harder, K., Nikaido, H. and Matsuhashi, M. (1981) Antimicrob. Agents Chemother. 20, 549–552.
12. Jaffe, A., Chabbert, Y.A. and Derlot, E. (1983) Antimicrob. Agents Chemother. 23, 622–625.
13. Medeiros, A.A., O'Brien, T.F., Rosenberg, E.Y. and Nikaido, H. (1987) J. Infect. Dis. 156, 751–757.
14. Nikaido, H., Rosenberg, E.Y. and Foulds, J. (1983) J. Bacteriol. 153, 232–240.
15. Young, J.D.E., Blake, M., Mauro, A. and Cohn, Z. (1983) Proc. Natl. Acad. Sci. USA 80, 3831–3835.
16. Sen, K., Hellman, J. and Nikaido, H. (1988) J. Biol. Chem. 1182–1187.
17. Nikaido, H. and Normark, S. (1987) Mol. Microbiol. 1, 29–36.
18. Nikaido, H. and Hancock, R.E.W. (1986) in: J.R. Sokatch (Ed.), The Bacteria, Vol. X, Academic Press, New York, pp. 145–193.
19. Nikaido, H., Nikaido, K. and Harayama, S. (1991) J. Biol. Chem. 266, 770–779.
20. Bellido, F., Martin, N.L., Siehnel, R.J. and Hancock, R.E.W. (1992) J. Bacteriol. 174, 5196–5203.
21. Duchene, M., Schweizer, A., Lottspeich, F., Grauss, G., Marget, M., Vogel, K., von Specht, B.-U. and Domdey, H. (1988) J. Bacteriol. 170, 155–162.
22. Sugawara, E. and Nikaido, H. (1992) J. Biol. Chem. 267, 2507–2511.
23. Angus, B.L., Carey, A.M., Caron, D.A., Kropinski, A.M.B. and Hancock, R.E.W. (1982) Antimicrob. Agents Chemother. 21, 299–309.
24. Yoshimura, F. and Nikaido, H. (1982) J. Bacteriol. 152, 636–642.
25. Pugsley, A.P. and Schnaitman, C.A. (1978) J. Bacteriol. 133, 1181–1189.
26. Burns, J.L., Mendelman, P.M., Levy, J., Stull, T.L. and Smith, A.L. (1986) Antimicrob. Agents Chemother. 27, 46–54.

27. Vachon, V., Lyew, D.J. and Coulton, J.W. (1985) J. Bacteriol. 162, 918–924.
28. Burns, J.L., Hedin, L.A. and Lien, D.M. (1989) Antimicrob. Agents Chemother. 33, 136–141.
29. Hirai, K., Aoyama, H., Irikura, T., Iyobe, S. and Mitsuhashi, S. (1986) 29, 535–538.
30. Luckey, M. and Nikaido, H. (1980) Proc. Natl. Acad. Sci. USA 77, 167–171.
31. Luckey, M. and Nikaido, H. (1980) Biochem. Biophys. Res. Commun. 93, 166–171.
32. Benz, R., Schmidt, A. and Vos-Scheperkeuter, G.H. (1987) J. Membr. Biol. 100, 21–29.
33. Maier, C., Bremer, E., Schmidt, A. and Benz, R. (1988) J. Biol. Chem. 263, 2493–2499.
34. Freundlieb, S., Ehmann, U. and Boos, W. (1988) J. Biol. Chem. 263, 314–320.
35. Quinn, J.F., Dudek, E.J., DiVincenzo, C.A., Lucks, D.A. and Lerner, S.A. (1986) J. Infect. Dis. 154, 289–294.
36. Trias, J. and Nikaido, H. (1990) Antimicrob. Agents Chemother. 34, 52–57.
37. Trias, J., Dufresne, J., Levesque, R.C. and Nikaido, H. (1989) Antimicrob. Agents Chemother. 33, 1201–1206.
38. Trias, J. and Nikaido, H. (1990) J. Biol. Chem. 265, 15680–15684.
39. Fukuoka, T., Masuda, N., Takenouchi, T., Sekine, N., Iijima, N. and Ohya, S. (1991) Antimicrob. Agents Chemother. 35, 529–532.
40. Nikaido, H. (1992) in: E. Galli, S. Silver and B. Witholt (Eds.), Pseudomonas, Molecular Biology and Biotechnology, American Society for Microbiology, Washington, DC, pp. 146–153.
41. Raimondi, A., Traverso, A. and Nikaido, H. (1991) Antimicrob. Agents Chemother. 35, 1174–1180.
42. Postle, K. (1990) Mol. Microbiol. 4, 2019–2025.
43. Kadner, R.J. (1990) Mol. Microbiol. 4, 2027–2033.
44. Rutz, J.M., Liu, J., Lyons, J.A., Goranson, J., Armstrong, S.K., McIntosh, M.A., Feix, J.B. and Klebba, P.E. (1992) Science 258, 471–475.
45. Reynolds, P.R., Mottur, G.P. and Bradbeer, C. (1980) J. Biol. Chem. 255, 4313–4319.
46. Matzanke, B.F., Ecker, D.J., Yang, T.-S., Huynh, B.H., Muller, G. and Raymond, K.N. (1986) J. Bacteriol. 167, 674–680.
47. Neilands, J.B. (1981) Annu. Rev. Biochem. 50, 715–731.
48. Pugsley, A.P., Zimmermann, W. and Wehrli, W. (1987) J. Gen. Microbiol. 133, 3505–3511.
49. Ohi, N., Aoki, B., Shinozaki, T., Moro, K., Noto, T., Nehashi, T., Okazaki, H. and Matsunaga, I. (1986) J. Antibiot. 39, 230–241.
50. Watanabe, N., Nagasu, T., Katsu, K. and Kitoh, K. (1987) Antimicrob. Agents Chemother. 31, 497–504.
51. Curtis, N.A.C., Eisenstadt, R.L., East, S.J., Cornford, R.J., Walker, L.A. and White, A.J. (1988) Antimicrob. Agents Chemother. 32, 1879–1886.
52. Mochizuki, H., Yamada, H., Oikawa, Y., Murakami, K., Ishiguro, J., Kosuzume, H., Aizawa, N. and Mochida, E. (1988) Antimicrob. Agents Chemother. 32, 1648–1654.
53. Hashizume, T., Sanada, M., Nakagawa, S. and Tanaka, N. (1990) J. Antibiot. 43, 1617–1620.
54. Nikaido, H. and Rosenberg, E.Y. (1990) J. Bacteriol. 172, 1361–1367.
55. Sisley, P., Griffiths, J.W., Monsey, D. and Harris, A.M. (1990) Antimicrob. Agents Chemother. 34, 1806–1808.
56. Southgate, R., Stachulski, A.V., Basker, M.J. and Knott, S.J. (1990) J. Antibiot. 43, 574–577.
57. Arisawa, M., Sekine, Y., Shimizu, S., Takano, H., Angehrn, P. and Then, R.L. (1991) Antimicrob. Agents Chemother. 35, 653–659.
58. Minnick, A.A., McKee, J.A., Dolence, E.K. and Miller, M.J. (1992) Antimicrob. Agents Chemother. 36, 840–850.
59. Hancock, R.E.W., Hantke, K. and Braun, V. (1977) Arch. Microbiol. 114, 231–239.
60. Stein, W.D. (1967) The Movement of Molecules across Cell Membranes, Academic Press, New York.
61. Takeuchi, Y. and Nikaido, H. (1981) Biochemistry 20, 523–529.
62. Nikaido, H., Takeuchi, Y., Ohnishi, S.-I. and Nakae, T. (1977) Biochim. Biophys. Acta 465, 152–164.
63. Labischinski, H., Barnickel, G., Bradaczek, H., Naumann, D., Rietschel, E.T. and Giesbrecht, P. (1985) J. Bacteriol. 162, 9–20.
64. Vaara, M., Plachy, W.Z. and Nikaido, H. (1990) Biochim. Biophys. Acta 1024, 152–158.
65. Plesiat, P. and Nikaido, H. (1992) Mol. Microbiol. 6, 1323–1333.
66. Nikaido, H. (1976) Biochim. Biophys. Acta 433, 118–132.

558

67. Yoshimoto, T., Higashi, H., Kanatani, A., Lin, X.-S., Nagai, H., Oyama, H., Kurazono, K. and Tsuru, D. (1991) J. Bacteriol. 173, 2173–2179.
68. Barthelemy, P., Autissier, D., Gerbaud, G. and Courvalin, P. (1984) J. Antibiot. 37, 1692–1696.
69. Russell, A.D. and Gould, G.W. (1988) J. Appl. Bacteriol. 65, 167S–195S.
70. Nikaido, H. and Thanassi, D.G. (1993) Antimicrob. Agents Chemother. 37, 1393–1399.
71. Takacs-Novak, K., Noszal, B., Hermecz, I., Kereszturi, G., Podanyi, B. and Szasz, G. (1990) J. Pharm. Sci. 79: 1023–1028.
72. Rigler, N.E., Bag, S.P., Leyden, D.E., Sudmeier, J.L. and Reilley, C.N. (1965) Anal. Chem. 37: 872–875.
73. McMurry, L.M., Cullinane, J.C. and Levy, S.B. (1982) Antimicrob. Agents Chemother. 22, 791–799.
74. Leive, L., Telesetsky, S., Coleman, Jr., W. G. and Carr, D. (1984) Antimicrob. Agents Chemother. 25, 539–544.
75. Vaara, M. and Vaara, T. (1983) Antimicrob. Agents Chemother. 24, 114–122.
76. Vaara, M. (1992) Microbiol. Rev. 56, 395–411.
77. Hancock, R.E.W. (1981) J. Antimicrob. Chemother. 8, 429–445.
78. Hancock, R.E.W. and Wong, P.G.W. (1984) Antimicrob. Agents Chemother. 26, 48–52.
79. Scherrer, R. and Gerhardt, P. (1971) J. Bacteriol. 107, 718–735.
80. Jarlier, V. and Nikaido, H. (1990) J. Bacteriol. 172, 1418–1423.
81. Trias, J., Jarlier, V. and Benz, R. (1992) Science 258, 1479–1481.
82. Nikaido, H., Kim, S.-H. and Rosenberg, E.Y. (1993) Mol. Microbiol. 8, 1025–1030.
83. McNeil, M., Daffe, M. and Brennan, P.J. (1991) J. Biol. Chem. 266, 13217–13323.
84. Minnikin, D.E. (1982) in: C. Ratledge and J. Stanford (Eds.), The Biology of the Mycobacteria, Academic Press, New York, pp. 95–184.
85. Nikaido, H. and Jarlier, V. (1991) Res. Microbiol. 142, 437–443.

Species Index

Subject Index

574